Atomic weights are based on carbon-12.
Numbers in parentheses are the mass
numbers of the most stable isotopes.

VIIIA or 0

| | | | | | | | 2 Helium He 4.00 |

		IIIA	IVA	VA	VIA	VIIA	
		5 Boron B 10.81	6 Carbon C 12.01	7 Nitrogen N 14.01	8 Oxygen O 16.00	9 Fluorine F 19.00	10 Neon Ne 20.18
		13 Aluminum Al 26.98	14 Silicon Si 28.09	15 Phosphorus P 30.97	16 Sulfur S 32.06	17 Chlorine Cl 35.45	18 Argon Ar 39.95

IB	IIB						
29 Copper Cu 63.546	30 Zinc Zn 65.38	31 Gallium Ga 69.72	32 Germanium Ge 72.59	33 Arsenic As 74.92	34 Selenium Se 78.96	35 Bromine Br 79.90	36 Krypton Kr 83.80
47 Silver Ag 107.87	48 Cadmium Cd 112.41	49 Indium In 114.82	50 Tin Sn 118.69	51 Antimony Sb 121.75	52 Tellurium Te 127.60	53 Iodine I 126.90	54 Xenon Xe 131.30
79 Gold Au 196.97	80 Mercury Hg 200.59	81 Thallium Tl 204.37	82 Lead Pb 207.2	83 Bismuth Bi 208.98	84 Polonium Po (209)	85 Astatine At (210)	86 Radon Rn (222)

64 Gadolinium Gd 157.25	65 Terbium Tb 158.93	66 Dysprosium Dy 162.50	67 Holmium Ho 164.93	68 Erbium Er 167.26	69 Thulium Tm 168.93	70 Ytterbium Yb 173.04	71 Lutetium Lu 174.97
96 Curium Cm (247)	97 Berkelium Bk (247)	98 Californium Cf (251)	99 Einsteinium Es (252)	100 Fermium Fm (257)	101 Mendelevium Md (258)	102 Nobelium No (259)	103 Lawrencium Lr (260)

SIXTH EDITION

Chemistry for the Health Sciences

George I. Sackheim

Associate Professor Emeritus
University of Illinois, Chicago

Dennis D. Lehman

Northwestern University

Macmillan Publishing Company

New York

Macmillan Publishing Company
866 Third Avenue, New York, New York 10022

Collier Macmillan Canada, Inc.

Library of Congress Cataloging in Publication Data

Sackheim, George I.
    Chemistry for the health sciences/George I. Sackheim. Dennis D. Lehman.—6th ed.
        p. cm.
    ISBN 0-02-405151-9
    1. Chemistry. 2. Medicine. I. Lehman, Dennis D. II. Title.
    [DNLM: 1. Chemistry. 2. Medicine.   QD 33 S121c]
    QD33.S13   1990
    540—dc20
    DNLM/DLC   89-13870
                                                                                      CIP

        Printing: 1 2 3 4 5 6 7 8   Year: 0 1 2 3 4 5 6 7 8 9

Credits for Chapter Opening Photographs

Ch. 1: Mettler Instrument Corporation
Ch. 2: Anthony Bouch
Ch. 3: Mark Antman/The Image Works
Ch. 4: Peter Menzel/Stock Boston
Ch. 5: Courtesy Tripos Corporation
Ch. 6: Anthony Bouch
Ch. 7: Anthony Bouch
Ch. 8: Joseph Schuyler/Stock, Boston
Ch. 9: John Schultz, PAR/NYC
Ch. 10: Ulrike Welsch
Ch. 11: Alan Carey/The Image Works
Ch. 12: Tom Pantages
Ch. 13: Anthony Bouch
Ch. 14: Peter G. Aitken/Photo Researchers, Inc.
Ch. 15: Ellis Herwig/The Picture Cube
Ch. 16: Courtesy Dow Chemical Company
Ch. 17: John Schultz, PAR/NYC, Inc.
Ch. 18: Ulrike Welsch/Photo Researchers, Inc.
Ch. 19: IBM Almaden Research Center
Ch. 20: Adam Hart-Davis/Science Photo Library, Photo Researchers, Inc.
Ch. 21: Anthony Bouch
Ch. 22: Ulrike Welsch
Ch. 23: Harriet Gans/The Image Works
Ch. 24: Anthony Bouch
Ch. 25: Peter Southwick/Stock, Boston
Ch. 26: Computer Graphics Laboratory, courtesy Regents of University of California, San Francisco
Ch. 27: Fred Lightfoot/George Washington University Medical Center
Ch. 28: Dr. Jeremy Burgess/Science Photo Library, Photo Researchers, Inc.
Ch. 29: Steve Takatsuno/The Picture Cube
Ch. 30: Christopher Morrow/Stock, Boston
Ch. 31: John Griffin/The Image Works
Ch. 32: AFIP 39251
Ch. 33: Tom Kelly
Ch. 34: Ulrike Welsch
Ch. 35: Christopher Morrow/Stock, Boston

Preface

This textbook of chemistry is designed primarily for first-year students in various health-related programs—nursing, dietetics, laboratory technology, inhalation therapy, dental hygiene, dental assisting, medical assisting, dental technology, and so on. Emphasis is placed on *practical* aspects of inorganic chemistry, organic chemistry, and biochemistry. Theoretic topics are dealt with only as an aid to understanding bodily processes in the human.

Organization of the Text

Part I, "Inorganic Chemistry," stresses relationships with the life processes that are the subject of Part III, "Biochemistry." Among these related topics and processes are

1. Acids, bases, salts, and electrolytes / acid–base balance and electrolyte balance in the body.
2. Oxidation–reduction / biologic oxidation–reduction reactions in the mitochondria of cells.
3. Solutions / the solvent action involved in digestion.
4. Colloids / the nature and properties of proteins, amino acids, and nucleic acids.
5. Covalent compounds / the bonds that must be broken and rearranged in the formation of high-energy phosphate bonds.
6. Emulsions / the need for emulsification of fats before digestion.
7. Nuclear chemistry and radioactivity / biologic effects of radiation on cells and organs.

Part II, "Organic Chemistry," introduces the various classes of organic compounds—hydrocarbons, alcohols, ethers, aromatic compounds, and heterocyclic compounds. In addition, the text discussions relate such compounds to carbohydrates, fats, proteins, vitamins, hormones, and nucleic acids.

Part III, "Biochemistry," deals with the chemical and molecular basis of life itself. The various chemical processes taking place in the body are described in terms of both normal and abnormal metabolism. The role of ATP, the principal direct source of energy for the body, is stressed throughout the chapters on metabolism. The formation and decomposition of this compound, and the energies involved, serve to indicate the complexity of "normal" processes. Also emphasized is the role of coenzymes, such as CoA, in metabolism. Discussions of excesses and deficiencies of vitamins and hormones are designed to demonstrate the involved interrelationships in the body's metabolic processes. In the chapter on heredity the combination of chemistry and the molecular basis of life is evidenced by the many recent advances in our understanding of DNA structure and the replication of DNA and RNA.

An outline at the beginning of each chapter indicates the topics to be discussed and, in general, how they are related. The Summary at the end of each chapter is also designed to help the reader identify the particularly important aspects of the subject matter. The Questions and Problems at the ends of the chapters may be used for oral review or assigned as homework. Finally, the Practice Test will aid students in checking their understanding of the topics of the chapter. Answers to odd-numbered questions and the practice test questions appear at the end of the book.

Changes in This Edition

- New to this edition: Chapters 31, "Immunology," and 35, "Clinical Chemistry."
- Topics introduced or given expanded treatment: absorption of iron, of vitamins · AIDS · allosteric regulation · amido sugars · biosynthesis of nonessential amino acids · black and white photography · blood group substances · breath-alcohol analyzer · cardiac hormones · chemiosmotic theory · cyclic GMP · daily requirements of vitamins · defects of carbohydrate digestion and absorption · desalinization · disposal of radioactive waste · enzyme inhibitors · fetal hemoglobin · fluoridation of water free rotation of bonds · genetic markers · glycogen storage diseases · heterocyclic amines in the brain · hormones of the pineal gland, of the blood · hormones that regulate calcium metabolism · immunoglobulines · interconversion of hexoses · irradiation of food · lipid storage diseases · medical use of $He-O_2$ mixture · mono-, di-, and triprotic acids · myocardial infarction · myoglobin · neurotransmitters · nutritionally essential and nonessential amino acids · oncogenes · phosphate esters · plasma lipoproteins · polarity of molecules · resonance · reverse osmosis.
- Many newly redrawn illustrations and new photographs.
- Expanded Glossary.
- A multiple choice Practice Test for each chapter.

Supplements

A comprehensive set of laboratory experiments (Sackheim and Lehman, *Laboratory Chemistry for the Health Sciences*, Macmillan Publishing Company) has been designed to supplement various topics in this textbook. Performing these experiments will aid the student in understanding the fundamental concepts involved. An Instructor's Manual, also available from Macmillan, contains answers to the text questions and a test bank.

Acknowledgments

Many reviewers have given us invaluable criticism and suggestions for the five earlier editions, and we thank them again. For this sixth edition specifically, we express our appreciation to Fay E. Bennett R. N., Director, *Update*; Ronald E. DiStefano, Northhampton Community College; David F. McCormick, Sinclair Community College; Valeria Meehan, City College of San Francisco; Donald Terpening, Ulster Community College; Barbara Varian, Pennsylvania State University–Behrend College.

G. I. S.
D. D. L.

Contents

Part III   Biochemistry

PART I

Inorganic Chemistry

1

Units of Measurement

Use of a laboratory balance.

The International System (SI)

We use measurements every day of our lives. We measure gasoline by the gallon and milk by the quart; we buy meat by the pound and pay for postage by the ounce. A recipe may call for a cup of flour, a tablespoonful of butter, and $\frac{1}{4}$ teaspoonful of salt. In chemistry, all measurements are made in the metric system or in its expanded modernized version, the SI system, from the French *Le Système International d'Unités*. The main advantage of the SI system is that it is a decimal system; that is, the units are all multiples of ten of larger or smaller units. In the English system there is no such common number. Recall that there are 3 feet in a yard, 4 quarts in a gallon, and 16 ounces in a pound.

Mass and Weight

The mass of an object is a measure of the amount of matter it contains. The weight of an object depends on the pull of gravity. Mass remains constant regardless of location, whereas weight can vary slightly from place to place on the surface of the Earth. For our purposes, however, we will use the terms mass and weight interchangeably.

Fundamental Units

The fundamental unit of length in the SI system is the meter. A meter is a little longer than a yard. The meter was originally defined as one ten-millionth of the distance of a meridian from the equator to the North Pole on a line passing through Paris. The standard meter was later defined as the length between two marks on a platinum-iridium bar kept at the temperature of melting ice and stored at the International Bureau of Weights and Measures near Paris. Subsequently the standard meter was redefined in terms of the orange-red wavelengths of light emitted by the element krypton-86. In October 1983 the General Conference on Weights and Measures defined the meter as the distance light travels in a vacuum in 1/299,792,458 second. The word meter is abbreviated as m.

The fundamental SI unit of mass (weight) is the kilogram. A kilogram is slightly greater than 2 pounds. The basic laboratory unit of mass (weight) is the gram. A gram is $\frac{1}{454}$ pound.

The fundamental SI unit of time is the second (s). Decimal-based multiples and submultiples can also be used, but the older units such as minutes, hours, and days are still in common use.

The fundamental SI unit of temperature is the kelvin (K). Note that the word degree and the symbol for degree are not used. In medical applications, however, temperature is usually measured in degrees Celsius (°C). We will discuss the relationship between temperatures in kelvins and degrees Celsius later in this chapter.

Table 1-1 Prefixes Used in the SI System

Prefix	Abbreviation	Decimal Expresion	Exponential Expression
exa	E	1,000,000,000,000,000,000	10^{18}
peta	P	1,000,000,000,000,000	10^{15}
tera	T	1,000,000,000,000	10^{12}
giga	G	1,000,000,000	10^{9}
mega	M	1,000,000	10^{6}
kilo	k	1,000	10^{3}
hekto	h	100	10^{2}
deka	da	10	10^{1}
—	—	1	10^{0}
deci	d	0.1	10^{-1}
centi	c	0.01	10^{-2}
milli	m	0.001	10^{-3}
micro	μ (or **mc**)	0.000001	10^{-6}
nano	n	0.000000001	10^{-9}
pico	p	0.000000000001	10^{-12}
femto	f	0.000000000000001	10^{-15}
atto	a	0.000000000000000001	10^{-18}

Derived Units

Volume is not a fundamental SI unit because it can be derived from length (the volume of a rectangular object can be calculated from the relationship volume = length × width × height). In the SI system, volume is expressed in the unit cubic meter (m^3). One cubic meter contains approximately 250 gallons. A more common unit of volume is the **liter** (often spelled out but abbreviated by chemists as **L** and by some others as **l**). A liter is slightly more than 1 quart.

The basic units most commonly used in medicine are the meter, the gram, and the liter. For this reason, these are the units used throughout this text.

Table 1-2 Metric Conversions

Mass (Weight)	
1 g = 1000 mg	one gram = one thousand milligrams
1 g = 100 cg	one gram = one hundred centigrams
1 g = 10 dg	one gram = ten decigrams
1 kg = 1000 g	one kilogram = one thousand grams
Length	
1 m = 1000 mm	one meter = one thousand millimeters
1 m = 100 cm	one meter = one hundred centimeters
1 m = 10 dm	one meter = ten decimeters
1 km = 1000 m	one kilometer = one thousand meters
Volume	
1 L = 1000 mL	one liter = one thousand milliliters
1 L = 100 cL	one liter = one hundred centiliters
1 L = 100 dL	one liter = ten deciliters
1 kL = 1000 L	one kiloliter = one thousand liters

In the SI system prefixes are used to designate various multiples or submultiples. The most commonly used prefixes are indicated in color in Table 1-1. For example, kilo- means one thousand and centi- means one-hundredth. Thus, a kilometer (km) is one thousand meters, or 1000 m, and a centigram (cg) is $\frac{1}{100}$ gram, or 0.01 g. Likewise, a milliliter (mL) is 0.001 L (see Table 1-2). Another unit of volume is the cubic centimeter (cm^3); 1 cm^3 is equal to 1 mL, so these two units may be used interchangeably. Although the term cubic centimeter is abbreviated as cm^3 by chemists, the older abbreviation cc is still seen at times. Likewise 1 MCi (1 megacurie) means 1 million Ci, and 1 μg or 1 mcg (1 microgram) means 0.000001 g. Note that the letter m indicates both meter and milli-; however, when used by itself, it indicates meter.

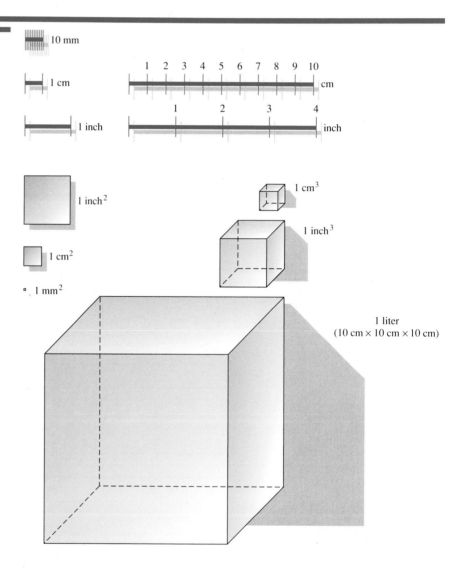

Figure 1-1   Comparison of English and SI units ($\frac{1}{2}$ scale).

Conversion Units

Table 1-3 English–SI Conversion Factors

1 in. = 2.54 cm
1 lb = 454 g
2.2 lb = 1 kg
1.06 qt = 1 L

Frequently we need to convert from one system of measurement to another (see Figure 1-1 and Table 1-3). This will no longer be necessary when the United States is fully converted to the SI system. The procedure is the same when converting from one system to another or within the English system. Consider, for example, the following English units.

$$12 \text{ in.} = 1 \text{ ft}$$

$$3 \text{ ft} = 1 \text{ yd}$$

$$1760 \text{ yd} = 1 \text{ mi}$$

Figure 1-2 Use of units in various medical examples. [Courtesy of Merck Sharp & Dohme (a); ICI Pharmaceuticals Group, a business unit of ICI Americas Inc. (b); Smith Kline & French Laboratories (c); Travenol Laboratories, Inc. (d); Becton-Dickinson, Division of Becton Dickinson and Company, Rutherford, NJ 07070 (e)]

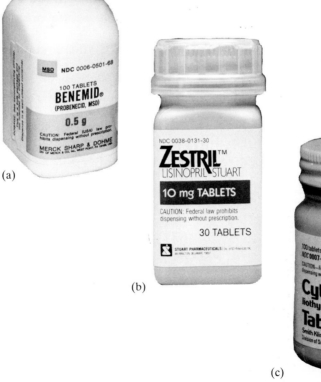

(a)

(b)

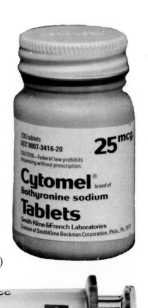

(c)

(d)

(e)

Factor Label Method

When asked to convert 6 ft to inches, many students merely multiply by 12. However, mathematically, the correct procedure is to multiply by the conversion unit 12 in./1 ft so that the result is correct both numerically and in units.

$$6 \text{ ft} = 6 \text{ ft} \times \frac{12 \text{ in.}}{1 \text{ ft}} = 72 \text{ in.}$$

Likewise, to convert 3.58 m to millimeters (mm), first recall that there are 1000 mm in a meter. The procedure is to multiply by 1000 mm/1 m. The unwanted units cancel, leaving the answer in millimeters, as desired.

$$3.58 \text{ m} = 3.58 \text{ m} \times \frac{1000 \text{ mm}}{1 \text{ m}} = 3580 \text{ mm}$$

(Note that if the 3.58 m were divided by the conversion unit of 1000 mm/1 m the units would not cancel, indicating an error in computation.)

When converting 20 in. to centimeters, the conversion factor 1 in. = 2.54 cm should be used (see Table 1-3). Thus

$$20 \text{ in.} = 20 \text{ in.} \times \frac{2.54 \text{ cm}}{1 \text{ in.}} = 50.8 \text{ cm}$$

Likewise, when converting 55 lb to kilograms, the conversion factor 2.2 lb = 1 kg should be used. Thus

$$55 \text{ lb} = 55 \text{ lb} \times \frac{1 \text{ kg}}{2.2 \text{ lb}} = 25 \text{ kg}$$

Density

The density of an object helps to characterize it physically and is defined as its mass (or weight) divided by its volume (see Figure 1-3). Density can be expressed by the formula

$$D = \frac{M}{V}$$

Figure 1-3   A comparison of the weights of 1 mL (or 1 cm^3) of
various substances.

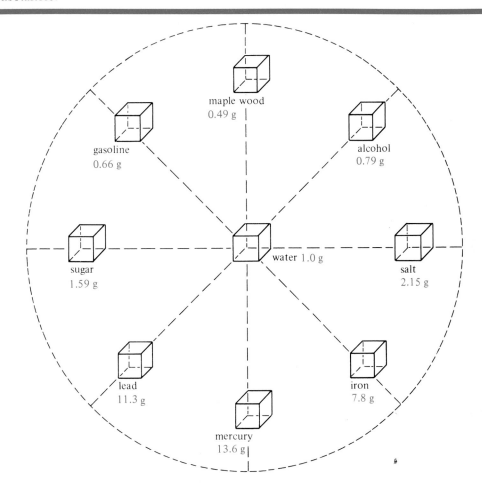

where D is the density, M is the mass, and V is the volume. Density must have units of mass per unit volume such as g/mL, g/cm^3, lb/ft^3, or mg/L.

Example 1-1   What is the density of the mercury in a thermometer if 31.2 g of it occupies 2.29 mL?

$$D = \frac{M}{V}$$

$$= \frac{31.2 \text{ g}}{2.29 \text{ mL}} = 13.6 \frac{\text{g}}{\text{mL}}$$

> **Example 1-2**   Alcohol has a density of 0.80 g/mL. How much will 100 mL of it weigh?
>
> $$D = \frac{M}{V}$$
>
> $$M = D \times V$$
>
> $$= 0.80 \; \frac{g}{mL} \times 100 \; mL = 80 \; g$$

Specific Gravity

Specific gravity (sp gr) is defined as the weight of an object compared to the weight of an equal volume of water, or as the density of an object compared to the density of water (which is 1 g/mL). Specific gravity can be expressed mathematically as

$$\text{specific gravity} = \frac{\text{density of object}}{\text{density of water}}$$

> **Example 1-3**   The density of mercury is 13.6 g/mL. What is its specific gravity?
>
> $$\text{specific gravity} = \frac{\text{density of object}}{\text{density of water}} = \frac{13.6 \; g/mL}{1 \; g/mL} = 13.6$$
>
> Note that the specific gravity and the density (in the metric system) have the same number, but density has units whereas specific gravity does not. Thus, if an object has a density of 1.5 g/mL, its specific gravity will be 1.5; likewise, an object of specific gravity 0.75 will have a density of 0.75 g/mL.

The specific gravity of normal human urine ranges from 1.003 to 1.030. Values outside this range are of diagnostic value. For example, the urine of a person suffering from diabetes mellitus contains large amounts of sugar and thus will have a specific gravity greater than normal. Conversely, a person suffering from diabetes insipidus will have a urine specific gravity close to 1.000 because of the large amount of water being excreted (see page 526).

Temperature Scales

Temperature is a measure of the availability of heat or cold or, in simpler terms, of how hot or cold a substance is. It can be measured by means of a thermometer. When we speak of body temperature as being 98.6 degrees or when we say that the outside temperature is 80 degrees or 40 degrees or even 2 degrees below zero, we are usually speaking in terms of degrees Fahrenheit (°F) even though we may not express the unit itself. In chemistry, the Celsius (C) and Kelvin (K) temperature scales are commonly used. Let us see first how the Celsius and Fahrenheit temperature scales compare with one another.

If a Fahrenheit and a Celsius thermometer are placed in a mixture of ice and water, the reading on the Fahrenheit scale will be 32° and the reading on the Celsius scale 0° (see Figure 1-4)—that is, 32 °F corresponds to 0 °C. These temperatures indicate the freezing point of water. They also indicate the melting point of ice.

Next, if the same two thermometers are placed in boiling water at 1 atmosphere pressure, the reading on the Fahrenheit scale will be 212° and the reading on the Celsius scale will be 100°. These temperatures indicate the boiling point of water.

Thus, there is a 180-degree difference between the boiling and the freezing points of water on the Fahrenheit scale (212° to 32°) and a 100-degree difference between the boiling and freezing points of water on the Celsius scale (100° to 0°). The difference between the freezing points of water on the Fahrenheit and Celsius scales is 32° (32° − 0°).

This information can be combined into the following formulas.

$$°F = \frac{9}{5}°C + 32° \quad \text{and} \quad °C = \frac{5}{9}(°F - 32°)$$

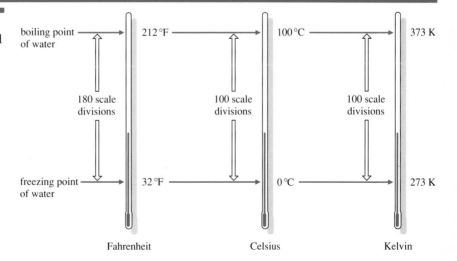

Figure 1-4 Comparison of Fahrenheit, Celsius, and Kelvin temperature scales.

The first formula is used to change Celsius temperatures to the corresponding Fahrenheit readings; the second formula is used to do the reverse, to change Fahrenheit readings to Celsius. Note that the first formula can be changed to the second by subtracting 32° from both sides and then dividing by $\frac{9}{5}$.

Example 1-4   Change 80 °C to °F.
Using the formula °F = $\frac{9}{5}$ °C + 32°,

$$°F = \tfrac{9}{5}(80°) + 32°$$
$$= 144° + 32°$$
$$= 176 \text{ °F}$$

Thus,

$$80 \text{ °C} = 176 \text{ °F}$$

Example 1-5   Change 50 °F to °C.
Using the formula °C = $\frac{5}{9}$(°F − 32),

$$°C = \tfrac{5}{9}(50° - 32°)$$
$$= \tfrac{5}{9}(18°)$$
$$= 10 \text{ °C}$$

Thus,

$$50 \text{ °F} = 10 \text{ °C}$$

A Kelvin thermometer will indicate the freezing point of water as 273 K and the boiling point of water as 373 K. Temperatures on the Kelvin scale are given as numbers only, without the degree sign. Why pick 273 K for the freezing point of water? Why not some even number? This temperature was selected so that zero on the Kelvin scale would represent the lowest temperature it is possible to reach. This temperature, 0 K, is also called **absolute zero.**

There are 100 units between the boiling and freezing points of water on the Kelvin scale (373 K − 273 K), just as there are on the Celsius scale. The difference between the freezing points of water on the Kelvin and Celsius scales is 273 (273 − 0). This information can be combined into the following formula.

$$K = °C + 273$$

> *Example 1-6*   Change 37 °C to K.
> Using the formula K = °C + 273,
>
> $$K = 37 + 273$$
> $$= 310$$
>
> Thus,
>
> $$37\,°C = 310\,K$$

Summary

The SI units of length and mass (weight) are the meter (m) and the kilogram (kg). Common units of length, weight, and volume are the meter (m), the gram (g), and the liter (L), respectively. Prefixes in common use are micro (1/1,000,000), milli (1/1000), centi (1/100), deci (1/10), kilo (1000), and mega (1,000,000).

Density is defined as mass divided by volume, or $D = M/V$.

Specific gravity is defined as the density of a substance compared to the density of water.

Although Fahrenheit and Celsius temperature scales are in common use, the Celsius (and Kelvin) temperature scales are the ones used almost exclusively in scientific measurements.

To convert Fahrenheit temperatures to Celsius, use the formula $°C = \frac{5}{9}(°F - 32°)$. To convert Celsius temperature to Fahrenheit, use the formula $°F = \frac{9}{5}°C + 32°$. To convert Celsius temperature to Kelvin temperature, use the formula $K = °C + 273$.

Questions and Problems

A

1. Body temperature is 98.6 °F. What is it in Celsius?
2. Which patient has a higher fever, one with a temperature of 101 °F or one with a temperature of 38.2 °C?
3. At the top of a certain mountain, water boils at 85 °C. What is the Fahrenheit boiling point? the Kelvin?
4. If the water in a pressure cooker boils at 250 °F, what will be the Celsius boiling point? the Kelvin?
5. Compare the following units: mL and cm^3.
6. What is the SI standard of length? the English?
7. What is the SI standard of mass? temperature?
8. What do the following prefixes indicate: milli-, deci-, kilo-, centi-?
9. What do the following abbreviations indicate: g, cm, mm, kL, m?

10. Convert the following:
    (a) 87.6 mm = _____ m
    (b) 2.5 L = _____ mL
    (c) 375 mL = _____ L
    (d) 1.76 g = _____ mg
    (e) 30.6 m = _____ dm
    (f) 394 cg = _____ g
    (g) 16.4 cm = _____ m
    (h) 0.54 kL = _____ L
    (i) 1.02 g = _____ kg
    (j) 500 cm^3 = _____ mL
11. Convert the following (answers to one decimal place):
    (a) 20 in. = _____ cm
    (b) 22 lb = _____ kg
    (c) 19.4 kg = _____ lb
    (d) 63.5 cm = _____ in.
    (e) 10 cm = _____ in.
    (f) 50 kg = _____ lb
    (g) 5 ft 4 in. = _____ cm
    (h) 176 lb = _____ kg

B

12. A patient is 5 ft 10 in. tall and weighs 160 lb. What is his height in centimeters and his weight in kilograms?

13. The following information was recorded for a patient: height 152.5 cm, weight 68 kg, temperature 37.2 °C. What would be the readings in inches, pounds, and degrees Fahrenheit?

14. What is the density of a medication if the contents of a filled 2-mL syringe weigh 2.50 g?

15. Normal urine has a density between 1.003 g/mL and 1.030 g/mL. Assuming a value of 1.020 g/mL, what will be the weight of a 250-mL sample of urine?

16. An object has a specific gravity of 3.06. What is its density in the metric system?

17. Mercury has a density of 13.6 g/mL. What volume will 200 g of it occupy?

18. An order for a medication reads: "Give 1.5 mg per kilogram of body weight." How much medication should be given to a patient weighing 165 lb?

19. What do the following mean in terms of known units: kilowatts, milliseconds, megacuries, microamperes?

20. A baby weighs 7000 g. How many kilograms does the baby weigh? How many pounds?

21. A rectangular object is 25 cm tall, 15 cm wide, and 10 cm deep. What is its volume in cubic centimeters, liters, and kiloliters?

22. Using Figure 1-3, what is the specific gravity of lead? salt?

23. Compare weight and mass.

24. An order for a medication reads: "Give 0.04 mL per kilogram of body weight." How much medication should be given to a patient weighing 110 lb?

25. An order for a medication reads: "Give 0.6 mg per kilogram of body weight." How much medication should be given to a patient weighing 131 lb? Give the answer to the closest whole number of milligrams.

Practice Test

1. 67.3 mL = _____ cm^3
   a. 0.673   b. 6.73
   c. 67.3   d. 673

2. 50 μg = _____ mg
   a. 0.050   b. 0.50
   c. 5.0   d. 50

3. 125 m= _____ cm
   a. 1.25   b. 12.5
   c. 1250   d. 12,500

4. 38 °C = _____ K
   a. 102   b. 235   c. 311   d. 380

5. An object of mass 100 g has a volume of 40 mL. Its density is _____ g/mL.
   a. 0.4   b. 2.5   c. 60   d. 140

6. An object has a specific gravity of 0.75 and a mass of 200 g. Its volume in milliliters is _____ .
   a. 150   b. 267   c. 275   d. 375

7. 95 °F = _____°C
   a. 35   b. 35.6   c. 36   d. 36.6

8. 165 lb = _____ kg
   a. 66   b. 75   c. 325   d. 363

9. 5'10" = _____ cm
   a. 24   b. 125   c. 154   d. 178

10. What does the prefix mega- mean?
    a. 1000   b. 1,000,000
    c. 1,000,000,000   d. 1,000,000,000,000

2

Properties of Matter

Liquid nitrogen (boiling point −196 °C) being poured from
a flask. Note that liquid nitrogen does not boil away
immediately.

What Is Matter?

Matter is anything that occupies space and has weight. Everything we see or feel is matter, as well as many things we cannot see or feel. Such things as trees, food, machinery, and soil are examples of matter we can see and feel. Air is an example of matter we cannot see, yet we know that it is all around us. Not all matter is of the same type. Matter can be classified as solid, liquid, or gas.

States of Matter

A piece of iron is an example of matter in the solid state. A bar of iron has a definite shape, a shape that cannot be easily changed. The volume of the piece of iron (the amount of space that it occupies) is a definite volume; it cannot easily be changed. Also, the piece of iron does not flow.

When a pint of water is poured from a container of one shape into a container of another shape, the water (a liquid) assumes the shape of the new container as far as it fills it. However, the volume of the water remains the same—1 pt—regardless of the shape of the container into which it is poured (see Figure 2-1).

When air is pumped into an empty bottle, the air occupies all of the space and also takes the shape of the container. Forcing more air into the bottle will increase the pressure, but the air will still occupy all of the space and take the shape of the container. If the air is transferred to another bottle of different shape and size, again the air will occupy all of the space and take the shape of the container.

In most solids, the particles are closely adhering and tightly packed in a highly ordered system. The motion of the particles is highly restricted (Figure 2-2a). These factors account for the fact that solids have a definite shape and resist changes in that shape. Tight packing also accounts for the incompressibility of solids. Heating allows the particles in a solid to move about slightly. Therefore, most solids expand on heating.

In liquids, the particles are moderately ordered and farther apart (but still in contact with each other) than the particles in solids (see Figure 2-2b). This loose structure accounts for the fact that liquids flow and have no definite shape. The particles in a liquid are close enough to resist compression. Liquids expand slightly on heating.

In gases, the motion of the particles is unrestricted so that the particles are independent of one another and are relatively very far apart (see Figure 2-2c). This independence allows gases to assume any shape or volume, to be compressed or expanded.

Some solids have a high density (gold, 19.3 g/mL), whereas others have a low density (cork, 0.2 g/mL). Many liquids have a relatively low density (water, 1 g/mL; gasoline, 0.66 g/mL), but some such as mercury (13.6 g/mL) have a high density. All gases

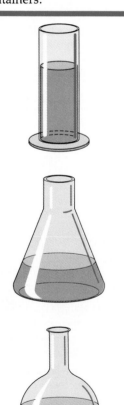

Figure 2-1 The same quantity (1 pint or approximately 500 mL) of water in variously shaped containers.

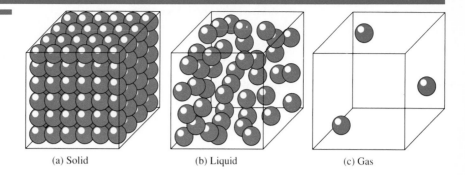

Figure 2-2 Rough representation of the particles of matter in (a) solid, (b) liquid, and (c) gas.

(a) Solid   (b) Liquid   (c) Gas

have a very low density, expressed in the units grams per liter (g/L) (air, 1.3 g/L; hydrogen, 0.09 g/L). In summary:

1. Solids have a definite shape and a definite volume, do not flow, and have particles that are closely adhering (interacting) and tightly packed in a highly ordered system. The motion of the particles in solids is highly restricted so that solids generally are incompressible. Most solids expand slightly when heated and may exhibit high or low density.
2. Liquids have no definite shape, do have a definite volume, flow, have particles that are relatively close to one another and are moderately ordered with some interaction, and may have high or low density. Liquids are incompressible and expand slightly when heated.
3. Gases have no definite shape and no definite volume, flow, and have particles whose motion is unrestricted so that they are independent and relatively far apart. Gases have a low density, are highly compressible, and expand greatly when heated.

Changes of State

Consider a piece of solid material. It consists of particles in a definite arrangement, as indicated in the preceding section. These particles are not motionless—they vibrate about fixed points. As heat is added to the solid, the temperature rises (see Figure 2-3, A). Because of the increased temperature, the rate of vibration increases and the particles move slightly farther apart. Eventually, a point is reached when the vibrating particles can no longer retain their orderly arrangement. When this occurs, the solid begins to melt and becomes a liquid. The temperature at which this occurs is called the **melting point** of the solid. As more heat is added, more of the solid melts. All of the heat being added is used to change the state from solid to liquid. This is indicated by plateau B in Figure 2-3, showing that there is a constant temperature while the solid is changing to a liquid (melting).

Figure 2-3 Temperature-time diagram. Section A indicates heating of the solid; section B, a change of state from solid to liquid; section C, heating of the liquid; section D, a change of state from liquid to gas; section E, heating of the gas.

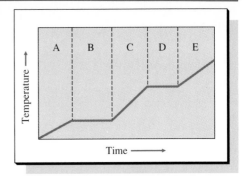

The number of calories required to change 1 g of solid to liquid at the melting point is called the **heat of fusion.** The heat of fusion of ice is 80 cal/g.

After all of the solid has melted, additional heat causes the temperature of the liquid to rise (Figure 2-3, C). The particles move faster and also move farther apart. Soon a point is reached at which the particles move far enough apart to become independent of one another. When this occurs, the liquid changes to a gas; it boils. The temperature at which this occurs is called the **boiling point** of the liquid. As more heat is added, more of the liquid boils, but all of the heat being added is used to change the state from liquid to gas. Thus the temperature remains constant during boiling (Figure 2-3, D).

The number of calories required to change 1 g of liquid to gas at the boiling point is called the **heat of vaporization.** The heat of vaporization of water is 540 cal/g. Water has a very high heat of vaporization. Steam will therefore cause a much more severe burn than boiling water because of its much higher heat content. Once all the liquid has boiled, additional heat increases the temperature of the gas (Figure 2-3, E).

The reverse steps are also possible. A gas can be cooled to become a liquid. Further cooling of that liquid changes it into a solid.

In general, then, matter can usually be changed from one state to another merely by changing its temperature. Such changes are called **physical changes.** A physical change is one in which no new substance is produced, although there may be a change of state or density, or both.

However, not all substances can be changed from solid to liquid to gas merely by changing the temperature. For example, a piece of wood bursts into flame when heated sufficiently; it does not become liquid. Dry ice changes from solid to gas without becoming a liquid.

Dissolving sugar in water, breaking a sheet of glass, and freezing water also bring about physical changes. No new substance is produced, and each substance retains its own properties. Note that when water freezes there is a change in both state and density, since the density of water is 1.00 g/mL and that of ice 0.917 g/mL.

The other type of change that matter can undergo is called a **chemical change.** A chemical change is one in which one or more substances disappear and a new substance or new substances are formed. The new substance or substances have entirely different properties from those of the original substance or substances.

Burning a piece of wood causes a chemical change. New substances—ash and smoke—are produced. These substances have properties different from those of the original piece of wood. Chemical changes are also involved in cooking food and in digestion.

Properties of Matter

One portion of matter can be distinguished from another by means of its properties. These distinguishing properties of matter can be classified into two main types: **physical properties** and **chemical properties.**

Physical Properties

Physical properties include color, odor, taste, solubility in water, density, hardness, melting point, and boiling point. These physical properties can serve to identify a substance, although not all of these properties may be necessary for the identification. For example, when we say that the color of a substance is white, we automatically eliminate all substances that are not white. Next, if we say that the white substance is odorless, we can eliminate all white objects that have an odor, leaving a smaller number of substances that are both white and odorless. If we continue to eliminate in this manner by using additional physical properties, such as density, hardness, melting point, and boiling point, eventually only one substance will fit all of these properties—the substance we are trying to identify.

Chemical Properties

Properties such as reacting (or not reacting) in air, reacting (or not reacting) with an acid, or burning (or not burning) in a flame are chemical properties. A substance can be identified by means of its chemical properties, but it is usually much simpler to do so by means of the physical properties.

Comparison of Physical and Chemical Properties

A physical property tells what a substance *is*—it is white or it is green; it is odorless or it has a sharp odor; it is hard or it is soft. A chemical property tells what a substance *does*—it burns or it does not burn; it reacts with an acid or it does not react with an acid; and so on.

Energy

Energy is defined as the ability to do work. The muscles in our bodies get their energy from the chemical reactions that take place in the muscle cells. The heat energy necessary to keep our bodies at a temperature of 98.6 °F or 37 °C comes from the oxidation of the foods we eat. The electric energy we use in our homes comes from burning a fuel or from atomic energy. Energy exists in several forms—heat, light, electric, mechanical, sound, chemical, and atomic. Energy can be classified into two categories: **kinetic energy** and **potential energy**.

Kinetic Energy

Kinetic energy is energy of motion; that is, energy that is doing something now, such as heat energy obtained from burning wood, light energy from an incandescent lamp, mechanical energy from a motor, and atomic energy from a nuclear reactor.

Potential Energy

Potential energy is stored energy, energy not associated with motion. Examples of potential energy are a dry cell (which can supply electric energy when it is connected to something), food (which supplies energy to our bodies when it is metabolized), and water at the top of a waterfall (which becomes kinetic energy that can supply mechanical energy as it falls to the bottom).

Chemical energy is a form of potential energy. Most chemical reactions involve changes in heat energy. If heat is given off during a chemical reaction, that reaction is said to be **exothermic**. If heat is absorbed during a chemical reaction, that reaction is said to be **endothermic**.

Transformation of Energy

Energy can be transformed from one form into another. Thus, burning a piece of coal changes its potential energy into heat (kinetic) energy. The heat energy thus produced might be used to boil water, which produces large amounts of steam. The steam might be used to drive a generator to produce electric energy. In turn, this electric energy might be used to drive a motor (mechanical energy), produce light in a fluorescent lamp (light energy), operate a radio (sound energy), or operate a toaster (heat energy).

The sun produces energy by nuclear reactions and radiates this energy to the Earth. Plants on the Earth pick up the light energy from the sun during the process of photosynthesis and produce compounds that contain chemical energy. When humans eat these compounds in the plants, their bodies convert the chemical energy into heat energy and mechanical energy.

Conservation of Energy and Matter

The **law of conservation of energy** states that energy is neither created nor destroyed during a chemical reaction. Energy can be changed from one form to another, but the total amount of energy remains the same regardless of what form the energy is changed into.

The **law of conservation of matter** states that during a chemical reaction matter is neither created nor destroyed. This means that the total weights of substances before they react should be the same as the weights of the products after the reaction. For example, if a candle is placed in a sealed container, a certain weight will be obtained for both the candle and the container. If the candle is lit and allowed to burn (inside the sealed container) and if it is weighed as it continues to burn, the weight will be found to remain the same even though part of the candle is disappearing and several gaseous products are being produced. The same is true for a camera flashbulb before and after it is fired. Other experiments performed under very carefully controlled conditions have produced similar results—that is, the sum of the weights after a reaction is the same as the sum of the weights before the reaction.

In the early twentieth century Albert Einstein stated that matter and energy were interchangeable. That is, under certain conditions, matter could be changed into energy or energy into matter. These changes do not occur under the conditions of an ordinary chemical reaction, so that the laws of conservation of energy and conservation of matter are still used. However, these laws can be combined into one overall law, which states that matter and energy cannot be created or destroyed but they can be converted from one to the other.

Measurement of Energy

Heat is the most common form of energy; all other forms of energy can be converted into heat energy. The unit of heat energy is the **calorie,** which is defined as the amount of heat required to raise the temperature of 1 g of water one degree Celsius. The calorie is abbreviated as cal. It is a rather small unit of heat.

A larger unit of heat, the kilocalorie,[†] is equal to 1000 cal. The kilocalorie is abbreviated as kcal. The kilocalorie is used when measuring the heat energy of the body and for nutritional values of foods.

Another unit of heat is the joule (abbreviated J); 1 cal equals 4.18 J. Calories are used primarily in medical work, whereas joules (and kilojoules) are used in chemical work.

[†]   In the past people in the field of nutrition used the words "large Calorie," which they abbreviated Cal for the kilocalorie. This usage is confusing and now obsolete. The word kilocalorie states clearly what the measurement is—1000 cal—and cannot be confused with the calorie by an oversight in capitalization.

Table 2-1 Caloric Values of Some Foods

Food	Portion	Kilocalories
Milk, whole	1 cup	160
Cheese, cheddar	1 ounce	115
Hamburger, broiled	3 ounces	245
Almonds, shelled, whole kernels	1 cup	850
Carrots, cooked	1 cup	91
Potato chips, 2-in. diameter	10 chips	115
Orange juice, fresh	1 cup	115
Bread, white, toasted	1 slice	70
Brownies, with nuts	1 brownie	95
Cola drink	12 ounces	145

There are three principal kinds of foods that produce energy in the body: carbohydrates, fats, and proteins. The oxidation of 1 g of carbohydrate produces 4 kcal, the oxidation of 1 g of fat produces 9 kcal, and that of 1 g of protein produces 4 kcal. Table 2-1 lists the caloric values of some common foods.

The number of calories produced by a chemical reaction can be calculated in terms of the amount of water that can be heated from one temperature to another by using the formula

$$\text{number of calories} = \text{number of grams of water}$$
$$\times \text{ change in temperature in } °C$$
$$\times \text{ specific heat of water}$$

where the specific heat of water is 1 cal/g·°C.

Example 2-1 How many calories are produced during oxidation (burning) of a piece of food if the heat is sufficient to warm 2000 g of water from 20 °C to 38 °C?

$$\text{number of calories} = \text{number of grams of water}$$
$$\times \text{ change in temperature in } °C$$
$$\times \text{ specific heat of water}$$

$$= 2000 \text{ g} \times (38 - 20)°C \times 1 \frac{\text{cal}}{\text{g} \cdot °C}$$

$$= 36,000 \text{ cal} = 36 \text{ kcal}$$

Example 2-2 How much carbohydrate must be oxidized in the body to produce 36 kcal? How much fat? How much protein?

Since the oxidation of 1 g of carbohydrate produces 4 kcal, it will require $\frac{36}{4}$ g, or 9 g, of carbohydrate to produce 36 kcal.

Since the oxidation of 1 g of fat produces 9 kcal, it will require $\frac{36}{9}$ g or 4 g of fat.

Since the oxidation of 1 g of protein produces 4 kcal, as did carbohydrate, it will require 9 g of protein.

Composition of Matter

All matter can be divided into three classes, depending upon the properties of the material being considered. These three classes are elements, compounds, and mixtures.

Elements

Elements are the building blocks of all matter. An element can be defined as a substance that cannot be broken down into any simpler substance by ordinary chemical means. There are more than 100 known elements. The names of some of these elements, such as oxygen, hydrogen, iron, carbon, copper, mercury, and uranium, are probably familiar. The names of some of the rarer elements, such as cesium, einsteinium, molybdenum, and xenon, may not be familiar. Many elements are in common industrial use. Consider iron in steel, germanium in transistors, and tungsten in incandescent light bulbs.

Because an element cannot be broken down into anything simpler by ordinary chemical means, it must contain only one type of substance.

Elements can be classified into two main types: metals and nonmetals. Each has its own specific properties.

Classification

Metals conduct heat and electricity. They have a luster (a shiny surface) similar to that of silver and aluminum. Metals reflect light. Some metals are ductile (they can be drawn out into a thin wire). Some metals are malleable (they can be pounded out into thin sheets). Some metals have a high tensile strength. Metals such as iron, copper, and silver are solid at room temperature, but mercury is a liquid.

Nonmetals usually do not conduct heat and electricity very well. They have little luster and seldom reflect light. Nonmetals frequently are brittle. They cannot be pounded into thin sheets (they are not malleable) and cannot be drawn into thin wires (they are not ductile). Nonmetals may be solids at room temperature (carbon and sulfur), liquid at room temperature (bromine), or gaseous at room temperature (oxygen and nitrogen). There are many more metallic elements than nonmetallic; see the periodic table inside the front cover. Among the nonmetals are the noble gases, which are gases at room temperature. They are relatively unreactive and were formerly

Table 2-2   Symbols for Some Elements

Symbol Using First Letter of Name	Symbol Using First Two Letters of Name	Symbol Using First Letter and One Other Letter in the Name	Symbol based on Latin Name
C (carbon)	Ca (calcium)	Cl (chlorine)	Cu (copper—cuprum)
H (hydrogen)	Al (aluminum)	Zn (zinc)	Na (sodium—natrium)
O (oxygen)	Ni (nickel)	Mg (magnesium)	Ag (silver—argentum)
N (nitrogen)	Ne (neon)	As (arsenic)	Au (gold—aurum)
I (iodine)	Br (bromine)	Mn (manganese)	Sn (tin—stannum)
S (sulfur)	Si (silicon)	Cr (chromium)	Fe (iron—ferrum)
P (phosphorus)	Co (cobalt)	Cd (cadmium)	Pb (lead—plumbum)

called inert or rare gases. Among the noble gases are the elements helium, neon, and krypton.

Symbols for the Elements

Each element can be identified by a symbol that represents that element. The symbol C stands for the element carbon, S for sulfur, and O for oxygen. In these instances, the symbol is the first letter of the name of the element. Because conflicts would arise if the first letter were to be used for every element, the first two letters are used for some elements, such as Ca for calcium and Al for aluminum. Or the symbol may use the first letter and one other letter to suggest a sound that is apparent in the name, such as Zn for zinc and Cl for chlorine. Note that when two letters are used to form a symbol, the first letter is capitalized and the second letter is not.

The symbols for some elements are based upon their Latin names. Ag, the symbol for silver, comes from the Latin word **argentum.** Fe, the symbol for iron, comes from the Latin word **ferrum.** Table 2-2 lists the symbols and names of various elements based on these categories.

Elements Present in the Human Body

Table 2-3 lists the elements necessary for life, the symbols for these elements, and their functions in the body.

Compounds

When an electric current is passed through a container of water, the water is decomposed into two gases, each having its own set of properties. One gas is hydrogen and the other is oxygen (see Figure 2-4). If sugar is placed in a test tube and heated strongly, drops of water will collect at the top of the test tube while pieces of carbon will remain at the bottom. Both of these changes are chemical changes; that is, the original substance has disappeared and some new substances have been formed. Both sugar and water are classified as compounds, as are such other substances as salt, boric acid,

Figure 2-4   Electrolysis of water.

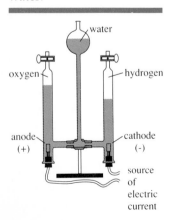

water

oxygen

hydrogen

anode (+)

cathode (-)

source of electric current

Table 2-3 Elements of Life

Element	Symbol	Function
Major components of molecules found in humans		
Oxygen	O	Required for water and organic compounds
Carbon	C	Required for organic compounds
Hydrogen	H	Required for water and organic compounds
Nitrogen	N	Required for many organic compounds, and for all proteins
Sulfur	S	Required for some proteins and some organic compounds
Nutritionally important elements required in amounts greater than 100 mg/day		
Calcium	Ca	Required for bones and teeth; necessary for certain enzymes, nerve muscle function, hormonal action, cellular motility, and clotting of the blood
Phosphorus	P	Required for bones and teeth; necessary for high-energy compounds, nucleoproteins, nucleic acids, phospholipids, and some proteins
Magnesium	Mg	Required for many enzymes; necessary for energy reactions requiring ATP
Sodium	Na	Principal positive extracellular ion
Potassium	K	Principal positive intracellular ion
Chlorine	Cl	Principal negative ion
Trace elements necessary for humans		
Iodine	I	Required for thyroid hormones
Fluorine	F	Required for bones and teeth; inhibitor of certain enzymes
Iron	Fe	Required for hemoglobin and many enzymes
Copper	Cu	Required for many oxidative enzymes, for the synthesis of hemoglobin, and for normal bone formation
Zinc	Zn	Required for many enzymes; related to action of insulin; essential for normal growth and reproduction and for nucleic acid metabolism
Manganese	Mn	Required for some enzymes acting in the mitochondria; essential for normal bone structure, reproduction, and normal functioning of central nervous system
Cobalt	Co	Required for vitamin B_{12}
Molybdenum	Mo	Required for some enzymes; essential for purine metabolism
Chromium	Cr	Related to action of insulin
Selenium	Se	Essential for growth and fertility of animals; closely related to action of vitamin E
Trace elements required for animal nutrition but having no known essential function in humans		
Arsenic	As	
Cadmium	Cd	
Nickel	Ni	
Silicon	Si	
Tin	Sn	
Vanadium	V	

carbon dioxide, and ether. Both water and sugar can be broken down into other substances by chemical means. Compounds, then, are substances that can be broken down into simpler substances by chemical means. Compare this definition with that of elements, which cannot be broken down into simpler substances by ordinary means.

Take a crystal of sugar from a sugar bowl. Examine it carefully. List its physical and chemical properties. Take another crystal of sugar from the sugar bowl and again list its physical and chemical properties. The properties are identical, as are the pieces of sugar. That is, the sugar in a bowl of sugar is homogeneous; it is the same throughout.

If a sample of sugar is analyzed, it will be found to contain carbon, hydrogen, and oxygen. The percentage of carbon in a portion of sugar can be calculated by weighing the piece of sugar as it is and again after heating it until nothing is left but carbon. Repeating this procedure with another portion of sugar will yield the same percentage of carbon. Likewise, analysis of several portions of sugar would give identical results for the percentage of oxygen and for the percentage of hydrogen. The analysis of many samples of water gives a constant percentage of oxygen and hydrogen. The result of examining many different compounds for the percentages of their components can be stated as the law of definite proportions—compounds have a definite proportion or percentage by weight of the substances from which they were made.

It has already been mentioned that compounds have properties that are entirely different from those of the substances from which they were made. That is, water, a compound, has entirely different properties from the oxygen and hydrogen it is composed of.

Compounds, then, have the following characteristics:

1. They can be separated into their component substances by chemical means.
2. They are homogeneous in composition.
3. They have a definite proportion by weight of the substances from which they were made.
4. They have different properties from those of the substances from which they were made.

Mixtures

If a few crystals of salt are dissolved in a cup full of water, a mixture of salt and water—or salt water—is produced. If a few more crystals of salt are added to that same salt water, a little stronger salt solution is produced. If a teaspoonful of salt is added to the cup and stirred until it is dissolved, an even stronger salt solution will be produced. The salt and water form a mixture when they are placed together and stirred. The salt waters made by dissolving different amounts of salt in water will have different compositions, depending upon how much salt was added to the water. Thus, one property of a mixture is that it can have a variable composition or variable proportions. More water, more salt, or both salt and water can be added to change the strength of the mixture, but in each case a salt water is produced.

The property of variable proportions is characteristic of any mixture. If sugar and sand are stirred together, a mixture is produced regardless of the amounts of each used. Likewise, sugar and iron filings can be mixed, and again, the proportion of iron filings to sugar does not alter the fact that it is a mixture.

The ingredients of any mixture can be separated from each other by such physical processes as evaporation and filtering. Salt can be separated from a salt–water mixture merely by evaporating the water. The sugar–sand mixture can be separated by placing it in water, stirring to dissolve the sugar, filtering the solution, and recovering the sand. The sugar can then be recovered from the filtered solution by evaporating the water. A sugar–iron filings mixture can be separated by passing a magnet over the mixture. The magnet will attract the iron filings, leaving the sugar behind. Mixtures can also be separated by a process known as chromatography (see page 424).

Evaporation, chromatography, and separation by means of a magnet are examples of a physical change—one in which no new substance is produced. In each of the mixtures, the individual substances retain their own properties. There is no evidence of a chemical reaction because no new substance is produced. Thus, in the salt–water mixture, the salt retains its own properties, as does the water. In the sugar–sand mixture each retains its own properties. The sugar and the sand can be recognized separately under a microscope.

In summary, then, mixtures have the following characteristics.

1. They have no definite proportion or composition.
2. They can be separated into their component substances by physical means.
3. They retain the properties of the individual substances from which they were made.

Mixtures can be either homogeneous or heterogeneous in composition. **Homogeneous** means the same composition throughout. **Heterogeneous** means different composition throughout. Let us consider two of the mixtures already discussed, the salt–water and the sugar–sand mixtures.

Samples of a salt–water mixture are found to be identical from the top, from the middle, and from the bottom of a container of salt water. That is, the salt–water mixture is homogeneous. It is of the same composition throughout.

Now consider a mixture of sugar and sand. Samples from different parts of the mixture, when examined carefully under a microscope or with a magnifying glass, would not appear to have the same composition. That is, a mixture of sugar and sand is heterogeneous.

A mixture of two or more solids is heterogeneous. One solid can be distinguished from another, even if they are ground together in making the mixture. On the other hand, a solution is a homogeneous mixture. When a substance is dissolved in a liquid such as water, the mixture becomes the same throughout; it is homogeneous.

Although both the salt–water and sugar–sand mixtures contain two different substances, mixtures can be made from any number of substances. Note the word *substances*. It can mean either elements or compounds. That is, a mixture can be composed of two (or more) elements (powdered iron and powdered sulfur), two (or more) compounds (sugar and salt), or both elements and compounds (iodine and water).

Summary

Matter is anything that occupies space and has weight. Matter can exist in the solid, the liquid, or the gaseous state. These states of matter are interchangeable and depend primarily upon temperature. Physical properties of matter describe what a substance is; chemical properties describe what a substance does.

Energy is the ability to do work. The two types of energy are kinetic, the energy of motion, and potential or stored energy. The law of conservation of energy states that energy can be transformed from one type to another but is not created or destroyed. The law of conservation of matter states that matter is not created or destroyed. These two laws can be combined into one stating that matter and energy are not created or destroyed but can be converted from one to the other.

The calorie is the unit of heat energy and is the amount of heat required to raise the temperature of 1 g of water 1 °C. One kilocalorie (kcal) equals 1000 cal.

Matter can be divided into three categories: elements, compounds, and mixtures.

Elements are substances that cannot be broken down into simpler substances by ordinary means. Elements are homogeneous in composition. Elements are either metals or nonmetals. Each element can be represented by a symbol.

Compounds can be separated into their component substances by chemical means, are homogeneous in composition, have a definite proportion by weight of their component substances, and have properties different from those of the substances from which they were made.

Mixtures have no definite proportions or composition, can be separated into their component substances by physical means, and retain the properties of the individual substances from which they were made. Mixtures can be made from two or more substances. Mixtures of solids are heterogeneous; mixtures made by dissolving a substance in a liquid are homogeneous in composition.

Questions and Problems

A

1. Compare the properties of solids, liquids, and gases as to
   (a) Definite shape
   (b) Definite volume
   (c) Density
   (d) Motion of particles
   (e) Compressibility
   (f) Expansion upon heating
2. Which of the following properties are physical, and which are chemical?
   (a) Odor   (b) Reactivity
   (c) Taste   (d) Boiling point
   (e) Flammability   (f) Density
   (g) Color   (h) Melting point
3. Define heat of vaporization; heat of fusion.
4. Which of the following substances are elements, which are compounds, and which are mixtures?
   (a) Milk   (b) Zinc
   (c) Mercury   (d) Paint
   (e) Carbon   (f) Water
   (g) Air   (h) Table salt
5. Distinguish between potential energy and kinetic energy.
6. State the laws of conservation of energy, conservation of matter, conservation of matter and energy.
7. How many calories are required to change the temperature of 500 g of water from 15 °C to 27 °C? How many grams of protein must be oxidized to produce this energy?
8. Which of the following processes involve a physical change, and which involve a chemical change?
   (a) Breaking glass
   (b) Burning wood
   (c) Boiling water
   (d) Digesting food
   (e) Distilling mercury
   (f) Electrolysis of water
   (g) Winding a clock
   (h) Souring of milk
9. Which *must* be homogeneous: elements, compounds, or mixtures? Which *may* be?
10. What elements do the following symbols represent?
    (a) O   (b) N   (c) Fe
    (d) K   (e) Na   (f) Cl
    (g) H   (h) F   (i) S
    (j) Ag   (k) Al   (l) Cr

11. What are the symbols for the following elements?
    (a) Magnesium   (b) Calcium
    (c) Iodine   (d) Carbon
    (e) Phosphorus   (f) Zinc
    (g) Copper   (h) Bromine
    (i) Mercury   (j) Boron
    (k) Silver   (l) Manganese
12. Why are gases compressible?
13. Can all substances be changed from solid to liquid to gas? Explain.
14. Which particular element is required for each of the following?
    (a) Clotting of the blood
    (b) Thyroid hormones
    (c) Formation of all protein
    (d) Vitamin B_{12}
15. Which element(s) are required for the following?
    (a) Action of insulin
    (b) Formation of bones and teeth
    (c) Synthesis of hemoglobin

B

16. Why does the temperature of a liquid remain constant while the liquid is boiling?
17. What happens to the particles when a solid melts?
18. What is an exothermic reaction? Give an example.
19. Why do gases and liquids flow whereas solids do not?
20. Why do solids have a definite shape whereas liquids and gases do not?
21. Which is larger, a joule or a calorie?
22. Which expands more when heated—a solid, liquid, or gas? Why?
23. Compare the properties of metals and nonmetals.

Practice Test

1. An example of a physical property is _____.
   a. height   b. weight
   c. color   d. reactivity
2. An example of a chemical property is _____.
   a. odor   b. density
   c. temperature   d. reactivity
3. Gases have _____.
   a. definite volume   b. definite shape
   c. compressibility   d. slow-moving particles

4. An example of an element is _____.
   a. carbon   b. water
   c. mercuric sulfide   d. sand
5. An example of a mixture is _____.
   a. water   b. air
   c. iron   d. zinc oxide
6. The number of calories required to raise the temperature of 200 g of water from 18 °C to 25 °C is _____.
   a. 218   b. 225   c. 1400   d. 3600

7. The symbol for the element potassium is _____.
   a. K   b. P   c. Fe   d. Po
8. Which is the symbol for magnesium?
   a. Mg   b. Mn   c. Ma   d. M
9. The element necessary for production of thyroid hormones is _____.
   a. F   b. Fe   c. I   d. Ca
10. Which must be homogeneous?
   a. elements   b. compounds
   c. mixtures   d. both a and b

3

Structure of Matter

A ball and stick model represents the three-dimensional structure of a molecule.

The Atom

The symbol of an element not only represents that element, it also represents one atom of that element. But what is an atom? We have all heard of atoms in connection with atomic bombs, "splitting the atom," and atomic power.

Consider a bar of iron. Iron is an element. It has certain properties. Cutting the bar in half produces two pieces of iron. Both pieces have the same properties as the original bar. Continued cutting produces smaller and smaller pieces, all with identical properties. In time, we could theoretically arrive at the smallest piece of iron attainable. This smallest piece of iron is an atom—an atom of iron. If this atom of iron were cut in two, particles with different properties would be produced. It would no longer be iron. Thus, an **atom** can be defined as the smallest portion of an element that retains all of the properties of the element.

A piece of iron is made up of many atoms of iron; a piece of copper, of many atoms of copper; and a piece of silver, of many atoms of silver. The atoms of one element differ from those of another and so give characteristic properties of each element. Atoms are called building blocks of the universe. A chemist uses different kinds of atoms to build chemical compounds just as we all use the different letters of the alphabet to form words. Since there are more than 100 elements, there are more than 100 different kinds of atoms.

Size

Although an atom is extremely small, its size can be accurately measured. An atom has a diameter of approximately one hundred-millionth of a centimeter (1/100,000,000 cm). Because an atom is so small, one hundred trillion of them (100,000,000,000,000) could be placed on the head of a pin.

Weight

An atom is extremely small; therefore, it is not surprising that an atom weighs very little. In fact, it would take 18 million billion billion (18,000,000,000,000,000,000,000,000) hydrogen atoms to weigh 1 oz.

Inside the Atom

The early chemists believed that the atom was solid and indivisible, that it could not be broken down into any simpler substances. Modern theory states that the atom is composed of a small, heavy nucleus with particles surrounding it at relatively great distances. Thus an atom is composed mostly of empty space, that space being between the nucleus and the surrounding particles.

If the nucleus of an atom could be expanded so that it was about 400 ft in diameter, the closest surrounding particle would be

Table 3-1

Particle	Symbol	Charge	Approximate Mass (Weight), amu	Location in the Atom
Proton	p or p$^+$	+1	1	Inside nucleus
Electron	e or e$^-$	−1	1/1837	Outside nucleus
Neutron	n or n^0	0	1	Inside nucleus

4000 miles away. Between that nucleus and the closest particle would be empty space, 4000 miles of it. Actually, over 99.9 percent of the volume of an atom is empty space.

Fundamental Particles

Atoms are considered to be made primarily of three fundamental particles, the **proton,** the **electron,** and the **neutron.**[†] The proton (p) has a charge of positive one (+1) and mass (weight) of approximately one atomic mass unit (amu).[‡] Protons are located inside the nucleus of the atom. The electron (e) has a negative charge (−1) and a mass (weight) of 1/1837 amu. Electrons are located outside the nucleus of the atom. The neutron (n) has no charge; it is neutral, as the name implies. It has a mass (weight) of approximately 1 amu. Neutrons are located inside the nucleus (see Table 3-1). In addition to these three fundamental particles there are many, many more particles—among them the positron, the meson, and the neutrino—but a discussion of these is beyond the scope of this book.

Atomic Weight

One atom cannot be weighed even on the most sensitive weighing device. However, the weights of individual atoms can be determined accurately by weighing large numbers of them. As would be expected, the weights of individual atoms are infinitesimal. But the chemist is not as interested in the exact weights of atoms as in their relative weights. The relative weight of an atom is called its **atomic weight.** The chemist uses atomic weights rather than exact weights.

[†] Theoretical physicists now believe that electrons are fundamental particles but that protons and neutrons are made of even smaller particles called quarks. However, in this discussion we shall assume that protons, neutrons, and electrons are all fundamental particles.

[‡] One atomic mass unit (amu) has a mass (weight) of 1.6605×10^{-24} g. A proton has a mass of 1.6726×10^{-24} g or 1.0073 amu. A neutron has a mass of 1.67495×10^{-24} g or 1.0087 amu. An electron has a mass of 9.1095×10^{-28} g or 5.5×10^{-4} amu. The actual masses of the fundamental particles are cumbersome to deal with, so we use the amu. For simplicity, as indicated in Table 3-1, we will assume that the proton and the neutron each have a mass (weight) of 1 amu and that the electron has a negligible mass.

Atomic weights are easier to use than the exact weights, and they are just as accurate because they can be determined very precisely.

What does the term *relative weight* mean? The chemist has arbitrarily given the carbon-12 atom (see page 37) a weight of 12.0000 amu.[†] The weights of atoms of all other elements can then be compared with this weight. Thus, if an atom of an element is exactly twice as heavy as the carbon-12 atom, that element is assigned a relative weight, or atomic weight, of 24.0000. For our purposes, we will usually use the atomic weights as whole numbers, ignoring the decimal values. Thus, the atomic weight of carbon-12 is 12, that of oxygen is 16, and that of sodium is 23. For precise work, however, the exact atomic weights must be used.

The atomic weights of all the elements are listed in the periodic table inside the front cover and also in the chart inside the back cover.

Atomic Number

Each element has a given atomic number that represents that element and no other. The atomic number indicates the number of protons in the nucleus of an atom of that element. However, since all atoms are electrically neutral, there must be as many electrons (negative charges) as protons (positive charges). Therefore, the atomic number also tells the number of electrons in the atom, these electrons being located outside the nucleus.

The atomic number is written as a subscript to the left of the symbol. Thus, $_6C$ indicates that the carbon atom has an atomic number of 6.

Mass Number

The mass number of a nucleus is equal to the total number of protons and neutrons in that nucleus. The mass number is always a whole number.

The mass number is written as a superscript to the left of the symbol. Thus $^{12}_6C$ indicates a carbon atom with an atomic number of 6 and a mass number of 12.

Number of Neutrons in the Nucleus

Because the weight of the electron is quite small (1/1837 that of the proton or the neutron) and because only the electrons are located outside the nucleus of the atom, practically all the weight of the atom

[†]   In 1961, by international agreement, the atomic weights of the elements were based upon the carbon-12 atom having a relative weight of 12.0000 amu. Previously, the atomic weights had been based upon oxygen at 16.0000 amu.

is located in its nucleus. The atomic number of an element indicates the number of protons in its nucleus. Knowing that each proton weighs 1 amu, you can calculate the total weight of these protons. The rest of the weight of the atom must be due to the neutrons in the nucleus. Knowing that each neutron weighs 1 amu, you can calculate the number of neutrons present in the nucleus.

For example, if the atomic number of an element is 5, there must be five protons in the nucleus of the atom and also five electrons outside the nucleus. If the mass number of that element is 11, the number of neutrons can be calculated as follows: since the five protons weigh 5 amu and the whole atom weighs 11 amu, the neutrons in it must weigh 6 amu. Knowing that each neutron weighs 1 amu, the number of neutrons must be 6. **The number of neutrons can be found by subtracting the atomic number of an element from its mass number.**

Structure of the Atom

What does the atom look like? The simplest atom, the hydrogen atom, has the atomic number 1 and a mass number of 1. The atomic number indicates one proton inside the nucleus of this atom; it also indicates one electron outside that nucleus. The number of neutrons can be calculated by subtracting the atomic number (1) from the mass number (1); thus, there are no neutrons in the nucleus of a hydrogen atom. The hydrogen atom can be represented as follows.

$$\begin{pmatrix} 1 \text{ p} \\ 0 \text{ n} \end{pmatrix} \quad 1 \text{ e}$$

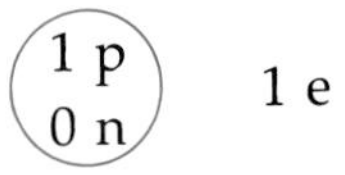

hydrogen, atomic number 1, mass number 1, $_1^1\text{H}$

The circle represents the nucleus of the atom, the letter p the protons, e the electrons, and n the neutrons.

The element helium—atomic number 2 and mass number 4—has two protons in the nucleus and two electrons outside that nucleus. The number of neutrons is two (atomic number subtracted from mass number). The helium atom can thus be represented as follows.

$$\begin{pmatrix} 2 \text{ p} \\ 2 \text{ n} \end{pmatrix} \quad 2 \text{ e}$$

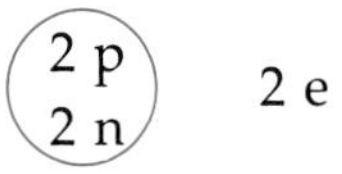

helium, atomic number 2, mass number 4, $_2^4\text{He}$

The sodium atom—atomic number 11 and mass number 23—has 11 protons in its nucleus, 11 electrons outside its nucleus, and 12 neutrons ($23 - 11$) in its nucleus.

sodium, atomic number 11, mass number 23, $^{23}_{11}$Na

The uranium atom—atomic number 92 and mass number 238—has in its nucleus 92 protons and 146 neutrons. Outside the nucleus are 92 electrons.

uranium, atomic number 92, mass number 238, $^{238}_{92}$U

Isotopes

The periodic chart at the front of the book indicates that the element chlorine has an atomic number of 17 and an atomic weight of 35.5 (to one decimal place). According to the discussion in the previous paragraphs, the chlorine atom should have 17 protons in its nucleus, 17 electrons outside its nucleus, and 18.5 neutrons (35.5 − 17) in its nucleus. However, a neutron is a fundamental particle so there can never be a fraction of a neutron. How can this problem be resolved?

The answer is that there are two types of chlorine atoms. One chlorine atom has a mass number of 35 and the other has a mass number of 37. (Chlorine with a mass number of 36 does not exist in nature.) Both of these types of atoms of chlorine have an atomic number of 17. These two varieties of chlorine are termed **isotopes.** Isotopes are defined as atoms of an element having the same atomic numbers but different mass numbers. The first isotope of chlorine—atomic number 17 and mass number 35—has 17 protons in its nucleus, 17 electrons outside its nucleus, and 18 neutrons (35 − 17) in its nucleus. The second isotope of chlorine—atomic number 17 and mass number 37—has 17 protons in its nucleus, 17 electrons outside its nucleus, and 20 neutrons (37 − 17) in its nucleus.

$^{35}_{17}$Cl

$^{37}_{17}$Cl

The atomic weight is the average weight of all the isotopes. If the two isotopes of chlorine, mass numbers 35 and 37, were present in equal amounts, the atomic (average) weight would be 36. However, since the atomic weight is listed as 35.5, the isotope of mass number 35 must be the predominant one because the atomic weight is closer to 35 than to 37.

The element carbon—atomic number 6—has three isotopes. Their mass numbers are 12, 13, and 14. They all have atomic number 6, which means that they all have six protons in their nucleus and six electrons outside their nucleus. The isotope of mass number 12 has six neutrons in its nucleus; the isotope of mass number 13 has seven neutrons in its nucleus; and the isotope of mass number 14 has eight neutrons in its nucleus. Carbon-12 is the most abundant since the atomic weight of carbon is 12.011, indicating small amounts of the other isotopes.

6 p 6 n	6 e	6 p 7 n	6 e	6 p 8 n	6 e
$^{12}_{6}C$		$^{13}_{6}C$		$^{14}_{6}C$	

three isotopes of carbon

Isotopes have been defined as atoms of an element having the same atomic number but different mass numbers; therefore isotopes of an element must have the same number of protons and electrons but different numbers of neutrons.

Most of the known elements have isotopes. Some have only two, whereas others have many more. In addition to the naturally occurring isotopes, there are many more artificially prepared isotopes. These isotopes will be discussed in the chapter on radioactivity (Chapter 4). In general, isotopes have identical chemical properties because they contain the same number of electrons as well as the same number of protons. However, isotopes have different physical properties.

Arrangement of the Electrons in the Atom

There is a definite order to the arrangement of electrons in atoms. The electrons are located in **energy levels.** An energy level represents a volume occupied by an electron cloud (Figure 3-1). For our purposes, we shall use the terms electron cloud and energy level interchangeably.

The maximum number of electrons in each energy level can be calculated from the formula $2n^2$, where n is the number of the energy level counting out from the nucleus. Thus, the first energy level holds a maximum of two electrons (if $n = 1$, $2n^2 = 2$); the second energy level holds a maximum of eight electrons (if $n = 2$, $2n^2 = 8$), and the third energy level has a maximum of 18 electrons. However, as will be discussed later in this chapter, the maximum number of electrons in any outer energy level is always eight.

The first energy level must be completely filled before electrons can begin filling the second energy level; the second energy level must be completely filled before electrons can begin filling the third energy level.

Figure 3-1   Electron clouds.

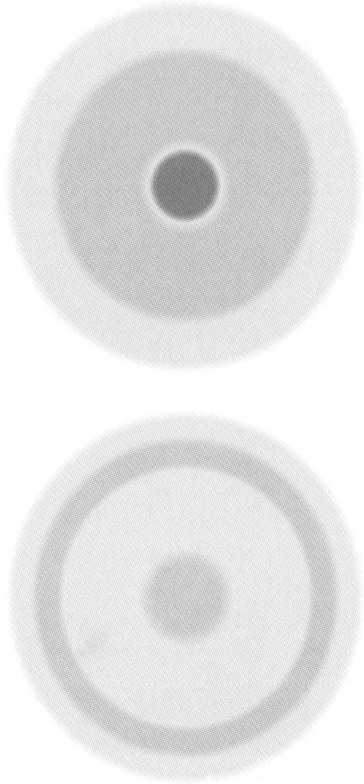

The element hydrogen—atomic number 1 and mass number 1—has one proton and no neutrons in its nucleus and one electron outside of its nucleus. This one electron can go into the first energy level, so the hydrogen atom can be represented as follows.

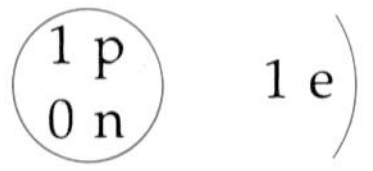

$$\left(\begin{array}{c} 1\,p \\ 0\,n \end{array}\right) \quad 1\,e\,\Big)$$

hydrogen, atomic number 1, mass number 1

The curved line indicates the first energy level.

The element helium—atomic number 2 and mass number 4—has two protons and two neutrons in its nucleus. The helium atom has two electrons outside its nucleus. These two electrons can go into the first energy level, which can hold a maximum of two electrons, as shown.

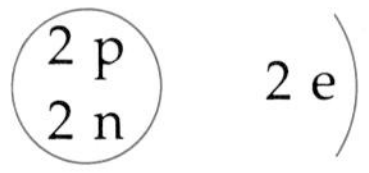

$$\left(\begin{array}{c} 2\,p \\ 2\,n \end{array}\right) \quad 2\,e\,\Big)$$

helium, atomic number 2, mass number 4

The lithium atom—atomic number 3 and mass number 7—has three protons and four neutrons in its nucleus. Outside of its nucleus it has three electrons. Two of these electrons can go into the first energy level. The third electron must go into the second energy level because the first energy level can hold only two electrons.

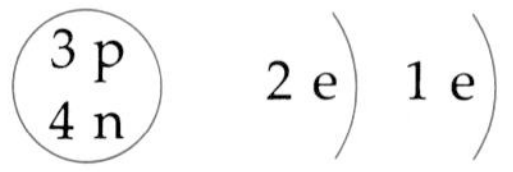

$$\left(\begin{array}{c} 3\,p \\ 4\,n \end{array}\right) \quad 2\,e\,\Big) \quad 1\,e\,\Big)$$

lithium, atomic number 3, mass number 7

The element sodium—atomic number 11 and mass number 23—has 11 protons and 12 neutrons in its nucleus and 11 electrons outside its nucleus. The first energy level can hold two electrons and the second energy level eight electrons, so the one remaining electron must go into the third energy level. Therefore, the structure of the sodium atom is as follows:

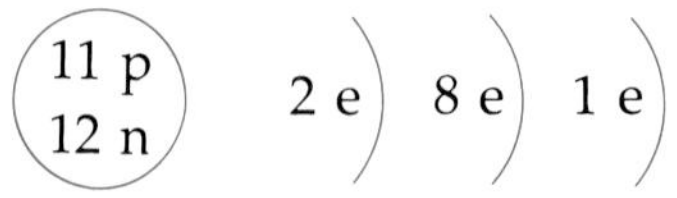

$$\left(\begin{array}{c} 11\,p \\ 12\,n \end{array}\right) \quad 2\,e\,\Big) \quad 8\,e\,\Big) \quad 1\,e\,\Big)$$

sodium, atomic number 11, mass number 23

Table 3-2 lists the progression of atomic numbers and the electron arrangement in the first 18 elements. The arrangement of the electrons for the heavier elements does not follow these simple rules and will not be discussed in this text.

Table 3-2 Electron Arrangements for the First 18 Elements

| Element | Symbol | Atomic Number | Electron Arrangement | | |
			First Energy Level	Second Energy Level	Third Energy Level
Hydrogen	H	1	1		
Helium	He	2	2		
Lithium	Li	3	2	1	
Beryllium	Be	4	2	2	
Boron	B	5	2	3	
Carbon	C	6	2	4	
Nitrogen	N	7	2	5	
Oxygen	O	8	2	6	
Fluorine	F	9	2	7	
Neon	Ne	10	2	8	
Sodium	Na	11	2	8	1
Magnesium	Mg	12	2	8	2
Aluminum	Al	13	2	8	3
Silicon	Si	14	2	8	4
Phosphorus	P	15	2	8	5
Sulfur	S	16	2	8	6
Chlorine	Cl	17	2	8	7
Argon	Ar	18	2	8	8

Energy Sublevels

An energy level is composed of sublevels, differing from one another in their spatial arrangement. The sublevels are composed of atomic orbitals, also called orbitals. The rules relating to sublevels are

1. Each energy level has one *s* sublevel containing one *s* orbital.
2. Beginning with the second energy level, each energy level contains a *p* sublevel, which consists of three *p* orbitals.
3. Beginning with the third energy level, each energy level contains a *d* sublevel, which consists of five *d* orbitals.
4. Beginning with the fourth energy level, each energy level contains an *f* sublevel, which consists of seven *f* orbitals.
5. Each orbital can contain no more than two electrons.
6. Each group of orbitals must be completely filled before electrons can begin to fill the next one.

This can be represented as follows.

Sublevels	*s*	*p*	*d*	*f*
Number of orbitals	1	3	5	7
Total number of electrons possible	2	6	10	14

If only the *s* sublevel is filled, there is a total of 2 electrons.
If the *s* and *p* sublevels are filled, there is a total of 8 electrons.
If the *s*, *p*, and *d* sublevels are filled, there is a total of 18 electrons.
If the *s*, *p*, *d*, and *f* sublevels are filled, there is a total of 32 electrons.

Recall that the maximum number of electrons in the first, second, third, and fourth energy levels is 2, 8, 18, and 32, respectively.

The orbitals are filled in a very definite sequence, as indicated in the following chart.

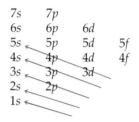

7s   7p
6s   6p   6d
5s   5p   5d   5f
4s   4p   4d   4f
3s   3p   3d
2s   2p
1s

The order of filling the orbitals can be found by reading diagonally upward in the direction indicated by the arrows. Thus the orbitals are filled in the order 1s, 2s, 2p, 3s, 3p, 4s, 3d, 4p, and so on.

Example 3-1   Hydrogen, $_1$H, has only one electron. This electron must fill the lowest energy sublevel, the 1s orbital, so the hydrogen atom is indicated as $1s^1$, which means one electron in the s orbital of the first energy level.

Example 3-2   Helium, $_2$He, has two electrons. Both of these electrons can go into the 1s orbital since this orbital can hold a total of two electrons. Thus, helium may be indicated as $1s^2$, meaning two electrons in the s orbital of the first energy level.

Example 3-3   Lithium, $_3$Li, has a total of three electrons. The 1s orbital can hold only two electrons, or $1s^2$. The remaining electron must go into the next available orbital, the 2s orbital, so the electron arrangement for lithium is $1s^2 2s^1$, where the $2s^1$ indicates one electron in the s orbital of the second energy level.

Example 3-4   For beryllium, $_4$Be, there is a total of four electrons. The first orbital, the 1s, can hold two electrons, or $1s^2$. The next orbital, the 2s, can hold two electrons, or $2s^2$, so the electron arrangement for beryllium is $1s^2 2s^2$.

Example 3-5   For boron, $_5$B, there is a total of five electrons. The first orbital, the 1s, can hold two electrons, or $1s^2$. The second orbital, the 2s, can also hold two electrons, or $2s^2$ The fifth electron must go into the next available orbital, the 2p, or $2p^1$, so the electron arrangement for boron is $1s^2 2s^2 2p^1$.

Example 3-6   For fluorine, $_9$F, there is a total of nine electrons. The first orbital can hold two electrons, or $1s^2$. The second orbital can also hold two electrons, or $2s^2$. The remaining five electrons go into the 2p orbitals since there are three of these 2p orbitals and each orbital can hold two electrons. Thus, the electron arrangement for fluorine is $1s^2 2s^2 2p^5$.

Example 3-7 For potassium, $_{19}K$, there is a total of 19 electrons. The $1s$ orbital can hold two electrons and the $2s$ orbital can also hold two electrons. The $2p$ orbitals can hold six electrons, making a total ten electrons so far. The next orbital to be filled is the $3s$, which can hold two electrons. Then comes the $3p$ orbitals, which can hold a total of six electrons, making a grand total of 18 electrons so far. The one remaining electron must go into the next orbital in order, the $4s$, so the electron arrangement for potassium is $1s^2 2s^2 2p^6 3s^2 3p^6 4s^1$.

Example 3-8 Cobalt, $_{27}Co$, has a total of 27 electrons. The $1s$ orbital can hold two electrons, or $1s^2$. The $2s$ orbital can also hold two electrons, or $2s^2$, and the $2p$ orbitals can hold six electrons, or $2p^6$. The $3s$ orbital can hold two electrons, $3s^2$, and the $3p$ orbitals can hold six electrons, or $3p^6$. The total number of electrons placed in orbitals so far is 18, with nine more to be placed. The next orbital in order is the $4s$, which can hold two electrons, or $4s^2$, followed by the $3d$ orbitals, which can hold up to 10 electrons. Therefore, the remaining seven electrons can go into the $3d$ orbitals, or $3d^7$, so the electron arrangement for cobalt is $1s^2 2s^2 2p^6 3s^2 3p^6 4s^2 3d^7$.

Noble Gases

The noble (inert) gases are unreactive because they have an outer energy level of eight electrons (except for helium, which requires only two electrons to fill its outer energy level). Eight electrons in the outer energy level corresponds to filled s and p orbitals, which in turn leads to great stability. Table 3-3 compares the electron arrangements of the noble gases.

The Periodic Table

Toward the end of the nineteenth century, chemists noted many similarities between elements and tried to group them into certain families with similar properties and reactions. This was the beginning of the periodic table.

Table 3-3 The Noble Gases

Symbol	Atomic Number	Electron Arrangement by Energy Level					
		1	2	3	4	5	6
He	2	2					
Ne	10	2	8				
Ar	18	2	8	8			
Kr	36	2	8	18	8		
Xe	54	2	8	18	18	8	
Rn	86	2	8	18	32	18	8

Figure 3-2   Abbreviated periodic table.

The modern periodic table places the elements according to the number and arrangement of the electrons in the atom. Look at the abbreviated periodic table shown in Figure 3-2 and the complete periodic table inside the front cover of this book. Note that the table is divided into horizontal rows and vertical columns. Each box in the table represents one element. There are over 100 known elements; thus, there are over 100 different kinds of atoms, each having its own place in the periodic table. The vertical columns are called **groups** or **families** and are labeled with Roman numerals. The horizontal rows are called **periods** and are indicated by Arabic numerals.

Each box on the chart contains a symbol with a number above and below that symbol. The name of the element is given above the symbol. Consider the box with the symbol O in it,

8

Oxygen

O

16.00

The symbol O represents the element oxygen. The atomic number of the element (8) is given above the name. Below the symbol is the atomic weight, 16.00.

Under group IA of the periodic chart are such elements as hydrogen (H), lithium (Li), and sodium (Na). The electron arrangement of these atoms is given in Table 3-4. Note that they all have one electron in their outermost or highest energy level. (A-group elements are commonly called main group elements.)

Consider the elements beryllium (Be) and magnesium (Mg). Each of these elements has two electrons in its outermost energy level; they are in group IIA. Likewise, in group IIIA the elements aluminum (Al) and boron (B) have three electrons in their outermost or highest energy level. In group VIIA the elements fluorine (F) and chlorine (Cl) each have seven electrons in their outermost or highest energy level.

In general, the A-group number corresponds to the number of electrons in the highest energy level. The principal exception to this rule is group 0 at the far right of the table. This group contains the

Table 3-4	First Energy Level	Second Energy Level	Third Energy Level
Hydrogen	1		
Lithium	2	1	
Sodium	2	8	1

noble gases. The noble gases all have eight electrons in their highest energy level except for helium, which contains only two (because the first energy level is filled when it contains two electrons).

The electrons in the highest energy level are called **valence electrons.** The elements in a group (family) have similar chemical properties because they have the same number of valence electrons. We will see that elements in the same group form similar compounds and frequently can substitute for each other.

The B-group elements are called "transition" elements. They are all metals and usually have two electrons in their highest energy level.

Reading horizontally, in period 1 there are only two elements, hydrogen and helium. Both of these elements have an electron or electrons in the first energy level only. In period 2 and period 3 there are eight elements, all of which have one or more electrons in their highest energy level, the second and third energy levels, respectively. In period 4 there are 18 elements, all having electrons in their fourth energy level.

Another generalization, then, is that the period corresponds to the number of energy levels in the atom. Thus, the element oxygen—group VIA and period 2—has six electrons in its highest energy level (from group VIA) and two energy levels (from period 2). The element vanadium (symbol V) is in group VB, period 4. Vanadium has two electrons in its highest energy level (B-group elements usually have two electrons in their highest energy level) and four energy levels (from period 4).

Summary

Atoms have an extremely small size and weight. The weights of all atoms are compared to that of the carbon-12 atom, which has been assigned a weight of 12.0000 amu. These relative weights are called atomic weights.

The atom is composed of a nucleus containing protons and neutrons and of electrons surrounding that nucleus. Both protons and neutrons each weigh 1 amu; electrons weigh practically nothing. Almost the entire weight of the atom is therefore in its nucleus.

The atomic number of an element indicates the number of protons inside the nucleus of an atom of that element and also the number of electrons outside that nucleus.

The number of neutrons can be found by subtracting the atomic number of an element from its mass number.

Isotopes are atoms having the same atomic number but different mass numbers.

Most elements have isotopes; some have two, many have several.

The electrons are located in electron clouds or energy levels. The first energy level holds a maximum of two electrons, the second a maximum of eight electrons, and the third energy level a maximum of 18 electrons.

The first electron energy level must be filled before electrons can begin filling the second. The second energy level must be filled before electrons can begin filling the third energy level.

Electron energy levels are composed of sublevels, which in turn consist of orbitals. Each orbital can hold a maximum of two electrons. The order of filling the orbitals is 1s, 2s, 2p, 3s, 3p, 4s, 3d, and so on.

The noble gases each have eight electrons in their highest energy level with the exception of helium, which has only two.

The periodic table lists all of the elements in the order of their atomic numbers. The vertical columns are called groups and the horizontal rows are called periods.

The A-group number indicates the number of electrons in the highest energy level. The B-group elements are called transition elements; they are all metals and usually have two electrons in their highest energy level. The period indicates the number of energy levels.

Questions and Problems

A

1. Diagram the following atoms showing protons, neutrons, and electrons in each energy level and sublevel.

Symbol	Atomic Number	Mass Number
N	7	14
Be	4	9
F	9	19
S	16	32
Ne	10	20
Si	14	28

2. Diagram the structures of the following isotopes showing electrons in each energy level.

Symbol	Atomic Number	Mass Number
H	1	1,2,3
O	8	16,17,18
Mg	12	24,25,26
N	7	13,14
Ne	10	20,22

3. Using the periodic table inside the front cover of this book, predict the number of electrons in the highest energy level and also the number of energy levels in the following elements.
   (a) Beryllium (Be)
   (b) Phosphorus (P)
   (c) Bromine (Br)
   (d) Calcium (Ca)
   (e) Rubidium (Rb)
   (f) Arsenic (As)
   (g) Radium (Ra)
   (h) Zirconium (Zr)
   (i) Manganese (Mn)
   (j) Actinium (Ac)
   (k) Radon (Rn)

4. How do isotopes differ from one another with respect to chemical properties? physical properties?

5. What is an atom?

6. State the mass, charge, and location in the atom of the three fundamental particles.

7. Where is the mass of the atom concentrated?

8. What are the vertical columns in the periodic table called?

9. What are the horizontal rows in the periodic table called?

10. In general the elements in the periodic table are listed in order of increasing atomic weights, but there are exceptions. Give two examples.

B

11. Are all A-group elements metals? all B-group elements?

12. Group 0 contains what type of elements?

13. In which part of the periodic chart are the metals located? the nonmetals? the noble gases?

14. Elements 93 through 106 have been prepared synthetically. Are they all transition elements?

15. What are valence electrons? noble gases?

16. Chlorine consists of two isotopes of mass numbers 35 and 37 amu. If both isotopes were present in equal amounts, what would be the atomic weight of chlorine? If 75 percent of the lighter isotope and 25 percent of the heavier isotope were present, what would be the atomic weight?

17. Why do elements in the same group have similar chemical properties?

18. An atom of an element contains 15 protons and 18 neutrons.
    (a) How many electrons does it have?
    (b) What is its mass number?
    (c) What is its atomic number?
    (d) In which group is it?
    (e) In which period is it?
    (f) Is it a metal or nonmetal?
    (g) What is its symbol?

19. An element is located in group VIA, period 4. What can you tell about that element?
20. As new elements are prepared synthetically, will they fit into the existing periodic chart? Explain.

Practice Test

1. Most of the mass of the atom is located in the _____ .
    a. protons   b. nucleus
    c. neutrons   d. electrons
2. The fundamental particle with a +1 charge is the _____ .
    a. proton   b. neutron
    c. electron   d. none of these
3. Electrons are found _____ .
    a. inside the nucleus
    b. outside the nucleus
    c. either inside or outside the nucleus
4. Isotopes differ in the number of _____ .
    a. protons   b. electrons
    c. neutrons   d. charged particles

5. All noble gases, except helium, have _____ electrons in their outermost or highest energy level?
    a. 2   b. 4   c. 6   d. 8
6. The maximum number of electrons in the third energy level is _____ .
    a. 8   b. 18   c. 32   d. 64
7. The $^{15}_{7}N$ atom contains _____ neutrons.
    a. 7   b. 8   c. 15   d. 22
8. Transition elements usually have _____ electrons in their outermost or highest energy level.
    a. 0   b. 2   c. 8   d. 18
9. The relative weight of an atom is called its _____ .
    a. atomic weight   b. nuclear weight
    c. isotopic weight   d. all of these
10. The atomic number of an element indicates the number of _____ in the atom.
    a. protons
    b. electrons
    c. neutrons
    d. protons and neutrons

4

Radioactivity

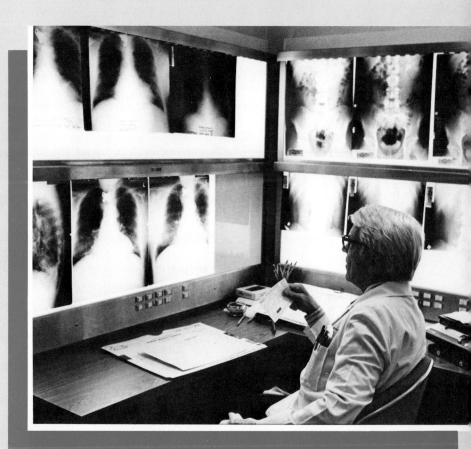

Radiologist examining X-rays.

Discovery of Radioactivity

In 1896 a French physicist, Henri Becquerel (1852–1908), found that uranium crystals had the property of "fogging" a photographic plate that had been placed near those crystals. This fogging took place even though the photographic plate was wrapped in black paper. By placing crystals of uranium on a photographic plate covered with black paper and then developing the plate, he obtained a self-photograph of the crystals. Becquerel concluded that the uranium gave off some kind of radiation or rays that affected the photographic plate. Figure 4-1 illustrates an *autoradiograph[†] of a radioactive bone section.

Substances like uranium that spontaneously give off radiation are said to be **radioactive.** Radioactivity is the property that causes an element to emit radiation. This radiation comes from the nucleus of the atom.

Types of Radiation Produced by a Radioactive Substance

The following experiment was performed to study the radiation produced by a radioactive element (Figure 4-2). A piece of radium was placed at the bottom of a thick lead well. The purpose of the lead was to absorb all the radiation except that going directly upward. The escaping radiation was allowed to fall on a photographic plate. When the radiation was passed through a strong electrostatic field, three different areas showed up on the photographic plate. This indicated that there were actually three different kinds of radiation. These were called alpha, beta, and gamma.

Alpha Particles

Alpha particles (α particles) are attracted toward the negative electrostatic field, which indicates that they are positively charged. Alpha particles consist of positively charged helium nuclei; that is, they consist of the nuclei of helium atoms (each of which contains two protons and two neutrons) and so they have a charge of +2. Alpha particles have a very low penetrating power. They can be stopped by a piece of paper or by a thin sheet of aluminum foil. Alpha particles are relatively harmless when they strike the body because they do not penetrate the outer layer of the skin. However, if a source of alpha particles is inhaled or ingested or gets into the body through an open wound, then those particles can cause damage to the cells and to the internal organs.

Alpha particles result from the radioactive decay of heavy elements such as uranium and radium.

[†] See the glossary for starred terms.

Figure 4-1 A section of bone from the body of a former radium watch-dial painter who, in order to maintain a fine tip on his brush, was in the habit of touching the tip with his tongue. (a) Darkened areas of damaged bone. (b) An auto-radiograph in which the bone "took its own picture" by being held against film, showing areas exposed by the radium alpha particles. Note that the areas of high alpha activity correspond to the areas of maximum damage in (a). [Reproduced from N. Frigerio, *Your Body and Radiation*, U.S. Atomic Energy Commission, Washington, DC, 1967. Photos from Argonne National Laboratory.]

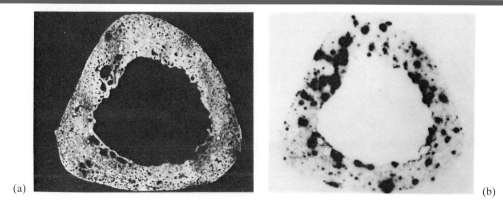

(a)   (b)

Beta Particles

Beta particles (β particles) are attracted toward the positive electrostatic field, which indicates that they consist of negatively charged particles. Beta particles consist of high-speed electrons that travel at a speed in excess of 100,000 miles per second. Note that beta particles (electrons) are produced in the nucleus by the transformation of a neutron into a proton and an electron. The electron is emitted as a beta particle and the proton remains in the nucleus. Beta particles (electrons) have a charge of -1.

Figure 4-2 Radium emits radiation that an electrostatic field separates into alpha (α) particles, beta (β) particles, and gamma (γ) rays.

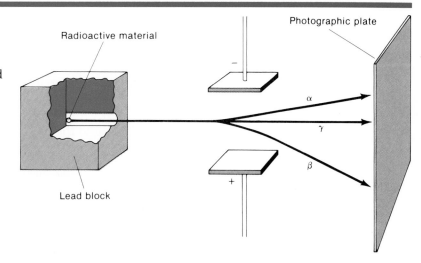

Photographic plate

Radioactive material

Lead block

Beta particles have a slight penetrating power. They pass through a sheet of paper but can be stopped by heavy clothing. When beta particles strike the body, they penetrate only a few millimeters and do not reach any vital organs. If a source of beta particles should be inhaled or ingested, those particles could also cause internal damage to the body cells and organs.

Note in Figure 4-2 that beta particles are deflected by the electrostatic field to a much greater extent than are the alpha particles. This indicates that beta particles have a much smaller mass than alpha particles.

Gamma Rays

Gamma rays (γ rays) are not affected by an electrostatic field because they have no charge. They are not particles at all. They have no mass. Gamma rays are a form of electromagnetic radiation similar to X-rays. Gamma rays are very penetrating; they will pass through the body, causing cellular damage as they travel through (see page 71). Gamma rays are often emitted along with alpha or beta particles. Gamma rays originate from unstable atoms releasing energy to gain stability.

Other Types of Radiation

X-rays are a form of electromagnetic radiation usually produced by machines, whereas gamma rays are emitted by radioactive substances.

Neutrons are released from elements that undergo spontaneous fission. Their relatively large mass gives them great energy, and because they have no charge they readily penetrate the body. Neutrons are used in the treatment of cancer.

Nuclear Reactions

When the nucleus of an atom emits a ray, its atomic number and mass number may change. Such changes in the nucleus are called nuclear reactions.

In writing equations for nuclear reactions the following symbols are used.

1. The atomic number and the mass number are written to the left of the symbol of the element.

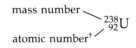

---

† Note that in the periodic table (inside front cover) the atomic number is above the symbol.

2. The alpha particle, which consists of a helium nucleus, is indicated as ^4_2He.
3. The beta particle, which is an electron, is indicated as $_{-1}^{0}\text{e}$. (The beta particle has arbitrarily been assigned an atomic number of -1.)
4. The gamma ray, which is a radiation only, is indicated as γ.
5. The neutron is indicated as ^1_0n.
6. The proton is indicated by ^1_1H.
7. The deuteron, which is the nucleus of a heavy hydrogen atom, is indicated by ^2_1H.

The following rules must be observed when writing nuclear reactions.

1. The sum of the atomic numbers on both sides of the equation must be the same.
2. The sum of the mass numbers on both sides of the equation must be the same.

When a uranium atom—atomic number 92 and mass number 238—loses an alpha particle, the nuclear reaction can be written as follows.

$$^{238}_{92}\text{U} \longrightarrow \alpha + \text{?} \qquad \text{or} \qquad ^{238}_{92}\text{U} \longrightarrow {}^4_2\text{He} + \text{X}$$

where X is the product other than an alpha particle produced during the decomposition. The atomic number on the left side of the equation is 92 and on the right side it is 2, so the atomic number of X must be 90. Likewise, the mass number on the left side of the equation is 238 and on the right side it is 4, so the mass number of X must be $238 - 4$ or 234. Therefore, the reaction is written as

$$^{238}_{92}\text{U} \longrightarrow {}^4_2\text{He} + {}^{234}_{90}\text{X}$$

Element 90 (from the periodic table in the front of the book) is thorium, symbol Th. The complete reaction therefore becomes

$$^{238}_{92}\text{U} \longrightarrow {}^4_2\text{He} + {}^{234}_{90}\text{Th}$$

indicating that when uranium gives off an alpha particle, it changes into a new element, thorium. If a thorium atom should emit an alpha particle, the reaction could be written as

$$^{234}_{90}\text{Th} \longrightarrow {}^4_2\text{He} + \text{Z}$$

The atomic number of element Z must be 88 in order for the sum of the atomic numbers to be the same on both sides of the equation.

Likewise, the mass number of element Z must be 230, so the equation would be

$$^{234}_{90}\text{Th} \longrightarrow {}^4_2\text{He} + {}^{230}_{88}\text{Z}$$

The element of atomic number 88 is radium, symbol Ra, so the actual reaction is

$$^{234}_{90}\text{Th} \longrightarrow {}^4_2\text{He} + {}^{230}_{88}\text{Ra}$$

Thus, when an atom emits an alpha particle, its mass number decreases by 4 and its atomic number decreases by 2.

If a thorium atom, $^{234}_{90}\text{Th}$, emits a beta particle, $^{0}_{-1}\text{e}$, the following reaction takes place.

$$^{234}_{90}\text{Th} \longrightarrow {}^0_{-1}\text{e} + {}^{234}_{91}\text{Pa}$$

Therefore, when a thorium atom emits a beta particle, an isotope of protactinium (Pa) is produced. Note that the sum of the atomic numbers is the same on both sides of the equation, as is the sum of the mass numbers.

When an atom emits a beta particle, its mass number remains the same, but its atomic number increases by 1.

When an atom emits a gamma ray, there is no change in the atomic number or mass number because the gamma ray is not a particle. It has no mass and no charge. However, even though the same element is produced in the reaction, it has a slightly lower energy because of the energy carried away by the gamma ray,

$$^{238}_{92}\text{U} \longrightarrow \gamma + {}^{238}_{92}\text{U}^*$$

where the * indicates a slightly lower energy.

Another way of indicating a loss of a gamma ray and the resulting change in energy is

$$^{99m}_{43}\text{Tc} \longrightarrow {}^{99}_{43}\text{Tc} + \gamma$$

where the "m" indicates a metastable (unstable) form of technetium (Tc). Thus, the unstable form of technetium emits a gamma ray and becomes a more stable (less energetic) isotope of technetium. ^{99m}Tc is used in various types of scans (see page 61).

Natural and Artificial Radioactivity

Some elements, such as uranium and radium, are naturally radioactive. Natural radioactivity can be defined as the spontaneous change of one element into another. Many of the heavier elements are naturally radioactive, whereas most of the lighter elements are naturally nonradioactive.

However, normally nonradioactive lighter elements can be changed into radioactive ones by bombardment with such particles as protons ($_1^1H$), neutrons ($_0^1n$), electrons ($_{-1}^0e$), or alpha particles ($_2^4He$). The radioactive substances so produced are said to be artificially radioactive. They may be produced in a nuclear reactor. If nitrogen ($_7^{14}N$) is bombarded with neutrons (**$_0^1n$**), the following nuclear reaction takes place:

$$_7^{14}N + {_0^1}n \longrightarrow {_6^{14}}C + {_1^1}H$$

The products are a radioactive isotope of carbon, ^{14}C, and a proton. Note that again the sum of the atomic numbers on both sides of the equation is the same; likewise the sum of the mass numbers.

Units of Radiation

Radiation is measured in terms of several different units, depending upon whether the measurement relates to a physical or a biologic effect.

The physical unit of radiation is a measure of the number of nuclear disintegrations occurring per second in a radioactive source. The standard unit is the *curie, which is defined as the number of nuclear disintegrations occurring per second in 1 g of radium; 1 curie (1 Ci) equals 37 billion disintegrations per second. Smaller units are the millicurie (1 mCi = 37 million disintegrations per second) and the microcurie (1 μCi = 37,000 disintegrations per second). These smaller units are frequently used in describing an amount of radioactive fallout. The curie is not useful in biologic work because it simply indicates the number of disintegrations per second regardless of the type of radiation and regardless of the effect of that radiation upon tissue.

The *roentgen (abbreviated R) is a unit of radiation generally applied to X-rays and gamma rays only. X-rays and gamma rays produce ionization in air and also in tissue. The roentgen is defined as the intensity of X-rays or gamma rays that produces 2 billion ion pairs (see page 71) in 1 mL of air. This is not the same for tissue as it is for air, so that the roentgen does not accurately indicate the amount of radiation on tissue.

The *rad (*r*adiation *a*bsorbed *d*ose) refers to the amount of radiation energy absorbed by tissue that has been radiated. One rad corresponds to the absorption of 100 ergs of energy per gram of tissue. An erg is a very small unit of energy. More than 40 million ergs are required to equal 1 cal. However, even though the erg is an extremely small unit of energy, the effect of 1 rad (100 ergs per gram) is important because of the ionization that the radiation produces in the cells. The SI unit for absorbed dosage of radiation is the *gray* (Gy). 1 Gy = 100 rad.

The *rem (*r*adiation *e*quivalent, *m*an) represents the amount of radiation absorbed by a human being. This unit of measure takes into consideration the difference in energy for various radioactive sources. A rem is the amount of ionizing radiation that, when absorbed by a human, has an effect equal to the absorption of 1 R. A smaller unit is the millirem, mrem.

Rem = rad × quality factor, with the quality factor being larger for types of radiation that produce greater numbers of ions along the path of the radiation. The production of a greater number of ions implies more damage to the individual cell.

LET (*l*inear *e*nergy *t*ransfer) is the amount of energy transferred to the atoms in a medium by radiation. Alpha particles with their large mass and slow speed, for example, impart much more energy over their path than do electrons. In general, the higher the LET of the radiation, the greater the injury for a given absorbed dose. Gamma rays, X-rays, and beta particles usually have low LETs, whereas alpha particles and neutrons have high LET values.

Detection and Measurement of Radiation

The problem of detecting and measuring radiation is very important in medical work, particularly in the protection of personnel. One device used to detect radiation is the Geiger counter. This device consists of a glass tube containing a gas at low pressure through which runs a wire connected to a high-voltage power supply. When the device is brought close to a radioactive substance, the radiation causes a momentary current to flow through the tube. A speaker is usually placed in the circuit to produce a click, indicating a momentary flow of current. Sometimes a counting device is connected to the tube to indicate the amount of radiation.

Scanners (see page 56 and Figure 4-4) use another type of device, called a **scintillation counter**, to detect and measure radiation. This detector consists of a crystal of sodium iodide containing a small amount of thallium iodide. When the crystal is hit by radiation, it gives off a flash of light, a scintillation; hence the name. A counting device records these scintillations, and the result is produced as a "scan."

X-ray technicians and others who work around radiation usually are required to wear film badges (see Figure 4-3). These badges indicate the accumulated amount of radiation to which they have been exposed. They contain a piece of photographic film whose darkening is directly proportional to the amount of radiation received. This film must be checked frequently to see how much radiation has been absorbed.

Figure 4-3   (a) Film
badges. (b) Film badge
worn by X-ray technician.
[Courtesy Tech/Ops
Landauer, Inc., Glenwood,
IL.]

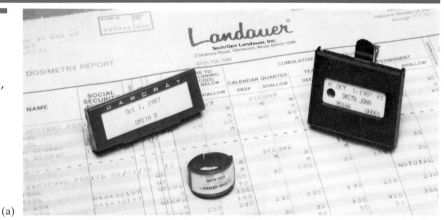

(a)

(b)

Radioisotopes

The isotopes produced artificially by bombardment with one of the
various particles are called radioactive isotopes, or radioisotopes.
Radioisotopes have the same chemical properties as nonradioactive
isotopes of the same element because chemical properties are based
upon electrons only, and isotopes of an element have identical
electron structures. Thus, an organism cannot distinguish between
normal carbon (^{12}C) and radioactive carbon (^{14}C).

   If a plant is exposed to carbon dioxide containing ^{14}C, the resulting carbohydrates will contain radioactive carbon. The movement of
the radioactive atoms as they proceed through the plant can be

Figure 4-4 A scanner, an instrument used for depicting the distribution of radioactive material in the human body. [Courtesy Picker International, Northford, CT.]

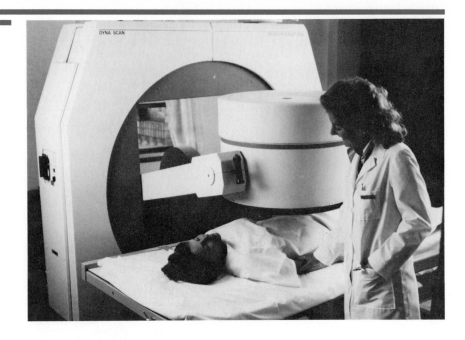

followed by means of a Geiger counter. Radioisotopes introduced into living organisms are called "tagged" atoms because their path can be followed readily as they move through the organism.

A radioisotope is usually indicated by its symbol and mass number only. The atomic number is not necessary because the symbol itself serves to identify the element. Some radioisotopes commonly used in medicine and in biochemistry are ^{131}I, ^{60}Co, ^{99m}Tc, ^{14}C, and ^{59}Fe (see page 59).

Radioisotopes are used medicinally in the diagnosis and treatment of various disorders of the human body. A medical tool called a scanner helps to locate malignancies (see Figure 4-4). A patient is given a selected radioisotope that will accumulate in the body area being studied. The scanner moves back and forth across that site and detects the difference in uptake between healthy and abnormal tissue. This information is fed to a receiver and recorder, which then produces a picture called a scan (see Figure 4-5). Proper interpretation of a scan can tell not only if there is a malignancy but also where it is located and its exact dimensions.

Positron Emission Tomography (PET)

Some synthetic radioactive substances give off positrons from their nucleus. Positrons are positively charged electrons and are produced as indicated in the following reaction.

$$\underset{\text{proton}}{^{1}_{1}\text{p}} \longrightarrow \underset{\text{neutron}}{^{1}_{0}\text{n}} + \underset{\text{positron}}{^{0}_{1}\text{e}}$$

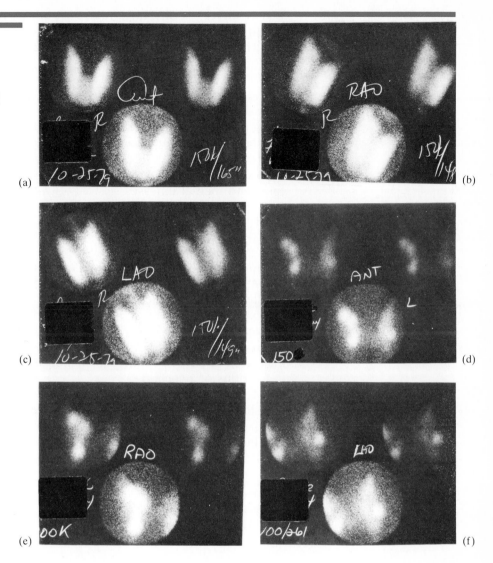

Figure 4-5 Thyroid scans. [Courtesy Department of Nuclear Medicine, Michael Reese Hospital and Medical Center, Chicago, IL.]

The positron thus emitted can exist for only a very short time before colliding with an electron. When this occurs, the electron (negatively charged) and the positron (positively charged) annihilate each other and in so doing convert their masses into two bursts of gamma rays.

$$_{-1}^{0}e \quad + \quad _{1}^{0}e \quad \longrightarrow \quad 2\,\gamma$$

$$\text{electron} \qquad \text{positron} \qquad \text{gamma rays}$$

Positron-emitting elements such as ^{11}C, ^{15}O, ^{13}N, and ^{18}F are made part of a molecule that, when administered to a patient, will travel to the section of the body being studied. There it will produce gamma rays inside the tissue (as opposed to radiation treatment from outside the body). Positron-emitting elements have a short half-life (see following section), ranging from 2 min to 2 hr. Because of this short half-life, large amounts of these materials can be given with relatively small exposure to the patient. In addition, because of the short half-life, repeated measurements can be made effectively.

The PET method allows glucose, made partially from positron-emitting ^{11}C instead of the normal ^{12}C, to be traced through the sensory sections of the brain. In one test, when volunteers were stimulated visually in only one eye, the region of the brain on the opposite side showed increased glucose usage. Similar tests are being performed for other senses, such as sound and smell. Not only can the change in glucose usage be determined but also the location of that usage can be pinpointed. PET is also being used in detecting such disorders as epilepsy, heart disease, stroke, Parkinson's disease, and mental illnesses.

Half-life

When a radioactive element gives off a particle, it changes or decays into another element. The rate of decay of all radioactive elements is not the same. Some elements decay rapidly, whereas others decay at an extremely slow rate. The half-life of a radioactive element is defined as the amount of time required for half of the atoms in a given sample to decay. Some radioactive elements have a half-life measured in terms of billions of years, whereas others are measured in fractions of a second.

The radioisotope ^{131}I has a half-life of approximately 8 days. Consider a 60-mg sample of ^{131}I. After 8 days (one half-life period of time) only half as much, or 30 mg, would be left. After another half-life period (a total of 16 days), only half of that amount, or

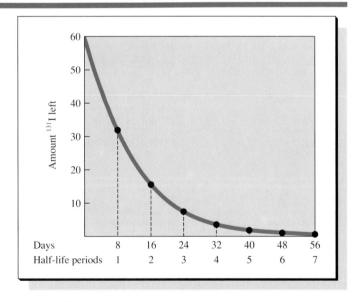

Figure 4-6 Decay curve for a radioactive substance, showing amount of ^{131}I versus time and half-life periods.

	Radioisotope	Half-life	Radiation
Table 4-1 Half-life Periods of Several Common Radioisotopes	^{99m}Tc	6 hr	γ
	^{59}Fe	45 days	$\beta + \gamma$
	^{198}Au	2.7 days	$\beta + \gamma$
	^{131}I	8.0 days	$\beta + \gamma$
	^{123}I	13 hr	$\beta + \gamma$
	^{32}P	14.3 days	β
	^{60}Co	5.3 yr	$\beta + \gamma$
	^{14}C	5760 yr	β
	^{3}H	12 yr	β
	^{67}Ga	78 hr	γ
	^{51}Cr	27.8 days	γ
	^{24}Na	14.8 hr	$\beta + \gamma$
	^{111}In	2.8 days	γ

15 mg, of ^{131}I would be left. During every half-life period, half of the remaining amount decays.

This can be shown diagrammatically in a decay curve in which the amount of radioactive material is plotted against elapsed time (Figure 4-6).

For medical work, a radioisotope must have a half-life such that it will remain in the body long enough to supply the radiation needed and yet not expose the body to excess radiation. This requires either a relatively short half-life or rapid elimination from the body.

The half-life periods of several commonly used radioisotopes are listed in Table 4-1.

Radioisotopes of long half-life are very dangerous to the body. For example, radium has a half-life of 1590 years; therefore, if it is taken into the body, it continues to give off its radiation during the lifetime of that person. Radium and strontium-90 are members of group IIA in the periodic chart. This indicates that they act like calcium. They are easily taken up in the bones and are not readily excreted.

Biologic half-life is the time taken for one-half of an administered radioactive substance to be lost through biologic processes.

Radioisotopes in Medicine

When a radioisotope is to be used for diagnostic purposes, it must meet several criteria. Among them are the following: The radioactive element must be contained in a compound that will tend to concentrate in the area under study or in certain abnormal tissues. Since the presence of radiation is usually determined by an external counter or by a scan, a radioisotope emitting alpha particles is not generally used because such particles have too low a penetrating power to be detected outside the body. Beta-emitting radioisotopes must be located very close to the skin to be detected. Radioisotopes with gamma radiation are preferred.

The radioisotope selected should have a short half-life and should be in the form of a compound that will be eliminated from the body shortly after its diagnostic use is completed. Thus, the body will receive a minimum amount of radiation after the test is completed. In addition, the amount of radioisotope used should be as small as practicable.

When a radioisotope is to be used for therapy, external measurement is not so necessary, so alpha as well as beta and gamma emitters can be used. In radiation therapy, selected cells or tissues are to be destroyed without damage to nearby healthy tissues. Thus, the given radioisotope should have the property of concentrating in the desired area and, preferably, should emit alpha or beta particles because these have limited penetrating power and will not damage adjacent tissues.

A radioimmunoassay technique measures the blood levels of an enzyme called CK BB. The blood levels of this enzyme increase after damage to the nervous system. This method allows rapid assessment of brain damage that can occur during a stroke.

Some of the many radioisotopes in common medical use are discussed in the sections that follow.

Iodine-131, Iodine-123

Iodine-131 and iodine-123 are used in the diagnosis and treatment of thyroid conditions. The thyroid gland requires iodine to function normally. A patient suspected of having a thyroid disorder is given a drink of water containing a small amount of ^{131}I (or ^{123}I) in the form of sodium iodide. If the thyroid is functioning normally, it should take up about 12 percent of the radioactive iodine within a few hours. This iodine uptake can be measured with a scan. If less than the normal amount of ^{131}I is taken up, the patient may have a hypothyroid condition. If the amount of ^{131}I taken up is greater than normal, a condition known as hyperthyroidism may exist.

For a typical thyroid scan, a dose of 50 μCi is administered orally. This amount of radiation gives 50 rads to the thyroid gland and 0.02 rad to the total body. Another method for a thyroid scan involves the use of ^{99m}Tc. This radioisotope has the following advantages: (1) 5 mCi of ^{99m}Tc administered intravenously gives a dose of only 1 rad to the thyroid gland and 0.02 rad to the whole body, and (2) the scan can be performed 20 min after injection rather than after a 24-hr wait as when ^{131}I is used.

In addition, ^{131}I is used for scans of the adrenal glands, the gallbladder, and the liver. ^{131}I is used to screen for impaired fat digestion and decreased intestinal absorption of fats. A normal result implies good pancreatic function. ^{131}I is employed *therapeutically in the treatment of hyperthyroidism and in cancer of the thyroid gland.

^{125}I injected intravenously in the form of radioisotope-labeled fibrinogen is being used to detect deep *thromboses.

Technetium-99m

^{99m}Tc (m for metastable) is one of the most widely used radioisotopes for various types of scans. Sodium pertechnate is used for brain scans and thyroid scans (see Figure 4-5); technetium sulfur colloid, for liver scans and bone marrow scans; technetium macroaggregates of albumin, for lung-perfusion scans; technetium pyrophosphate, for bone scans; technetium diethylenetriaminepentaacetic acid (DTPA), for renal scans; and technetium-labeled red blood cells, for pericardial studies. Technetium albuminate is used to scan the placenta for bleeding.

^{99m}Tc is also used to measure blood volume. A sample of this radioisotope is injected into the bloodstream. A short while later a sample of blood is withdrawn and its radioactivity is measured. By comparing the radioactivity of the ^{99m}Tc and the radioactivity present in the sample withdrawn, an accurate determination of the total blood volume can be obtained.

Scanners (CT, PET, and MRI) for the detection of tumors and cancerous tissue have greatly reduced the use of ^{99m}Tc for these purposes.

Cobalt-60, Cobalt-52

^{60}Co gives off powerful gamma rays as well as beta particles. For treatment purposes, the beta particles are shielded out and only the gamma rays are used. ^{60}Co has a half-life of 5.3 years and is used as a substitute for radium because it is much cheaper and easier to handle. This radioisotope is employed in the treatment of many different types of cancer. Hospitals use large machines containing a cobalt "bomb" to supply gamma radiation directly to the cancer site (see Figure 4-7).

^{52}Co is used in the diagnosis of vitamin B_{12} malabsorption, a condition that might occur in pernicious anemia.

Other Radioisotopes

Among the many other radioisotopes in medical use are xenon-133 (^{133}Xe) and krypton-81m (^{81m}Kr) for ventilation lung scans, selenium-75 (^{75}Se) for pancreas scans, and gallium-67 (^{67}Ga) for whole-body scans for tumors. ^{3}H is used to determine the total amount of water present in the body; ^{14}C has been employed to study the path of carbohydrates, fats, and proteins in the body and their conversion

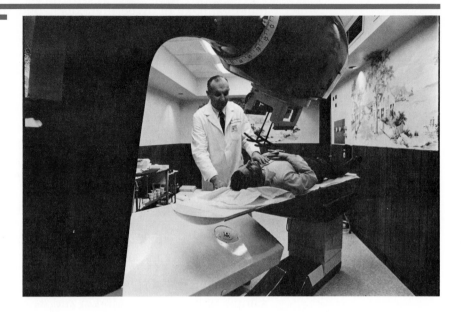

Figure 4-7 Cobalt machine in use. [Courtesy University of Illinois Hospitals, Chicago, IL.]

into other substances; and gold-198 (^{198}Au) has been employed for the treatment of pleural and peritoneal *metastases.

Unlike ^{99m}Tc, which concentrates in damaged heart cells, thallium-201 (^{201}Tl) concentrates in normal heart muscle but not in abnormal tissue. Thus, damaged heart muscle areas show up on a scan as "cold spots." Cesium-129 (^{129}Cs) is used for scans on patients suspected of having acute *myocardial infarction.

Iron-59 (^{59}Fe) is used to measure the rate of disappearance of iron from the plasma, the plasma iron turnover rate, bone marrow function, and the utilization of iron in red blood cell production and to determine anemias caused by iron deficiency or chronic infection. The iron is usually injected as ferrous citrate, with blood samples being withdrawn at various intervals depending on the diagnostic test.

Strontium-85 (^{85}Sr) is used for bone scans. When used in conjunction with X-rays, it is possible to diagnose fractures, metastatic bone diseases, osteomyelitis, and *neoplasm. Since ^{85}Sr remains in the bone for a long period of time, it is possible to repeat the bone scan a month after the original test.

Chromium-51 (^{51}Cr), in the form of sodium chromate that has been tagged to albumin, is used to determine protein loss through the gastrointestinal tract. It is also used to determine the size, shape, and location of the spleen and can be used in blood studies including plasma volume, red blood cell mass, and red blood cell survival time.

Phosphorus-32 (^{32}P) provides a useful means of determining eye tumors. It is not possible to perform a biopsy of the eye without loss of vision and destruction of the eye. Since ^{32}P uptake is greater in ocular tumors than in other areas of the eye, the location of a tumor can be pinpointed. ^{32}P is also used in the treatment of leukemia.

Rubidium-82 (^{82}Rb) shows promise for diagnosing heart defects. This radioisotope, with a half-life of 75 sec, is extracted from coronary blood by the heart muscle as if it were potassium.

Radioimmunoassay

Radioimmunoassay (RIA) is a method of measuring the concentration of substances that are present in very small amounts in the blood. The procedure is based upon the body's ability to provide immunity against disease.

When a foreign substance enters the bloodstream, the body produces antibodies to react with and neutralize the invading material. The foreign body, which causes the production of antibodies, is called an *antigen. The reaction is

$$\text{antigen + antibody} \longrightarrow \text{antigen–antibody complex}$$

For each antigen there is a specific antibody. An antibody that protects against one disease will, in general, have no effect on other diseases.

Suppose an RIA is ordered for renin, an enzyme of the kidney. The laboratory will produce an artificially radioactive form of renin and also the specific antibody for that substance. As indicated in the previous paragraph, the antigen reacts with the antibody to form an antigen–antibody complex. However, in this case there are two antigens, the one present in the blood and the artificially radioactive one. Both can react with the antibody. The two antigen–antibody complexes are separated from the rest of the material. When the ratio of the radioactive complex to the nonradioactive complex is measured, the concentration of the antigen being determined can be calculated.

This method is capable of determining protein concentration as low as 10^{-9} g/mL. Concentrations of lower molecular weight compounds can be determined to within 10^{-12} g/mL.

Dying heart muscle releases MB isoenzyme, which is different from that produced by other types of cells. Radioimmunoassay for this isoenzyme can help diagnose heart attack in persons with chest pains.

X-ray Therapy

X-rays are a penetrating type of radiation, similar to gamma rays but of a lower energy. The amount of radiation and how deep it penetrates the tissue are adjustable in an X-ray machine, whereas these factors are generally fixed in a radioactive source such as Ra or ^{60}Co. X-rays can be used for treatment of superficial skin conditions by adjusting the voltage of the machine so that it produces a "soft" or

Figure 4-8 Dental X-rays. (a) Root canal. (b) Bone destruction with calculus formation on root surfaces. (c) Periapical abscess. (d) Carious lesion. (e) Retained bone fragment.

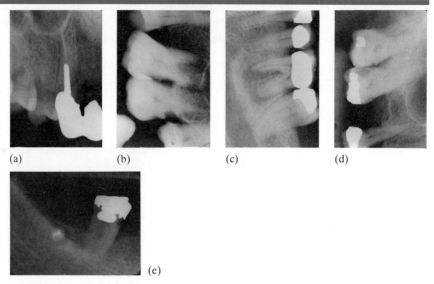

(a)   (b)   (c)   (d)

(e)

nonpenetrating radiation. Although more penetrating X-rays have been used for the treatment of deep-seated malignancies, the cobalt machine (see Figure 4-7) is now widely used for the treatment of different types of cancer.

Another form of X-ray therapy employs oxygen at pressures higher than atmospheric. This treatment is based upon the fact that cancer cells are almost three times as sensitive to destruction by X-rays when the tissues are under 3 atmospheres (atm) pressure as they are at normal atmospheric pressure. This is not true for normal cells. The patient is placed in a chamber containing air at 3 atm pressure and then given X-ray treatments. This type of treatment is called hyperbaric cancer radiation.

X-rays will not pass through bone and teeth as easily as through tissue, so dental X-rays show the presence of cavities, advanced bone destruction, and abscesses as well as the positions of normal and impacted teeth (see Figure 4-8).

Radiopaque substances are compounds that absorb X-rays, thereby allowing body parts to become visible on film. One such radiopaque substance, barium sulfate ($BaSO_4$), is used to detect abnormalities in the stomach and esophagus. Barium sulfate, which is highly insoluble in water and is also nontoxic, is administered orally in the form of a suspension.

Diagnostic Instrumentation

X-ray Scanners

In addition to X-ray machines, another type of instrument used is the CT scanner (computerized tomography). This type of scanner rotates in a circle around the body and makes sharp, detailed records

Figure 4-9 (a) CT scan of the abdomen demonstrating the liver, spleen, stomach, and aorta and the upper pole of the left kidney. (b) CT scan of the abdomen reveals the liver, including caudate and quadrate lobes, right adrenal gland, inferior vena cava, and aorta. (c, d) CT scans of the abdomen at the level of the kidneys. [EMI Medical Inc., Northbrook, IL. Scans courtesy Edward Mallinckrodt Institute of Radiology.]

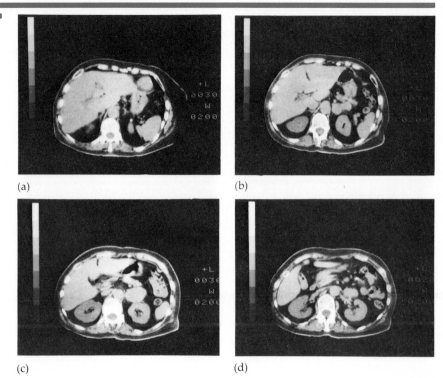

(a) (b)

(c) (d)

of narrow strips of a cross section of the body, each record yielding thousands of bits of information. All of the information is fed into a computer, which then produces a picture of the body section scanned. The results are available in minutes on television screens in black and white (see Figure 4-9) or in color. Even though the scanner makes many passes, the patient is exposed to about the same amount of radiation as with traditional X-ray equipment.

The use of the CT scanner makes possible a detailed examination of any part of a patient's body for the detection of tumors of the brain, breast, kidney, lung, or pancreas. It can also detect abnormal cavities in the spinal cord and enlargements of such organs as the liver, spleen, and heart. In addition, this equipment allows examinations to be performed on an outpatient basis, since no preparation is necessary. The examination can even be performed with the patient wearing clothing.

CT scanners are also used to distinguish between benign and malignant lung tumors. A lung nodule that contains calcium absorbs more X-radiation than the surrounding tissues and so appears darker on the "scan." It has been found that calcium-containing lung nodules are almost always benign.

Xeroradiography

In xeroradiography, conventional X-ray equipment is used to make the exposure. However, the image is produced by an entirely dif-

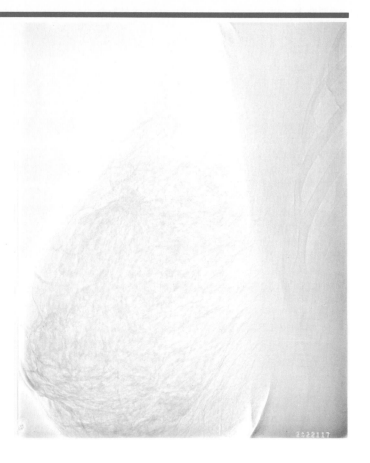

Figure 4-10   A mammogram showing cancer of the breast. [Courtesy Dr. Leonard Berlin, Rush North Shore Medical Center, Skokie, IL.]

ferent process. In X-ray work, the image is produced on a piece of film by a photochemical process. In xeroradiography the image is produced on opaque paper by a photoelectric process. In contrast to X-rays, the xeroradiographic process is dry and requires no darkroom.

Xeroradiography is applied principally to *mammography for early detection of breast cancer. It produces better resolution and makes the interpretation of soft tissue studies easier and more accurate (see Figure 4-10). Xeroradiography is also useful in the search for foreign bodies, such as plastic, wood, and glass, that may not show up on a regular X-ray.

Ultrasonography

Ultrasonography employs very high frequency sound waves in place of X-rays. Normal sound waves have a frequency range from 20 to 20,000 *hertz (Hz). Ultrasound frequencies are those above 20,000 Hz. Those used in diagnostic ultrasound range from 1 million to 15 million Hz (1 to 15 MHz).

Modern ultrasound equipment uses a transducer, a device that both emits and receives sound waves. When connected to a computer, such a device can produce images of internal organs. Photos can also be taken (see Figure 4-11). This technique is not useful for bones and gases, which conduct sound waves poorly, but it is quite useful for liquids and so may be used on human tissue. Advantages of this technique are that it produces no radiation and also that it is relatively inexpensive.

Ultrasound can be used to monitor fetal growth. A special form

Figure 4-11 (a) Ultrasound of a pregnant woman being observed. (b) Lithotripter in use on a gallbladder. [Courtesy Rush North Shore Medical Center, Skokie, IL.]

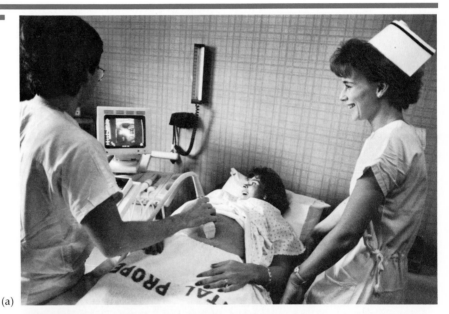

(a)

(b)

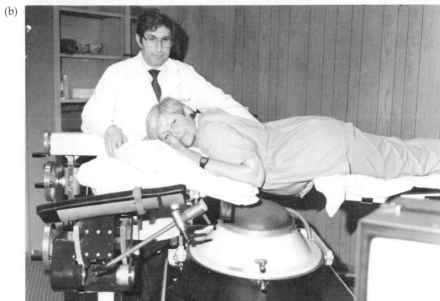

of ultrasound imaging called echocardiography is used to monitor fetal heart problems.

A generator called a lithotripter produces short, intense bursts of *ultrasonic waves, which are used to break up kidney stones.

MRI

MRI (magnetic resonance imaging) is a noninvasive method of following biochemical reactions in both cells and entire organs under normal physiological conditions. The principle of MRI, in general, is that a sample is placed in a strong magnetic field and subjected to radio waves of a frequency appropriate to the nucleus of the element being studied. For biochemical work, ^{31}P and ^{13}C are frequently used. In chemical work, this method is called NMR, nuclear magnetic resonance.

With MRI it is possible to follow changes in phosphate compounds during skeletal muscle contraction and even in heart muscle

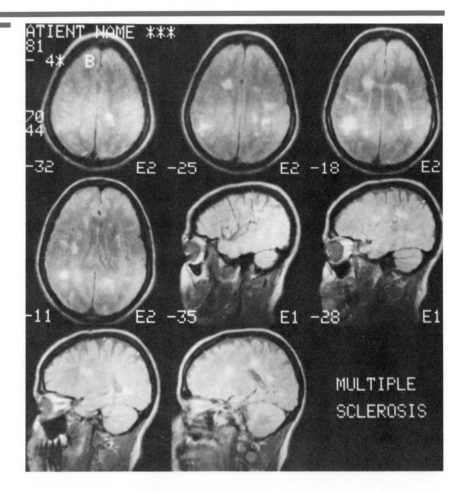

Figure 4-12 MRI scans, multiple sclerosis. [Courtesy Dr. Leonard Berlin, Rush North Shore Medical Center, Skokie, IL.]

itself. Changes in heart muscle caused by oxygen deprivation can now be studied.

MRI is used to scan for brain damage in newborns that have suffered asphyxia or similar problems. MRI can detect brain damage due to strokes and heart attacks within minutes after an attack rather than the hours or days needed in a CT scan. However, MRI poses a potential hazard to patients with pacemakers and artificial metal joints.

Biologic Effects of Radiation

Externally, alpha and beta particles are relatively harmless to humans, for they have slight penetrating power (see Figure 4-13). Gamma rays, with their great penetrating power, have a very definite effect upon the body. If a radioactive substance is taken inside the body, it is the alpha particles that are most harmful.

Protection

Shielding, distance, and limiting exposure are the only effective preventive methods against radiation exposure.

Exposure to external radiation can be controlled by increasing the distance between the body and the source of the radiation. The amount of radiation received varies inversely as the square of the distance; therefore, doubling the distance from a radioactive source permits the body to receive only one-fourth as much radiation. Shielding material, such as lead, when placed between the body and the radioactive source will also protect the body against radiation. This is the reason that dentists and dental assistants stand behind a

Figure 4-13 Changes in pigmentation of skin due to beta burns. (a) A burn area on the neck of a Rongelap native one month after accidental exposure to radiation in 1954. (b) The same burn area one year after the accident, showing complete recovery. [Reproduced from N. Frigerio, *Your Body and Radiation*. U.S. Atomic Energy Commission, Washington, DC, 1967. Photographs from Brookhaven National Laboratory.]

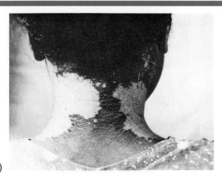
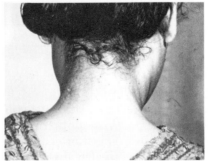

(a)   (b)

lead shield when taking X-rays. They also should shield patients by using lead aprons across their bodies. Clothing or plastic will offer protection against beta emitters.

Example 4-1   A nurse receives an exposure of 20 mrem when standing 3 ft from a radioactive source. What will be the exposure at a distance of (a) 6 ft? (b) 10 ft?

Since exposure to radiation varies inversely with the square of the distance, we can use the relationship

$$\frac{\text{exposure at distance 1}}{\text{exposure at distance 2}} = \frac{(\text{distance 2})^2}{(\text{distance 1})^2}$$

(a) Exposure at distance 1 is 20 mrem; distance 1 is 3 ft and distance 2 is 6 ft. Exposure at distance 2 is unknown, or x.

$$\frac{20 \text{ mrem}}{x} = \frac{(6 \text{ ft})^2}{(3 \text{ ft})^2}$$

$$\frac{20 \text{ mrem}}{x} = \frac{36 \text{ ft}^2}{9 \text{ ft}^2}$$

Canceling ft^2 and cross-multiplying, we have

$$36x = 180 \text{ mrem}$$

$$x = 5 \text{ mrem}$$

(b) Distance 2 is now 10 ft, so

$$\frac{20 \text{ mrem}}{x} = \frac{(10 \text{ ft})^2}{(3 \text{ ft})^2}$$

$$\frac{20 \text{ mrem}}{x} = \frac{100 \text{ ft}^2}{9 \text{ ft}^2}$$

$$100x = 180 \text{ mrem}$$

$$x = 1.8 \text{ mrem}$$

Radiologists, and others working with radioactive material, must wear protective gloves and handle the material with long tongs to avoid direct contact with the body.

Another factor to consider is the length of time of the exposure. The shorter the time that any individual is exposed to radiation, the smaller the dose that will be absorbed.

Radioactive wastes must be disposed of in specially marked containers and should be handled with care.

Persons in regular contact with radioactive material or radiation must wear film badges or other types of detectors (see page 54), and these must be checked periodically for exposure to radiation.

Radiation can strike the molecules within a cell. Since water is the most abundant cellular molecule, we will consider the effects of radiation on a water molecule. Radiation may knock an electron from a water molecule, or it may remove a hydrogen ion, as indicated in the following equations:

$$H_2O \xrightarrow{\text{radiation}} e^- + H_2O^+$$

$$H_2O \xrightarrow{\text{radiation}} H^+ + OH^-$$

In each case, a pair of ions, an ion pair, is produced (see page 53). That is, radiation causes ionization within the cells. Ionization within the cell can disrupt the chemical processes going on inside that cell. Ionization can alter DNA. As a result, some cells may die or fail to multiply, but other cells will continue to live and reproduce. Because of the alteration in DNA, there may be genetic changes (mutations) that can show up in future generations. Large amounts of radiation can also produce cataracts, sterility, and leukemia. Exposure to radiation lessens life expectancy. Rapidly dividing tissue is highly susceptible to radiation. For this reason, unnecessary irradiation of growing children and pregnant women should be avoided. However, since cancerous cells are also rapidly dividing, they too should be highly susceptible to radiation. Hence, irradiation is the form of treatment for some cancers.

Exposure to large amounts of radiation can cause "radiation sickness." The symptoms are gastrointestinal disturbances (nausea, vomiting, diarrhea, general body weakness), a drop in red and white blood cell counts, loss of hair, extensive skin damage, and ulcerative sores that are difficult to heal. Extremely large doses of radiation (over 5000 rads) are fatal.

The lethal effects of radiation can be expressed in terms of the dose that will kill 50 percent of an exposed population within 30 days. The dose is abbreviated LD_{50}^{30}. Typical values of LD_{50}^{30} for various organisms are listed in Table 4-2. Note the inverse relationship between the complexity of an organism and its tolerances to damage by radiation. Table 4-3 indicates the biologic effects of increasing amounts of radiation on rats, and Table 4-4 indicates the symptoms in humans.

The National Council on Radiation Protection and Measurement and the International Commission on Radiological Protection have set the following radiation standards:

1. A dose not exceeding 0.5 rem (500 mrem) per year of whole-body exposure for individual members of the general population.

Table 4-2 LD_{50}^{30} Values (in rems)

Dog	300
Human	450
Monkey	600
Rabbit	750
Rat	800
Bacteria	100,000
Viruses	1,000,000

Table 4-3 Effect of Radiation on Rats	Number of Roentgens	Time Period	Results
	10	—	Genetic changes
	50	—	Shortened life span
	100	30 days	Cataracts; leukemia
	500	10–14 days	Destruction of spleen, sternum, bone marrow; disappearance of leukocytes and platelets; decrease in gamma globulin
	1,000	4–6 days	Failure of intestinal system; loss of intestinal mucosa; bacterial invasion; no antibodies present; no immunity
	10,000	1–2 days	Severe body burns
	100,000	1–2 hr	Destruction of central nervous system

Table 4-4 Symptoms in Humans Due to Exposure to Acute Whole-Body Radiation	Dose (rads)	Symptoms (for 50% of population)
	50–100	Decrease in circulating lymphocytes
	120	Anorexia
	170	Nausea
	210	Vomiting
	242	Diarrhea

2. An average dose to the general population not exceeding 0.17 rem (170 mrem) per year, whole-body radiation.
3. A dose not exceeding 5 rem per year, whole-body exposure for radiation workers.

It is also recommended that actual exposures to radiation be kept as low as possible.

Sources of Radiation

The body receives radiation externally from three principal sources: natural background radiation, medical radiation, and fallout and radioactive wastes. Background radiation comes from space and from radioactive material present in the soil, in the air, in water, and in building materials. The average natural background radiation in the United States is 120 to 150 mrem per year. Global fallout contributes about 4 mrem per year, and radiation from occupational sources about 1 to 2 mrem per year, which is about the same as that received from a television set.

The amount of diagnostic medical radiation varies with the type and frequency of medical treatment. The average in the United

Table 4-5 Amount of Radiation Received During Various Types of Medical Treatment

Type of Treatment	Average Exposure (in mrem)
Chest X-ray	27
Abdomen X-ray	620
Diagnostic upper GI series	1,970
Dental X-ray	910
^{131}I thyroid scan	10,000–20,000 (to thyroid)
^{99m}Tc thyroid scan	1,500 (to thyroid)
Radium implant treatment	3,000–8,000

States is 70 mrem per year. However, X-ray photographs of specific parts of the body may give that body part a very high amount of radiation. Fluoroscopy produces an even greater amount of radiation. Table 4-5 indicates the amount of radiation received during selected types of medical treatment.

Irradiation of Food

The Food and Drug Administration (FDA) has recently allowed irradiation treatment of food. Irradiation consists of exposing food to some form of ionizing radiation such as gamma rays or X-rays to kill insects and microorganisms and also to halt the ripening of fruit. ^{60}Co is most commonly used for this purpose. Irradiation lengthens the shelf life of the food and reduces the need for preservatives, some of which have toxic effects. Table 4-6 gives FDA-allowed limits for irradiation of foods.

Nuclear Energy

Some people believe that nuclear power is the answer to the world's energy problems. Others say that it is too dangerous to use and should be outlawed. What is nuclear power? How can the nucleus of an atom produce such tremendous amounts of energy? We shall discuss these topics under two general headings: nuclear fission and nuclear fusion.

Table 4-6 Radiation Limits for Foods Set by FDA

Food	Dose Limit (kGy)[a]
Fruits and vegetables	1
Dehydrated herbs, seeds, teas	30
Pork	0.3–1
White potatoes	50–140
Wheat, wheat flour	200–500

[a] 1 kilogray = 100,000 rad.

Nuclear Fission

When bombarded with neutrons, the nuclei of several heavy elements split into smaller pieces. This process is called *fission. ^{235}U can be split by a neutron into smaller pieces, such as strontium and xenon, accompanied by the release of more neutrons and a tremendous amount of energy. When this reaction takes place, the sum of the masses of the products is less than the masses of the reactants. That is, nuclear reactions do not obey the law of conservation of mass. They do, however, obey the combined law of conservation of mass and energy. The amount of mass that disappears is converted into an equivalent amount of energy. This amount of energy can be calculated by using Einstein's equation

$$E = mc^2$$

where E is the amount of energy, m is the loss in mass, and c is a constant, the speed of light. When m is in kilograms and c is in meters per second, E is in joules.

Consider the fission of ^{235}U, written in equation form, with weights of reactants and products also being indicated.

$$^1_0n \; + \; ^{235}_{92}U \; \longrightarrow \; ^{94}_{38}Sr \; + \; ^{139}_{54}Xe \; + \; 3\,^1_0n \; + \; energy$$

(kg)   1.0087   234.9934   93.9154   138.9179   3(1.0087)

Total mass of reactants	236.0021 kg
Total mass of products	235.8594 kg
Loss in mass	0.1427 kg

Using Einstein's equation, $E = mc^2$, we find

$$E = 0.1427 \text{ kg} \times (3.00 \times 10^8 \text{ m/s})^2$$
$$= 1.28 \times 10^{16} \text{ joules}$$

This is equivalent to 3 million million kcal, a tremendous amount of energy.

Chain Reactions

Note in the above reaction that when a neutron strikes a uranium nucleus, three more neutrons are produced. If each of these three neutrons strikes another uranium nucleus, then nine more neutrons will be produced. This process keeps on building up as more and more uranium nuclei undergo fission. Such a reaction, called a *chain reaction, can sustain itself, producing more and more energy (see Figure 4-14). However, a minimum amount of uranium nuclei (called the critical mass) must be present for a chain reaction to occur.

Figure 4-14 Chain reaction.

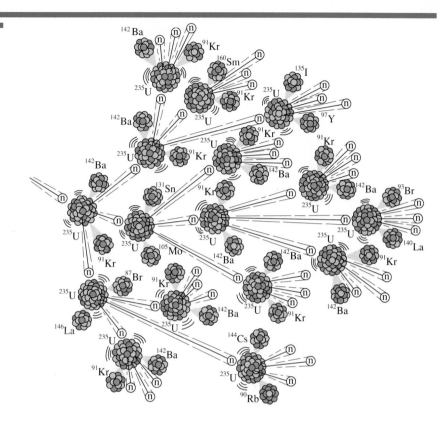

In an atomic bomb, the chain reaction is uncontrolled, producing a tremendous explosion. If the chain reaction is controlled, large amounts of useful energy are available. Devices that control nuclear chain reactions are called **reactors**.

Nuclear Reactors

A nuclear reactor (Figure 4-15) uses a fissionable isotope such as ^{235}U or ^{239}Pu as its fuel. Control rods made of cadmium or boron steel, which absorb neutrons, are used to control the reaction. When inserted into the reactor, these rods absorb neutrons and slow down the rate of the reaction. When pulled partway out, they allow the rate of the reaction to increase by absorbing fewer neutrons. Surrounding the fissionable material is a moderator such as water or graphite, which slows down the neutrons produced during fission.

The energy produced during fission is primarily in the form of heat. A heat transfer system removes heat from the core of the reactor and transfers this heat to a steam-generating unit.

Naturally, because of the intense radiation in the reactor, there must be sufficient shielding to protect personnel from radiation. Also, safety equipment is designed to shut down the reactor in case of a malfunction or radiation leakage.

Figure 4-15   Diagram of a nuclear power plant.

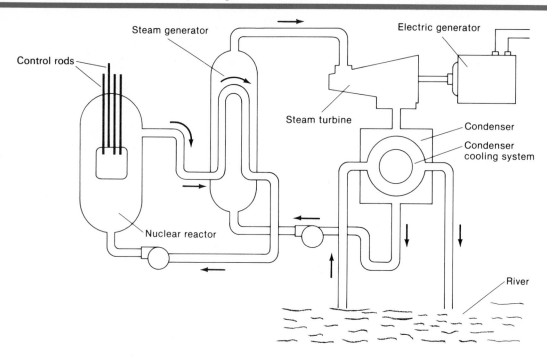

Fusion

The sun and other stars produce their energy by a process called *fusion. In this process, small nuclei are combined to form larger ones. In the sun, the overall reaction is the combination of hydrogen to form helium, with the subsequent release of incredible amounts of energy. Note that fusion is the opposite of fission. Fusion is the combination of small nuclei to make larger ones; fission is the splitting of a larger nucleus into smaller ones. It has been said that fusion is the energy source of the future. Fusion produces more energy than a comparable fission reaction, and with less radioactive by-products to be disposed of. The fuel for a fusion reaction is deuterium, an isotope of hydrogen, which comes from seawater and is relatively plentiful. What has prevented the production of fusion reactors?

The answer lies in the fact that in order for a fusion reaction to work, temperatures in excess of 100 million° C must be produced. Also, suitable containers for the fusion reaction have not been designed in a workable form.

When a fusion reaction is produced in the laboratory, it will probably be the following.

$$\underset{\substack{\text{isotopes of} \\ \text{hydrogen}}}{{}^2_1H + {}^3_1H} \longrightarrow {}^4_2He + {}^1_0n + energy$$

The explosion of a hydrogen bomb is an example of an uncontrolled fusion reaction.

Disposal of Radioactive Waste

Waste disposal in the United States has become a problem because of the diminishing amount of landfill available. A larger problem is the disposal of radioactive materials, not only from nuclear power plants and nuclear weapons production but also from medical applications.

From the beginning of the nuclear age until 1982 there was no permanent method of disposal of radioactive waste. It was all placed in temporary storage. Much of the waste will remain very radioactive for 20,000 years or longer. The potential problems of genetic mutation and cancer due to radioactive materials make the safe permanent disposal of such substances a matter of immediate concern.

The Nuclear Waste Policy Act of 1982 mandated the choosing and preparing of deep underground disposal sites for nuclear waste materials. Of great importance in the site selection are the stability of the rock formation in which the waste is to be placed; the absence of volcanic or earthquake activity; and the presence of rocks impermeable to water. With proper disposal, the amount of radioactivity will gradually diminish with little or no effect upon the environment.

Summary

Radioactivity is the property of emitting radiation from the nucleus of an atom. The three types of radiation are alpha, beta, and gamma. Alpha particles are positively charged helium nuclei. Beta particles are high-speed electrons and are negatively charged. Gamma rays are a high-energy form of electromagnetic radiation and have no charge or mass.

In nuclear reactions both the sum of the atomic numbers and the sum of the mass numbers are the same on both sides of an equation. In addition to naturally occurring radioactive substances, artificially radioactive substances can be prepared by bombardment with particles such as protons, neutrons, and alpha particles.

A scanner helps to locate malignancies by moving back and forth across the site being studied and detecting the radiation in each area over which it travels. The radiation comes from radioisotopes administered to the patient, which are selected to accumulate at the desired body part.

The half-life of a radioisotope is the amount of time required for half of its atoms to decay. For medical work, a radioisotope must have a half-life long enough to give the body part the radiation it needs and short enough that it will not give the patient too much radiation during the period it remains in the body.

Radioisotopes are used in the diagnosis and treatment of various disorders in the body — ^{131}I for thyroid conditions, ^{32}P for eye tumors, ^{99m}Tc for scans, and ^{60}Co for radiation therapy.

X-rays are a type of radiation similar to gamma rays. The penetrating powers of X-rays can be controlled, whereas those of gamma rays cannot.

X-ray treatment combined with high-pressure oxygen has had some success in the treatment of cancer.

The units of radiation are the curie, the gray, the roentgen, the rad, and the rem. The roentgen applies primarily to X-rays. The rem is the unit most commonly used in relation to the body.

Radiation produces ionization within the cells, causing some type of damage. Small amounts of radiation produce genetic changes; larger amounts cause a shortened life span. Very large amounts of radiation can cause death within a short period of time.

X-rays are harmful to the body because of the effects produced by the radiation, so care should be taken to avoid unnecessary exposure. Fluoroscopy produces even more radiation than X-ray photographs.

Nuclear fission involves the splitting of a large nucleus into smaller ones with large amounts of energy being produced. Nuclear fusion involves the combining of small nuclei into larger ones, with the release of tremendous amounts of energy.

The disposal of radioactive waste is an ongoing problem, with the effect upon the environment being an item of top priority.

Questions and Problems

A

1. What are alpha particles? beta particles? gamma rays?
2. Define the term *half-life.*
3. Balance the following equations.
   (a) $^9_4Be + {}^4_2He \longrightarrow {}^1_0n + ?$
   (b) $^{27}_{13}Al + {}^1_0n \longrightarrow {}^4_2He + ?$
   (c) $^{30}_{15}P \longrightarrow {}^0_{-1}e + ?$
4. What is artificial radioactivity?
5. What are radioisotopes? Give one use for (a) ^{14}C, (b) ^{131}I, and (c) ^{59}Fe.
6. If 100 mg of ^{32}P is present on a certain day, approximately how much will be present 2 months later?
7. If 2 mg of ^{99m}Tc is present at 8 A.M. Monday, approximately how much was present at 2 A.M. on the preceding Sunday?
8. How can radiation be detected?
9. What are X-rays used for?
10. What are the units of radiation?
11. List some of the physiologic effects of radiation.
12. What can be done to minimize the effects of radiation on the body?
13. From where does the body receive external radiation?
14. Describe the use of a scanner. Name one radioisotope used in scans, and indicate where it can be used.
15. How can blood volume be determined?
16. What is xeroradiography? For what purposes is it used?

17. What is ultrasonography?
18. Compare criteria for selection of radioisotopes for diagnostic and therapeutic uses.
19. How does the CT scanner work?
20. What are the values of radiation dose for the general population? for radiation workers?
21. Explain changes in atomic number and mass number when an atom emits an alpha particle, a beta particle, a gamma ray.

B

22. What does the "m" in ^{99m}Tc stand for?
23. Compare nuclear fission with nuclear fusion.
24. How does a nuclear power plant produce energy?
25. Where does the energy of nuclear fission come from? How can the amount of energy of this type of reaction be calculated?
26. What is a chain reaction?
27. What is PET? How is it used medically?
28. What is a scintillation counter?
29. What is radioimmunoassay? For what purposes is it used?
30. What is a radiopaque substance? Where is it used?
31. How does radiation produce ion pairs? What effect can they have on cells?
32. What do the numbers in LD^{30}_{50} indicate?
33. If a nurse standing 2 ft from a patient with a radium implant receives an exposure of 32 mrem, what would be the exposure level 8 ft away?

Practice Test

1. The charge on an alpha particle is _____.
   a. +1   b. −1   c. +2   d. −2
2. The type of radiation carrying a negative charge is the _____.
   a. alpha particle   b. beta particle
   c. gamma ray   d. none of these
3. The most penetrating type of radiation is _____.
   a. alpha   b. beta
   c. gamma   d. delta
4. In the reaction $^{234}_{90}U \longrightarrow {}^{4}_{2}He + X$ the atomic number of element X is _____.
   a. 88   b. 90   c. 91   d. 93
5. The mass number of element X in question 4 is _____.
   a. 230   b. 232   c. 234   d. 236
6. An element has a half-life of 1 week. If 100 mg is present today, how many milligrams will be present 3 weeks from now?
   a. 100   b. 50   c. 25   d. 12.5
7. A device to detect radiation is the _____.
   a. scintillation counter
   b. Geiger counter
   c. film badge
   d. all of these
8. The unit of radiation directly relating to human tissue is the _____.
   a. curie
   b. roentgen
   c. rad
   d. rem
9. Radiation _____.
   a. decreases life expectancy
   b. may cause nausea
   c. affects rapidly dividing cells
   d. all of these
10. If a person receives a certain amount of radiation from a source at a distance of 10 ft, what fraction of that radiation will that person receive at a distance of 20 ft?
    a. $\frac{3}{4}$ b. $\frac{1}{2}$ c. $\frac{1}{4}$ d. 0

5

Chemical Bonding

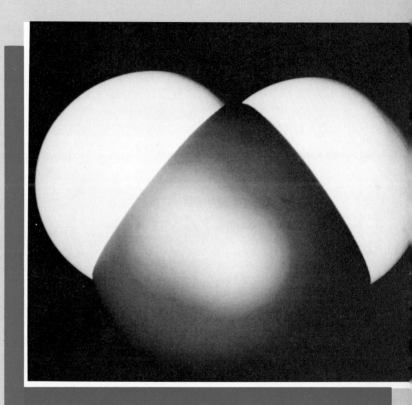

Computer-generated molecular model of water (H_2O).

Molecules

A molecule is a combination of two or more atoms. These atoms may be of the same elements, as in the oxygen molecule (O_2), or of different elements, as in the hydrogen chloride molecule (HCl). A more complicated molecule is that of glucose, $C_6H_{12}O_6$. What holds the atoms together in a molecule? Atoms are held together by bonds that can be classified into two main types: ionic and covalent.

Stability of the Atom

Most atoms are considered stable (nonreactive) when their highest (outer) energy level has eight electrons in it and is therefore filled. The noble gases neon, argon, krypton, xenon, and radon all have eight electrons in their highest energy level. They are stable. One exception to this rule of eight (the octet rule) is the lightest noble gas, helium, which is stable even though it has only two electrons in its highest energy level because that energy level is the first and can hold only two electrons.

Atoms that do not have eight electrons in their highest energy level may lose, gain, or share their valence electrons with other atoms in order to reach a more stable structure with lower chemical potential energy. This process of rearrangement of the valence electrons is responsible for chemical reactions between atoms.

Symbols and Formulas

A symbol not only identifies an element but also represents one atom of that element. Thus, the symbol Cu designates the element copper and also indicates one atom of copper (the number 1 being understood and not written). Two atoms of copper are designated as 2 Cu.

A formula consists of a group of symbols that represent the elements present in a substance. It also indicates one molecule of that substance. Thus the formula NaCl indicates that the compound (sodium chloride) consists of one atom of sodium (Na) and one atom of chlorine (Cl).

If there is more than one atom of an element present in a compound, numerical subscripts are used to indicate how many atoms of each element are present. In the compound HNO_3 (nitric acid) there are one atom of hydrogen (H), one atom of nitrogen (N), and three atoms of oxygen (O), all of which make up one molecule of HNO_3. In the compound $K_2Cr_2O_7$ (potassium dichromate) there are two atoms of potassium (K), two atoms of chromium (Cr), and seven atoms of oxygen making up one molecule.

To designate more than one molecule of that substance, a number (a **coefficient**) is placed in front of the formula for that substance. For example, 2 HNO_3 indicates two molecules of HNO_3; 6 $K_2Cr_2O_7$ indicates six molecules of $K_2Cr_2O_7$.

The formula O_2 indicates one molecule of oxygen with the 1 being understood. This molecule consists of two atoms of oxygen. The formula H_2 indicates one molecule of hydrogen, which consists of two atoms of hydrogen. Both O_2 and H_2 are called **diatomic molecules** because they are each made up of two atoms. Other examples of diatomic molecules are N_2, F_2, Cl_2, Br_2, and I_2. Water (H_2O) is a triatomic molecule—it contains two atoms of hydrogen and one atom of oxygen.

Molecules may also be **monatomic**; that is, they can consist of only one atom. Examples of monatomic molecules are neon (Ne), and argon (Ar). Molecules of other elements such as sulfur, S_8, are **polyatomic**; they contain several atoms in their molecules.

Be very careful in distinguishing between 2 O and O_2. The 2 O represents two atoms of oxygen that are not combined; they are separate, independent atoms; O_2 represents one molecule of oxygen, which consists of two atoms of oxygen that are chemically combined with a covalent bond between them. (Two molecules of oxygen would be shown as 2 O_2.) This note of caution applies to other diatomic molecules as well.

Electron-Dot Structures

The electron-dot structure of an atom (also called a Lewis structure) is an abbreviated representation for the structure of that atom. In this system, the nucleus and all of the energy levels except the highest one are represented by the symbol for that element. Each valence electron is indicated by a dot. For example, the element sodium (symbol Na, atomic number 11) has its nucleus surrounded by 11 electrons— 2 in the first energy level, 8 in the second energy level, and 1 in the third (highest) energy level. The electron-dot structure for the sodium atom is Na·, with the dot representing the one valence electron and the symbol Na representing the remainder of the atom. Carbon, atomic number 6, has the electron configuration 2e) 4e). The electron-dot representation for carbon is

$$\cdot \overset{\displaystyle \cdot}{\underset{\displaystyle \cdot}{C}} \cdot$$

Argon, atomic number 18, has the electron configuration of 2e) 8e) 8e). The electron-dot structure for argon is

$$: \overset{\displaystyle \cdot \cdot}{\underset{\displaystyle \cdot \cdot}{Ar}} :$$

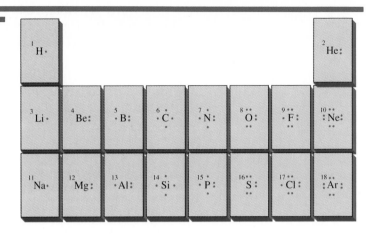

Figure 5-1 Electron-dot structures of the first 18 elements.

Formation of Ions

Figure 5-1 shows electron-dot structures of the elements in the first three periods of the periodic chart. Consider the sodium atom with the electron structure 2e) 8e) 1e). If the sodium atom loses its one outer electron, it will reach a noble gas structure of eight electrons in its outer energy level. A noble gas structure has great stability.

When a sodium atom loses an electron, it becomes a positively charged particle called a sodium ion. This reaction may be written as

$$\mathrm{Na} \cdot - e^- \longrightarrow \mathrm{Na}^+$$

or, more appropriately,

$$\mathrm{Na} \longrightarrow \mathrm{Na}^+ + e^-$$

where the positive sign indicates a charge of +1 on the sodium ion. (Note that the number 1 is understood and not written.) The charge on the sodium ion is positive because the sodium ion still has eleven protons in its nucleus but now has only ten electrons outside that nucleus.

Likewise, the aluminum atom, which has the electron structure 2e) 8e) 3e), loses all three outer electrons when it forms an aluminum ion with a charge of +3 (written to the upper right of the symbol as 3+).

$$\cdot \mathrm{Al} \colon \longrightarrow \mathrm{Al}^{3+} + 3\,e^- \quad \text{or simply} \quad \mathrm{Al} \longrightarrow \mathrm{Al}^{3+} + 3\,e^-$$

A metal that has one valence electron forms an ion with a 1+ charge; a metal with two valence electrons forms an ion with a 2+ charge, and so on. The positive charge on a metallic ion is equal to the number of electrons lost by the metal.

Elements that have six or seven electrons in their highest energy level tend to gain electrons to reach a stable configuration of eight. Such elements are called nonmetals. (Most elements having four or five outer electrons are also nonmetals. These will be discussed separately under covalent bonds.)

Consider the element chlorine, 2e) 8e) 7e). The chlorine atom will tend to gain one electron to bring its highest energy level to eight, reaching a stable (noble gas) structure. Chlorine will thus form an ion with a charge of 1−, or, omitting the dots,

$$Cl + e^- \longrightarrow Cl^-$$

Since the ion has one more electron than the atom, it will have a charge of 1−, again with the 1 being understood and not written.

Likewise, the sulfur atom, 2e) 6e), can gain two electrons to form an ion with a charge of 2−.

$$S + 2e^- \longrightarrow S^{2-}$$

The S^{2-} ion has eight electrons in the highest energy level, a noble gas structure.

An atom that has either lost or gained electrons in its highest energy level is called an ion. Ions formed from a metal will have a positive charge equal to the number of electrons lost. Ions formed from nonmetals will have a negative charge equal to the number of electrons gained.

Positive ions are attracted toward a negatively charged electrode called a cathode. Such ions are called *cations. Likewise, negative ions are attracted toward a positively charged electrode, an anode. These ions are called *anions. Common cations in body fluids are the sodium ion, Na^+, the potassium ion, K^+, and the calcium ion, Ca^{2+}. The chloride ion, Cl^-, is the most common anion in body fluids.

Size of Ions

When a metal loses an electron (or electrons), the positive charge on the nucleus is greater than the negative charge in the electron energy levels, so the nucleus pulls in the electrons and thus decreases the size of the ion. That is, for metals, the ionic radius is less then the atomic radius.

When a nonmetal gains an electron (or electrons), the positive charge on the nucleus is less than the negative charge in the electron energy levels; thus the nucleus cannot hold the electrons as tightly as before. Therefore, for nonmetals, the ionic radius is greater than the atomic radius.

Figure 5-2 illustrates the relative sizes of several atoms and their corresponding ions.

Figure 5-2   Atomic and ionic sizes of some elements.

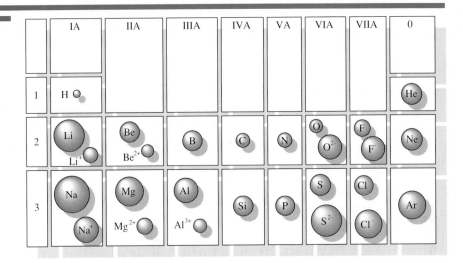

Ionic Bonds

When a sodium atom (Na) combines with a chlorine atom (Cl) to form a sodium chloride molecule (NaCl), the sodium atom loses one electron to form a positively charged sodium ion (Na^+). At the same time the chlorine atom gains that one electron to form a negatively charged chloride ion (Cl^-).

The reaction is

$$Na\cdot + :\ddot{\underset{\cdot\cdot}{Cl}}: \longrightarrow Na^+ + :\ddot{\underset{\cdot\cdot}{Cl}}:^- \qquad \text{or} \qquad Na + Cl \longrightarrow Na^+ + Cl^-$$

The positively charged sodium ion and the negatively charged chloride ion will be attracted to each other and will be held together by the electrostatic attraction of their charges (opposite charges attract each other). This type of bonding is called an **ionic bond.** An ionic bond results from the transfer of an electron or electrons from one atom to another with the formation of ions that attract one another.

Another example of a transfer of electrons from a metal to a nonmetal is in the reaction between magnesium (Mg) and two chlorine atoms.

$$Mg: \left\langle \begin{array}{c} :\ddot{\underset{\cdot\cdot}{Cl}}: \\[1em] + \\[1em] :\ddot{\underset{\cdot\cdot}{Cl}}: \end{array} \right. \longrightarrow Mg^{2+} + \begin{array}{c} :\ddot{\underset{\cdot\cdot}{Cl}}:^- \\[1em] \\[1em] :\ddot{\underset{\cdot\cdot}{Cl}}:^- \end{array}$$

or

$$Mg + 2\,Cl \longrightarrow Mg^{2+} + 2\,Cl^-$$

where the positively charged magnesium ion and the negatively charged chloride ions are held together by ionic bonds. (Again each ion has a completed highest energy level of eight.)

Polyatomic Ions

A group of atoms that stay together and act as a unit in a chemical reaction is called a **polyatomic ion.** A polyatomic ion acts as if it were a simple ion. Table 5-1 indicates the name, formula, and charge of several common polyatomic ions.

Naming Ionic Compounds

Compounds that contain ions are called ionic compounds or electrolytes. As will be discussed later, ionic compounds fall into three categories—acids, bases, and salts.

Ionic compounds that contain only two types of elements are called **binary** compounds. To name binary compounds, the following system is used.

name of positive ion followed by stem of negative ion + *ide*

Table 5-1 Common Polyatomic Ions

Name	Formula and Charge
Sulfate	SO_4^{2-}
Nitrate	NO_3^-
Phosphate	PO_4^{3-}
Carbonate	CO_3^{2-}
Hydroxide	OH^-
Bicarbonate	HCO_3^-
Ammonium	NH_4^+

Note that the names of all binary compounds end in the letters *ide*. The stem is the first part of the name of the element forming the negative ion. Stems for some common elements are listed in Table 5-2.

Table 5-2 Stems for Some Common Elements

Element	Stem
Oxygen	ox
Chlorine	chlor
Bromine	brom
Iodine	iod
Nitrogen	nitr
Sulfur	sulf
Carbon	carb
Phosphorus	phosph

Example 5-1 What is the name of the compound $CaBr_2$?
The name of the positive ion, which is always written first in an ionic compound, is calcium. The negative ion comes from Br, the element bromine. The stem for bromine is **brom**. The ending is **ide.** Thus the name of $CaBr_2$ is **calcium bromide.**

Example 5-2 What is the name of the compound KCl?
The name of the positive ion is potassium. The stem of the negative ion, from the element chlorine, Cl, is **chlor.** The ending is **ide.** The name is **potassium chloride.**

For ionic compounds containing polyatomic ions, the following system of naming is used.

name of positive ion followed by name of polyatomic ion

> *Example 5-3*   Name the compound $MgSO_4$.
>   The positive ion is **magnesium**. The name of the SO_4 polyatomic ion, from Table 5-1, is **sulfate**. Thus, $MgSO_4$ is called **magnesium sulfate**.

One exception to the rule for naming binary compounds is the compound NH_4Cl, where the NH_4 polyatomic ion is treated as if it were a simple positive ion. Thus the compound NH_4Cl is treated as if it were a binary compound and is called ammonium chloride.

Covalent Bonds

Ionic bonding results from the loss or gain of electrons. However, there is another method by which atoms can be bonded together. This is by the sharing of electrons (covalent bonding).

In the chlorine molecule, Cl_2, each of the two chlorine atoms has seven outer electrons. In this case both atoms will share electrons so that each will have a completed outer energy level of eight electrons. Diagram (5-1) shows two chlorine atoms with their electrons so situated that each has eight electrons around it.

$$:\ddot{\underset{..}{Cl}}\cdot \; + \; \cdot\ddot{\underset{..}{Cl}}: \; \longrightarrow \; \left(:\ddot{\underset{..}{Cl}}\!:\!\ddot{\underset{..}{Cl}}:\right) \qquad (5\text{-}1)$$

Note that each atom has a noble gas structure, with eight outer electrons. Each of the chlorine atoms is sharing one electron with the other. The bond that holds these two atoms together is called a **covalent bond**. Note that in the chlorine molecule, Cl_2, there has been no electron loss or gain and so there are no chloride ions present. This is one of the primary differences between ionic and covalent bonds. In compounds containing ionic bonds, ions are present, whereas in compounds containing covalent bonds no ions are present.

The covalent bond between the two chlorine atoms can be indicated by a short line joining the atoms, Cl—Cl, with the electrons being understood and not written.

Covalent bonds can also be formed between atoms of different elements. In compounds containing covalent bonds, each atom usually has eight electrons around it, since eight electrons in the outer energy level represent a stable structure. An exception to this rule (see page 82) is hydrogen, which in compounds has only two electrons around it (Recall that the first energy level can hold only two electrons.)

The compound carbon tetrachloride, CCl_4, can be diagrammed in either of the ways shown at (5-2).

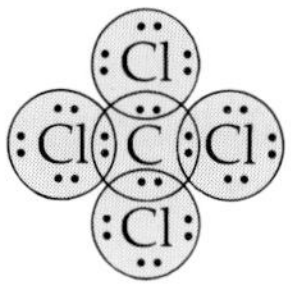

$$Cl—C—Cl \quad (5\text{-}2)$$

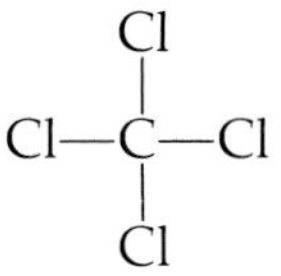

The compound ammonia, NH_3, can be represented as

$$H—N—H \quad (5\text{-}3)$$

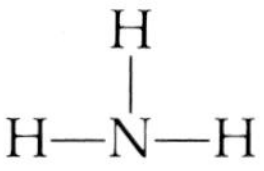

In carbon tetrachloride there are four covalent bonds, one between each of the chlorines and the carbon. In ammonia there are three covalent bonds. Note that in these structures each element has eight electrons around it, except for hydrogen, which has only two.

Since metals tend to lose electrons, they usually do not form covalent compounds. Thus, we can say that most covalent compounds are formed between nonmetals.

The compound carbon dioxide, CO_2, can be represented as

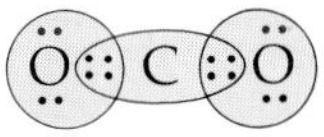

$$O=C=O \quad (5\text{-}4)$$

There are two double covalent bonds present in carbon dioxide. In the previous examples, single covalent bonds were present, representing one shared pair of electrons. Single covalent bonds are also called **single bonds**. A double covalent bond (also called a **double bond**) represents two shared pairs of electrons. A triple covalent bond (**triple bond**) represents three shared pairs of electrons, as in the compound nitrogen, N_2,

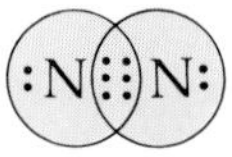

$$N≡N \quad (5\text{-}5)$$

Note that in CO_2 and in N_2 there are eight electrons around each atom.

Nonpolar and Polar Covalent Bonds

Consider the compound Cl_2 or Cl—Cl. There is a single bond representing a pair of shared electrons between the two chlorine atoms. The two chlorines are identical, and so the electrons should be shared equally between them. Such a type of bond is called a **nonpolar covalent bond**.

In the compound HCl, or H—Cl, the single bond again represents a shared pair of electrons. However, this pair of electrons is not shared equally. Let us see why.

Electronegativity is the attraction of an atom for electrons. The greater the electronegativity, the greater the attraction for electrons;

Table 5-3 Electronegativities	F	4.0	Br	2.8	H	2.1
of Various Elements	O	3.5	C, S	2.5	Ca	1.0
	N,Cl	3.1	I	2.4	Na	0.9

the lower the electronegativity the less the attraction for electrons. Table 5-3 indicates the electronegativities of some elements.

Again consider the HCl molecule. Chlorine is more electronegative than hydrogen and so attracts the shared electrons more strongly. Thus, the shared pair of electrons will be closer to the chlorine than to the hydrogen, or

$$H:\overset{..}{\underset{..}{Cl}}:\qquad \text{which can also be represented as}\qquad \overset{\delta+}{H}\ \overset{\delta-}{Cl}$$

where the δ sign (Greek delta) indicates a partial charge. So, one end of the HCl molecule has a partial positive charge and the other end has a partial negative charge. Note, however, that there are no ions formed, only partial charges. Such a type of bond is called a **polar covalent bond.** Thus, with equally shared electrons, as between like atoms or between atoms of equal electronegativity, nonpolar covalent bonds are formed. When atoms of different electronegativity form a covalent bond, the bond is always polar.

Another example of a compound containing a polar covalent bond is water, H_2O. Its structure is

$$H:\overset{..}{\underset{..}{O}}:\qquad \text{or}\qquad \overset{\delta+}{H}-\overset{\delta-}{O}\diagdown_{H^{\delta+}}$$
$$\underset{H}{}$$

Because oxygen is more electronegative than hydrogen, it becomes the negative side of the molecule. The hydrogens are at the positive side. The polar nature of the water molecule will be discussed in more detail in Chapter 10.

However, a nonpolar molecule may contain polar bonds. Consider the compound CCl_4 whose partial charges are indicated in structure (5-6).

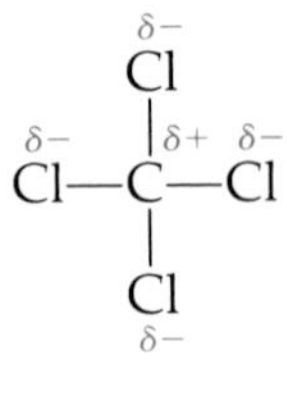

$$\begin{array}{c} \overset{\delta-}{Cl} \\ | \\ \overset{\delta-}{Cl}-\overset{\delta+}{C}-\overset{\delta-}{Cl} \\ | \\ \underset{\delta-}{Cl} \end{array} \qquad (5\text{-}6)$$

The bonds between the carbon and each chlorine are polar. But there is no negative end to the molecule and no positive end. Rather, the outer part is negative and the inner part positive (partially), so the molecule is nonpolar. In general, symmetrical molecules are nonpolar even though they may contain polar bonds.

Resonance

Sometimes more than one electron-dot structure can be drawn for the same group of atoms. Consider the compound SO_3. Three different electron-dot structures can be drawn, all of which satisfy the requirements of eight electrons around each atom.

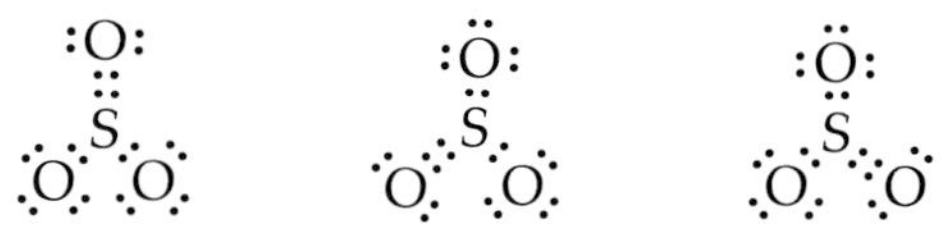

or

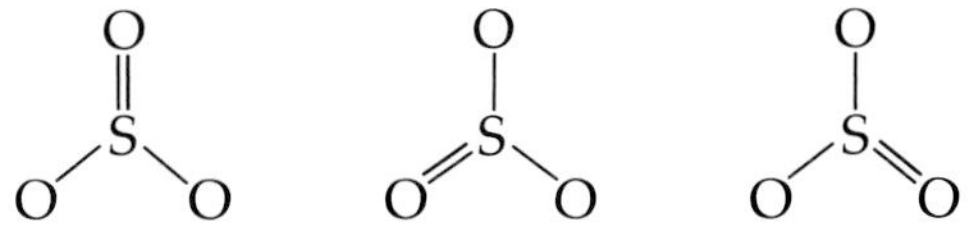

According to these structures, there are two different types of bonds between the sulfur and the oxygen atoms—single covalent bonds and a double covalent bond. However, experimental evidence shows that all the bonds in the SO_3 molecule are the same.

The above three electron-dot structures are said to be in **resonance** and are so indicated by a double-headed arrow as follows.

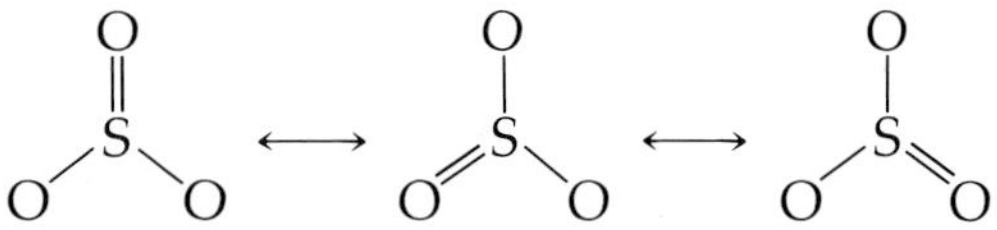

Resonance occurs when more than one electron-dot structure can be drawn for a given molecule or ion. The resulting structures are called **resonance structures.**

Resonance occurs because electrons are not fixed objects near a given atom. Instead they move around the entire molecule. Therefore the correct structure lies somewhere between the various resonance structures. For most purposes in inorganic chemistry, the electron-dot structures are satisfactory, but for organic compounds based on benzene (see Chapter 19), resonance structures are important and must always be considered.

Naming Covalent Compounds

Covalent binary compounds have names ending in *ide*, as do ionic binary compounds. To name covalent binary compounds, the following system is used.

prefix + name of first element followed by
prefix + stem of second element + *ide*

Note that this system is similar to that for ionic compounds except that prefixes are used for covalent compounds, whereas no prefixes are used in naming ionic compounds. The prefixes in common use are

mono-, 1   di-, 2   tri-, 3   tetra-, 4   penta-,5

However, the prefix mono- is usually understood and not written.

Example 5-4   Name the covalent compound CCl_4.
   The name of the first element is carbon. There is only one carbon atom indicated in the formula, so the prefix is mono- (understood and not written). The second element is chlorine, stem chlor (see Table 5-2). Since there are four chlorines indicated in the formula, the prefix is tetra-. The ending as with all binary compounds is ide; so the name is carbon tetrachloride.

Example 5-5   Name the covalent compound P_2O_3.
   By following the previous example and using the prefixes di- and tri-, respectively, the name is found to be diphosphorus trioxide.

Strengths of Bonds

Consider Table 5-4 of the melting points of several ionic and covalent compounds. In general, compounds containing ionic bonds have higher melting points than compounds containing covalent bonds. That is, it takes more energy (heat) to separate the particles in ionic compounds than it takes to separate those in covalent compounds. Although the bonds holding ionic compounds together are generally weaker than those holding covalent compounds together, ionic compounds contain many more bonds than covalent compounds.

Table 5-4   Melting Points of Various Substances

Ionic Compound	Melting Point (°C)	Covalent Compound	Melting Point (°C)
Sodium chloride, NaCl	800	Glucose, $C_6H_{12}O_6$	146
Calcium chloride, $CaCl_2$	782	Carbon tetrachloride, CCl_4	−23
Zinc oxide, ZnO	1975	Urea, NH_2CONH_2	133

However, covalent substances, such as diamond, that contain a network of covalent bonds also have extremely high melting points.

Water solutions of ionic compounds conduct electricity (they are electrolytes) because they contain ions. Water solutions of covalent compounds do not contain ions and do not conduct electricity.

Most ionic compounds are soluble in polar solvents such as water. Most covalent compounds are insoluble in polar solvents.

Most ionic compounds are insoluble in nonpolar solvents such as benzene. Many covalent compounds are soluble in nonpolar solvents.

Molten ionic compounds conduct electricity because they contain ions. Molten covalent compounds contain no ions and do not conduct electricity.

Oxidation Numbers

For ionic compounds, the oxidation number, sometimes called the *charge* of an element, is equal to the number of electrons lost or gained and therefore is the same as the charge on the ion. That is, in sodium chloride, Na^+Cl^-, the oxidation number of sodium is +1 and that of chlorine is −1. In the compound $MgBr_2$, where the magnesium ion has a charge of 2+ and each bromide ion a charge of 1−, the oxidation number of magnesium is +2 and that of each bromine −1.

For covalent compounds, where electrons are shared and not transferred, oxidation numbers are assigned to elements using the following rules.

1. All elements in their free state have an oxidation number of zero.
2. The oxidation number of oxygen is −2 (except in peroxides, where it is −1).
3. The oxidation number of hydrogen is +1 (except in metal hydrides, where it is −1).
4. The sum of oxidation numbers in all compounds must equal zero. (That is, all compounds are electrically neutral.)
5. All elements in group IA have an oxidation number of +1.
6. All elements in group IIA have an oxidation number of +2.

Table 5-5 lists the oxidation numbers of various elements.

Calculating Oxidation Numbers from Formulas

As has been previously mentioned, the sum of the oxidation numbers in any compound must equal zero. Let us find the oxidation number of zinc in zinc oxide, ZnO. Note that the oxidation number of oxygen, as listed in Table 5-5, is −2. Writing the formula of the compound with the known oxidation number above,

$$? + -2 = 0$$
$$Zn \quad O$$

Positive Oxidation Numbers		Negative Oxidation Numbers	
Name and Symbol	Oxidation Number	Name and Symbol	Oxidation Number
Hydrogen H^+	+1	Chloride Cl^-	−1
Sodium Na^+	+1	Bromide Br^-	−1
Potassium $\cdot K^+$	+1	Iodide I^-	−1
Silver Ag^+	+1	Sulfide S^{2-}	−2
Ammonium $NH_4{}^+$	+1	Oxide O^{2-}	−2
Calcium Ca^{2+}	+2		
Magnesium Mg^{2+}	+2		
Aluminum Al^{3+}	+3		
Iron Fe^{2+} and Fe^{3+}	+2 and +3		
Copper Cu^+ and Cu^{2+}	+1 and +2		
Tin Sn^{2+} and Sn^{4+}	+2 and +4		

Table 5-5   Oxidation Numbers of Some Elements and Ammonium Ion[a]

[a] Note that some elements, such as copper, tin, and iron, have more than one oxidation number.

we see that the oxidation number of the Zn must be +2 in order for the sum of the oxidation numbers to be zero.

Next let us find the oxidation number of Mn in potassium permanganate, $KMnO_4$. From the table we see that the oxidation number of K is +1 and that of O is −2. Therefore, four oxygens will have a total oxidation number of 4(−2), or

$$+1 + (?) + 4(-2) = 0$$
$$\text{K   Mn   } O_4$$

In order for the sum of the oxidation numbers to be zero, the oxidation number of Mn must be +7.

Now consider the compound diarsenic trisulfide, As_2S_3. The oxidation number of S from the table is −2. Therefore, three S's will have a total oxidation number of 3(−2).

$$? + 3(-2) = 0$$
$$\text{As}_2 \text{   S}_3$$

In order for the sum of the oxidation numbers to be zero, the total oxidation number of the As atoms must be +6. However, this value of +6 applies to two As atoms, so the oxidation number of each As is +3.

Writing Formulas from Oxidation Numbers

To write the formula of a compound formed between calcium and chlorine, look up the oxidation numbers of these elements in

Table 5-5 and write them above the symbols

$$\overset{+2\,-1}{\text{CaCl}}$$

Note that the sum of the oxidation numbers is not zero, so this is not the correct formula for the compound. An easy method for obtaining the correct formula is to use a system of crisscrossing the oxidation numbers:

$$\overset{+2}{\text{Ca}}\quad\overset{-1}{\text{Cl}}$$

Thus the formula of the compound between calcium and chlorine is $CaCl_2$. (Note that the subscript 1 is always understood and never written.) In this compound the sum of the oxidation numbers ($+2$ for the calcium and -2 for the two chlorines) does equal zero.

To write the formula for the compound formed between magnesium and the phosphate ion, we first write the oxidation numbers above the symbols and then crisscross them,

$$\overset{+2}{\text{Mg}}\quad\overset{-3}{\text{PO}_4}$$

so that the formula is $Mg_3(PO_4)_2$, where the sum of the oxidation numbers now equals zero ($+6$ from three Mg's and -6 from two PO_4's). The parentheses around the PO_4^{3-} ion indicate that the ion occurs more than once in the formula. If the ion occurs only once in a compound, parentheses are not necessary. Thus, in the compound between the sodium ion (oxidation number $+1$) and the nitrate ion (oxidation number -1), the formula is simply written as $NaNO_3$.

When the positive and negative oxidation numbers are equal, the formula is correctly written without subscripts. The compound formed between calcium (oxidation number $+2$) and the sulfate ion (oxidation number -2) is $CaSO_4$ since the sum of the oxidation numbers is already zero.

Occasionally, when both positive and negative oxidation numbers are even numbers, the formula of the compound can be simplified by dividing by 2. Thus, the compound formed between tin (oxidation number $+4$) and the sulfate ion (oxidation number -2) can be written as

$$\overset{+4}{\text{Sn}}\quad\overset{-2}{\text{SO}_4}$$

or $Sn_2(SO_4)_4$, which should be simplified to $Sn(SO_4)_2$.

Molecular Weight

The molecular weight or formula weight of any compound is the sum of the atomic weights[†] of all of the atoms present in one molecule of that compound. The molecular weight of sodium bromide, NaBr, is 103, which represents the sum of the atomic weight of sodium (23) plus that of bromine (80). (See Table of Atomic Weights inside back cover.)

Calculating the Molecular Weight (Formula Weight) of a Compound

To find the molecular weight (MW) of a compound, add the atomic weights of all of the atoms that are present in that compound. In the compound H_2O, the molecular weight can be calculated by adding the weight of two atoms of hydrogen and one atom of oxygen.

$$
\begin{array}{ll}
\text{2 hydrogen atoms (at. wt 1)} & 2 \times 1 \;=\; 2 \\
\text{1 oxygen atom (at. wt 16)} & 1 \times 16 = \underline{16} \\
& \text{molecular weight} = 18
\end{array}
$$

The molecular weight of glucose, $C_6H_{12}O_6$, can be calculated as follows:

$$
\begin{array}{ll}
\text{6 carbon atoms (at. wt 12)} & 6 \times 12 = 72 \\
\text{12 hydrogen atoms (at. wt 1)} & 12 \times 1 = 12 \\
\text{6 oxygen atoms (at. wt 16)} & 6 \times 16 = \underline{96} \\
& \text{molecular weight} = 180
\end{array}
$$

The molecular weight of calcium phosphate, $Ca_3(PO_4)_2$, can be calculated as follows:

$$
\begin{array}{ll}
\text{3 calcium atoms (at. wt 40)} & 3 \times 40 = 120 \\
\text{2 phosphorus atoms (at. wt 31)} & 2 \times 31 = 62 \\
\text{8 oxygen atoms (at. wt 16)} & 8 \times 16 = \underline{128} \\
& \text{molecular weight} = 310
\end{array}
$$

Percentage Composition

The percentage composition of a compound can be calculated from the relative atomic weights of the elements present in that compound. Consider the compound $Ca_3(PO_4)_2$, calcium phosphate, whose molecular weight was found to be 310 (see previous paragraph). Of this weight, 120 is calcium, 62 phosphorus, and 128 oxygen. Then

---

[†]   Atomic weights are usually rounded off to the closest whole number.

$$\% \ Ca = \frac{\text{weight of calcium in compound}}{\text{weight of compound}} \times 100 = \frac{120}{310} \times 100 = 38.7\%$$

$$\% \ P = \frac{\text{weight of phosphorus in compound}}{\text{weight of compound}} \times 100 = \frac{62}{310} \times 100 = 20.0\%$$

$$\% \ O = \frac{\text{weight of oxygen in compound}}{\text{weight of compound}} \times 100 = \frac{128}{310} \times 100 = 41.3\%$$

$$\text{total} = \overline{100\%}$$

The Mole

We measure distances on earth in terms of miles. However, this unit is far too small for distances to the stars. We need a much larger unit, the light year. So, too, chemists need a larger unit to weigh molecules, since individual molecules are too small to measure even with the most sensitive equipment. Such a unit, the **mole** (abbreviated mol) is defined as the number of atoms in 12.000 g of ^{12}C. One mole[†] of anything—molecules, atoms, ions, electrons—always contains the same number of particles, 6.02×10^{23} (602,000,000,000,000,000,000,000). This large number is called **Avogadro's number**.

One mole of any substance has a weight, in grams, equal to its molecular (or atomic) weight. That is,

1 mol of water, H_2O,   contains 6.02×10^{23} molecules and weighs   18 g
1 mol of carbon, C,   contains 6.02×10^{23} atoms   and weighs   12 g
1 mol of glucose, $C_6H_{12}O_6$, contains 6.02×10^{23} molecules and weighs 180 g

Note that H_2O and $C_6H_{12}O_6$ are molecules, so we deal with Avogadro's number of molecules. Carbon, C, represents an atom, so we deal with Avogadro's number of atoms.

Example 5-6 How much does 1 mol of NH_3, ammonia, weigh? How many molecules are present in 1 mol of ammonia?

The molecular weight of ammonia is 17 (N = 14 and each H = 1). So 1 mol of NH_3 weighs 17 g. By definition, 1 mol of any substance contains Avogadro's number of particles, so 1 mol of NH_3 contains 6.02×10^{23} molecules.

Example 5-7 How much will 2 mol of NH_3 weigh?

If 1 mol of NH_3 weighs 17 g, 2 mol will weigh 34 g. We can write this mathematically as follows, using the conversion factor 1 mol NH_3 = 17 g:

[†] Note that *mole* is a "counting word" like dozen.

$$2 \text{ mol NH}_3 \times \frac{17 \text{ g NH}_3}{1 \text{ mol NH}_3} = 34 \text{ g NH}_3$$

Example 5-8   A container holds 45 g of sugar, $C_6H_{12}O_6$ (molecular weight 180). How many moles of sugar are present?

Since the molecular weight of sugar is 180, 1 mol of sugar weighs 180 g. Using this latter figure as a conversion factor,

$$45 \text{ g sugar} \times \frac{1 \text{ mol sugar}}{180 \text{ g sugar}} = 0.25 \text{ mol sugar}$$

Example 5-9   How many molecules are present in 27 g of water, H_2O (molecular weight 18)?

We know that 1 mol of any substance contains Avogadro's number of particles, 6.02×10^{23}. So first we have to convert grams of water to moles of water (1 mol of water weighs 18 g) and then moles of water to molecules of water (1 mol of water contains 6.02×10^{23} molecules).

$$27 \text{ g H}_2\text{O} \times \frac{1 \text{ mol H}_2\text{O}}{18 \text{ g H}_2\text{O}} = 1.5 \text{ mol H}_2\text{O}$$

$$1.5 \text{ mol H}_2\text{O} \times \frac{6.02 \times 10^{23} \text{ molecules H}_2\text{O}}{1 \text{ mol H}_2\text{O}}$$
$$= 9.03 \times 10^{23} \text{ molecules H}_2\text{O}$$

Empirical and Molecular Formulas

An **empirical (simplest) formula** represents the relative number of each type of atom present in each molecule of a given compound.

A **molecular formula** represents the actual number of atoms present in each molecule of a given compound.

The empirical formula for both acetylene and benzene is CH, indicating one atom of carbon for each atom of hydrogen in both compounds. The molecular formula for acetylene is C_2H_2, and that of benzene is C_6H_6. Note that in both of these compounds the ratio of carbons to hydrogens is 1:1.

The empirical formula does not always represent the actual number of atoms present and so cannot represent the molecular weight. The molecular formula is always a simple integral multiple (1, 2, 3, etc.) of the empirical formula.

Example 5-10   A compound contains 11.2 percent hydrogen and 88.8 percent oxygen. What is its empirical formula? If its molecular weight is 18, what is its molecular formula?

Step 1   *Assume that 100 g of the compound is present.* If there is 100 g of compound present, there will be 11.2 g of hydrogen (11.2 percent of 100 g) and also 88.8 g of oxygen (88.8 percent of 100 g).

Step 2   *Convert grams to moles.*

$$\text{For hydrogen:}\quad 11.2 \text{ g} \times \frac{1 \text{ mol}}{1.0 \text{ g}} = 11.2 \text{ mol}$$

$$\text{For oxygen:}\quad 88.8 \text{ g} \times \frac{1 \text{ mol}}{16 \text{ g}} = 5.6 \text{ mol}$$

Step 3   *Find the empirical formula by dividing by the smaller number of moles.*

$$\text{For hydrogen:}\quad \frac{11.2 \text{ mol}}{5.6 \text{ mol}} = 2$$

$$\text{For oxygen:}\quad \frac{5.6 \text{ mol}}{5.6 \text{ mol}} = 1$$

The ratio of 2 mol of hydrogen to 1 mol of oxygen gives the empirical formula of H_2O.

Step 4   *Find the molecular formula by dividing the weight of the empirical formula into the molecular weight and multiplying that number by the empirical formula.*
    The weight of H_2O = 18 (H = 1, O = 16). The molecular weight is 18, so the ratio of 18/18 = 1, and thus in this case the empirical formula is also the molecular formula.

Example 5-11   A compound of molecular weight 56 contains 85.6 percent carbon and 14.4 percent hydrogen. What is its empirical formula? its molecular formula?

Step 1 $$\text{Carbon:}\quad 85.6 \text{ g} \times \frac{1 \text{ mol}}{12 \text{ g}} = 7.1 \text{ mol}$$

$$\text{Hydrogen:}\quad 14.4 \text{ g} \times \frac{1 \text{ mol}}{1 \text{ g}} = 14.4 \text{ mol}$$

Step 2 $$\text{For carbon:}\quad \frac{7.1 \text{ mol}}{7.1 \text{ mol}} = 1$$

$$\text{For hydrogen:}\quad \frac{14.4 \text{ mol}}{7.1 \text{ mol}} = 2$$

Note that the number of moles is rounded off to whole numbers because in step 1 the exact atomic weights are not used.

Step 3   The empirical formula is CH_2.

Step 4   The weight of CH_2 is 14. The molecular weight was given as 56. Since $56/14 = 4$, the molecular formula is $(CH_2)_4$ or C_4H_8.

Example 5-12   A compound of molecular weight 270 contains 17.0 percent sodium, 47.4 percent sulfur, and 35.6 percent oxygen. What is its molecular formula?

Step 1

Sodium: $17.0 \text{ g} \times \dfrac{1 \text{ mol}}{23 \text{ g}} = 0.74 \text{ mol}$

Sulfur: $47.4 \text{ g} \times \dfrac{1 \text{ mol}}{32 \text{ g}} = 1.5 \text{ mol}$

Oxygen: $35.6 \text{ g} \times \dfrac{1 \text{ mol}}{16 \text{ g}} = 2.2 \text{ mol}$

Step 2

For sodium: $\dfrac{0.74 \text{ mol}}{0.74 \text{ mol}} = 1$

For sulfur: $\dfrac{1.5 \text{ mol}}{0.74 \text{ mol}} = 2$

For oxygen: $\dfrac{2.2 \text{ mol}}{0.74 \text{ mol}} = 3$

Step 3   The empirical formula is NaS_2O_3.

Step 4   The weight of the empirical formula is 135 $(23 + 2 \times 32 + 3 \times 16)$. The given molecular weight is 270. $270/135 = 2$, so the molecular formula is $Na_2S_4O_6$.

Summary

Molecules are combinations of two or more atoms. Atoms are held together in molecules by ionic or covalent bonds.

A symbol for an element not only identifies that element but also represents one atom of that element. A formula consists of a group of symbols that represent the elements present in a substance.

The number of outer electrons determines the chemical properties of the atom. Atoms are most stable when they have eight electrons in their highest (outer) energy level.

The electron-dot structure of an element uses the symbol of that element to represent the nucleus and all of the electrons except those in the highest energy level. Each electron in that highest energy level is represented by a dot placed near the symbol.

Metals have one, two, or three electrons in their highest energy level and tend to lose all those electrons to form positively charged ions. Nonmetals with six or seven electrons in their highest energy level tend to gain electrons to bring that energy level to eight, thereby forming ions with a negative charge. Positively charged ions are called cations; negatively charged ions are called anions.

The ionic radius of a metal is less than that of the corresponding atom;

the ionic radius of a nonmetal is greater than that of the corresponding atom.

When a metal loses an electron to form a positively charged ion and a nonmetal gains that electron to form a negatively charged ion, these ions are held together by the attraction of their charges. This type of bonding is called ionic bonding.

A polyatomic ion is a group of atoms that acts as a unit in a chemical reaction.

Ionic binary compounds are named by writing the name of the positive ion and then the stem of the negative ion with the ending *ide*. Ionic compounds containing polyatomic ions are named by writing the name of the positive ion and then the name of the polyatomic ion.

Nonmetals may also share electrons to complete their highest energy level with eight. Such a bond is called a covalent bond. In a covalent bond no ions are formed. When a covalent bond is formed, each element has eight electrons around it, except for hydrogen, which has only two.

Covalent compounds in which a pair of electrons is shared either between two identical atoms or between two atoms of equal electronegativity contain nonpolar covalent bonds. Covalent compounds containing bonds between atoms of different electronegativity contain polar covalent bonds.

Covalent compounds are named by writing a prefix and the name of the first element and then writing a prefix, the stem of the second element, and the ending *ide*.

For ionic compounds, the oxidation number is the same as the charge on the ion. For covalent compounds the oxidation number is an arbitrary number based on unequal sharing of electrons. The oxidation number of oxygen is -2, and that of hydrogen is $+1$. In all compounds the sum of the oxidation numbers must equal zero.

In writing the formula of a compound from a table of oxidation numbers, the values of the oxidation numbers are crisscrossed. Care must be taken to note when the oxidation numbers are divisible by 2 so that the formula can be simplified.

The molecular weight of a compound is equal to the sum of the atomic weights of the atoms present in that compound. One mole of any substance contains Avogadro's number (6.02×10^{23}) of particles.

An empirical (simplest) formula represents the relative number of each type of element present in a given compound.

A molecular formula represents the actual number of atoms present in each molecule of a given compound.

Resonance occurs when more than one electron-dot structure can be drawn for a molecule or ion.

Questions and Problems

A

1. What is a molecule?
2. What are valence electrons? What effect do they have on the properties of an atom?
3. What is meant by the term *electron-dot structure?*

4. Give the electron-dot structures for the following elements (use periodic table).
   (a) Phosphorus   (b) Oxygen
   (c) Hydrogen   (d) Neon
   (e) Nitrogen   (f) Aluminum
5. What electron configuration does an element usually need to reach maximum stability?

6. What is an ion? What type of elements form positively charged ions? negatively charged ions?

7. What type of bond consists of ions held together by the attraction of their charges?

8. What is an anion? a cation? Give an example of each.

9. Name the following ionic compounds.
   (a) KI   (b) $CaSO_4$ (c) ZnS
   (d) $NaNO_3$ (e) $AlCl_3$

10. Define *covalent bond*. Compare ionic and covalent bonds.

11. Draw the electron-dot structures for the following covalent compounds.
    (a) H_2 (b) HCl   (c) H_2S
    (d) PH_3 (e) CH_4 (f) N_2

12. What is a polyatomic ion? Give two examples.

13. What is a single covalent bond? double bond? triple bond? Give an example of each.

14. What is a nonpolar covalent bond? a polar covalent bond? Give an example of each.

15. Name the following covalent compounds.
    (a) PCl_3 (b) SO_2 (c) N_2O_5
    (d) CS_2 (e) ICl

16. Write the formulas for the compounds formed from the following ion pairs.
    (a) H^+ and $SO_4{}^{2-}$ (b) Fe^{2+} and Cl^-
    (c) Cu^+ and I^- (d) Ca^{2+} and $PO_4{}^{3-}$
    (e) Ba^{2+} and $HCO_3{}^-$ (f) Mg^{2+} and $NO_3{}^-$
    (g) Ca^{2+} and S^{2-} (h) Al^{3+} and $CO_3{}^{2-}$

17. Write the formulas for the compounds formed between the following.
    (a) silver ions and sulfate ions
    (b) potassium ions and bicarbonate ions
    (c) ammonium ions and sulfide ions
    (d) hydrogen ions and nitrate ions

18. Calculate the oxidation number for *each* atom of the underlined element (consult Table 5-1 for charges and Table 5-5 for oxidation numbers).
    (a) K$\underline{Cl}O_4$ (b) Na$\underline{N}O_2$ (c) $\underline{Fe}_2O_3$
    (d) $\underline{As}_2S_5$ (e) $H_4\underline{P}_2O_7$ (f) $K_2\underline{Cr}_2O_7$
    (g) $\underline{Ra}(HCO_3)_2$ (h) $(\underline{N}H_4)_2S$

19. What do the following symbols or formulas indicate: H_2, $2H$, CO_2, CO?

20. Calculate the molecular weight of each of the following compounds (use atomic weights as whole numbers).
    (a) $NaNO_3$ (b) KBr
    (c) $Ca(HCO_3)_2$ (d) $C_{12}H_{22}O_{11}$
    (e) $C_{57}H_{110}O_6$

21. What is the difference between an empirical formula and a molecular formula? Can they ever be the same?

22. Calculate the empirical formula for a compound containing 31.9 percent potassium, 28.9 percent chlorine, and 39.2 percent oxygen.

23. Which is larger, a metallic atom or its corresponding ion? Why? Which is larger, a nonmetallic atom or its corresponding ion? Why?

24. Why would water solutions of ionic compounds be expected to conduct electricity?

B

25. Explain how a nonpolar molecule can contain polar bonds.

26. Why is fluorine the most electronegative element?

27. What is the predominant cation in extracellular body fluids? in intracellular body fluids?

28. What is the predominant anion in extracellular body fluids? in intracellular body fluids?

29. How can you tell whether a covalent bond will be polar or nonpolar?

30. Why does diamond have such a high melting point? What type of bonds does it have?

31. Why is the oxidation number of sodium +1 rather than +2 or -1?

32. Can an element have more than one oxidation number? Explain.

33. Can both elements in a binary compound have a positive oxidation number? Explain.

34. When writing the formula of a binary compound, is the first element always a metal? a nonmetal? Is the second element always a metal? a nonmetal?

35. Do all covalent bonds consist of atoms having eight electrons around them? Explain.

36. What is the empirical formula and the molecular formula for a compound of molecular weight 284 that contains 43.7 percent phosphorus and 56.3 percent oxygen?

Practice Test

1. The electron-dot structure for sodium is _____.
   a. Na   b. Na·   c. Na^+ d. Na

2. The formula for aluminum chloride is _____.
   a. $AlCl_3$ b. $AlClO_3$
   c. $Al(ClO_3)_3$ d. AlOCl

3. In a single covalent bond, how many electrons are shared?
   a. 2   b. 4   c. 6   d. 8

4. The name of CS_2 is _____.

a. carbon sulfate   b. carbon disulfide
c. carbon disulfate   d. carbon sulfide

5. Which of the following represents an empirical formula?

a. CH   b. C_2H_2
c. C_3H_6 d. C_5H_{10}

6. Metals tend to lose electrons to form _____ .

a. positively charged ions
b. negatively charged ions
c. ions that are larger than the original atoms
d. ions that are heavier than the original atoms

7. The molecular weight of $C_6H_{12}O_6$ (C = 12, H = 1, O = 16) is _____ .

a. 24   b. 90   c. 180   d. 360

8. An example of a polar compound is _____ .

a. H_2O b. CCl_4 c. CO_2 d. H_2

9. The oxidation number of S in H_2SO_4 is _____ .

a. 2   b. 4   c. 6   d. 8

10. What does the sign $\leftrightarrow$ indicate?

a. equilibrium   b. resonance
c. uncertain reaction   d. none of these

6

Chemical Equations and Reactions

Student observing magnesium burning.

Chemical Equations

When an electric current (energy) is passed through water (a process known as **electrolysis**), hydrogen gas and oxygen gas are produced. The chemist uses symbols and formulas in a chemical equation to describe this chemical reaction.

$$H_2O(l) + energy \longrightarrow H_2(g) + O_2(g)$$

The arrow is used instead of an equal sign and is read as "yields" or "produces." The plus sign on the right-hand side of the equation is read as "and." A plus sign on the left-hand side of the equation is read as "reacts with." The materials that react are called the **reactants.** The reactants are written on the left-hand side of the equation. The substances that are produced are called the **products.** They are written on the right-hand side of the equation. The (l) indicates a liquid, (g) a gas. A solid, or precipitate (an insoluble solid), is indicated by (s). The energy involved in the reaction can be written in words on the left (or right) side of the equation; it may be indicated above the arrow; or it may be indicated by the symbol Δ (Greek delta). The use of a catalyst can be shown above or below the arrow.

A chemical equation must be balanced before any specific interpretation can be made about that reaction. For example, if we wish to know how much oxygen will be required to metabolize a given amount of a particular carbohydrate, we must first set up a balanced equation for the reaction.

Balancing Chemical Equations

Note that the chemical equation shown below does not contain the same numbers of hydrogen and oxygen atoms on both sides of the arrow. To be balanced, a chemical equation must contain the same number of atoms of each element on both sides. Thus equation (6-1) is *not* balanced.

$$H_2O(l) + energy \longrightarrow H_2(g) + O_2(g) \qquad \text{(unbalanced)} \quad \text{(6-1)}$$

In balancing a chemical equation, you must not change the subscripts (the small numbers to the right of the symbols) because doing so would change either the reactants or the products, thus changing the meaning of the reaction. Instead, place coefficients in front of the symbols and formulas to indicate how many atoms or molecules of each are needed.

In the equation $H_2O \rightarrow H_2 + O_2$, there are two hydrogen atoms on each side of the equation. There are one oxygen atom on the left side and two oxygen atoms on the right side of the equation. To get two atoms of oxygen on the left side of the equation (to balance the two on the right side), place a 2 in front of the H_2O. The 2 cannot be placed as a subscript after the O in H_2O because then another substance would be represented, not water. The 2 cannot be placed between the H and the O because this also would change the meaning of the formula. Therefore, place the 2 in front of the H_2O.

$$2\,H_2O(l) + energy \longrightarrow H_2(g) + O_2(g) \qquad \text{(unbalanced)}$$

However, now there are four hydrogen atoms (two H_2's) on the left side of the equation. In order to get four hydrogen atoms on the right side of the equation, place a 2 in front of the H_2. There are already two oxygen atoms on each side of the equation. Thus the balanced equation is as follows.

$$2\,H_2O(l) + energy \longrightarrow 2\,H_2(g) + O_2(g) \qquad \text{(balanced)}$$

This balanced equation now shows that two molecules of water, on electrolysis, yield two molecules of hydrogen gas and one molecule of oxygen gas.

When aluminum metal reacts with sulfuric acid, the products are hydrogen gas, H_2, and aluminum sulfate, $Al_2(SO_4)_3$. The unbalanced equation for this reaction is

$$Al + H_2SO_4 \longrightarrow Al_2(SO_4)_3 + H_2(g) \qquad \text{(unbalanced)} \qquad \text{(6-2)}$$

Equation (6-2) is not balanced because there are more aluminum atoms on the right side of the equation than on the left side. The same is true for the sulfur and oxygen atoms. To balance an equation of this type, pick out the most complicated-looking formula and assume that one molecule of it is present. The most complicated-looking formula in equation (6-2) is $Al_2(SO_4)_3$. Assuming that one molecule of it is produced, then there are two atoms of aluminum on the right side of the equation. To balance this, place a 2 in front of the Al on the left side of the equation. (For simplicity, we will omit the Δ sign from all the equations that follow.)

$$2\,Al + 3\,H_2SO_4 \longrightarrow Al_2(SO_4)_3 + H_2(g) \qquad \text{(unbalanced)}$$

Next note that there are three SO_4 groups in the molecule of $Al_2(SO_4)_3$. There must then be three SO_4 groups on the left side of the equation. To get these three groups place a 3 in front of the H_2SO_4.

$$2\,Al + 3\,H_2SO_4 \longrightarrow Al_2(SO_4)_3 + H_2(g) \qquad \text{(unbalanced)}$$

To complete the equation, note that there are now six hydrogen atoms on the left side of the equation (in the three H_2's). Therefore, there must be six hydrogen atoms on the right side, so place a 3 in front of the H_2.

$$2 \text{ Al} + 3 \text{ H}_2\text{SO}_4 \longrightarrow \text{Al}_2(\text{SO}_4)_3 + 3 \text{ H}_2(g) \qquad \text{(balanced)}$$

Now the equation is balanced. There are two aluminums, six hydrogens, three sulfurs, and twelve oxygens (or three SO_4's) on each side of the equation.

When sulfur is burned in excess oxygen, sulfur trioxide is produced. When this reaction is written in equation form, it becomes equation (6-3).

$$S(s) + O_2(g) \longrightarrow SO_3(g) \qquad \text{(unbalanced)} \qquad (6\text{-}3)$$

Following the balancing procedure, pick out the most complicated compound and take one molecule of it. Thus, in equation (6-3), take one molecule of SO_3. This molecule contains one atom of sulfur. There is already one atom of sulfur on the left side of the equation. There are three atoms of oxygen on the right side of the equation and only two on the left side. However, there is no *whole* number that can be placed in front of the O_2 to make three oxygen atoms on that side of the equation. If a 2 is placed there, there will be four atoms of oxygen. In this case, then, instead of selecting one molecule of the most complicated compound, select two molecules of it.

$$S(s) + O_2(g) \longrightarrow 2 SO_3(g) \qquad \text{(unbalanced)}$$

Then, in order to balance two sulfur atoms on the right side of the equation, start with two sulfur atoms on the left side.

$$2 S(s) + O_2(g) \longrightarrow 2 SO_3(g) \qquad \text{(unbalanced)}$$

Next, the right side of the equation contains six oxygen atoms and so must the left side. Place a 3 in front of the O_2 in order to have six oxygen atoms on that side of the equation. The equation then is balanced.

$$2 S(s) + 3 O_2(g) \longrightarrow 2 SO_3(g) \qquad \text{(balanced)}$$

Types of Chemical Reactions

Chemical reactions can be divided into combination reactions, decomposition reactions, single replacement or substitution reactions, and double displacement or metathesis reactions. These are illustrated in Table 6-1. Note that the example equations are written as balanced equations.

Other types of chemical reactions include equilibrium reactions (see following paragraphs) and oxidation–reduction reactions (see Chapter 9).

Table 6-1   Chemical Reactions

Type	General Equation	Example
Combination	$A + B \rightarrow AB$	$C + O_2 \rightarrow CO_2$ $2\,S + 3\,O_2 \rightarrow 2\,SO_3$
Decomposition	$AB \rightarrow A + B$	$2\,HgO \rightarrow 2\,Hg + O_2(g)$ $2\,KClO_3 \rightarrow 2\,KCl + 3\,O_2(g)$
Single replacement	$A + BC \rightarrow AC + B$ or $A + BC \rightarrow BA + C$	$Zn + CuSO_4 \rightarrow ZnSO_4 + Cu$ $Cl_2 + 2\,NaBr \rightarrow 2\,NaCl + Br_2$
Double displacement	$AB + CD \rightarrow AD + CB$	$Na_2SO_4 + BaCl_2 \rightarrow 2\,NaCl + BaSO_4(s)$ $FeS + 2\,HCl \rightarrow FeCl_2 + H_2S(g)$

Equilibrium Reactions

Often, when two or more reactants unite to form a certain number of products, these products themselves unite to re-form the original reactants. Reactions of this type are called **reversible reactions.** They are indicated by double arrows $\rightleftharpoons$ showing that the reaction may proceed in either direction depending upon the conditions that exist.

If we start with a mixture of N_2 and H_2, at a given temperature and pressure (with a catalyst), we will soon have some NH_3 formed. As more NH_3 is formed, it will begin to decompose into N_2 and H_2, or

$$N_2 + 3\,H_2 \rightleftharpoons 2\,NH_3$$

When the rates of formation and decomposition become equal, a chemical **equilibrium** exists. This does not mean that all reaction has stopped; it merely means that the rate of decomposition is the same as the rate of formation so the composition remains constant. An equilibrium can be defined as a dynamic state in which the **rate of the forward reaction is equal to the rate of the reverse reaction.**

Two examples of equilibrium reactions in the body are

$$HCO_3^- + H^+ \rightleftharpoons H_2CO_3 \rightleftharpoons CO_2 + H_2O$$

and

$$\text{hemoglobin} + \text{oxygen} \rightleftharpoons \text{oxyhemoglobin}$$

Equilibrium Constant

Consider the following general equilibrium reaction:

$$A + B \rightleftharpoons C + D$$

The **law of mass action** states that the rate of a chemical reaction is proportional to the concentration of the reacting substances. So, for the above reaction, rate forward $= k_1[A][B]$, where k_1 is a proportionality constant and the brackets, [ ], indicate concentrations in the units moles per liter. Likewise, the rate of the reverse reaction equals $k_2[C][D]$, where k_2 is another proportionality constant.

At equilibrium, the rate of the forward reaction is equal to the rate of the reverse reaction, so

$$k_1 \times [A] \times [B] = k_2 \times [C] \times [D]$$

from which we have

$$\frac{[C] \times [D]}{[A] \times [B]} = \frac{k_1}{k_2} = K_{eq}$$

where K_{eq} is the equilibrium constant (since the ratio of two constants k_1/k_2 is itself another constant).

In general, the equilibrium constant, K_{eq}, equals the product of the concentrations of the products divided by the product of the concentrations of the reactants, each concentration raised to the power indicated by its coefficient in the equation. So, for the reaction $4\,A + 3\,C \rightleftharpoons 2\,D + F$, we have

$$K_{eq} = \frac{[D]^2[F]}{[A]^4[C]^3}$$

Example 6-1   In the conversion of glucose to vitamin C, the following equilibrium reaction takes place:

$$\underset{\text{gluconic acid}}{C_5H_{11}O_5COOH} \rightleftharpoons \underset{\text{hydrogen ion}}{H^+} + \underset{\text{gluconate ion}}{C_5H_{11}O_5COO^-}$$

If the equilibrium concentrations in moles per liter are $C_5H_{11}O_5COOH$, 0.10; H^+, 3.7×10^{-3}; $C_5H_{11}O_5COO^-$, 3.7×10^{-3}, calculate the value of K_{eq}.

Using the formula,

$$K_{eq} = \frac{[H^+][C_5H_{11}O_5COO^-]}{[C_5H_{11}O_5COOH]}$$

$$K_{eq} = \frac{(3.7 \times 10^{-3})(3.7 \times 10^{-3})}{0.10}$$

$$= 1.4 \times 10^{-4}$$

Example 6-2   In the manufacture of wood alcohol (page 284), the following equilibrium reaction occurs: $CO + 2\,H_2 \rightleftharpoons$

CH₃OH. At equilibrium, the concentrations in moles per liter are CO, 0.025; H₂, 0.050; CH₃OH, 0.12. Calculate the value of K_{eq}.

$$K_{eq} = \frac{[CH_3OH]}{[CO][H_2]^2} = \frac{0.12}{(0.025)(0.050)^2} = 1.9 \times 10^3$$

In general, a large value of K_{eq} indicates an equilibrium that has been shifted far to the right, whereas a small value indicates one shifted to the left.

Le Châtelier's Principle

Le Châtelier's principle states that if a stress is applied to a reaction at equilibrium, the equilibrium will be displaced in such a direction as to relieve that stress. Thus, if we apply a stress such as a change in concentration or temperature, we should be able to predict the results of such a stress upon the given equilibrium.

Effect of Concentration

Let us see what the effect upon the equilibrium and upon the equilibrium constant will be if we add more A to a mixture of A, B, C, and D. The reaction is

$$A + B \rightleftharpoons C + D$$

Increasing the concentration of a reactant increases the number of collisions between reactant molecules so that the rate of the forward reaction is greater than the rate of the reverse reaction. The system is no longer in equilibrium. However, as the reactant molecules are used up, the rate of the forward reaction will decrease and the rate of the reverse reaction will increase until a new equilibrium is established. That is, the addition of more A will cause a stress that the reaction will tend to oppose by using up more A, thus shifting the equilibrium to the right. Since the equilibrium constant depends upon temperature only, the addition of more A will have no effect upon it. Thus, the addition of more A increases the amount of products; likewise for the addition of more B to the equilibrium mixture. In accordance with Le Châtelier's principle, removal of a product (or products) should also shift the equilibrium to the right.

Consider the enzyme-catalyzed reaction

$$CO_2 + H_2O \rightleftharpoons H_2CO_3$$

Carbon dioxide is a waste product produced in the cells. As the CO_2 flows into the blood, its concentration increases and so, according to

Le Châtelier's principle, the equilibrium shifts toward the right, producing more H_2CO_3. In the lungs, as carbon dioxide is exhaled, the equilibrium is shifted to the left, again according to Le Châtelier's principle.

Effect of Temperature

In the reaction

$$4\ HCl + O_2 \rightleftharpoons 2\ H_2O + 2\ Cl_2 + heat$$

heat is liberated when the reaction proceeds to the right and is absorbed when it proceeds to the left.

If we raise the temperature of this reaction mixture at equilibrium, the reaction will tend to go in a direction to relieve this stress. That is, as the temperature is raised, the equilibrium will shift to the left, favoring the reaction that tends to absorb this heat. Conversely, as the temperature is lowered, the reaction speeds more to the right for the production of more heat to relieve the new stress.

Effect of a Catalyst

The addition of a catalyst will speed up the forward reaction but will speed up the reverse reaction equally. Thus, all that is accomplished is that the system reaches equilibrium much sooner. The equilibrium reached is the same as would have been reached if no catalyst had been used. Again we note that temperature is the only thing that affects the equilibrium constant, so naturally a catalyst will have no effect on it.

Interpreting Chemical Equations

A balanced chemical equation specifies a great deal of quantitative information. Consider the balanced equation

$$2\ H_2 + O_2 \longrightarrow 2\ H_2O \tag{6-4}$$

Equation (6-4) indicates that two molecules of H_2 react with one molecule of O_2 to produce two molecules of H_2O. Note that the numbers are the same as the coefficients (numbers in front of the substances) in the balanced equation. That is why the equation must be balanced before any type of calculation can be carried out.

Equation (6-4) can also be interpreted in terms of moles: 2 mol of H_2 react with 1 mol of O_2 to produce 2 mol of H_2O.

We can substitute molecular weights in grams for moles in equation (6-4). Thus the equation can be represented as

$$2 H_2 + \quad O_2 \quad \longrightarrow \quad 2 H_2O$$

$$\text{2 molecules} + \text{1 molecule} \longrightarrow \text{2 molecules}$$

$$\text{2 mol} + \quad \text{1 mol} \quad \longrightarrow \quad \text{2 mol}$$

$$2(2 \text{ g}) + \quad 32 \text{ g} \quad \longrightarrow \quad 2(18 \text{ g})$$

Note that the law of conservation of mass must be satisfied.

Example 6-3 Given the balanced equation

$$C_6H_{12}O_6 + 6 O_2 \longrightarrow 6 CO_2 + 6 H_2O + 686 \text{ kcal}$$

how many moles of CO_2 will be produced from 3.5 mol of $C_6H_{12}O_6$?

We note from the balanced equation that 1 mol of $C_6H_{12}O_6$ yields 6 mol of CO_2. Using this as a conversion factor,

$$3.5 \text{ mol } C_6H_{12}O_6 \times \frac{6 \text{ mol } CO_2}{1 \text{ mol } C_6H_{12}O_6} = 21 \text{ mol } CO_2$$

Example 6-4 Using the balanced equation in Example 6.3, how many grams of water will be produced from the reaction of 0.50 mol of $C_6H_{12}O_6$?

The balanced equation indicates that 1 mol of $C_6H_{12}O_6$ yields 6 mol of H_2O, so

$$0.50 \text{ mol } C_6H_{12}O_6 \times \frac{6 \text{ mol } H_2O}{1 \text{ mol } C_6H_{12}O_6} = 3 \text{ mol } H_2O$$

Then, recalling that 1 mol of H_2O weighs 18 g, we obtain

$$3 \text{ mol } H_2O \times \frac{18 \text{ g } H_2O}{1 \text{ mol } H_2O} = 54 \text{ g } H_2O$$

Example 6-5 Using the balanced equation in Example 6-3, how much energy will be produced from the complete combustion of 0.5 mol of $C_6H_{12}O_6$?

The balanced equation indicates that complete oxidation of 1 mol of $C_6H_{12}O_6$ yields 686 kcal, so

$$0.50 \text{ mol } C_6H_{12}O_6 \times \frac{686 \text{ kcal}}{1 \text{ mol } C_6H_{12}O_6} = 343 \text{ kcal}$$

Reaction Rates

Some chemical reactions proceed at a slow rate. Iron, for example, rusts very slowly. Wood takes years to decay. On the other hand, some chemical reactions proceed more rapidly. Coal burns steadily and quickly. Concrete begins to set within a few hours. Some chemical reactions not only occur rapidly, they take place almost instantaneously. Consider the violent explosion of dynamite. Within a fraction of a second, the complete reaction has taken place.

In order for a chemical reaction to occur between two substances, their molecules must collide (interact) with sufficient energy to overcome any force of repulsion between the electron clouds.

The minimum amount of energy required for a collision to produce a successful reaction is called the *activation energy (Figure 6-1).

What determines the speed of a chemical reaction? The speed of a chemical reaction depends upon several factors: (1) the nature of the reacting substances, (2) the temperature, (3) the concentration of the reacting substances, (4) the presence of a catalyst, and (5) the surface area and the intimacy of contact of the reacting substances.

Nature of Reacting Substances

When a solution of sodium sulfate (Na_2SO_4) is mixed with a solution of barium chloride ($BaCl_2$), a white precipitate of barium sulfate ($BaSO_4$) is formed immediately.

$$Na_2SO_4 + BaCl_2 \longrightarrow 2\,NaCl + BaSO_4(s) \qquad (6\text{-}5)$$

Equation (6-5) can be rewritten to show the ions of which these salts consist.

Figure 6-1   A catalyst increases the rate of a reaction by lowering the activation energy so more reactant molecules collide with enough energy to react.

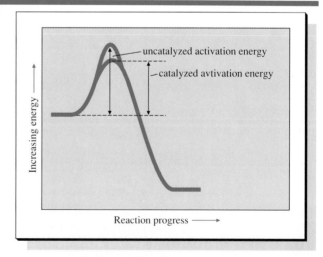

$$2\,Na^+ + SO_4^{2-} + Ba^{2+} + 2\,Cl^- \longrightarrow 2\,Na^+ + 2\,Cl^- + BaSO_4(s)$$

Next, as in any algebraic equation, cancel the sodium ions and the chloride ions from both sides of equation (6-5), leaving the net equation

$$Ba^{2+} + SO_4^{2-} \longrightarrow BaSO_4(s)$$

This is an example of an ionic reaction—the reaction between ions. Many of the reactions taking place in the body are of this type.

Consider, however, the reaction between hydrogen (H_2) and oxygen (O_2) to form water (H_2O). This reaction proceeds very slowly, even at a temperature of 200 °C, unless a spark is introduced into the mixture.

$$2\,H_2 + O_2 \xrightarrow{\text{spark}} 2\,H_2O$$

In this reaction, it is necessary for the bonds between the hydrogen atoms in the hydrogen molecules to be broken. Also, the bonds between the oxygen atoms must be broken before the reaction can occur. This is an example of a reaction in which covalent bonds must be broken and new ones formed. In general, such reactions proceed much more slowly than ionic reactions.

Temperature

As the temperature rises, the speed of a chemical reaction increases because at higher temperatures the molecules move faster and therefore collide more often. Thus even a slight change in temperature can affect the speed of a reaction with noticeable results. Every 10 °C rise in temperature doubles the rate, or speed, of the reaction. In other words, the reaction would occur in one-half the time. For a 20 °C rise, the reaction would occur four times faster, or in one-fourth the time.

A patient who has a fever of only a few degrees has an increased pulse rate and also an increased respiratory rate. Reactions taking place throughout the body proceed at an accelerated rate.

When the temperature of the human body drops, the various metabolic processes slow down considerably. This fact is of great importance, for example, during open-heart surgery when the temperature of the body is lowered considerably.

A drop in body temperature of 2 to 3 °F causes uncontrolled shivering as the body tries to generate heat by muscle activity. Death from exposure occurs when the body temperature drops below 78 °F. Victims who fall overboard in water whose temperature is near freezing may survive only a short time unless they are rescued or are wearing the proper type of insulated coverings. A person who has been "chilled" in such an accident should be given dry clothing and warmed as soon as possible. Warm liquids are helpful in increasing body temperature, as are "sugars." Contrary to popular belief, alcohol is not helpful in such a situation because it dilates blood vessels, allowing cold blood to flow more rapidly to the vital organs.

Concentration

The concentration of a reactant is the amount present in a given unit of volume. The more of a given material present in a certain volume, the greater its concentration. Greater concentration produces faster reactions because there are more molecules that can react.

A patient with a respiratory disease can breathe more easily when using a nasal catheter with oxygen because the concentration of the oxygen in the lungs is increased. This increased concentration increases the speed of oxygen uptake, making breathing easier for the patient.

Catalyst

When a protein substance is placed in water and heated, a hydrolysis reaction (see page 414) proceeds at an extremely slow speed. If a strong acid is added to the mixture, the reaction proceeds at a much faster rate. The acid is not used up (that is, it is not changed chemically). Its presence merely increases the speed of the reaction. Any substance that increases the speed of a reaction without itself being changed chemically is called a **catalyst.**

Catalysts function to increase the rate of a reaction because they provide a pathway with a lower activation energy (see Figure 6-1). Many of the chemical reactions used in industry would not be practical without a catalyst. They would take too long to be of commercial use.

The body uses catalysts to enable its chemical reactions to proceed at a rapid pace. Those catalysts present in the body are called **enzymes.** During digestion, for example, the food undergoes many chemical changes, each under the influence of a specific enzyme. There are also **inhibitors** that slow down rather than speed up chemical reactions.

Surface Area

The speed of a chemical reaction also depends upon the amount of surface area present in the reacting substances. Although a pile of flour is quite harmless, the same flour in the form of dust can cause a dangerous explosion. This effect is due to the tremendous amount of surface area of the dust. This large surface area can react rapidly with the oxygen in the air to cause an explosion.

Many medications are given in the form of finely divided suspended solids. In this manner, more surface area means more rapid absorption in the body.

Increasing the concentration of the reacting substances can also be considered as increasing the amount of surface area.

Summary

A chemical equation uses symbols and formulas to represent a chemical reaction. The substances on the left side of the equation are called reactants and those on the right side products. A (g) indicates a gas; an (s) indicates a precipitate, or solid, and (l) a liquid.

To balance a chemical equation, pick out the most complicated-looking compound and assume that one molecule of it is present. Then proceed back and forth adding coefficients in front of the reactants and products until the number of atoms of each type is the same on both sides of the equation.

The four types of chemical reactions and their general equations are

Combination	$A + B \longrightarrow AB$
Decomposition	$AB \longrightarrow A + B$
Single replacement	$A + BC \longrightarrow AC + B$
	or
	$A + BC \longrightarrow BA + C$
Double displacement	$AB + CD \longrightarrow AD + CB$

An equilibrium reaction is a dynamic state in which the rate of the forward reaction is equal to the rate of the reverse reaction.

Le Châtelier's principle states that if a stress is applied to a reaction at equilibrium, the equilibrium will be displaced in such a direction as to relieve that stress.

The speed of a chemical reaction depends upon the nature of the reacting substances, the temperature, the concentration of the reacting substances, the presence of a catalyst, and the surface area of the reacting substances.

Questions and Problems

A

1. Balance the following equations:
   (a) $Mg + O_2 \longrightarrow MgO$
   (b) $Zn + HCl \longrightarrow ZnCl_2 + H_2$
   (c) $C + O_2 \longrightarrow CO_2$
   (d) $NaCl + AgNO_3 \longrightarrow AgCl + NaNO_3$
   (e) $ZnSO_4 + NaOH \longrightarrow Zn(OH)_2 + Na_2SO_4$
   (f) $Fe + O_2 \longrightarrow Fe_3O_4$
   (g) $Mg + AgNO_3 \longrightarrow Mg(NO_3)_2 + Ag$
   (h) $Al(OH)_3 + H_2SO_4 \longrightarrow Al_2(SO_4)_3 + H_2O$
2. Label each of the reactions in question 1 as to type.
3. What factors determine the rate of a chemical reaction? Indicate at least one practical application of each factor.

4. Calculate the percentage composition of each compound in question 1 (answers to one decimal place).
5. What is a mole? What does 1 mol of $C_{12}H_{22}O_{11}$ weigh?
6. How many moles are present in 500 g of $CaCO_3$? How many molecules?
7. Given the balanced equation $Zn + H_2SO_4 \rightarrow ZnSO_4 + H_2$, how many moles of H_2 can be produced from 2.5 mol of Zn?
8. How many grams of $ZnSO_4$ can be produced from 0.54 mol of H_2SO_4 according to the equation in question 7?
9. Define equilibrium. What does Le Châtelier's principle state?
10. Calculate the value of K_{eq} for the reaction $A + 2 B \rightleftharpoons 3 C$ if the molar concentrations are A, 1.0; B, 1.5; and C, 0.80.

B

11. How much will 6.02×10^{23} atoms of Ne weigh?
12. How many moles of $Fe_2(SO_4)_3$ are present in 6.00×10^{-5} g?
13. How many molecules are present in 5 mol of H_2O? How many atoms?
14. Why must an equation be balanced before weight calculations can be made?
15. How many grams of S will react with oxygen to yield 2.5 mol of SO_3?
$$2\,S + 3\,O_2 \longrightarrow 2\,SO_3$$
16. Calculate K_{eq} for the reaction $N_2 + 3\,H_2 \rightleftharpoons$ $2\,NH_3 + $ heat if the molar concentrations are N_2, 5.0; H_2, 2.0; NH_3, 50.0.
17. How may the reaction in question 16 be shifted to the right? to the left?
18. What effect will a change of temperature have on the equilibrium in question 16?
19. Why does an increase in the concentration of a reactant increase the rate of the forward reaction?
20. What effect does a catalyst have upon an equilibrium? Why?
21. Is a reaction at equilibrium static or dynamic? Explain.

Practice Test

Use the following equation for questions 1 and 2.

$$Zn + HCl \longrightarrow ZnCl_2 + H_2(g)$$

1. The (g) indicates _____.
   a. a precipitate   b. a gas
   c. a liquid   d. a solid
2. When the equation is balanced, the number in front of the HCl is _____.
   a. 1   b. 2   c. 2   d. 4

Use the following balanced equation for questions 3–7.

$$2\,N_2 + 3\,O_2 \rightleftharpoons 2\,N_2O_3 + \text{heat (all gases)}$$

3. An increase in N_2 will shift the equilibrium _____.
   a. to the right
   b. to the left
   c. not at all
4. An increase in pressure will shift the equilibrium _____.
   a. to the right
   b. to the left
   c. not at all
5. An increase in temperature will shift the equilibrium _____.
   a. to the right
   b. to the left
   c. not at all
6. The addition of a catalyst will speed up the _____.
   a. forward reaction
   b. the reverse reaction
   c. both reactions equally
7. The forward reaction is an example of _____.
   a. a combination reaction
   b. a decomposition reaction
   c. a single replacement reaction
   d. a double displacement reaction

Use the following balanced equation for questions 8–10.

$$2\,H_2 + O_2 \longrightarrow 2\,H_2O$$

8. How many moles of water will be produced from 6 moles of H_2?
   a. 2   b. 4   c. 6   d. 8
9. How many grams of water will be produced from 2 moles of H_2?
   a. 9   b. 18   c. 36   d. 72
10. How many moles of H_2 are required to react completely with 6.5 moles of O_2?
    a. 3.25   b. 6.5   c. 13   d. 19.5

1. Public health: Diabetes is an international problem affecting all peoples and all age groups. *[Centers for Disease Control]*

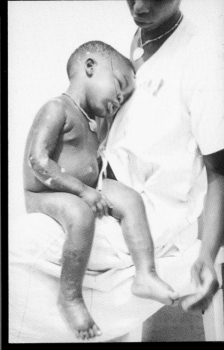

2. Illness: Treatment of illnesses is a primary concern of all health personnel. *[Michael Latham, Cornell University]*

3. Food: A well-balanced diet is essential for the well-being of all individuals. *[Dick Durrance/Woodfin Camp, Inc.]*

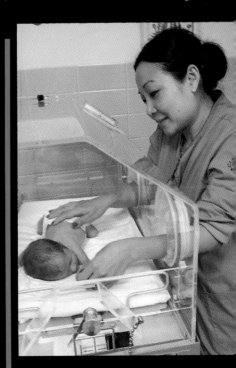

4. Patient care: Care of an infant

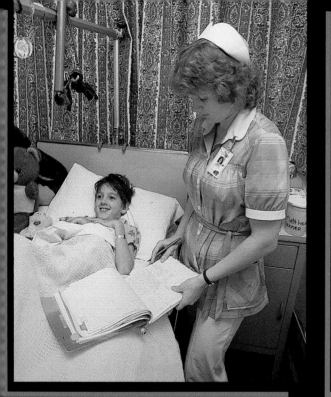

5. Patient care: A child in a hospital. *[Tom Kelly]*

6. Patient care: Care of a patient in a wheelchair. *[Catherine Kamow/Woodfin Camp, Inc.]*

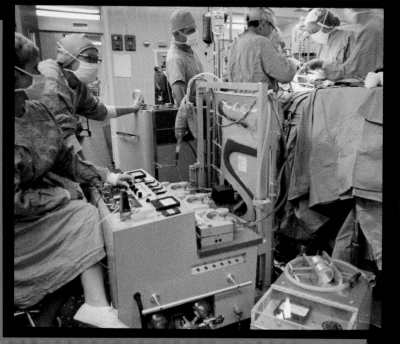

7. Blood: Use of blood in operating room. *[AP-Wide World Photos]*

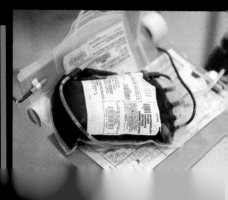

9. Laboratory: Scientist at work in maximum containment laboratory at Centers for Diseases Control where little-known contagious viruses are studied. *[Centers for Disease Control]*

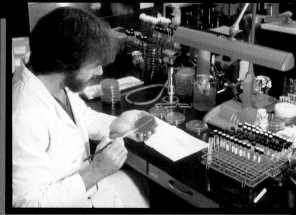

10. Laboratory: Preparing a culture of suspected pathogen (e.g., bacteria). *[Centers for Disease Control]*

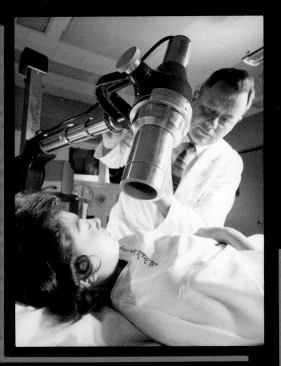

1. Medical testing: Use of Geiger counter in diagnosis of thyroid condition. *[Brookhaven National Laboratory]*

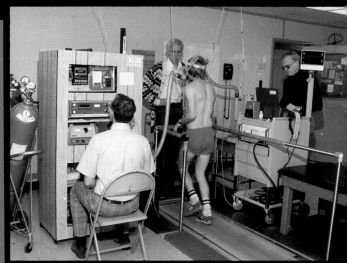

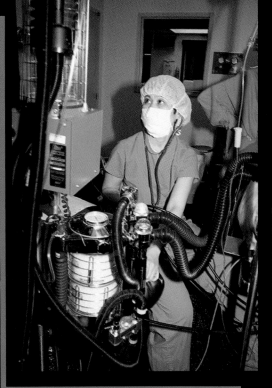

13. Anesthesiology: Anesthesiologist in operating room. *[Ulrike Welsch]*

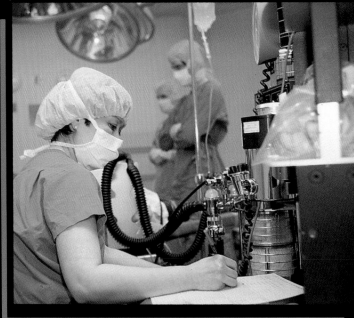

14. Anesthesiology: Anesthesiologist in operating room. *[Ulrike Welsch]*

15. Anesthesiology: Checking vital signs during surgery. *[Robert Severi Woodfin Camp, Inc.]*

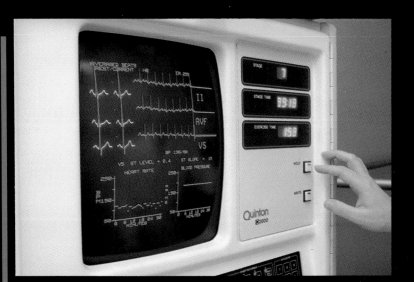

7

The Gaseous State

Technician delivering gas sample to gas chromatograph.

General Properties

In Chapter 2 we saw that gases have no definite shape, no definite volume, and a low density. Gases have a much greater volume than an equal mass of solid or liquid. For example, 1 g of liquid water occupies 1 mL and 1 g of solid water (ice) occupies nearly 1 mL, but 1 g of water vapor (at 0 °C and 1 atm pressure) occupies nearly 1250 mL. These and other properties of gases can be explained in terms of the kinetic molecular theory.

The Kinetic Molecular Theory

The principal assumptions of the kinetic molecular theory are as follows.

1. Gases consist of tiny particles called molecules.
2. The distance between the molecules of a gas is very great compared to the size of the molecules themselves (that is, the volume occupied by a gas is mostly empty space).
3. Gas molecules are in rapid motion and move in straight lines, frequently colliding with each other and with the walls of the container.
4. Gas molecules do not attract each other.
5. When molecules of a gas collide with each other or with the walls of the container, they bounce back with no loss of energy. Such collisions are said to be perfectly elastic.
6. The average kinetic energy of the molecules is the same for all gases at the same temperature. The average kinetic energy increases as the temperature increases and decreases as the temperature decreases.

Let us see how the various properties of gases can be explained in terms of this theory.

1. If the distance between molecules is very great compared to the size of the molecules themselves, the molecules will occupy only a small fraction of that volume so the density will be very low.
2. If the molecules are moving rapidly in all directions, they can fill any size container. They can keep on moving until they hit a wall of the container or until they hit each other and bounce back; that is, the gas will have no definite volume and no definite shape.
3. If the molecules of a gas strike the walls of a container, they should exert a pressure on each wall. And, since the molecules are moving in all directions, they should exert a pressure equally in all directions. Gases do just this.
4. If a bottle of ether is opened in a room, the odor is soon apparent in all parts of that room; that is, the molecules of ether gas diffuse into the air (a mixture of gases). According to the kinetic

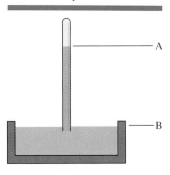

Figure 7-1 Pressure is measured by a barometer.

molecular theory, there is a great deal of empty space between the molecules of a gas, so the ether molecules (or any other gas) can diffuse into the spaces between the air molecules.

5. We can show that the collisions must be perfectly elastic by means of a reverse type of reasoning. Suppose that the collisions between molecules were not perfectly elastic; that is, suppose some energy was lost upon collision with other molecules and with the walls of the container. Eventually, the gas molecules would have so little energy left that they would settle to the bottom of the container. However, gases never settle. Therefore, the collisions between the molecules themselves and with the walls of the container must be perfectly elastic.

6. Since gas molecules are so far apart, it should be possible to force them closer together by increasing the pressure; that is, gases should be compressible, as indeed they are.

Measurement of Pressure

Pressure is measured by a barometer, one form of which is shown in Figure 7-1. A glass tube about 1 yd long is filled with mercury and placed open end down in a dish of mercury. The mercury falls in the tube, leaving a vacuum above it, until the pressure exerted by the air just balances the mercury column in the tube. Thus the atmospheric pressure is expressed as being equal to so many millimeters of mercury (mm Hg) or the height of the column from A to B.

One unit of pressure is the torr, named after the Italian scientist Torricelli, who invented the barometer. One torr is the pressure exerted by 1 mm of mercury at sea level. The SI unit of pressure is the pascal (Pa).

Standard pressure is 1.00 atm and will support a column of mercury 760 mm tall. Standard pressure can be expressed in many different units. Among these are the following.

760 mm Hg	29.92 in. Hg	1.013×10^5 pascals (Pa)
76.0 cm Hg	14.7 lb/in.^{2}	101.3 kilopascals (kPa)
760 torr		

The Gas Laws

Boyle's Law

If the volume of a gas is reduced, the molecules will have less space in which to move. Therefore, they will strike the walls of the container more often and cause a greater pressure.

The relationship between the volume of a given quantity of a gas and its pressure is expressed by **Boyle's law,** which states that the volume occupied by a gas is inversely proportional to the pressure if the temperature remains constant.

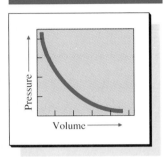

This may be illustrated by Figure 7-2, a graph of pressure versus volume. Note that as the pressure increases the volume of the gas decreases, and as the pressure decreases the volume of the gas increases.

In normal breathing, the diaphragm moves downward, allowing the lungs to expand. The increased volume in the lungs causes the pressure in the lungs to drop slightly. Since air always flows from an area of high pressure to one of lower pressure, air flows into the lungs. When the diaphragm moves upward, the volume of the lungs decreases and the pressure of the air inside the lungs increases. Now the air flows out of the lungs (again from an area of high pressure to one of lower pressure).

A direct application of Boyle's law is seen in a chest respirator (see Figure 7-3), a machine used in the treatment of patients with respiratory difficulties. When the pressure inside the respirator is decreased, the air in the lungs expands, forcing the diaphragm down. When the pressure in the respirator is increased, the volume of air in the lungs is decreased, allowing the diaphragm to move upward again. This alternate increase and decrease in pressure enables the patient to breathe even though he or she cannot control the movement of the diaphragm muscles.

Another example of Boyle's law is in the use of a sphygmomanometer, a device used to measure blood pressure. When the rubber bulb is squeezed, the volume of air in that bulb is decreased and its pressure is increased. This increased pressure is transmitted to the cuff (see Figure 7-4).

Boyle's law can be stated mathematically as

Figure 7-3 Chest respirator in use. [Courtesy LIFECARE, 5505 Central Ave., Boulder, CO 80301.]

$$P_1V_1 = P_2V_2 \qquad \text{(at constant temperature)}$$

where P_1 and P_2 are the initial and final pressures, and V_1 and V_2 are the initial and final volumes, respectively. Pressures are usually expressed in millimeters of mercury (mm Hg). Recall that 760 mm Hg equals 1 atm pressure.

Example 7-1 What volume will 500 mL of gas initially at 25 °C and 750 mm Hg occupy when conditions change to 25 °C and 650 mm Hg?

Note first that the temperature is constant so that we can use Boyle's law.

Note also that P_1 is 750 mm Hg, V_1 is 500 mL, and P_2 is 650 mm Hg. Then, using $P_1V_1 = P_2V_2$, we have

$$750 \text{ mm Hg} \times 500 \text{ mL} = 650 \text{ mm Hg} \times V_2$$

Dividing by 650 mm Hg, we have

$$\frac{750 \text{ mm Hg} \times 500 \text{ mL}}{650 \text{ mm Hg}} = V_2 \quad \text{and} \quad V_2 = 577 \text{ mL}$$

Example 7-2 A gas exerts a pressure of 858 mm Hg when confined in a 5-L container. What will be the pressure if the gas is confined in a 10-L container at constant temperature?

Again using $P_1 V_1 = P_2 V_2$,

$$858 \text{ mm Hg} \times 5 \text{ L} = P_2 \times 10 \text{ L}$$

$$P_2 = 429 \text{ mm Hg}$$

Charles's Law

When gases are heated, they expand; when gases are cooled, they contract. The relationship between volume and temperature is expressed by Charles's law, which states that the volume of a fixed quantity of a gas is directly proportional to its Kelvin temperature if the pressure remains constant. Recall that Kelvin (absolute) temperature is Celsius temperature plus 273 (see page 12).

Charles's law can be explained in terms of the kinetic molecular theory. As the temperature of a gas is increased, the molecules move faster and strike the walls of the container more often. However, if the pressure is to be kept constant, then the volume must increase; that is, at constant pressure, the higher the temperature, the greater the volume of a gas, and vice versa.

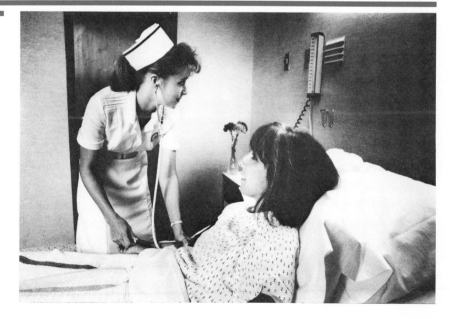

Figure 7-4 Sphygmomanometer, an instrument that makes use of Boyle's law to measure blood pressure. [Saint Francis Hospital School of Nursing, Evanston, IL. ©Jean Clough, 1988.]

A direct application of Charles's law can be seen in such equipment as an incubator. When air comes into contact with the heating element, it expands and becomes lighter. This lighter air rises, causing a circulation of warm air throughout the incubator.

Charles's law can be expressed mathematically as

$$\frac{V_1}{T_1} = \frac{V_2}{T_2} \qquad \text{(at constant pressure)}$$

where V_1 and V_2 are the initial and final volumes and T_1 and T_2 the initial and final temperatures, respectively, all temperatures Kelvin.

Example 7-3   A sample of gas occupies 368 mL at 27°C and 600 mm Hg. What will be the volume of that gas at 127 °C and 600 mm Hg?

We note that pressure is constant so we can use Charles's law. Also note that V_1 is 368 mL, T_1 is 27 + 273, or 300 K, and T_2 is 127 + 273, or 400 K. Then, substituting into

$$\frac{V_1}{T_1} = \frac{V_2}{T_2}$$

we have

$$\frac{368 \text{ mL}}{300 \text{ K}} = \frac{V_2}{400 \text{ K}}$$

$$V_2 = 491 \text{ mL}$$

Example 7-4   At constant pressure 200 mL of gas at 35 °C is cooled to −20 °C. What will be its new volume?

Here V_1 is 200 mL, T_1 is 35 + 273, or 308 K; and T_2 is −20 + 273, or 253 K. Then

$$\frac{V_1}{T_1} = \frac{V_2}{T_2}$$

$$\frac{200 \text{ mL}}{308 \text{ K}} = \frac{V_2}{253 \text{ K}}$$

$$V_2 = 164 \text{ mL}$$

Pressure – Temperature Relationship

Another relationship can be expressed between the pressure exerted by a gas and its temperature. These two factors are directly pro-

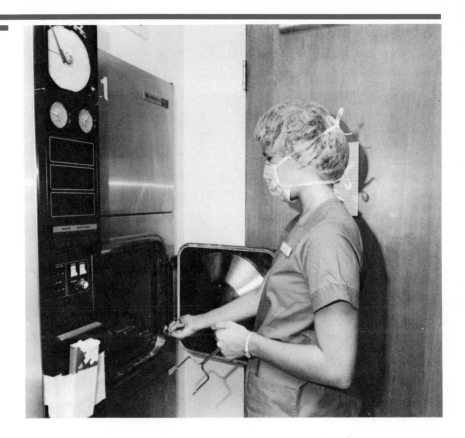

Figure 7-5 Autoclave in use. [Courtesy Methodist Medical Center of Illinois School of Nursing, Peoria, IL.]

portional. That is, as the temperature of a gas increases, the pressure increases, and vice versa, if the volume remains constant.

A common application of this relationship is the autoclave, a device used in hospitals for sterilization (see Figure 7-5). The normal temperature of steam is 100 °C, but in an autoclave it can rise as high as 120 °C because of the increased pressure. This higher temperature is sufficient to destroy any microorganisms that may exist in the material being autoclaved.

Combined Gas Laws

Boyle's law refers to the volume of a fixed quantity of a gas at constant temperature; Charles's law refers to the volume of such a gas at constant pressure. However, frequently neither of these factors is constant. In such a case we use the combined gas laws, which can be stated mathematically as

$$\frac{P_1 V_1}{T_1} = \frac{P_2 V_2}{T_2}$$

Example 7-5   What volume will 250 mL of gas at 27 °C and 800 mm Hg occupy at STP? (STP means standard temperature and pressure, 0 °C and 760 mm Hg or 760 torr, where the units "mm Hg" and "torr" are commonly used interchangeably.)
   Using

$$\frac{P_1 V_1}{T_1} = \frac{P_2 V_2}{T_2}$$

we have

$$\frac{800 \text{ mm Hg} \times 250 \text{ mL}}{300 \text{ K}} = \frac{760 \text{ mm Hg} \times V_2}{273 \text{ K}}$$

$$V_2 = 239 \text{ mL}$$

Dalton's Law

Dalton's law refers to a mixture of gases rather than to a pure gas. Dalton's law states that in a mixture of gases, each gas exerts a partial pressure proportional to its concentration. For example, if air contains 21 percent oxygen, then 21 percent of the total air pressure is exerted by the oxygen. Normal air pressure of 1 atm will support a mercury column 760 mm high. The partial pressure of the oxygen in the air would be 21 percent of 760 mm Hg, or $0.21 \times 760 = 160$ mm Hg.

   Dalton's law can be explained in terms of the kinetic molecular theory. Since there is no attraction between gas molecules, each kind of molecule strikes the walls of the container the same number of times per second as if it were the only kind of molecule present. That is, the pressure exerted by each gas (its partial pressure) is not affected by the presence of other gases. Each gas exerts a partial pressure proportional to the number of molecules of that gas (proportional to its concentration).

   Gases always diffuse from an area of higher partial pressure to one of lower partial pressure. An example of the diffusion of gases caused by a difference in partial pressures is found in our own bodies (see Figure 7-6). The partial pressure of oxygen in the inspired air is 158 mm Hg. The partial pressure of oxygen in the *alveoli is 104 mm Hg. Therefore, oxygen passes from the lungs into the alveoli (from a higher partial pressure to a lower one). From the alveoli the oxygen diffuses into the venous blood (from a partial pressure of 104 mm Hg to one of 40 mm Hg). This diffusion of oxygen into the venous blood in the lungs changes the venous blood into arterial blood, in which the partial pressure of oxygen is 95 mm Hg. When the arterial blood reaches the tissues, where the partial pressure of oxygen is 40 mm Hg, oxygen diffuses to those

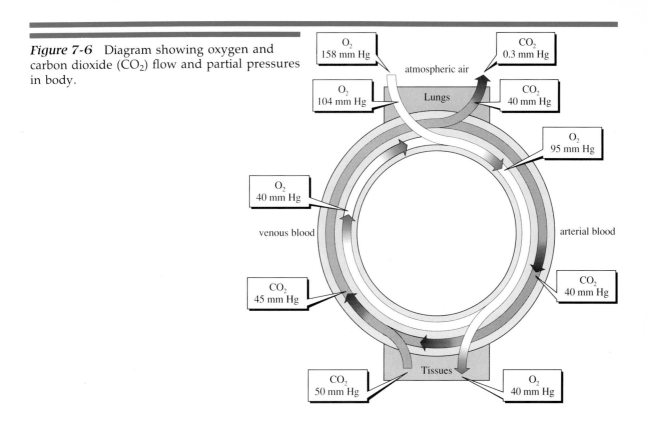

Figure 7-6 Diagram showing oxygen and carbon dioxide (CO_2) flow and partial pressures in body.

O_2 158 mm Hg

CO_2 0.3 mm Hg

atmospheric air

O_2 104 mm Hg

Lungs

CO_2 40 mm Hg

O_2 95 mm Hg

O_2 40 mm Hg

venous blood

arterial blood

CO_2 45 mm Hg

CO_2 40 mm Hg

CO_2 50 mm Hg

Tissues

O_2 40 mm Hg

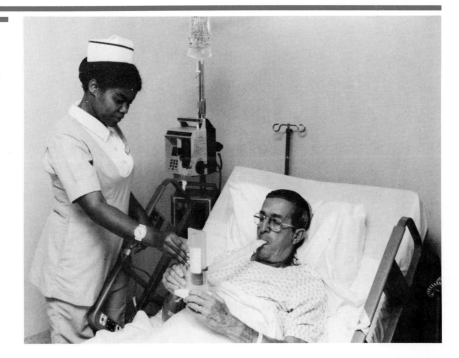

Figure 7-7 Intermittent partial pressure breathing apparatus. [Courtesy Methodist Medical Center of Illinois School of Nursing, Peoria, IL.]

tissues (again from a higher partial pressure to a lower one). When the arterial blood loses oxygen to the tissues, its oxygen partial pressure drops to 40 mm Hg, and it becomes venous blood, which returns to the lungs to begin the cycle anew.

Conversely, in the tissues the partial pressure of the carbon dioxide is 50 mm Hg, and in the arterial blood it is 40 mm Hg, so that carbon dioxide diffuses out of the tissues into the blood. When the arterial blood picks up carbon dioxide (and at the same time loses oxygen) it becomes venous blood with a carbon dioxide partial pressure of 45 mm Hg. This venous blood, in turn, loses carbon dioxide to the alveoli, where the carbon dioxide partial pressure is 40 mm Hg. From the alveoli the carbon dioxide passes into the lungs (from a partial pressure of 40 mm Hg to one of 0.3 mm Hg) and then is exhaled.

Use is made of Dalton's law in the intermittent partial pressure breathing apparatus, which fills the patient's lungs by increasing the partial pressure of the gas being inhaled (see Figure 7-7).

Graham's Law

If a container of ammonia and a container of ether are opened simultaneously, a person standing at the far end of the room will notice the odor of ammonia before that of the ether. Both gases will diffuse through the air, but ammonia has a molecular weight of 17 and ether a molecular weight of 74. Lighter molecules diffuse faster than heavier ones. This relationship was stated by Graham as, "The rates of diffusion of gases are inversely proportional to the square roots of the molecular weights" (or inversely proportional to the square roots of the densities).

$$\frac{\text{rate of diffusion gas}_1}{\text{rate of diffusion gas}_2} = \sqrt{\frac{\text{molecular weight gas}_2}{\text{molecular weight gas}_1}}$$

Henry's Law

Carbonated drinks are made by dissolving carbon dioxide, under pressure, in a given liquid. When the container is opened, the pressure is released and bubbles of carbon dioxide escape.

This relationship is expressed in terms of Henry's law, which states, "The solubility of a gas in a liquid at a given temperature is proportional to the pressure of that gas." That is, more gas dissolves at higher pressures and less at lower pressures.

Oxygen should be more soluble in blood at pressures above normal. Hyperbaric chambers, in which air pressures are two to three times normal, are used in the treatment of cancer (see page 64) and for victims of scuba diving accidents (see Figure 7-8). Decompression sickness can occur in any individual who has been breathing compressed gas at depths greater than 30 ft. Flying within

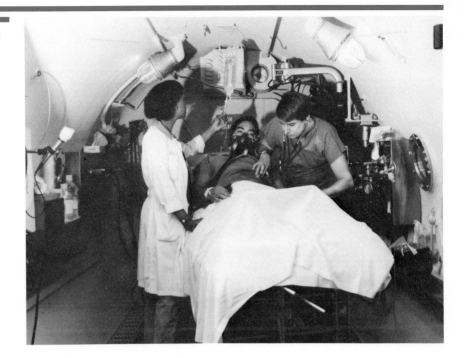

Figure 7-8 Treatment in hyperbaric chamber. [Courtesy intensive care unit, Edgewater Hospital, Chicago, IL.]

24 hr of diving can precipitate illness after an otherwise safe decompression.

Deep-sea divers are subject to pressures far above normal and so can have greater amounts of dissolved gases in their bloodstream. This condition might lead to the "bends" (see page 144).

Molar Gas Volumes

The volume occupied by 1 mol of any gas at STP (*s*tandard *t*emperature and *p*ressure, 0 °C and 760 mm Hg) is 22.4 L. However, this volume may be used only at STP. If a gas is at some other temperature or pressure, its volume must first be converted to STP before the molar gas volume can be used.

Example 7-6 How many moles of O_2 gas are present in 5.6 L at STP? Since 1 mol of a gas at STP occupies 22.4 L, then

$$5.6 \text{ L} \times \frac{1 \text{ mol}}{22.4 \text{ L}} = 0.25 \text{ mol } O_2$$

Example 7-7 What volume will 1.50 mol of CH_4 gas occupy at STP?

$$1.50 \text{ mol} \times \frac{22.4 \text{ L}}{1 \text{ mol}} = 33.6 \text{ L at STP}$$

Example 7-8   At STP, 2.50 g of a gas occupies 500 mL. What is the molecular weight of that gas?

First we must convert the volume to liters.

$$\frac{2.50 \text{ g}}{500 \text{ mL}} \times \frac{1000 \text{ mL}}{1 \text{ L}} = \frac{5.00 \text{ g}}{1 \text{ L}}$$

Then using the conversion factor 1 mol occupies 22.4 L,

$$\frac{5.00 \text{ g}}{1 \text{ L}} \times \frac{22.4 \text{ L}}{1 \text{ mol}} = \frac{112 \text{ g}}{1 \text{ mol}}$$

That is, 1 mol of the gas weighs 112 g, so the molecular weight is 112.

Air Pollution

Air pollutants can be gases such as sulfur dioxide, nitrogen oxides, ozone, hydrocarbons, and carbon monoxide, or they can be particulate matter such as smoke particles, asbestos, and lead *aerosols.

Concentrations of pollutants in the air are often expressed as parts per million (ppm), where 1 ppm corresponds to one part pollutant to 1 million parts air.

Concentrations of pollutants at levels far below 1 ppm can have an adverse effect upon human life. For example, 0.2 ppm sulfur dioxide in the atmosphere leads to an increased death rate, and 0.02 ppm peroxybenzoyl nitrate (a constituent of smog) causes severe eye irritation.

The word *smog* is derived from the words *smoke* and *fog*. Smog is characterized by air that contains lung and eye irritants along with reduced visibility. Smog is formed frequently during a thermal inversion. Normally, warm air near the ground surface rises and carries away pollutants. However, during a thermal inversion, the air near the surface is cooler than the air above it and so remains at the surface, keeping the air pollutants down at that level.

Gases

The principal source of three gaseous pollutants—carbon monoxide, hydrocarbons, and nitrogen oxides—is the automobile. Tobacco smoke is a source of carbon monoxide as well as particulate matter (ash): fossil-fuel-powered electrical generating stations are also a major source of air pollutants such as sulfur dioxide.

Each pollutant poses a different threat to human life.

1. **Carbon monoxide** is a deadly poison that interferes with the transportation of oxygen by the blood by competing with oxygen for the iron-binding sites in hemoglobin. Low concentrations of this gas,

(a)

(b)

Figure 7-9 (a) Normal lung tissue and lung tissue darkened by deposits from smoking and air pollution. (b) The darkened area shows the distention of small airways, which is characteristic of emphysema. [Courtesy Chicago Lung Association, photo by Milton Regier; lung tissue provided by David W. Cugell, M.D., Northwestern University.]

such as are found in automobiles, garages, downtown streets during rush hours, and space-heated rooms, cause impairments of judgment and vision. Evidence has shown that intermittent exposure to carbon monoxide at low levels of concentration can cause strokes and hypertension in susceptible individuals. High concentrations of carbon monoxide cause headache, drowsiness, coma, respiratory failure, and death.

2. **Nitrogen oxides** are just as dangerous a pollutant as carbon monoxide, even though environmental groups do not stress them equally. The first effects of nitrogen oxides upon humans is an irritation of the eyes and respiratory passages. Concentrations of 1.6 to 5 ppm of nitrogen dioxide for a 1-hr exposure cause increased airway resistance and diminish diffusing capacity of the lungs. Concentrations of 25 to 100 ppm cause acute but reversible bronchitis and *pneumonitis. Concentrations above 100 ppm are usually fatal, with death resulting from pulmonary *edema.

Nitrogen dioxide, NO_2, can react with water to form nitric acid, a constituent of "acid rain."

3. **Oxides of sulfur** cause acute airway spasm and poor airway clearance in all people and exert a deadly effect upon patients already disabled by lung disease. Concentrations of 8 to 10 ppm cause immediate throat irritation, and concentrations of 20 ppm cause immediate coughing.

Particulate matter absorbs sulfur dioxide, and the resulting tiny particles can enter the small air passages in the lungs, causing spasms and destruction of cells. Particles of the smallest size penetrate deepest into the lungs and remain there the longest.

Sulfur dioxide, SO_2, is oxidized in the air to sulfur trioxide, SO_3. Sulfur trioxide can react with the moisture in the air to form sulfuric acid, another constituent of acid rain.

Why is acid rain so detrimental? It affects trees and fruits and vegetables with considerable commercial consequences. Also it can react with calcium carbonate, which is found in marble and in mortar. Many famous marble sculptures around the world are rapidly deteriorating because of the acid in the air. The mortar between the bricks is also affected by acid rain and is slowly being dissolved, leaving the bricks with very little to hold them together.

4. **Tobacco smoke** is one of the most dangerous types of air pollution. Smokers inhale large amounts of carbon monoxide, tars, and particulate matter. People around smokers are also exposed to these same pollutants. Evidence has shown that expectant mothers who smoke have smaller babies and babies with a higher infant

mortality rate than those of nonsmokers. Smokers also have a greater chance of developing lung cancer, *emphysema, and cardiovascular disease (see Figure 7-9).

5. Ozone is formed by the action of sunlight on oxygen and is normally present in the atmosphere in extremely small amounts. Ozone reacts with hydrocarbon emissions and with oxides of nitrogen to form peroxyacyl nitrates, which are the eye irritants of smog. Since ozone is produced by the action of sunlight, its levels in the air are usually lower at night than during the day. Concentrations as low as 0.15 part per million can, within 1 hr, adversely affect such plants as tomatoes and corn. While low concentrations of ozone cause eye irritation, higher concentrations can cause pulmonary edema and hemorrhaging.

Los Angeles-type smog is primarily a photochemical phenomenon because many of the changes that take place in the air pollutants are due to the action of sunlight upon the substances mentioned previously.

The Ozone Layer

Ozone is found in a layer 20 to 50 km (12 to 30 miles) above the Earth's surface. Ozone is formed by the reaction of the sun's high-energy ultraviolet radiation upon oxygen molecules in the upper air.

$$3\,O_2 \underset{\text{radiation}}{\overset{\text{ultraviolet}}{\rightleftharpoons}} 2\,O_3$$

This layer, called the *ozone layer*, filters out most of the sun's high-energy ultraviolet radiation. This type of radiation is detrimental to life, and therefore the ozone layer serves a vital function. If the ozone layer were reduced or eliminated, the effect on life would be catastrophic.

The two principal threats to the ozone layer are

1. Chlorofluorocarbons, used in aerosol spray cans, in refrigeration, and in air conditioners.
2. Oxides of nitrogen.

Chlorofluorocarbons are widely used in air conditioning and refrigeration units and in aerosol spray cans because they are *volatile, are chemically unreactive, and liquefy under pressure. In the atmosphere, they gradually diffuse upward into the stratosphere where they can react with the ozone layer and reduce the amount of ozone present. Even if the ozone layer were depleted by only a small amount, the result would be a marked increase in skin cancer, caused by the dangerous ultraviolet radiation from the sun reaching the Earth's surface.

Oxides of nitrogen such as NO and NO_2 are formed in the exhaust of automobiles and planes. In the lower levels of the atmosphere, although these compounds are detrimental to human health, they pose no threat to the ozone layer because they do not remain in the atmosphere long enough to rise into the stratosphere. However, supersonic and military planes do inject large amounts of oxides of nitrogen into the ozone layer because they fly at such great altitudes. This is one reason the use of supersonic planes has not been greatly expanded.

Summary

The properties of gases can be explained in terms of the kinetic molecular theory, which states that (1) gases are composed of tiny particles called molecules, (2) the distances between molecules of a gas are very great compared to the size of the molecules themselves, (3) gas molecules move rapidly in straight lines, (4) gas molecules do not attract each other, (5) collisions between molecules and between the molecules and the walls of the container are perfectly elastic, and (6) the average kinetic energy of the molecules is the same for all gases at the same temperature.

Pressure is measured by means of a barometer. Standard pressure is 1 atm and will support a column of mercury 760 mm tall.

Boyle's law states that the volume of a fixed quantity of a gas is inversely proportional to the pressure if the temperature remains constant. Boyle's law can be expressed mathematically as $P_1V_1 = P_2V_2$.

Charles's law states that the volume of a fixed quantity of a gas is directly proportional to its Kelvin temperature if the pressure remains constant. Charles's law can be stated mathematically as $V_1/T_1 = V_2/T_2$.

The combined gas laws can be stated mathematically as $P_1V_1/T_1 = P_2V_2/T_2$.

Dalton's law states that in a mixture of gases each gas exerts a partial pressure proportional to its concentration.

Gases diffuse from an area of high partial pressure to one of lower partial pressure.

Graham's law states that the rates of diffusion of gases are inversely proportional to the square roots of their molecular weights.

The molar gas volume is 22.4 L at STP. It is the volume occupied by 1 mol of any gas under those conditions.

Air pollutants may be gases such as carbon monoxide, ozone, nitrogen oxides, and sulfur oxides, and they also may be particulate matter such as smoke particles, asbestos, and lead aerosols. Each pollutant poses a different threat to human life.

Questions and Problems

A

1. Explain in terms of the kinetic molecular theory (a) why gases have a low density, (b) why gases are compressible, (c) why gases do not settle, (d) why gases diffuse into each other, (e) why gases have no definite shape or volume, (f) Charles's law, and (g) Boyle's law.

2. What volume will 800 mL of neon gas initially at 25 °C and 720 mm Hg occupy at 25 °C and 680 mm Hg?

3. Calculate the pressure at which 1.50 L of oxygen at 13 °C and 675 mm Hg will occupy 1.75 L at 13 °C.

4. What would be the change in volume of 200 mL of oxygen at 27 °C and 600 mm Hg when the temperature rises to 227 °C at 600 mm Hg?

5. When 3.00 ft³ of hydrogen at 40 °C and 1.15 atm pressure is cooled to −40 °C and 1.15 atm, what will the volume be?
6. What volume will 250 mL of carbon dioxide gas at 15 °C and 900 mm Hg occupy at STP (0 °C and 760 mm Hg)?
7. Explain how a chest respirator enables a person whose diaphragm is paralyzed to breathe.
8. Explain how an autoclave works in terms of the appropriate gas law.
9. Why does carbon dioxide pass from the blood to the lungs instead of vice versa?
10. Explain why oxygen diffuses to the tissues and carbon dioxide diffuses from the tissues instead of vice versa.
11. Which gas will diffuse faster: methane, CH_4, or ammonia, NH_3?
12. Arrange the following gases in order of increasing rates of diffusion: H_2, O_2, NH_3, CO_2, SO_2, C_2H_6, Ar, and He.
13. What are the principal sources of air pollution?
14. What are the hazards of cigarette smoking?
15. What threats to life are caused by (a) oxides of nitrogen, (b) ozone, (c) carbon monoxide, (d) oxides of sulfur, (e) particulate matter?

B

16. Explain in terms of the kinetic molecular theory why gases exert a pressure.
17. Why do gases exert pressure equally in all directions?
18. Why are collisions between molecules elastic?
19. Explain how a sphygmomanometer works.
20. Explain why air flows into and out of the lungs.
21. Explain in terms of the kinetic molecular theory why gases exert a partial pressure proportional to the concentration.
22. Define Henry's law. What use is made of this law in the treatment of cancer?
23. Explain "the bends" in terms of Henry's law.
24. What is "smog"? What are its effects?
25. Explain how a barometer works.
26. What does a "falling barometer" indicate in terms of weather conditions?
27. How many moles of CO_2 are present in 2.8 L at STP?
28. What volume will 0.75 mol of C_2H_2 gas at STP occupy?
29. What is "acid rain," and why is it harmful?
30. At STP, 1.65 g of a gas occupy 224 mL. What is the molecular weight of that gas?

31. Why is the ozone layer so important? What substances might affect it?

Practice Test

1. Pressure is measured by means of a _____.
   a. thermometer   b. barometer
   c. sphygmomanometer   d. hydrometer
2. Boyle's law applies when which of the following remains constant?
   a. temperature   b. pressure
   c. volume   d. none of these
3. Charles's law applies when which of these remains constant?
   a. temperature   b. pressure
   c. volume   d. none of these
4. 500 mL of CO_2 gas at 27 °C and 600 mm Hg pressure will occupy what volume at −73 °C and 1200 mm Hg pressure?
   a. 166 mL   b. 333 mL
   c. 1500 mL   d. 2000 mL
5. The partial pressure of oxygen is higher _____.
   a. in the air than in the lungs
   b. in the tissues than in the bloodstream
   c. in venous blood than in arterial blood
   d. in the tissues than in the alveoli
6. The molar gas volume, in liters, is _____.
   a. 11.2   b. 22.4
   c. 44.8   d. none of these
7. Gases are compressible because the molecules _____.
   a. are in rapid motion
   b. do not attract one another
   c. rebound without loss of energy
   d. are far apart
8. How many moles of CO_2 are present in 11.2 L at STP?
   a. 0.1   b. 0.5
   c. 1.0   d. 11.2
9. 200 mL of gas at 0 °C and 900 mm Hg pressure will occupy what volume at 0 °C and 450 mm Hg pressure?
   a. 100 mL   b. 200 mL
   c. 300 mL   d. 400 mL
10. As the pressure of a gas increases, its solubility in water _____.
    a. increases
    b. decreases
    c. is unaffected

8

Oxygen and Other Gases

Oxygen therapy at home.

Oxygen

Occurrence

Oxygen is the most abundant element on the earth's surface. Air is about 21 percent free oxygen. The oceans and lakes on the earth's surface consist of about 80 percent oxygen in the combined state. The Earth's crust consists of about 50 percent oxygen, combined mostly with silicon. Oxygen compounds comprise most of the weight of plants and animals.

Properties

Physical Properties

At room temperature oxygen is a colorless, odorless, tasteless gas. It is slightly heavier than air and is slightly soluble in water.

The method of preparation illustrated in Figure 8-1 shows that oxygen does not dissolve appreciably in water. If it were soluble in water, it could not be collected by this method. However, a small amount of oxygen does dissolve in water. This amount, though small, is of very definite importance to marine life. It is this small amount of dissolved oxygen that enables fish and other aquatic animals to "breathe."

The density of oxygen is 1.43 g/L. The density of air is 1.29 g/L. Therefore, oxygen is slightly heavier than air. When oxygen is collected in the laboratory, it is kept in covered bottles with the mouth upward.

When cooled sufficiently, oxygen forms a pale blue liquid that boils at −182.5 °C. Further cooling produces a pale blue solid when the oxygen freezes at −218.4 °C.

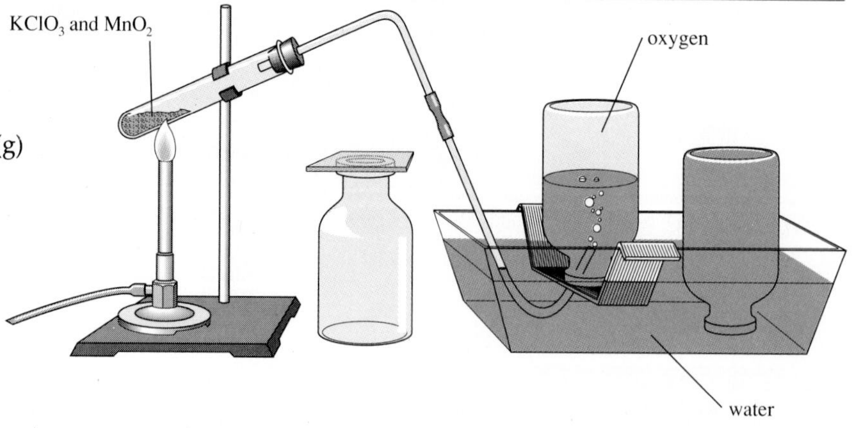

Figure 8-1 Laboratory preparation of oxygen. The reaction is

$$2\ KClO_3 \xrightarrow[\text{heat}]{MnO_2} 2\ KCl + 3\ O_2(g)$$

KClO$_3$ and MnO$_2$

oxygen

water

Chemical Properties

Oxygen is a moderately active element at room temperature but is extremely active at higher temperatures. It combines with almost all elements to produce a class of compounds called oxides.

$$2\,Mg \;+\; O_2 \longrightarrow 2\,MgO$$

magnesium   oxygen   magnesium oxide

$$C \;+\; O_2 \longrightarrow CO_2$$

carbon   oxygen   carbon dioxide

The reaction between oxygen and some other substance is an example of **oxidation.** Common examples of oxidation are the rusting of iron, the burning of a candle, and the decay of wood. Oxidation also occurs in living plant and animal tissues. These oxidation reactions are able to occur rapidly at relatively low temperatures because of the presence of specific catalysts called enzymes. An example of such a reaction occurring in the human body is the oxidation of glucose, a simple sugar, to carbon dioxide and water.

$$C_6H_{12}O_6 + 6\,O_2 \xrightarrow{\;enzymes\;} 6\,CO_2 + 6\,H_2O + energy$$

glucose   oxygen   carbon   water
dioxide

(This reaction is greatly oversimplified. As will be discussed later, when glucose is oxidized to carbon dioxide and water in the body, there are many intermediate steps involved, each with its own particular enzyme.)

Combustion

Wood, coal, and gas burn, that is, undergo combustion, in the presence of oxygen. Combustion can be defined as a rapid oxidation in which heat and light are produced, usually accompanied by a flame. Oxygen supports combustion (that is, substances burn in oxygen), but oxygen itself does not burn.

Combustion is usually thought of in terms of a reaction with oxygen. However, oxygen is not absolutely necessary for a combustion reaction. When powdered iron and sulfur are heated together in a test tube, a rapid reaction occurs in which heat and light are given off. This is also a combustion reaction even though no oxygen is involved.

Spontaneous Combustion

When iron combines with the oxygen of the air, it rusts, continuously liberating a small amount of heat as the reaction continues.

The total amount of liberated heat, however, will be the same as that which would have been liberated had the iron been burned in oxygen. That is, the total amount of heat produced is the same regardless of whether the reaction proceeds rapidly or slowly.

This is the principle underlying spontaneous combustion, which can be defined as a slow oxidation that develops by itself into combustion. How is such a process possible? If rags containing a "drying oil" such as boiled linseed oil are placed in an open dry container without adequate ventilation, the oil will slowly combine with the oxygen in the air, liberating a small amount of heat during the process. Without ventilation the heat will not be dissipated, especially since rags are such poor conductors of heat. As the oxidation proceeds, more and more heat will be liberated until a sufficient amount accumulates to start the rags burning. This is spontaneous combustion.

If such rags are placed in a closed container, the oxidation will not continue after the oxygen supply is used up. Likewise, if the rags are hung in a place where freely circulating air can carry away the heat, no spontaneous combustion can occur.

Fire Prevention and Control

Care should be taken in handling and storing flammable liquids such as ether and alcohol. Where smoking is allowed, nonflammable receptacles should be provided for the butts. One frequent cause of fires is lighted cigarettes thrown into a wastebasket. Oily rags and mops should be stored in well-ventilated fireproof lockers to avoid the danger of spontaneous combustion.

When a fire does occur, how can it be extinguished? There are two methods of putting out a fire: removing the oxygen from the burning material or lowering the temperature of the burning substance below its kindling point.

One type of fire extinguisher contains carbon dioxide gas under pressure (see Figure 8-2). This type of extinguisher has several advantages. The carbon dioxide is extremely cold and lowers the temperature of the burning substance. A large amount of carbon dioxide, which is heavier than oxygen, surrounds the burning area, keeping out the oxygen. For these reasons the carbon dioxide should be directed at the base of the flame. This type of extinguisher is recommended for electrical fires because the carbon dioxide does not conduct electricity. It is also used for oil and gasoline fires because it shuts off the oxygen supply and at the same time lowers the temperature. When a fire is extinguished, the carbon dioxide disappears into the air so that it does not have to be cleaned up along with the material that was burning. A cylinder of carbon dioxide holds a large volume of the gas under pressure and thus has a great capability in fire fighting.

Carbon tetrachloride (CCl_4) was formerly used in fire extinguishers, but its use for this purpose has been outlawed because (1)

Figure 8-2 Carbon dioxide fire extinguisher. [Courtesy The Ansul Co., Marinette, WI.]

its vapors are toxic and (2) a hot fire converts it to phosgene ($COCl_2$), a poisonous gas.

Preparation

Laboratory Methods

One very common method for preparing oxygen in the laboratory is by heating potassium chlorate, $KClO_3$ (see Figure 8-1). The reaction is

$$2\ KClO_3 \xrightarrow{\text{heat}} 2\ KCl + 3\ O_2(g)$$

It takes a considerable amount of heat to produce oxygen by this method because the potassium chlorate must be heated to its melting point (370 °C) before it gives off oxygen.

However, when manganese dioxide (MnO_2) is added to the potassium chlorate, oxygen is evolved from the heated mixture at a much lower temperature and at a more rapid rate. The manganese dioxide acts as a catalyst in this reaction; a catalyst increases the speed of the reaction but does not take part in it. The presence of a catalyst in a reaction is indicated over the arrow in the equation for that reaction.

$$2\ KClO_3 \xrightarrow[\text{heat}]{MnO_2} 2\ KCl + 3\ O_2(g)$$

Oxygen can also be produced in the laboratory by the electrolysis of water.

$$2\ H_2O \xrightarrow{\text{electricity}} 2\ H_2(g) + O_2(g)$$

Commercial Method

The commercial source of oxygen is the air, an inexhaustible supply as long as we have trees and other green plants, both on land and in the water. When air is cooled to a low enough temperature under compression, it becomes a liquid. Liquid air, like ordinary air, consists mostly of nitrogen and oxygen. Liquid nitrogen boils at −196 °C. Liquid oxygen boils at −182.5 °C. When liquid air is allowed to stand, the nitrogen boils off first (because of its lower boiling point), leaving almost pure oxygen behind. Liquid oxygen is stored in steel cylinders under high pressure.

Uses

Medical Uses

Oxygen is necessary to life. When oxygen is taken into the lungs, it combines with the hemoglobin of the blood to form a compound called oxyhemoglobin (see page 551). The blood carries the oxyhemoglobin to the tissues where oxygen is released. This oxygen then reacts with the food products in the cells, producing energy. At the same time, carbon dioxide is produced and is carried back to the lungs where it is exhaled. Blood going to the tissues (arterial blood) contains oxyhemoglobin and has a characteristic bright red color. Blood coming from the tissues (venous blood) does not contain oxyhemoglobin and has a characteristic reddish purple color.

Patients with lung diseases such as pneumonia frequently do not have enough functioning lung tissue to pick up sufficient oxygen from the air. These patients are given a mixture of oxygen and air to breathe instead of air alone. Then, because of the higher partial pressure of oxygen, a small functioning area of the lung can pick up more oxygen than it could if air alone were breathed in. This enables the patient to breathe and live until the diseased area is cured and returns to normal. The oxygen can be administered by nasal cannula or mask (see Figures 8-3 and 8-4). Oxygen is given to patients with lung cancer to help them adjust to a decreased lung area.

Living tissues require oxygen. Without it they soon die. However, if the temperature is lowered sufficiently, tissues can survive with very little oxygen. At normal body temperature, 98.6 °F, the brain is extremely sensitive to a lack of oxygen. However, it too can live for a longer period of time without oxygen if its temperature is

Figure 8-3 Nasal oxygen cannula. [Courtesy Puritan-Bennett Corporation.]

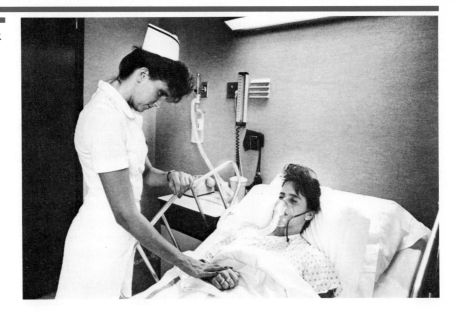

Figure 8-4 Aerosol mask in use. [Saint Francis Hospital School of Nursing, Evanston, IL. © Jean Clough, 1988.]

lowered. But how low is low when we are talking about body temperature?

In one form of surgery used to repair ruptured blood vessels in the brain, the body temperature is dropped from normal to 86 °F by means of an ice bath. At this temperature the heart can be stopped for about 15 min without damage to the body. However, brain surgery requires more time than the 15 min allowed. After the body temperature is lowered to 86 °F and after the heart is stopped, a saltwater solution at 32 °F is pumped directly into the main artery that feeds the brain. This cold solution lowers the brain's temperature to approximately 60 °F; at this temperature brain surgery can be completed without damage to either the brain or the body. After surgery, the heart is restarted and blood (carrying oxygen) again flows to the brain.

Hypothermia is also used for open-heart surgical procedures. The patient is placed on a plastic pad through which an iced solution is run in order to lower body temperature to 25 or 20 °C (77 or 68 °F). However, the main cooling is done through the bypass machine during the surgery. The heart itself is cooled by perfusion of coronary arteries with an iced solution or packed in an ice solution. The temperature of the heart can be lowered to 5 °C (41 °F). Hypothermia of the organs is also necessary during transplant surgery.

Cooling the scalp during chemotherapy reduces the amount of drug absorbed by hair follicles. Such a treatment has been effective in preventing hair loss in many patients without reducing the anticancer effect of the drug being absorbed.

Hyperthermia is a technique of using microwave radiation to raise the temperature of target cancer cells without harming nearby normal tissue.

Since oxygen supports combustion, an object such as a candle, which burns slowly in air, will burn very vigorously in oxygen. Therefore, patients and others must not smoke or use matches in a room where oxygen is in use. The following precautions should also be taken.

1. Electric devices such as radios, televisions, electric shavers, and so on are likewise banned because of the danger that a spark from the equipment could cause a fire.
2. Electric signal cords should be replaced by a hand bell because of the danger of a spark.
3. Patients should not be given backrubs with alcohol or oil because of the danger of a fire. Instead, lotion or powder should be used.
4. Oil or grease should never be applied to any part of the oxygen equipment. Nurses should take care not to have oil on their hands when manipulating the regulator of the oxygen tank.

*Hypoxia is a condition in which the body does not receive enough oxygen. In cases of hypoxia, oxygen must be administered to permit the body to function normally. Newborn babies who have difficulty breathing are given oxygen containing a small amount of carbon dioxide. The carbon dioxide stimulates the respiratory center of the brain so that the rate of breathing increases and the oxygen is picked up more rapidly and carried to the tissues.

During dental surgery when nitrous oxide is used as the anesthetic, oxygen must be administered along with the nitrous oxide to prevent *asphyxiation (lack of oxygen). Firefighters who breathe large quantities of smoke may suffer from asphyxiation. They are treated by breathing from an oxygen mask. Death by smoke inhalation during a fire is often the result of breathing in noxious material such as hydrogen cyanide, HCN, or phosgene produced by the burning of plastics.

A person who is under water for several minutes and becomes unconscious because of lack of oxygen is given oxygen in an effort to build up the oxygen concentration in the blood rapidly. Pilots and astronauts breathe oxygen through a face mask because of the decrease or lack of oxygen in the atmosphere around them, as do miners, deep-sea divers, and smoke fighters.

In the hospital, oxygen under pressure (hyperbaric treatment) is administered in the treatment of cancerous tissues (see page 64), gangrene, and carbon monoxide poisoning.

Another use of oxygen in the hospital is in the determination of the basal metabolic rate (BMR). Basal metabolism tests measure the energy production of the body by measuring the amount of oxygen a patient breathes during a specified period of time (usually 6 min). The BMR has been used as an indicator of thyroid function, but its use for this purpose has been largely supplanted by tests using ^{131}I (radioactive iodine).

One measure of the body's use of oxygen (pulmonary function)

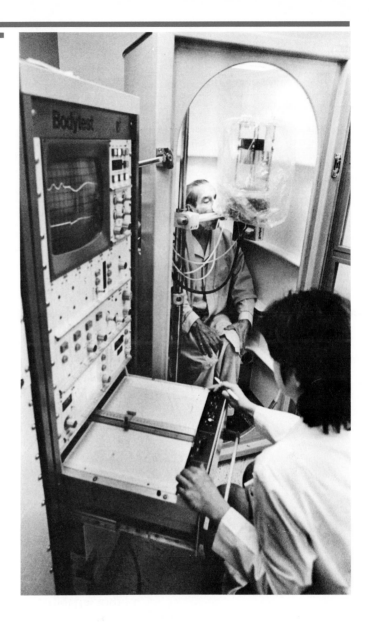

Figure 8-5 A technician tests a man in the pulmonary function laboratory. [Courtesy University of Illinois at Chicago.]

is the MET, derived from the word metabolism. One MET corresponds to the consumption of 1 mL of O_2 per minute for each kilogram of body weight (see Figure 8-5). Table 8-1 compares MET values for various activities.

Formerly, all premature infants were routinely given oxygen until respiratory sufficiency had been established (see Figure 8-6). Now it is known that when the concentration of oxygen rises above 40 percent in the inspired air, premature infants develop retrolental fibroplasia, a disease that affects the eyes. This disease produces complete or nearly complete blindness due to separation and fibrosis of the retina.

Table 8-1 MET values

	MET
At rest	1
Walking 2 mph	3
Running 7 mph	12
Walking upstairs	15
Shoveling snow	7

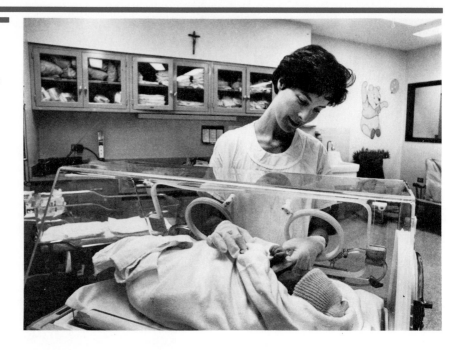

Figure 8-6 Infant in isolette. [Saint Francis Hospital School of Nursing, Evanston, IL. © Jean Clough, [1988.]

Commercial Uses

Deep-sea divers who work at great depths formerly were supplied air under pressure to enable them to breathe. However, this process introduced a difficulty in that the nitrogen in the air became much more soluble in the blood under the higher pressure. When the diver was taken out of the water too rapidly, this dissolved nitrogen became less soluble because of the decrease in pressure. When this happened, bubbles of nitrogen formed in the blood, causing a condition called the "bends." Unless the diver could be placed immediately in a compression chamber under high pressure, which then was gradually reduced to normal pressure, the results were generally fatal. Recall the similar effect of tiny bubbles being formed in a liquid when a bottle of warm carbonated beverage is opened.

A method enabling the diver to breathe a mixture of oxygen and helium is now used. Helium is less soluble in the blood than nitrogen and so decreases the chances of the bends occurring.

A mixture of helium and oxygen can flow more easily through a partially obstructed air passageway than can a mixture of nitrogen and oxygen. In small children, the airways are extremely tiny, so even a partial obstruction can cause major problems. Administration of a $He-O_2$ mixture to an infant suffering from an infection that results in the swelling of the air passageways allows the passage of oxygen to the body until the antibiotics can take effect and cure the infection.

The $He-O_2$ mixture is also useful in buying time for adult patients who have a tumor partially blocking the air passageways and so cannot get enough oxygen into their bodies. Normally, such patients cannot survive long enough for radiation to shrink the tumor. With the He–O_2 mixture, they can. They are then gradually weaned off the He–O_2 mixture. Such a treatment is not a cure but merely a method of buying time until the breathing difficulty can be corrected.

Industrially, oxygen is used in the operation of furnaces for production of iron and steel. It is also used in welding torches. These torches frequently use acetylene as their fuel. When acetylene is burned with oxygen, the heat produced is high enough to melt steel. Liquid oxygen (LOX) is used in rockets for launching satellites.

Hydrogen Peroxide

Another product besides water (H_2O) is possible between the elements hydrogen and oxygen. This compound is hydrogen peroxide, H_2O_2, produced by means of the following reactions:

$$Ba + O_2 \longrightarrow BaO_2$$

$$BaO_2 + H_2SO_4 \longrightarrow BaSO_4 + H_2O_2$$

Hydrogen peroxide is a pale blue oily liquid that boils at 150 °C and freezes at -0.41 °C. It is very unstable and decomposes according to the following reaction:

$$2\ H_2O_2 \longrightarrow 2\ H_2O + O_2$$

This reaction is induced by light, and therefore hydrogen peroxide is usually stored in dark-colored bottles. A 3 percent solution of H_2O_2 is used as an antiseptic; a 30 percent solution is used as a bleach and as an oxidizing agent in chemistry laboratories.

The oxidation state of oxygen in water and in oxides is -2. The oxidation state of oxygen in the free state is 0. In peroxides, the oxidation state of the oxygen (the oxidation number) is -1.

Hydrogen peroxide is an intermediate product in the reduction of O_2 to H_2O in many cellular reduction reactions. Since hydrogen peroxide is toxic to living cells, it must be removed quickly before it can cause harm. The enzyme that catalyzes such a reaction is catalase (see page 442).

Superoxides

When a metal reacts with oxygen, an oxide is formed.

$$2\ Al + 3\ O_2 \longrightarrow 2\ Al_2O_3 \quad \text{(aluminum oxide)}$$

When an active metal reacts with oxygen, a peroxide is formed.

$$2 \, Na + O_2 \longrightarrow Na_2O_2 \quad \text{(sodium peroxide)}$$

When a very active metal reacts with oxygen, a superoxide is formed.

$$K + O_2 \longrightarrow KO_2 \quad \text{(potassium superoxide)}$$

In superoxides, the oxidation number of oxygen is $-\frac{1}{2}$. Superoxide ions exist in the human body and in some microorganisms. In the production of uric acid (page 528), superoxide ions are produced. Superoxide ions are formed during the oxidation of Fe^{2+} to Fe^{3+} in the hemoglobin molecule. Superoxide ions are also involved in the immune response (see page 577) and possibly in the aging process.

Both the superoxide ion (O_2^-) and hydrogen peroxide (H_2O_2) are very reactive substances and extremely dangerous to living cells; therefore, their removal is of great importance to the well-being of the organism. The superoxide ion is rapidly converted to O_2 and H_2O_2 by the enzyme superoxide dismutase. The H_2O_2 thus produced is decomposed into H_2O and O_2 by the enzyme catalase (page 442).

Ozone

Oxygen occurs most frequently as the diatomic molecule O_2. However, it also exists as a triatomic molecule, O_3, called ozone (Figure 8-7). These two different forms of the element oxygen are called allotropic forms. **Allotropic forms** of an element have different physical and chemical properties. Other elements that exist in allotropic forms are carbon, sulfur, phosphorus, tin, and lead.

Figure 8-7 Models of oxygen and ozone molecules.

Oxygen molecule

Ozone molecule

Preparation

When air or oxygen is passed between two electrically charged plates, the volume decreases and a pale blue gas with a strong odor is formed. This gas is ozone. An equation for the formation of ozone is

$$3 \, O_2 \rightleftharpoons 2 \, O_3$$

The double arrow indicates an **equilibrium reaction,** one that proceeds in both directions.

Ozone is formed by the action of ultraviolet light from the sun upon the oxygen in the air. It is also formed in the air by electric discharges (lightning). Ozone can also be found around high-voltage machinery where sparks convert some of the oxygen of the air into ozone.

The ozone layer in the upper atmosphere helps protect the Earth from the harmful effect of the sun's ultraviolet radiation (see page 132).

Properties and Uses

Ozone is a colorless (or sometimes slightly blue) gas at room temperature. It has a very pungent odor similar to that of garlic. It is heavier than oxygen and more soluble in water. Liquid ozone is blue. Ozone is a powerful oxidizing agent. The use of ozone is rather limited because it is poisonous and unstable. Ozone is irritating to the mucous membranes, and it is quite toxic to the body except in extremely small amounts.

Ozone is not tolerated in industrial establishments in concentrations of more than 1 part per million (ppm), whereas carbon monoxide (CO) can be tolerated in strengths as high as 100 ppm. In other words, ozone is 100 times as poisonous as carbon monoxide. It is also 100 times as poisonous as hydrogen sulfide (H_2S) and approximately 10 times as poisonous as hydrogen cyanide (HCN).

Ozone is not used medically for any purpose because it reaches a fatal concentration long before it can act in any useful way. However, the industrial use of ozone is increasing rapidly. It is a more powerful oxidizing agent than oxygen itself. Its usefulness as a *bactericide, a decolorizer, and a deodorizer is great when it is employed carefully and in sufficient strengths. Ozone is used in the rapid aging of wood, for rapid drying of varnishes and inks, in the treatment of water, and in the disinfection of swimming-pool water, but it should only be handled with great care by trained personnel.

Nitrous Oxide

Nitrous oxide, N_2O, is frequently used as a general anesthetic. It is a colorless gas with almost no odor or taste. It is heavier than air and has a relatively low solubility in blood. N_2O does not combine with hemoglobin and is carried dissolved in the bloodstream. It is excreted, unchanged, primarily through the lungs, but a small fraction may escape through the skin. It has been found that nitrous oxide can be useful in relieving the pain in heart attack victims by reducing the strain on the heart muscles.

Although N_2O is not flammable, it does support combustion, as does oxygen, so care must be taken with its use.

The highest concentration of N_2O that can be safely given for the maintenance of anesthesia is 70 percent. Above this concentration hypoxia develops. However, at a concentration of 70 percent or less, N_2O is not potent enough for most patients, so additional drugs such as halothane or methoxyflurane are used to complete the anesthesia.

Nitrous oxide can react with the benzpyrene (page 323) in the air to produce *mutagenic substances. Large doses of vitamin E seem to protect mice and rats from these substances; however, humans do not benefit in the same way.

Noxious Gases

Carbon monoxide, CO, is a colorless, odorless, tasteless gas. It is nonirritating to the body but has a great affinity for hemoglobin. CO is poisonous because the hemoglobin forms such a strong bond with the gas that the blood is unable to carry sufficient oxygen to the tissues.

Vapors of **carbon tetrachloride,** CCl_4, are very toxic to the body because of the damaging effects upon both the liver and the kidneys. CCl_4 has been used as an anesthetic, as a cleaning fluid, in fire extinguishers, and for the management of hookworm infestations; however, because of its great toxicity and its *carcinogenic properties, its use has been banned by the U.S. Food and Drug Administration. Today carbon tetrachloride has no clinical use at all.

Chlorine is a greenish yellow gas with a pungent odor. In neutral or acid solutions chlorine acts not only as a bactericide but also as a *virucide and *amoebicide. It is less effective as a germicide in basic solution. Chlorine kills microorganisms in water and so is used for water purification. In large amounts, chlorine vapors are toxic because they react with and destroy lung tissues.

Sulfur dioxide is released into the air during the combustion of fossil fuels. Catalytic converters in automobiles change SO_2 to SO_3, which can then form sulfuric acid and other sulfates that are extremely harmful to the health of humans and aquatic creatures.

Hydrogen sulfide, H_2S, is characterized by its distinctive odor of rotten eggs. H_2S is extremely poisonous, even more so than hydrogen cyanide (HCN), but its odor warns a person of its presence long before its concentration reaches lethal proportions. This is not true for HCN.

Summary

Oxygen is the most abundant element on the Earth's surface. It is colorless, odorless, tasteless, slightly heavier than air, and very slightly soluble in water.

Oxygen combines with most elements to form a class of compounds called oxides. The reaction between oxygen and some other substances is an example of oxidation. Oxidation also occurs in plant and animal tissues in the presence of specific catalysts called enzymes.

Combustion is a rapid reaction with oxygen, one in which heat and light are produced, usually accompanied by a flame. Spontaneous combustion is a slow oxidation that develops by itself into combustion.

Fire extinguishers accomplish their work either by removing the oxygen from the burning substance or by lowering its temperature below its kindling point, or both.

Oxygen can be prepared in the laboratory by heating such compounds as potassium chlorate ($KClO_3$) or by the electrolysis of water.

Oxygen is prepared commercially from liquid air.

Oxygen is used medically in the treatment of various respiratory diseases. Oxygen is inhaled and carried to the tissues by the hemoglobin in the blood. Care must be taken to prevent fires in the room of a patient receiving oxygen therapy. This means no smoking or use of electric devices, no alcohol or oil backrubs.

Asphyxiation (hypoxia) can be overcome by the administration of oxygen. Oxygen is used in the treatment of cancer for some patients and in the determination of basal metabolic rate.

Industrially, oxygen is used in welding, in the production of steel, and in rockets.

When oxygen reacts with an active metal, a peroxide is formed; when oxygen reacts with a very active metal, a superoxide is formed.

Ozone (O_3) is an allotropic form of oxygen (O_2). Ozone is colorless gas with a pungent odor. It is very toxic to the body. Ozone is not used medically but has many industrial uses; it must be employed under carefully controlled conditions by trained personnel.

Nitrous oxide is used as a general anesthetic, usually in conjunction with another drug such as halothane or methoxyflurane.

Among the noxious gases are carbon monoxide, the vapor of carbon tetrachloride, chlorine, sulfur dioxide, and hydrogen sulfide.

Questions and Problems

A

1. After oxygen is collected in the laboratory, why is the bottle kept mouth upward?
2. List several physical properties of oxygen.
3. When elements combine with oxygen, what type of compound is produced?
4. What is combustion? How does it differ from spontaneous combustion?
5. What precautions should a nurse take in preventing fires?
6. Describe the operation of one type of fire extinguisher.
7. How can oxygen be prepared in the laboratory? commercially?
8. How is oxygen carried to the tissues?
9. Discuss the medical uses of oxygen and also the hazards involved in oxygen therapy.
10. What causes retrolental fibroplasia? How can this disease be prevented?
11. Why is carbon dioxide sometimes given along with oxygen to newborn babies?
12. What is hypoxia?
13. What is the "bends," and how can it be overcome?
14. What is ozone? List some properties of this substance. How can it be prepared?
15. List the general properties of nitrous oxide. What precautions should be taken with its use?
16. Why must nitrous oxide be used at concentrations less than 70 percent? Why is it used in conjunction with other drugs?
17. Why is carbon monoxide poisonous?
18. Why is carbon tetrachloride dangerous?
19. Under what conditions is a solution of chlorine an effective germicide?
20. Why are chlorine vapors toxic to the body?
21. Compare the reactions of metals, active metals, and very active metals with oxygen.
22. Why is hydrogen peroxide usually stored in dark bottles?
23. What are the oxidation states of oxygen in free oxygen, an oxide, a peroxide, a superoxide?
24. Where are superoxides formed in the body? How are they removed?
25. How is hydrogen peroxide removed from living cells?

B

26. Compare the reaction involving the oxidation of glucose with photosynthesis.
27. Do all oxidation reactions require oxygen? Explain.
28. Why are fire extinguishers containing carbon tetrachloride no longer in use?

29. What causes the red color of blood?
30. Compare the use of a nasal catheter with the use of an oxygen tent.

Practice Test

1. Oxygen is _____.
   a. heavier than air and insoluble in water
   b. lighter than air and insoluble in water
   c. lighter than air and slightly soluble in water
   d. heavier than air and slightly soluble in water
2. Commercially, oxygen is produced from _____.
   a. air   b. water
   c. $KClO_3$ d. metal oxides
3. The molecular formula for ozone is _____.
   a. O   b. O_2 c. O_3 d. O_4
4. Oxides are compounds formed by the reaction of oxygen with _____.
   a. metals only
   b. nonmetals only
   c. either metals or nonmetals
5. A gas frequently used for general anesthesia is _____.
   a. chlorine
   b. carbon tetrachloride
   c. ozone
   d. nitrous oxide
6. Spontaneous combustion can be prevented by _____.
   a. use of a closed container
   b. addition of oxygen
   c. poor ventilation
   d. none of these
7. Oxygen can be produced in the laboratory by _____.
   a. heating any metal oxide
   b. heating any nonmetal oxide
   c. electrolysis of water
   d. all of these
8. If a patient is using oxygen, which of the following precautions should be taken?
   a. ban electric devices
   b. ban electric signal cord
   c. avoid smoking
   d. all of these
9. Oxygen can be mixed with _____ gas for use in diving equipment.
   a. H_2 b. He   c. CO_2 d. N_2
10. An example of a noxious gas is _____.
    a. CCl_4 b. CO
    c. Cl_2 d. all of these

9

Oxidation–Reduction

A pool maintenance worker places a capsule of chlorinated disinfectant into the water supply of a whirlpool bath.

Oxidation

Oxidation can be defined as "a loss of electrons." Consider the electron-dot structures in reaction (9-1). The sodium atom has one outer electron. When the sodium atom loses this one electron, it forms a sodium ion with a + 1 charge. This loss of an electron is defined as oxidation. Therefore, the sodium atom was oxidized.

$$Na\cdot + :\ddot{\underset{\cdot\cdot}{Cl}}: \longrightarrow Na^+ + :\ddot{\underset{\cdot\cdot}{Cl}}:^-$$ (9-1)

A second definition of oxidation states that it is an increase in oxidation number. Consider reaction (9-2).

$$2\,Na + Cl_2 \longrightarrow 2\,NaCl$$ (9-2)

An uncombined element has an oxidation number of zero. (See page 93 for a discussion of oxidation numbers.) The oxidation number of sodium in sodium chloride (NaCl) is +1, and that of chlorine is −1. Therefore, reaction (9-2) can be written as follows:

$$\overset{0}{2\,Na} + \overset{0}{Cl_2} \longrightarrow \overset{+1}{2\,Na} + \overset{-1}{2\,Cl}$$

where the upper numbers indicate the respective oxidation numbers of the substances. The sodium has changed in oxidation number from zero to +1, a gain. This is oxidation. The sodium atom was oxidized.

The cells in the body "burn" glucose, producing carbon dioxide, water, and energy.

$$\overset{0}{C_6H_{12}O_6} + \overset{0}{6\,O_2} \longrightarrow \overset{+4\,-2}{6\,CO_2} + \overset{-2}{6\,H_2O} + energy$$

glucose

The oxidation number of each carbon atom in glucose is zero. The oxidation number of the carbon atom in carbon dioxide (CO_2) is +4.[†] Therefore, the carbon atom increased in oxidation number. A gain in oxidation number is oxidation; therefore, the carbon atom in glucose was oxidized, or it can be said that the glucose, which contains the carbon atom, was oxidized.

A third definition of oxidation is "addition of oxygen." Consider reaction (9-3), which involves the oxidation of formaldehyde to formic acid (this type of reaction will be discussed in Chapter 18).

[†] Refer to Chapter 5.

$$\underset{\underset{C_6}{} \quad \underset{H_{12}}{} \quad \underset{O_6}{}}{6(0) + 12(+1) + 6(-2) = 0} \quad \text{and} \quad \underset{\underset{C}{} \quad \underset{O_2}{}}{+4 + 2(-2) = 0}$$

$$2 \text{ HCHO} + \text{O}_2 \longrightarrow 2 \text{ HCOOH} \qquad (9\text{-}3)$$

formaldehyde   formic acid

Note that this reaction involves the addition of oxygen.

A fourth definition of oxidation involves the "removal of hydrogen." In reaction (9-4),

$$2 \text{ CH}_3\text{CH}_2\text{OH} + \text{O}_2 \longrightarrow 2 \text{ CH}_3\text{CHO} + 2 \text{ H}_2\text{O} \qquad (9\text{-}4)$$

ethanol   acetaldehyde

ethanol is changed to acetaldehyde. This process is called oxidation because it involves the loss of hydrogen.

The following oxidation reactions take place in the body. They will be discussed in detail in the appropriate chapters on carbohydrates, fats, and proteins.

$$\text{carbohydrate} + \text{O}_2 \longrightarrow \text{CO}_2 + \text{H}_2\text{O} + \text{energy}$$

$$\text{fat} + \text{O}_2 \longrightarrow \text{CO}_2 + \text{H}_2\text{O} + \text{energy}$$

$$\text{protein} + \text{O}_2 \longrightarrow \text{CO}_2 + \text{H}_2\text{O} + \text{urea} + \text{energy}$$

Thus, oxidation can be defined as

1. An increase in oxidation number.
2. A loss of electrons.
3. A gain of oxygen.
4. A loss of hydrogen.

Reduction

Oxidation is defined as a loss of electrons and also as an increase in oxidation number. Reduction is the opposite of oxidation—a gain of electrons and, therefore, a decrease in oxidation number. **Oxidation can never take place without reduction** because something must be able to pick up the electrons lost by the oxidized atom, ion, or compound. Free electrons cannot exist by themselves for very long.

In the reactions of sodium with chlorine,

$$\overset{0}{2 \text{ Na}} + \overset{0}{\text{Cl}_2} \longrightarrow \overset{+1}{2 \text{ Na}^+} + \overset{-1}{2 \text{ Cl}^-}$$

the sodium increased in oxidation number from 0 to +1. It was oxidized. At the same time, the chlorine decreased in oxidation number from 0 to −1. Therefore, the chlorine was reduced. Consider the structures in reaction (9-1). The chlorine atom has seven outer electrons. It gains one electron to form the choride ion with a charge of −1. This gain of an electron is called reduction. Thus, by either definition, the chlorine was reduced.

In the reaction of glucose with oxygen,

$$\overset{0}{C_6H_{12}O_6} + 6\,\overset{0}{O_2} \longrightarrow 6\,\overset{+4\,-2}{CO_2} + 6\,\overset{-2}{H_2O} + \text{energy}$$

glucose is oxidized because the carbon atoms change in oxidation number from 0 to +4. However, at the same time, the oxygen changes from 0 to −2, a decrease in oxidation number. Therefore, the oxygen is reduced.

Since reduction is the opposite of oxidation, reduction can also be defined as a loss of oxygen or as a gain of hydrogen. Equations (9-5) and (9-6) illustrate reduction reactions that will be discussed in Chapter 18.

$$CH_3COOH + H_2 \longrightarrow CH_3CHO + H_2O \qquad (9\text{-}5)$$
$$\text{acetic acid} \qquad\qquad\qquad \text{ethanal}$$

$$CH_3COCH_3 + H_2 \longrightarrow CH_3CH(OH)CH_3 \qquad (9\text{-}6)$$
$$\text{acetone} \qquad\qquad\qquad \text{isopropyl alcohol}$$

Reaction (9-5) is reduction because it involves a loss of oxygen; reaction (9-6) is reduction because it involves a gain of hydrogen.
Thus, reduction is

1. A decrease in oxidation number.
2. A gain of electrons.
3. A loss of oxygen.
4. A gain of hydrogen.

Oxidizing Agents and Reducing Agents

In reaction (9-7),

$$H_2 + PbO \longrightarrow Pb + H_2O \qquad (9\text{-}7)$$

$$\overset{0}{H_2} + \overset{+2\,-2}{PbO} \longrightarrow \overset{0}{Pb} + \overset{2(+1)\,-2}{H_2O}$$

the lead decreases in oxidation number from +2 to 0. Therefore, the lead was reduced. What reduced it? What supplied the electrons that it had to gain in order to decrease in oxidation number? The answer is that the hydrogen reduced it. The hydrogen supplied the electrons so that the lead could be reduced; therefore, hydrogen is called the **reducing agent**. The substance that causes the reduction of an element or compound is known as a reducing agent.

At the same time, the hydrogen gained in oxidation number from 0 to +1. Therefore, the hydrogen was oxidized. What oxidized it? What picked up the electrons that the hydrogen must have lost in being oxidized? The answer is that the PbO picked up these electrons; thus, the PbO is called the **oxidizing agent**. An oxidizing agent is defined as a substance that causes the oxidation of some element or compound.

Another way of stating the same thing is to say that whatever is oxidized is the reducing agent and whatever is reduced is the oxidizing agent.

Rewriting equation (9-7) and indicating what was oxidized, what was reduced, what the oxidizing agent was, and what the reducing agent was, we have the following:

$$H_2 \; + \; PbO \; \longrightarrow \; Pb + H_2O$$

oxidized	reduced
(reducing	(oxidizing
agent)	agent)

Balancing Oxidation–Reduction Reactions

To balance a redox equation use the following rules:

1. Write the complete equation listing the oxidation numbers of all the elements (see page 94), and locate those elements that are changing their oxidation number.
2. Calculate the change in oxidation number, and indicate this change below the equation.
3. Multiply the change in oxidation number for each atom by the total number of those atoms that are changing in oxidation number.
4. Find a least common multiple for the total increase and decrease in oxidation number.
5. Equate the increase and decrease in oxidation numbers, since they must always be equal.
6. Complete the remainder of the equation by adding the proper coefficients to make sure that there are the same number of each type of atom on both sides of the equation.
7. Note that any element in the free state, that is, uncombined, has an oxidation number of zero.

Example 9-1   Balance $H_2 + O_2 \longrightarrow H_2O$

Step 1   *Write in all the oxidation numbers.*

$$\overset{0}{H_2} + \overset{0}{O_2} \longrightarrow \overset{2(+1)-2}{H_2O}$$

(Since the molecules of hydrogen and oxygen are uncombined, their atoms have an oxidation number of zero; from Table 5-5 we know that the oxidation number of oxygen in water is -2 and that of each hydrogen is $+1$).

Step 2   *Show the increase and decrease in oxidation numbers.*
Since the hydrogen is changing from an oxidation number of 0 to $+1$, there is an increase of 1 in oxidation number. The oxygen, in changing from an oxidation number of 0 to -2, decreases by 2 in oxidation number. These changes can be indicated in the equation as

$$\overset{0}{H_2} + \overset{0}{O_2} \longrightarrow \overset{2(+1)-2}{H_2O}$$

↑1 for   ↓2 for
| each | each
| atom   ↓ atom

where the upward-pointing arrow indicates an increase in oxidation number and the downward-pointing arrow indicates a decrease in oxidation number.

Step 3   *Find the total change in oxidation number for each of the elements that is changing.*

The total increase in oxidation number for the hydrogen is 2 (1 for each of the hydrogens). The total decrease in oxidation number for the oxygen is 4 (2 for each oxygen). These total changes in oxidation number can be indicated as

$$\textit{Total changes:} \quad \overset{0}{H_2} + \overset{0}{O_2} \longrightarrow \overset{2(+1)-2}{H_2O}$$

↑2   ↓4

Step 4   *The least common mutiple (LCM) for an increase of 2 and a decrease of 4 is 4.*

Step 5   *Equate the increase in oxidation number with the decrease.*

The increase in oxidation number for the hydrogen (2) goes into the LCM twice, so we place a 2 in front of the H_2. The decrease in oxidation number for the O_2 (4) goes into the LCM once, so we place a 1 in front of the O_2 (no number in front of a formula designates 1). Thus, we have

$$2 H_2 + O_2 \longrightarrow H_2O$$

where the total increase in oxidation number is now equal to the total decrease.

Step 6   *Complete the equation.*

Since we now have a total of four H's on the left side of the equation, we must have four H's on the right side, so we place a 2 in front of the H_2O and then we have four H's on each side of the equation, or

$$2 H_2 + O_2 \longrightarrow 2 H_2O$$

We note that there are also two O's on each side of the equation, and it is therefore completely balanced.

After you have tried several redox equations you will be able to combine many of these steps and save time in balancing.

Let us see what was *oxidized* and what was *reduced*. Since the definition of oxidation is an increase in oxidation number, the H_2 must have been oxidized because its oxidation number did increase (from 0 to 1 each). Also, the O_2 must have been reduced because it decreased in oxidation number (from 0 to -2 each). The H_2 was oxidized. What oxidized it? The O_2; so the O_2 is the oxidizing agent. The O_2 was reduced. What reduced it? The H_2; so the H_2 is the reducing agent.

$$H_2 \qquad\qquad O_2$$

oxidized   reduced
(reducing agent)   (oxidizing agent)

Example 9-2 Balance $KClO_3 \longrightarrow KCl + O_2$.
Listing the oxidation numbers we have

$$\overset{\overset{\displaystyle 3\times}{\displaystyle +1+5-2}}{KClO_3} \longrightarrow \overset{+1-1}{KCl} + \overset{0}{O_2}$$

Note that the Cl and the O are changing in oxidation number. The Cl, in changing from $+5$ to -1, exhibits a decrease of 6 in oxidation number. The O, in changing from -2 to 0, exhibits an increase of 2 in oxidation number of each of the O atoms present. Since there are three O atoms present, the total increase in oxidation number is 6. These changes may be indicated as

$$Total\ changes:\quad \overset{\overset{\displaystyle 3\times}{\displaystyle +5-2}}{KClO_3} \longrightarrow \overset{-1}{KCl} + \overset{0}{O_2}$$

Because the increase in oxidation number here does equal the decrease, we need no LCM and so leave the left side of the equation as is.

Now to complete the equation. Since we have one $KClO_3$ on the left side of the equation we can have only one KCl on the right because the number of K's on each side of the equation must be the same and so must the number of Cl's. Since we have three O's on the left side of the equation we must have three O's on the right side; we get these by placing a $1\frac{1}{2}$ in front of the O_2 ($1\frac{1}{2} \times O_2 = 3\,O$). Thus we have

$$KClO_3 \longrightarrow KCl + 1\tfrac{1}{2}\,O_2$$

However, we can can never have fractions of atoms or molecules in a chemical equation, so we multiply the whole equation by 2 to get

$$2\,KClO_3 \longrightarrow 2\,KCl + 3\,O_2$$

which is balanced because it contains the same number of each type of atom on each side of the equation.

What was reduced? An error made by many students is the statement that the chlorine was reduced because it decreased in oxidation number. It is true that a decrease in oxidation number is reduction, but the chlorine did not decrease in oxidation number, the Cl^{+5} decreased. This makes quite a difference as we shall see in the following paragraph.

What was oxidized? The O^{2-} was oxidized because it increased in oxidation number. If we had said that the oxygen was oxidized, that would have been incorrect because oxygen is a product of the reaction and not one of the original reactants.

Also the Cl^{5+}, since it was reduced, is the oxidizing agent, and the O^{2-}, which was oxidized, is the reducing agent.

Example 9-3   Balance $P + HNO_3 + H_2O \longrightarrow NO + H_3PO_4$.
We will combine a few of the steps to save space. Therefore, we have

$$\begin{array}{ccccccc} & & 3\times & & & 3\times & 4\times \\ 0 & +1+5-2 & 2(1)-2 & & +2-2 & +1+5-2 \\ \textit{Total changes:}\quad P & + HNO_3 & + H_2O & \longrightarrow & NO & + H_3PO_4 \end{array}$$

$\uparrow 5 \qquad \downarrow 3$

The LCM between an increase of 5 and a decrease of 3 is 15. The increase in oxidation number by the P (5) goes into the LCM three times, so we place a 3 in front of the P; the decrease in oxidation number by the N (3) goes into the LCM five times, so we place a 5 in front of the HNO_3 (to give us five N's) or

$$3\,P + 5\,HNO_3 + (\ )\,H_2O \longrightarrow (\ )\,NO + (\ )\,H_3PO_4$$

Three P's on the left side of the equation must give three P's on the right, so we put a 3 in front of the H_3PO_4 to give us a total of three P's there. Five HNO_3 on the left side of the equation contain five N's, so we must have five N's on the right side, we get these by placing a 5 in front of the NO.

$$3\,P + 5\,HNO_3 + (\ )\,H_2O \longrightarrow 5\,NO + 3\,H_3PO_4$$

Now to complete the equation, let us balance the H's. There are nine H's in the H_3PO_4's on the right side of the equation. On the left side there are five H's in the HNO_3's, so we need four more. We get these by taking two H_2O's (four H's), or

$$3\,P + 5\,HNO_3 + 2\,H_2O \longrightarrow 5\,NO + 3\,H_3PO_4$$

To check, let us see if the O's balance. There are $(5 \times 3) + (2 \times 1) = 17$ O's on the left side and $(5 \times 1) + (3 \times 4) = 17$ O's on the right side, so the equation is correctly balanced.

What was oxidized? The P, since it increased in oxidation number. Thus it is the reducing agent.

What was reduced? The N^{5+} (or we could say the HNO_3) was reduced because it decreased in oxidation number, and therefore it is the oxidizing agent.

Example 9-4   Balance

$$FeSO_4 + K_2Cr_2O_7 + H_2SO_4 \rightarrow Fe_2(SO_4)_3 + Cr_2(SO_4)_3 + K_2SO_4 + H_2O$$

Since the equation is a little more complex than those we have studied so far, we will solve it carefully and slowly to see why we take each step. First, we will list the oxidation numbers, giving the whole (SO_4) polyatomic ion a charge of -2 instead of giving the separate oxidation numbers of the S and the O because we see that SO_4 occurs unchanged on both sides of the equation. So we have

$$\overset{+2\ -2}{FeSO_4} + \overset{\overset{2\times\ 2\times\ 7\times}{}}{\underset{+1\ +6\ -2}{K_2Cr_2O_7}} + \overset{\overset{2\times}{}}{\underset{+1\ -2}{H_2SO_4}} \rightarrow \overset{\overset{2\times\ 3\times}{}}{\underset{+3\ -2}{Fe_2(SO_4)_3}} + \overset{\overset{2\times\ 3\times}{}}{\underset{+3\ -2}{Cr_2(SO_4)_3}} + \overset{\overset{2\times}{}}{\underset{+1\ -2}{K_2SO_4}} + \overset{\overset{2\times}{}}{\underset{+1\ -2}{H_2O}}$$

We note that the Fe is changing from $+2$ to $+3$, which is an increase of 1, and the Cr is changing from $+6$ to $+3$, which is a decrease of 3 in oxidation number for each Cr or a total decrease of 6 in oxidation number for the two Cr's, or

$$\overset{+2}{FeSO_4} + \overset{2(+6)}{K_2Cr_2O_7} + H_2SO_4 \rightarrow \overset{2(+3)}{Fe_2(SO_4)_3} + \overset{2(+3)}{Cr_2(SO_4)_3} + K_2SO_4 + H_2O$$

$\uparrow 1 \qquad\qquad \downarrow 6$

The LCM for an increase of 1 and a decrease of 6 in oxidation number is 6. Therefore, an increase of 1 (by the Fe) goes into the LCM six times, so we place a 6 in front of the $FeSO_4$; a decrease of 6 in oxidation number (by the Cr's in $K_2Cr_2O_7$) goes into the LCM once, so we place a 1 (understood) in front of the $K_2Cr_2O_7$, or

$$6\ FeSO_4 + K_2Cr_2O_7 + (\ )\ H_2SO_4 \longrightarrow$$
$$(\ )\ Fe_2(SO_4)_3 + (\ )\ Cr_2(SO_4)_3 + (\ )\ K_2SO_4 + (\ )\ H_2O$$

Now to complete the rest of the equation. Since there are six Fe's on the left side there must be six on the right; we get these by placing a 3 in front of the $Fe_2(SO_4)_3$ (three Fe_2 = six Fe). Since there are two K's on the left side of the equation, there must be two on the right, and we get these by taking one K_2SO_4. Since there are two Cr's on the left side, there must be two on the right, and so we place a 1 in front of the $Cr_2(SO_4)_3$, or

$$6\ FeSO_4 + K_2Cr_2O_7 + (\ )\ H_2SO_4 \longrightarrow$$
$$3\ Fe_2(SO_4)_3 + Cr_2(SO_4)_3 + K_2SO_4 + (\ )\ H_2O$$

Now let us count the SO_4 groups on the right side of the equation. There are nine in the three $Fe_2(SO_4)_3$ plus three in the $Cr_2(SO_4)_3$ plus one in the K_2SO_4, or a total of 13 SO_4's. We started with six SO_4's in the 6 $FeSO_4$, so we need seven more; we get these by taking seven H_2SO_4. These seven H_2's on the left side must give seven H_2's on the right, so we have seven H_2O's, or

$$6\ FeSO_4 + K_2Cr_2O_7 + 7\ H_2SO_4 \longrightarrow$$
$$3\ Fe_2(SO_4)_3 + Cr_2(SO_4)_3 + K_2SO_4 + 7\ H_2O$$

To check if the equation is correctly balanced, let us add up the O's on each side and see if the totals are the same.

	Left Side		Right Side	
6 $FeSO_4$	24 O's	3 $Fe_2(SO_4)_3$	36 O's
1 $K_2Cr_2O_7$	7 O's	$Cr_2(SO_4)_3$	12 O's
7 H_2SO_4	28 O's	K_2SO_4	4 O's
		7 H_2O	7 O's
	59 O's		59 O's

And so we see that the equation is correctly balanced.

What is the oxidizing agent? We know that the oxidizing agent is reduced, which means that it decreases in oxidation number. The Cr^{6+}, in going from +6 to +3, decreases in oxidation number, therefore, it is the oxidizing agent; or we might say that the $K_2Cr_2O_7$ is the oxidizing agent, since the Cr^{6+} is a part of this compound and the rest of it does not change in oxidation number. Likewise, the reducing agent is the Fe^{2+} or the $FeSO_4$ (which is oxidized).

Oxidation and reduction reactions produce the energy the body needs to carry out its normal functions. Oxidation–reduction in the body involves either oxygen or hydrogen, or both.

Enzymes involved in oxidation–reduction reactions in the body are called oxidoreductases (see page 441). Many of these enzymes are present in the mitochondria (page 442).

Importance of Oxidation–Reduction

Antiseptic Effects

Because they are oxidizing agents, many *antiseptics have the property of killing bacteria. Among these is chlorine, which oxidizes organic matter and bacteria and so is used in the treatment of water to make it potable (see page 177). Calcium hypochlorite, $Ca(OCl)_2$, another commonly used oxidizing agent and bleaching powder, is used as a disinfectant for clothes and hospital beds. Table 9-1 lists some of the common antiseptics that are oxidizing agents.

Formaldehyde and sulfur dioxide are two reducing agents used in disinfecting rooms formerly occupied by patients with contagious diseases.

Effects on Hair Protein

Oxidizing and reducing agents denature protein by affecting the disulfide bonds (see page 416) of the amino acid cysteine. Use is made of this effect in "home permanents." Hair protein is primarily

	Formula	Name	Use
Table 9-1 Antiseptic Agents	3% H_2O_2	Hydrogen peroxide	Minor cuts and scratches
	$KMnO_4$	Potassium permanganate	Treatment of infection in urethra and bladder
	$KClO_3$	Potassium chlorate	Treatment of sore throat
	I_2 in H_2O	Lugol's solution	Treatment of minor cuts
	NaOCl	Sodium hypochlorite (Dakin's solution)	Treatment of wounds

keratin, and keratin contains a large amount of cysteine. During the treatment, a reducing agent is used first. This substance breaks the disulfide bonds in the hair protein. The hair is then shaped with rollers. The new shape is "set" by using an oxidizing agent, which forms new disulfide bonds in the desired places. The hair will retain its new shape only until new hair grows out. Then the entire process has to be repeated.

Stain Removal

Oxidizing agents and reducing agents are used to remove most stains that cannot otherwise be removed. Table 9-2 lists some of the common stain removers and indicates where they can be used safely.

Black-and-White Photography

Black-and-white photographic film contains an emulsion of silver bromide, AgBr, spread over a transparent surface. Exposure to light activates the silver in the silver bromide. When the film is *developed*, by reacting it with a mild oxidizing agent, the activated silver ions are reduced to elemental silver, which appears black. The greater the

Table 9-2 Stain Removers

Substance	Name	Function	Use
H_2O_2	Hydrogen peroxide	Oxidizing agent	Blood stains on cotton or linen
$KMnO_4$	Potassium permanganate	Strong oxidizing agent	Almost all stains on white fabrics except rayon ($KMnO_4$ stain must be removed, usually with oxalic acid)
$(COOH)_2$	Oxalic acid	Reducing agent	Rust spots and $KMnO_4$ stains
NaOCl	Javelle water	Oxidizing agent	Effective on almost all stains on cotton and linen (not to be used on wool or silk)
$Na_2S_2O_3$	Sodium thiosulfate	Reducing agent	Iodine and silver stains

original exposure to light, the larger the amount of activated silver bromide and hence the darker the area on the developed film.

Once the film is developed, it is *fixed* by placing it in a solution of sodium thiosulfate, *hypo*. The reaction is

$$AgBr + 2\,S_2O_3^{2-} \longrightarrow Ag(S_2O_3)_2^{3-} + Br^-$$

unreacted   thiosulfate   silver thiosulfate   bromide
silver bromide   ion   ion   ion

The resulting silver product, $Ag(S_2O_3)_2^{3-}$, is soluble and is washed away, leaving the elemental silver behind. The resulting image is called a *negative*. Examples of such negatives are X-ray films.

To produce a print, a *positive*, light is passed through a negative onto a fresh piece of photographic film. Then the film is developed, fixed, and printed. The result is a black-and-white photograph where the dark areas on the negative show up white and the white areas of the negative are dark.

Breath-Alcohol Analyzer

Reactions involving oxidation–reduction are used to measure the amount of alcohol in a driver's breath. A sample of the driver's breath is blown through an orange-colored solution of acidified potassium dichromate. If alcohol, which is a reducing agent, is present, it causes the following reaction to take place.

$$3\,C_2H_5OH + 2\,K_2Cr_2O_7 + 8\,H_2SO_4 \longrightarrow$$

alcohol   potassium   sulfuric
dichromate   acid

$$3\,CH_3COOH + 2\,Cr_2(SO_4)_3 + 2\,K_2SO_4 + 11\,H_2O$$

acetic   chromic   potassium   water
acid   sulfate   sulfate

The chromic sulfate thus produced is green. The greater the amount of alcohol in a driver's breath, the greater the change from orange to green. The actual alcohol content can be determined by comparing the color produced with that of a standardized chart.

Summary

Oxidation is defined as a loss of electrons. Oxidation is also an increase in oxidation number, a combination with oxygen, or a loss of hydrogen.

Reduction is a gain of electrons. Reduction is also a decrease in oxidation number, a gain of hydrogen, or a loss of oxygen. Oxidation can never take place without reduction and vice versa.

Whatever is oxidized is called a reducing agent.

Whatever is reduced is called an oxidizing agent.

Oxidizing and reducing agents are useful as antiseptics and also as stain removers.

Questions and Problems

A

1. In the following equations, indicate the oxidizing agent and the reducing agent:
   (a) $Cl_2 + 2 KI \longrightarrow 2 KCl + I_2$
   (b) $2 Al + 3 H_2SO_4 \longrightarrow Al_2(SO_4)_3 + 3 H_2$
   (c) $Cu + 2 AgNO_3 \longrightarrow Cu(NO_3)_2 + 2 Ag$
   (d) $2 H_2 + O_2 \longrightarrow 2 H_2O$

2. In the following equations, what was oxidized and what was reduced?
   (a) $4 Fe + 3 O_2 \longrightarrow 2 Fe_2O_3$
   (b) $2 S + 3 O_2 \longrightarrow 2 SO_3$
   (c) $Zn + 2 HCl \longrightarrow ZnCl_2 + H_2$
   (d) $CuO + H_2 \longrightarrow Cu + H_2O$
   (e) $MnO_2 + 4 HCl \longrightarrow MnCl_2 + Cl_2 + 2 H_2O$

3. Why can oxidation never take place without reduction?

4. List several substances that can be used to remove stains.

B

5. Give an example of oxidation involving (a) loss of electrons, (b) gain of oxygen, (c) loss of hydrogen.

6. Consider the reaction

$$2 H_2 + O_2 \rightleftharpoons 2 H_2O$$

Does it involve oxidation–reduction in both directions? Are all oxidation–reduction reactions reversible?

7. Discuss oxidation–reduction in terms of disulfide bonds.

8. Balance the following oxidation–reduction reactions:
   (a) $Zn + HCl \longrightarrow ZnCl_2 + H_2$
   (b) $P + HNO_3 + H_2O \longrightarrow NO + H_3PO_4$
   (c) $MnCl_2 + NaOH + Br_2 \longrightarrow$
   $\qquad\qquad MnO_2 + NaCl + NaBr + H_2O$
   (d) $Mn(NO_3)_2 + NaBiO_3 + HNO_3 \longrightarrow$
   $\qquad\qquad HMnO_4 + Bi(NO_3)_3 + NaNO_3 + H_2O$

9. In each of the equations in question 8, indicate what was oxidized, what was reduced, the oxidizing agent, and the reducing agent.

10. Is an X-ray a "true" picture of a body part? Explain.

Practice Test

1. Oxidation may involve a _____.
   a. gain of electrons   b. gain of oxygen
   c. gain of hydrogen
   d. decrease in oxidation number

2. Reduction _____.
   a. always involves oxygen
   b. always involves hydrogen
   c. can never take place without oxidation
   d. occurs in nuclear reactions

3. A reduction agent _____.
   a. is oxidized   b. is reduced
   c. does not change in oxidation number
   d. none of these

Use the following balanced oxidation–reduction reaction for questions 4 and 5.

$$Cu + 4 HNO_3 \longrightarrow Cu(NO_3)_2 + 2 NO_2 + 2 H_2O$$

4. The oxidizing agent is _____.
   a. Cu   b. $Cu(NO_3)_2$
   c. HNO_3 d. NO_2

5. The substance oxidized is _____.
   a. Cu   b. $Cu(NO_3)_2$
   c. HNO_3 d. NO_2

6. Oxidizing agents can act as _____.
   a. analgesics   b. antiseptics
   c. antipyretics   d. all of these

7. Oxidizing and reducing agents denature protein by affecting which type of bond?
   a. hydrogen   b. oxygen
   c. disulfide   d. triphosphate

8. The reaction $2 Na + Cl_2 \longrightarrow 2 NaCl$ is an example of _____.
   a. oxidation   b. reduction
   c. both oxidation and reduction
   d. neither oxidation nor reduction

9. Black-and-white photography uses an emulsion containing the compound _____.
   a. $BaSO_4$ b. KCl
   c. NH_4NO_3 d. AgBr

10. The use of a breath-alcohol analyzer is based on the fact that alcohol acts as a(n) _____.
    a. oxidizing agent   b. reducing agent
    c. antiseptic   d. analgesic

10

Water

Evaporation of water from the skin produces a cooling effect.

The Importance of Water

Water is one of the most important chemicals known. Without it neither animal nor plant life would exist. Humans can live for a few weeks without food but for only a few days without water. Almost three-fourths of the Earth's surface is covered by water. Rain, snow, sleet, hail, fog, and dew are manifestations of the water vapor present in the air.

Water is present on the solid part of the Earth's surface as lakes, streams, waterfalls, and glaciers. The human body is approximately 50 percent water. Water is essential in the processes of digestion, circulation, elimination, and the regulation of body temperature. Indeed, every activity of every cell in the body takes place in a watery environment. Normally, in the body, water intake equals water output. If water intake is greater than water output, a condition known as *edema results (see Figure 10-1). If water intake is less than output, dehydration occurs.

Water is important as a solvent. Many substances dissolve in water; sugar, salt, and alcohol are examples.

Physical Properties of Water

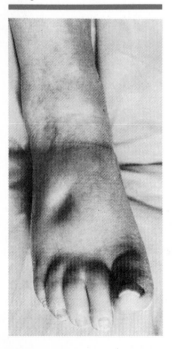

Figure 10-1 Patient's leg showing pitting edema. [Courtesy Evanston Hospital, Evanston, IL.]

Pure water is colorless, odorless, and tasteless. Tap water owes its taste to dissolved gases and minerals. Large bodies of water such as lakes and oceans appear blue owing to the presence of finely divided solid material and also to the reflection of the sky.

When water at room temperature is cooled, its volume contracts. However, water is an unusual liquid in that, after it is cooled to 4 °C, further cooling causes an expansion in volume. When water freezes at 0 °C, it expands even more, increasing in volume by almost 10 percent and decreasing in density. These changes on freezing explain why ice floats and why pipes may burst when the water in them freezes. Expansion of water during frostbite explains the damage to the cells.

The density of water changes with the temperature. At 4 °C water has its smallest volume and its maximum density, 1 g/mL. However, for all practical purposes, the density of water is given as 1 g/mL regardless of temperature, since the variation in density between 0 and 100 °C is quite small.

Pure water boils at 100 °C at 1 atm pressure. At lower pressures it boils at a lower temperature. In certain mountainous localities, water boils at 80 °C because of the lower pressure. If more heat is applied to the water, it merely boils faster. The boiling point does not increase. When water boils at 80 °C, there may not be sufficient heat to cook food. In this case, the food must be heated in a pressure cooker. This device increases pressure and so increases the boiling point of the water. The same principle holds in the autoclave, where the increased pressure raises the boiling point.

For water to evaporate, a certain amount of heat is necessary. This amount of heat is approximately 540 cal/g. When water is placed on the skin, the heat it needs to make it evaporate comes from that skin. Therefore, the skin loses heat and so is cooled (see page 169). The evaporation of perspiration also cools the skin.

For ice to melt, the amount of heat required is 80 cal/g. When an ice pack is placed on the skin, the heat needed to melt the ice comes from the body, thus lowering the temperature of the body.

Physical Constants Based on Water

The freezing point of water (0 °C or 32 °F) and the boiling point of water at 1 atm pressure (100 °C or 212 °F) are the standard reference points for the measurement of temperature.

The weight of 1 mL of water at 4 °C (its maximum density) is the gram. Specific gravity is based upon water. It is defined as the weight of a substance compared with the weight of an equal volume of water.

The calorie is defined as the amount of heat required to change the temperature of 1 g of water by one degree Celsius.

Structure of the Water Molecule

Pure water does not conduct electricity. This indicates that water is a covalent compound in which the atoms share electrons. Thus each hydrogen atom of the water molecule shares its one electron with the oxygen atom.

We might expect the oxygen and the hydrogen atoms in the water molecule to be arranged in a straight line such as HOH. However, laboratory evidence indicates that atoms in the water molecule are arranged in a nonlinear manner with the angle between the hydrogen atoms being approximately 105 ° (see page 90).

$$H \overset{..}{\underset{\times}{\times}} \overset{..}{O} :$$
$$\overset{..}{\underset{\times}{H}}$$

The ×'s represent the electrons of the hydrogen atom, and the dots the electrons from the oxygen atom. However, the oxygen atom has a greater attraction for electrons than the hydrogen atom has. Therefore, the electrons will spend more of their time closer to the oxygen atom than to the hydrogen atom. This shifting of the electrons toward the oxygen atom will tend to give the oxygen atom a slight negative charge, whereas the hydrogen atoms have a slight positive charge. The structure and distribution of relative charges in the water molecule can be represented as

$$^{\delta+}H \quad \overset{..}{\underset{\times}{\times}} \overset{..}{O} :^{\delta-}$$
$$105° \searrow H^{\delta+}$$

Figure 10-2   Model of the dissolution of an ionic solid such as sodium chloride (Na^+Cl^-) in water.

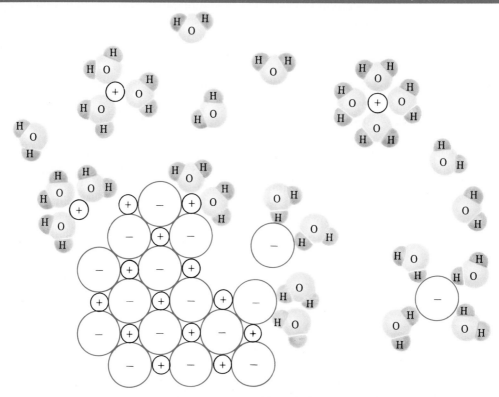

Molecules in which there is an unequal or uneven *distribution* of charges are called **polar molecules.** Water is a polar molecule. The polar nature of the water molecule is responsible for its property of dissolving many materials. When the ionic compound sodium chloride (NaCl) is placed in water, it dissolves partly because of the water molecule attracting the ions and pulling them apart (see Figure 10-2). Sodium chloride is a polar compound because of the uneven distribution of charges, Na^+Cl^-, within the molecule. A general rule is that polar compounds dissolve in polar liquids, and nonpolar compounds dissolve in nonpolar liquids. Nonpolar compounds generally do not dissolve in polar liquids. Carbon tetrachloride (CCl_4) and benzene (C_6H_6) are nonpolar liquids. Water is a polar liquid. As should be expected, carbon tetrachloride dissolves in benzene but not in water. Likewise, benzene does not dissolve in water (see page 186).

Hydrogen Bonding

Since water molecules are polar, we might expect the positive (hydrogen) side of one water molecule to attract the negative (oxygen) side of one or more other water molecules. This type of attraction is

Figure 10-3 Hydrogen bonding in (a) water and (b) amino acids.

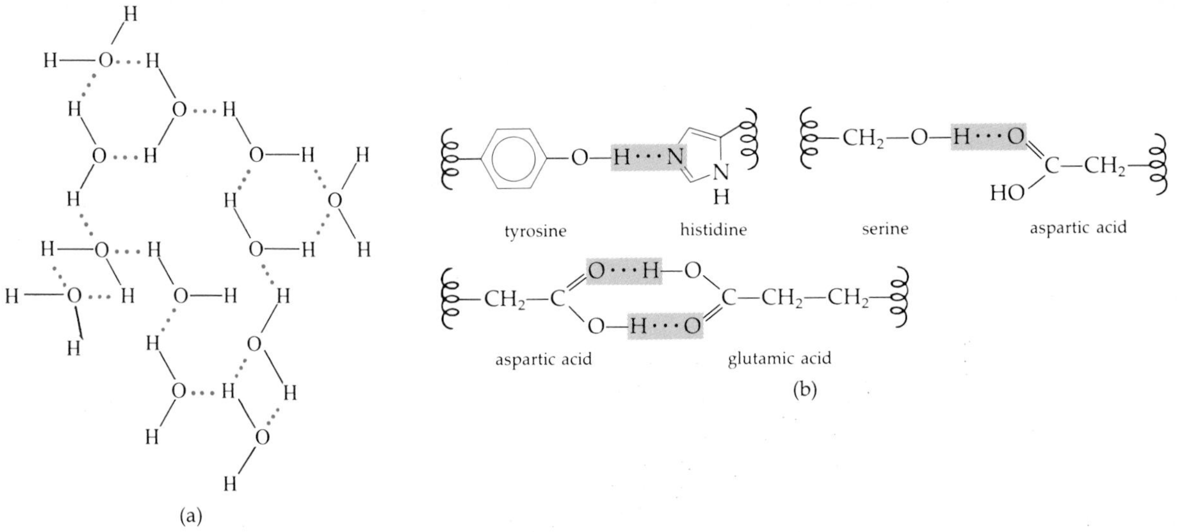

called **hydrogen bonding** and is a weak type of bonding that, as the name implies, always involves hydrogen (see Figure 10-3). The three-dimensional shapes of proteins as well as DNA and RNA are partially due to the presence of hydrogen bonds.

Hydrogen bonds also account for many of the unusual properties of water. In the following section we will see how this type of bonding can account for the abnormally low vapor pressure of water.

Evaporation

The particles in a liquid, such as water, are not held as tightly as those in a solid (see Figure 2-2), so they have some freedom of motion. When some of the surface molecules escape completely from the liquid, the process is called evaporation. As the temperature of the liquid rises, the molecules move faster, so more of them can escape from the surface. Thus, the rate of evaporation increases as the temperature increases.

Evaporation requires energy. When water evaporates, it takes energy (540 cal/g) either from its surroundings or from the remaining water. When water is placed on the skin, it evaporates by taking heat from the body. That is, evaporation is a cooling process. Alcohol evaporates faster than water, so it has a greater cooling effect when placed on the skin. Thus, alcohol sponge baths lower body temperature faster than water sponge baths.

Consider two containers of water, one open to the air and the other sealed (Figure 10-4). The water in the open container will continue to evaporate until none is left. However, the situation is different in the sealed container. As more and more of the surface

Figure 10-4 Evaporation takes place from the open container; equilibrium is established in the closed container, after which the amounts of evaporation and condensation become equal.

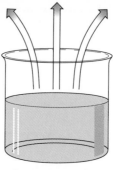
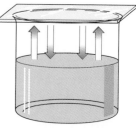

Open water container   Sealed water container

molecules escape into the air above the liquid, some of those molecules will return to the liquid. Soon the rate of evaporation will equal the rate at which the gaseous molecules return to the liquid. This condition is known as an *equilibrium.* The pressure exerted by the gaseous form of the liquid under equilibrium conditions is known as the vapor pressure of the liquid.

Since the rate of evaporation increases as the temperature increases, the vapor pressure also increases as the temperature increases. Table 10-1 indicates the vapor pressure of water at various temperatures. Table 10-2 indicates the vapor pressures of various liquids at room temperature (20 °C).

Because alcohol has a higher vapor pressure than water at the same temperature, alcohol will evaporate faster than water. Therefore, medications containing alcohol should be tightly closed when stored. If not, some of the alcohol will evaporate, thus changing the strength of that medication.

Table 10-1 Vapor Pressure of Water

Temperature (°C)	Pressure (mm Hg)	Temperature (°C)	Pressure (mm Hg)
20	18	60	149
30	32	70	234
40	55	80	355
50	93	90	526
		100	760

Table 10-2 Vapor Pressures of Various Liquids at Room Temperature (20 °C)

Liquid	Pressure (mm Hg)
Water	18
Alcohol (ethyl)	44
Benzene	75
Acetone	177
Ether (ethyl)	442

Water has a relatively low vapor pressure. Why? Recall that water molecules are held together by hydrogen bonds. In order for water to evaporate, not only must the surface molecules have enough energy to escape but they must also have enough energy to break the hydrogen bonds holding them together. Therefore the tendency to evaporate is low, and hence the vapor pressure is low.

Boiling Point

Both boiling and evaporation involve a change from the liquid to the gaseous state. In boiling, however, heat is applied directly, whereas in evaporation heat is taken from the surroundings.

The boiling point of a liquid is defined as the temperature at which the vapor pressure of the liquid is equal to atmospheric pressure. For water, the temperature at which the vapor pressure equals 1 atm (760 mm Hg) is 100 °C (see Table 10-1). Thus, at 1 atm pressure water boils at 100 °C. If the atmospheric pressure is lower than 1 atm, water will boil at a correspondingly lower temperature. Conversely, at pressures above 1 atm, water boils at a temperature higher than 100 °C. An autoclave (see page 125) uses this principle.

Chemical Properties

Electrolysis

When water undergoes electrolysis—that is, when an electric current is passed through it—hydrogen gas (H_2) and oxygen gas (O_2) are formed. The volume of hydrogen produced is twice that of oxygen.

$$2\,H_2O \xrightarrow{\text{electric current}} 2\,H_2(g) + O_2(g)$$

Stability

When water is heated to 100 °C at 1 atm pressure, it boils and turns into a gas, steam. Even when the steam is heated to a high temperature, it does not decompose. The water molecule is extremely stable. There is very little decomposition of the water molecule even at a temperature of 1600 °C.

Reaction with Metal Oxides

Water reacts with soluble metal oxides to form a class of compounds called **bases** (see Chapter 13). For example,

$$CaO + H_2O \longrightarrow Ca(OH)_2$$

calcium   calcium hydroxide,
oxide   a base

$$Na_2O + H_2O \longrightarrow 2\,NaOH$$

sodium   sodium hydroxide,
oxide   a base

Reaction with Nonmetal Oxides

Water reacts with soluble nonmetal oxides to form a class of compounds called **acids** (see Chapter 13). For example,

$$CO_2 + H_2O \longrightarrow H_2CO_3$$

carbon   carbonic
dioxide   acid

$$SO_3 + H_2O \longrightarrow H_2SO_4$$

sulfur   sulfuric
trioxide   acid

Reaction with Active Metals

When an active metal such as sodium or potassium is placed in water, a vigorous reaction takes place, with the rapid evolution of hydrogen gas. At the same time, a base is formed.

$$2\,Na + 2\,H_2O \longrightarrow 2\,NaOH + H_2(g)$$

sodium   sodium hydroxide

Formation of Hydrates

When water solutions of some soluble compounds are evaporated, the subtances separate as crystals that contain the given compound combined with water in a definite proportion by weight. Crystals that contain a definite proportion of water as part of their crystalline structure are called **hydrates**. The water contained in a hydrate is called **water of hydration** or **water of crystallization**.

Barium chloride crystallizes from solution as a hydrate containing two molecules of water. The formula for this hydrate is $BaCl_2 \cdot 2H_2O$. It is called barium chloride dihydrate. The dot in the middle of the compound indicates a loose association between two molecules of water and the molecule of barium chloride. When the water of hydration is removed from a hydrate, the resulting compound is said to be **anhydrous** (without water). When barium chloride dihydrate is heated, the water of hydration is driven off, leaving anhydrous barium chloride.

$$BaCl_2 \cdot 2H_2O \xrightarrow{\text{heat}} BaCl_2 + 2\,H_2O(g)$$

barium chloride   anhydrous
dihydrate   barium
chloride

Substances that lose their water of hydration on exposure to air are said to be **efflorescent.** An example of an efflorescent hydrate is washing soda ($Na_2CO_3 \cdot 10H_2O$), sodium carbonate decahydrate. Substances that pick up moisture from the air are said to be **hygroscopic.** An example of a hygroscopic compound is calcium chloride ($CaCl_2$). A hygroscopic compound can be used as a drying agent because it will pick up and remove moisture from the air.

A hydrate of particular importance in the medical field is plaster of paris. When plaster of paris is mixed with water, it froms a hard crystalline compound called gypsum.

$$(CaSO_4)_2 \cdot H_2O \ + 3 \ H_2O \longrightarrow 2(CaSO_4 \cdot 2H_2O)$$

plaster of paris (soft)   gypsum (hard)

Plaster of paris was formerly used extensively in preparing surgical casts (see Figure 10-5). It was spread on crinoline to form a bandage. When the bandage was to be used, it was placed in water, wrung out, and quickly applied. Plaster of paris expands on setting, so the cast may be comfortable when applied but become too tight as it sets. Therefore, it is very important for the nurse to check the circulation to the body part to which a plaster cast has been applied.

Another method of preparing surgical casts involves the use of knitted fiberglass fabric impregnated with a water-activated urethane resin. Exposure to moisture or water initiates a chemical reaction that results in a rigid tape within a few minutes. These casts have the advantages that they are lighter than comparable

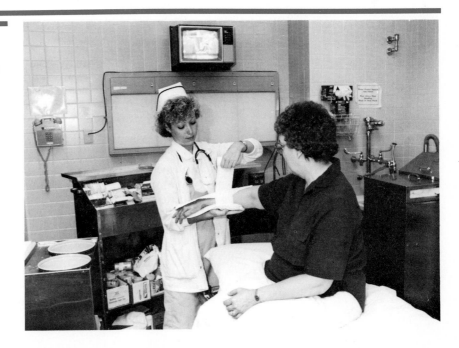

Figure 10-5 Cast being applied. [Courtesy Methodist Medical Center of Illinois School of Nursing, Peoria, IL.]

	Physical	Chemical
Table 10-3 Properties of Water	Colorless, except in deep layer, when it is greenish blue	Stable compound, not easily decomposed; at 1600 °C about 0.3% dissociates
	Odorless	Reacts violently with sodium and potassium
	Tasteless	
	At 1 atm, water freezes and ice melts at 0 °C or 32 °F	Reacts with metal oxides—CaO, for example—to form bases
	At 1 atm, water boils and steam condenses at 100 °C or 212 °F	Reacts with nonmetal oxides—SO_2, for example—to form acids
	Water expands when it freezes to ice	When crystals are formed from aqueous solutions of certain substances, water and the solute combine to build crystals, which are called hydrates
	Density at 4 °C, 1.0 g/mL	
	Specific heat, 1.0 cal/g · °C	
	Heat of vaporization, 540 cal/g	
	Heat of fusion, 80 cal/g	

plaster casts, they can be immersed in water, they are porous and allow free air circulation, and they are resistant to breakage.

Other hydrates are $MgSO_4 \cdot 7H_2O$, magnesium sulfate heptahydrate (commonly known as Epsom salts), and $Na_2SO_4 \cdot 10H_2O$, sodium sulfate decahydrate (commonly known as Glauber's salts). Both of these hydrates are used as *cathartics.

Hydrolysis

When some salts are placed in water, hydrolysis occurs. Hydrolysis is the reaction of a compound with water. When ammonium chloride, a salt, is placed in water, some hydrolysis occurs according to the following equilibrium reaction.

$$NH_4Cl \; + H_2O \; \rightleftharpoons \; NH_4OH \; + \quad HCl$$

| ammonium chloride | water | ammonium hydroxide | hydrochloric acid |

Note that hydrolysis in this example is the reverse of neutralization.

Hydrolysis occurs during the process of digestion of foods. For example, sucrose is hydrolyzed into glucose and fructose through the action of an enzyme.

$$C_{12}H_{22}O_{11} + H_2O \xrightarrow{\text{enzyme}} C_6H_{12}O_6 + C_6H_{12}O_6$$

| sucrose | | glucose | fructose |

Note that both glucose and fructose have the same formula,

$C_6H_{12}O_6$. How two different compounds can have the same formula will be discussed in Chapter 21.

The hydrolysis of fats in the body yields fatty acids and glycerol. The hydrolysis of proteins yields amino acids. Specific enzymes are necessary for these hydrolytic reactions. They will be discussed in the appropriate chapters on fats and protein.

Purification of Water

Impurities Present in Water

Natural water contains many dissolved and suspended materials. Rainwater contains dissolved gases—oxygen, nitrogen, and carbon dioxide—plus air pollutants (see page 130), suspended dust particles, and other particulate matter. Groundwater contains minerals dissolved from the soil through which the water has passed. It also contains some suspended materials. Seawater contains over 3.5 percent dissolved matter, the principal compound being sodium chloride. Both sea- and groundwater also contain dissolved and undissolved pollutants.

Lake water or river water may appear clear when a glass full of it is held up to the light, or it may at first contain suspended clay or mud, which tends to settle slowly, leaving what appears to be pure water. However, either of these "clear" waters may contain bacteria and other microorganisms that can be quite harmful to the body. Their destruction or removal is necessary for the proper purification of water. Water can be purified by several processes. The most common are distillation, boiling, filtration, and aeration.

Distillation

Distillation is a process of converting water to steam and then changing the steam back to water again. In the laboratory, a "still" is used to prepare distilled water. In the still shown in Figure 10-6, impure water is placed in the flask at the left and then heated to boiling. As the water boils and changes into steam, it passes into a condenser, which consists of a long glass tube surrounded by another glass tube through which cold water runs. The cold water, by absorbing heat, causes the steam to condense back into liquid water, which then runs out of the end of the tube into the receiving vessel at the right. The suspended and dissolved solids (including the bacteria) that were present in the impure water remain behind in the flask. They do not pass over into the condenser with the steam. The dissolved gases originally present in the impure water, however, do pass over with the steam. The usual practice is to discard the first few milliliters of distilled water coming from the condenser, since they will contain most of the dissolved gases.

Although distilled water is pure water, it is too expensive and the process too slow for large-scale use. The principal use of distilled water in the hospital is in the preparation of sterile solutions.

Figure 10-6 A laboratory distillation apparatus.

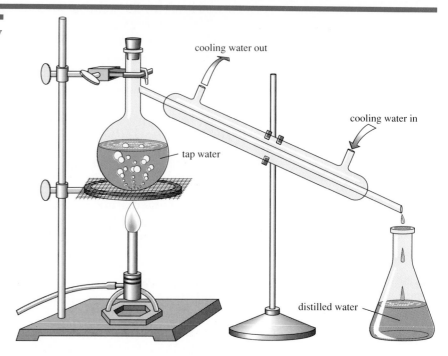

cooling water out

cooling water in

tap water

distilled water

Boiling

Groundwater or contaminated water can usually be made safe for drinking by boiling it for at least 15 min. The boiling does not remove the dissolved impurities but does kill any bacteria that might be present. Freshly boiled water has a flat taste because of the loss of dissolved gases. The taste may be brought back to normal by pouring the water back and forth from one clean vessel to another. This process, called aeration, allows air to dissolve in the water again.

Sedimentation and Filtration

For large-scale use, water is first allowed to stand in large reservoirs where most of the suspended dirt, clay, and mud settle out. This process is called sedimentation. However, sedimentation is a very slow process. Put some finely divided clay in a graduated cylinder full of water, shake it, and see how slowly the clay settles out. Commercially, a mixture of aluminum sulfate and lime is added to the water. These two chemicals combine to form aluminum hydroxide which precipitates as a gelatinous (sticky) substance. As the sticky aluminum hydroxide settles out, it carries down with it most of the suspended material. The main advantage of this material is that it settles much more rapidly than does the suspended material by itself and so increases the rate of sedimentation.

After sedimentation, the water is filtered through several beds of sand and gravel to remove the rest of the suspended material. The water then is essentially free of suspended material. However, this process does not remove the dissolved material or much of the bacteria originally present. The water must then be treated with chlorine to kill the bacteria before it is fit to drink.

Aeration

Water can be purified by exposing it to air for a considerable period of time. The oxygen in the air dissolves in the water and destroys the bacteria by the process of oxidation. The oxygen also oxidizes the dissolved organic material in the water so that the bacteria have no source of food. However, this process is slow and expensive because of the long time involved in exposing water to the air. Commercially, aeration is accomplished by spraying filtered chlorinated water into the air. This additional process also removes objectionable odors from the water.

Hard and Soft Water

When a small amount of soap is added to soft water, it forms copious suds. When a small amount of soap is added to hard water, it forms a precipitate or scum and no lather. What is the difference between soft water and hard water? Hard water contains dissolved compounds of calcium and magnesium. Soft water may contain other dissolved compounds, but these compounds do not cause hardness. The calcium and magnesium compounds (which cause the hardness) react with soap to form a precipitate, thus removing the soap from the water. More and more soap must be added until all the hardness-causing compounds are removed. Only then will the soap cause a lather. The reactions involved are as follows.

$$Ca^{2+} + Na(soap) \longrightarrow Ca(soap)(s) + 2\,Na^+$$

$$Mg^{2+} + Na(soap) \longrightarrow Mg(soap)(s) + 2\,Na^+$$

hardness   hardness   no
(a precipitate)   hardness

The precipitated soap adheres to washed materials, making them rough and irritating to tender skin, or to washed hair, making it sticky and gummy (see Figure 10-7). Food cooked in hard water is likely to be tougher than that cooked in soft water because of the presence of additional minerals. When hard water is boiled, some of the salts form a deposit on the inside of the container in which it is heated. Look inside an old teakettle at home and see the "boiler scale." Hard water must never be used to sterilize surgical instruments because the precipitated salts will dull the cutting edges. (If iron compounds are present in water, they also will cause hardness.)

Detergents have replaced soaps for washing clothes because they do not precipitate in hard water.

Figure 10-7 When the hair is shampooed in hard water, curd clings to the hair strands, dulling their natural luster and interfering with their ability to reflect light. The hair strands at top are stringy and not clean because of the clinging hard water curd. Those at bottom, washed in soft water, are radiant and clean. [Courtesy Culligan Water Institute, Northbrook, IL.]

Home Methods for Water Softening

Temporary hardness is hardness that can be removed by boiling. It is caused by the bicarbonates of calcium and magnesium. Boiling converts these to insoluble carbonates and carbon dioxide gas, thus removing part of the hardness.

$$Ca(HCO_3)_2 \xrightarrow{\text{heat}} CaCO_3(s) + CO_2(g) + H_2O$$

Other soluble compounds of calcium and magnesium cause permanent hardness. Permanent hardness is not affected by boiling.

Ammonia (ammonium hydroxide) is frequently used to soften water used in washing clothes and windows because it precipitates all the ions that cause temporary hardness and some of those that cause permanent hardness. Borax, sodium tetraborate ($Na_2B_4O_7$), is also frequently used in the home as a laundry water softener. Its effect is similar to that of ammonia. Washing soda ($NaCO_3$), frequently used as a home-laundry water softener, removes both temporary and permanent hardness from the water. Trisodium phosphate, TSP (Na_3PO_4), which is another home-laundry water softener, has an action similar to that of washing soda. However, TSP is no longer recommended for laundry use because of the effects of phosphates on our lakes and streams.

Commercial Water Softeners

The preceding methods for water softening are practical for use in a home, but they are too expensive for large-scale commercial use. One commercial method used is the lime-soda process. In this process lime, $Ca(OH)_2$, and then soda, Na_2CO_3, are added to water to remove both temporary and permanent hardness.

Complex silicates called zeolites are also used commercially to remove both types of hardness from water. These zeolites, in the form of large granules, are placed in a cylindrical container and hard water is allowed to flow through that container (see Figure 10-8). As the water passes over the zeolites, the sodium ions in the zeolite replace the calcium and magnesium ions in the water. This process, called **ion exchange,** takes the calcium and magnesium ions out of the water and replaces them with sodium ions. Sodium ions do not cause hardness. The reactions involved are

$$Ca^{2+} + Na_2Z \longrightarrow CaZ + 2\,Na^+$$
$$\text{sodium zeolite} \qquad \text{calcium zeolite}$$

$$Mg^{2+} + Na_2Z \longrightarrow MgZ + 2\,Na^+$$
$$\text{sodium zeolite} \qquad \text{magnesium zeolite}$$

The zeolite in the cylinder will be gradually used up as more and more water is softened. However, the used zeolite can be regener-

Figure 10-8 Zeolite tank. [Courtesy Culligan Water Institute, Northbrook, IL.]

ated and used over and over again. Many hospitals use this process for water softening. However, this type of softened water must be avoided by patients on low sodium diets.

Water that has been softened by the zeolite process still has dissolved minerals in it—the sodium compounds obtained by ion exchange. To prepare water comparable to distilled water, all soluble minerals must be removed. This can be accomplished by using a set of two different resins, one to remove the soluble metal (positive) ions and the other to remove the negative ions present in the water. If a sample of hard water is assumed to contain calcium chloride, $CaCl_2$, the dissolved mineral can be removed as follows:

$$CaCl_2 + 2\,HY \longrightarrow 2\,HCl + CaY$$
$$\text{resin}$$

The HCl then reacts with the second resin, XNH_2,

$$HCl + XNH_2 \longrightarrow XNH_3Cl$$
$$\text{resin}$$

so that no ions (aside from those of the water itself) are left in the water. Such water is called deionized water. Although this process is inexpensive and quick, it will not remove bacteria or dissolved nonelectrolytes.

Fluoridation of Water

Approximately one part per million (1 ppm) of fluoride ion added to drinking water in the form of NaF reduces dental caries. One of the major constituents of bones and teeth is hydroxyapatite $[Ca_5(PO_4)_3OH]$. The presence of the hydroxide ion (OH^-) makes this substance susceptible to attack by acidic substances such as soft drinks with a low pH (see page 233). Replacement of the hydroxide ion with a fluoride makes teeth more resistant to acid attack and hence more resistant to decay.

Fluorides are also used in toothpastes and gels in the form of stannous fluoride (SnF_2).

Water as a Moderator

Many nuclear reactors use water as a moderator, a substance that slows down neutrons. Why slow down neutrons? Because slower-moving neutrons can react with ^{235}U more effectively than faster-moving neutrons (see page 75).

Water Pollution

What is polluted water? Strictly speaking, it is any water that is not pure. However, tap water contains many dissolved and suspended substances. It is not pure, yet it is not called polluted water. Any substance that prevents or prohibits the normal use of water is termed a pollutant. The signs of polluted water are usually quite obvious—oil and dead fish floating on the surface of a body of water or deposited along the shores, a bad taste to drinking water, a foul odor along a waterfront, unchecked growth of aquatic weeds along the shore, or tainted fish that cannot be eaten.

Water pollutants can be classified into several categories.

1. Oxygen-demanding wastes. Dissolved oxygen is required for both plant and animal life in a body of water. Anything that tends to decrease the supply of this vital element endangers the survival of the life forms. Oxygen-demanding wastes are acted upon by bacteria in the presence of oxygen, thus leading to a depletion of the dissolved oxygen. Oxygen-demanding pollutants include sewage and wastes from papermills, food-processing plants, and other industrial processes that discharge organic materials into the water.

2. Disease-causing agents. Among the diseases that can be caused by pathogenic microorganisms present in polluted water are typhoid fever, cholera, infectious hepatitis, and poliomyelitis.

3. Radioactive material. Low-level radioactive wastes from nuclear power plants are sealed in concrete and buried underground. High-level radioactive wastes are initially stored as liquids in large underground tanks and later converted into solid form for burial in concrete. In either case, leakage can lead to pollution of nearby water supplies.

4. Heat. Although heat is not normally considered a pollutant of waterways, it does have a detrimental effect on the amount of dissolved oxygen. Thermal pollution results when water is used as a coolant for industrial plants and nuclear reactors and then returned to its source.

In addition to decreasing the amount of dissolved oxygen, thermal pollution also causes an increase in the rate of chemical reactions. The metabolic processes of fish and microorganisms are speeded up, increasing their need for oxygen, at a time when the supply of oxygen is diminishing. Higher water temperatures can also be fatal to certain forms of marine life.

5. Plant nutrients. Nutrients stimulate the growth of aquatic plants. This may lead to lower levels of dissolved oxygen. It may also lead to disagreeable odors when the large amount of plant material decays. Excessive plant growth is often unsightly and interferes with recreational use of water. Excess phosphorus in sewage comes from phosphate detergents and is one cause of this type of pollution.

6. Synthetic organic chemicals. In this category of pollutants are such substances as surfactants in detergents, pesticides, plastics, and food additives.

7. Inorganic chemicals and minerals. These pollutants come from industrial wastes as well as from runoff water from urban areas. One example of such a pollutant is mercury. It was once believed that metallic mercury was inert and settled to the bottom of a lake. It is now known that anaerobic bacteria in bottom muds are capable of

converting this mercury into compounds that are poisonous. Another pollutant in this category is sulfuric acid, which is formed by the reaction of sulfur-containing ores with water and oxygen in the air. Salt is also a pollutant, which can occur when brine from oil wells is released into fresh water.

Summary

Water is one of the most important chemicals known. Pure water is a colorless, odorless, flat-tasting liquid that freezes at 0 °C at 1 atm pressure. Water has its maximum density at 4 °C. When water freezes, its volume increases by almost 10 percent.

Water is used as a solvent for many substances. Many chemical reactions take place only when the reactants are dissolved in water. The evaporation of water or perspiration from the skin is a cooling process because the skin provides (loses) the heat required to change the liquid to the vapor state. Water is the standard of reference for such physical constants as the temperature scale, specific gravity, and the calorie.

The water molecule is a covalent one with a hydrogen–oxygen–hydrogen angle of about 105 °. Water is a polar liquid. Polar liquids usually dissolve polar compounds, and nonpolar liquids dissolve nonpolar compounds.

The electrolysis of water yields hydrogen and oxygen gases. Water does not otherwise appreciably decompose into these gases, even when heated to 1600 °C.

Water reacts with certain metal oxides to form bases; water reacts with certain nonmetal oxides to form acids. Water reacts with sodium and potassium to form hydrogen gas and a base. Water reacts with certain salts to form hydrates. Hydrates that lose their water of hydration on standing are said to be efflorescent. Substances that pick up moisture from the air are said to be hygroscopic.

Hydrolysis of a salt is the reaction of that salt with water whereby the water molecule is split.

Water can be purified by distillation, by boiling, by sedimentation and filtration, and by aeration.

Hardness in water is caused by ions of calcium and magnesium (and iron). Temporary hardness is due to bicarbonates of calcium and magnesium. Materials causing temporary hardness can be removed by boiling. Materials causing permanent hardness cannot be removed by boiling. Hard water can be softened by using ammonia, washing soda, trisodium phosphate, lime-soda, or zeolite. Deionized water contains no contaminating ions and is frequently used in place of distilled water.

Water pollution can be caused by oxygen-demanding wastes, disease-causing agents, radioactive materials, heat, plant nutrients, synthetic organic chemicals, and inorganic chemicals and materials.

Questions and Problems

A

1. Why is pure water flat tasting?
2. Why do lakes and oceans appear blue?
3. Why does ice float?
4. At what temperature does water have its maximum density?
5. Why use a pressure cooker at high altitudes?
6. How much heat is required to evaporate 10 g of water from the skin?
7. Why is evaporation a cooling process?
8. Water is used as the standard of reference for which physical constants?
9. Diagram the structure of the water molecule. Why is the molecule polar?

10. Why does salt dissolve in water?
11. Will carbon tetrachloride (CCl_4), a nonpolar liquid, dissolve in water?
12. What type of compound is produced when water reacts with a metal oxide?
13. What type of compound is produced when water reacts with a nonmetal oxide?
14. What type of reaction occurs when a salt reacts with water?
15. Why purify water?
16. Describe the preparation of distilled water.
17. What effect does boiling have on water as far as purification is concerned? as far as the removal of impurities is concerned?
18. Of what importance are filtration and aeration of water?
19. What is a hydrate?
20. What reaction occurs when a hydrate is heated?
21. What is efflorescence?
22. What is a hygroscopic compound? For what purpose can it be used?
23. What causes temporary hardness in water? How can it be removed?
24. What causes permanent hardness in water? How can it be removed?
25. Describe the zeolite process for water softening.
26. What is deionized water, and how can it be prepared?
27. What is polluted water? What signs indicate such a condition?
28. What diseases can be caused by polluted water?
29. A depletion of the oxygen content of water can be caused by what factors?
30. How are radioactive wastes disposed of? Why are precautions necessary?
31. What might be the effects of thermal pollution? What might cause this type of pollution?
32. Indicate several synthetic organic water pollutants. Where do they come from?

B

33. What causes the vapor pressure of a liquid? How is vapor pressure affected by temperature?
34. What effect does hydrogen bonding have on the boiling point of water? Explain.
35. How is the vapor pressure of a liquid related to its boiling point?
36. How is the boiling point of a liquid related to the atmospheric pressure?

37. Why is hydrolysis important in the human body?
38. Is hard water safe to drink? Explain.
39. Should zeolite-softened water be used by all persons? Explain.

Practice Test

1. Water _____.
   a. expands on freezing
   b. boils at 100 °C at 1 atm pressure
   c. has a maximum density at 4 °C
   d. all of these
2. An example of a cooling process is_____.
   a. distillation   b. condensation
   c. solvation   d. evaporation
3. Water _____.
   a. is a polar compound
   b. is a nonpolar compound
   c. dissolves all substances
   d. dissolves only nonpolar substances
4. Water reacts with metal oxides to form _____.
   a. acids   b. bases
   c. salts   d. hydrates
5. One process for purifying water is _____.
   a. distillation   b. condensation
   c. solvation   d. evaporation
6. Hardness in water is due to the presence of _____.
   a. Na^+ b. K^+ c. Ca^{2+} d. Cu^+
7. Water used as a moderator _____.
   a. slows down neutrons
   b. absorbs neutrons
   c. activates neutrons
   d. has no effect on neutrons
8. The reaction of water with a salt is called _____
   a. hydrolysis   b. acidification
   c. efflorescence   d. neutralization
9. As the atmospheric pressure decreases, the boiling point of water _____.
   a. increases
   b. decreases
   c. is unaffected
10. Water has a relatively low vapor pressure because of _____.
    a. hydrogen bonds   b. dissolved minerals
    c. dissolved gases   d. ionic bonds

11

Liquid Mixtures

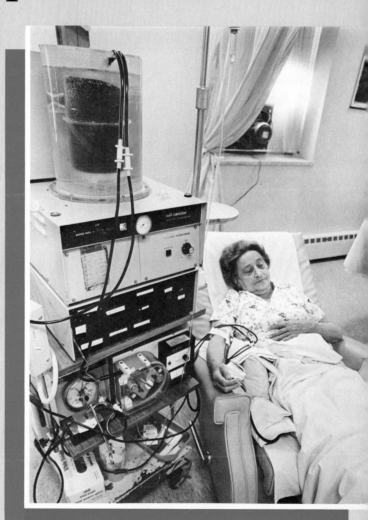

Woman in hemodialysis in a hospital.

Liquid mixtures can be divided into four types: solutions, suspensions, colloids, and emulsions. Each type has its own specific properties and uses.

Solutions

General Properties

A solution is a homogenous mixture of two or more substances evenly distributed in each other. A liquid solution consists of two parts: the solid, liquid, or gaseous material that has dissolved, the solute, and a liquid material in which it has dissolved, the solvent.

When a crystal of salt is placed in water, which is then stirred, the crystal dissolves and a clear solution is formed. When more salt is added to this salt–water solution, it too dissolves, making the solution more concentrated than the previous one. Even more salt can be dissolved in the water to make it a much more concentrated salt solution. Thus, one of the properties of solutions is that they have a variable composition. Varying amounts of salt and water can be mixed to form various concentrations of salt-water.

When salt is dissolved in water to make salt water, the solution formed is clear and colorless. When sugar is dissolved in water, again a clear, colorless solution is formed. When copper sulfate is dissolved in water, it also forms a clear solution. However, the solutions formed with the salt and the sugar are colorless, whereas that formed with copper sulfate is blue. Solutions are always clear. They may or may not have a color. Clear merely means that the solution is transparent to light.

When a salt solution is examined under a high-power microscope, it appears to be homogeneous. The same is true for a sugar solution. In general, all solutions are homogeneous. The solute cannot be distinguished from the solvent in a solution.

When a solution is allowed to stand undisturbed for a long period of time, no crystals of solute settle out, provided the solvent is not allowed to evaporate. This is another property of solutions—the solute does not settle out.

The salt in a salt–water solution can be recovered by allowing the water to evaporate; the same is true of the sugar in a sugar solution. In general, solutions can be separated by physical means.

If a solution (such as salt water) is poured into a funnel containing a piece of filter paper, the solution will pass through unchanged. That is, the particles in solution must be smaller than the openings in the filter paper.

The properties of solutions are summarized as follows. Solutions

1. Consist of a soluble material or materials (the solute) dissolved in a liquid (the solvent).

2. Have a variable composition.
3. Are clear.
4. Are homogeneous.
5. Do not settle.
6. Can be separated by physical means.
7. Pass through filter paper.

Solvents Other Than Water

Solvents other than water are also used. One common solvent used in hospitals is alcohol. An alcohol solution used medicinally is called a *tincture. Tincture of iodine contains iodine dissolved in alcohol. Tincture of green soap contains potassium soap dissolved in alcohol. Ether is an excellent solvent for fats and oils.

Factors Affecting Solubility of a Solute

Temperature

Most solid solutes are more soluble in hot water than in cold water. Figure 11-1 shows that KNO_3 becomes much more soluble as the temperature increases; however, $Ce_2(SO_4)_3$ becomes less soluble with an increase in temperature, and $NaCl$ shows little change in

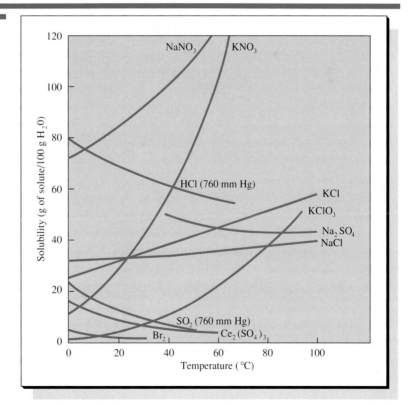

Figure 11-1 Variation in the solubilities of several substances in water as a function of temperature.

solubility. Gases such as HCl and SO_2 become less soluble with increasing temperature. The solubility of Br_2, a liquid, is practically unaffected by temperature.

Pressure

A change in pressure has no noticeable effect on the solubility of a solid or liquid solute in a given solvent but will affect the solubility of a gaseous solute. The greater the pressure, the greater the solubility of a gas in a liquid.

Surface Area

Although surface area does not affect the amount of solute that will dissolve, it does affect the rate of dissolution. The greater the amount of surface area, the quicker a solute will dissolve in a solvent. Thus, to make a solid solute dissolve faster we frequently powder it, thereby increasing the surface area.

Stirring

The rate at which a solute dissolves can also be increased by stirring the mixture. The process of stirring brings fresh solvent into contact with the solute and so permits more rapid solution.

Therefore, to dissolve most solid solutes rapidly, the solute should be powdered and the mixture should be heated while it is stirred.

Nature of Solvent

In general, polar liquids dissolve polar compounds and nonpolar liquids dissolve nonpolar compounds. Water is a polar liquid (see page 168) and dissolves polar compounds such as sodium chloride (NaCl).

Sometimes, in spite of all attempts, a substance does not appreciably dissolve in water. If the substance is nonpolar, then it should not be expected to be soluble in water. But it should be soluble in a nonpolar liquid. Sometimes polar materials do not dissolve in water. These are called insoluble salts (see page 238).

Table 11-1 lists several polar and nonpolar liquids. Using this table we could predict that potassium iodide (KI), a polar compound, should be soluble in water and that most likely it would not be soluble in a nonpolar liquid such as benzene. Likewise we could predict that oils and waxes, which are nonpolar compounds, should be insoluble in water but soluble in ether (or other nonpolar liquids).

Table 11-1 Some Polar and Nonpolar Liquids

Polar Liquids		Nonpolar Liquids	
Name	Formula	Name	Formula
Water	H_2O	Benzene	C_6H_6
Methyl alcohol	CH_3OH	Carbon tetrachloride	CCl_4
Ethyl alcohol	C_2H_5OH	Ether	$C_2H_5OC_2H_5$

In addition to liquid solutions, there are also solid solutions. They usually consist of one metal dissolved in another while both are in the molten (liquid) state. Such a solid solution is called an *alloy. Some alloys are used in preparing replacements for such body parts as hip bones and knee joints.

A special type of alloy consists of a metal dissolved in mercury, a liquid metal. An alloy containing mercury is called an **amalgam**. If silver is dissolved in mercury, a silver amalgam is produced. This substance was used in dental work to fill a cavity in a tooth.

Importance of Solutions

During digestion, foods are changed to soluble substances so that they can pass into the bloodstream and be carried to all parts of the body. At the same time the waste products of the body are dissolved in the blood and carried to other parts of the body where they can be eliminated. Plants obtain minerals from the groundwater in which those minerals have dissolved.

Many chemical reactions take place in solution. When solid silver nitrate is mixed with solid sodium chloride, no reaction takes place because the movement of the ions in the solid state is highly restricted. However, when a solution of silver nitrate is mixed with a solution of sodium chloride, a precipitate of silver chloride is formed instantaneously. This reaction occurs because the ions in the solution are free to move and react with other ions.

Many medications are administered orally, subcutaneously, or intravenously as solutions.

Drugs must be in solution before they can be absorbed from the gastrointestinal (GI) tract. As you might expect, when drugs are taken in solution, such as syrups and elixirs, they are absorbed more rapidly than drugs in a solid form, such as tablets and capsules.

Strength of Solutions

Dilute and Concentrated

When a few crystals of sugar are placed in a beaker of water, a *dilute* sugar solution is produced. As more and more sugar is added to the water, the solution becomes more *concentrated*. But when does the solution change from dilute to concentrated? There is no sharp dividing line. "Dilute" merely means that the solution contains a small amount of solute in relation to solvent. "Concentrated" merely means that the solution contains a large amount of solute in relation to solvent.

However, both dilute and concentrated are relative terms. A dilute sugar solution may contain 5 g of sugar per 100 mL of solution, whereas 5 g (the same amount) of boric acid per 100 mL of solution will produce a concentrated boric acid solution. That is, the terms dilute and concentrated usually have no specific quantitative meaning and so are not generally used for medical applications.

Saturated

Suppose a small amount of salt is placed in a beaker of water. When the mixture is stirred, all the salt will dissolve. If more and more salt is added with stirring, a point will soon be reached where some of the salt settles to the bottom of the beaker. This excess salt does not dissolve even upon more rapid agitation. This type of solution is called a *saturated solution*. Some of the crystals are continually dissolving and going into solution, but at the same time, the same amount of solute crystallizes out of the solution. This type of interchange, as you may remember, is called *equilibrium*.

Thus, a saturated solution can be defined as one in which there is an equilibrium between the solute and the solution. A saturated solution can also be defined as one that contains all the solute that it can hold under the given conditions.

Unsaturated

An unsaturated solution contains less of a solute than it could hold under normal conditions. Suppose that a saturated solution of glucose in a certain amount of water contains 25 g of glucose. An unsaturated glucose solution would be one that contained less than 25 g of glucose in the same amount of water. In an unsaturated solution, no equilibrium exists because there is no undissolved solute.

Like dilute and concentrated, the terms saturated and unsaturated are relative terms. The same amount of two different solutes may produce entirely different types of solutions. For glucose, 5 g in 100 mL of water produces an unsaturated solution, whereas 5 g of boric acid in 100 mL of water produces a saturated solution. Therefore, the terms saturated solution and unsaturated solution are not used for medical applications.

Supersaturated

Under certain conditions a solvent can be made to dissolve more solute than its saturated solution can hold under the same conditions. A supersaturated solution can be prepared by adding excess solute to a saturated solution, heating that mixture, filtering off the excess solute, and then allowing the liquid to cool slowly. If this is done carefully, some excess solute will remain dissolved, thus forming a supersaturated solution. However, such a solution is very unstable. If one crystal of solute is added, or if the liquid is shaken, the excess solute will crystallize out immediately and a saturated solution will remain. The formation of gallstones involves the precipitation of cholesterol from a supersaturated solution.

Percentage Solutions

The **weight–volume method** expresses the weight of solute in a given volume of solvent, usually water. A 10 percent glucose solution will contain 10 g of glucose per 100 mL of solution. A 0.9 percent saline solution will contain 0.9 g of sodium chloride per

100 mL of solution. The percentage indicates the number of grams of solute per 100 mL of solution.

Example 11-1   Prepare 500 mL of 2 percent citric acid solution.

First let us calculate how much citric acid will be required. A 2 percent citric acid solution contains 2 g citric acid per 100 mL of solution. Therefore, in 500 mL of solution there should be

$$500 \text{ mL} \times \frac{2 \text{ g citric acid}}{100 \text{ mL}} = 10 \text{ g citric acid}$$

To prepare the solution, proceed as follows.

1. Weigh out exactly 10 g of citric acid.
2. Dissolve the 10 g of citric acid in a small amount of water contained in a 500-mL graduated cylinder.
3. Add water to the 500-mL mark and stir.

Note that the 10 g of citric acid was dissolved in water and then diluted to the required volume—500 mL. It was *not* dissolved directly in 500 mL of water. (If it had been, the final volume would have been more than 500 mL.)

Example 11-2   A patient is given 1000 mL of 0.9 percent NaCl intravenously. How many grams of NaCl did the patient receive?

$$0.9\% \text{ means } \frac{0.9 \text{ g NaCl}}{100 \text{ mL solution}}$$

$$1000 \text{ mL} \times \frac{0.9 \text{ g NaCl}}{100 \text{ mL}} = 9 \text{ g NaCl}$$

In clinical work involving dilute solutions, concentrations are sometimes expressed in terms of milligram percent (mg %), which indicates the number of milligrams of solute per 100 mL of solution. Milligram percent is also referred to as milligrams per deciliter (mg/dL). Table 30-1 lists the values of blood components in these units.

Parts per Million

Low concentrations may be expressed in the units *milligrams per liter* (mg/L). Another unit for expressing low concentrations is parts per million (ppm). One part per million is equivalent to 1 mg/L. That is, if a solution has a concentration of 40 mg/L, its concentration can also be expressed as 40 ppm.

Parts per million are used to indicate the hardness of water and also to show the concentration of both common substances and pollutants in water and in air.

Extremely low concentrations of pollutants are expressed in the units **parts per billion**, which is equivalent to milligrams per 1000 L (mg/1000 L).

Ratio Solutions

Another method of expressing concentration is a ratio solution. A 1:1000 merthiolate solution contains 1 g of merthiolate in 1000 mL of solution. A 1:10,000 $KMnO_4$ solution contains 1 g of $KMnO_4$ in 10,000 mL of solution. The first number in the ratio indicates the number of grams of solute, and the second number gives the number of milliliters of solution. As with percentage solutions, the solute is dissolved in a small amount of solvent (water) and then diluted to the desired volume.

Percentage and ratio solutions are frequently used by doctors, nurses, and pharmacists.

Molar Solutions

Molar solutions are used most frequently by chemists. A molar solution is defined as one that contains 1 mol (see page 97) of solute per liter of solution. A 1 molar (1 M) solution of glucose ($C_6H_{12}O_6$) will contain 1 mol of glucose (180 g) in 1 L of solution. As before, the solute is dissolved in a small amount of water and then diluted to the desired volume.

Example 11-3 Prepare 3 L of 2 M (2 molar) KCl (molecular weight 74.5).

The problem calls for the preparation of 3 L of 2 M KCl or 3 L × 2 M KCl. Recall that molarity means moles per liter. So,

$$3 \text{ L} \times 2 \text{ M KCl} = 3 \text{ L} \times \frac{2 \text{ mol KCl}}{1 \text{ L}} = 6 \text{ mol KCl}$$

Then, since 1 mol KCl weighs 74.5 g,

$$6 \text{ mol KCl} = 6 \text{ mol KCl} \times \frac{74.5 \text{ g KCl}}{1 \text{ mol KCl}} = 447 \text{ g KCl}$$

So, we take 447 g KCl, dissolve it in water, and dilute to a total of 3 L.

Example 11-4 Prepare 500 mL of 0.1 M NaOH (molecular weight 40).

The problem calls for 500 mL of 0.1 M NaOH, or 500 mL × 0.1 M NaOH. Changing molarity to moles per liter and also changing milliliters to liters, we have

$$500 \text{ mL} \times \frac{1 \text{ L}}{1000 \text{ mL}} = \frac{0.1 \text{ mol NaOH}}{1 \text{ L}} = 0.05 \text{ mol NaOH}$$

Then, changing moles of NaOH to grams of NaOH,

$$0.05 \text{ mol NaOH} \times \frac{40 \text{ g NaOH}}{1 \text{ mol NaOH}} = 2 \text{ g NaOH}$$

Thus, we should dissolve 2 g of NaOH in water and dilute to 500 mL.

Example 11-5 How many grams of glucose are present in 0.5 L of 2.0 M glucose solution? The molecular weight of glucose is 180.

Changing molarity to moles per liter, we have

$$0.5 \text{ L} \times \frac{2.0 \text{ mol glucose}}{1 \text{ L}} = 1 \text{ mol glucose}$$

Then,

$$1 \text{ mol glucose} \times \frac{180 \text{ g glucose}}{1 \text{ mol glucose}} = 180 \text{ g glucose}$$

Normal Solutions

A 1 normal (1 N) solution contains one gram equivalent weight of solute per liter of solution. The gram equivalent weight of an acid can be calculated by dividing the weight of 1 mol of that acid by the number of replaceable hydrogens that it contains. For example, hydrochloric acid,

$$HCl \longrightarrow H^+ + Cl^-$$

contains one replaceable hydrogen. The gram equivalent weight of HCl is the weight of 1 mol (36.5 g) divided by 1, or 36.5 g.

The gram equivalent weight of sulfuric acid (H_2SO_4),

$$H_2SO_4 \longrightarrow 2 H^+ + SO_4^{2-}$$

is the weight of 1 mol of H_2SO_4 (98 g) divided by 2, or 49 g.

The gram equivalent weight of a base can be calculated by dividing the weight of 1 mol of that base by the number of OH^- groups it contains. The gram equivalent weight of sodium hydroxide (NaOH),

$$NaOH \longrightarrow Na^+ + OH^-$$

is the weight of 1 mol of NaOH (40 g) divided by 1, or 40 g.

The gram equivalent weight of calcium hydroxide [$Ca(OH)_2$]

$$Ca(OH)_2 \longrightarrow Ca^{2+} + 2\,OH^-$$

is the weight of 1 mol of $Ca(OH)_2$ (74 g) divided by 2, or 37 g.

Example 11-6   Prepare 2 L of 1.5 N H_2SO_4.

The problem calls for the preparation of 2 L of 1.5 N H_2SO_4, or 2 L × 1.5 N H_2SO_4. Recall that N means gram equivalent weights per liter. So,

$$2\,L \times 1.5\,N\ H_2SO_4 = 2\,\cancel{L} \times \frac{1.5\ \text{gram equivalent weights } H_2SO_4}{\cancel{L}}$$

$$= 3 \text{ gram equivalent weights } H_2SO_4$$

According to the previous calculation, the gram equivalent weight of H_2SO_4 is 49 g, so

$$3\ \cancel{\text{gram equivalent weights } H_2SO_4}$$

$$\times \frac{49 \text{ g } H_2SO_4}{\cancel{1 \text{ gram equivalent } H_2SO_4}} = 147 \text{ g } H_2SO_4$$

So we take 147 g of H_2SO_4, dissolve it in water, and dilute to a total of 2 L.

Example 11-7   Prepare 100 mL of 0.2 N NaOH (molecular weight of 40).

The problem calls for 100 mL of 0.2 N NaOH or 100 mL × 0.2 N NaOH. Changing normality to gram equivalent weights per liter and also changing milliliters to liters, we have

$$100\ \cancel{mL} \times \frac{1\ \cancel{L}}{1000\ \cancel{mL}} \times \frac{0.2 \text{ gram equivalent weight NaOH}}{1\ \cancel{L}}$$

$$= 0.02 \text{ gram equivalent weight NaOH}$$

Then, changing gram equivalent weights of NaOH to grams of NaOH (1 gram equivalent weight of NaOH weighs 40 g),

$$0.02 \; \cancel{\text{gram equivalent weights NaOH}}$$
$$\times \; \frac{40 \text{ g NaOH}}{1 \; \cancel{\text{gram equivalent weight NaOH}}} = 0.8 \text{ g NaOH}$$

Thus we should dissolve 0.8 g of NaOH in water and dilute to 100 mL.

Milliequivalents per Liter

Normality is a very common concentration unit used in a procedure called titration whereby the strength of an unknown acid or base can be determined (see page 231).

Concentrations of ions in body fluids are frequently expressed in the units milliequivalents per liter, mEq/L (see page 542). The number of equivalents, of an ion is determined by multiplying the number of moles of ions by the value of the charge the ion carries. That is, 1 mol of sodium ions (Na^+) contains 1 Eq of sodium ions. One mole of calcium ions Ca^{2+}, contains 2 Eq of calcium ions, and 1 mol of CO_3^{2-} ions contains 2 Eq of carbonate ions.

One milliequivalent (mEq) is 1/1000 of an equivalent.

Example 11-8 A solution contains 0.045 mol of Na_2CO_3/L. How many equivalents per liter and milliequivalents per liter of each ion are present?

$$Na_2CO_3 \longrightarrow 2\,Na^+ + CO_3^{2-}$$

According to the equation, 1 mol of Na_2CO_3 yields 2 mol of Na^+ and 1 mol of CO_3^{2-}. So,

$$0.045 \text{ mol/L } Na_2CO_3 \longrightarrow 2 \times 0.045 = 0.090 \text{ mol } Na^+/L$$

$$0.090 \text{ mol/L } Na^+ = 0.090 \text{ Eq/L } Na^+$$

$$0.090 \; \cancel{\text{Eq}}/\text{L } Na^+ = 0.090 \; \cancel{\text{Eq}}/\text{L } Na^+ \times \frac{1000 \text{ mEq}}{1 \; \cancel{\text{Eq}}}$$

$$= 90 \text{ mEq/L } Na^+$$

Likewise,

$$0.045 \text{ mol/L } Na_2CO_3 \longrightarrow 0.045 \text{ mol/L } CO_3^{2-}$$

$$0.045 \text{ mol/L } CO_3^{2-} = 0.090 \text{ Eq/L } CO_3^{2-}$$

$$0.090 \; \cancel{\text{Eq}}/\text{L } CO_3^{2-} \times \frac{1000 \text{ mEq}}{1 \; \cancel{\text{Eq}}} = 90 \text{ mEq/L } CO_3^{2-}$$

Dilution of Solutions

It is often necessary to prepare a weaker solution from a stronger (stock) solution. To do so, we must add water to the stock solution. But how much water must be added? To answer this question we use the relationship

$$\frac{\text{initial}}{\text{volume}} \times \frac{\text{initial}}{\text{concentration}} = \frac{\text{final}}{\text{volume}} \times \frac{\text{final}}{\text{concentration}}$$

where the initial volume is the amount of stock solution to be used, the initial concentration is the strength of the stock solution, the final volume is the amount of dilute solution to be prepared, and the final concentration is the strength of the dilute solution to be prepared.

Example 11-9   Prepare 100 mL of 0.9 percent saline solution from 10 percent saline solution.

Using the above formula where initial volume = x, final volume = 100 mL, initial concentration = 10 percent, and final concentration = 0.9 percent,

$$x \times 10\% = 100 \text{ mL} \times 0.9\%$$

$$x = \frac{100 \text{ mL} \times 0.9\%}{10\%} = 9 \text{ mL}$$

Thus, 9 mL of 10 percent saline solution should be diluted to a volume of 100 mL to prepare 100 mL of 0.9 percent saline solution. The solution could also be prepared by adding 91 mL of water to the 9 mL of 10 percent saline solution.

Osmolarity

The osmotic pressure of a solution is expressed in terms of the osmolarity of a solution. *Osmolarity (osmol) is a measure of the number of particles in solution and will be discussed later in this section under Osmotic Pressure.

Special Properties of Solutions

Effect of Solute on Boiling Point and Freezing Point

Whenever a nonvolatile solute is dissolved in a solvent, the boiling point of the solution thus prepared is always greater than that of the pure solvent. Pure water boils at 100 °C at 1 atm pressure. A solution of salt in water or a solution of sugar in water will boil at a temperature above 100 °C.

Likewise, when a nonvolatile solute is dissolved in a solvent,

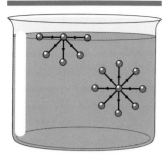

Figure 11-2 Surface tension of a liquid.

the freezing point of the solution is always less than the freezing point of the pure solvent. The freezing point of water is 0 °C. The freezing point of salt solution or sugar solution is always less than 0 °C.

Use is made of these facts in the cooling system of automobiles. Antifreeze is added to the water in the car radiator to lower the freezing point so that the liquid will not freeze when the temperature drops below 32 °F. The same material is used in some areas to raise the boiling point of the liquid in the automobile radiator—to prevent it from boiling over when the temperature rises.

Surface Tension

Consider a water molecule in the center of a beaker of water (see Figure 11-2). This water molecule will be attracted in all directions by the water molecules around it. Next, consider a molecule at the surface of the water. This molecule is attracted sideways and downward, but it is not attracted very much by the air molecules above it. Therefore there is a net downward attraction on the surface water molecules. This downward pull on the surface molecules causes them to form a surface film. Surface tension can be defined as the force that causes the surface of a liquid to contract. Surface tension also is the force necessary to break this surface film. All liquids exhibit surface tension; the surface tension of water is higher than that of most liquids.

Surface tension is responsible for the formation of drops of water on a greasy surface. The surface film holds the drop in a spherical shape rather than letting it spread over the surface as a sheet of water.

Some medications designed for use on the tissues in the throat contain a very special surface-active agent—one that will reduce the surface tension of the water. This surface-active agent, called a **surfactant,** lowers the surface tension of the liquid so that it spreads rapidly over the tissues rather than collecting in the form of droplets with less "active" surface area. Soaps and detergents are *surfactants. Bile, which is secreted by the liver, contains bile salts, which act as surfactants. These surfactants help in the digestion of fats (page 456).

A lack of a surfactant in the lungs of premature infants causes respiratory distress syndrome (RDS), formerly called hyaline membrane disease. The surfactant is necessary to form a coating on the inner lining of the small air sacs (alveoli) in the lungs. If the surfactant is present in low amounts or is not present at all, the surface tension in the alveoli rises, causing portions of the lung to collapse and producing respiratory distress.

The attraction of like molecules for each other is known as **cohesion;** the attraction of unlike molecules is known as **adhesion.** Water rises in a capillary tube because of surface tension and also because of adhesion of the water molecules to the glass. This effect, known as **capillary action,** is used when drawing blood samples.

Table 11-2 Relative Viscosities of Various Liquids at 20 °C

Alcohol (ethyl)	1.20
Blood, male	4.71
Blood, female	4.46
Ether (ethyl)	0.23
Glycerol	1490
Water	1.00

Capillary action also accounts for the absorption of water by absorbent cotton and by paper towels.

Viscosity

Some liquids flow readily (water) whereas others (molasses) do not. A measure of the resistance to flow is called *viscosity (see page 412). Table 11-2 indicates the viscosities of several liquids compared to water, which is assigned a value of 1.00. Note the viscosity of blood compared to that of water.

As the temperature increases, liquids tend to flow more readily, and so the viscosity decreases.

Diffusion

Diffusion, also known as passive transport, is the process whereby a substance moves from an area of its higher concentration to a region where it is less concentrated. That is, the molecules of the substance move from an area where they are crowded together and where molecular collisions are frequent to a region where they are less crowded and collisions will occur less often. The greater the difference in concentration between the two areas (the greater the concentration gradient), the faster will be the rate of diffusion. During diffusion, no external source of energy is required.

For example, when a crystal of copper sulfate pentahydrate (a blue crystalline substance) is dropped into a cylinder of water, the blue color is soon observed in the water surrounding the crystal. After a while, the blue color can be seen extending upward from the crystal. The liquid at the bottom of the cylinder will be darker blue, and the liquid above it will be lighter blue. After several hours the entire contents of the cylinder will be uniformly blue. That is, the solute particles from the crystal are uniformly distributed into all parts of the solution.

Gases will diffuse into one another. When a bottle of ether is opened the odor can soon be detected at a distance. Diffusion into a gas takes place more rapidly than diffusion into a liquid.

Another example of diffusion is the loss of perspiration from the body. The moisture flows from an area of high concentration (the skin) to one of lower concentration (the air). The higher the moisture content of the air (the lower the concentration gradient), the slower the rate of evaporation of moisture from the skin.

Passive transport (diffusion) occurs along a concentration gradient from an area of higher concentration to one of lower concentration with no energy other than kinetic molecular energy being required.

Active transport occurs when a substance is moved against the concentration gradient, that is, from an area of low concentration to one of higher concentration. In this process energy is required. Active transport is responsible for the high concentration of potassium ions inside the cells compared to a much lower potassium ion concentration outside the cells. Active transport is also responsible for the low concentration of sodium ions inside the cells compared to a much higher sodium ion concentration outside the cells. The amount of energy required for active transport could be as much as 35 percent of the energy output of a resting cell.

Osmosis

When the diffusing substance is water and when the diffusion takes place through a *semipermeable (properly, selectively permeable) membrane, the process is called osmosis.

Osmosis can be defined as the diffusion of water (solvent) through a semipermeable membrane from a weaker solution (one containing less dissolved solute) to a stronger solution (one containing more dissolved solute). Osmosis can also be defined as the diffusion of water (solvent) down a concentration gradient from an area of high solvent concentration (a weak solution) to a region of low solvent concentration (a stronger solution).

Consider two salt solutions (one dilute and the other concentrated) separated by a semipermeable membrane. The two solutions will tend to equalize in concentration. That is, the dilute one will tend to become more concentrated, whereas the concentrated one will tend to become more dilute. How can they do this? There are two possibilities. First, osmosis can take place. That is, the solvent, water, can diffuse through the membrane from the weaker to the stronger solution (or from an area of high solvent concentration to a region of lower solvent concentration). This process will continue until the two solutions have the same concentration.

The second possibility is that of diffusion. The solute will diffuse through the membrane from the stronger to the weaker solution (down the concentration gradient) until the two solutions have equal concentrations. Both osmosis and diffusion can occur at the same time, but not at the same rate. In general, osmosis occurs more rapidly than diffusion.

An example of osmosis that is quite common in the home can be observed by placing a dried prune in water (see Figure 11-3). The skin of the prune acts as a semipermeable membrane. Inside the prune are rather concentrated juices. The water surrounding the prune is certainly dilute in comparison to the juices inside. Thus there are two different concentrations of a solution separated by a semipermeable membrane, and osmosis can take place. In which direction? In osmosis the diffusion of solvent is from the weaker to the stronger solution. Therefore the water will diffuse into the prune, causing it to swell.

Another example of osmosis can be seen when a cucumber is placed in a strong salt solution (Figure 11-4). The skin of the cucumber acts as a semipermeable membrane. The liquid inside the cucumber is quite dilute in comparison to that of the salt solution. Therefore, osmosis takes place between solutions of two different

Figure 11-3 Osmosis: prune in water.

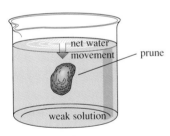

Figure 11-4 Osmosis: cucumber in saltwater.

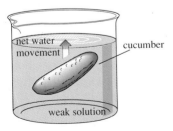

concentrations. The flow of water, again from dilute to concentrated, is from the cucumber into the solution. Thus the cucumber shrinks and becomes a pickle, again by the process of osmosis.

Sailors lost at sea die of dehydration if they drink saltwater. In this case, the seawater has a higher concentration of salts than the body fluids, so water diffuses out of the tissues and dehydration results.

Osmotic Pressure

Consider a beaker of water into which is placed a thistle tube with a semipermeable membrane over its end as in Figure 11-5. Assume that the thistle tube contains sugar solution. In which direction will osmosis take place? Osmosis will take place with the water diffusing into the thistle tube (from weaker to stronger solution). As the water diffuses into the thistle tube, the water level in the tube will rise. The rising water in the thistle tube will exert a certain amount of pressure, as does any column of liquid. The pressure exerted during osmotic flow is called osmotic pressure. This concept is of great importance in the regulation of fluid and electrolyte balance in the body (see page 558).

The osmotic pressure of a solution can be expressed in terms of *osmolarity and depends upon the number of particles in solution. The unit of osmolarity is the osmol. For dilute solutions, osmolarity is

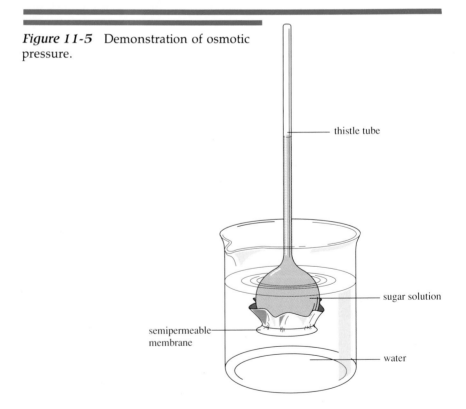

Figure 11-5   Demonstration of osmotic pressure.

thistle tube

sugar solution

semipermeable membrane

water

expressed in milliosmols (mosmol). Body fluids have an osmolarity of about 300 mosmol.

Osmolarity can be calculated from molarity as follows.

$$\text{osmolarity} = \text{molarity} \times \frac{\text{number of particles}}{\text{molecule of solute}}$$

Example 11-10 What is the osmolarity of a 1 M glucose solution?

Glucose is a nonelectrolyte (see page 216); it is undissociated. That is, it produces no ions in solution. Therefore, one molecule of glucose yields one particle in solution. Thus,

$$\text{osmolarity} = \text{molarity} \times \frac{\text{number of particles}}{\text{molecule of solute}}$$

$$= 1 \text{ M} \times 1 = 1 \text{ osmol}$$

Example 11-11 What is the osmolarity of a 1 M NaCl solution?

Each NaCl yields two ions, Na^+ and Cl^-, in solution (see page 217). Thus,

$$\text{osmolarity} = 1 \text{ M} \times 2 = 2 \text{ osmol}$$

Therefore we should expect a 1 M NaCl solution to have twice the osmotic pressure of a 1 M glucose solution, as it does.

Example 11-12 What is the osmolarity of a 0.1 M Na_2SO_4 solution?

$$Na_2SO_4 \longrightarrow 2 Na^+ + SO_4^{2-}$$

Since each molecule of Na_2SO_4 yields three ions in solution,

$$\text{osmolarity} = 0.1 \text{ M} \times 3 = 0.3 \text{ osmol}$$

Example 11-13 What is the osmolarity of physiologic saline solution whose concentration is 0.9 percent NaCl? (0.9 percent NaCl is equivalent to 0.15 M NaCl).

Since each NaCl yields two ions, Na^+ and Cl^-, osmolarity = 0.15 M × 2 = 0.30 osmol.

Then, changing osmol to milliosmol (1 osmol = 1000 mosmol),

$$0.30 \text{ osmol} = 0.30 \text{ \underline{osmol}} \times \frac{1000 \text{ mosmol}}{1 \text{ \underline{osmol}}} = 300 \text{ mosmol}$$

Example 11-14   What is the osmolarity of a 5.5 percent glucose solution? (A 5.5 percent glucose solution is 0.305 M.) Since glucose yields no ions in solution,

$$osmolarity = 0.305 \text{ M} \times 1 = 0.305 \text{ osmol}$$

$$= 0.305 \text{ osmol} \times \frac{1000 \text{ mosmol}}{1 \text{ osmol}} = 305 \text{ mosmol}$$

A cardioplegic solution is used for cardiac instillation during open heart surgery. This buffered solution contains the following electrolytes.

	mEq/L		mEq/L
Ca^{2+}	2.4	Na^+	120
Mg^{2+}	32	Cl^-	160
K^+	16	HCO_3^-	10

The osmolar concentration of this solution is 280 mosmol/L, and the pH[†] is approximately 7.8.

Reverse Osmosis

As was indicated in the previous paragraphs, when water and a salt solution are separated by a semipermeable membrane, osmosis takes place. There is net water diffusion from the weaker to the stronger solution with the result that osmotic pressure is produced. However, if a pressure greater than the osmotic pressure is applied to the salt solution side of the membrane (Figure 11-6), the entire process will be reversed. That is, water will diffuse from the salt solution side to the water side. This process is called **reverse osmosis** and is used in the desalination of seawater, which is not drinkable because of its high salt content.

Figure 11-6 Reverse osmosis.

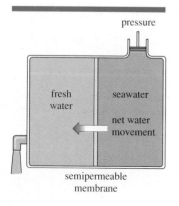

pressure

fresh water

seawater

net water movement

semipermeable membrane

Isotonic Solutions

Two solutions that have the same solute concentration are said to be **isotonic**. The normal salt concentration of the blood is approximately equal to that of a 0.9 percent sodium chloride solution. The common name for 0.9 percent sodium chloride solution is **physiologic saline solution**. The blood and physiologic saline solution are isotonic—they have the same salt concentration. A 5.5 percent glucose solution is also approximately isotonic with body fluids.

Suppose that a red blood cell were placed in a small amount of physiologic saline solution. What would happen? The red blood cell is surrounded by its cell wall, which acts as a semipermeable mem-

† pH will be discussed in Chapter 13.

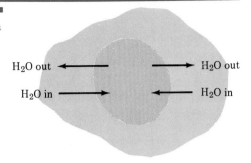

Figure 11-7 Red blood cell in isotonic solution. There is no change in the shape of a cell since water enters and leaves at the same rate.

H$_2$O out ← → H$_2$O out

H$_2$O in → ← H$_2$O in

brane. Will osmosis take place? The answer is no, because there is no difference in concentration on either side of the semipermeable membrane. Actually, the water molecules move in both directions equally with no net change (see Figure 11-7). Thus, physiologic saline solution can be given intravenously to a patient without any effect on the red blood cells (see Figure 11-8). Physiologic saline is administered under the following conditions:

Figure 11-8 Intravenous infusion with physiologic saline solution. [Stock, Boston]

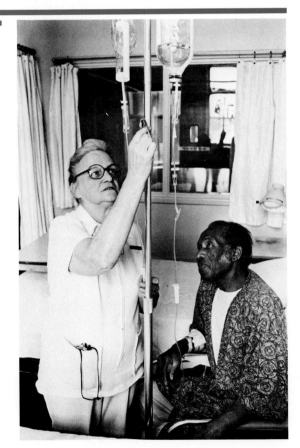

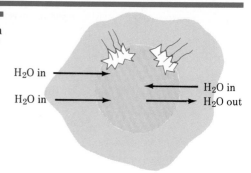

Figure 11-9   Red blood cell in hypotonic solution. The rate of water passing into the cell exceeds the rate of loss of water. As a result the cell distends and may burst.

1. When the patient has become dehydrated.
2. When the patient has lost considerable fluid, as in the case of hemorrhage.
3. To prevent postoperative shock.

Hypotonic Solutions

A **hypotonic** solution is one that contains a lower solute concentration than that of another solution. Distilled water and tap water are hypotonic compared with blood.

Suppose that a red blood cell is placed in water (a hypotonic solution). What will happen? The salt concentration in the red blood cell is higher than that of the water. Therefore, osmosis will take place, with the water diffusing into the red blood cell (from dilute to concentrated solution) (see Figure 11-9). The red blood cell thus enlarges until it bursts. This bursting of a red blood cell because of a hypotonic solution is called *hemolysis. During hemolysis the blood is said to be *laked. Thus a hypotonic solution is *not* usually used for transfusions.

Hypertonic Solutions

A *hypertonic solution is one that contains a higher solute concentration than that of another solution. A 5 percent sodium chloride solution or a 10 percent glucose solution is an example of a hypertonic solution when compared with blood.

Suppose a red blood cell is placed in a hypertonic solution. What will happen? The salt concentration in the red blood cell is less than that in the hypertonic solution. Therefore, osmosis will take place with the water diffusing out of the red blood cell (from dilute to concentrated) (see Figure 11-10). The red blood cell thus shrinks. This shrinking of the red blood cell in a hypertonic solution is called *plasmolysis.

Usually only isotonic solutions can be safely introduced into the bloodstream. Hypotonic solutions can cause hemolysis, and hypertonic solutions can cause plasmolysis.

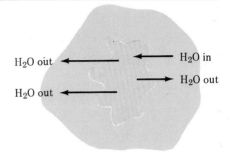

Figure 11-10 Red blood cell in hypertonic solution. The rate at which water leaves the cell exceeds the rate of water entering the cell. This causes the cell to shrink.

Saline *cathartics such as magnesium sulfate, milk of magnesia, and magnesium citrate are absorbed from the large intestine slowly and incompletely. When these substances are ingested, a hypertonic solution is produced in the large intestine and water will diffuse from the tissue spaces into the intestinal tract until the solution is again isotonic with the body fluids. This additional water in the large intestine produces a watery stool that is easily evacuated. Therefore, the continual use of cathartics may cause a patient to become dehydrated. On the same basis, cathartics have been used to rid the body of excess fluid, although diuretics are now preferred for this purpose.

Suspensions

Suppose that some powdered clay is placed in water and vigorously shaken. A suspension of clay in water will be produced. This suspension will not be clear; it will be opaque. Upon standing, the clay will slowly settle. The composition of the suspension is actually changing as the clay settles out, so it is a heterogeneous mixture.

The clay is not dissolved in the water; it is merely suspended in it. When the suspension is poured into a funnel lined with filter paper, only the water passes through the filter paper; the clay does not. Evidently the suspended clay particles are too large to pass through the holes in a piece of filter paper. Undoubtedly they will also be too large to pass through a membrane that has even finer openings (see Figure 11-11).

Properties of suspensions are summarized as follows. Suspensions

1. Consist of an insoluble substance dispersed in a liquid.
2. Are heterogeneous.
3. Are not clear.
4. Settle.
5. Do not pass through filter paper.
6. Do not pass through membranes.

Some medications, such as milk of magnesia, are administered as a suspension. Many bottles of medication state on the label "shake before using." Most suspensions use water as the suspending medium, but procaine penicillin G, for example, is usually administered as an oil suspension.

A mist is a suspension of a liquid in a gas. Water droplets suspended in air are one example of a mist.

Patients with a decreased water content in their lungs usually have thickened bronchial secretions. To increase the water content of the lungs requires a higher-than-normal water content in the inspired air. A nebulizer is a device that generates an aerosol mist consisting of large water particles that can penetrate into the trachea and large bronchi. An ultrasonic nebulizer produces a supersaturated mist by means of ultrasonic sound waves. Such a mist is very effective in inhalation therapy because the particle size is small enough to reach the smallest *bronchioles. Also, the concentration of the mist is greater than that produced by other means.

Care must be taken in the use of a nebulizer to avoid (1) bacterial contamination in the water reservoir of the equipment (such bacteria will go directly into the bronchi) and (2) excess water being added to the patient's lungs.

Colloids

The third class of liquid mixtures, called colloids, consists of tiny particles suspended in a liquid. We might ask, if these particles are suspended in a liquid, why aren't they suspensions? The answer is that these colloids behave quite differently from ordinary suspensions and have an entirely different set of properties.

Size

When a colloid (or a colloidal dispersion as it is frequently called) is poured into a funnel lined with filter paper, the colloid passes through the filter paper. This indicates that the colloidal particles are smaller than the openings in a filter paper. When a colloid is placed in a membrane, the colloidal particles do not pass through. Thus, colloidal particles are larger than the openings in the membrane.

We know that solutions pass through filter paper. They can also pass through certain types of membranes, whereas suspensions pass through neither. Therefore colloidal particles must be intermediate in size between solution particles and suspension particles. Colloidal particle sizes are measured in nanometers (nm) (1 nanometer is 0.000000001 m or one-millionth of a millimeter). Colloidal particles range in size from 1 to 100 nm. Solution particles are smaller than 1 nm, whereas suspension particles are larger than 100 nm (see Figure 11-11).

Colloids have a tremendous amount of surface area because

Figure 11-11 Relative sizes of solution, colloid, and suspension particles. Particles in solution are ions or molecules; colloidal particles consist of small clumps of molecules; suspension particles consist of large clumps of molecules.

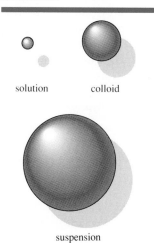

solution   colloid

suspension

they consist of so many tiny particles. Consider a cube 1 cm on an edge. The volume of such a cube is 1 cubic centimeter (cm^3) and the surface area is 6 square centimeters (cm^2). (There are six faces to a cube, and each face has an area of 1 cm^2.

If the cube is cut in half, the total volume of both pieces is still 1 cm^3. However, the total surface area has now increased to 8 cm^2 (cutting introduces two new faces each having an area of 1 cm^2). Cutting in half again still retains the volume of 1 cm^3. The surface area now is 10 cm^2. Continued cutting further increases the surface area.

When the size of the individual particle reaches colloidal size— approximately 10 nm—the total amount of surface area is about 60,000,000 cm^2. This tremendous surface area gives colloids one of their most important properties—adsorption.

Adsorption

Adsorption is defined as the property of holding substances to a surface. Colloidal charcoal will adsorb tremendous amounts of gas. It has a selective adsorption, as do most colloids. Coconut charcoal is used in gas masks because it selectively adsorbs poisonous gases from the air. It does not adsorb ordinary gases from the air.

Several commonly used medications owe their use to their adsorbent properties. Charcoal tablets are administered to patients to aid digestion. Kaolin, a finely divided aluminum silicate, is administered for the relief of diarrhea. Colloidal silver adsorbed on protein (Argyrol) has been used as a germicide.

Electric Charge

Almost all colloidal particles have an electric charge—either positive or negative. Why? Colloids selectively adsorb ions on their surface. If a colloid selectively adsorbs negative ions, it becomes a negatively charged colloid. If it adsorbs positive ions, it becomes a positively charged colloid.

Consider two negatively charged colloidal particles suspended in water. What would happen if these particles were to come close together? They would repel each other because of the repulsion of their like charges. Therefore, these colloidal particles will have little tendency to form large particles that would then settle out.

How can colloids be made to settle? Colloidal particles have an electric charge that will repel all other similarly charged colloidal particles. However, these charged particles *will* attract particles of opposite charge. Therefore, a negative colloid can be made to coagulate (begin to settle out) by adding to it positively charged particles.

Bichloride of mercury ($HgCl_2$) is a poisonous substance. When swallowed, it forms a positive colloid in the stomach. The antidote for this type of poisoning is egg white, which is a negative colloid. These two oppositely charged colloids neutralize each other and coagulate in the stomach. The stomach must then be pumped out to remove the coagulated material. If this is not done, the stomach will digest the egg white, exposing the body once again to the poisonous substance.

Protein can also be coagulated by heat. Egg white—a colloidal substance—is quickly coagulated when heated.

The charge of a colloid can be determined by placing it in a U tube containing two electrodes. When a current is passed through the U tube, each electrode will attract particles of opposite charge. A negative colloid will begin to accumulate around the positive electrode, and a positive colloid around the negative electrode. The movement of electrically charged suspended particles toward an oppositely charged electrode is called **electrophoresis.** Electrophoresis is a slow process because the charged particles are not soluble and are much larger than the particles in solution (see page 548).

Tyndall Effect

When a strong beam of light is passed through a colloidal dispersion, the beam becomes visible because the colloidal particles reflect and scatter the light. This phenomenon is called the Tyndall effect (see Figure 11-12). The Tyndall effect can also be observed when a beam of sunlight passes through a darkened room. The dust particles in the air scatter the light so that the sun's rays become visible. The blue color of the sky is due to scattered light, as is the blue color of the ocean. However, we actually do not see the colloidal particles; we merely see the light scattered by them.

When a strong beam of light is passed through a solution, no Tyndall effect is observed because the solution particles are too small to scatter the light. Thus, the Tyndall effect is a way of distinguishing between solutions and colloids.

Brownian Movement

When colloidal particles are observed with a transmission electron microscope, the particles are seen to move in a haphazard irregular motion called Brownian movement (see Figure 11-13).

What causes this irregular motion? It cannot be caused by the vibration of the slide on a microscope stand because the same motion can be observed when the microscope is mounted on a concrete pillar sunk deep into the earth. Brownian movement can be observed during the day or night, in the city and the country, in warm weather and cold weather, at high altitudes and low altitudes. The strangest characteristic of all is that this particular motion *never* ceases. It can be observed in containers that have been sealed for years. It has also been observed in liquid *occlusions in quartz samples that have been undisturbed for perhaps millions of years.

Careful investigation by scientists established the fact that Brownian movement is not due to the colloidal particles themselves but

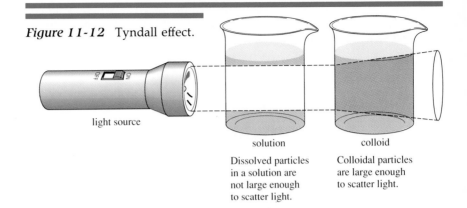

Figure 11-12 Tyndall effect.

light source

solution

colloid

Dissolved particles in a solution are not large enough to scatter light.

Colloidal particles are large enough to scatter light.

rather to their bombardment by the molecules of the suspending medium. The molecules of the suspending medium are in continuous random motion. When these rapidly moving molecules of the medium strike the colloidal particles, they cause these particles to have the random motion characteristic of Brownian movement.

Gels and Sols

Colloidal dispersions can be subdivided into two classes. Where there is a strong attraction between the colloidal particle and the suspending liquid (water), the system is said to be *hydrophilic (water loving). Systems of this type are called **gels.** Gelatin in water is an example of such a system. Gels are semisolid and semirigid; they do not flow easily.

A colloidal system where there is little attraction between the suspended particles and the suspending water is *hydrophobic (water hating). Systems of this type are called **sols.** They pour easily. A

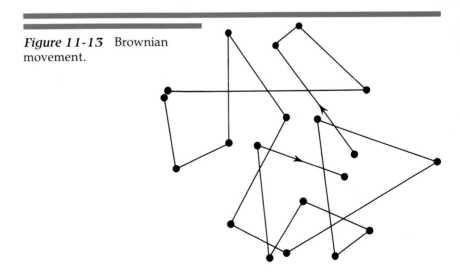

Figure 11-13 Brownian movement.

small amount of starch in water forms a sol. A hydrosol is a colloidal dispersion in water, while an aerosol is a colloidal dispersion in air.

If a gel, such as gelatin, is heated, it turns into a sol but returns to its original gel state on cooling. Protoplasm has the ability to change gel (in membranes) into sol and vice versa.

Dialysis

Dialysis is the separation of solute particles from colloidal particles by means of a semipermeable membrane. Recall that solute particles can pass through semipermeable membranes but colloidal ones cannot. Suppose that a colloidal starch suspension and sodium chloride solution are placed inside that type of membrane, which in turn is placed in a beaker of water. The starch is a colloid and cannot pass through the membrane. The salt is in solution and does pass through the membrane. This is an example of diffusion. The salt will continue to pass through the membrane until the salt concentration inside the membrane is the same as that in the water surrounding the membrane.

When this happens, no more salt will be removed from the mixture inside the membrane bag. However, if the bag is suspended in running water, soon all the salt will be removed from the inside, leaving behind only the starch—the colloid. This is dialysis—the separation of a solute from a colloid by means of a membrane (see Figure 11-14). *Antitoxins are prepared by this method. The impure material is placed inside a container made of membrane suspended in running water. The soluble impurities diffuse out through the membrane, leaving the pure antitoxin behind. The same process is used to prepare low-sodium milk. Milk enclosed in a membrane is

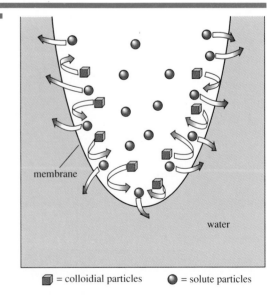

Figure 11-14 Dialysis.

membrane

water

▣ = colloidial particles   ● = solute particles

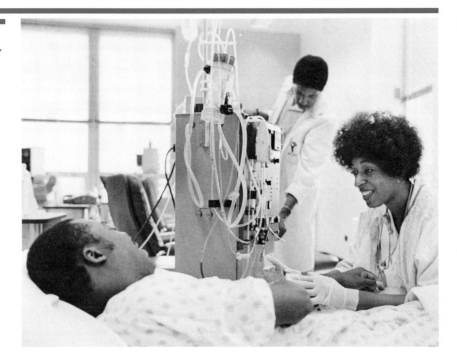

Figure 11-15 Artificial kidney machine. [Courtesy Michael Reese Hospital and Medical Center, Chicago, IL.]

suspended in running water. The soluble salt (sodium chloride) leaves the milk, so the remaining liquid is practically free of sodium compounds.

In the body the membranes in the kidneys allow the soluble waste material to pass through. The same membranes do not allow protein to pass through, since proteins are colloids. In an average 150-lb adult, approximately 180 L of blood is purified daily.

Hemodialysis

Hemodialysis refers to the removal of soluble waste products from the bloodstream by means of a membrane. Purification of the blood can be accomplished in this way because soluble particles can diffuse through dialyzing membranes, whereas blood cells and plasma proteins cannot. When a patient has problems related to renal excretion, an artificial kidney machine can be used (see Figure 11-15). This machine applies the principles of hemodialysis.

The artificial kidney machine consists of a long cellophane tube wrapped around itself to form a coil and immersed in a temperature-controlled solution whose chemical composition is carefully regulated according to the needs of the patient. The patient's blood is pumped through the coil, and the soluble end products of protein catabolism, water, and *exogenous poisons are removed from the blood. At the same time the blood cells and plasma proteins remain in the blood (recall that proteins, which are colloids, cannot pass through a membrane, and here the cellophane acts as the membrane). If the solution in which the coil is immersed has the same concentration of sodium ions as the blood, no net diffusion of sodium ions will take place. The same applies to other soluble substances in the bloodstream. That is, by regulating the composition of the solution, the waste products and unwanted material can be removed from the bloodstream.

A patient may remain on the artificial kidney machine 4 to 7 hr or even more. During this time the solution must be changed at intervals to avoid accumulation of waste products. Otherwise, the waste products would no longer continue to diffuse out of the blood. After passing through the coil, the blood is returned to the patient's veins.

As we might expect, the dialysate is mostly water. The kind of water used in its preparation is of utmost importance to the patient. If the water has been softened by the zeolite process (see page 178), it may contain too high a concentration of sodium ions (and so add sodium to the blood; see page 561). Improper levels of calcium and magnesium ions in the water can lead to metabolic bone diseases. Traces of copper in the water (from copper tubing in the equipment copper pipes in the water supply, or copper sulfate used to kill algae in the water treatment plant) have caused fatalities in dialysis patients.

Emulsions

When two liquids are mixed, either they dissolve in each other or they do not. Two liquids that are soluble in each other are **miscible**. Two liquids that do not dissolve in each other are **immiscible**.

Suppose two immiscible liquids such as oil and water are poured together and then shaken vigorously. The oil forms tiny drops, which are suspended in the water. After a while the tiny drops come together to form larger drops; these soon rise to the top and separate from the water. This type of liquid mixture is called a **temporary emulsion**—emulsion because it consists of a liquid colloidally suspended in a liquid, and temporary because it separates.

What are the properties of a temporary emulsion? First, as was just noted, it separates. Since it separates, it must be heterogeneous; its composition continues to change. Temporary emulsions, such as oil in water, are not clear. If placed in a funnel containing a piece of filter paper, they do not pass through the filter paper. Neither do they pass through any membrane.

A temporary emulsion such as oil and water separates because the oil drops attract one another to form larger drops, which soon rise to the top. However, suppose that each drop of oil were given a negative charge. Then the drops would not attract one another. They would not settle out; they would form a **permanent emulsion**. To give the oil drops an electric charge we need an emulsifying agent. An emulsifying agent is a protective colloid that coats the suspended oil drops and prevents them from coming together. Thus the surface area of oil is greatly increased by the addition of an emulsifying agent. Permanent emulsions are not clear; they are homogeneous. They do not settle; they do not pass through either filter paper or membranes.

An example of a permanent emulsion is mayonnaise, which is an emulsion of oil in vinegar with egg yolk as an emulsifying agent. Milk is an emulsion of butterfat in water, with casein acting as the emulsifying agent. Soap acts as an emulsifying agent on grease and oils in water. Many medications are given in the form of emulsions, gum acacia being the most common emulsifying agent for the dispersion of fats.

Summary

Solutions consist of a solute and a solvent. Solutions are clear and homogeneous, have a variable composition, do not settle, may be separated by physical means, and pass through filter papers.

Factors affecting the solubility of a solute are temperature, pressure, surface area, agitation, and nature of the solvent.

Many chemical reactions take place in solution. Many medications are administered as solutions.

Solutions may be labeled as dilute or concentrated, as saturated or unsaturated, all of which are relative terms and do not indicate any definite amount of solute and solvent.

A percentage solution indicates the number of grams of solute per 100 mL of solution. A ratio solution indicates the number of grams of solute and the number of milliliters of solution. A molar solution indicates the number of moles of solute per liter of solution. A normal solution indicates the number of gram equivalent weights of solute per liter of solution.

Concentrations of ions in body fluids are frequently expressed in the units milliequivalents per liter (mEq/L).

When a nonvolatile solute is dissolved in a solvent, the boiling point is elevated and the freezing point is depressed.

All liquids exhibit surface tension, which causes the surface molecules to form a surface film.

The movement of solute into a solvent or through a solution is called diffusion. The flow of solvent through a semipermeable membrane is called osmosis. The pressure exerted during osmosis is called osmotic pressure. Osmotic pressure is expressed in terms of the osmolarity of the solution.

An isotonic solution has the same salt concentration as blood and is used for transfusions.

A hypotonic solution has a solute concentration less than that of blood. If injected into the bloodstream, a hypotonic solution can cause hemolysis—the bursting of the red blood cells.

A hypertonic solution has a solute concentration greater than that of blood. If injected into the bloodstream, a hypertonic solution may cause plasmolysis—the shrinking of the red blood cells.

Suspensions consist of a nonsoluble solid suspended in a liquid medium. Suspensions are not clear; they settle out; they are heterogeneous; they do not pass through a filter paper; and they do not pass through a membrane.

Colloids consist of tiny particles suspended in a liquid. Colloids do not settle; they pass through filter paper but not through membranes; they adsorb (hold) particles on their surface; they have electric charges, owing to the adsorption of charged particles (ions); they exhibit the Tyndall effect and Brownian movement.

Colloidal dispersions can be subdivided into two classes—sols and gels.

Dialysis is the separation of a solute from a colloid by means of a semipermeable membrane.

Hemodialysis refers to the removal of soluble waste products from the bloodstream by means of a membrane. When a patient has problems related to renal excretion, an artificial kidney machine may be used.

An emulsion consists of a liquid suspended in a liquid. An emulsion that settles is called a temporary emulsion When an emulsifying agent is added to a temporary emulsion, it becomes a permanent emulsion.

Questions and Problems

A

1. Give specific directions for preparing
   (a) 100 mL of 5 percent boric acid solution.
   (b) a 1:300 KMnO$_4$ solution.
   (c) 2 L of 1 M NaCl solution.
   (d) 500 mL of 0.9 percent NaCl solution.
2. Define (a) osmosis; (b) dialysis; (c) adsorption; (d) osmotic pressure; (e) hypertonic solution; (f) diffusion; (g) saturated solution.
3. Why must an isotonic solution be used during a blood transfusion?
4. What factors affect the rate of solution of a solid solute?
5. List the general properties of solutions.
6. List the general properties of suspensions.
7. List the general properties of colloids.
8. List the general properties of emulsions.
9. What is the Tyndall effect?
10. What is Brownian movement?
11. What is surface tension?
12. How do colloids obtain their electric charge?
13. What effect does a solute have on the boiling point and on the freezing point of a solution?
14. Explain how an artificial kidney machine functions.
15. Why must the solution in an artificial kidney machine be changed at intervals during use over a long time?
16. Why does diffusion take place more rapidly in a gas than in a liquid?
17. How can a temporary emulsion be changed into a permanent emulsion?
18. How can a colloid be made to settle? Explain.

19. Why do colloids pass through filter paper but not through membranes?
20. What is the size range of colloidal particles? solution particles? suspension particles?
21. What is meant by the term *tincture*?
22. Does pressure always affect the amount of a solute that will dissolve in a given solvent? Explain.
23. How can you tell whether a solution is saturated or unsaturated?
24. What is a nebulizer? For what purposes is it used in the hospital? What precautions should be taken with its use?
25. Why is the control of water important in hemodialysis? Would you expect control of temperature also to be important? Explain.
26. What is the difference between a sol and a gel? Give an example of each.
27. How can a supersaturated solution be prepared? Is it stable? Explain.
28. A patient is given 1500 mL of 10 percent glucose solution intravenously in a 24-hr period. How many grams of glucose did the patient receive per hour?
29. How many grams of NaCl (MW 58.5) are present in 100 mL of 0.5 M solution?
30. A 0.125 M solution of Al$_2$(SO$_4$)$_3$ will contain how many milliequivalents per liter of each ion?

$$Al_2(SO_4)_3 \longrightarrow 2\ Al^{3+} + 3\ SO_4{}^{2-}$$

31. How would you prepare 1.5 L of 1.2 N KOH solution?
32. How would you prepare 50 mL of 0.25 N HNO$_3$ solution?

B

33. Which substance of those shown in Figure 11-1 has the greatest increase in solubility with an increase in temperature?

34. From Figure 11-1 calculate the number of grams of $NaNO_3$ that will dissolve in 500 g of water at 20 °C.

35. What is an alloy? an amalgam? Give one medical use for each.

36. How does a percent solution compare with a milligram percent solution?

37. Give specific directions for preparing
    (a) 500 mL of 2 percent KI solution from 10 percent stock KI solution.
    (b) 200 mL of 1:100 solution from 1:10 stock solution.

38. What is a surfactant? Explain why surfactants are needed in the body.

39. What is adhesion? cohesion? capillary action? viscosity?

40. What is the purpose of a blood extender?

41. What is reverse osmosis? How does it work?

42. Explain how drinking saltwater can cause dehydration.

43. Calculate the osmolarity of the following solutions.
    (a) 0.5 M sucrose solution
    (b) 0.1 M K_3PO_4 solution
       $(K_3PO_4 \rightarrow 3 K^+ + PO_4^{3-})$.
    (c) 10 percent glucose solution (0.55 M).

44. Upon what basis can the solubility of a solute in a solvent be predicted? Explain.

Practice Test

1. A 10 percent glucose (MW 180) solution will contain how many grams of glucose per 100 mL?

    a. 1   b. 5   c. 10   d. 18

2. When a nonvolatile solute is dissolved in water, the boiling point _____.
   a. increases
   b. decreases
   c. is unaffected

3. A solution that has the same salt concentration as blood is said to be _____.
   a. hypotonic   b. hypertonic
   c. isotonic   d. normal

4. A type of solution that can cause hemolysis is _____.
   a. hypertonic   b. hypotonic
   c. isotonic   d. normal

5. Colloids _____.
   a. settle
   b. pass through filter paper
   c. pass through membranes
   d. all of these

6. The separation of a solution from a colloid by means of a semipermeable membrane is called _____.
   a. osmosis   b. hemolysis
   c. electrolysis   d. dialysis

7. How many grams of glucose (MW 180) are required to prepare 500 mL of a 0.1 M solution?
   a. 1   b. 9   c. 18   d. 36

8. How many grams of NaOH (MW 40) are required to prepare 2 L of a 0.1 N solution?
   a. 4   b. 8   c. 40   d. 80

9. How many milliliters of 10 percent NaCl solution will be used to prepare 400 mL of a 1 percent solution?
   a. 10   b. 40   c. 100   d. 400

10. Colloids exhibit _____.
    a. Brownian movement
    b. Tyndall effect
    c. both of these effects
    d. neither of these effects

12

Ionization

Spreading salt on an icy sidewalk causes the ice to melt.

Conductivity of Solutions

Figure 12-1 illustrates an apparatus designed to show whether a liquid will conduct electricity or not. When two metal plates called electrodes are immersed in a liquid and the current is turned on, the bulb will light if the liquid conducts a current. If a liquid does not conduct a current, the bulb will not glow at all.

It can be shown experimentally that the only water solutions that conduct electricity are those of acids, bases, and salts. The substances whose water solutions conduct electricity are called **electrolytes.** Substances whose water solutions do not conduct electricity are called nonelectrolytes. Pure water is a nonelectrolyte, as is alcohol. Sugar solutions are also nonelectrolytes.

Effect of Electrolytes on Boiling Points and Freezing Points

Electrolytes have an unusual effect on the boiling point and the freezing point of a solution compared to the boiling point and the freezing point of the solvent itself. When a solid compound is dissolved in water, the resulting solution has a boiling point above 100 °C. That is, solid solutes raise the boiling point of water. However, the results are considerably exaggerated if the solute is an electrolyte rather than a nonelectrolyte.

Figure 12-1 Conductivity apparatus.

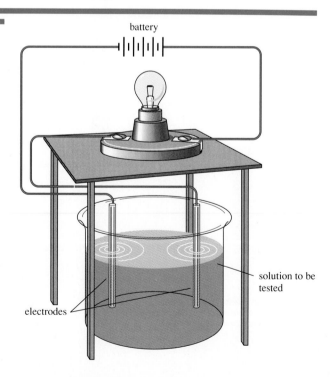

Consider a 1 M solution of an electrolyte such as sodium chloride and a 1 M solution of a nonelectrolyte such as sugar. The increase in the boiling point of the water containing the electrolyte sodium chloride will be approximately twice that for the water containing the nonelectrolyte sugar.

Likewise, the freezing point of the sodium chloride solution will be lowered approximately twice as much as that of the sugar solution.

Why do electrolytes conduct electricity? Why do they have an effect on the boiling point and freezing point of the solution?

Theory of Ionization

In 1887, a Swedish chemist, Svante Arrhenius, proposed a theory to explain the behavior of electrolytes in solution. The main points of his theory are as follows:

1. When electrolytes are placed in water, the molecules of the electrolyte break up into particles called ions. This process he called ionization.
2. Some of the ions have a positive charge, others a negative charge.
3. The sum of the positive charges is equal to the sum of the negative charges. That is, the original molecules were neutral, so the sum of the charges making up this molecule must also be neutral.
4. The conductance of electricity by solutions of electrolytes is due to the presence of ions.
5. Nonelectrolytes do not conduct electricity because of the absence of ions.
6. The effect of electrolytes on the boiling point and on the freezing point of a solution is due to the increased number of particles (ions) present in the solution.

Arrhenius proposed that an equilibrium exists between the ions and the nonionized molecules. This equilibrium can be represented by the equation

$$\underset{\substack{\text{sodium chloride} \\ \text{molecule}}}{NaCl} \rightleftharpoons \underset{\substack{\text{sodium} \\ \text{ion}}}{Na^+} + \underset{\substack{\text{chloride} \\ \text{ion}}}{Cl^-}$$

Arrhenius also proposed that the ionization should increase as the solution becomes more dilute. That is, in extremely dilute solutions, the electrolyte molecule would be almost completely dissociated (ionized).

Arrhenius's theory had to be modified as more information became available. Yet the modern theory of ionization retains most of his principles.

Arrhenius believed that the molecules of an electrolyte such as sodium chloride broke up into ions when placed in water. Later evidence proved conclusively that sodium chloride exists as ions even in the solid state. When sodium chloride is formed from its elements, the sodium atom loses one electron to form a positively charged sodium ion. At the same time the chlorine atom gains one electron to form a negatively charged chloride ion. Thus, sodium chloride exists in an ionic state even in the solid state. When sodium chloride is dissolved in water, the ions are free to move around.

Solid sodium chloride consists of sodium ions and chloride ions bonded together by ionic bonds in a definite crystalline structure. These ions are held tightly in place and cannot move about to any great extent. Thus they are not free to conduct an electric current, and solid sodium chloride does not conduct electricity. If sodium chloride is heated until it melts (about 800 °C), the resulting liquid will conduct electricity because the ions have some freedom to move around.

Conductivity of Solutions of Electrolytes

When sodium chloride crystals are placed in water, the ions present in the crystal are free to move about in the solution. These positively and negatively charged ions will be attracted to oppositely charged electrodes. Thus, if two electrodes are connected to a battery and are placed in a sodium chloride solution, the solution will conduct electricity.

The positively charged sodium ions will be attracted to the negative electrode, the cathode. Ions attracted to a cathode are called **cations.** At the same time, the negatively charged chloride ions will be attracted to the positive electrode, the anode. Ions attracted toward an anode are called **anions.** This movement of ions through the solution consists of a flow of current. If a nonelectrolyte is placed in water, no ions are formed and no conductance takes place.

Strong and Weak Electrolytes

Acids are electrolytes. However, all acids do not behave the same when placed in water. A dilute solution of hydrochloric acid (HCl) is a strong electrolyte. When a beaker containing it is placed under a conductivity apparatus, the bulb glows brightly. However, when a beaker containing a dilute solution of acetic acid ($HC_2H_3O_2$) is placed under the conductivity apparatus, the bulb glows dimly. This indicates that the acetic acid is a weak electrolyte. Likewise, sodium hydroxide (NaOH) solution is a strong electrolyte, whereas ammonium hydroxide (NH_4OH) solution is a weak electrolyte.

How can one acid or base be a strong electrolyte and another one be weak? What accounts for the difference between the types of electrolytes?

When hydrochloric acid or sodium hydroxide is placed in water, it dissociates (breaks up) almost completely into ions, as indicated by the arrows pointing in one direction only.

$$HCl \longrightarrow H^+ + Cl^-$$
$$NaOH \longrightarrow Na^+ + OH^-$$

Because these substances are just about completely ionized, they are called strong electrolytes.

However, some acids and bases ionize only to a limited extent when they are placed in water. They remain primarily as nonionized molecules. These substances maintain an equilibrium between the nonionized molecule and the ions, with the equilibrium being far to the left, as indicated by the arrows.

$$HC_2H_3O_2 \rightleftharpoons H^+ + C_2H_3O_2^-$$
$$NH_4OH \rightleftharpoons NH_4^+ + OH^-$$

These substances are called **weak electrolytes.**

Thus the original Arrhenius theory is true for weak electrolytes but not for strong ones.

Other Evidence of Ionization

It has already been mentioned that the conductivity of solutions of electrolytes indicates the presence of ions. Likewise, the effect of electrolytes on the boiling point and the freezing point can be explained by the presence of ions. The change in the boiling point or freezing point depends on the number of particles present in solution. Sugar ($C_6H_{12}O_6$) contributes only one particle per molecule because it is not ionized. Sodium chloride (NaCl) contributes two particles—the two ions—and so should have twice the effect on the boiling point and the freezing point.

Another factor indicating the presence of ions is the instantaneous reaction of solutions of electrolytes. When a solution of sodium chloride (NaCl) is mixed with a solution of silver nitrate ($AgNO_3$), a white precipitate of silver chloride (AgCl) is formed instantaneously.

$$Na^+ + Cl^- + Ag^+ + NO_3^- \longrightarrow AgCl(s) + Na^+ + NO_3^-$$

The silver chloride is written as a molecule rather than as ions. This is because that substance precipitates, thus removing the ions from the solution.

When an acid reacts with a base, a salt and water are formed. If potassium hydroxide, KOH, is reacted with nitric acid, HNO_3, potassium nitrate and water are formed. The ionic reaction is

$$K^+ + OH^- + H^+ + NO_3^- \longrightarrow K^+ + NO_3^- + H_2O$$

Potassium nitrate is a soluble salt, an electrolyte, which is ionized. Water is a nonelectrolyte; it is not ionized and is therefore written as a molecule.

Hydrolysis of salts (see page 239) can also be explained on the basis of ionization. When sodium acetate, $NaC_2H_3O_2$, is placed in water, the acetate ion reacts with the water according to the equation

$$Na^+ + \underset{\text{acetate ion}}{C_2H_3O_2^-} + H_2O \rightleftharpoons \underset{\text{acetic acid}}{HC_2H_3O_2} + Na^+ + OH^- \quad (12\text{-}1)$$

The acetic acid, being a weak acid, does not ionize appreciably to furnish any hydrogen ions and so is written as a molecule in equation (12-1). If the ions common to both sides of equation (12-1) (the sodium ions) are eliminated, the net reaction becomes

$$C_2H_3O_2^- + H_2O \rightleftharpoons HC_2H_3O_2 + OH^- \quad (12\text{-}2)$$

That is, the acetate ion (which came from a weak acid) hydrolyzes (reacts with water) to form nonionized acetic acid, a weak electrolyte, leaving hydroxide ions in solution. Therefore, when sodium acetate is placed in water, hydroxide ions are formed, causing the solution to be basic. Note that ions from a strong acid or a strong base (see page 239) do not hydrolyze. Thus, the sodium ion in reaction (12-1) does not react with water and so can be eliminated from both sides of the equation, giving equation (12-2).

Importance of Ions in Body Chemistry

Ions play the chief role in the various processes that take place in the body. Many of the body's vital processes take place in the ionic state within the cell. Table 12-1 lists some of the more important ions that are found in the body.

In addition, ions are necessary as part of the blood buffer system. They cause osmotic pressure in the cells and are necessary to control the contraction and relaxation of muscles. Ions are necessary to carry nerve impulses and help regulate the digestive processes (see pages 554 and 561).

Summary

Substances whose water solutions conduct electricity are called electrolytes. Soluble acids, bases, and salts are electrolytes. Solutions that do not conduct electricity are nonelectrolytes.

Electrolytes have an effect on the boiling and freezing points of a solution. This effect is due to the presence of ions.

Ionization (the formation of ions) accounts for the electrical conductivity of a solution. When a substance ionizes, the sum of the positive charges equals the sum of the negative charges. The positive ions, cations, are

Table 12-1 Ions Found in the Body	Calcium ion	Ca^{2+}	Necessary for clotting of the blood; for formation of milk curd during digestion in the stomach; for formation of bones and teeth; for action of muscle, including heart
	Iron ion	Fe^{2+}	Necessary for formation of hemoglobin and cytochromes
	Sodium ion	Na^+	Principal extracellular positive ion
	Potassium ion	K^+	Principal intracellular positive ion
	Chloride ion	Cl^-	Intracellular and extracellular negative ion
	Bicarbonate ion	HCO_3^-	Extracellular negative ion
	Iodide ion	I^-	Present in thyroid hormones
	Ammonium ion	NH_4^+	Plays a role in maintaining body's acid-base balance
	Phosphate ion	PO_4^{3-}	Plays an important role, along with calcium ions, in the formation of bones and teeth
	Magnesium ion	Mg^{2+}	An important activator for many enzyme systems (See Table 2-3, page 25, for other ions necessary for enzymes.)

attracted toward the cathode (negative electrode), whereas the negative ions, the anions, are attracted toward the anode (positive electrode).

The modern theory of ionization considers that ions are already present in a crystalline salt and that these ions are released to move about when that substance is placed in solution.

Most salts are strong electrolytes because they are completely (strongly) ionized. Acids and bases that are strong electrolytes are highly ionized. Acids and bases that are poor electrolytes are weakly ionized.

The presence of ions is of great importance in maintaining the electrolyte balance of body fluids.

Questions and Problems

A

1. Describe a laboratory experiment to determine whether or not a substance is an electrolyte.
2. What effect does an electrolyte have on the boiling point of a solution? Why?
3. What effect does an electrolyte have on the freezing point of a solution? Why?
4. What effect does a nonelectrolyte have on the freezing point of a solution? Why?
5. State the main ideas of Arrhenius's theory of ionization.
7. How does an electrolyte conduct a current?
8. What is an anion? a cation?
9. What is a strong electrolyte? Give an example.
10. What is a weak electrolyte? Give an example.
11. List several ions necessary for the proper functioning of the body. What purpose does each ion serve?

B

12. When an insoluble salt is placed in water, will the mixture conduct electricity? Explain.
13. Compare the osmolarity of equal molar solutions of glucose and sodium chloride. What effect will these solutions have on the boiling point?
14. Explain why solid sodium chloride does not conduct electricity whereas molten sodium chloride does.
15. Compare reversible and irreversible reactions in terms of strong and weak electrolytes.

Practice Test

1. A solution that conducts electricity is called a(n) _____.
   a. normal solution
   b. molar solution
   c. electrolyte
   d. hydrolyte

2. Ions attracted to a positively charged electrode are called _____ .
   a. anions
   b. cations
   c. electrolytes
   d. hydrolytes

3. An ion necessary for the proper functioning of the thyroid gland is _____ .
   a. K^+ b. Cl^- c. Na^+ d. I^-

4. An electrolyte has what effect upon the boiling point of a solution?
   a. increases it
   b. decreases it
   c. has no effect

5. An ion required for the formation of hemoglobin is _____ .
   a. Ca^{2+}
   b. K^+
   c. Fe^{2+}
   d. HCO_3^-

6. Ions attracted to a negatively charged electrode are called _____ .
   a. anions
   b. cations
   c. pions
   d. muons

7. A solution that conducts electricity does so because of the presence of _____ .
   a. hydrogen bonds
   b. polarization
   c. hydration
   d. ions

8. The freezing point of a solution of an electrolyte is _____ that of a solution of a nonelectrolyte of the same concentration.
   a. higher than
   b. lower than
   c. the same as

9. An ion necessary for the formation of bones and teeth is _____ .
   a. Ca^{2+} b. Mg^{2+} c. Fe^{2+} d. I^-

10. An ion necessary for the formation of bones and teeth is _____ .
    a. Cl^- b. PO_4^{3-} c. HCO_3^- d. SO_4^{2-}

13

Acids and Bases

Titration of an acid in the flask with a base in the buret.

Acids

Acids can be defined as compounds that yield or donate hydrogen ions (H^+) in a water solution. Brønsted (a Danish chemist) **defined an acid as a substance that donates protons.** Note that a hydrogen ion results when a hydrogen atom loses its only electron, leaving a single proton, so both definitions say essentially the same thing. That is, a hydrogen ion is a proton.

It is the hydrogen ions that are responsible for the particular properties of acids. Most people think of acids as being liquids. However, there are many solid acids; boric acid and citric acid are two common examples. The names and formulas of the most common acids are given in Table 13-1.

Properties of Acids

Acids yield hydrogen ions when placed in water solution.

$$HCl \longrightarrow H^+ + Cl^-$$

hydrochloric   hydrogen   chloride
acid   ion   ion

$$HNO_3 \longrightarrow H^+ + NO_3^-$$

nitric   hydrogen   nitrate
acid   ion   ion

$$HC_2H_3O_2 \rightleftharpoons H^+ + C_2H_3O_2^-$$

acetic   hydrogen   acetate
acid   ion   ion

Recall that strong acids are almost completely ionized in solution, whereas weak acids are only partially ionized.

Hydrochloric, nitric, and acetic acids are called *monoprotic acids* because each molecule yields one hydrogen ion in solution.

Sulfuric acid (H_2SO_4) is an example of a *diprotic acid*; it yields two hydrogen ions per molecule.

$$H_2SO_4 \longrightarrow H^+ + HSO_4^-$$

sulfuric   hydrogen   bisulfate
acid   ion   ion

$$HSO_4^- \rightleftharpoons H^+ + SO_4^{2-}$$

bisulfate   hydrogen   sulfate
ion   ion   ion

Table 13-1 Common Acids

HCl	Hydrochloric acid
H_2SO_4	Sulfuric acid
HNO_3	Nitric acid
H_2CO_3	Carbonic acid
H_3PO_4	Phosphoric acid
$HC_2H_3O_2$	Acetic acid

Sulfuric acid is a strong electrolyte, as indicated by the single arrow, which shows complete ionization. HSO_4^- is a weak acid, as indicated by its equilibrium reaction.

Phosphoric acid (H_3PO_4) is a *triprotic acid*, as the following equations indicate.

$$H_3PO_4 \rightleftharpoons H^+ + H_2PO_4^-$$

phosphoric acid ・ hydrogen ion ・ dihydrogen phosphate ion

$$H_2PO_4^- \rightleftharpoons H^+ + HPO_4^{2-}$$

dihydrogen phosphate ion ・ hydrogen ion ・ monohydrogen phosphate ion

$$HPO_4^{2-} \rightleftharpoons H^+ + PO_4^{3-}$$

monohydrogen phosphate ion ・ hydrogen ion ・ phosphate ion

Hydrogen ions are too reactive to exist in solution by themselves. They react with water to form *hydronium ions* (H_3O^+).

$$H^+ + H_2O \rightleftharpoons H_3O^+$$

While this is the more correct representation of the presence of hydrogen ions in solution, for simplicity we will use the term H^+ in this book.

Solutions of acids have a sour taste. Lemon and grapefruit juices owe their sour taste to citric acid. Vinegar owes its sour taste to acetic acid. Sour milk owes its taste partly to lactic acid.

When acids react with certain compounds, these compounds change in color. Substances that change in color in the presence of acids are called **indicators.** One of the most common indicators for acids is litmus. Blue litmus turns red in the presence of an acid (in the presence of hydrogen ions). Another common indicator—phenolphthalein—turns from red to colorless in the presence of an acid.

Acids react with metal oxides and hydroxides to form water and a salt, for example,

$$2\,HCl + MgO \longrightarrow H_2O + MgCl_2$$

acid ・ metal oxide ・ water ・ salt

$$H_2SO_4 + 2\,NaOH \longrightarrow 2\,H_2O + Na_2SO_4$$

acid ・ metal hydroxide ・ water ・ salt

The reaction of acids with certain hydroxides (called bases) is termed **neutralization.** That is, acids neutralize bases to form water and a salt.

If we rewrite the above neutralization reaction in ionic form, we have

$$2\,H^+ + SO_4^{2-} + 2\,Na^+ + 2\,OH^- \longrightarrow 2\,H_2O + 2\,Na^+ + SO_4^{2-}$$

Canceling ions that appear in the same quantities on both sides of the equation,

Table 13-2   The Activity
Series of the Metals

K	Potassium
Ca	Calcium
Na	Sodium
Mg	Magnesium
Al	Aluminum
Zn	Zinc
Fe	Iron
Sn	Tin
Pb	Lead
H	Hydrogen
Cu	Copper
Hg	Mercury
Ag	Silver
Au	Gold

$$2\,H^+ + \cancel{SO_4^{2-}} + \cancel{2\,Na^+} + 2\,OH^- \longrightarrow 2\,H_2O + \cancel{2\,Na^+} + \cancel{SO_4^{2-}}$$

we have the equation

$$2\,H^+ + 2\,OH^- \longrightarrow 2\,H_2O$$

which simplifies to

$$H^+ + OH^- \longrightarrow H_2O$$

This final reaction is called a *net equation*. It indicates that a hydrogen ion (H^+) from any acid reacts with a hydroxide ion (OH^-) from any base to form nonionized water.

The activity series of metals (Table 13-2) lists metals in order of decreasing activity. Note that hydrogen is classified with the metals. **Any metal above hydrogen in this series will displace the hydrogen from an acid.** The farther above hydrogen a metal is in the activity series, the greater will be its tendency to displace hydrogen from an acid.

Acids react with any metal above hydrogen in the activity series to produce hydrogen gas and a salt, for example,

$$Zn + H_2SO_4 \longrightarrow \underset{\text{salt}}{ZnSO_4} + H_2(g)$$

$$Mg + 2\,HCl \longrightarrow \underset{\text{salt}}{MgCl_2} + H_2(g)$$

These are examples of single-replacement reactions—the metal replaces the hydrogen in the acid. Thus acids cannot be stored in containers made of these active metals. Iron is above hydrogen in the activity series and should replace hydrogen in an acid. Therefore, acids should not be allowed to come into contact with surgical or dental instruments, which are frequently made of stainless steel. Acids are usually stored in glass or plastic containers. Reactions involving acids are usually carried out in glass or plastic vessels.

Since any metal above hydrogen in the activity series will replace the hydrogen from an acid, any metal below hydrogen in the series should not be able to replace a hydrogen from an acid. That is, we should expect the mixing of copper with hydrochloric acid to produce no hydrogen, and it does not.

$$Cu + HCl \longrightarrow \text{no reaction}$$

Acids react with carbonates and bicarbonates to form carbon dioxide, water, and salts. For example,

$$\underset{\text{acid}}{2\,HCl} + \underset{\text{carbonate}}{CaCO_3} \longrightarrow \underset{\text{salt}}{CaCl_2} + \underset{\substack{\text{carbonic} \\ \text{acid}}}{H_2CO_3}$$

and

$$H_2CO_3 \longrightarrow CO_2(g) + H_2O$$

$$\underset{\text{acid}}{HNO_3} + \underset{\text{bicarbonate}}{NaHCO_3} \longrightarrow \underset{\text{salt}}{NaNO_3} + \underset{\substack{\text{carbonic} \\ \text{acid}}}{H_2CO_3}$$

and

$$H_2CO_3 \longrightarrow H_2O + CO_2(g)$$

The stomach normally secretes hydrochloric acid (HCl), which is required for the digestion of protein (page 460). Emotional stress can lead to hyperacidity, too great an acid concentration in the stomach. Commercial **antacids** are available to react with the excess stomach acid. Among these are

Tums, which contains calcium carbonate ($CaCO_3$), magnesium carbonate ($MgCO_3$), and magnesium trisilicate ($Mg_2Si_3O_8$).

Maalox, which contains magnesium hydroxide [$Mg(OH)_2$] and aluminum hydroxide [$Al(OH)_3$] in suspension.

Rolaids, which contains aluminum sodium dihydroxycarbonate [$AlNa(OH)_2CO_3$].

All of these antacid substances can react with the hydrochloric acid present in the stomach.

Alka-Seltzer, another commercial antacid, contains calcium dihydrogen phosphate [$Ca(H_2PO_4)_2$], sodium bicarbonate ($NaHCO_3$), citric acid, and aspirin. When placed in water, the bicarbonate ions react with the acidic components to produce carbon dioxide gas (CO_2), which provides the "fizz" or effervescence. The net ionic reaction is

$$H^+ + HCO_3^- \longrightarrow CO_2(g) + H_2O$$

(The prefix *bi-* is sometimes used to indicate hydrogen in a compound. Sodium bicarbonate is also known as sodium hydrogen carbonate.) Sodium bicarbonate ($NaHCO_3$) can also be used to remove excess stomach acidity. However, its continued use may interfere with the normal digestive processes in the stomach. In addition, the continued use of $NaHCO_3$ adds a considerable amount of sodium ions to the body.

Strong acids will attack clothing. Vegetable fibers such as cotton and linen, animal fibers such as wool and silk, and synthetic fibers are rapidly destroyed by strong acids. All these effects are actually due to the hydrogen ions present in the acids.

Strong acids (acids with a high hydrogen ion concentration) also have an effect on tissues. Concentrated nitric acid (HNO_3) and concentrated sulfuric acid (H_2SO_4) are extremely corrosive to the skin; therefore, great care must be exercised in handling them. The yellowing of the skin by nitric acid is a test that is specific for protein (see page 424). If a strong acid is spilled on the skin, a serious burn may result. The area should be washed copiously with water. Then it should be treated with sodium bicarbonate to neutralize any remaining acid. Dilute acids are not as corrosive to the tissues. A few may even be used internally in the body.

Drugs are absorbed from the gastrointestinal tract more rapidly when they are nonionized. If a weakly acidic drug is swallowed, it will remain nonionized in the stomach because of the low pH (high acid content; see page 233) of the stomach. Since the drug is nonionized, it will be absorbed rapidly. Conversely, a weakly basic drug will be highly ionized in the stomach and thus will be poorly absorbed there.

Aspirin, acetylsalicylic acid (see page 328), is over 90 percent nonionized in a strongly acidic solution, whereas it is about 1 percent nonionized in a neutral solution. Thus we should expect aspirin to be absorbed rapidly from the stomach where the pH is near 2 and more slowly from the small intestines where the pH is above 7.

Uses of Acids

Acids such as hydrochloric acid (HCl, commercially known as muriatic acid) are used industrially and in laboratory work in large amounts. Several acids are also used medically; among these are hydrochloric, nitric, hypochlorous, boric, acetylsalicylic, and ascorbic acids.

Hydrochloric acid, normally found in the gastric juices, is necessary for the proper digestion of proteins in the stomach. Patients who have a lower than normal amount of hydrochloric acid in the stomach, a condition called hypoacidity, are given dilute hydrochloric acid orally before meals to overcome this deficiency.

Nitric acid (HNO_3) is used to test for the presence of albumin in urine because it will coagulate protein. Nitric acid has been used to remove warts, but dichloroacetic acid (bichloracetic acid) and trichloroacetic acid are now commonly used for this purpose.

Hypochlorous acid (HClO) is used as a disinfectant for floors and walls in the hospital.

Boric acid (H_3BO_3) has had extensive use as a germicide. Although boric acid has been used in eyewashes, its use in solutions or as a powder on extensive inflamed surfaces or in body cavities is now practically obsolete. Containers of boric acid should have a label reading "poison."

Acetylsalicylic acid (aspirin) is widely used as an *analgesic and as an *antipyretic. Aspirin is frequently taken by people with a cold to relieve headache, muscle pain, and fever. However, the aspirin does not remove the source of infection or effect a cure. Aspirin also may interfere with the normal clotting of the blood and may cause bleeding in the stomach of some individuals (see page 328).

Ascorbic acid (vitamin C) is normally found in citrus fruits and is used in the prevention and treatment of scurvy.

Bases

Bases can be defined as substances that yield hydroxide (OH^-) ions in a water solution. That is, bases increase the hydroxide ion concentration in water. If we write the ionization reactions (see Chapter 12) of the bases sodium hydroxide and potassium hydroxide, we have

$$NaOH \longrightarrow Na^+ + OH^-$$

$$KOH \longrightarrow K^+ + OH^-$$

A more general definition of a base is that of Brønsted: **a base is a substance that accepts protons.** According to this definition, substances that yield hydroxide ions in solution are bases, as are substances that yield bicarbonate ions, because these ions are then free to combine with or accept protons (hydrogen ions) as indicated in the following reactions:

$$OH^- + H^+ \longrightarrow H_2O$$

$$\underset{\text{base}}{HCO_3^-} + \underset{\text{proton}}{H^+} \longrightarrow H_2CO_3$$

Strong bases are highly ionized and have a great attraction for protons. Weak bases are slightly ionized and have a weak attraction for protons. Ammonia (NH_3) is a base according to the Brønsted definition because it accepts a proton from water or acids.

$$NH_3 + H_2O \rightleftharpoons NH_4^+ + OH^-$$

C_2H_5OH (ethyl alcohol) is not a base because it does not accept a proton; it does not ionize in water.

Table 13-3 indicates several commonly used bases. Note that they consist of a metal ion ionically bonded to an OH^- ion. The only exception to this rule is the base ammonium hydroxide (NH_4OH) in which the ammonium ion (NH_4^+) is considered to act as a metal ion. Bases are produced when metallic oxides are dissolved in water, for example,

$$\underset{\substack{\text{metal} \\ \text{oxide}}}{CaO} + \underset{\text{water}}{H_2O} \longrightarrow \underset{\substack{\text{calcium hydroxide,} \\ \text{a base}}}{Ca(OH)_2}$$

Table 13-3 Commonly Used Bases

Name	Formula
Sodium hydroxide	$NaOH$
Ammonium hydroxide	NH_4OH
Potassium hydroxide	KOH
Calcium hydroxide	$Ca(OH)_2$
Magnesium hydroxide	$Mg(OH)_2$

Properties of Bases

Solutions of bases have a slippery, soapy feeling and a biting, bitter taste. Like acids, bases also react with indicators. Bases turn red litmus blue, turn methyl orange from red to yellow, and turn phenolphthalein from colorless to red.

Bases neutralize acids to form water and a salt, for example,

$$Ca(OH)_2 + H_2SO_4 \longrightarrow 2\,H_2O + CaSO_4$$

base   acid   water   salt

Strong bases react with certain metals to produce hydrogen gas, for example,

$$2\,Al\ +\ 6\,NaOH\ +\ 6\,H_2\dot{O} \longrightarrow 3\,H_2(g)\ +\ 2\,Na_3Al(OH)_6$$

aluminum   sodium hydroxide,   water   hydrogen   sodium aluminate,
a strong base   a soluble compound

Thus a strong base such as lye (NaOH) should never be used or stored in an aluminum container because it will rapidly react with and dissolve the container.

Strong bases have a high hydroxide ion concentration. They have a corrosive effect on tissues because of their ability to react with proteins and fats. If a strong base is spilled on the skin, a serious burn may result. The procedure in this case is to apply copious amounts of water.

Strong laundry soaps are quite basic and should not be used for washing woolen clothing because the hydroxide ion will attack the fibers and cause them to shrink. Particular care must be taken not to use strong soap on diapers because, if it is not thoroughly removed, the basic soap can cause severe sores on the tender skin of a baby.

Uses of Bases

Sodium hydroxide (NaOH), commonly known as lye, is used to remove fats and grease from clogged drains. It is quite caustic, and care must be exercised in handling this substance. Sodium hydroxide is also used in the conversion of fat to soap (see page 387).

Calcium hydroxide solution [$Ca(OH)_2$], commonly known as lime water, is used to overcome excess acidity in the stomach. It is also used medicinally as an antidote for oxalic acid poisoning because it reacts with the oxalic acid to form an insoluble compound, calcium oxalate.

Magnesium hydroxide [$Mg(OH)_2$] is commonly known as milk of magnesia. In dilute solutions it is used as an antacid for the stomach. In the form of a suspension of magnesium hydroxide in water, it is used as a laxative.

Spirits of ammonia, which contains ammonium hydroxide

(NH₄OH) and ammonium carbonate [(NH₄)₂CO₃], is used as a heart and respiratory stimulant. Ammonium hydroxide, also known as household ammonia, is used as a water softener for washing clothes.

Acid–Base Titration

When an acid reacts with a base (neutralization), a salt and water are produced. This reaction can be carried out in the laboratory by a process called **titration** (see Figure 13-1). In this process, a buret is filled with an acid of known concentration. A buret is a cylindrical glass tube with graduated markings, so that the exact volume of liquid withdrawn can be easily determined. A flask placed below the buret contains a measured volume of a base of unknown concentration. An indicator is added to the flask. The indicator is one that will change in color when the acid has reacted with all of the base present. This is called the *endpoint* or equivalence point of the titration.

The acid is added drop by drop to the flask until the indicator just begins to change color. The concentration of the base (or acid if

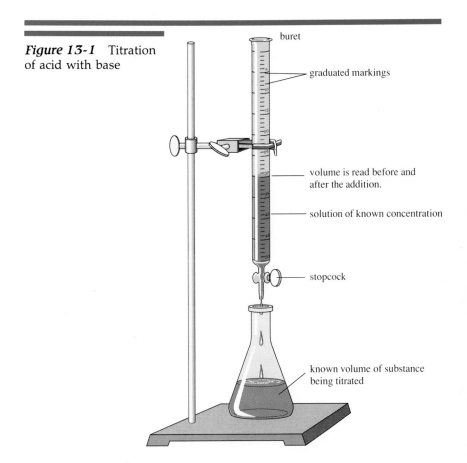

Figure 13-1 Titration of acid with base

buret

graduated markings

volume is read before and after the addition.

solution of known concentration

stopcock

known volume of substance being titrated

the buret contains a base and the flask an acid) can be determined by using the following formula.

$$\text{volume}_{acid} \times \text{normality}_{acid} = \text{volume}_{base} \times \text{normality}_{base}$$

Example 13-1   27.8 mL of 0.1 N HCl solution reacts with 20.0 mL of NaOH solution. What is the concentration of the NaOH solution?
  Using the above formula, we have

$$27.8 \text{ mL} \times 0.1 \text{ N} = 20.0 \text{ mL} \times x$$

$$x = 0.139 \text{ N NaOH}$$

Example 13-2   35.2 mL of 0.15 N KOH reacts with 40.0 mL of H_2SO_4 solution. What is the normality of the acid?

$$35.2 \text{ mL} \times 0.15 \text{ N} = 40.0 \text{ mL} \times x$$

$$x = 0.132 \text{ N } H_2SO_4$$

Ionization of Water

Even though pure water does not conduct electricity, very exact measurements show that a very slight amount of ionization does take place. The reaction is

$$H_2O \rightleftharpoons H^+ + OH^-$$

At 25 °C, only one water molecule out of 10 million ionizes (dissociates). The concentration of hydrogen ions in pure water is 10^{-7} M, as is the hydroxide ion concentration. Since the hydrogen and hydroxide ion concentrations are equal, pure water is neutral.
  The equilibrium constant for water (page 109) can be written as

$$K_{eq} = \frac{[H^+][OH^-]}{[H_2O]}$$

where the [ ] indicates concentration in moles per liter. However, the concentration of water remains constant, so the equilibrium constant can be rewritten as

$$K_{eq} \times [H_2O] = [H^+][OH^-] = K_w$$

where K_w, the ion product of water, is itself the product of two constants, K_{eq} and $[H_2O]$.

At equilibrium at 25 °C, pure water contains 10^{-7} M H^+ and 10^{-7} M OH^-, so

$$K_w = [H^+][OH^-] = 10^{-7} \times 10^{-7} = 10^{-14}$$

Since $[H^+][OH^-]$ always equals 10^{-14}, if $[H^+]$ increases, the $[OH^-]$ must decrease, and vice versa. A solution in which the $[H^+]$ is greater than the $[OH^-]$ is acidic, and a solution in which the $[H^+]$ is less than the $[OH^-]$ is basic. For example, in 0.01 M HCl, $[H^+]$ is 10^{-2}. Therefore, since $[H^+][OH^-] = 10^{-14}$, $(10^{-2})[OH^-] = 10^{-14}$ and $[OH^-] = 10^{-12}$. The solution is acidic because the $[H^+]$ is greater than $[OH^-]$; likewise, in a 0.1 M NaOH solution, $[OH^-] = 10^{-1}$, so $[H^+] = 10^{-13}$. The solution is basic since $[H^+]$ is less than $[OH^-]$.

pH

A few drops of concentrated hydrochloric acid in water produce a dilute acid solution. A few more drops produce another solution, still dilute but a little stronger than the previous one. If a piece of blue litmus paper is placed in either of these two solutions it will turn red, indicating that the solution is acidic. However, it will not tell which one is more strongly acidic. Likewise if a piece of red litmus paper turns blue when placed in a solution, it merely indicates that the solution is basic or alkaline. It does not indicate how strongly basic the solution is. The term pH is used to indicate the exact strength of an acid or a base. The pH indicates the hydrogen ion concentration in a solution.

Mathematically, pH is defined as the negative logarithm (log) of the hydrogen ion concentration, or

$$pH = -\log [H^+]$$

A logarithm is an exponent. Therefore, the logarithm of 10^{-2} is -2 and $\log 10^{-12}$ is -12. Thus, a solution that has $[H^+] = 10^{-4}$ has a pH of 4; that is,

$$pH = -\log [H^+] = -\log 10^{-4} = -(-4) = 4$$

A pH of 7 indicates a neutral solution because $[H^+] = [OH^-]$. pH values below 7 indicate an acidic solution: pHs between 5 and 7 indicate a weakly acidic solution; values between 2 and 5, a moderately acidic solution; and pHs between 0 and 2, a strongly acidic solution.

Likewise, pHs above 7 indicate a basic solution: pH values between 7 and 9 indicate a weakly basic solution; those between 9

Table 13-4   pH Values[a]

pH	Strength of Acid ([H$^+$] in mol/L)	Strength of Base ([OH$^-$] in mol/L)	
0	10^0	10^{-14}	Strong acid
1	10^{-1}	10^{-13}	
2	10^{-2}	10^{-12}	Moderate acid
3	10^{-3}	10^{-11}	
4	10^{-4}	10^{-10}	
5	10^{-5}	10^{-9}	Weak acid
6	10^{-6}	10^{-8}	
7	10^{-7}	10^{-7}	Neutral
8	10^{-8}	10^{-6}	Weak base
9	10^{-9}	10^{-5}	
10	10^{-10}	10^{-4}	Moderate base
11	10^{-11}	10^{-3}	
12	10^{-12}	10^{-2}	
13	10^{-13}	10^{-1}	Strong base
14	10^{-14}	10^0	

[a] See Appendix for an explanation of the use of negative exponential numbers in scientific notation.

and 12, a moderately basic solution; and pHs between 12 and 14, a strongly basic solution. This is summarized in Table 13-4.

The pHs of some common body fluids and some household substances are listed in Table 13-5. From these values it can be seen that blood is a slightly basic liquid, the gastric juices strongly acidic, bile weakly basic, and urine and saliva both weakly acidic or weakly basic. The pH of pure water is 7.0.

A difference of 1 in pH value represents a tenfold difference in strength. That is, an acid of pH 4.5 is ten times as strong as one of pH 5.5. Likewise, a base of pH 10.7 is ten times as strong as one of pH 9.7, and 100 times as strong (10 × 10) as one of pH 8.7. Therefore, a small change in pH indicates a definite change in acid or base strength.

To measure pH in a laboratory, a pH meter (Figure 13-2) can be used. This instrument is standardized by placing the electrodes into a solution of known pH to see that it is functioning and recording properly. Then the electrodes are placed in a solution of unknown pH and the pH is determined by reading the value on the pH meter.

Table 13-5   pH Values of Body Fluids and Common Household Substances

Fluid	pH Range	Substance	pH Range
Blood	7.35–7.45	Black coffee	4.8–5.2
Gastric juices	1.6–1.8	Eggs	7.6–8.0
Bile	7.8–8.6	Lemon juice	2.8–3.4
Urine	5.5–7.5	Milk	6.3–6.6
Saliva	6.2–7.4	Tap water	6.5–8.0

Figure 13-2 Digital pH meter. [Courtesy Corning Glass Works, Corning, N.Y.]

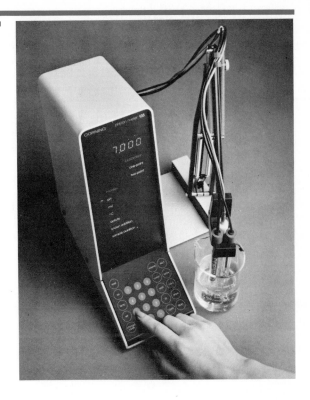

A quicker but less accurate method is to touch a drop of the liquid to a specially prepared piece of indicator paper and then determine the pH by comparison with pH color scale.

Summary

Acids are compounds that yield hydrogen ions or protons in solution. The general properties of acids are sour taste, effect on colors of indicators, reaction with metal oxides and hydroxides (neutralization), reaction with metals to yield hydrogen gas, reaction with carbonates and bicarbonates to produce carbon dioxide gas, effect on clothing, and effect on tissues.

Antacids are used to neutralize excess acidity in the stomach.

Acids and bases are used industrially, pharmaceutically, and in the hospital and laboratory.

Bases are compounds yielding hydroxide (OH^-) ions in solution. Bases are also compounds that accept protons. Bases have a slippery, soapy feeling; they affect the colors of indicators, neutralize acids, react with certain metals to produce hydrogen, and affect tissue and clothing.

Titration is a method for determining the strength of an unknown acid (or base) by using a buret containing a base (or acid) of known concentration and an indicator.

The ion product of water is $K_w = [H^+] \times [OH^-] = 10^{-14}$.

The pH of a solution indicates numerically the acid (or base) strength of the solution in terms of its hydrogen ion concentration. A pH of 7 indicates a neutral solution; a pH below 7, an acid solution; a pH above 7, a basic solution.

Questions and Problems

A

1. Define (a) acid; (b) base; (c) neutralization; (d) pH.
2. List the general properties of acids.
3. List the general properties of bases.
4. What effect does an acid have on litmus paper?
5. What effect does a base have on litmus paper?
6. Will aluminum replace the hydrogen from an acid? Explain.
7. Why are acids usually stored in a container made of glass rather than metal?
8. What substances might be used to lower the acidity of the contents of the stomach?
9. What treatment should be given if a strong acid is accidentally spilled on the skin?
10. Name four acids used in the hospital, and give one medical use for each.
11. Name three bases used in the hospital, and give one medical use for each.
12. How can pH be measured?
13. What is the pH of blood? of urine? of saliva?
14. Indicate whether each of the substances in question 13 is acidic, basic, or neutral?

B

15. Compare the general and Brønsted definitions of an acid and of a base.
16. Explain how antacids work.
17. Compare the ionization of strong and weak acids and strong and weak bases.
18. Discuss the absorption of aspirin from the gastrointestinal system in terms of pH.
19. Why is KOH a base and C_2H_5OH not a base?
20. If a solution has a $[H^+]$ of 10^{-3}, what is the $[OH^-]$? What is the pH? Is the solution acidic or basic?
21. According to Table 13–5, the pH of water is 6.5 to 8.0. Why isn't the pH 7.0?

22. Why does lemon juice have such a low pH?
23. 29.7 mL of what strength HNO_3 will neutralize 38.6 mL of 0.25 N KOH?
24. 31.8 mL of what strength NaOH will neutralize 46.5 mL of 0.40 N H_2SO_4?
25. How many milliliters of 0.15 N $Ca(OH)_2$ will 29.5 mL of 0.10 N HCl neutralize?

Practice Test

1. An example of a diprotic acid is _____.
   a. HCl
   b. H_2SO_4
   c. H_2O
   d. CaH_2
2. When an acid reacts with a base, the reaction is called _____.
   a. equilibrium
   b. hydrolysis
   c. hydration
   d. neutralization
3. Acids react with _____.
   a. all metals
   b. all bases
   c. all salts
   d. all of these
4. An acid found in the stomach is _____.
   a. HCl
   b. HNO_3
   c. H_2SO_4
   d. H_3PO_4
5. An example of a strong base is _____.
   a. $Mg(OH)_2$
   b. $Ca(OH)_2$
   c. NH_4OH
   d. KOH
6. 50 mL of 0.10 N HCl will neutralize _____ milliliters of 0.20 N NaOH?
   a. 25   b. 50   c. 75   d. 100
7. An acid is a substance that _____.
   a. donates protons
   b. donates electrons
   c. donates neutrons
   d. none of these
8. Which pH indicates a strong acid?
   a. 10.2   b. 7.2   c. 5.2   d. 1.2
9. The pH of freshly distilled water is _____.
   a. 1.0   b. 5.0   c. 7.0   d. 10.0
10. An example of an antacid is _____.
    a. $CaCO_3$
    b. $Mg(OH)_2$
    c. $AlNa(OH)_2CO_3$
    d. all of these

14

Salts

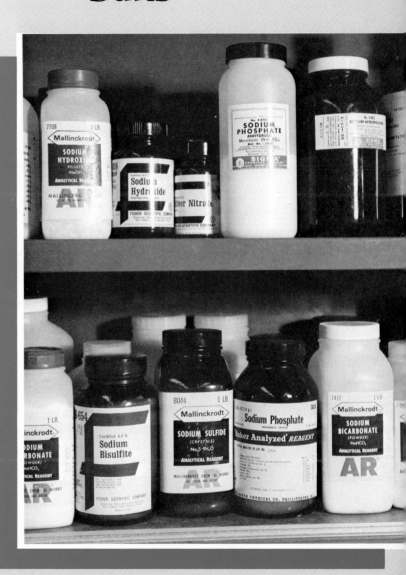

Shelves with sodium hydroxide and various salts.

Acids have one ion in common, the hydrogen ion (H^+). Consider the dissociation of the following salts

$$NaCl \longrightarrow Na^+ + Cl^-$$

$$K_2SO_4 \longrightarrow 2\,K^+ + SO_4^{2-}$$

$$Mg(NO_3)_2 \longrightarrow Mg^{2+} + 2\,NO_3^-$$

Salts have no common ion. Salts in solution yield a positive ion and a negative ion.

Salts are formed by the reaction of an acid and a base, for example,

Acid		Base		Salt		Water
HCl	+	KOH	$\longrightarrow$	KCl	+	H_2O
H_2SO_4	+	$Mg(OH)_2$	$\longrightarrow$	$MgSO_4$	+	$2\,H_2O$
$2\,HNO_3$	+	$Zn(OH)_2$	$\longrightarrow$	$Zn(NO_3)_2$	+	$2\,H_2O$

Recall that the reaction of an acid with a base is called neutralization.

Solubility of Salts

Some salts are quite soluble in water. Others are classified as slightly soluble or insoluble. Table 14-1 indicates the solubility of most

Table 14-1 Solubility of Common Salts and Bases

Soluble	Insoluble
Sodium salts	Carbonates (except sodium, potassium, ammonium)
Potassium salts	Phosphates (except sodium, potassium, ammonium)
Ammonium salts	Sulfides (except sodium, potassium, ammonium)
Acetates	
Nitrates	
Chlorides (except silver, lead, and mercury +1)	Hydroxides (except sodium, potassium, ammonium)
Sulfates (except calcium, barium, and lead)	

Table 14-2 Water Solubility of Some Salts

Name of Salt	Formula	Solubility
Sodium chloride	$NaCl$	Soluble
Silver chloride	$AgCl$	Insoluble
Sodium sulfate	Na_2SO_4	Soluble
Zinc nitrate	$Zn(NO_3)_2$	Soluble
Barium sulfate	$BaSO_4$	Insoluble
Calcium phosphate	$Ca_3(PO_4)_2$	Insoluble
Magnesium carbonate	$MgCO_3$	Insoluble

common salts, and Table 14-2 indicates the solubilities in water of selected common salts as predicted by using Table 14.1.

Reactions of Salts

Hydrolysis

Hydrolysis is the reaction of a compound with the hydrogen ion or the hydroxide ion derived from water. Ions derived from a weak acid or base hydrolyze (react with water) to form the corresponding acid or base. Ions derived from a strong acid or base do not hydrolyze. For simplicity we will assume that the acids and bases listed in Table 14-3 are strong and that all other commonly used acids and bases are weak. (For a discussion of strong and weak acids and bases see page 218.)

Since salts are produced by the reaction of an acid with a base, salts must contain parts of each. The positive ion of a salt is derived from a base, and its negative ion is derived from an acid.

Let us consider the hydrolysis of the salt sodium cyanide (NaCN). We first write the formula of the salt in ionic form, sodium ion (Na^+) and cyanide ion (CN^-). Directly below these ions we write the ionized formula for water, with the OH^- ion below the positive ion of the salt and the H^+ ion below the negative ion of the salt, or

$$\begin{array}{c|c} Na^+ & CN^- \\ OH^- & H^+ \end{array} \tag{14-1}$$

The sodium ion is derived from one of the strong bases (NaOH) and does not hydrolyze. The CN^- ion is not derived from one of the strong acids and does hydrolyze. Reading upward on the right side of the line in setup (14-1), we have the essentially nonionized compound HCN. Thus the products of hydrolysis of NaCN are the two ions on the left side of the line—Na^+ and OH^-—and the nonionized compound on the right side of the line—HCN. The hydrolysis reaction for NaCN can be written as

$$Na^+ + CN^- + H_2O \rightleftharpoons Na^+ + OH^- + HCN$$

Since the hydrolysis reaction produces OH^- ions, the solution will be basic. This is what we should expect from the hydrolysis of NaCN, which is derived from a strong base and a weak acid.

Table 14-3 Strong Acids and Bases in Common Use	Strong Acids		Strong Bases	
	HCl	Hydrochloric acid	NaOH	Sodium hydroxide
	HNO_3	Nitric acid	KOH	Potassium hydroxide
	H_2SO_4	Sulfuric acid	$Ca(OH)_2$	Calcium hydroxide

Table 14-4 pH of Solution Produced During Hydrolysis of a Salt	Substances from Which Salt Was Derived	Predicted pH	Example
	Strong acid, weak base	Below 7	$Al_2(SO_4)_3$
	Weak acid, strong base	Above 7	KCN
	Weak acid, weak base	Not easily predicted	$NH_4C_2H_3O_2$

Now consider the hydrolysis of NH_4Cl. If we write this compound in ionic form, with water in its ionized form below it, and place the OH^- ion below the positive ion of the salt and the H^+ ion of the water below the negative ion of the salt, we have setup (14-2).

$$\begin{array}{c|c} NH_4^+ & Cl^- \\ OH^- & H^+ \end{array} \qquad (14\text{-}2)$$

The NH_4^+ ion is not derived from one of the strong bases and does hydrolyze. The nonionized compound formed, as indicated by reading down on the left side of the line, is NH_4OH. The Cl^- ion is derived from one of the strong acids and does not hydrolyze. The right side of the line in (14-2) indicates the ions remaining in solution, Cl^- and H^+. Thus the hydrolysis reaction is

$$NH_4^+ + Cl^- + H_2O \rightleftharpoons NH_4OH + H^+ + Cl^-$$

The resulting solution will be acidic because of the presence of hydrogen ions. This is what should be expected, since the compound NH_4Cl is derived from a weak base and a strong acid.

A prediction of the pH of the solution produced by the hydrolysis of a salt is shown in Table 14-4.

Reaction with Metals

Some metals react with salt solutions to form another salt and a different metal. Refer to the activity series of metals (page 226) and recall that any metal can replace a metal ion below it in the activity series. Thus zinc can replace the copper from the copper sulfate solution because the zinc is higher in the activity series than copper. This is an example of a single replacement reaction.

$$Zn + CuSO_4 \longrightarrow ZnSO_4 + Cu$$

Reaction with Other Salts

Two different salts in solution can react by double displacement. Consider reactions (14-3) and (14-4).

$$K_2SO_4 + Ba(NO_3)_2 \longrightarrow 2\ KNO_3 + BaSO_4(s) \qquad (14\text{-}3)$$

$$K_2SO_4 + Zn(NO_3)_2 \longrightarrow \text{no reaction} \qquad (14\text{-}4)$$

Why does reaction (14-3) proceed whereas (14-4) does not? Review the solubility rules (page 238). Note that one of the products formed in reaction (14-3), $BaSO_4$, is insoluble in water. Therefore reaction (14-3) will proceed. In reaction (14-4), the products, if formed, would be $ZnSO_4$ and KNO_3. Both of these salts are soluble and ionized; therefore, no reaction will take place. In order for a reaction to take place between solutions of two salts, at least one of the products must be insoluble in water, a weak electrolyte, or a nonionized compound.

Reaction with Acids and Bases

Salts react with acids or bases to form other salts and other acids and bases. These reactions will proceed if one of the products is insoluble in water or is an insoluble gas.

$$CaCO_3 + 2\ HCl \longrightarrow CaCl_2 + H_2O + CO_2(g) \quad \text{(a gas is formed)}$$

$$AlCl_3 + 3\ NaOH \longrightarrow 3\ NaCl + Al(OH)_3(s) \quad \text{(a precipitate is formed)}$$

Uses of Salts

Salts are necessary for the proper growth and metabolism of the body. Iron salts are necessary for the formation of hemoglobin; iodine salts for the proper functioning of the thyroid gland; calcium

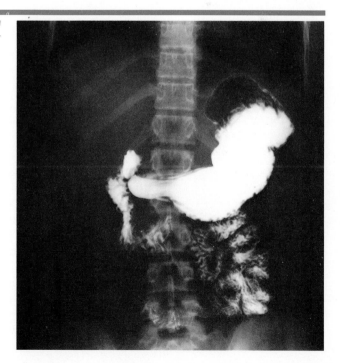

Figure 14-1 X-ray of the stomach. [Courtesy Dr. Leonard Berlin, Director of Radiology, Rush Northshore Medical Center, Skokie, IL.]

Table 14-5   Common Salts and Their Uses

Classification	Formula	Chemical Name	Common Name
Antacid	$CaCO_3$	Calcium carbonate	Precipitated chalk
	$NaHCO_3$	Sodium bicarbonate	Baking soda
Cathartics	Na_2SO_4	Sodium sulfate	Glaubers salt
	$MgSO_4 \cdot 7H_2O$	Magnesium sulfate	Epsom salts
	$MgCO_3$	Magnesium carbonate	—
	$MgHC_6H_5O_7 \cdot 5H_2O$	Magnesium citrate	Citrate of magnesia
	$KNaC_4H_4O_6 \cdot 4H_2O$	Potassium sodium tartrate	Rochelle salt
Diuretic	NH_4Cl	Ammonium chloride	Sal ammoniac
Expectorants	NH_4Cl	Ammonium chloride	Sal ammoniac
	KI	Potassium iodide	—
Germicide	$AgNO_3$	Silver nitrate	Lunar caustic
Miscellaneous Uses			
X-ray work	$BaSO_4$	Barium sulfate	Barium
Caries reduction	NaF	Sodium fluoride	—
	SnF_2	Stannous fluoride	—
For casts	$(CaSO_4)_2 \cdot H_2O$	Calcium sulfate hydrate	Plaster of paris
Treatment of anemia	$FeSO_4$	Ferrous sulfate	—
Decrease of blood clotting time	$CaCl_2$	Calcium chloride	—
Physiologic saline solution used for irrigation and as IV replacement fluid	$NaCl$	Sodium chloride	Table salt
Thyroid treatment	KI	Potassium iodide	—
	NaI	Sodium iodide	—
Prevent clotting of stored blood	$Na_3C_6H_5O_7$	Sodium citrate	—

and phosphorus salts for the formation of bones and teeth; sodium and potassium salts help regulate the acid–base balance of the body. Salts regulate the irritability of nerve and muscle cells and the beating of the heart. Salts help maintain the proper osmotic pressure of the cells.

Many salts have specific uses. Barium sulfate ($BaSO_4$) is used for X-ray work. Even though barium compounds are poisonous, barium sulfate is insoluble in body fluids and so has no effect on the body. Barium sulfate is opaque to X-rays and, when swallowed, it can be used to outline the gastrointestinal (GI) system for X-ray photographs (see Figure 14-1).

Table 14-5 lists some common salts and their specific uses in medicine.

Buffer Solutions

The pH of pure water (a neutral solution) is 7.0. If an acid is added to water, the pH goes down. How far below 7.0 it goes depends on how much acid and how strong an acid is added. When a base is added to pure water, the pH rises above 7.0.

However, when small amounts of acid or base are added to a buffer solution, the pH does not change appreciably. A buffer solution is defined as a solution that will resist changes in pH upon the addition of small amounts of either acid or base.

Buffer solutions, or buffers, are found in all body fluids and are responsible for helping maintain the proper pH of those fluids. The normal pH range of the blood is 7.35 to 7.45. Even a slight change in pH can cause a very definite pathologic condition. When the pH falls below 7.35, the condition is known as *acidosis. *Alkalosis is the condition under which the pH of the blood rises above 7.45 (see page 554).

What does a buffer solution consist of, and how does it work? A buffer system usually consists of a weak acid and a salt of a weak acid. There are several buffer systems in the blood. One of these consists of carbonic acid (H_2CO_3), a weak acid, and sodium bicarbonate ($NaHCO_3$), the salt of a weak acid.

Suppose that an acid such as hydrochloric acid (HCl) enters the bloodstream. The HCl reacts with the $NaHCO_3$ part of the buffer according to the reaction

$$HCl + NaHCO_3 \longrightarrow NaCl + H_2CO_3$$

The NaCl produced is neutral; it does not hydrolyze (see page 239). The H_2CO_3 produced is part of the original buffer system and is only slightly ionized. In the body, acids (hydrogen ions) are produced by various metabolic processes. When these acids enter the bloodstream, they are removed by this reaction or a similar reaction with other buffers.

A base such as sodium hydroxide (NaOH) would react with the carbonic acid part of the buffer,

$$NaOH + H_2CO_3 \longrightarrow H_2O + NaHCO_3$$

forming water, a harmless neutral normal metabolite, and sodium bicarbonate, part of the original buffer. Thus, when either acid or base is added to the buffer, something neutral (NaCl or water) is formed plus more of the buffer system (H_2CO_3 or $NaHCO_3$). Therefore, the pH of the blood should not change because the buffer system has reduced the number of free hydrogen or hydroxide ions.

The principal intracellular buffer is the phosphate buffer system, which consists of a mixture of monohydrogen phosphate ions (HPO_4^{2-}) and dihydrogen phosphate ions ($H_2PO_4^{-}$). Excess acid (H^+) reacts with the monohydrogen phosphate ions,

$$H^+ + HPO_4^{2-} \longrightarrow H_2PO_4^{-}$$

Excess base (OH^-) reacts with the dihydrogen phosphate ions,

$$OH^- + H_2PO_4^- \longrightarrow HPO_4^{2-} + H_2O$$

In each case, more buffer is produced and the pH remains constant.

In addition to the carbonate and phosphate buffers there are also several organic buffer systems. These will be discussed in Chapter 30.

If there is an overproduction of acid in the tissues and if these acids cannot be excreted rapidly enough, a condition known as acidosis results during which the buffers are unable to handle the excess acid. Acidosis can occur in certain diseases, such as diabetes mellitus and during starvation (see page 554).

Prolonged vomiting can result in alkalosis because of the continued loss of the acid contents of the stomach.

Summary

Salts are formed by the reaction of an acid with a base. Salts yield ions other than hydrogen or hydroxide.

Some salts are soluble and others are insoluble in water. The solubility of most common salts can be determined from the solubility rules.

Hydrolysis is a double displacement reaction in which water is a reactant.

Salts react with some metals to yield other salts and other metals. Salts can react with other salts by a double displacement reaction. Salts can react with acids or bases to form other salts and other acids or bases.

Salts serve a definite purpose in the various metabolic processses of the body.

Buffer solutions do not change in pH upon the addition of small amounts of acid or base. Buffer solutions help maintain the proper pH of the body fluids.

Questions and Problems

A

1. Define (a) salt; (b) hydrolysis; (c) buffer solution.
2. Name five soluble salts.
3. Name ten insoluble salts.
4. Indicate whether the hydrolysis of the following salts will produce an acidic, a basic, or a neutral solution: (a) sodium carbonate; (b) magnesium chloride; (c) ammonium nitrate; (d) lead nitrate; (e) potassium sulfate.
5. Give an example of the reaction of a salt with a metal.
6. Give an example of the reaction of a salt with another salt.
7. Give an example of the reaction of a salt with an acid.
8. Give an example of the reaction of a salt with a base.
9. Name ten salts commonly used in the hospital, and indicate what each is used for.
10. What is acidosis? alkalosis?
11. Explain how a buffer solution works.

B

12. Why do ions derived from weak acids or bases hydrolyze whereas those from strong acids or bases do not?
13. Why are buffer solutions important in the body?
14. Is silver chloride soluble in water? Will it hydrolyze in water?
15. When the following compounds are placed in water, will the resulting solution have a pH of 7? above 7? below 7?
    (a) KCl   (b) NaClO   (c) NH_4NO_3

Practice Test

1. An example of an insoluble salt is _____.
   a. $CaCl_2$ b. $CaCO_3$
   c. $Ca(NO_3)_2$ d. none of these

2. The hydrolysis of which of the following salts will produce an acidic solution?.
   a. KCl
   b. NaCN
   c. NH_4NO_3
   d. KI
3. A salt used in X-ray work is _____.
   a. $CaSO_4$
   b. $BaSO_4$
   c. NH_4NO_3
   d. KI
4. A solution used to maintain a constant pH in body fluids is a _____.
   a. hydrate
   b. substrate
   c. hydrolyzate
   d. buffer
5. An example of a soluble salt is _____.
   a. $CaCO_3$
   b. $CaCl_2$
   c. $Ca_3(PO_4)_2$
   d. none of these
6. The hydrolysis of which salt will produce a basic solution?

   a. KCN
   b. $MgSO_4$
   c. NaCl
   d. $AlPO_4$
7. A salt used in caries reduction is _____.
   a. $CaCl_2$
   b. $MgSO_4$
   c. $AgNO_3$
   d. NaF
8. A salt used as a germicide is _____.
   a. $Ca(NO_3)_2$
   b. $Mg(NO_3)_2$
   c. $AgNO_3$
   d. $NaNO_3$
9. Which of the following pHs could indicate alkalosis?
   a. 6.95
   b. 7.15
   c. 7.35
   d. 7.55
10. The principal intracellular buffer is a _____.
   a. sulfate
   b. phosphate
   c. carbonate
   d. chloride

PART

II

Organic

Chemistry

15

Introduction to Organic Chemistry

Computer-generated model of a portion of a DNA molecule.

In the eighteenth century it was believed that a "vital force" was needed to make the compounds produced by living cells, which were classified as organic compounds. However, this belief was overthrown by a German chemist, Friedrich Wöhler, in 1828. He prepared urea, a compound normally found in the blood and urine, by heating a solution of ammonium cyanate, an inorganic compound.

$$NH_4CNO \xrightarrow{heat} NH_2-\overset{\overset{\textstyle O}{\|}}{C}-NH_2$$

ammonium cyanate   urea

After Wöhler's work many other organic compounds were produced in the laboratory. This led to the subdivision of chemistry into two-parts—inorganic and organic. Organic chemistry was defined as the chemistry of carbon compounds. Why have one category for the element carbon and place all other elements in the other category? The answer is that although there are tens of thousands of inorganic compounds known today, *millions* of organic compounds are known.

Importance of Organic Chemistry

Organic chemistry is important in that it is the chemistry associated with all living matter in both plants and animals. Carbohydrates, fats, proteins, vitamins, hormones, enzymes, and many drugs are organic compounds. Wool, silk, cotton, linen, and such synthetic fibers as nylon, rayon, and Dacron contain organic compounds. So do perfumes, dyes, flavors, soaps, detergents, plastics, gasolines, and oils.

Comparison of Organic and Inorganic Compounds

Organic compounds differ from inorganic compounds in many ways. The most important of these are listed below.

1. Most organic compounds are flammable. Most inorganic compounds are nonflammable.
2. Most organic compounds have low melting points. Most inorganic compounds have high melting points.
3. Most organic compounds have a low boiling point. Most inorganic compounds have high boiling points.
4. Most organic compounds are soluble in nonpolar liquids. Most inorganic compounds are insoluble in nonpolar liquids.
5. Most organic compounds are insoluble in water. Many inorganic compounds are soluble in water.
6. Organic compounds are held together by covalent bonds. Many inorganic compounds contain ionic bonds.

7. Organic reactions usually take place between molecules. Inorganic reactions usually take place between ions.
8. Organic compounds generally contain many atoms. Inorganic compounds usually contain relatively few atoms.
9. Organic compounds have a complex structure. Inorganic compounds have a simpler structure.

Bonding

Organic compounds—compounds of carbon—are held together by covalent bonds. Recall that covalent bonds are formed by sharing electrons. In organic chemistry the term **bond** is used to designate a shared pair of electrons. Thus, the statement is made that carbon forms four bonds; it has an oxidation number of −4. Bonds are usually represented by a short, straight line connecting the atoms.

Each carbon atom in the following compounds forms four bonds.

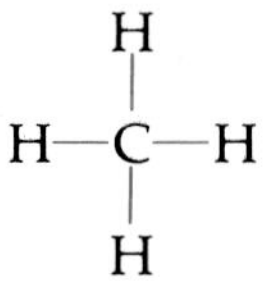

$$
\begin{array}{ccc}
& \text{H} \\
& | \\
\text{H}-\text{C}-\text{H} \\
& | \\
& \text{H}
\end{array}
\qquad
\begin{array}{c}
\text{H} \quad \text{H} \\
| \quad\; | \\
\text{H}-\text{C}=\text{C}-\text{H}
\end{array}
\qquad
\text{H}-\text{C}\equiv\text{C}-\text{H}
$$

| four single bonds to carbon | double bond between carbons | triple bond between carbons |

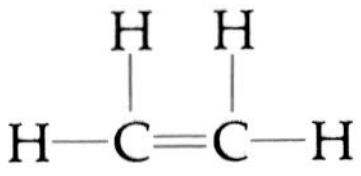

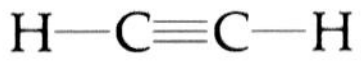

Likewise, since the oxygen atom has an oxidation number of −2, it forms two bonds, as shown in the following compounds.

$$
\begin{array}{c}
\text{H} \\
| \\
\text{H}-\text{C}-\text{O}-\text{H} \\
| \\
\text{H}
\end{array}
\qquad
\begin{array}{c}
\text{H} \\
| \\
\text{H}-\text{C}=\text{O}
\end{array}
$$

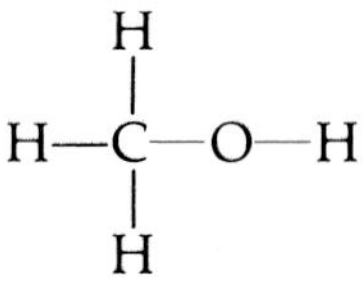

| two single bonds to oxygen | double bond to oxygen |

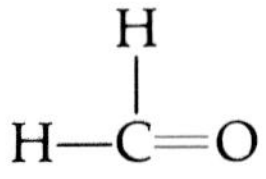

Nitrogen, with an oxidation number of −3, forms three bonds.

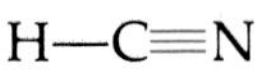

$$
\text{H}-\text{C}\equiv\text{N}
\qquad
\begin{array}{c}
\text{H} \quad \text{H} \\
| \quad\; | \\
\text{H}-\text{C}-\text{N}-\text{H} \\
| \\
\text{H}
\end{array}
\qquad
\begin{array}{c}
\text{H} \\
| \\
\text{H}-\text{C}-\text{N}-\text{N}=\text{O} \\
| \quad\; | \\
\text{H} \quad \text{H}
\end{array}
$$

| triple bond to nitrogen | three single bonds to nitrogen | one single and one double bond to nitrogen |

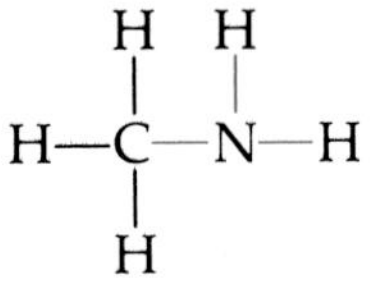

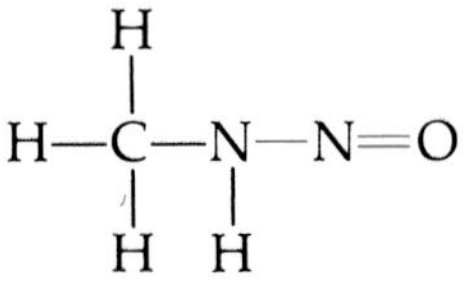

Note that in all the preceding examples, hydrogen, with an oxidation number of +1, forms only one bond. The halogens (fluorine, chlorine, bromine, and iodine), all with an oxidation number of -1, also form only one bond.

Structural Formulas

As will be seen later, organic compounds are often written using a structural rather than a molecular formula. What is the difference between a structural formula and a molecular formula? And why use the former and not the latter?

Consider an organic compound with the molecular formula C_2H_6O. In inorganic chemistry a formula of this type designates a specific compound. However, in organic chemistry this is not always true; one formula can designate more than one compound. Let us see how this is possible.

If the carbons, hydrogens, and oxygens in C_2H_6O are arranged in such a manner that each carbon atom has four bonds attached to it, each hydrogen atom has one bond, and the oxygen has two bonds, there are two possible structures, both having the formula C_2H_6O:

$$
\begin{array}{ccc}
\begin{array}{c}
\ \ \text{H}\ \ \text{H} \\
\ \ |\ \ \ | \\
\text{H}-\text{C}-\text{C}-\text{O}-\text{H} \\
\ \ |\ \ \ | \\
\ \ \text{H}\ \ \text{H}
\end{array}
& \text{and} &
\begin{array}{c}
\ \ \text{H}\ \ \ \ \ \ \text{H} \\
\ \ |\ \ \ \ \ \ \ | \\
\text{H}-\text{C}-\text{O}-\text{C}-\text{H} \\
\ \ |\ \ \ \ \ \ \ | \\
\ \ \text{H}\ \ \ \ \ \ \text{H}
\end{array}
& (15\text{-}1)
\end{array}
$$

Each compound contains two carbons, six hydrogens, and one oxygen, so each satisfies all the bond requirements. However, these two compounds have different structures and different properties and are actually different compounds. They are isomers of each other. The first compound is called ethyl alcohol, the second dimethyl ether. This difference in the structure of compounds having the same molecular formula illustrates the importance of using structural rather than molecular formulas for organic compounds.

In diagram (15-1) the lines represent bonds or shared electrons. Using electron dots, these structures are

$$
\begin{array}{ccc}
\begin{array}{c}
\text{H H} \\
\text{H}\!:\!\ddot{\text{C}}\!:\!\ddot{\text{C}}\!:\!\ddot{\text{O}}\!:\!\text{H} \\
\text{H H}
\end{array}
& \text{and} &
\begin{array}{c}
\text{H}\ \ \ \ \text{H} \\
\text{H}\!:\!\ddot{\text{C}}\!:\!\ddot{\text{O}}\!:\!\ddot{\text{C}}\!:\!\text{H} \\
\text{H}\ \ \ \ \text{H}
\end{array}
\end{array}
$$

It is simpler to write a structural formula using the bond notation.

Isomers

Isomers are defined as compounds having the same molecular formula but different structural formulas. Thus, the compound C_2H_6O has two isomers. The compound C_6H_{14} has five isomers, as illus-

Figure 15-1   Isomers of C_6H_{14}.

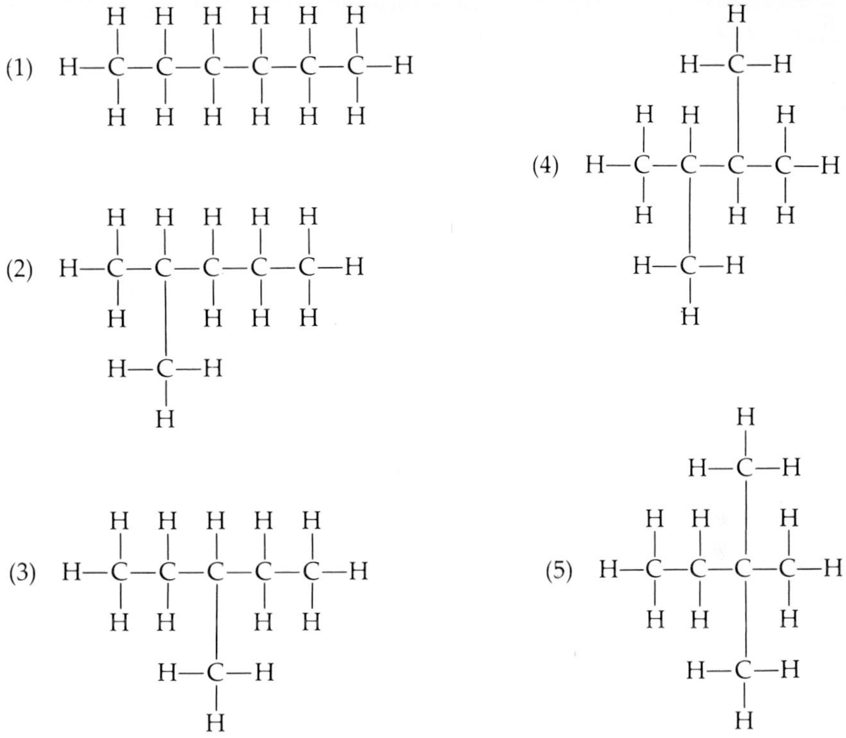

(1)
$$
\begin{array}{cccccc}
H & H & H & H & H & H \\
| & | & | & | & | & | \\
H-C-&C-&C-&C-&C-&C-H \\
| & | & | & | & | & | \\
H & H & H & H & H & H
\end{array}
$$

(2)
$$
\begin{array}{ccccc}
H & H & H & H & H \\
| & | & | & | & | \\
H-C-&C-&C-&C-&C-H \\
| & | & | & | & | \\
H & & H & H & H
\end{array}
$$
$$
\begin{array}{c}
H-C-H \\
| \\
H
\end{array}
$$

(3)
$$
\begin{array}{ccccc}
H & H & H & H & H \\
| & | & | & | & | \\
H-C-&C-&C-&C-&C-H \\
| & | & | & | & | \\
H & H & & H & H
\end{array}
$$
$$
\begin{array}{c}
H-C-H \\
| \\
H
\end{array}
$$

(4)
$$
\begin{array}{c}
H \\
| \\
H-C-H
\end{array}
$$
$$
\begin{array}{cccc}
H & H & & H \\
| & | & | & | \\
H-C-&C-&C-&C-H \\
| & | & | & | \\
H & & H & H
\end{array}
$$
$$
\begin{array}{c}
H-C-H \\
| \\
H
\end{array}
$$

(5)
$$
\begin{array}{c}
H \\
| \\
H-C-H
\end{array}
$$
$$
\begin{array}{cccc}
H & H & & H \\
| & | & | & | \\
H-C-&C-&C-&C-H \\
| & | & | & | \\
H & H & & H
\end{array}
$$
$$
\begin{array}{c}
H-C-H \\
| \\
H
\end{array}
$$

trated in Figure 15-1. $C_6H_{12}O_6$ is usually called glucose, but this molecular formula actually represents 16 different compounds or isomers, each of which is a different sugar (see page 359).

Three-Dimensional Arrangement of the Bonds in the Carbon Atom

It should be realized that the compounds represented by structural formulas are three-dimensional and not planar. For example, each carbon atom has four bonds attached to it. If these bonds are symmetrically arranged, the planar representation of the structure is

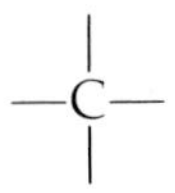

$$
\begin{array}{c}
| \\
-C- \\
|
\end{array}
$$

What does it actually look like in three dimensions? The four bonds of the carbon atom are arranged in a *tetrahedral shape. The carbon atom lies at the center of the tetrahedron; the angle between the bonds is 109.5°. Therefore the compound CH_4, methane, can be represented as shown in Figure 15-2.

Figure 15-2 Tetrahedral structure of methane, CH_4

H
C
H
H
H
H
C
H
H
H

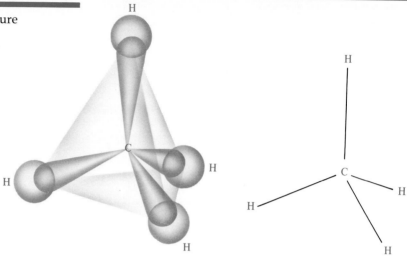

Bonding Ability of Carbon

What is unique about the element carbon that it forms so many compounds? The other elements form relatively few compounds. The answer is that the carbon atom has the ability to bond other carbon atoms to itself to form very large and complex molecules.

Figure 15-3 Types of carbon compounds.

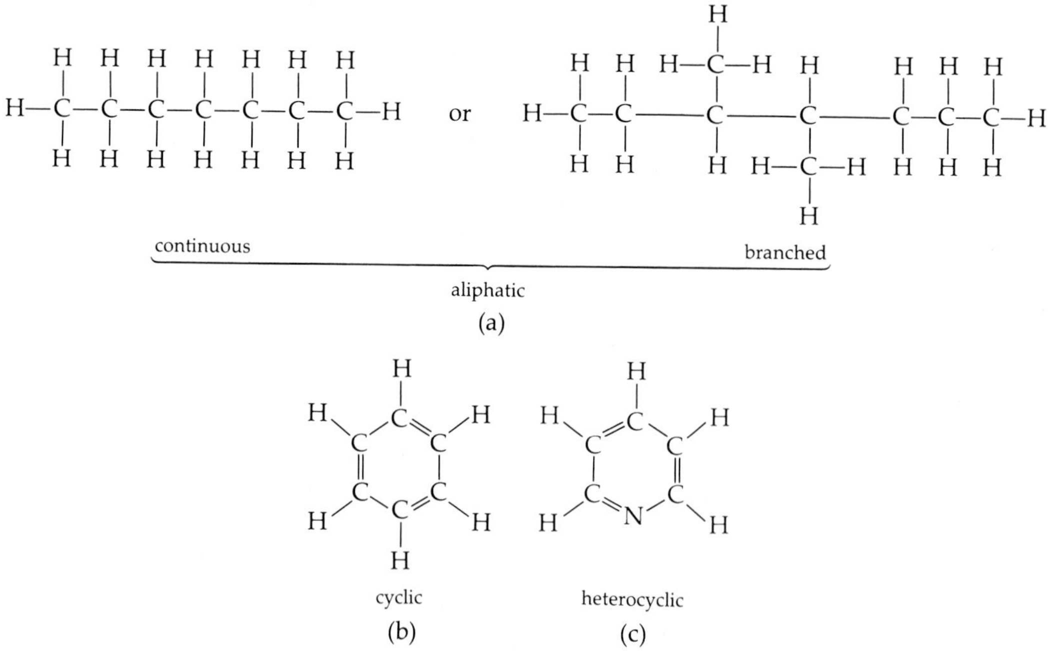

or
continuous
branched
aliphatic
(a)
cyclic
(b)
heterocyclic
(c)

Carbon atoms can join together to form continuous or branched chains of carbon atoms. Compounds of this type are called **aliphatic compounds** (Figure 15-3a). Carbon compounds can also bond together in the shape of rings to form **cyclic** (also called **aromatic**) compounds (Figure 15-3b). A third type of organic compounds are the **heterocyclic compounds,** which also have a ring structure. However, this ring structure contains some element other than carbon in the ring (Figure 15-3c). In this particular case the other element is nitrogen.

Carbon Compounds with Other Elements

Carbon also forms compounds with other elements besides hydrogen and oxygen. If a chlorine atom is bonded to a carbon atom, the compound chloromethane or methyl chloride is formed.

$$
\begin{array}{c}
\text{H} \\
| \\
\text{H}-\overset{\displaystyle}{\underset{\displaystyle |}{\text{C}}}-\text{Cl} \\
| \\
\text{H}
\end{array}
$$

Examples of carbon bonded to oxygen and nitrogen appear earlier in this chapter.

Summary

Organic chemistry is defined as the chemistry of carbon compounds. Most organic compounds differ from inorganic compounds as follows: they are combustible; they have lower melting points; they are insoluble in water; reaction takes place between molecules rather than between ions; the molecules contain many atoms; the molecules have a complex structure.

In organic chemistry the term *bond* is used rather than oxidation number. A bond is indicated by a short line and represents a pair of shared electrons. The carbon atom always has four bonds associated with it; the oxygen atom, two; the hydrogen atom, one.

Structural formulas are used for organic compounds rather than molecular formulas because the same molecular formula can often represent more than one structural formula.

Isomers are compounds having the same molecular formula but different structural formulas.

The four single bonds of the carbon atom are arranged in three dimensions in a tetrahedral structure with the carbon atom at the center and each bond pointing to a corner.

Organic compounds can be divided into three categories: aliphatic compounds, aromatic compounds, and heterocyclic compounds. Heterocyclic compounds contain elements other than carbon in the ring.

Carbon also forms compounds with other elements besides hydrogen and oxygen.

Questions and Problems

A

1. What was Wöhler's best-known contribution to organic chemistry?
2. Compare the properties of organic and inorganic compounds.
3. What is a "bond" in an organic compound?
4. How are the bonds arranged around the carbon atom in methane?
5. Why are structural formulas rather than molecular formulas used for organic compounds?
6. What are isomers? Do all organic compounds have isomers?
7. The compound C_5H_{12} has three isomers. Draw their structures.
8. What are the three major types of organic compounds?

B

9. Phosphorus is in group VA of the periodic chart. How many bonds should be attached to a phosphorus atom?
10. How many bonds should be attached to sulfur? bromine?
11. What is the tetrahedral angle?
12. Is it possible for a heterocyclic compound to contain oxygen in the ring? hydrogen?

Practice Test

1. Carbon forms how many bonds?
   a. 1   b. 2   c. 3   d. 4
2. Hydrogen forms how many bonds?.
   a. 1   b. 2   c. 3   d. 4

3. Organic compounds consisting of continuous or branched chains are called _____.
   a. aliphatic   b. aromatic
   c. heterocyclic   d. cyclic
4. The bonds of a carbon atom are arranged in what shape?.
   a. planar   b. trigonal
   c. tetrahedral   d. octahedral
5. Organic compounds having the same molecular formula but different structural formulas are called _____.
   a. polymers   b. tetramers
   c. cyclic   d. isomers
6. Organic compounds have _____.
   a. complex structures
   b. many atoms
   c. low melting points
   d. all of these
7. Most organic compounds are _____.
   a. flammable
   b. insoluble in water
   c. held together by covalent bonds
   d. all of these
8. An organic bond represents _____.
   a. a pair of shared electrons
   b. a pair of shared protons
   c. a transfer of electrons
   d. two pairs of shared electrons
9. An aromatic compound consists of a _____.
   a. chain of carbon atoms
   b. ring of carbon atoms
   c. halogenated organic compound
   d. none of these
10. Carbon can bond to _____.
    a. O   b. H
    c. N   d. all of these

16

Hydrocarbons

Engineers monitor the production of Saran wrap brand plastic film.

Alkanes

As the name implies, hydrocarbons are compounds that contain carbon and hydrogen only.

Consider the hydrocarbon with only one carbon atom. Since a carbon atom must have four bonds, four hydrogen atoms can be attached to that carbon atom. The hydrocarbon thus formed is called methane and has the structural formula shown in Figure 16-1.

If two carbon atoms are bonded together, six hydrogen atoms can be joined to them. This hydrocarbon is called ethane. The molecular formula is C_2H_6, and the structure is shown in Figure 16-2.

The hydrocarbon of three carbon atoms needs eight hydrogens to satisfy all the bonds. This compound, C_3H_8, is called propane (see Figure 16-3).

These compounds are called alkanes, and they are said to be saturated; that is, they have single covalent bonds between carbon atoms.

Table 16-1 lists several hydrocarbons. Note that the names of all alkanes end in -ane. The names of the first four compounds are not

Figure 16-1 Methane.

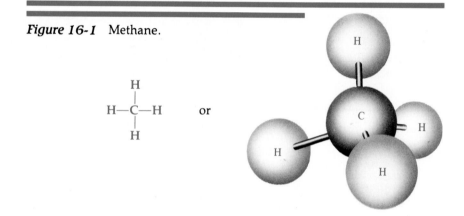

Figure 16-2 Ethane.

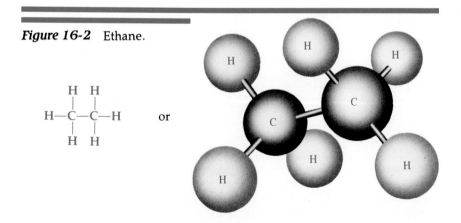

Figure 16-3   Propane.

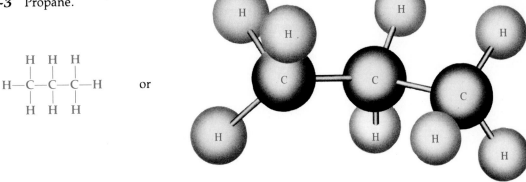

H—C—C—C—H   or

Table 16-1   Alkanes

Number of Carbon Atoms	Name	Molecular Formula	Structural Formula	Condensed Structural Formula
1	Methane	CH_4	H—C—H (with H above and H below)	CH_4
2	Ethane	C_2H_6	H—C—C—H	CH_3—CH_3 or CH_3CH_3
3	Propane	C_3H_8	H—C—C—C—H	CH_3—CH_2—CH_3 or $CH_3CH_2CH_3$
4	Butane	C_4H_{10}	H—C—C—C—C—H and H—C—C—C—H (with H—C—H below)	CH_3—CH_2—CH_2—CH_3 or $CH_3(CH_2)_2CH_3$ CH_3—CH—CH_3 with CH_3 or $CH_3CH(CH_3)CH_3$
5	Pentane	C_5H_{12}	3 isomers[a]	
6	Hexane	C_6H_{14}	5 isomers	
7	Heptane	C_7H_{16}	9 isomers	
8	Octane	C_8H_{18}	18 isomers	
9	Nonane	C_9H_{20}	35 isomers	
10	Decane	$C_{10}H_{22}$	75 isomers	

[a] The number of isomers increases rapidly as the number of carbons in the compound increases. $C_{40}H_{82}$ has 62,491,178,805,831 possible isomers.

systematic; they were named before a system of nomenclature was devised. Beginning with the hydrocarbon containing five carbon atoms, however, the names follow a definite pattern. The name **pentane** was derived from the prefix **penta-,** meaning five; pentane contains five carbon atoms. The name **hexane** is derived from the prefix **hexa-,** which means six; and so on through the series.

Structural formulas can be condensed as indicated in Table 16-1 and discussed on the next page.

General Formula

The general formula for alkanes is

$$C_nH_{2n+2}$$

where n represents the number of carbon atoms. That is, if a compound contains n carbon atoms, the number of hydrogen atoms is twice n plus two more.

Butane contains four carbon atoms. The number of hydrogens should be $(2 \times 4) + 2$ or 10, which agrees with the molecular formula for butane listed in Table 16-1. Likewise, octane has eight carbon atoms and so should have $(2 \times 8) + 2$, or 18, hydrogen atoms, which it does.

Continuing with a larger number of carbon atoms, the alkane containing 16 carbon atoms would have $(2 \times 16) + 2$, or 34, hydrogen atoms, giving it the formula $C_{16}H_{34}$.

Carbon atoms connected by single bonds exhibit *free rotation* about those bonds. That is, the carbon atoms can rotate with respect to each other. Thus the molecular formula C_3H_8, which represents a continuous chain of three carbon atoms, can be diagrammed in several ways, such as

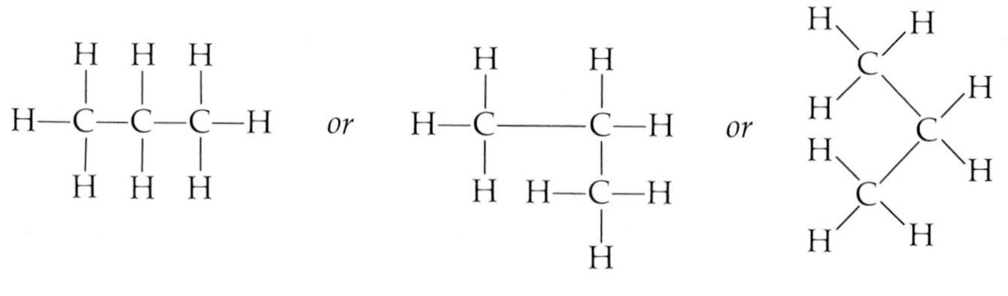

However, they all represent the same compound, which consists of three carbon atoms connected in one continuous chain.

In larger molecules, bonding between carbon atoms usually produces twisted shapes because of molecular interactions. However, for simplicity we will consider such compounds as having straight chains.

In the case of butane, C_4H_{10}, we can draw several models for four carbon atoms connected continuously

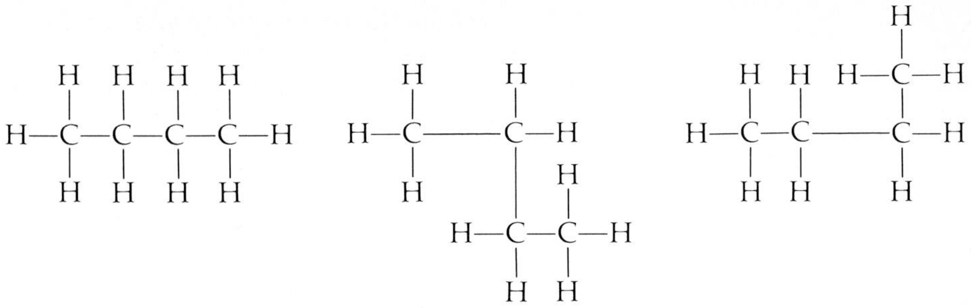

However, because of free rotation of the carbon-to-carbon bonds, they all represent the same compound, consisting of four continuously connected carbon atoms.

But the structure

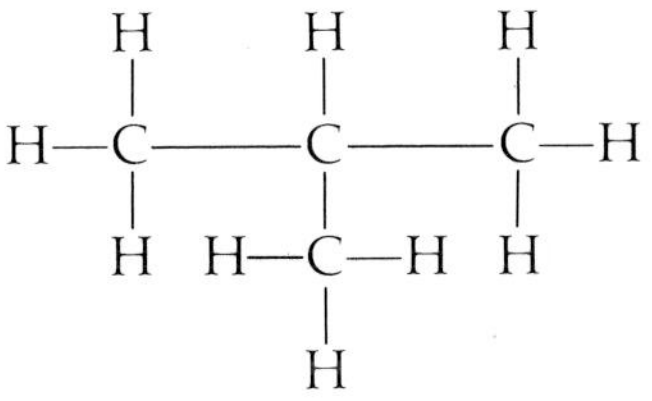

which also has the molecular formula C_4H_{10}, cannot rotate its bonds to form a continuous chain of four carbon atoms. Therefore it is an isomer—it has the same molecular formula as the preceding compound but a different structural formula.

Condensed Structural Formulas

It is usually too cumbersome to write a structural formula showing all the bonds and all the individual hydrogen atoms. So we frequently condense the formula, writing the hydrogens attached to a carbon atom after that carbon atom.

There are several different types of condensed structural formulas that can be used according to choice. For example,

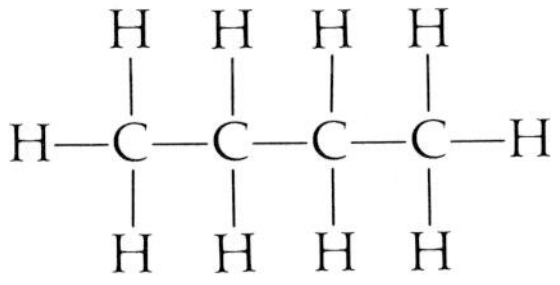

can be condensed to

$$CH_3—CH_2—CH_2—CH_3$$

It can be condensed even further by omitting the bonds and assuming that they are present. Thus

$$CH_3CH_2CH_2CH_3$$

By grouping identical arrangements of atoms and indicating the number of these arrangements with a subscript, another condensed formula for the same compound can be written as

$$CH_3(CH_2)_2CH_3$$

For a branched compound, any group attached to the chain is indicated by parentheses placed around that group following the carbon to which it is attached. For example, the formula of the compound

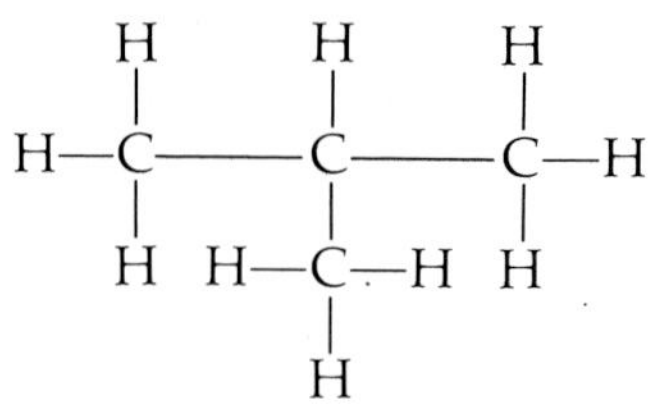

$$\begin{array}{ccccc}
& H & & H & & H \\
& | & & | & & | \\
H- & C & - & C & - & C-H \\
& | & & | & & | \\
& H & H- & C-H & H \\
& & & | & \\
& & & H &
\end{array}$$

can be condensed to

$$CH_3-CH(CH_3)-CH_3 \quad or \quad CH_3CH(CH_3)CH_3$$

Sometimes carbon atoms are grouped together to form a condensed structural formula. Thus

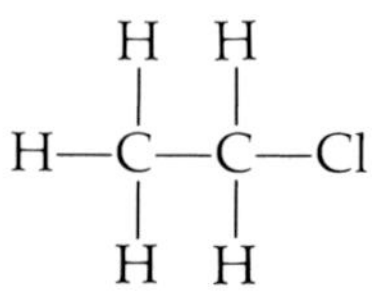

$$\begin{array}{ccc}
& H & H \\
& | & | \\
H- & C- & C-Cl \\
& | & | \\
& H & H
\end{array}$$

can be condensed to CH_3-CH_2-Cl or CH_3CH_2Cl *or* C_2H_5Cl.

Alkyl Groups

When a hydrogen atom is removed from an alkane, an alkyl group is formed. The names of the alkyl groups are obtained by changing the ending of the name from *-ane* to **-yl.** The alkyl group of one carbon atom formed from the alkane methane is called the **methyl** group. The alkyl group of two carbon atoms formed from ethane is called the **ethyl** group. The alkyl group of three carbon atoms formed from propane is called propyl (see Table 16-2).

The compound CH_3Cl, or

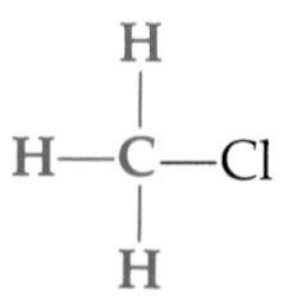

$$\begin{array}{c}
H \\
| \\
H-C-Cl \\
| \\
H
\end{array}$$

	Name of Alkyl Group	Condensed Structural Formula
Table 16-2 Simple Alkyl Groups	Methyl	CH_3—
	Ethyl	CH_3—CH_2— *or* C_2H_5—
	Propyl	CH_3—CH_2—CH_2— *or* C_3H_7—
	Butyl	CH_3—CH_2—CH_2—CH_2— *or* C_4H_9—

is made up of a methyl group (CH_3—) attached to a chlorine atom. It is called methyl chloride. The compound C_2H_5I consists of an ethyl group (C_2H_5—) bonded to an iodine. It is called ethyl iodide.

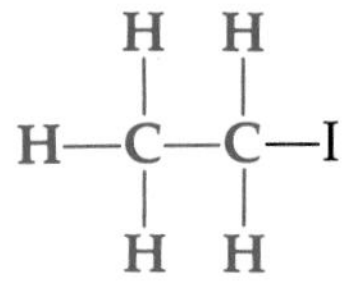

$$H—\overset{\displaystyle H}{\underset{\displaystyle H}{C}}—\overset{\displaystyle H}{\underset{\displaystyle H}{C}}—I$$

Naming Alkanes

An international system of nomenclature for organic compounds has been devised and is recognized and used by chemists all over the world. This system was devised and approved by the International Union of Pure and Applied Chemistry and is frequently designated by the initials *IUPAC. The rules of the IUPAC system are

1. Pick out the longest continuous chain of carbon atoms.
2. Identify that chain as an alkane.
3. Pick out the alkyl groups attached to that chain.
4. Number the carbons in the chain, starting at whichever end of the chain will give the smallest numbers to the carbons to which the alkyl groups are attached. Continue the numbering of this carbon chain in the same direction from one end to the other.
5. List the numbers and the names of the alkyl groups in alphabetical order.
6. Use commas between numbers and a dash between a number and a letter.

> *Example 16-1* Name the following compound. 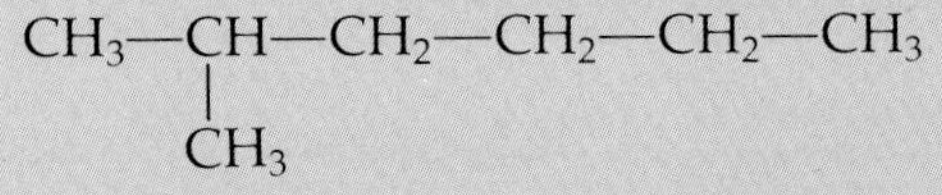
>
> $$CH_3—\underset{\displaystyle \overset{|}{CH_3}}{CH}—CH_2—CH_2—CH_2—CH_3$$
>
> The longest chain contains six carbon atoms; therefore, this compound is a *hexane*. To identify the alkyl group attached to the chain and also to identify the chain itself, it is sometimes

easier to highlight the chain. Then whatever is attached to the chain, the alkyl groups, will be outside the shading and can be easily noticed.

$$CH_3\text{—}CH\text{—}CH_2\text{—}CH_2\text{—}CH_2\text{—}CH_3$$
$$\underset{CH_3}{|}$$

Attached to the chain (sticking out) is an alkyl group of one carbon atom, the CH_3—or methyl group. Thus, this compound is a methylhexane.

The next step calls for numbering the carbons in the chain. They can be numbered in either direction, from left to right or from right to left. It should be observed that the methyl group is on the second carbon atom from the left end or on the fifth carbon atom from the right end. The rule states that the numbering should be such that the carbon to which the alkyl group is attached has the smallest number. Therefore the correct name of this compound is 2-methylhexane, which indicates that there is a methyl group on the second carbon from the end in a chain consisting of six carbon atoms. If the methyl group had been pointing upward instead of down, the compound would still be the same and so would the name.

Example 16-2   Name the following compound.

$$\overset{\displaystyle CH_3}{\underset{\displaystyle CH_3}{CH_3\text{—}CH_2\text{—}CH\text{—}CH\text{—}CH_3}}$$

The longest chain contains five carbon atoms. This compound is a *pentane*. There are two methyl groups attached to the chain, one on the second carbon and one on the third carbon. This time the numbering is from right to left in order to obtain the lowest numbers for the alkyl groups.

$$\overset{\displaystyle CH_3}{\underset{\displaystyle CH_3}{\overset{5}{C}H_3\text{—}\overset{4}{C}H_2\text{—}\overset{3}{C}H\text{—}\overset{2}{C}H\text{—}\overset{1}{C}H_3}}$$

This compound is called 2,3-dimethylpentane, where the prefix *di-* indicates that there are two identical groups. Dimethyl means two methyl groups, and the numbers tell us on which carbon atoms they are located.

Whenever an alkyl group appears more than once in a compound, a prefix is used to designate how many of these alkyl groups are present in that compound. The most commonly used prefixes are

di-	which means two
tri-	which means three
tetra-	which means four
penta-	which means five

Example 16-3 Name the following compound.

$$CH_3-CH-CH-CH_2-CH-CH-CH_2-CH_3$$

with CH_3 and CH_3 above, and CH_3 and C_2H_5 below.

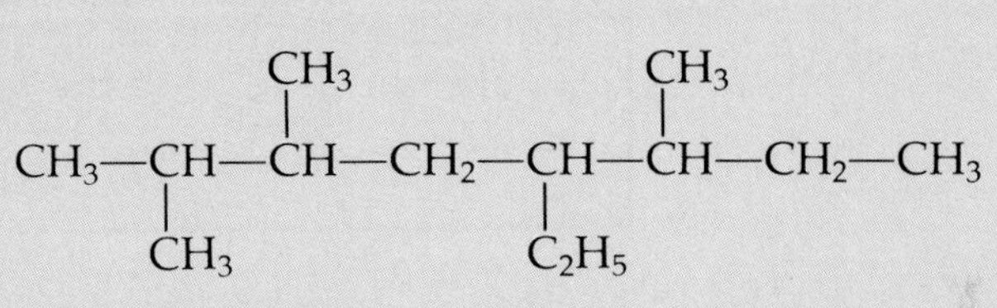

We first highlight the longest chain. This chain contains eight carbon atoms, so the compound is an *octane*.

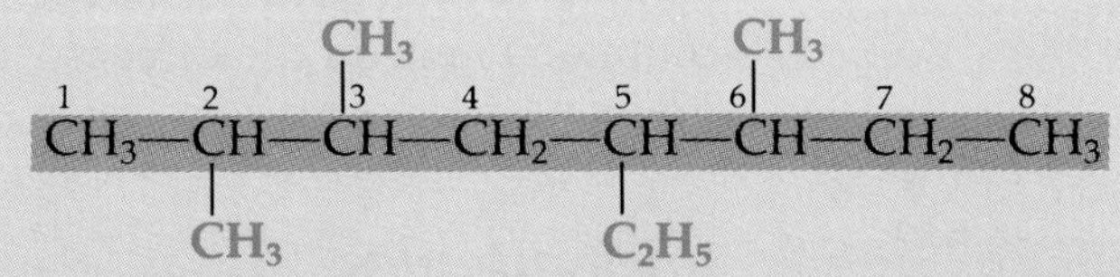

$$\overset{1}{CH_3}-\overset{2}{CH}-\overset{3}{CH}-\overset{4}{CH_2}-\overset{5}{CH}-\overset{6}{CH}-\overset{7}{CH_2}-\overset{8}{CH_3}$$

with CH_3 above carbons 3 and 6, and CH_3, C_2H_5 below.

Attached to the chain are three methyl groups and one ethyl group. The chain should be numbered from left to right in order to obtain alkyl groups of the lowest numbers. There are methyl groups on carbons numbered 2, 3, and 6 and an ethyl group on carbon number 5 of the eight carbon chain.

The correct name of this compound is **5-ethyl-2,3,6-trimethyloctane.** Note that the alkyl groups are named in alphabetical order. Also note that the prefix *tri-* is used to indicate three alkyl groups of the same type.

Example 16-4 Name the following compound.

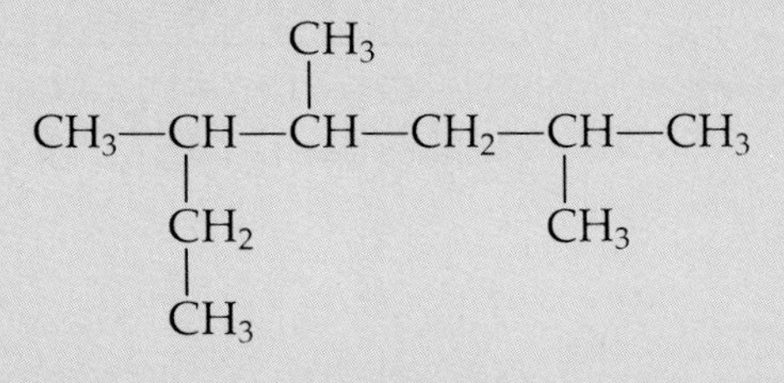

$$CH_3-CH-CH-CH_2-CH-CH_3$$

with CH_3 above, and CH_2, CH_3 below, and CH_3 below the CH_2.

First highlight the longest chain. Note that this chain contains seven carbon atoms so that this compound is a *heptane*.

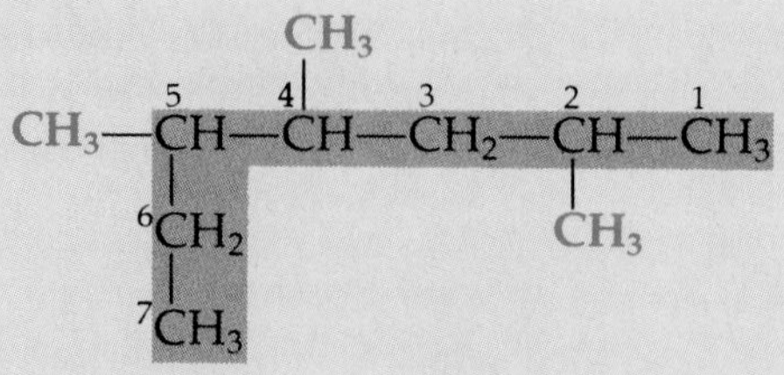

$$CH_3-\overset{5}{C}H-\overset{4}{C}H-\overset{3}{C}H_2-\overset{2}{C}H-\overset{1}{C}H_3$$

with CH₃ attached above carbon 4, CH₃ below carbon 2, and ⁶CH₂—⁷CH₃ below carbon 5.

Attached to the chain are three methyl groups at carbons numbered 2, 4, and 5. Thus the name of this compound is 2,4,5-trimethylheptane. Recall rule number 1 (page 263), which says "pick out the longest continuous chain of carbon atoms." If you had picked a chain of six carbon atoms (straight across) this would not have been the longest continuous chain of carbon atoms.

Many simple hydrocarbons and hydrocarbon derivatives have common names that were in use before the IUPAC system was adopted. Many of these names are still in use although IUPAC naming is preferred. Table 16-3 indicates the IUPAC name and the common name for the three isomers of pentane, C_5H_{12}. Note that the prefix *n*- (for normal) indicates a straight-chain compound.

Cycloalkanes

Alkanes also exist in the shape of a ring. Such structures are called cycloalkanes. They are named by placing the prefix cyclo- before the name of the corresponding straight-chain alkane. Thus, the cyclic alkane of three carbons is called cyclopropane. Its structure is

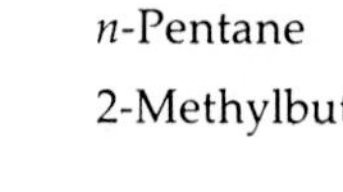

$$\begin{array}{c} CH_2 \\ \diagup \quad \diagdown \\ H_2C \underline{\qquad} CH_2 \end{array}$$

	Structure	IUPAC Name	Common Name
Table 16-3 Names of Simple Hydrocarbons	$CH_3-CH_2-CH_2-CH_2-CH_3$	*n*-Pentane	Pentane

	$CH_3-CH-CH_2-CH_3$
	      $\mid$
	     CH_3
		2-Methylbutane	Isopentane

	$\quad\;\;CH_3$ $\quad\;\;\mid$ CH_3-C-CH_3 $\quad\;\;\mid$ $\quad\;\;CH_3$	2,2-Dimethylpropane	Neopentane

Cyclopropane is used medicinally as a general anesthetic. Both induction and recovery time are short. Muscular relaxation is greater than with nitrous oxide but less than with ether. As with ether the danger of explosion is great, so care must be taken in its use.

The structures of cycloalkanes can also be abbreviated by using a geometric shape in which each line represents a bond between carbon atoms and each corner a carbon atom and its hydrogens. The following structures represent cyclopropane and cyclobutane, respectively.

cyclopropane   cyclobutane

When alkyl groups are attached to a cycloalkane, the compound is numbered in such a manner as to give the groups the lowest numbers. When only one group is attached to a cycloalkane, the number 1 is understood and not written.

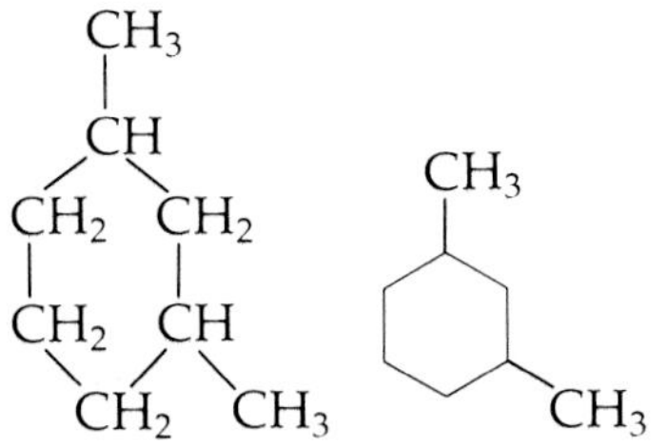

1,3-dimethylcyclohexane 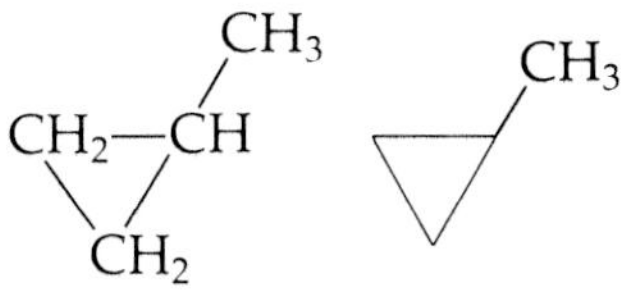   methylcyclopropane

For an additional method of naming cycloalkanes, see page 354.

Haloalkanes

Alkanes may have one or more *halogen atoms replacing hydrogens. Such compounds are called haloalkanes. The simplest chloroalkane is

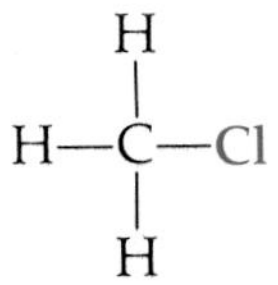

$$H-\underset{\underset{H}{|}}{\overset{\overset{H}{|}}{C}}-Cl$$

chloromethane
(methyl chloride)

The IUPAC name is given first followed by the common name in parentheses.

Further substitution of chlorines for hydrogen leads to

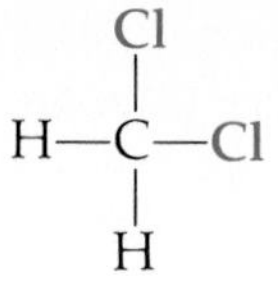

$$
\begin{array}{ccc}
\text{Cl} & \text{Cl} & \text{Cl} \\
| & | & | \\
\text{H—C—Cl} & \text{H—C—Cl} & \text{Cl—C—Cl} \\
| & | & | \\
\text{H} & \text{Cl} & \text{Cl}
\end{array}
$$

dichloromethane	trichloromethane	tetrachloromethane
(methylene chloride)	(chloroform)	(carbon tetrachloride)

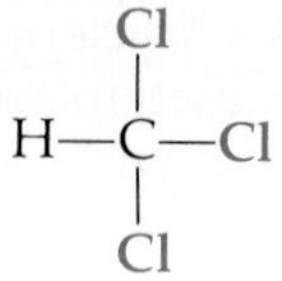

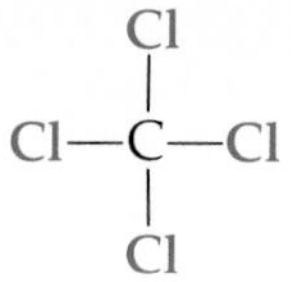

Similar compounds can be formed with bromine and iodine. The common name for $CHCl_3$ is chloroform. Likewise, $CHBr_3$ is bromoform, and CHI_3 is iodoform.

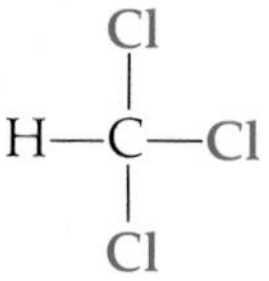

$$
\begin{array}{ccc}
\text{Cl} & \text{Br} & \text{I} \\
| & | & | \\
\text{H—C—Cl} & \text{H—C—Br} & \text{H—C—I} \\
| & | & | \\
\text{Cl} & \text{Br} & \text{I}
\end{array}
$$

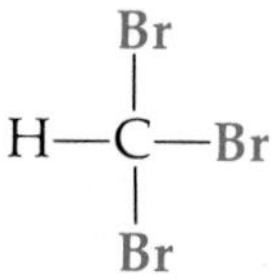

trichloromethane	tribromomethane	triiodomethane
(chloroform)	(bromoform)	(iodoform)

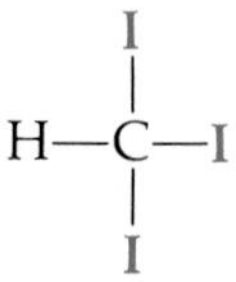

Since there is only one carbon atom, no numbers are necessary.

Table 16-4 indicates both the IUPAC and common names of some simple haloalkanes.

A hydrocarbon derivative used as an anesthetic is halothane, 2-bromo-2-chloro-1,1,1-trifluoroethane. (Note that the groups attached to the carbons are named in alphabetical order.)

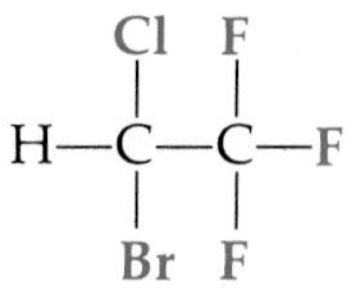

$$
\begin{array}{cc}
\text{Cl} & \text{F} \\
| & | \\
\text{H—C—C—F} \\
| & | \\
\text{Br} & \text{F}
\end{array}
$$

Its main advantage over other anesthetics formerly used is that it is nonflammable and is not irritating to the respiratory passages. Halothane is usually used in conjunction with nitrous oxide (see page 147) and with muscle relaxants to provide general anesthesia for surgery of all types, but it can also be administered alone.

A similar compound, chloral hydrate, induces sleep and prevents convulsions. A combination of ethyl alcohol and chloral hydrate is known as a "Mickey Finn" or "knockout drops."

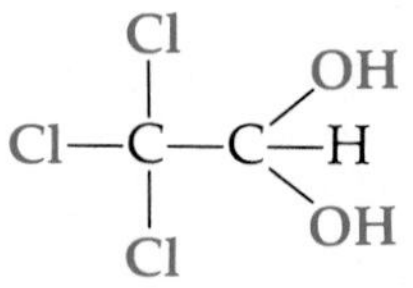

$$
\begin{array}{cc}
\text{Cl} & \\
| & \text{OH} \\
\text{Cl—C—C—H} & \\
| & \text{OH} \\
\text{Cl} &
\end{array}
$$

chloral hydrate

In veterinary medicine, chloral hydrate is used as a narcotic and anesthetic for cattle, horses, and poultry.

Other widely used haloalkanes include ethyl chloride, a fast-

Table 16-4   IUPAC and Common Names of Some Haloalkanes

Structure	IUPAC Name	Common Name
$CH_3—CH_2—CH_2—CH_2Br$	1-Bromobutane	*n*-Bromobutane
$CH_3—CH_2—CH—CH_3$ $\quad\quad\quad\;\;\mid$ $\quad\quad\quad\;\;Br$	2-Bromobutane	*sec*-Bromobutane[a]
$CH_3—CH—CH_2Br$ $\quad\quad\;\mid$ $\quad\quad\;CH_3$	1-Bromo-2-methylpropane	Isobutyl bromide
$\quad\quad\;CH_3$ $\quad\quad\;\mid$ $CH_3—C—CH_3$ $\quad\quad\;\mid$ $\quad\quad\;Br$	2-Bromo-2-methylpropane	*tert*-Bromopropane[a]

[a] *sec* stands for secondary and *tert* for tertiary.

acting topically applied local anesthetic; Freon-12, used in aerosol containers and in refrigeration and air conditioning units (see page 132); Teflon and poly(vinyl chloride) (PVC), which are polymers. Teflon is used as a coating material for cooking utensils, and PVC is used in making plastic containers. Their structures are as follows:

$$CH_3—CH_2—Cl \qquad H—\overset{\displaystyle Cl}{\underset{\displaystyle Cl}{\overset{\mid}{\underset{\mid}{C}}}}—\overset{\displaystyle F}{\underset{\displaystyle F}{\overset{\mid}{\underset{\mid}{C}}}}—H$$

ethyl chloride   Freon-12

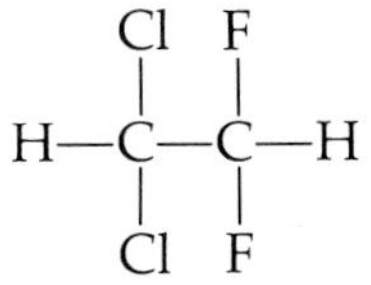

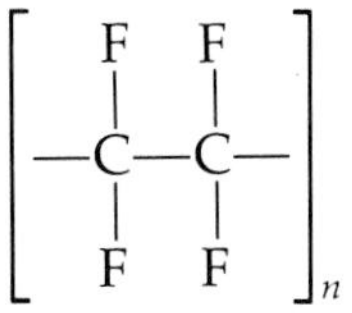

$$\left[\; \overset{\displaystyle F}{\underset{\displaystyle F}{\overset{\mid}{\underset{\mid}{C}}}}—\overset{\displaystyle F}{\underset{\displaystyle F}{\overset{\mid}{\underset{\mid}{C}}}}— \;\right]_n \qquad \left[\; —CH_2—\overset{\displaystyle Cl}{\underset{\mid}{\overset{\mid}{C}H_2}} \;\right]_n$$

Teflon 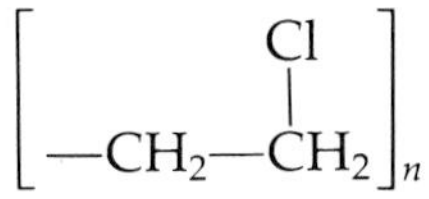   poly(vinyl chloride)

Alkenes

Alkanes have a single bond between the carbon atoms. Alkenes have a double bond (two bonds) between two of the carbon atoms.

$$C{=\!=}C$$

Consider two carbon atoms connected by a double bond. Since this double bond uses four electrons from both carbons, a total of only four hydrogen atoms will satisfy all of the remaining bonds. Recall that a single bond represents a pair of shared electrons; a double bond represents two pairs of shared electrons.

This compound thus becomes

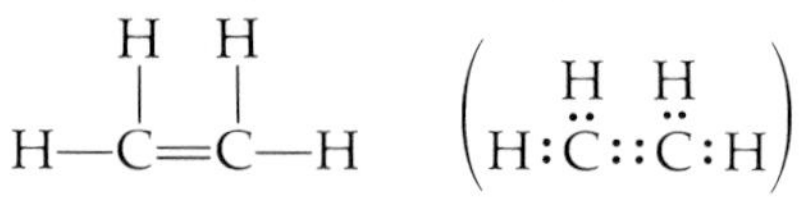

$$
\underset{\substack{|\\H}}{H}-\overset{\substack{H\\|}}{C}=\overset{\substack{H\\|}}{C}-H
\qquad
\left(H:\overset{..}{C}::\overset{..}{C}:H\right)
$$

and has the molecular formula C_2H_4. It is called **ethene.**

When three carbon atoms are arranged in a chain with a double bond between two of the carbon atoms, C=C—C, how many hydrogen atoms must be connected to these carbon atoms in order to satisfy all of the bond requirements? The answer is six, and the structure becomes

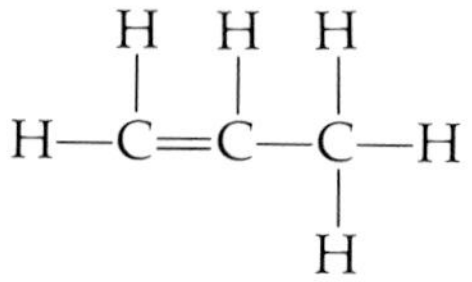

$$
H-\overset{\substack{H\\|}}{C}=\overset{\substack{H\\|}}{C}-\underset{\substack{|\\H}}{\overset{\substack{H\\|}}{C}}-H
$$

The molecular formula is C_3H_6. The name of this compound is **propene.**

Note that the names of these compounds end in **-ene.** This is true of all alkenes. It should be noted also that the names of these compounds are similar to those of the alkanes except for the ending, which is *-ene* instead of *-ane*.

Compare the structures of ethane (C_2H_6) and ethene (C_2H_4).

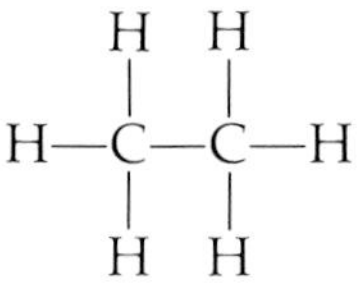

$$
\underset{\substack{|\\H}}{\overset{\substack{H\\|}}{H-C}}-\underset{\substack{|\\H}}{\overset{\substack{H\\|}}{C}}-H
\qquad\qquad
H-\overset{\substack{H\\|}}{C}=\overset{\substack{H\\|}}{C}-H
$$

$$
\text{ethane} \qquad\qquad\qquad \text{ethene}
$$

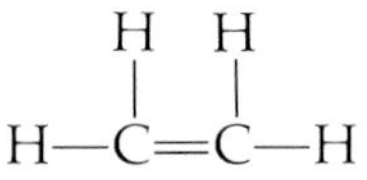

The three-carbon alkane is propane, and the three-carbon alkene is propene. Likewise the four-carbon alkane is called butane while the corresponding alkene is called butene. The names and formulas of some of the alkenes are listed in Table 16-5.

The general formula for alkenes is

$$
C_nH_{2n}
$$

There are twice as many hydrogen atoms as carbon atoms in every alkene. Thus octene has 8 carbon atoms and 16 hydrogen atoms, and the formula of an alkene of 15 carbon atoms would be $C_{15}H_{30}$.

The IUPAC names for alkenes are similar to those for alkanes except that the longest chain must contain the double bond. Consider the two isomers of butene listed in Table 16-5. The first is called 1-butene, indicating that the double bond is between carbons 1 and 2. The second compound is called 2-butene, indicating that the double bond lies between carbons 2 and 3. Note that the double bond is always given the lowest number, even if other groups are present.

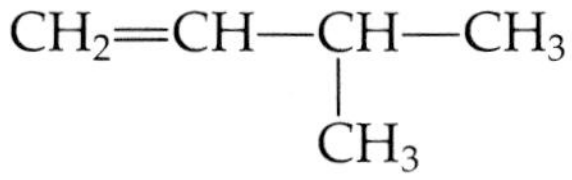

$$CH_2{=}CH{-}\underset{\underset{\displaystyle CH_3}{|}}{CH}{-}CH_3$$

3-methyl-1-butene

cyclobutene

(1 is understood, so it is assumed that the double bond is between carbon atoms 1 and 2)

Table 16-5 Some Simple Alkenes

No. of Carbon Atoms	Name	Molecular Formula	Structural Formula	Condensed Structural Formula
2	Ethene	C_2H_4	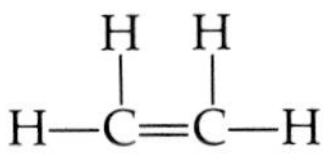	$CH_2{=}CH_2$
3	Propene	C_3H_6	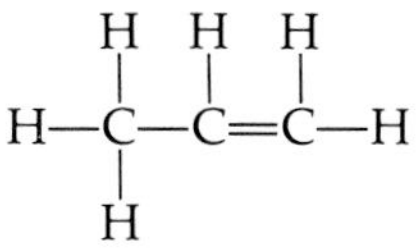	$CH_3CH{=}CH_2$
4	Butene	C_4H_8	*or*	$CH_2{=}CHCH_2CH_3$ *or* $CH_3CH{=}CHCH_3$
5	Pentene	C_5H_{10}	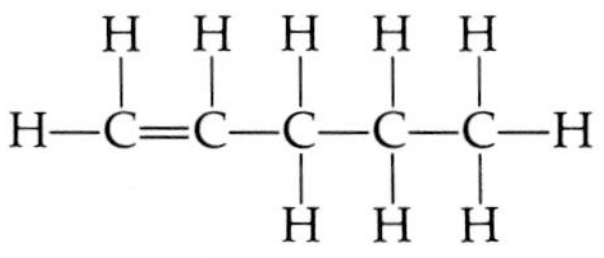 *or 5 other isomers*	$CH_2{=}CHCH_2CH_2CH_3$
6	Hexene	C_6H_{12}	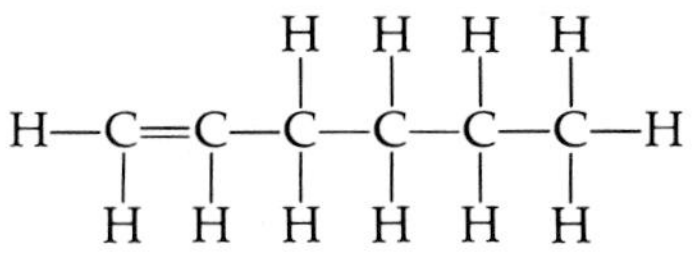 *or 14 other isomers*	$CH_2{=}CHCH_2CH_2CH_2CH_3$
7	Heptene	C_7H_{14}	30 isomers	
8	Octene	C_8H_{16}	66 isomers	

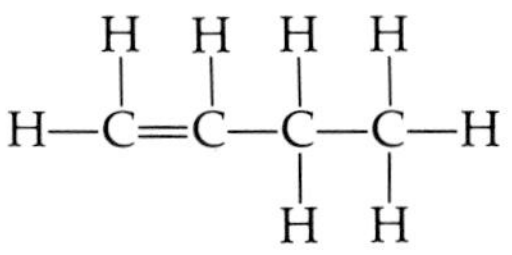

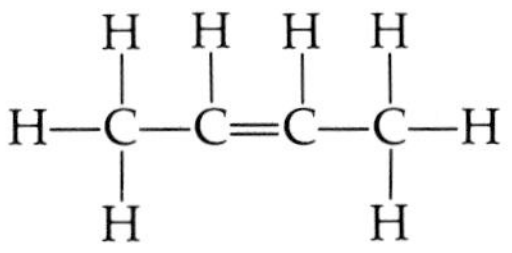

The following cycloalkene is numbered to show the double bond having the number 1 (understood) and the methyl group having the lowest possible number. If the numbering had been in the other direction, then the methyl group would have a larger number.

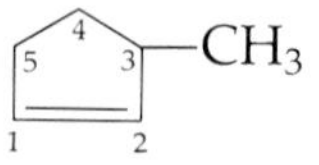

3-methylcyclopentene

The vinyl group has the structure

$$\begin{array}{cc} H & H \\ | & | \\ H-C{=}C- \end{array}$$

Vinyl chloride is used in the manufacture of such consumer products as floor tile, raincoats, phonograph records, fabrics, and furniture coverings. However, evidence has shown that several workers exposed to vinyl chloride during their work have died from a very rare form of liver cancer. In addition, exposure to vinyl chloride is suspected to be responsible for certain types of birth defects.

The structure of vinyl chloride is

$$\begin{array}{cc} H & H \\ | & | \\ H-C{=}C-Cl \end{array}$$

Alkynes

Consider two carbon atoms connected by a triple bond.

$$C{\equiv}C$$

How many hydrogen atoms must be connected to these two carbon atoms in order to satisfy all the bond requirements? The answer is two, so the molecular formula of this compound is C_2H_2. This compound is called **ethyne.** Its structure is

$$H-C{\equiv}C-H \quad (H{:}C{::}C{:}H)$$

Ethyne is commonly called acetylene. However, this name is not preferred because the ending *-ene* denotes a double bond, whereas this compound actually has a triple bond between the carbon atoms.

If three carbon atoms are placed in a chain with a triple bond between two of them,

$$C{\equiv}C-C$$

only four hydrogen atoms can be placed around these carbons to satisfy all the bonds. The compound then becomes

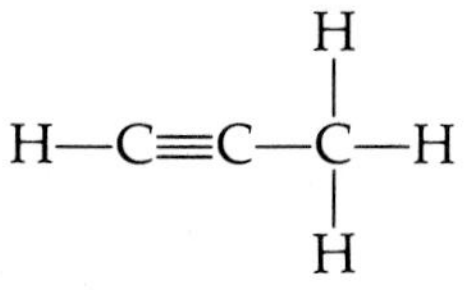

$$H-C\equiv C-\overset{\displaystyle H}{\underset{\displaystyle H}{\vphantom{|}C}}-H$$

with the molecular formula C_3H_4. This compound is called **propyne**.

These two compounds ethyne and propyne, are **alkynes**. They have a triple bond between two of the carbon atoms. All their names end in **-yne.** The general formula for alkynes is

$$C_nH_{2n-2}$$

Thus hexyne, which has six carbon atoms, has the formula $C_6H_{(2\times6)-2}$, or C_6H_{10}. Likewise, octyne has the molecular formula C_8H_{14}.

Alkynes are relatively rare compounds and do not normally occur in the human body.

As with alkenes, alkynes are named with the triple bond having the smallest number.

$$HC\equiv C-CH_2-CH_2-CH_3 \qquad CH_3-C\equiv C-CH_2-CH_3$$

1-pentyne   2-pentyne

$$CH_3-C\equiv C-\overset{\displaystyle}{\underset{\displaystyle CH_3}{\vphantom{|}CH}}-CH_2-CH_3$$

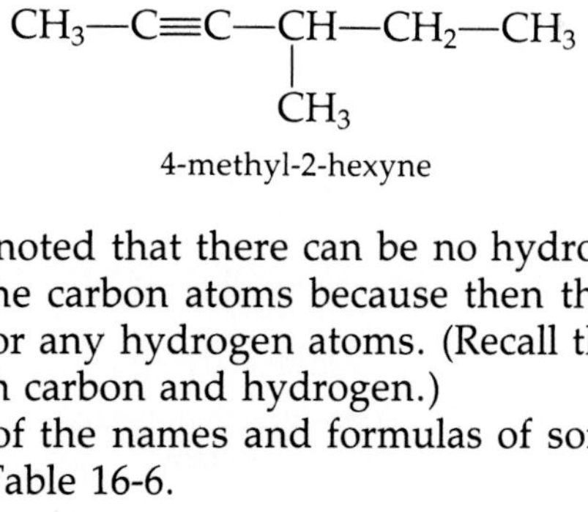

4-methyl-2-hexyne

It should be noted that there can be no hydrocarbon with four bonds between the carbon atoms because then there would be no bonds available for any hydrogen atoms. (Recall that hydrocarbons must contain both carbon and hydrogen.)

A summary of the names and formulas of some hydrocarbons can be found in Table 16-6.

Table 16-6 Hydrocarbons

No. of Carbon Atoms	Alkanes	Alkenes	Alkynes
1	Methane CH_4	—	—
2	Ethane C_2H_6	Ethene C_2H_4	Ethyne C_2H_2
3	Propane C_3H_8	Propene C_3H_6	Propyne C_3H_4
4	Butane C_4H_{10}	Butene C_4H_8	Butyne C_4H_6
General formula	C_nH_{2n+2}	C_nH_{2n}	C_nH_{2n-2}

Saturated and Unsaturated Hydrocarbons

Saturated hydrocarbons are those that have only single bonds between the carbon atoms. Alkanes are saturated compounds.

Unsaturated hydrocarbons contain at least one double bond or triple bond. Both alkenes and alkynes are unsaturated compounds.

Reactions of Saturated Hydrocarbons

Saturated hydrocarbons react by a process known as substitution. When ethane reacts with chlorine, Cl_2 or Cl—Cl, one of the chlorine atoms substitutes for one of the hydrogen atoms in the saturated compound.

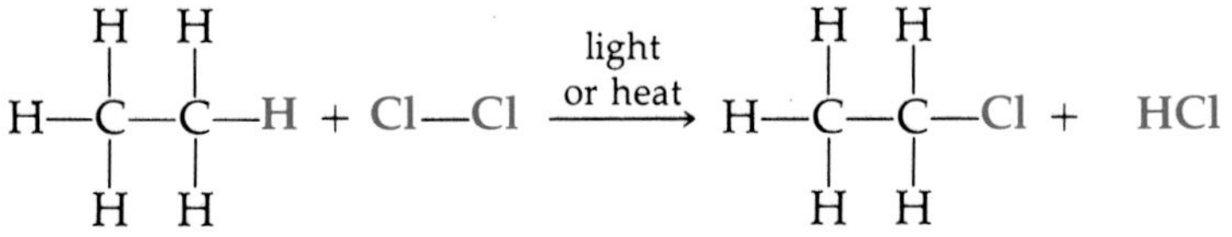

$$
\underset{\text{ethane}}{H-\overset{\displaystyle H}{\underset{\displaystyle H}{C}}-\overset{\displaystyle H}{\underset{\displaystyle H}{C}}-H} + \underset{\text{chlorine}}{Cl-Cl} \xrightarrow[\text{or heat}]{\text{light}} \underset{\substack{\text{chloroethane}\\\text{(ethyl chloride)}}}{H-\overset{\displaystyle H}{\underset{\displaystyle H}{C}}-\overset{\displaystyle H}{\underset{\displaystyle H}{C}}-Cl} + \underset{\substack{\text{hydrogen}\\\text{chloride}}}{HCl}
$$

Further chlorination may substitute additional chlorines for hydrogens.

Alkanes can be completely oxidized to yield carbon dioxide, water, and energy, as indicated in the following reactions.

$$
\underset{\text{methane}}{CH_4} + 2\,O_2 \longrightarrow CO_2 + 2\,H_2O + \text{energy}
$$

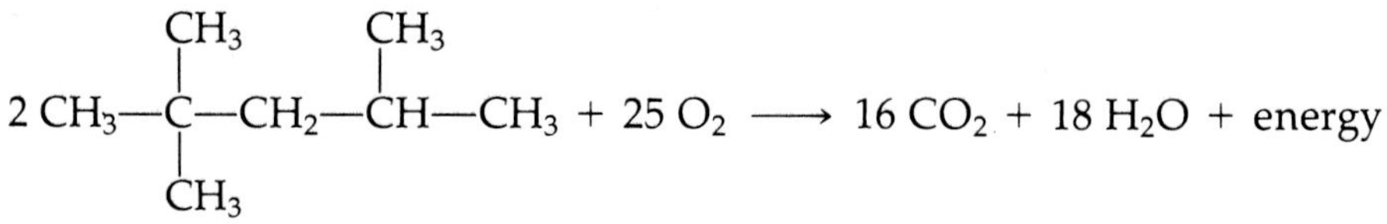

$$
2\ CH_3-\overset{\displaystyle CH_3}{\underset{\displaystyle CH_3}{C}}-CH_2-\overset{\displaystyle CH_3}{CH}-CH_3 + 25\,O_2 \longrightarrow 16\,CO_2 + 18\,H_2O + \text{energy}
$$

2,2,4-trimethylpentane
(isooctane)

Incomplete oxidation (combustion) of alkanes yields carbon monoxide in place of carbon dioxide. The equation for the incomplete oxidation of methane is

$$
2\,CH_4 + 3\,O_2 \longrightarrow 2\,CO + 4\,H_2O + \text{energy}
$$

Incomplete combustion of isooctane, a major component of gasoline, also yields carbon monoxide. This carbon monoxide is released into the air and becomes a deadly pollutant. Carbon monoxide is poisonous because it combines with hemoglobin and prevents that compound from carrying oxygen to the cells (see pages 544 and 551).

Reactions of Unsaturated Hydrocarbons

When ethene reacts with hydrogen (H_2) it reacts by a process known as addition. That is, the hydrogen atoms add to the double bond, making a single bond out of it. Chlorine (Cl_2) reacts similarly with propene.

$$
\begin{array}{ccc}
\underset{\text{ethene}}{H-\overset{\displaystyle H}{\underset{}{C}}=\overset{\displaystyle H}{\underset{}{C}}-H} \; + & \underset{\text{hydrogen}}{H_2} & \longrightarrow \quad \underset{\text{ethane}}{H-\overset{\displaystyle H}{\underset{\displaystyle H}{C}}-\overset{\displaystyle H}{\underset{\displaystyle H}{C}}-H}
\end{array}
$$

$$
\underset{\text{propene}}{H-\overset{\displaystyle H}{\underset{\displaystyle H}{C}}-\overset{\displaystyle H}{\underset{}{C}}=\overset{\displaystyle H}{\underset{}{C}}-H} \; + \; \underset{\text{chlorine}}{Cl-Cl} \; \longrightarrow \; \underset{\text{1,2-dichloropropane}}{H-\overset{\displaystyle H}{\underset{\displaystyle H}{C}}-\overset{\displaystyle H}{\underset{\displaystyle Cl}{C}}-\overset{\displaystyle H}{\underset{\displaystyle Cl}{C}}-H}
$$

Like alkenes, alkynes undergo addition reactions. Ethyne (acetylene) reacts with chlorine by a process of addition to form a single bond between the carbon atoms.

$$
\underset{\text{ethyne}}{H-C\equiv C-H} \; + \; \underset{\text{chlorine}}{2\,Cl-Cl} \; \longrightarrow \; \underset{\text{1,1,2,2-tetrachloroethane}}{H-\overset{\displaystyle Cl}{\underset{\displaystyle Cl}{C}}-\overset{\displaystyle Cl}{\underset{\displaystyle Cl}{C}}-H}
$$

In general the addition reaction takes place more rapidly and under milder conditions than the substitution reaction. Note that in these examples an unsaturated reactant has been changed into a saturated product. A practical application of this will be discussed in the hydrogenation of fats and oils (see page 387).

Sources of Hydrocarbons

The chief sources of hydrocarbons are petroleum and natural gas. Petroleum is a very complex mixture of solid, liquid, and gaseous hydrocarbons plus a few compounds of other elements. Natural gas is primarily a mixture of alkanes of one to four carbon atoms. In general, alkanes are gaseous if the compounds contain between 1 and 4 carbon atoms, liquid if they contain between 5 and 16 carbon atoms, and solid if they contain over 16 carbon atoms. The various hydrocarbons present in petroleum are isolated by a process known as fractional distillation.

One of the chief products of the fractional distillation of petroleum is gasoline. One measure of the quality of a gasoline is *octane number*. The octane number of a gasoline refers to how smoothly it burns. A high octane gasoline delivers power smoothly to the pistons. A low octane fuel burns too rapidly and causes the engine to "knock." Isooctane, 2,2,4-trimethylpentane, which has excellent combustion properties, was arbitrarily assigned an octane rating of 100. Normal heptane, *n*-heptane (C_7H_{16}), which causes considerable engine knocking, was assigned an octane rating of 0. Mixtures of these two compounds were burned in test engines to establish an "octane scale." That is, if a sample of gasoline has the same amount of knock as an 87 percent isooctane–13 percent heptane mixture, then that gasoline has an octane rating of 87.

$$CH_3-\underset{\underset{\displaystyle CH_3}{|}}{\overset{\overset{\displaystyle CH_3}{|}}{C}}-CH_2-\underset{\overset{\displaystyle CH_3}{|}}{CH}-CH_3 \qquad\qquad CH_3-CH_2-CH_2-CH_2-CH_2-CH_2-CH_3$$

isooctane   *n*-heptane

Until the mid-1960s automotive gasolines commonly contained the additive tetraethyllead [$Pb(C_2H_5)_4$]. This substance reduced knocking and increased the octane ratings of gasolines. However, it caused large amounts of lead compounds to be discharged into the air, with resulting hazards to human health. For that reason unleaded fuels must now be used in all modern cars.

Properties of Hydrocarbons

In general, as the number of alkane carbon atoms increases, the boiling point and density increase. Table 16-7 indicates the changes in these properties with increasing numbers of carbon atoms.

Table 16-7   Physical Properties of Some Alkanes

Condensed Structural Formula	Name	Boiling Point (°C)	Density (g/mL)
CH_4	Methane	−162	gas
CH_3CH_3	Ethane	−89	gas
$CH_3CH_2CH_3$	Propane	−42	gas
$CH_3(CH_2)_2CH_3$	Butane	−0.5	gas
$CH_3(CH_2)_3CH_3$	Pentane	36	0.626
$CH_3(CH_2)_4CH_3$	Hexane	69	0.659
$CH_3(CH_2)_5CH_3$	Heptane	98	0.684
$CH_3(CH_2)_6CH_3$	Octane	126	0.703
$CH_3(CH_2)_7CH_3$	Nonane	151	0.718
$CH_3(CH_2)_8CH_3$	Decane	174	0.730

Uses of Hydrocarbons

Methane is used for heating and cooking purposes both in the laboratory and in the home. Mineral oil is a mixture of saturated hydrocarbons and is used extensively in the hospital for lubricating purposes and also as a laxative. Rubber is also a hydrocarbon. Ethene (ethylene) and propene (propylene) have been used extensively to make plastics called polyethylene and polypropylene, respectively. Ethylene has been used as an anesthetic, but its popularity has declined since the introduction of neuromuscular blocking drugs. Acetylene has been employed by European surgeons as an anesthetic; however, the main use of acetylene (ethyne) in North America is for welding.

Polymers

When many small like or similar organic molecules are joined together to form a much larger molecule, the process is called **polymerization.** The product is called a polymer. Naturally occurring polymers include starch, cellulose, rubber, and protein. Synthetic polymers include plastics and such fibers as nylon, rayon, and Dacron (see page 286). Medical uses of polymers include synthetic heart valves and blood vessels, surgical mesh, disposable syringes, and drug containers (see Figure 16-4).

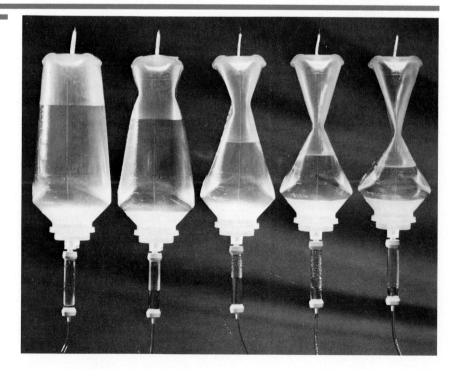

Figure 16-4 Plastic container in use for an intravenous infusion. [Courtesy American Hospital Supply Corporation, Evanston, IL.]

Table 16-8 Polymers in Commercial Use

Parent Compound	IUPAC Name (Common Name)	Polymer	Use
$CH_2{=}CH_2$	Ethene (Ethylene)	Polyethylene	Packaging materials
$CH_2{=}CHCH_3$	Propene (Propylene)	Polypropylene	Carpeting, clothing
$CH_2{=}CHC_6H_5$	Phenylethene (Styrene)	Polystyrene	Styrofoam products
$CH_2{=}CHCl$	Chloroethene (Vinyl chloride)	Poly(vinyl chloride) (PVC)	Tubing
$CH_2{-}CCl_2$	1,1-Dichloroethene (1,1-Dichloroethylene)		Plastic wrapping
$CF_2{-}CF_2$	Tetrafluoroethene (Tetrafluoroethylene)	Polytetrafluoroethylene	Teflon

The equation that follows indicates the formation of polyethylene from ethene (ethylene). Note that an initiator is required. This substance starts the reaction, which continues until a long chain is obtained. The number of units of ethylene joined together varies between 100 and 1000.

$$CH_2{=}CH_2 \xrightarrow[\text{initiator}]{\text{heat, pressure}} -(CH_2{-}CH_2){-}(CH_2{-}CH_2){-} \text{ etc.} \quad or \quad (-CH_2{-}CH_2{-})_n$$

ethene
(ethylene)

polyethylene

If the starting material is propene (propylene), then the polymer produced is called polypropylene. If styrene is the starting material, polystyrene is produced. Likewise, poly(vinyl chloride) (PVC) is produced from vinyl chloride.

Table 16-8 lists several polymers and their commercial use.

Summary

Hydrocarbons are organic compounds that contain the elements carbon and hydrogen only. The simplest hydrocarbon, the hydrocarbon containing only one carbon atom, is methane (CH_4). The hydrocarbon of two carbon atoms is called ethane; that of three carbon atoms, propane. These compounds are called alkanes. The general formula for alkanes is C_nH_{2n+2}, where n is the number of carbon atoms. The names of all alkanes end in *-ane*. Beginning with pentane, the prefixes indicate the number of carbon atoms present.

When an alkane loses a hydrogen atom, it forms an alkyl group whose name ends in *-yl*. The alkyl group of one carbon atom, derived from methane, is called the methyl group ($CH_3{-}$). The alkyl group of two carbon atoms, derived from ethane, is called the ethyl group ($C_2H_5{-}$).

To identify and name a hydrocarbon compound, pick out the longest continuous chain of carbon atoms and name that chain. Then pick out the alkyl groups attached to that chain. Number the carbons, starting at whichever end of the chain will give alkyl groups with the smallest numbers. List the numbers and names of the alkyl groups, using prefixes to designate alkyl groups occurring more than once.

Alkanes that occur in the shape of a ring are called cycloalkanes.

Haloalkanes have one or more hydrogens of an alkane replaced by a halogen.

Alkenes have a double bond between two of the carbon atoms. The names of all alkenes end in *-ene*. The general formula for alkenes is C_nH_{2n}.

Alkynes have a triple bond between two of the carbon atoms and have the general formula C_nH_{2n-2}. The names of all alkynes end in *-yne*.

A saturated hydrocarbon has only single bonds between the carbon atoms. An unsaturated hydrocarbon has double or triple bonds between its carbon atoms. Saturated hydrocarbons react by a process known as substitution. Unsaturated hydrocarbons react by a process known as addition. The addition reaction is usually much more rapid than the substitution reaction.

The chief sources of hydrocarbons are petroleum and natural gas.

Polymers consist of many small molecules joined together. Examples are polyethylene and polypropylene.

Questions and Problems

A

1. Name the following compounds.
   (a) Alkane of five carbon atoms; of two carbon atoms.
   (b) Alkyne of four carbon atoms; of three carbon atoms.
   (c) Alkene of eight carbon atoms; of two carbon atoms.

2. Classify the following hydrocarbons as alkanes, alkenes, or alkynes.
   (a) C_3H_6 (b) C_9H_{20} (c) $C_{14}H_{26}$
   (d) C_3H_8 (e) $C_{22}H_{44}$ (f) $C_{45}H_{88}$
   (g) C_5H_8 (h) C_5H_{10}

3. What is an alkyl group? Name and draw the structure of the alkyl group derived from ethane.

4. Name the following compounds.

   (a) CH_3—CH_2—CH—CH_2—CH_3
   　　　　　　　　|
   　　　　　　　CH_3

   (b)
   　　　　　　CH_3
   　　　　　　　|
   CH_3—CH—CH—CH_2—CH_2—CH_3
   　　　　　|
   　　　　CH_3

   (c)
   　　　　　　　　　　　　　CH_3
   　　　　　　　　　　　　　　|
   CH_3—CH—CH_2—CH—CH_2—CH—CH_2—CH_3
   　　　　|　　　　　　|
   　　　CH_3　　　　CH_3

5. What does the term saturated refer to in hydrocarbons?

6. What does the term unsaturated mean in terms of hydrocarbons?

7. How do saturated hydrocarbons react? Give an example of such a reaction.

8. How do unsaturated hydrocarbons react? Give an example of such a reaction.

9. What are the chief sources of hydrocarbons?

10. List several uses of hydrocarbons.

B

11. What is the IUPAC name for isooctane? Write the equation for its complete oxidation.

12. Draw the structures for (a) ethylcyclobutane, (b) 1,3-dichlorocyclobutane, (c) cyclopentane.

13. Draw structures for (a) 2-chloropropane, (b) *n*-pentane, (c) 2-iodo-2-methylbutane, (d) 1,3-dichloropropane.

14. What does the prefix *sec* refer to? Give an example.

15. Name the following compounds.

    (a)
    　　　　H　H　H　H
    　　　　|　|　|　|
    　　H—C—C—C—C—H
    　　　　|　|　|　|
    　　　　H　Br　Br　H

    (b)
    　　　　H　H　H　H
    　　　　|　|　|　|
    　　H—C=C—C—C—H
    　　　　　　　|　|
    　　　　　　Br　H

(c) 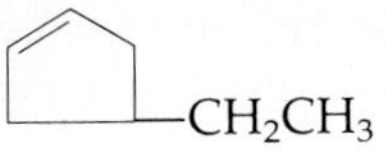 [structure: cyclopentene ring with —CH₂CH₃ substituent]

(d) 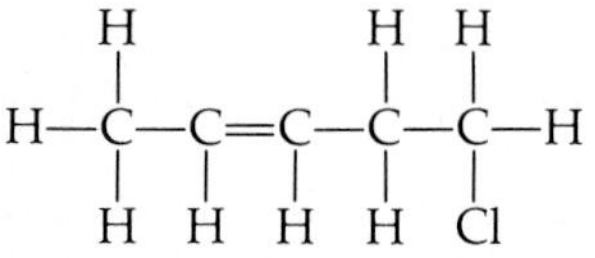 $H-\overset{\underset{|}{H}}{C}-\overset{\underset{|}{H}}{C}=\overset{\underset{|}{H}}{C}-\overset{\underset{|}{H}}{C}-\overset{\underset{|}{H}}{C}-H$ with Cl on last carbon

16. Compare complete and incomplete combustion of alkanes.
17. How are polymers formed? Give examples of uses of three different polymers.
18. Discuss the advantages and the disadvantages of the use of unleaded gasoline versus "ethyl" gas.

Practice Test

1. Which of the following is an alkane?
   a. C_6H_{12} b. C_6H_{14} c. C_6H_5 d. C_8H_{14}
2. Which of the following is an alkyne?
   a. C_7H_{14} b. C_6H_{14} c. C_6H_5 d. C_8H_{14}
3. The name of the following compound is _____ .

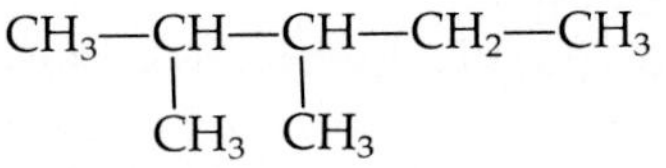

$$CH_3-\underset{\underset{CH_3}{|}}{CH}-\underset{\underset{CH_3}{|}}{CH}-CH_2-CH_3$$

    a. 2-methylpentane
    b. 2,3-dimethylpentane
    c. 2,3-methylpentane
    d. 4-methylpentane

4. The structure ☐ represents _____ .
   a. cyclopropane   b. cyclobutane
   c. cyclopentane   d. cubane
5. Alkanes react primarily by _____ .
   a. addition   b. substitution
   c. chlorination   d. oxidation
6. Alkenes react primarily by _____ .
   a. addition   b. substitution
   c. chlorination   d. oxidation
7. Acetylene is a common name for _____ .
   a. ethane   b. ethene
   c. ethyne   d. propane
8. An alkane used primarily for heating purposes is _____ .
   a. methane   b. ethane
   c. octane   d. decane
9. Which of the following condensed structures represents ethyl chloride?
   a. CH_3Cl b. C_2H_5Cl
   c. CH_2Cl_2 d. $C_2H_4Cl_2$
10. Which of the following is saturated?
   a. an alkane   b. an alkene
   c. an alkyne   d. none of these

17

Alcohols and Ethers

The alcohol ethylene glycol is used in antifreeze.

Alcohols

A functional group is a particular arrangement of a few atoms that imparts certain characteristic properties to an organic molecule. Alcohols are derivatives of hydrocarbons in which one or more of the hydrogen atoms has been replaced by a hydroxyl (—OH) functional group. Other functional groups, which will be discussed in Chapter 18, are those of aldehydes, ketones, esters, acids, amines, and amides (see the inside back cover).

In the IUPAC system, alcohols are named for the longest continuous chain containing the —OH group, with that functional group having the lowest numbers. The ending -e is changed to -ol to indicate the alcohol functional group. Thus, the alcohol derived from methane is methanol, and the alcohol derived from ethane is ethanol.

In the common system, alcohols are named by taking the name of the alkyl group and adding the word *alcohol*.

Examples of the names of simple alcohols in these two systems are

$$CH_3—OH \qquad CH_3—CH_2—OH \qquad CH_3—\underset{\underset{OH}{|}}{CH}—CH_3$$

| methanol | ethanol | 2-propanol |
| (methyl alcohol) | (ethyl alcohol) | (isopropyl alcohol) |

More complex alcohols are named according to the IUPAC system.

Example 17-1 Name the following compound.

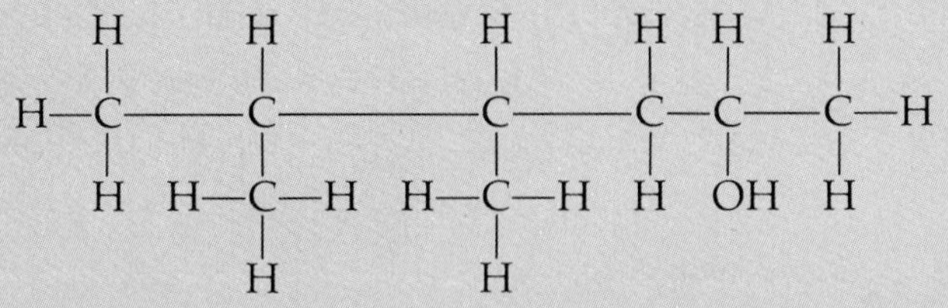

If this were a hydrocarbon, the name of the longest chain would be hexane. Since the chain has an —OH group, it is called hexanol. The —OH is on carbon number 2 (the lowest number when counting from the right side), and so the compound is a 2-hexanol. Methyl groups are attached at carbons number 4 and number 5, so the IUPAC name is 4,5-dimethyl-2-hexanol.

Example 17-2 Name   OH CH₃

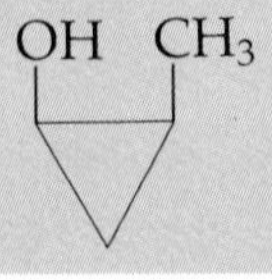

This is a cyclic alcohol of three carbon atoms, so it is a cyclopropanol. Because the —OH functional group is given the lowest number, it is assumed to be at carbon number 1 (understood). Therefore the methyl group must be at carbon number 2. Thus the name of the compound is 2-methylcyclopropanol.

Example 17-3   Name

$$H-\underset{\underset{H}{|}}{\overset{\overset{H}{|}}{C}}=\underset{\underset{H}{|}}{\overset{\overset{H}{|}}{C}}-\underset{\underset{H}{|}}{\overset{\overset{H}{|}}{C}}-\underset{\underset{H}{|}}{\overset{\overset{H}{|}}{C}}-OH$$

This compound is called 3-butenol because it is a four-carbon alcohol with a double bond between carbons number 3 and number 4. Since no number is given for the —OH group, it is assumed to be at carbon number 1.

Example 17-4   Name

Using the same system as before, the compound is called 2-cyclobutenol with the number 1 for the —OH group being understood. However, some chemists do indicate the position of the —OH group and call this compound 2-cyclobuten-1-ol.

Although the IUPAC system is preferred by chemists, medical personnel usually refer to alcohols by their common names. We will do so in the following sections.

The general formula for an alcohol is **ROH**, where the R signifies an alkyl group attached to an —OH functional group.

Alcohols contain an —OH group that does not ionize. Because alcohols do not ionize, their reactions are much slower than those of inorganic bases, which contain a hydroxide ion (OH⁻). Solutions of alcohols are nonelectrolytes; they are not bases. However, alcohols do react with acids to form compounds called esters, which will be discussed in Chapter 18.

Since oxygen is more electronegative than either carbon or hydrogen, alcohols are polar compounds. This polarity gives rise to hydrogen bonding between alcohol molecules and accounts for their relatively high boiling points.

The presence of an —OH group increases the ability of a substance to dissolve in water. Thus methyl alcohol and ethyl alcohol are quite soluble in water. However, as the length of the carbon chain increases, the molecules become more and more like alkanes and less and less like alcohols, so the solubility in water decreases.

Uses

Methyl Alcohol

Methyl alcohol (methanol), CH_3OH, is commonly known as wood alcohol. It is used as a solvent in many industrial reactions. Methyl alcohol should never be applied directly to the body nor should the vapors be inhaled because this substance can be absorbed both through the skin and through the respiratory tract. Ingestion of as little as 15 mL of methyl alcohol can cause blindness and 30 mL can cause death.

Methyl alcohol can be prepared from the distillation of wood. It is prepared commercially from carbon monoxide.

$$CO + 2 H_2 \xrightarrow[\text{heat}]{\text{Pt catalyst}} CH_3OH$$

Ethyl Alcohol

Ethyl alcohol (ethanol), CH_3CH_2OH, is known commonly as grain alcohol. In the hospital the word *alcohol* means ethyl alcohol.

One important property of ethyl alcohol is its ability to denature protein (see page 422). Because of this property, ethyl alcohol is widely used as an antiseptic. As an antiseptic, 70 percent alcohol is preferred to a stronger solution. It would seem that if 70 percent alcohol is a good antiseptic then 100 percent alcohol would be even better; however, the reverse is true. The 70 percent alcohol is actually a better antiseptic than the 100 percent alcohol.

Pure alcohol coagulates protein on contact. Suppose that pure alcohol is poured over a single-celled organism. The alcohol will penetrate the cell wall of that organism in all directions, coagulating the protein just inside the cell wall, as shown in Figure 17-1. This ring of coagulated protein would then prevent the alcohol from penetrating farther into the cell so that no more coagulation would take place. At this time the cell would become dormant, but not dead. Under the proper conditions the organism could again begin to function. If 70 percent alcohol is poured over a single-celled organism, the diluted alcohol also coagulates the protein, but at a slower rate, so that it penetrates all the way through the cell before coagulation can block it. Then all the cell proteins are coagulated, and the organism dies (see Figure 17-2).

Alcohol (ethyl) can also be used for sponge baths to reduce the fever of a patient. When alcohol is placed on the skin it evaporates rapidly. In order to evaporate, alcohol requires heat. This heat comes from the patient's skin. Thus an alcohol sponge bath will remove heat from the patient's skin and so lower the body temperature. A water sponge bath will do the same thing, but water evaporates more slowly than alcohol so the heat is removed more slowly. However, water sponge baths are in common use in many hospitals because they are cheaper. Since alcohol is flammable it cannot be used in a room where oxygen is in use.

Alcohol is used as a solvent for many substances. Alcohol solu-

Figure 17-1 Effect of 100 percent alcohol on bacteria.

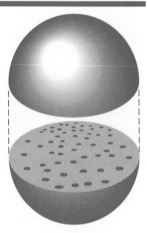

Bacteria before using alcohol

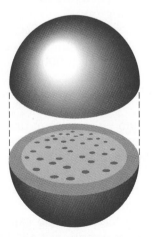

Bacteria after using 100% alcohol. Note that only outer layer is coagulated.

Figure 17-2 Effect of 70 percent alcohol on bacteria.

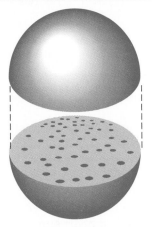

Bacteria before using alcohol

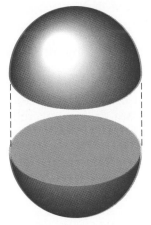

Bacteria after using 70% alcohol. Note that bacterial protein is coagulated all the way through.

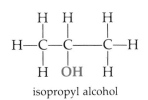

isopropyl alcohol

ethylene glycol

tions are called tinctures. Tincture of iodine consists of iodine dissolved in alcohol.

Ethyl alcohol is also used as a beverage. The concentration of alcohol in alcoholic beverages is expressed as "proof." The proof is twice the percentage of alcohol in the solution. Thus a beverage marked "100 proof" contains 50 percent alcohol. Alcohol slows reaction time, so driving under the influence of alcohol can be very dangerous. Alcohol is not a stimulant. It actually depresses the nervous system and can remove the normal inhibitions of a person. Excessive use of alcohol may cause the destruction of the liver, a condition known as cirrhosis.

Alcohol has a high tax rate because of its use as a beverage. For use in industry, substances are added to make the alcohol unfit for drinking. Such an alcohol is said to be **denatured.** It is not fit to drink but is still useful for industrial purposes. Common denaturants are wood alcohol, formaldehyde, and gasoline.

Alcohol (ethyl) can be prepared from the fermentation of blackstrap molasses, the residue that results from the purification of cane sugar. The principal constituent of molasses is sucrose ($C_{12}H_{22}O_{11}$). The fermentation reaction is brought about by the enzymes present in yeast.

$$C_{12}H_{22}O_{11} + H_2O \xrightarrow{\text{enzymes}} 4\,C_2H_5OH + 4\,CO_2$$
$$\text{sucrose} \qquad\qquad \text{ethyl alcohol}$$

The starches present in grains can be converted into sugar by malt. The sugar thus produced can be fermented under the influence of the enzymes in yeast to yield ethyl alcohol. Hence ethyl alcohol is also known as grain alcohol.

Ethyl alcohol can also be prepared synthetically from ethene.

$$CH_2{=}CH_2 + H_2O \xrightarrow{\text{catalyst}} C_2H_5OH$$
$$\text{ethene} \qquad\qquad \text{ethyl alcohol}$$

Isopropyl Alcohol

Isopropyl alcohol has the structural formula shown at the left. The IUPAC name for this compound is 2-propanol, indicating that the —OH functional group is on the second carbon of a three-carbon chain. Isopropyl alcohol is toxic and should not be taken internally. Since it is not absorbed through the skin, it is commonly used as rubbing alcohol and as an *astringent.

Ethylene Glycol

All the previously mentioned alcohols have one —OH functional group. They are called *monohydric alcohols.* Glycols are compounds with two —OH functional groups. They are examples of *dihydric alcohols.*

Ethylene glycol is used in preparations to moisten the skin. It is also used as a permanent antifreeze in car radiators and as a raw material in the manufacture of the polymer Dacron. The IUPAC name for ethylene glycol is 1,2-ethanediol.

Glycerol

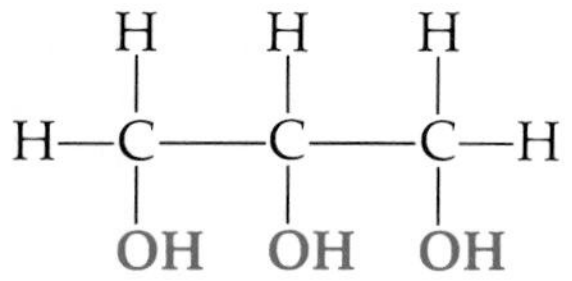

glycerol

Glycerol (sometimes called glycerin) is a trihydric or trihydroxy alcohol; it contains three —OH groups. The IUPAC name for glycerol is 1,2,3-propanetriol, since there is an —OH functional group on each carbon atom of the three-carbon chain.

Glycerol is an important alcohol in terms of body chemistry, especially as a constituent of fats (see Chapter 22). It is a byproduct of the manufacture of soap (see page 387) and is used in the preparation of cosmetics and hand lotions and also in suppositories. Glycerol is used in the laboratory as a lubricant for rubber tubing and stoppers. When treated with nitric acid, glycerin forms nitroglycerin, an explosive. Medicinally, nitroglycerin is used as a heart stimulant. It causes a dilation of the coronary arteries, thus increasing the supply of blood to the heart muscles.

Other Alcohols

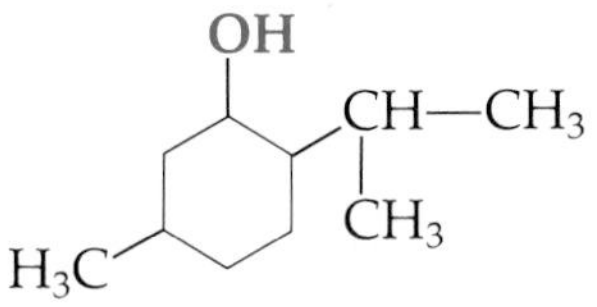

menthol

Menthol is an example of a cyclic alcohol. It has a cooling, refreshing feeling when rubbed on the skin and so is a frequently used ingredient in cosmetics and shaving lotions. Menthol is used in cough drops and nasal sprays.

Other alcohols of biological importance are cholesterol, retinol (vitamin A), and tocopherol (vitamin E). These will be discussed later in appropriate chapters.

Types

Primary Alcohols

A primary (1°) alcohol is one that contains an —OH functional group attached to a carbon that has one or no carbon atoms attached to it. All the structures in diagram (17-1) are primary alcohols.

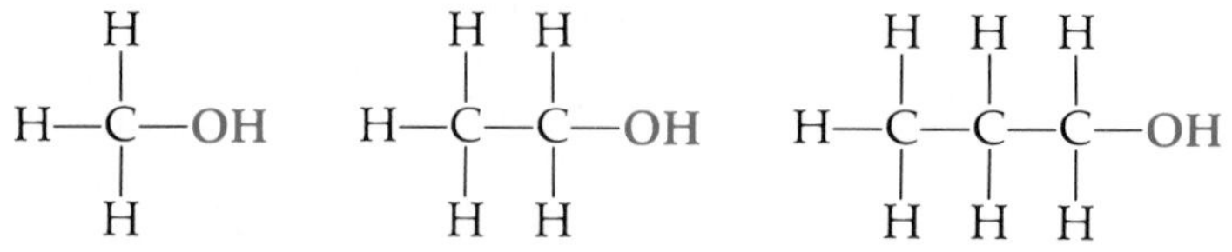

methyl alcohol   ethyl alcohol   propyl alcohol   (17-1)

Methyl alcohol (methanol) is an example of a primary alcohol in which the —OH functional group is attached to a carbon atom with no other carbons attached to it. Ethyl alcohol (ethanol) and propyl alcohol (propanol) are examples of primary alcohols in which the —OH functional group is attached to a carbon atom with one carbon atom attached to it. Note that in a primary alcohol the functional group (—OH) is at the end of the chain.

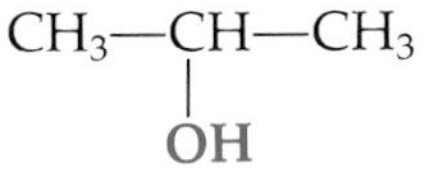

$$CH_3—CH—CH_3$$
$$|$$
$$OH$$

isopropyl alcohol
(2-propanol)

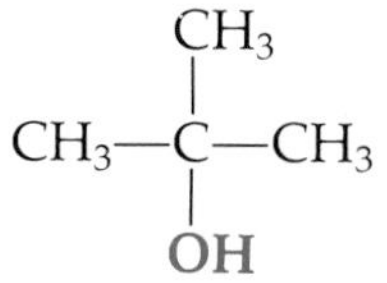

Secondary Alcohols

A secondary (2°) alcohol is one in which the —OH is attached to a carbon atom having two other carbon atoms attached to it. Isopropyl alcohol is an example of a secondary alcohol. Note that in a secondary alcohol the functional group is not at the end of the chain.

$$CH_3$$
$$|$$
$$CH_3—C—CH_3$$
$$|$$
$$OH$$

2-methyl-2-propanol, commonly known as tertiary butyl alcohol

Tertiary Alcohols

A tertiary (3°) alcohol is one in which the —OH is attached to a carbon atom that has three carbon atoms attached to it.

Reactions

Dehydration

In the presence of a dehydrating agent, such as H_2SO_4, alcohols can be dehydrated to form alkenes.

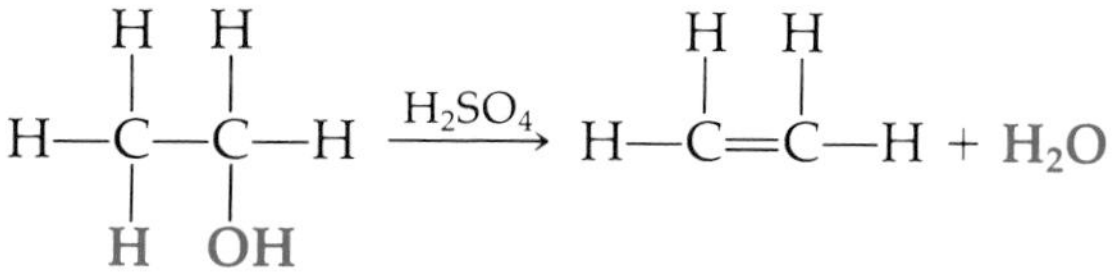

ethanol   ethene

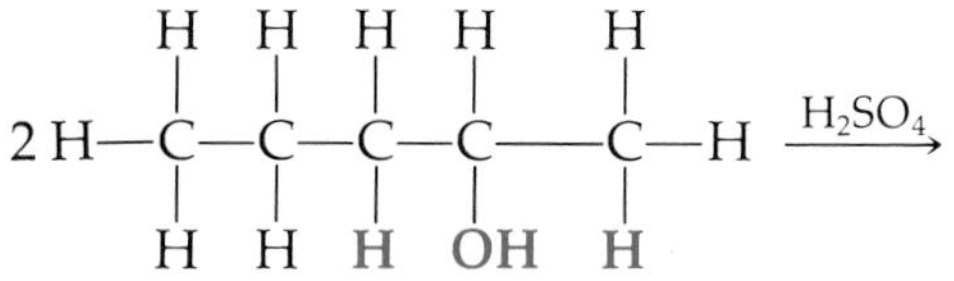

2-pentanol

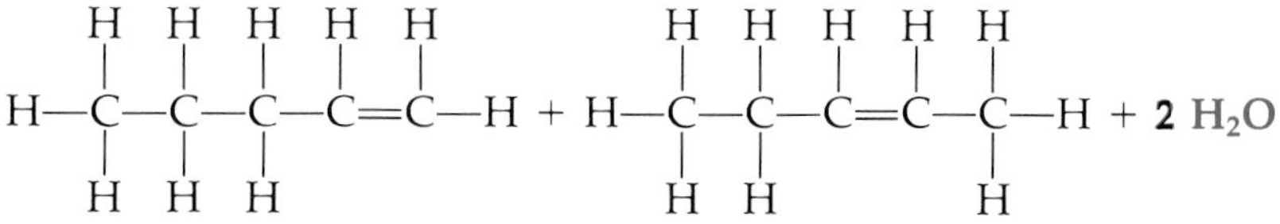

1-pentene   2-pentene

Formation of Ethers

As will be shown in the next section, alcohols can also react in the presence of H_2SO_4 to form ethers.

Oxidation

Primary alcohols can be oxidized to form aldehydes and then further oxidized to form acids and eventually CO_2 and H_2O. These reactions will be discussed in the next chapter. Some of these oxidation products are responsible for the toxicity of various alcohols.

$$ROH \xrightarrow{\text{oxidation}} RCHO \xrightarrow{\text{oxidation}} RCOOH$$

primary alcohol   aldehyde   acid

Secondary alcohols can be oxidized to yield ketones (see next chapter).

$$ROH \xrightarrow{\text{oxidation}} RCOR$$

secondary alcohol   ketone

Tertiary alcohols cannot be oxidized under ordinary conditions. Alcohols also react with organic acids to yield compounds called esters (see page 305).

Thiols

$$CH_3—SH$$

methanethiol
(methyl mercaptan)

$$CH_3—\underset{\underset{\displaystyle SH}{|}}{CH}—CH_3$$

2-propanethiol
(isopropyl mercaptan)

Thiols are sulfur analogs of alcohols and contain an —SH functional group in place of an —OH group. The IUPAC names of thiols are formed by adding the ending *-thiol* to the name of the parent hydrocarbon. Note that the *-e* ending of the parent compound is not deleted. The common names of thiols are formed by first naming the alkyl group and then adding the word mercaptan.

Many thiols are found in nature, and they all have a disagreeable odor. When an onion is cut, 1-propanethiol is released; garlic owes its odor to the presence of thiols; and thiols are responsible for the odor given off by skunks. Since natural gas used for heating and cooking is odorless, thiols are added so that a gas leak can be easily detected.

Unlike alcohols, thiols do not exhibit hydrogen bonding. Therefore they have lower boiling points than the corresponding alcohols. Also, since they do not exhibit hydrogen bonding, thiols are less soluble in water than alcohols with the same number of carbon atoms.

Thiols can be prepared by heating alkyl halides with sodium hydrogen sulfide, NaHS.

$$CH_3—CH_2—CH_2—Cl + Na\textbf{HS} \longrightarrow CH_3—CH_2—CH_2—\textbf{SH} + NaCl$$

Thiols are easily oxidized to disulfides.

$$R—CH_2—SH \xrightarrow{O_2} R—CH_2—S—S—CH_2—R$$

a thiol   a disulfide

An example of such an oxidation is the conversion of cysteine to cystine, a reaction that takes place when hair is given a "permanent" (see page 160).

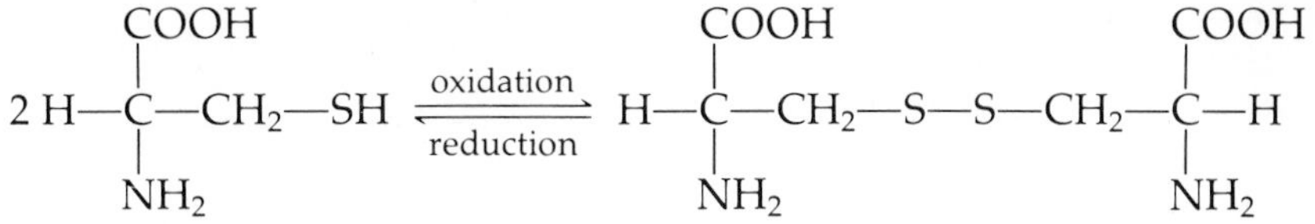

$$2\ \underset{\underset{NH_2}{|}}{\overset{\overset{COOH}{|}}{H-C}}-CH_2-SH \underset{\text{reduction}}{\overset{\text{oxidation}}{\rightleftharpoons}} \underset{\underset{NH_2}{|}}{\overset{\overset{COOH}{|}}{H-C}}-CH_2-S-S-CH_2-\underset{\underset{NH_2}{|}}{\overset{\overset{COOH}{|}}{C}}-H$$

cysteine   cystine

Disulfide bonds are also involved in the formation of some proteins (see page 416).

Thiols (also called sulfhydryls) are also important in enzymes involved in carbohydrate metabolism.

Ethers

An ether is formed during the dehydration of an alcohol. In this reaction the sulfuric acid can be considered a dehydrating agent that removes water from two molecules of alcohol. Consider the following reaction, where R indicates an alkyl group and hence ROH indicates an alcohol.

$$R-OH + HO-R \xrightarrow{H_2SO_4} R-O-R + H_2O$$

alcohol   alcohol   ether

This equation indicates that two molecules of alcohol (with the second molecular formula written backwards) react in the presence of sulfuric acid to form water and an ether. When methyl alcohol is reacted with sulfuric acid, methyl ether (also called dimethyl ether) is formed.

$$CH_3-OH + HO-CH_3 \xrightarrow{H_2SO_4} CH_3-O-CH_3 + H_2O$$

methyl alcohol   methyl alcohol   methyl ether
                                         (dimethyl ether)

When ethyl alcohol is treated with sulfuric acid, ethyl ether (also called diethyl ether) is formed.

$$CH_3-CH_2-OH + HO-CH_2-CH_3 \xrightarrow{H_2SO_4} CH_3-CH_2-O-CH_2-CH_3 + H_2O$$

ethyl alcohol   ethyl alcohol   ethyl ether
                                               (diethyl ether)

The ether is named for the alcohol from which it is made: ethyl ether from ethyl alcohol and methyl ether from methyl alcohol. The general formula for an ether is **ROR**.

Under the IUPAC system, the —OCH_3 group is called methoxy and the —OCH_2CH_3 group is called ethoxy. Thus, methyl ether, CH_3-O-CH_3, is called methoxymethane and ethyl ether, $CH_3-CH_2-O-CH_2-CH_3$, is called ethoxyethane.

$CH_3-O-CH_2-CH_3$

methoxyethane
(methyl ethyl ether)

CH₃—O—CH—CH₃
          |
         CH₃

2-methoxypropane
(methyl isopropyl ether)

If two different alcohols are reacted, a mixed ether is formed. Mixed ethers are named as alkoxy derivatives of hydrocarbons with the shorter chain being named as the alkoxy group and the longer chain as the alkane.

Under the common system, mixed ethers are named by listing the alkyl groups followed by the word ether.

Recall from the previous section that alcohols react with H_2SO_4 to form alkenes. Here it is indicated that alcohols also react with H_2SO_4 to form ethers. What causes two different reactions to take place? When alcohols react with H_2SO_4 at a temperature of 140 °C, ethers are formed. At temperatures above 150 °C, alkenes are formed.

Ethers have a low boiling point because the molecules do not form hydrogen bonds (see page 168). Ethers are good solvents because they are inert—they do not react with the solute. Ethyl ether is frequently used to extract organic material from naturally occurring substances. The low boiling point of the ether allows it to be easily removed and recovered. However, ethyl ether is very flammable and must be used with care. Ethers that remain in a laboratory for a long period of time may contain organic peroxides that are extremely explosive. Care must be taken that ethers are not allowed to stand undisturbed for a long period of time.

Ether as an Anesthetic

Ethyl ether, commonly known as ether, has been used quite extensively as a general anesthetic. It is very easy to administer, is an excellent muscular relaxant, and has very little effect upon the rate of respiration, blood pressure, or pulse rate. However, the disadvantages of ether outweigh its advantages. It is very flammable, it is irritating to the membranes of the respiratory tract; and it has an aftereffect of nausea. Today ether is infrequently employed as a general anesthetic except in laboratory work. It has been replaced by such nonflammable anesthetics as nitrous oxide (see page 268) and halothane (see page 147).

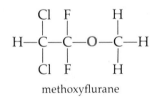

methoxyflurane

Methoxyflurane, 2,2-dichloro-1,1-difluoroethyl methyl ether, has also been used as a general anesthetic, but its use has been discontinued in several hospitals because of its effects on renal function.

     H   H   H   H
     | | | |
H—C=C—O—C=C—H

divinyl ether

Divinyl ether has been used as a surgical anesthetic but is now obsolete for this purpose.

Summary

Alcohols are derivatives of hydrocarbons with one or more of the hydrogen atoms replaced by an —OH group. The —OH (hydroxyl) group is a functional group that imparts to alcohol its particular properties.

The general formula for an alcohol is ROH, where the R represents an alkyl group attached to the —OH group.

Alcohols do not ionize; they are not bases; their reactions are slower than those of inorganic hydroxides.

The simplest alcohol is methyl alcohol, CH_3OH, also known as methanol. This alcohol is poisonous and should never be used internally or externally.

Ethyl alcohol, C_2H_5OH, is also known as ethanol. It is used as a disinfectant because it has the property of coagulating protein. It can also be used for sponge baths to reduce body temperature. Solutions of medications in alcohol are called tinctures.

Isopropyl alcohol is used primarily as rubbing alcohol. It is toxic and should never be used internally.

Glycerol or glycerin is a trihydric alcohol; it has three —OH groups in its molecule. Glycerol is a constituent of fats.

Alcohols can be divided into three categories: primary, where the —OH group is attached to a carbon atom having one or no carbon atoms attached to it; secondary, where the —OH group is attached to a carbon atom having two carbon atoms attached to it; and tertiary, where the —OH is attached to a carbon atom having three carbon atoms attached to it.

Thiols are sulfur analogs of alcohols and contain an —SH group.

Ethers are produced by the dehydration of an alcohol. Ethers are named according to the alcohol or alcohols from which they were produced.

Ethyl ether has been used as a general anesthetic. It is an excellent muscular relaxant and has little effect on the rate of respiration or pulse rate. However, ether is irritating to the membranes of the respiratory tract, it may cause nausea, and it is very flammable. It has been replaced by nonflammable, nonirritating anesthetics such as halothane and nitrous oxide.

Questions and Problems

A

1. What is the general formula for an alcohol?
2. Indicate several general properties of alcohols.
3. Draw the structure and give the IUPAC name of
   (a) methyl alcohol   (b) ethyl alcohol
   (c) isopropyl alcohol   (d) propyl alcohol
   (e) glycerin
4. Why should methyl alcohol never be used medicinally?
5. Explain why 70 percent alcohol is a better disinfectant than 100 percent alcohol.
6. Why is alcohol often used for sponge baths?
7. Why may water be substituted for alcohol in a sponge bath?
8. What is a tincture?
9. What is denatured alcohol? Where is it used?
10. What does the word "proof" mean in terms of alcohol?
11. How can alcohol be prepared commercially?
12. What is a common use for isopropyl alcohol?
13. Why is glycerol important in the body?
14. What is nitroglycerin used for medically? How can it be prepared?

15. List the three types of alcohols, and give an example of each.
16. How can an ether be prepared? What reagents are needed?
17. Write the equation for the formation of ethyl ether from ethyl alcohol.
18. What is a mixed ether? How can it be prepared?
19. List the disadvantages of ether as a general anesthetic.

B

20. What is a functional group?
21. How is methyl alcohol produced commercially? ethyl alcohol?
22. Why does the dehydration of 2-pentanol produce two different products?
23. Complete the following reactions:

$$CH_3CH_2CH_2CH_2OH \xrightarrow[\text{low temp}]{H_2SO_4}$$

$$CH_3CH_2CH_2CH_2OH \xrightarrow[\text{high temp}]{H_2SO_4}$$

24. Name the following compounds according to both the IUPAC and the common systems.
    (a) $CH_3{-}O{-}CH_2CH_3$

(b) CH_3CH_2—O—$CH_2CH_2CH_3$

(c) CH_3CH_2—O—$CHCH_2CH_3$
$\qquad\qquad\qquad\quad |$
$\qquad\qquad\qquad\quad CH_3$

(d) CH_3—O—CH_3

(e) CH_3—CH_2—SH

(f)

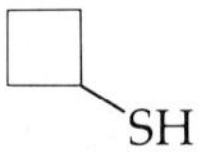

SH

Practice Test

1. The general formula for an alcohol is _____.
   (a) ROH   (b) ROR
   (c) RCHO   (d) C_nH_{2n+2}
2. The general formula for an ether is _____.
   (a) ROR   (b) ROH
   (c) RCHO   (d) C_nH_{2n}
3. An example of a trihydric alcohol is _____.
   (a) ethylene glycol
   (b) glycerol
   (c) ethyl alcohol
   (d) isopropyl alcohol
4. CH_3OH is called _____.
   (a) methanol   (b) ethanol
   (c) propanol   (d) isopropanol
5. CH_3OCH_3 is called _____.
   (a) dimethyl ether

(b) diethyl ether
(c) methyl ethyl ether
(d) ethyl ether

6. Which substance does not exhibit hydrogen bonding?
   (a) an alcohol
   (b) a thiol
   (c) water

Refer to the following structures for questions 7 and 8.

(a) CH_3OH

(b) CH_3—CH—CH_3
$\qquad\qquad\quad |$
$\qquad\qquad\quad OH$

(c) CH_3—O—CH_2—CH_3

(d) CH_3—$\overset{\displaystyle OH}{\underset{\displaystyle CH_3}{\overset{|}{\underset{|}{C}}}}$—$CH_3$

7. Which is a primary alcohol?
8. Which is a secondary alcohol?

9. 100 proof alcohol contains _____ percent alcohol.
   (a) 25   (b) 50   (c) 75   (d) 100
10. An alcoholic solution of a medication is called a(n) _____.
    (a) isotope   (b) tincture
    (c) emulsion   (d) mercaptan

18

Other Organic Compounds

A bee sting injects formic acid into the skin.

The oxidation of primary, secondary, and tertiary alcohols gives different types of products. If we consider oxidation as the removal of hydrogens (see page 153), then the oxidation of an alcohol can be said to involve the removal of one hydrogen from the —OH group of the alcohol and of a second hydrogen from the carbon atom to which the —OH group is attached. The oxidation of a primary alcohol can be written as

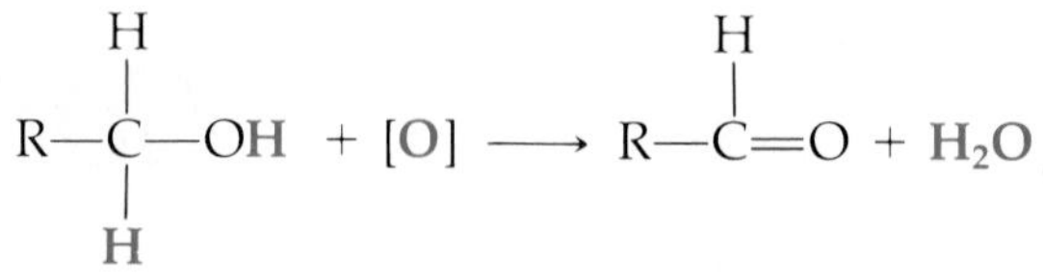

$$\underset{\text{primary alcohol}}{R-\overset{\displaystyle H}{\underset{\displaystyle H}{C}}-OH} + [O] \longrightarrow \underset{\text{aldehyde}}{R-\overset{\displaystyle H}{C}=O} + H_2O$$

The oxidation of an alcohol requires the use of some oxidizing agent such as $KMnO_4$, $K_2Cr_2O_7$, or CuO. However, for the sake of simplicity the oxidizing agent in the reactions shown in this chapter is simply listed as [O], which stands for any substance that will yield the oxygen needed for the reaction.

Aldehydes

Preparation by Oxidation of a Primary Alcohol

Recall that a primary alcohol has the —OH functional group bonded to a carbon with one or no other carbon atom attached to it. The following equation represents the oxidation of methyl alcohol (CH_3OH), a primary alcohol.

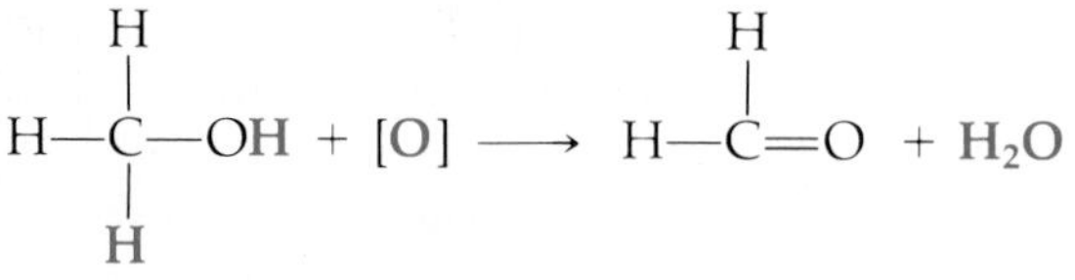

$$\underset{\text{methyl alcohol}}{H-\overset{\displaystyle H}{\underset{\displaystyle H}{C}}-OH} + [O] \longrightarrow \underset{\text{formaldehyde}}{H-\overset{\displaystyle H}{C}=O} + H_2O$$

Observe that during the oxidation one H was removed from the —OH group and another H from the carbon to which the —OH group was attached (the only carbon in this compound). Water is one product of this reaction; the other product is a new kind of compound called an aldehyde. In this example, the product is called formaldehyde. The formula for formaldehyde can also be written as HCHO.

The following reaction indicates the oxidation of ethyl alcohol (also a primary alcohol).

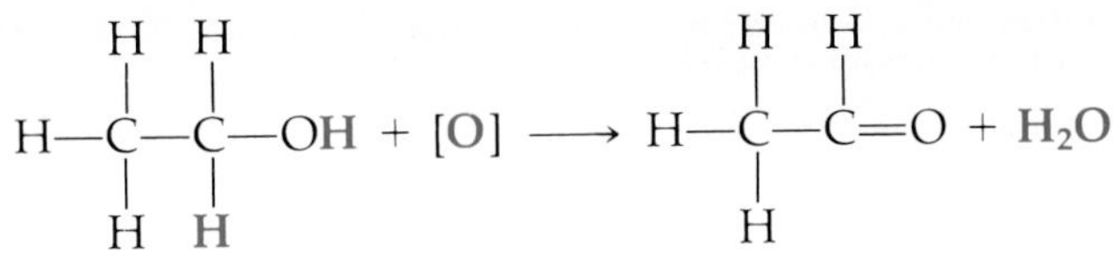

$$\underset{\text{ethyl alcohol}}{H-\overset{\displaystyle H}{\underset{\displaystyle H}{C}}-\overset{\displaystyle H}{\underset{\displaystyle H}{C}}-OH} + [O] \longrightarrow \underset{\text{acetaldehyde}}{H-\overset{\displaystyle H}{\underset{\displaystyle H}{C}}-\overset{\displaystyle H}{C}=O} + H_2O$$

The oxidation of ethyl alcohol, a primary alcohol, yields acetaldehyde, whose formula can also be written as CH_3CHO. In general,

$$\text{primary alcohol} \xrightarrow{\text{oxidation}} \text{aldehyde}$$

Aldehydes all have the —CHO group at the end of the chain. The general formula for an aldehyde is **RCHO**, which indicates that some alkyl group (R) is attached to a —CHO group at the end of the molecule.

The colored part of the following —CHO is called a **carbonyl group**. This group is also present in other types of compounds, as will be discussed in the following sections.

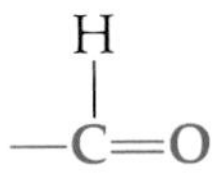

$$-\overset{\displaystyle H}{C}=O$$

Naming Aldehydes

As we have seen, the oxidation of methyl alcohol, a primary alcohol of one carbon atom, yields an aldehyde of one carbon atom, HCHO, formaldehyde. The oxidation of ethyl alcohol, a primary alcohol of two carbon atoms, yields an aldehyde of two carbon atoms, CH_3CHO, acetaldehyde. Note that the term *aldehyde* comes from the words *al*cohol *dehyd*rogenation.

The IUPAC names for all aldehydes end in **-al**. To name an aldehyde according to the IUPAC system, take the name of the longest chain containing the aldehyde group, drop the ending *-e*, and replace it with the ending *-al*. Thus, the following aldehyde, which contains four carbons, is called butanal.

$$CH_3-CH_2-CH_2-CHO$$

The aldehyde group is always at the end of the chain, at carbon 1, with that number being understood and not written, as indicated in the following compound.

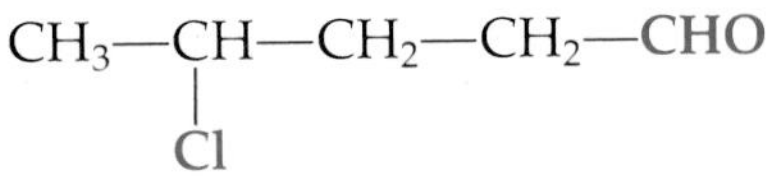

$$CH_3-\underset{\underset{\displaystyle Cl}{|}}{CH}-CH_2-CH_2-CHO$$

4-chloropentanal

Table 18-1 Comparison of the Names of Some Aldehydes

Condensed Structural Formula	IUPAC Name	Common Name
$HCHO$	Methanal	Formaldehyde
CH_3CHO	Ethanal	Acetaldehyde
$CH_3—CH_2—CHO$	Propanal	Propionaldehyde
$CH_3—CH—CHO$ $\quad\quad\;\; \mid$ $\quad\quad\;\; CH_3$	2-Methylpropanal	Isobutyraldehyde

Table 18-1 compares the IUPAC and common names for some simple aldehydes.

Uses of Aldehydes

Formaldehyde is a colorless gas with a very sharp odor. It is used in the laboratory as a water solution containing about 40 percent formaldehyde. The 40 percent solution, commonly known as formalin, is an effective germicide for the disinfection of excreta, rooms, and clothing. Formalin hardens protein, making it very insoluble in water. It is used in embalming fluids and also as a preservative for biologic specimens. Formaldehyde solutions should not be used directly on the patient or even in the room with the patient because of irritating fumes.

Formaldehyde and its oxidation product, formic acid, are primarily responsible for the systemic toxicity of methyl alcohol.

Glutaraldehyde is superior to formaldehyde as a sterilizing agent and is gradually replacing it. Glutaraldehyde is microcidal against all microorganisms, including spores and many viruses. Glutaraldehyde does not have the disagreeable odor that formaldehyde does, and it is less irritating to the eyes and skin.

Paraldehyde is formed by the polymerization (joining) of three molecules of acetaldehyde. Paraldehyde depresses the central nervous system. It is used as a hypnotic, a sleep producer. Paraldehyde is also used in the treatment of alcoholism (see following paragraph). In therapeutic dosages it is nontoxic; it does not depress heart action or respiration. Its disadvantages are its disagreeable taste and its unpleasant odor.

Acetaldehyde (ethanal) is responsible for many of the unpleasant side effects of ethyl alcohol consumption. The drug Antabuse, used to treat alcoholics, functions by increasing the concentration of acetaldehyde in the body.

Another aldehyde, glyceraldehyde, is an important component in the metabolism of carbohydrates (see page 480).

$$CHO$$
$$\mid$$
$$(CH_2)_3$$
$$\mid$$
$$CHO$$

glutaraldehyde

$$CHO$$
$$\mid$$
$$H—C—OH$$
$$\mid$$
$$H—C—OH$$
$$\mid$$
$$H$$

glyceraldehyde
(2,3-dihydroxypropanal)

paraldehyde

Tests for Aldehydes

In general, aldehydes are good reducing agents. Laboratory tests for the presence of aldehydes are based on their ability to reduce copper(II) (cupric) ions to form copper(I) (cuprous) oxide. When an aldehyde is heated with Benedict's or Fehling's solution or treated with a Clinitest tablet (all of which contain Cu^{2+} complex ion), a red precipitate of copper(I) oxide (Cu_2O) is formed. This is actually the test for glucose (sugar) in urine, since glucose is an aldehyde (see page 362).

Another laboratory test for the presence of an aldehyde involves the use of Tollens' reagent, which contains an Ag^+ complex ion. In this test, the presence of an aldehyde causes the formation of a bright, shiny mirror on the inside of the test tube. Hence the name "silver mirror test" (see pages 363–64).

Reactions of Aldehydes

Oxidation

Aldehydes can be oxidized to form acids, a type of reaction that will be discussed later in this chapter.

$$\underset{\text{aldehyde}}{RCHO} + [O] \longrightarrow \underset{\text{acid}}{RCOOH}$$

Note that this type of oxidation involves the addition of an oxygen (see page 152).

Aldehydes can be oxidized by $Cu(OH)_2$, according to the following reaction.

$$RCHO + Cu(OH)_2 \longrightarrow RCOOH + Cu_2O(s) + H_2O$$

This is the basis of the test for an aldehyde discussed above.

Reduction

Aldehydes can be reduced to the corresponding primary alcohols.

$$\underset{\text{aldehyde}}{RCHO} + [H] \longrightarrow \underset{\text{primary alcohol}}{RCH_2OH}$$

This is the reverse of the reaction whereby a primary alcohol was oxidized to yield an aldehyde.

Biologic oxidation–reduction in the body is carried out by substances called coenzymes (see page 438). One coenzyme, nicotinamide adenine dinucleotide (NAD^+), acts as an oxidizing agent and in turn is reduced to nicotinamide adenine dinucleotide hydride (NADH), as shown in the following reaction.

$$\underset{\substack{\text{oxidized form} \\ \text{of coenzyme}}}{NAD^+} + \underset{\text{ethyl alcohol}}{CH_3CH_2OH} \rightleftharpoons \underset{\substack{\text{reduced form} \\ \text{of coenzyme}}}{NADH} + \underset{\text{acetaldehyde}}{CH_3CHO} + H^+$$

The preceding oxidation–reduction reaction can also be written as

$$CH_3CH_2OH \quad\longleftarrow\quad\longrightarrow\quad NAD^+$$

$$CH_3CHO \quad\longleftarrow\quad\longrightarrow\quad NADH$$

Ketones

Preparation by Oxidation of a Secondary Alcohol

Recall that a secondary alcohol is one in which the —OH group is bonded to a carbon atom that is also bonded to two carbon atoms. Isopropyl alcohol is an example of a secondary alcohol.

$$CH_3-\overset{\displaystyle H}{\underset{\displaystyle OH}{C}}-CH_3$$

isopropyl alcohol
(2-propanol)

The oxidation of isopropyl alcohol is indicated by the equation

$$CH_3-\overset{\displaystyle H}{\underset{\displaystyle OH}{C}}-CH_3 + [O] \longrightarrow CH_3-\overset{\displaystyle }{\underset{\displaystyle O}{C}}-CH_3 + H_2O$$

isopropyl alcohol   acetone
(2-propanol)   (propanone)

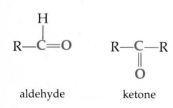

R—$\overset{\displaystyle H}{C}$=O   R—$\overset{\displaystyle }{\underset{\displaystyle O}{C}}$—R

aldehyde   ketone

As before, the oxygen atom from the oxidizing agent reacts with the H from the —OH group and with the H attached to the same carbon as the —OH group, forming water and a new class of compounds called ketones.

The oxidation of a secondary alcohol yields a ketone, of the general formula **RCOR**. That is, a ketone has two alkyl groups attached to a $>$C=O, **carbonyl, group.** This carbonyl group is present in both aldehydes and ketones. However, the carbonyl group is at the end of the chain in an aldehyde and not at the end in a ketone.

Naming Ketones

In the IUPAC system, the names of ketones end in **-one.** To name a ketone according to this system, take the name of the longest

	Structure	IUPAC	Common Name
Table 18-2 Names of Some Simple Ketones	CH$_3$—C—CH$_3$ $\overset{\|}{O}$	Propanone	Dimethyl ketone (commonly known as acetone)
	CH$_3$—C—CH$_2$—CH$_3$ $\overset{\|}{O}$	Butanone	Methyl ethyl ketone
	CH$_3$—C—CH$_2$—CH$_2$—CH$_3$ $\overset{\|}{O}$	2-Pentanone	Methyl propyl ketone
	CH$_3$—CH$_2$—C—CH$_2$—CH$_3$ $\overset{\|}{O}$	3-Pentanone	Diethyl ketone

CH$_3$—C—CH$_2$—CH$_3$
$\overset{\|}{O}$

butanone

alkane containing the carbonyl group, drop the ending -*e*, and add -*one*. Thus, the four-carbon ketone is called butanone.

In the common system for naming ketones, each alkyl group attached to the carbonyl group is named and the word ketone is added afterward. Thus the name of the preceding compound, according to the common system, is methyl ethyl ketone, since there is a methyl group attached to one end of the carbonyl group and an ethyl group attached to the other end.

Table 18-2 gives both common and IUPAC names of several ketones.

Uses of Ketones

Acetone (propanone) is the simplest ketone. Acetone is a good solvent for fats and oils. It is also frequently used in fingernail polish and in polish remover. Acetone is normally present in small amounts in the blood and urine. In diabetes mellitus it is present in larger amounts in the blood and urine and even in the expired air. Dihydroxyacetone is an intermediate in carbohydrate metabolism.

$$\overset{O}{\overset{\|}{HO—CH_2—C—CH_2—OH}}$$

dihydroxyacetone

Reactions of Ketones

What happens when a ketone is oxidized? Consider the formula for the ketone acetone.

$$\overset{}{\underset{\overset{\|}{O}}{CH_3—C—CH_3}}$$

acetone

There are no hydrogen atoms on the carbon atom of the carbonyl group. Therefore, ketones are not easily oxidized. They are normally unreactive. Ketones can be reduced, however, to the corresponding secondary alcohol. Ketones give a negative test with such oxidizing agents as Benedict's solution or Clinitest tablets. Recall that aldehydes give a positive test with these reagents.

The test for acetone and ketone bodies makes use of the reaction between sodium nitroprusside and ketones or ketone bodies to produce a lavender color.

Hemiacetals and Hemiketals

The reaction of an aldehyde or a ketone with an alcohol yields compounds known as *hemiacetals or *hemiketals, respectively.

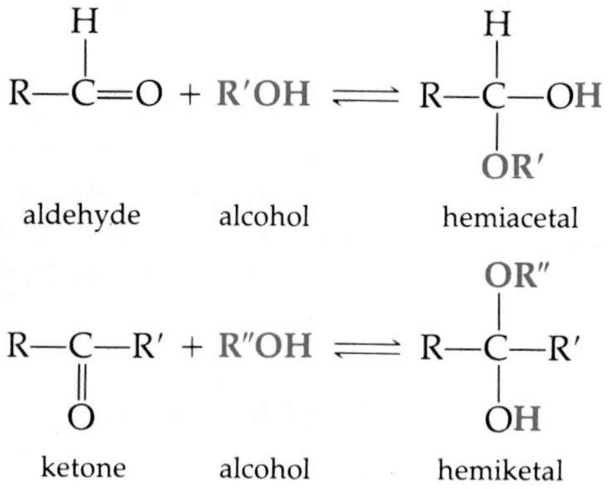

$$
\begin{array}{ccc}
& & \text{H} \\
& & | \\
\text{R—C}{=}\text{O} + \text{R'OH} \;\rightleftharpoons\; & & \text{R—C—OH} \\
& & | \\
& & \text{OR'} \\
\text{aldehyde} \qquad \text{alcohol} & & \text{hemiacetal}
\end{array}
$$

$$
\begin{array}{ccc}
& & \text{OR''} \\
& & | \\
\text{R—C—R'} + \text{R''OH} \;\rightleftharpoons\; & & \text{R—C—R'} \\
\| & & | \\
\text{O} & & \text{OH} \\
\text{ketone} \qquad \text{alcohol} & & \text{hemiketal}
\end{array}
$$

These types of compounds are important in discussing the structures of monosaccharides (see page 358).

If a hemiacetal or hemiketal reacts with a second molecule of alcohol, an *acetal or *ketal, respectively, is formed. These structures are important in disaccharides and polysaccharides (pages 367–69 and 371–72).

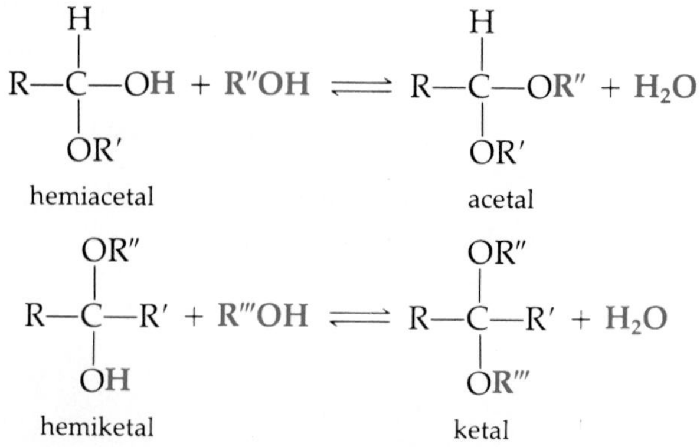

$$
\begin{array}{ccc}
\text{H} & & \text{H} \\
| & & | \\
\text{R—C—OH} + \text{R''OH} \;\rightleftharpoons\; & & \text{R—C—OR''} + \text{H}_2\text{O} \\
| & & | \\
\text{OR'} & & \text{OR'} \\
\text{hemiacetal} & & \text{acetal}
\end{array}
$$

$$
\begin{array}{ccc}
\text{OR''} & & \text{OR''} \\
| & & | \\
\text{R—C—R'} + \text{R'''OH} \;\rightleftharpoons\; & & \text{R—C—R'} + \text{H}_2\text{O} \\
| & & | \\
\text{OH} & & \text{OR'''} \\
\text{hemiketal} & & \text{ketal}
\end{array}
$$

Organic Acids

Preparation by Oxidation of an Aldehyde

The oxidation of a primary alcohol yields an aldehyde. Aldehydes in turn can be easily oxidized. When an aldehyde is oxidized, the reaction is

$$
\underset{\text{acetaldehyde}}{CH_3\overset{H}{\underset{|}{C}}=O} + [O] \longrightarrow \underset{\text{acetic acid}}{CH_3\overset{OH}{\underset{|}{C}}=O} \quad \begin{array}{l}\text{(also written as}\\ CH_3COOH)\end{array}
$$

The resulting compound is acidic because it yields hydrogen ions in solution. (Note that this reaction involved oxidation because of a gain in oxygen; see page 152.)

The functional group of an organic acid is —COOH, so the oxidation of an aldehyde to an acid can be written functionally as

$$
\underset{\text{aldehyde}}{R-CHO} \xrightarrow{[O]} \underset{\text{acid}}{R-COOH}
$$

The oxidation of methyl alcohol, a primary alcohol, to an aldehyde and then to an acid is illustrated in the following equation.

$$
\underset{\substack{\text{methyl alcohol}\\\text{(methanol)}}}{H-\overset{H}{\underset{H}{\underset{|}{\overset{|}{C}}}}-OH} \xrightarrow{[O]} \underset{\substack{\text{formaldehyde}\\\text{(methanal)}}}{H-\overset{}{\underset{H}{\underset{|}{C}}}=O} \xrightarrow{[O]} \underset{\substack{\text{formic acid}\\\text{(methanoic acid)}}}{H-\overset{}{\underset{OH}{\underset{|}{C}}}=O} \quad \begin{array}{l}\text{(also written}\\\text{as HCOOH)}\end{array}
$$

Primary alcohols can also be oxidized directly to acids.

$$
\underset{\substack{\text{ethyl alcohol}\\\text{(ethanol)}}}{CH_3CH_2OH} \xrightarrow{[O]} \underset{\substack{\text{acetic acid}\\\text{(ethanoic acid)}}}{CH_3\overset{}{\underset{OH}{\underset{|}{C}}}=O}
$$

Naming Organic Acids

$$
\underset{\text{ethanoic acid}}{CH_3-\overset{}{\underset{O}{\underset{\|}{C}}}-OH} \quad (18\text{-}1)
$$

The IUPAC names for organic acids end in *-oic acid*. To name an acid according to the IUPAC system, take the longest alkane containing the acid group, drop the ending *-e* and add *-oic acid*. Thus, formula (18-1) shows ethanoic acid.

The common names of acids are derived from the names of the aldehyde from which they may be prepared. Thus formula (18-1), derived from acetaldehyde, is called acetic acid.

Table 18-3 Names of Common Organic Acids

Formula	IUPAC Name	Common Name
HCOOH	Methanoic acid	Formic acid
CH_3COOH	Ethanoic acid	Acetic acid
CH_3CH_2COOH	Propanoic acid	Propionic acid
$CH_3CH_2CH_2COOH$	Butanoic acid	Butyric acid

The general formula for an acid is **RCOOH.** All organic acids contain at least one —COOH group. This group is called the **carboxyl group,** and it is this group that yields hydrogen ions.

Table 18-3 gives the common and IUPAC names of several common acids.

Organic acids containing two carboxyl groups are called *dicarboxylic acids.* Those containing three carboxyl groups are called *tricarboxylic acids.* Table 18-4 gives the names and structures of some common dicarboxylic acids.

Properties and Reactions of Organic Acids

Most organic acids are relatively weak acids since they ionize only slightly in water.

Table 18-4 Some Common Dicarboxylic Acids

Structure	Common Name	IUPAC Name
COOH | COOH	Oxalic acid	Ethanedioic acid
COOH | CH_2 | COOH	Malonic acid	Propanedioic acid
COOH | CH_2 | CH_2 | COOH	Succinic acid	Butanedioic acid
COOH | CH_2 | CH_2 | CH_2 | COOH	Glutaric acid	Pentanedioic acid

$$CH_3COOH \rightleftharpoons H^+ + CH_3COO^-$$

acetic acid   hydrogen   acetate ion
ion

Organic acids react with bases to form salts and water. The general reaction of an organic acid with a base to form a salt and water can be written as follows.

$$RCOOH + NaOH \rightleftharpoons RCOONa + H_2O$$

organic acid   base   organic salt   water

$$CH_3COOH + NaOH \rightleftharpoons CH_3COONa + H_2O$$

acetic acid   sodium   sodium acetate   water
hydroxide   (a salt)

Organic acids also react with bicarbonates and carbonates.

$$CH_3COOH + NaHCO_3 \longrightarrow CH_3COONa + CO_2(g) + H_2O$$

acetic acid   sodium bicarbonate   sodium acetate

$$2\ HCOOH + Na_2CO_3 \longrightarrow 2\ HCOONa + CO_2(g) + H_2O$$

formic acid   sodium carbonate   sodium formate

Organic acids containing few carbon atoms are soluble in water. As the length of the carbon chain increases, the solubility in water decreases.

Organic acids also react with alcohols to form a class of compounds called esters, which will be dealt with later in this chapter.

Medically Important Organic Acids

Formic acid (HCOOH) is a colorless liquid with a sharp irritating odor. Formic acid is found in the sting of bees and ants and causes the characteristic pain and swelling when it is injected into the tissues. It is one of the strongest organic acids.

Acetic acid (CH_3COOH) is one of the components of vinegar, where it is usually found as a 4 to 5 percent solution. Anhydrous acetic acid freezes at 17 °C and then looks like a chunk of ice. For this reason it is called *glacial acetic acid*. Acetic acid can be made by the

$$H-\overset{\displaystyle H}{\underset{\displaystyle H}{\overset{|}{\underset{|}{C}}}}-COOH$$
$$HO-\overset{|}{\underset{|}{C}}-COOH$$
$$H-\overset{|}{\underset{|}{C}}-COOH$$
$$H$$

citric acid

oxidation of ethyl alcohol. The acetyl group, $CH_3\overset{\displaystyle O}{\overset{\|}{C}}-$, derived from acetic acid, is very important in metabolic reactions.

Citric acid is found in citrus fruits. Its formula indicates that it is an alcohol as well as an acid. Citric acid contains one alcohol (—OH) group and three acid (—COOH) groups. It is an example of a tricarboxylic acid.

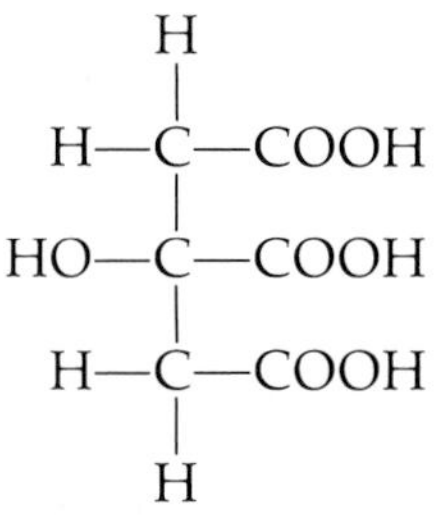

Magnesium citrate, a salt of citric acid, is used as a cathartic (a medication for stimulating the evacuation of the bowels). Sodium citrate, another salt of citric acid, is used as a blood anticoagulant. (It removes Ca^{2+} from the blood.)

Lactic acid is found in sour milk. It is formed in the fermentation of milk sugar, lactose. Its formula is

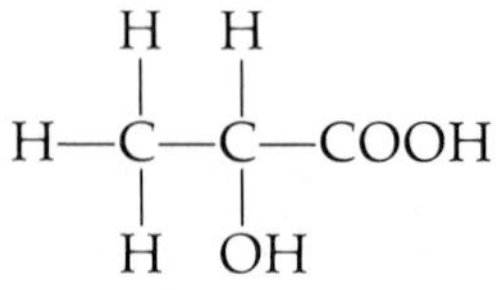

$$
\begin{array}{c}
\ \ \ \ \ \ \ \ \text{H}\ \ \ \text{H} \\
\ \ \ \ \ \ \ \ |\ \ \ \ \ | \\
\text{H—C—C—COOH} \\
\ \ \ \ \ \ \ \ |\ \ \ \ \ | \\
\ \ \ \ \ \ \ \ \text{H}\ \ \ \text{OH}
\end{array}
$$

lactic acid

Lactic acid is also both an acid and an alcohol. It is formed whenever the body produces energy anaerobically (see page 478).

Oxalic acid is another one of the strong, naturally occurring organic acids. Its formula is

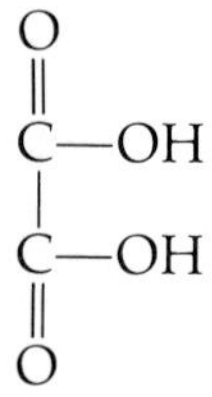

$$
\begin{array}{c}
\text{O} \\
\|\\
\text{C—OH} \\
|\\
\text{C—OH} \\
\|\\
\text{O}
\end{array}
$$

oxalic acid

Oxalic acid is used to remove stains, particularly rust and potassium permanganate stains, from clothing. It is poisonous when taken internally. Oxalate salts also prevent clotting by removing Ca^{2+} from the blood. However, oxalate can be used only for blood samples that are to be analyzed in the laboratory because these salts are poisonous and cannot be added directly to the bloodstream.

Pyruvic acid is produced during the anaerobic oxidation of glucose (see page 478). It is a keto acid.

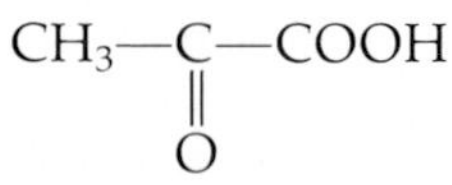

$$
\begin{array}{c}
\text{CH}_3\text{—C—COOH} \\
\|\\
\text{O}
\end{array}
$$

pyruvic acid

In muscle, pyruvic acid is reduced to lactic acid. In the tissues, pyruvic acid is changed to acetyl coenzyme A, which then enters the Krebs cycle (page 482).

Tartaric acid is another organic acid that is both an acid and an alcohol. Its formula is written as

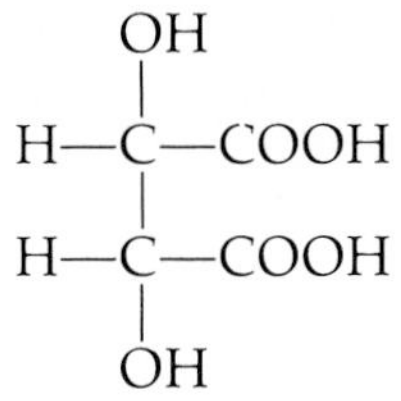

$$\begin{array}{c} \text{OH} \\ | \\ \text{H}-\text{C}-\text{COOH} \\ | \\ \text{H}-\text{C}-\text{COOH} \\ | \\ \text{OH} \end{array}$$

tartaric acid

Tartaric acid is found in several fruits, particularly grapes. Potassium hydrogen tartrate, an acid salt called cream of tartar, is used in making baking powders. Rochelle salts, or potassium sodium tartrate, is used as a mild cathartic. Antimony potassium tartrate, known as tartar emetic, is used in the treatment of schistosomiasis (a disease caused by a type of parasitic flatworm), but its use is inadvisable in the presence of severe hepatic, renal, or cardiac insufficiency.

Stearic acid is a solid greaselike acid that is insoluble in water. It is an example of a fatty acid. Its formula is $C_{17}H_{35}COOH$. The sodium salt of stearic acid, sodium stearate, is a commonly used soap.

$$C_{17}H_{35}COOH + NaOH \longrightarrow C_{17}H_{35}COONa + H_2O$$

stearic acid   sodium stearate

Esters

Esters are produced by the reaction of an organic acid with an alcohol and have the general formula.

$$RCOOR' \quad \text{or} \quad R\overset{\overset{\displaystyle O}{\|}}{C}OR'$$

The general reaction of an alcohol with an acid is illustrated by the following equation (where R and R' may be the same or different alkyl groups).

$$RCOOH + R'OH \rightleftharpoons RCOOR' + H_2O$$

acid   alcohol   ester   water

Note that the reaction is written with a double arrow, indicating that it is an equilibrium reaction; that is, the reverse reaction also takes place. Thus, esters hydrolyze to form organic acids and alcohols. Esters do not readily ionize in water solution.

When an organic acid reacts with an alcohol, the name of the ester is determined as follows.

IUPAC Name	*Common Name*
1. Write the name of the alkyl group of the alcohol	1. Write the name of the alcohol
2. Write the name of the acid minus the ending *-ic*	2. Write the name of the acid minus the ending *-ic*
3. Add the ending *-ate* to the name of the acid	3. Add the ending *-ate* to the name of the acid

An example is given in reaction (18-2).

$$
\underset{\substack{\text{ethanoic acid} \\ \text{(acetic acid)}}}{CH_3-\overset{\overset{\textstyle O}{\|}}{C}-OH} + \underset{\substack{\text{methanol} \\ \text{(methyl alcohol)}}}{HOCH_3} \rightleftharpoons \underset{\substack{\text{methyl ethanoate} \\ \text{(methyl acetate)}}}{CH_3-\overset{\overset{\textstyle O}{\|}}{C}-OCH_3} + H_2O \quad (18\text{-}2)
$$

Consider reaction (18-3) between ethyl alcohol and formic acid.

$$
\underset{\text{formic acid}}{H-\overset{\overset{\textstyle O}{\|}}{C}-OH} + \underset{\text{ethyl alcohol}}{HO-CH_2-CH_3} \rightleftharpoons \underset{\text{ethyl formate}}{H-\overset{\overset{\textstyle O}{\|}}{C}-O-CH_2-CH_3} + H_2O \quad (18\text{-}3)
$$

In reactions (18-2) and (18-3), the acid loses the —OH group and the alcohol loses an —H. This is indicated in diagram (18-4) of the structure of an ester.

$$
\underset{\substack{\text{acid} \\ \text{part}}}{R-\overset{\overset{\textstyle O}{\|}}{C}}\Big|\underset{\substack{\text{alcohol} \\ \text{part}}}{O-R'} \quad (18\text{-}4)
$$

Table 18-5   Names of Several Common Esters

Ester	Acid Part	Alcohol Part	IUPAC Name	Common Name
$H-\overset{\overset{\textstyle O}{\|}}{C}-O-CH_2CH_3$	$H-\overset{\overset{\textstyle O}{\|}}{C}-OH$	CH_3CH_2OH	Ethyl methanoate	Ethyl formate
$CH_3-\overset{\overset{\textstyle O}{\|}}{C}-O-CH_3$	$CH_3-\overset{\overset{\textstyle O}{\|}}{C}-OH$	CH_3OH	Methyl ethanoate	Methyl acetate
$CH_3CH_2-\overset{\overset{\textstyle O}{\|}}{C}-O-CH_2CH_3$	$CH_3CH_2-\overset{\overset{\textstyle O}{\|}}{C}-OH$	CH_3CH_2OH	Ethyl propanoate	Ethyl propionate
$CH_3CH_2CH_2-\overset{\overset{\textstyle O}{\|}}{C}-O-CH_2CH_2CH_3$	$CH_3CH_2CH_2-\overset{\overset{\textstyle O}{\|}}{C}-OH$	$CH_3CH_2CH_2OH$	Propyl butanoate	Propyl butyrate

Table 18-6 Esters and Synthetic Flavors

Ester	Flavor
Amyl acetate	Banana
Ethyl butyrate	Pineapple
Amyl butyrate	Apricot
Isoamyl acetate	Pear
Octyl acetate	Orange

Table 18-5 shows the derivation of several common esters and compares their common and IUPAC names.

Esters are important as solvents, perfumes, and flavoring agents. Table 18-6 indicates some of the esters used in preparing synthetic flavors. Many esters are also used medicinally (see Table 18-7).

The hydrolysis of esters takes place in the body when fats and oil are digested. The reaction, in general, is

$$RCOOR' + H_2O \xrightarrow{\text{enzymes}} RCOOH + R'OH$$

This type of reaction will be discussed in Chapter 22.

Esters also undergo saponification, a reaction with NaOH (see page 387).

Thioesters

Thioesters have the general formula

$$\overset{\displaystyle O}{\underset{\displaystyle \|}{R-C}}-S-R'$$

where the prefix *thio-* indicates the presence of a sulfur atom (see page 288). Many thioesters are found in biologic systems. The most important of these is acetyl coenzyme A, abbreviated acetyl CoA. Acetyl CoA is a key compound in metabolic reactions in the body. It is involved in the formation of fats (page 502), ketone bodies (page 499), and amino acids.

The structure of acetyl CoA is abbreviated as

$$\overset{\displaystyle O}{\underset{\displaystyle \|}{CH_3-C}}-S-CoA$$

The complete structure of acetyl CoA is given on page 438.

Table 18-7 Some Medical Uses of Esters

Esters	Use
Ethyl aminobenzoate (benzocaine)	Local anesthetic
Glyceryl trinitrate (nitroglycerin)	Used to dilate coronary arteries and lower blood pressure
Methyl salicylate (oil of wintergreen)	Used as a flavoring agent and also as a counterirritant in many liniments
Phenyl mercuric acetate	Used to disinfect instruments and as an antiseptic on cutaneous and mucosal surfaces

Phosphate Esters

When an organic compound containing an —OH group reacts with phosphoric acid (H_3PO_4), phosphate esters are formed. Phosphoric acid is a triprotic acid,

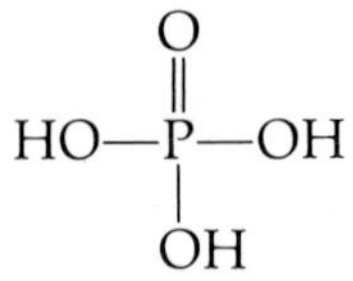

$$
\begin{array}{c}
O \\
\parallel \\
HO-P-OH \\
\mid \\
OH
\end{array}
$$

and so it can form various esters such as

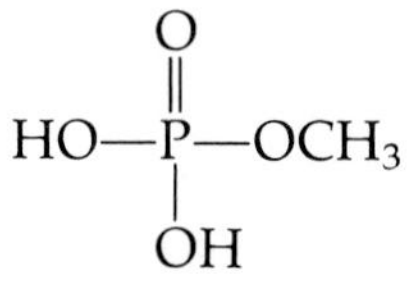

$$
\begin{array}{ccc}
O & O & O \\
\parallel & \parallel & \parallel \\
HO-P-OCH_3 & CH_3O-P-OCH_3 & CH_3O-P-OCH_3 \\
\mid & \mid & \mid \\
OH & OH & OCH_3
\end{array}
$$

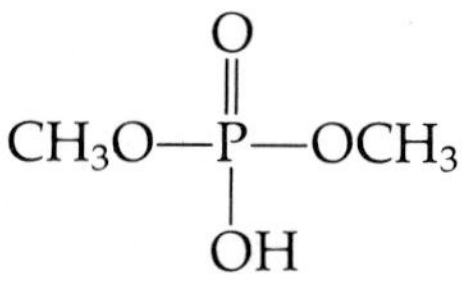

methyl dihydrogen   dimethyl hydrogen   trimethyl phosphate
phosphate   phosphate

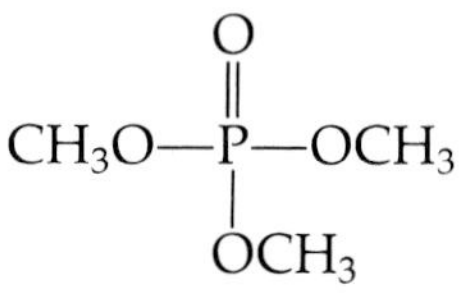

Phosphoric acid can also form more complex esters such as glyceraldehyde 3-phosphate and fructose 6-phosphate.

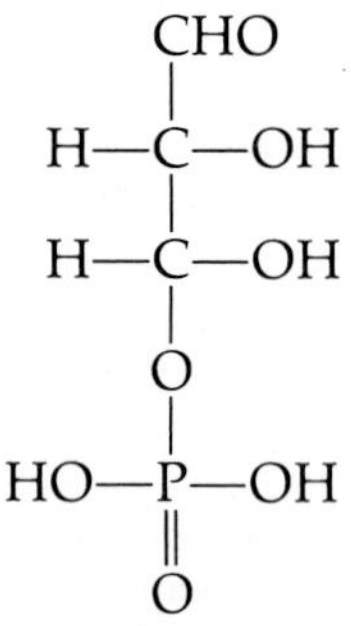

$$
\begin{array}{cc}
 & CH_2OH \\
 & \mid \\
 & C=O \\
 & \mid \\
 & HO-C-H \\
CHO & \mid \\
\mid & H-C-OH \\
H-C-OH & \mid \\
\mid & H-C-OH \\
H-C-OH & \mid \\
\mid & H-C-H \\
O & \mid \\
\mid & O \\
HO-P-OH & \mid \\
\parallel & HO-P-OH \\
O & \parallel \\
 & O
\end{array}
$$

glyceraldehyde 3-phosphate 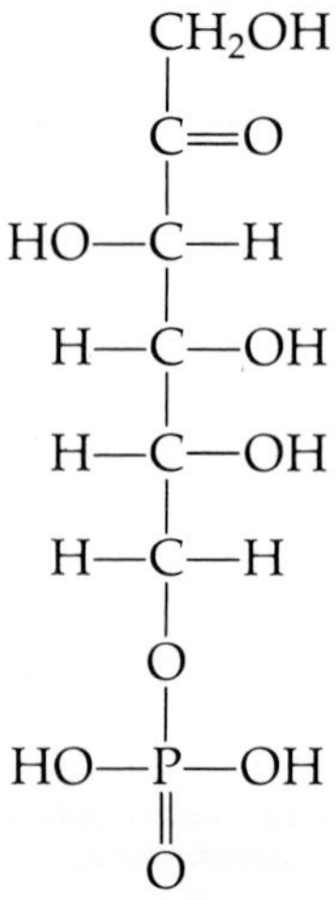   fructose 6-phosphate

Before sugars can be metabolized, they must first be converted into phosphate esters (see page 475). Phosphate esters serve many important functions in the body. They are present in

1. Adenosine triphosphate (ATP), adenosine diphosphate (ADP), and adenosine monophosphate (AMP), the body's energy compounds.

2. Cyclic adenosine monophosphate (cAMP), a chemical messenger that regulates cellular enzymes.
3. Deoxyribonucleic acid (DNA) and ribonucleic acid (RNA), the hereditary material of the cell.
4. Phospholipids, which are found in cell membranes and also in all tissue, particularly brain, liver, and spinal.
5. Phosphorylases, enzymes involved in the adding of phosphate groups to carbohydrates.

These uses of phosphate esters will be discussed later in appropriate chapters.

Amines

Amines are organic compounds derived from ammonia. There are three classes of amines: primary, secondary, and tertiary. Note that in relation to amines, the terms primary, secondary, and tertiary refer directly to the number of hydrogen atoms of ammonia that have been replaced by alkyl groups. In the case of alcohols, the terms primary, secondary, and tertiary refer to the number of carbon atoms attached to the carbon having the —OH group on it.

Primary amines are those in which one of the hydrogen atoms of ammonia (NH_3) has been replaced by an alkyl group. Primary amines have the general formula RNH_2

Secondary amines are those in which two of the hydrogen atoms of the ammonia have been replaced by alkyl groups. Secondary amines have the general formula R_2NH.

Tertiary amines are those in which all three hydrogen atoms of the ammonia have been replaced by alkyl groups. Tertiary amines have the general formula R_3N.

The common system names amines by placing the name of the alkyl group before the suffix *-amine*. If two or three identical alkyl groups are present, the prefix di- or tri- is used, as indicated in Table 18-8. The common system is used primarily for simple compounds. For more complex compounds, the IUPAC system is used. With this system, the prefix **amino-** is used to denote the presence of an —NH_2 group.

The following compounds illustrate the IUPAC names for some amino compounds (common names in parentheses).

$$CH_3—CH—CH_3 \qquad\qquad CH_2—\overset{\overset{\displaystyle O}{\|}}{C}—OH$$
$$\underset{NH_2}{|} \qquad\qquad\qquad\quad \underset{NH_2}{|}$$

2-aminopropane   2-aminoethanoic acid
(isopropylamine)   (aminoacetic acid)

Among the amines that are found in the body or are used medicinally are histamines (page 518), barbiturates (page 339),

Table 18-8   Ammonia and Simple Amines

Name	Formula	Example
Ammonia	NH_3 *or* 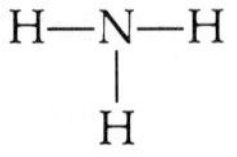 H—N—H$\;$(with H below)	
Primary amine	RNH_2	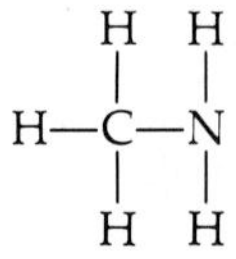 H—C—N (with H's) *or* CH_3NH_2 (methylamine)
Secondary amine	R_2NH	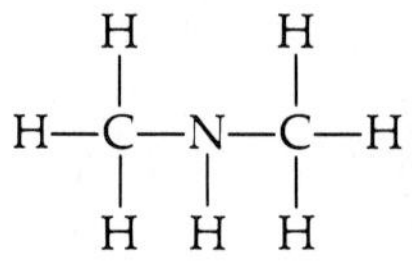 H—C—N—C—H (with H's) *or* $(CH_3)_2NH$ (dimethylamine)
Tertiary amine	R_3N	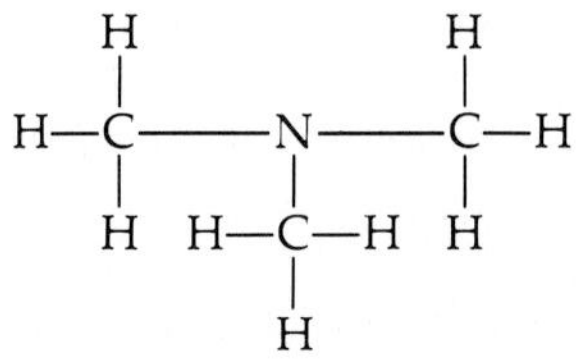 H—C———N———C—H (with H's and H—C—H) *or* $(CH_3)_3N$ (trimethylamine)

psychedelics (page 341), amphetamines (page 340), and nucleic acids (page 646).

Amines produced during the decay of once-living matter include putrescine and cadaverine (page 518).

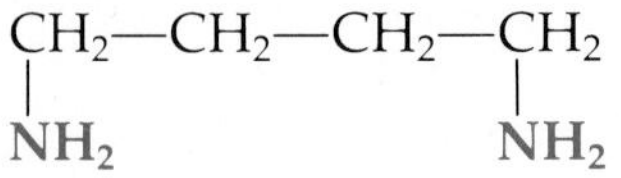

$$CH_2{-}CH_2{-}CH_2{-}CH_2 \qquad\qquad CH_2{-}CH_2{-}CH_2{-}CH_2{-}CH_2$$
$$|\qquad\qquad\qquad\qquad |\qquad\qquad\quad |\qquad\qquad\qquad\qquad\qquad\qquad |$$
$$NH_2\qquad\qquad\qquad NH_2\qquad\;\; NH_2\qquad\qquad\qquad\qquad\qquad NH_2$$

putrescine
(1,4-diaminobutane)

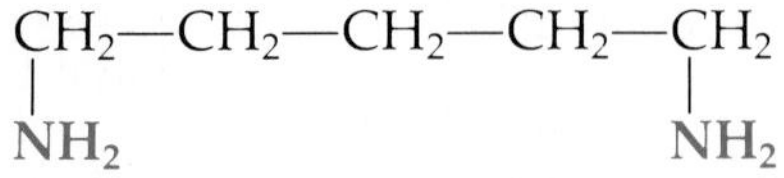

cadaverine
(1,5-diaminopentane)

Preparation of Amines

Amines can be produced by the reaction of alkyl halides with ammonia or with another amine. These reactions can be abbreviated, the inorganic by-products being ignored, as follows.

$$RCl \xrightarrow{\;NH_3\;} RNH_2$$

$$RCl \xrightarrow{\;R'NH_2\;} RNHR'$$

Examples of such reactions are

$$CH_3CH_2Cl \xrightarrow{\;NH_3\;} CH_3CH_2NH_2$$

$$CH_3CH_2Cl \xrightarrow{\;CH_3NH_2\;} CH_3CH_2NHCH_3$$

Reactions of Amines

When amines react with an acid such as hydrochloric acid, ammonium salts are produced. These ammonium salts are frequently named by placing the word hydrochloride after the name of the amine. Examples are thiamine hydrochloride (vitamin B_1) and procaine hydrochloride (Novocain, a local anesthetic). The hydrochlorides are administered rather than the amines themselves because they are more soluble in water.

Amines in general are basic compounds. They react with inorganic acids to form salts. Recall that one definition of a base is "a substance that accepts protons (H^+)" (see page 229). Compare the following reactions of ammonia and amines with HCl to the reaction of a base with an acid to form a salt.

$$NH_3 + HCl \longrightarrow NH_4^+ + Cl^-$$

$$RNH_2 + HCl \longrightarrow RNH_3^+ + Cl^-$$

$$R_2NH + HCl \longrightarrow R_2NH_2^+ + Cl^-$$

$$R_3N + HCl \longrightarrow R_3NH^+ + Cl^-$$

Quaternary ammonium salts are formed by the action of tertiary amines with organic halogen compounds. An example of this type of reaction is that between trimethylamine and methyl iodide. Tetramethylammonium iodide is formed.

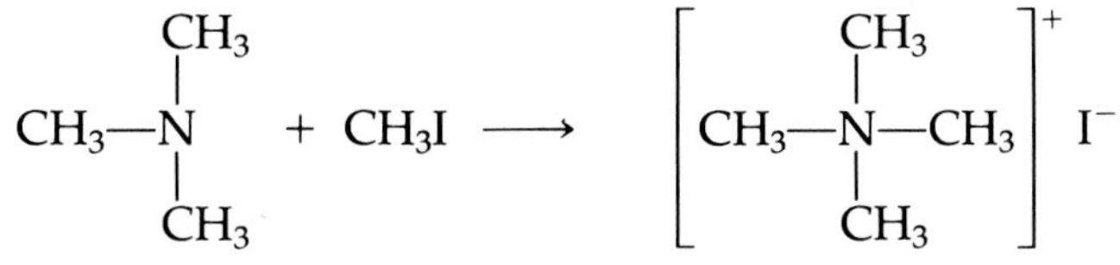

trimethylamine methyl iodide tetramethylammonium iodide

Some quaternary ammonium salts have both a detergent action and antibacterial activity. They are used medicinally as antiseptics. Benzalkonium chloride (zephiran chloride) is used in 0.1 percent solution for storage of sterilized instruments. A 0.01 to 0.02 percent solution is applied as a wet dressing to denuded areas. A 0.005 percent solution is used for irrigations of the bladder and the urethra.

Amino Acids

Amino acids are organic acids that contain an amine group. Examples of amino acids are

$$
\begin{array}{ccc}
\underset{\substack{|\\NH_2}}{\overset{\substack{H\\|}}{H-C-COOH}} &
\underset{\substack{|\ \ |\\H\ \ NH_2}}{\overset{\substack{H\ \ H\\|\ \ |}}{H-C-C-COOH}} &
\underset{\substack{|\ \ \ \ |\\NH_2\ \ H}}{\overset{\substack{H\ \ \ \ H\\|\ \ \ \ |}}{H-C--C-COOH}}
\end{array}
$$

glycine,   an α-amino acid   a β-amino acid
an α-amino acid

Note that Greek letters are used to designate the position of the amino group in the chain. The carbon atom next to the acid group, the —COOH group, is called the alpha (α) carbon. Then next in order come the beta (β), the gamma (γ), and the delta (δ) carbon atoms.

Amino acids contain an acid group, —COOH, which naturally is acidic. Amino acids also contain an —NH$_2$ group, which is basic (see page 311). Thus, amino acids exhibit both acidic and basic properties; amino acids can react with either acids or bases. Compounds that can act as, or react with, either acids or bases are called **amphoteric compounds.**

α-Amino acids are the building blocks of proteins. That is, proteins are polymers of α-amino acids (see Chapter 23).

Amides

Formation of Amides

Organic acids can react with ammonia or with amines to form a class of compounds called amides. Amides have a bond between the carbonyl group (—C=O) and the nitrogen.

$$
\underset{\text{acid}}{\overset{\overset{\textstyle O}{\|}}{R-C-OH}} + \underset{\text{ammonia}}{\overset{\overset{\textstyle H}{|}}{H-N-H}} \xrightarrow{\text{heat}} \underset{\text{amide}}{\overset{\overset{\textstyle O}{\|}}{R-C-NH_2}} + H_2O
$$

Amides are named as follows.

IUPAC System	*Common System*
1. Drop ending *-oic* and the word acid from the name of the acid.	1. Drop ending *-ic* and the word acid from the name of the acid.
2. Add *-amide*.	2. Add *-amide*.

For example,

$$
\underset{\substack{\text{ethanoic acid}\\\text{(acetic acid)}}}{\overset{\overset{\textstyle O}{\|}}{CH_3-C-OH}} + \underset{\substack{\text{ammonia}\\\text{(ammonia)}}}{\overset{\overset{\textstyle H}{|}}{H-N-H}} \xrightarrow{\text{heat}} \underset{\substack{\text{ethanamide}\\\text{(acetamide)}}}{\overset{\overset{\textstyle O}{\|}}{CH_3-C-NH_2}} + H_2O
$$

If an organic acid reacts with a primary or secondary amine, a substituted amide is formed.

$$R-\overset{\overset{\displaystyle O}{\|}}{C}-OH + H-\overset{\overset{\displaystyle H}{|}}{N}-R' \longrightarrow R-\overset{\overset{\displaystyle O}{\|}}{C}-\overset{\overset{\displaystyle H}{|}}{N}-R' + H_2O$$

acid   primary   a substituted
amine   amide

$$H-\overset{\overset{\displaystyle O}{\|}}{C}-OH + H-\overset{\overset{\displaystyle H}{|}}{N}-CH_2-CH_3 \longrightarrow H-\overset{\overset{\displaystyle O}{\|}}{C}-\overset{\overset{\displaystyle H}{|}}{N}-CH_2-CH_3 + H_2O$$

methanoic acid   ethylamine   *N*-ethylmethanamide
(formic acid)   (*N*-ethylformamide)

$$CH_3-\overset{\overset{\displaystyle O}{\|}}{C}-OH + H-\overset{\overset{\displaystyle CH_3}{|}}{N}-CH_3 \longrightarrow CH_3-\overset{\overset{\displaystyle O}{\|}}{C}-\overset{\overset{\displaystyle CH_3}{|}}{N}-CH_3 + H_2O$$

ethanoic acid   dimethylamine   *N,N*-dimethylethanamide
(acetic acid)   (*N,N*-dimethylacetamide)

Alkyl groups attached to the nitrogen are indicated by the letter *N*.

Amides can also be prepared by reacting an acid with thionyl chloride ($SOCl_2$) to form a compound called an acyl chloride, which in turn reacts with ammonia to form an amide. The reactions are

$$RCOOH + SOCl_2 \longrightarrow RCOCl$$

acid   thionyl chloride   acyl chloride

$$RCOCl + NH_3 \longrightarrow RCONH_2$$

acyl chloride   amide

Properties of Amides

Amides are neutral compounds compared to amines, which are basic. Acid hydrolysis of simple amides yields a carboxylic acid, whereas basic hydrolysis yields the salt of a carboxylic acid.

$$\text{Acid hydrolysis} \quad CH_3CONH_2 \xrightarrow[\text{H}_2\text{O}]{\text{HCl}} CH_3COOH$$

$$\text{Basic hydrolysis} \quad CH_3CONH_2 \xrightarrow[\text{H}_2\text{O}]{\text{NaOH}} CH_3COONa$$

If two amino acids combine, with the acid part of one reacting with the amine part of the other, the following type of reaction can occur.

$$R-\underset{\underset{\displaystyle NH_2}{|}}{CH}-\overset{\overset{\displaystyle O}{\|}}{C}-OH + H-\overset{\overset{\displaystyle H}{|}}{N}-\underset{\underset{\displaystyle R}{|}}{CH}-COOH \longrightarrow R-\underset{\underset{\displaystyle NH_2}{|}}{CH}-\overset{\overset{\displaystyle O}{\|}}{C}-\overset{\overset{\displaystyle H}{|}}{N}-\underset{\underset{\displaystyle R}{|}}{CH}-COOH + H_2O$$

amino acid   amino acid   a peptide

These two amino acids are said to be linked by a peptide (amide) bond. Compounds of this type will be discussed in Chapter 23.

Niacin, one of the B vitamins, is administered as an amide, niacinamide. Urea, one of the metabolic products of protein metabolism, can be considered the diamide of carbonic acid.

$$\underset{\text{carbonic acid}}{HO-\overset{\overset{\textstyle O}{\|}}{C}-OH} \qquad \underset{\text{urea}}{H_2N-\overset{\overset{\textstyle O}{\|}}{C}-NH_2}$$

Summary

During the oxidation of an alcohol, the oxygen atom reacts with the H from the —OH group and with a hydrogen attached to the same carbon that has the —OH group.

The oxidation of a primary alcohol yields an aldehyde. Aldehydes contain the —CHO group and have the general formula RCHO. The aldehyde of one carbon atom is known as formaldehyde or methanal; that of two carbon atoms, as acetaldehyde or ethanal. A water solution of formaldehyde, known as formalin, is commonly used as a germicide. Paraldehyde, formed by polymerizing molecules of acetaldehyde, depresses the central nervous system.

Aldehydes are good reducing agents. When an aldehyde is heated with $Cu(OH)_2$, a red precipitate of Cu_2O is formed.

The oxidation of a secondary alcohol yields a ketone. A ketone cannot be further oxidized without decomposition. Likewise, tertiary alcohols are not easily oxidized.

The oxidation of an aldehyde yields an acid. The general formula for an acid is RCOOH. The —COOH group, called the carboxyl group, furnishes hydrogen ions and so causes the acidic properties. Organic acids react with bases to form organic salts.

Formic acid (HCOOH) is found in the sting of bees. Acetic acid (CH_3COOH) is one of the components of vinegar. Citric acid is found in citrus fruits, lactic acid in milk.

When an organic acid reacts with an alcohol, a compound called an ester is produced. Esters have the general formula RCOOR′. Esters are named according to the alcohol and the acid from which they were made.

Amines are organic compounds in which an alkyl group has replaced one or more of the hydrogen atoms of ammonia, NH_3. Amines are basic compounds and readily form salts.

Amino acids are organic acids that contain an amine group. α-Amino acids are the building blocks of protein.

Amides are formed by the reaction of an organic acid with ammonia or with an amine. Amino acids are held together by peptide (amide) bonds.

Questions and Problems

A

1. Write the equation for the oxidation of methyl alcohol to an aldehyde. What product is formed?
2. Write the equation for the formation of acetaldehyde from ethyl alcohol.
3. What is the general formula for an aldehyde?
4. What does the suffix -al indicate?
5. What does the suffix -one indicate?
6. Indicate some uses for formaldehyde.
7. What is paraldehyde? What is it used for?
8. Describe the test for aldehydes.
9. Write the reaction for the oxidation of isopropyl alcohol.

10. The oxidation of a primary alcohol yields what type of compound?
11. The oxidation of a secondary alcohol yields what type of product?
12. Describe the oxidation of a tertiary alcohol; of a ketone.
13. Write the equation for the oxidation of formaldehyde, and name the product.
14. Write the equation for the oxidation of acetaldehyde and name the product.
15. What is the general formula for an acid?
16. Write the equation for the reaction between formic acid and potassium hydroxide.
17. Name several organic acids, and indicate where each can be found.
18. What is an ester? Indicate its general formula.
19. Write the equation for the reaction between formic acid and ethyl alcohol. Name the product.
20. Name two esters and give their medical uses.
21. What is an amine? Compare the structures of primary, secondary, and tertiary amines.
22. How do the terms *primary, secondary*, and *tertiary* for amines compare with the use of these same terms for alcohols?
23. What is an amino acid? What properties does it have?
24. For what purposes does the body use amino acids?
25. Why are amino acids amphoteric?
26. What is the type structure for an amide?
27. Write the reaction when formic acid and ammonia are heated. What is the name of the product?
28. What is a peptide bond? Where is it found?
29. Why is urea called a diamide?

B

30. In what type of compounds is a carbonyl group present?
31. Name the following compounds according to the IUPAC system.

(a) $CH_3-CH-CH_2-\overset{\displaystyle H}{\underset{\displaystyle Cl}{\overset{|}{C}}}=O$

(b) $Cl-\overset{\displaystyle Cl}{\underset{\displaystyle Cl}{\overset{|}{\underset{|}{C}}}}-\overset{\displaystyle O}{\overset{\|}{C}}-OH$

(c) $CH_3-CH-CH_2-\overset{\displaystyle}{\underset{\displaystyle H}{\overset{|}{C}}}=O$
$\overset{|}{CH_3}$

(d) $CH_3-\overset{\displaystyle CH_3}{\underset{\displaystyle O}{\overset{|}{\underset{\|}{C}}}}-CH-CH_3$

(e) $CH_3-\overset{\displaystyle}{\underset{\displaystyle NH_2}{\overset{|}{CH}}}-\overset{\displaystyle O}{\overset{\|}{C}}-OH$

(f) $CH_3-CH_2-\overset{\displaystyle O}{\overset{\|}{C}}-\overset{\displaystyle}{\underset{\displaystyle CH_3}{\overset{|}{CH}}}-CH_2-CH_3$

(g) $H-\overset{\displaystyle O}{\overset{\|}{C}}-\overset{\displaystyle}{\underset{\displaystyle H}{\overset{|}{N}}}-CH_2-CH_3$

32. Reduction of an aldehyde yields what type of compound? reduction of a ketone?
33. Why do aldehydes give a positive test with Clinitest, whereas ketones give a negative test?
34. What is a hemiacetal? an acetal? Why are they important in the body?
35. What is a hemiketal? a ketal? Why are they important in the body?
36. Why does the removal of calcium ions from the blood prevent clotting?

Practice Test

1. The general formula for an aldehyde is _____.
   a. ROH
   b. RCOR
   c. RCHO
   d. RCOOR
2. The general formula for an acid is _____.
   a. ROH
   b. RCOR
   c. RCHO
   d. RCOOH
3. The general formula for a primary amine is _____.
   a. RNH_2
   b. R_2NH
   c. R_3N
   d. R_4NCl
4. The oxidation of a secondary alcohol yields a(n) _____.
   a. acid
   b. aldehyde
   c. ketone
   d. ester
5. The reaction of an acid with an alcohol yields a(n) _____.
   a. amide
   b. amine
   c. ester
   d. amino acid

6. The name of the following compound is _____.

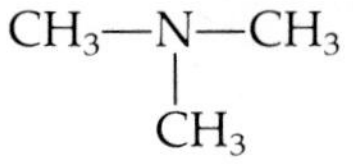

$$CH_3—N—CH_3$$
$$|$$
$$CH_3$$

   a. methylamine   b. dimethylamine
   c. trimethylamine   d. ethylmethylamine

7. An example of an amphoteric compound is a(n) _____.
   a. ester   b. amino acid
   c. substituted amine   d. alcohol

8. What type of compound yields a positive test with Clinitest tablets?
   a. alcohols   b. aldehydes
   c. ketones   d. esters

9. When an aldehyde reacts with excess alcohol, the product is called a(n) _____.
   a. acetal   b. hemiacetal
   c. ketal   d. hemiketal

10. An example of a basic compound is a(n) _____.
   a. ester   b. amine
   c. amide   d. alcohol

19

Aromatic Compounds

A picture of ring-shaped benzene molecules generated by a scanning tunneling microscope.

The term *aromatic* originally referred to certain compounds that had a pleasant odor and similar chemical and physical properties. Further studies of these compounds showed that they all had a ring-shaped structure. However, many compounds have a pleasant odor and do not have this ring-shaped structure. The term **aromatic** is now usually used to designate compounds whose bonding has features in common with that of benzene.

Benzene

Structure

Benzene has the formula C_6H_6. Since there is one hydrogen atom for every carbon atom, we might expect benzene to have several double bonds in order to fulfill the bonding requirements for carbon, thus making it unsaturated and predictably very reactive. On the contrary, however, benzene is quite stable. The structure for the benzene molecule was first deduced by Kekulé in 1865. He stated that the six carbon atoms were arranged in a ring with alternate single and double bonds, each carbon atom having one hydrogen atom attached to it. Kekulé also suggested that the position of the double and single bonds could change, producing two structures that represent benzene.

resonance structures of benzene

The actual structure of benzene is intermediate between these two Kekulé structures. This is indicated by the double-headed arrow. The two Kekulé structures are called *resonance structures. The

Figure 19-1 Electron-cloud picture of benzene.

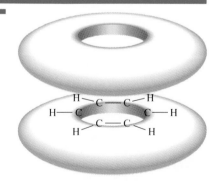

electron-cloud picture (Figure 19-1) of the benzene molecule shows two continuous doughnut-shaped electron clouds, one above and the other below the plane of the atoms.

The benzene structure is the basis of many thousands of organic compounds. The chemist has devised an abbreviated version of the benzene formula.

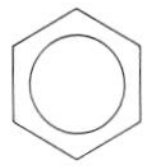

Abbreviated resonance structures for benzene found in other texts are

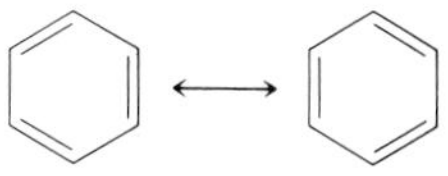

In all these representations it is assumed that a hydrogen atom is present on each carbon or at each corner unless otherwise indicated.

If one of the hydrogen atoms is replaced by a methyl group (CH_3—), it is indicated as follows.

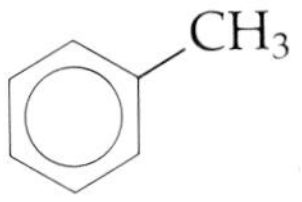

Properties

Benzene is a colorless liquid with a distinct gasolinelike odor. It is insoluble in water but soluble in alcohol and ether. Benzene is toxic when taken internally. Contact with the skin is harmful, and continued inhalation of benzene vapors decreases red and white blood cell counts. Benzene is now considered to be mildly carcinogenic, and care must be taken with its use. For this reason, the use of benzene in laboratory experiments by students has, or should be, discontinued. Also, benzene, commonly known as naphtha, has been banned as an ingredient in most consumer products.

Halogen Derivatives

Although benzene has three double bonds, it does not easily undergo addition reactions as do alkenes. Instead, like alkanes, benzene usually undergoes substitution reactions. This difference is due to resonance in the benzene structure.

When benzene is treated with chlorine, chlorobenzene is produced.

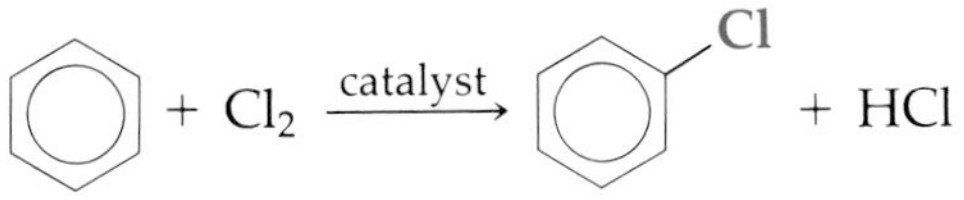

The structure of chlorobenzene indicates that one of the hydrogen atoms in the benzene ring has been replaced by a chlorine atom. Since all six carbon atoms and all six hydrogen atoms are equivalent in the benzene ring, there is only one possible *mono*substitution product. That is, all of the structures shown in diagram (19-1) are identical—they represent the same compound, chlorobenzene.

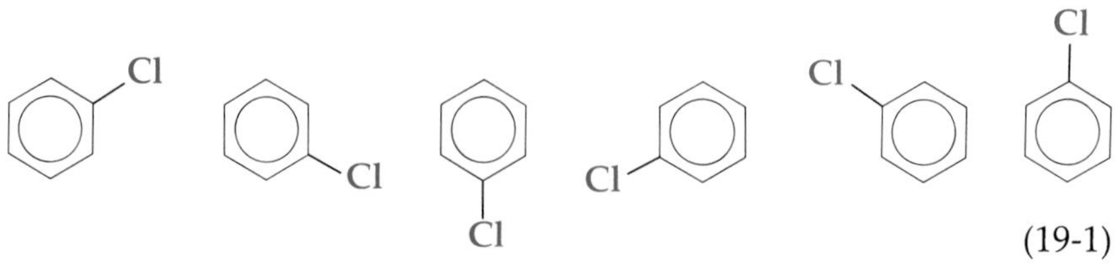

(19-1)

However, when two of the hydrogen atoms in the benzene ring are replaced, more than one possible *di*substitution product is possible. One of the disubstitution products obtained upon the reaction of chlorine with chlorobenzene is indicated as follows.

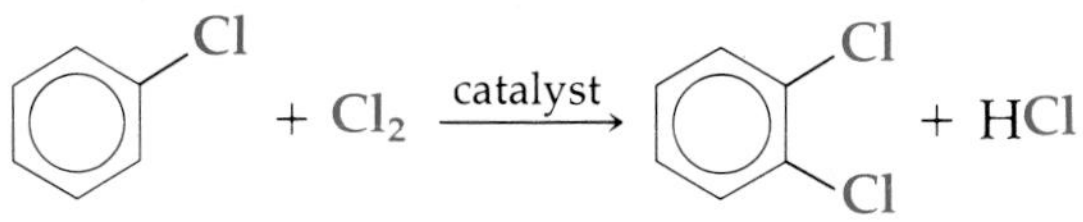

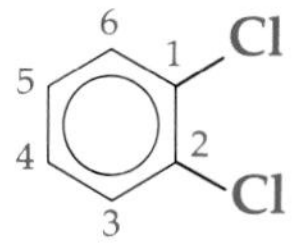

(19-2)   For the IUPAC name, the benzene ring is numbered from 1 to 6. The numbers must be such that the substituents attached to the ring have the lowest possible numbers, as shown in structure (19-2). The name of this compound is 1,2-dichlorobenzene, which indicates two chlorines on a benzene ring in positions 1 and 2.

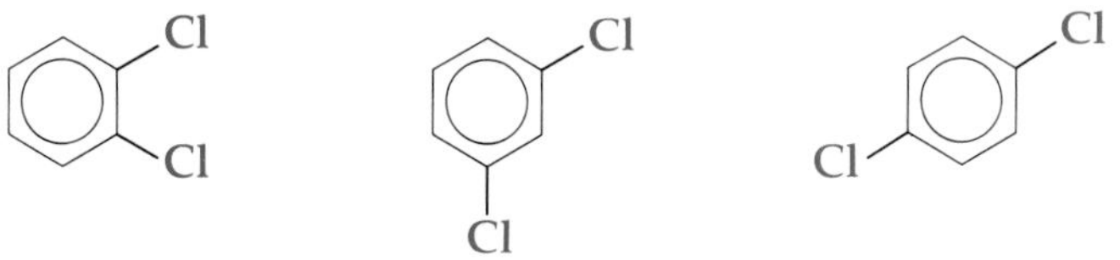

(19-3)

1,2-dichlorobenzene   1,3-dichlorobenzene   1,4-dichlorobenzene

Actually, when chlorine reacts with chlorobenzene, three different disubstitution products (isomers) are obtained. Their structures and names are indicated in (19-3).

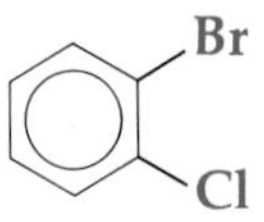

(19-4)   If the disubstitution products on the benzene ring are different, the same system of naming may be used. Consider compound (19-4) It may be numbered with either the bromine atom at position 1 and the chlorine atom at position 2 or vice versa, as shown in (19-5).

(19-5)

Thus the compound may be named either 1-bromo-2-chlorobenzene or 1-chloro-2-bromobenzene. The preferred name lists the substituents in alphabetical order, but pharmaceutical companies frequently use several different systems for naming their drugs.

The common system for naming disubstituted benzene compounds is based upon the use of prefixes rather than numbers to designate positions in the benzene ring. The prefix **ortho-** indicates substances on the benzene ring in positions next to each other. The compound shown in (19-6a) is called *ortho*-dichlorobenzene, or simply *o*-dichlorobenzene.

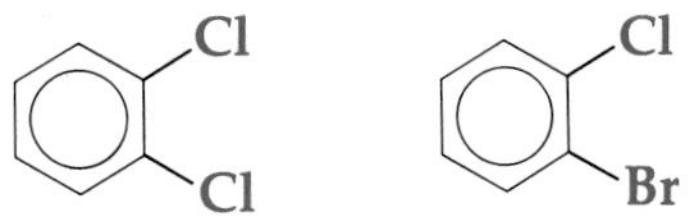

(19-6)

(a)   (b)

The compound shown in (19-6b) may be called either *o*-chlorobromobenzene or *o*-bromochlorobenzene; the alphabetic sequence is preferred.

When substituents on the benzene ring are separated by one carbon atom (in positions 1 and 3), the prefix used is **meta-**. The compound shown in (19-7a) is called *meta*-dichlorobenzene or *m*-dichlorobenzene.

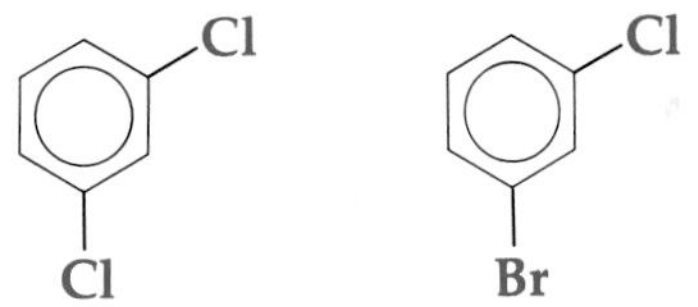

(19-7)

(a)   (b)

Structure (19-7b) is called *m*-bromochlorobenzene.

When two substituents on the benzene ring are separated by two carbon atoms (in positions 1 and 4) the prefix used is **para-**. The compound indicated in (19-8a) is called *para*-dichlorobenzene or *p*-dichlorobenzene. It is a modern version of mothballs.

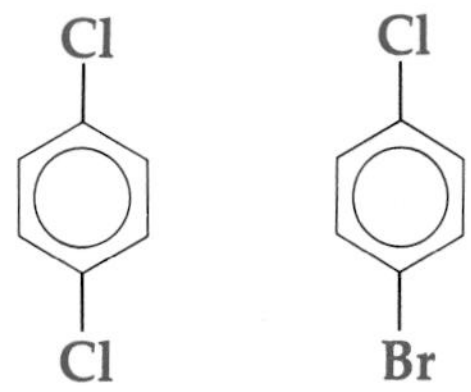

(19-8)

(a)   (b)

The structure of *p*-bromochlorobenzene is shown in (19-8b).

Other Derivatives

Toluene

The methyl derivative of benzene is commonly called toluene. Its IUPAC name is methylbenzene. Its structure is

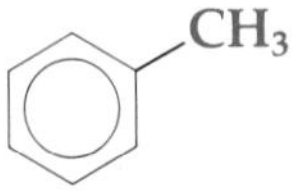

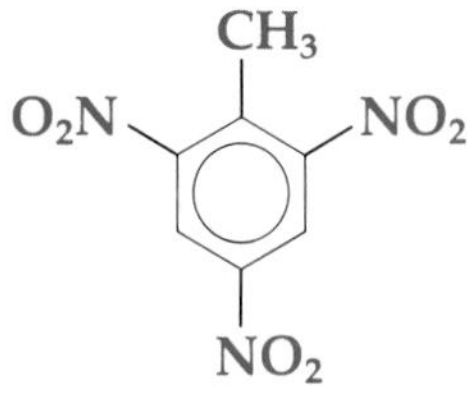

Toluene is a colorless liquid with a benzenelike odor. It is insoluble in water and soluble in alcohol and ether. Toluene is used as a preservative for urine specimens and in the preparation of dyes and explosives. Toluene is one ingredient in airplane glue. When "sniffed," it can produce blurred vision and a lack of coordination and may even be fatal. Toluene, however, is much less toxic than benzene and now is frequently being substituted for benzene.

A derivative of toluene, trinitrotoluene (TNT), is a powerful explosive.

TNT

Xylene

Xylene is a dimethylbenzene. Again, with two substitutions on the benzene ring, there are three possible structures of xylene.

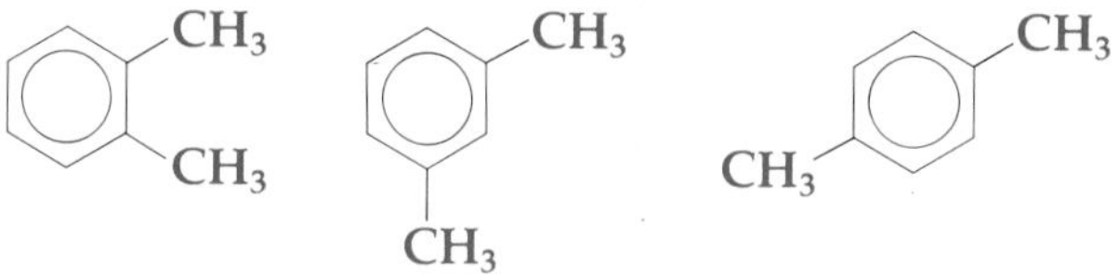

o-xylene   *m*-xylene   *p*-xylene

Xylenes are good solvents for oils and are used in cleaning lenses in microscopes.

Naphthalene

Naphthalene, $C_{10}H_8$, is an aromatic compound containing two benzene rings. These two rings are attached to each other in such a manner that they share two carbon atoms.

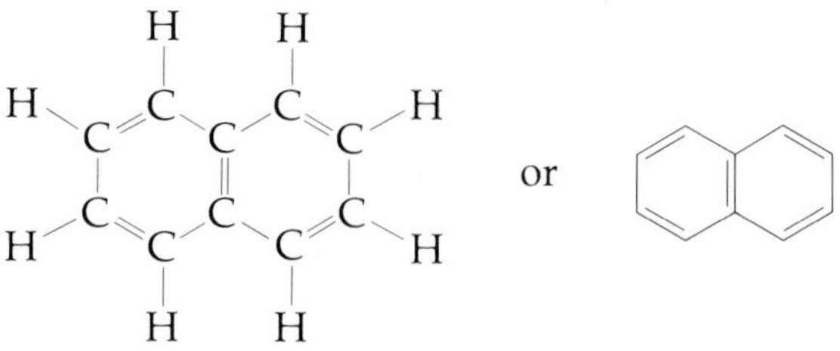

or

Naphthalene is a white crystalline solid obtained from coal tar.

Naphthalene crystals are frequently used in the home under the name mothballs. Functional groups may be attached to naphthalene in either of two positions, called alpha (α) and beta (β). For example, the structures of α- and β-naphthols are

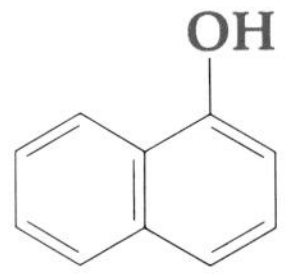

OH

and

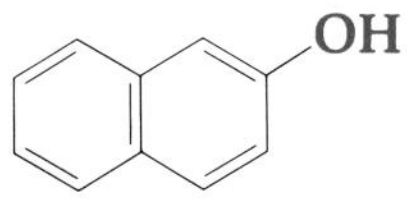

OH

α-naphthol β-naphthol

Note that naphthalene has four α positions and also four β positions.

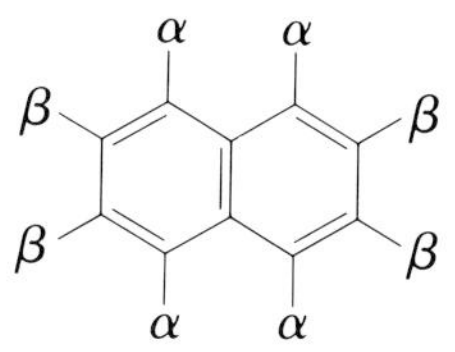

$$\alpha \quad \alpha$$
$$\beta \qquad\qquad \beta$$
$$\beta \qquad\qquad \beta$$
$$\alpha \quad \alpha$$

Anthracene and Phenanthrene

Anthracene and phenanthrene are aromatic compounds containing three benzene rings joined together. Their structures are

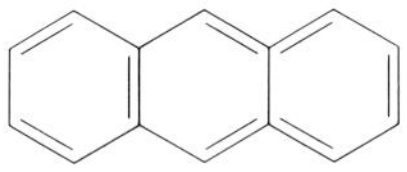

anthracene 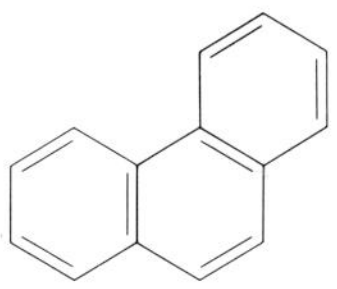   phenanthrene

Anthracene is used commercially in the manufacture of dyes. Phenanthrene is an isomer of anthracene. It also contains three benzene rings but in a different structural arrangement. Phenanthrene has the basic structure of many biologically and medically important compounds. Among these are the male and female sex hormones, vitamin D, cholesterol, bile acids, and some alkaloids.

Other Polycyclic Aromatic Compounds

benzpyrene

When coal or wood is burned, compounds containing several benzene rings are obtained. Such compounds are carcinogenic (cancer producing). One of the most active of this group of carcinogens is benzpyrene. In the seventeenth century chimney sweeps in England were the first people known to develop cancer from such compounds, which are also found in the "tar" of cigarettes. Industrial nations emit large amounts of benzpyrene into the air. This dangerous substance is also found in well-done charbroiled meats and smoked fish.

Other Aromatic Compounds

Phenols

When an —OH group is attached to a benzene ring, a class of compounds known as phenols is formed. Generally, phenols are like alcohols but have been placed in a class by themselves because phenols are weak acids and alcohols are not.

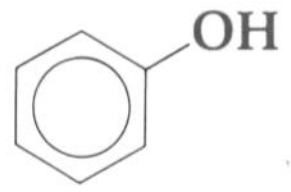

phenol

Phenol reacts with aqueous sodium hydroxide solution to form a water-soluble salt, sodium phenoxide. Alcohols such as ethanol and cyclohexanol do not react with basic solutions.

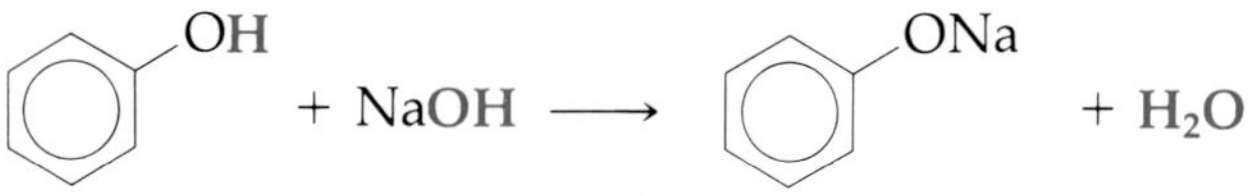

phenol   sodium phenoxide

$$CH_3CH_2OH + NaOH \longrightarrow \text{no reaction}$$
ethanol

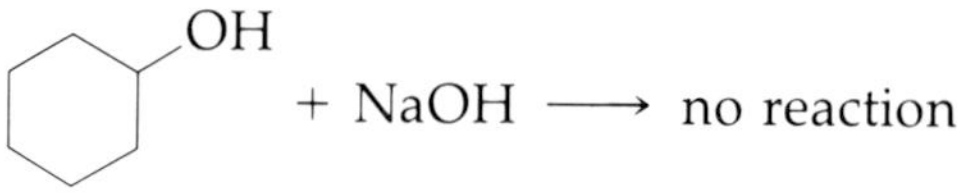

cyclohexanol

Pure phenol is a white crystalline solid with a low melting point, 41 °C. However, on exposure to light and air phenol turns reddish. Phenol is poisonous if taken internally, and externally it causes deep burns and blisters on the skin. If phenol should accidentally be spilled on the skin, it should be removed as quickly as possible with 50 percent alcohol, glycerin, sodium bicarbonate solution, or water.

Phenol was the original Lister *antiseptic and is still used as a disinfectant for surgical instruments and utensils, clothing and bed linens, floors, toilets, and sinks. Phenol is used commercially in the manufacture of dyes and plastics.

Phenol is the standard of reference for germicidal activity of disinfectants; that is, their activity is compared to that of phenol. If a 1 percent solution of a germicide kills organisms in the same time that a 5 percent solution of phenol does, then that germicide is said to have a **phenol coefficient** of 5.

Phenol Derivatives

The methyl derivatives of phenol are called cresols. There are three different cresols—*ortho-*, *meta-*, and *para-*cresols.

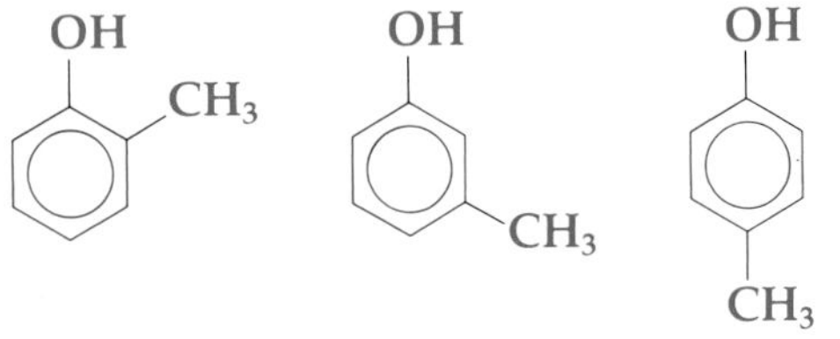

o-cresol   *m*-cresol   *p*-cresol

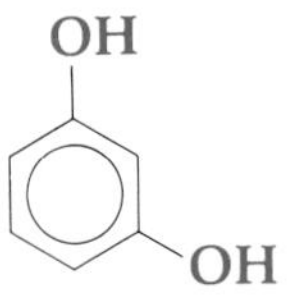

resorcinol

Usually, cresol is a mixture of all three of these isomers. Cresol is a better antiseptic than phenol and is also less toxic. One commonly used disinfectant, Lysol, is a mixture of the three cresols in water with soap added as an emulsifying agent. Even though cresols are less toxic than phenol, they are still poisonous and should be used for external purposes only.

Resorcinol is *m*-dihydroxybenzene. It is also an antiseptic but is not as good as phenol.

A resorcinol derivative, hexylresorcinol, is a much better antiseptic and germicide than resorcinol. It is commonly used in mouthwashes. Its structure is

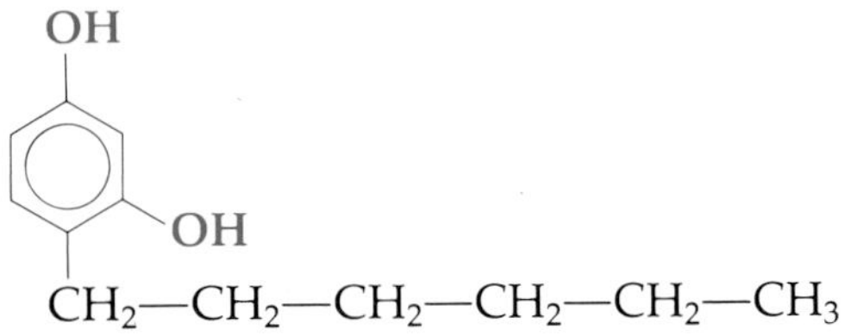

$$CH_2—CH_2—CH_2—CH_2—CH_2—CH_3$$

hexylresorcinol

Hexachlorophene has marked antibacterial activity. Its phenol coefficient is 125. Its activity is retained in the presence of soaps, so soaps containing hexachlorophene are used for surgical scrubs and also by food handlers and dentists. Hexachlorophene has also been employed in deodorants and cleansing creams, but its use for all these purposes has largely been discontinued because of the possibility of adverse effects on brain tissue, particularly in premature infants. The structure of hexachlorophene is

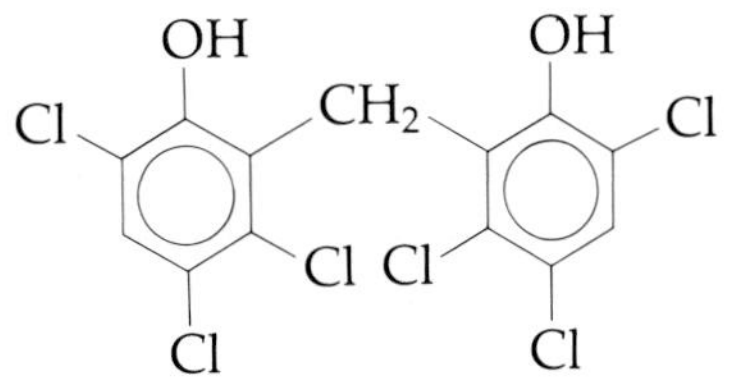

Two compounds commonly found in foods (listed as food preservatives) are BHA (butylated hydroxyanisole) and BHT (butylated hydroxytoluene).

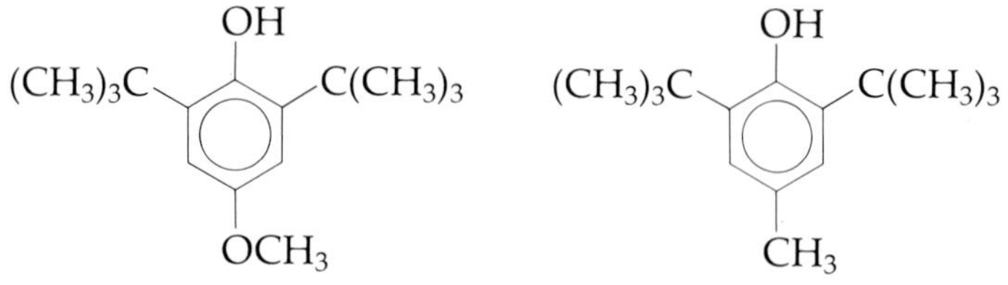

BHA

BHT

Hydroquinone is 1,4-dihydroxybenzene. Hydroquinone is easily oxidized to quinone. Quinone, in turn, is easily reduced to hydroquinone.

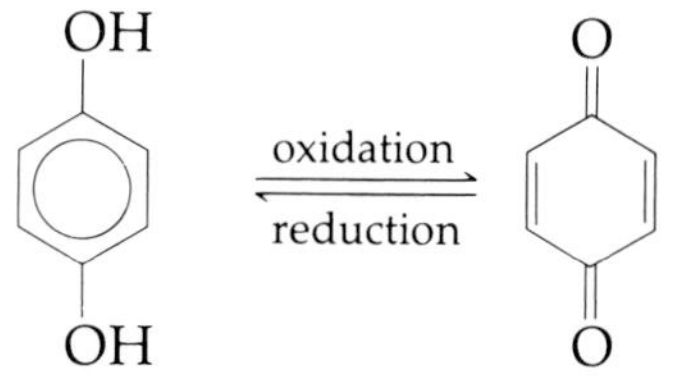

hydroquinone   quinone

Hydroquinone and quinones are important in the respiratory system. One such compound, ubiquinone, is also known as coenzyme Q (see page 439).

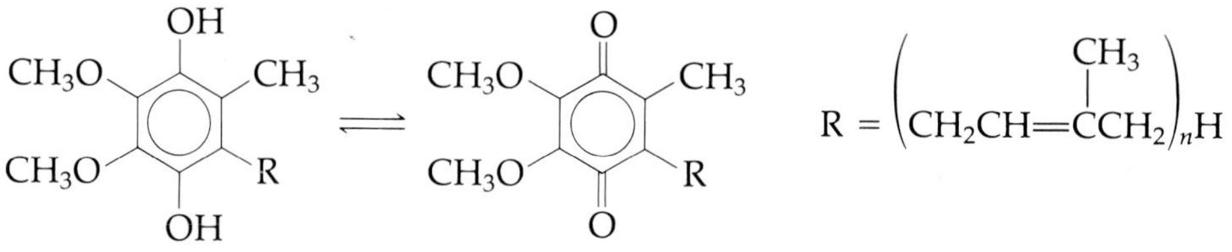

ubiquinone (coenzyme Q)
reduced form   oxidized form

Aldehydes

Aromatic aldehydes have the general formula **ArCHO,** where Ar stands for an aromatic ring. The simplest aromatic aldehyde is benzaldehyde, which consists of an aldehyde group attached to a benzene ring. Benzaldehyde is prepared by the mild oxidation of toluene, as shown in equation (19-11). Note that the side chain, the methyl group, is more susceptible to oxidation than the fairly stable benzene ring.

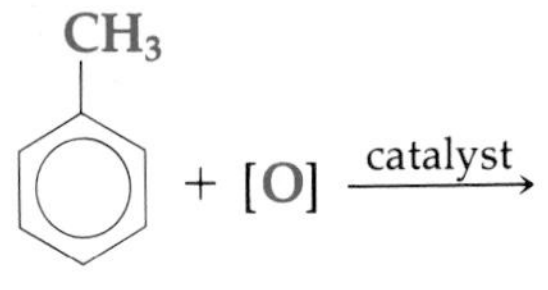

$$\underset{\text{toluene}}{\text{CH}_3\text{—}\bigcirc} + [O] \xrightarrow{\text{catalyst}} \underset{\text{benzaldehyde}}{\text{CHO—}\bigcirc} \qquad (19\text{-}9)$$

Benzaldehyde is a colorless oily liquid with an almondlike odor. It is used in the preparation of flavoring agents, perfumes, drugs, and dyes.

Vanillin occurs in vanilla beans and gives the particular taste and odor to vanilla extract. It also has an aldehyde structure. Cinnamic aldehyde (cinamaldehyde) is present in oil of cinnamon, an oil found in cinnamon bark. Both vanillin and cinnamic aldehyde can be prepared synthetically, and both are used as flavoring agents.

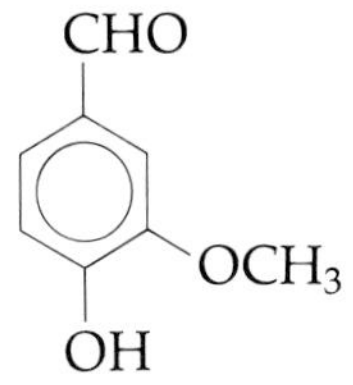

vanillin 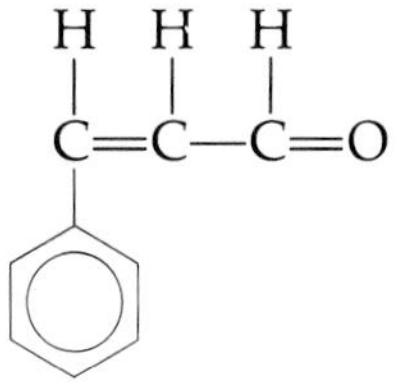   cinnamic aldehyde

Ketones

Aromatic ketones have the general formula **ArCOAr′** or **ArCOR**. The simplest aromatic ketone is acetophenone.

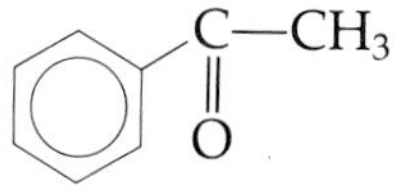

acetophenone 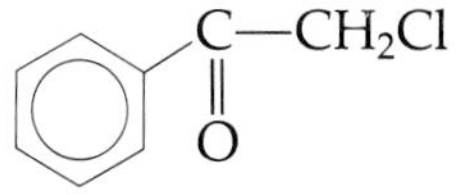   chloracetophenone

Acetophenone has been used as a hypnotic but has been supplanted for this purpose by newer and safer drugs.

Chloracetophenone is a *lacrimator and is used as a tear gas.

Among the aromatic ketones in the body are the sex hormones estrone, progesterone, testosterone, and androsterone (see pages 638–40).

Acids

Aromatic acids have the —COOH group just as aliphatic acids do. Aromatic acids are represented by the general formula **ArCOOH**.

The simplest aromatic acid is benzoic acid, which consists of a —COOH attached to a benzene ring. Benzoic acid can be produced by the oxidation of toluene.

CH$_3$ → COOH

$$\xrightarrow[\text{[O]}]{\text{catalyst}}$$

toluene   benzoic acid

Benzoic acid is a white crystalline compound slightly soluble in cold water and more soluble in hot water. It is used medicinally as an antifungal agent. The sodium salt of benzoic acid, sodium benzoate, is used as a preservative.

Salicylic acid is both an alcohol and an acid, as can be seen from its structure.

salicylic acid

Salicylic acid is a white crystalline compound with properties similar to those of benzoic acid. It is used in the treatment of fungal infections and also for the removal of warts and corns.

Commonly used compounds of salicylic acid are the salt sodium salicylate and the ester methyl salicylate.

COONa   OH   COOCH$_3$ OH

sodium salicylate   methyl salicylate

Sodium salicylate is used as an antipyretic (to reduce fever) and also to relieve pain of arthritis, bursitis, and headache. Methyl salicylate is a liquid with a pleasant odor, that of wintergreen. It is used topically to relieve pain in muscles and joints.

The acetyl derivative of salicylic acid (the acetyl group is CH$_3$CO—) is acetylsalicylic acid, more commonly known as aspirin. Aspirin is used as an *analgesic, as an *antipyretic, and for the treatment of colds, headaches, minor aches, and pains. Aspirin is also used in the treatment of rheumatic fever. Over 50 tons of aspirin is used daily in the United States. Aspirin is contraindicated after surgery because it interferes with the normal clotting of the blood and can induce hemorrhaging. Aspirin can cause bleeding of the stomach and therefore should not be taken on an "empty" stomach. About 1 person in 10,000 is allergic to aspirin.

COOH   OCOCH$_3$

acetylsalicylic acid (aspirin)

The action of aspirin is related directly to that of the prostaglandins (see page 398). Aspirin stimulates respiration directly, and overdoses can cause serious acid–base balance disturbances.

Recent evidence appears to indicate that aspirin can prevent blood clots from forming by interfering with the action of the blood platelets. There is also evidence that one aspirin every other day helps prevent heart attacks. While aspirin can prevent sickling of red blood cells (see page 659) in test tubes, clinical evidence of this effect in humans is far from complete.

Acetaminophen (Tylenol) has been used as a substitute for aspirin because it does not cause gastrointestinal bleeding and does not affect blood clotting. However, overdoses can lead to hepatic damage, as can be the case with aspirin itself. Also, it is not very anti-inflammatory.

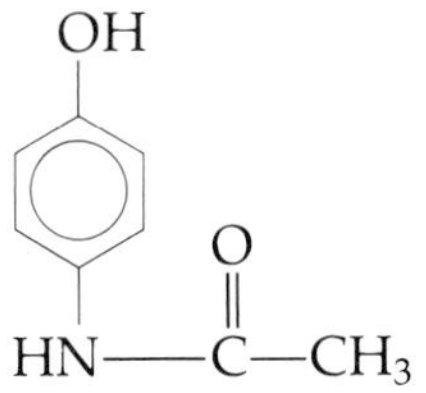

OH

HN——C—CH$_3$

acetaminophen (Tylenol)

Amines

The simplest aromatic amine is aniline, which consists of an amine group attached to the benzene ring. It is prepared by the reduction of nitrobenzene.

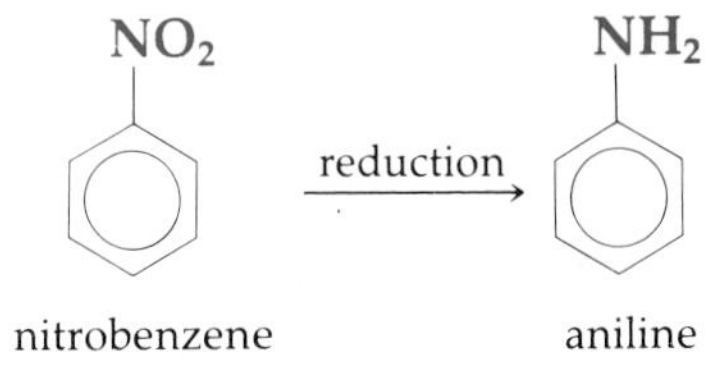

NO$_2$ $\xrightarrow{\text{reduction}}$ NH$_2$

nitrobenzene   aniline

Aniline is used commercially in the preparation of many dyes and drugs. When aniline is reacted with acetic acid, acetanilide is produced.

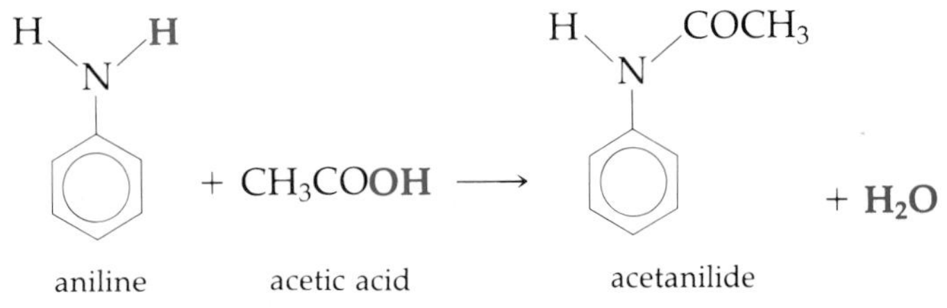

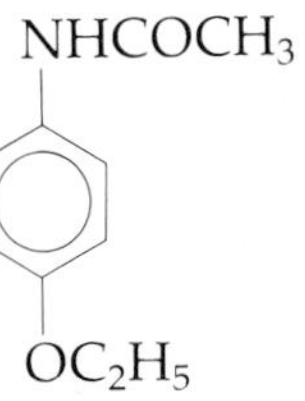

H H
  N   H COCH$_3$
                    N

+ CH$_3$COOH $\longrightarrow$ + H$_2$O

aniline   acetic acid   acetanilide

NHCOCH$_3$

OC$_2$H$_5$

phenacetin

Acetanilide has been used as an antipyretic and as an analgesic. A related compound, phenacetin, has also been used for similar purposes, but its use has largely been discontinued because of liver and kidney toxicity.

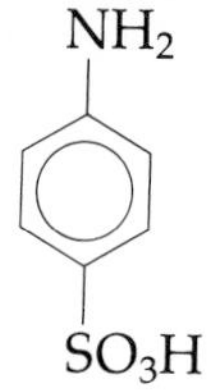

NH$_2$

SO$_3$H

sulfanilic acid

When aniline is heated with concentrated sulfuric acid, it forms *p*-aminobenzenesulfonic acid, commonly known as sulfanilic acid.

In 1936 a derivative of sulfanilic acid, sulfanilamide, was discovered to have definite therapeutic effects against such diseases as pneumonia, diarrhea, and streptococcal infections. However, further investigation showed that sulfanilamide had several disadvantages. It caused nausea, dizziness, anemia, and other toxic reactions in the body. Pharmaceutical chemists then developed a group of related sulfanilamide compounds with equal or better therapeutic value and with greatly decreased toxic effects. These were called *the sulfa drugs.*

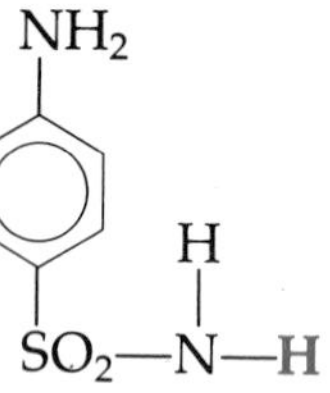

All of the sulfa drugs contain the *p*-aminobenzenesulfonamide group, which gives them their characteristic properties. Some other sulfa drugs are shown below.

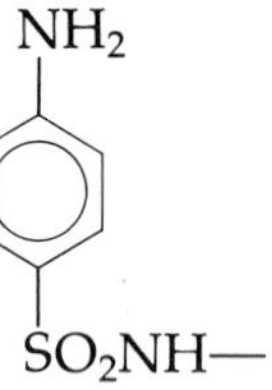

NH$_2$

SO$_2$—N—H
      H

sulfanilamide

NH$_2$

SO$_2$NH—

p-aminobenzenesulfonamide
group

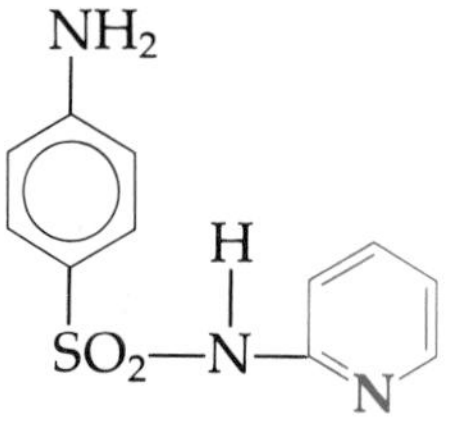

NH$_2$

SO$_2$—N
      H

sulfapyridine

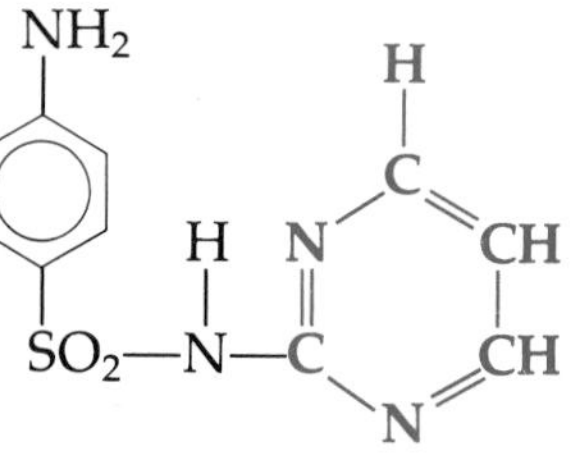

NH$_2$

SO$_2$—N—C
      H   N

sulfadiazine

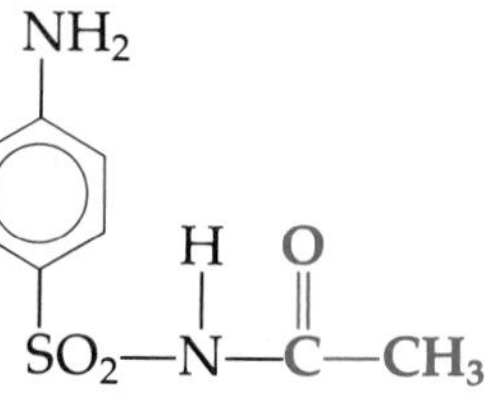

NH$_2$

SO$_2$—N—C—CH$_3$
      H   O

sulfacetamide

NH$_2$

SO$_2$—N
      H

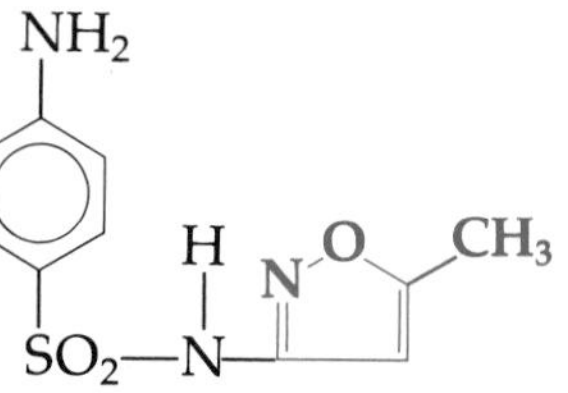

sulfamethoxazole

Since 1935, when the antibacterial action of Prontosil was discovered by Domagk, over 5000 sulfonamide (sulfa-type) compounds have been prepared and tested. Of these only a few have clinical use. The different groups attached to the parent substance determine the potency and toxicity. Most of the sulfa drugs are white crystalline substances that are insoluble in water. Their sodium compounds, however, are readily soluble.

The sulfa drugs are classified as *bacteriostatic and not *bactericidal. They appear to weaken or inhibit the growth of susceptible bacteria and make them more vulnerable to the action of phagocytes in the bloodstream of the host.

It is believed that the sulfa drugs interfere and compete with the use of *p*-aminobenzoic acid (PABA) for the formation of folic acid by bacteria (see page 446). Note the similarity of structures of sulfa drugs and PABA.

NH$_2$

COOH

PABA

p-Aminobenzoic acid is also used in suntan lotions to prevent the dangerous ultraviolet rays from reaching the skin.

Amides

The simplest aromatic amide is benzamide, which is formed by the reaction of benzoic acid with ammonia

benzoic acid   ammonia   benzamide

Aromatic amides have the general formula $ArCONH_2$. The amide of niacin, niacinamide, is one of the B vitamins (see page 606).

niacin   niacinamide

Summary

Benzene (C_6H_6) is a symmetrical six-sided ring compound with a hydrogen at each corner. One or more of the hydrogen atoms of benzene may be replaced by a halogen yielding halogen derivatives.

There is only one monosubstitution product of benzene for any substituent because all six positions on the ring are equivalent.

There are three possible disubstitution products of benzene for any pair of substituents. These compounds can be named numerically. If both the substituents are chlorine, the disubstitution products are 1,2-dichlorobenzene, 1,3-dichlorobenzene, and 1,4-dichlorobenzene.

Prefixes are also used to designate positions on the benzene ring. The prefix *ortho-* corresponds to positions 1,2; *meta-* to 1,3; and *para-* to 1,4.

Naphthalene contains two benzene rings joined together. Anthracene and phenanthrene contain three benzene rings joined together.

The simplest ring alcohol is phenol, a benzene ring with an —OH group attached. Phenol is used as a disinfectant of surgical instruments.

Cresols are methyl derivatives of phenol. Cresol is a better antiseptic than phenol and is also less toxic.

Benzoic acid consists of a benzene ring with a carboxyl (—COOH) group attached. Benzoic acid can be produced by the oxidation of toluene.

Acetylsalicylic acid, commonly known as aspirin, is an analgesic and an antipyretic and is used in the treatment of colds, headaches, minor aches, and pains. Aspirin is contraindicated after surgery.

Aniline consists of an amine group attached to a benzene ring. Aniline is the basic compound for many dyes and drugs. Among these are the sulfa drugs.

Questions and Problems

A

1. Indicate the resonance structures for benzene.
2. What is the abbreviation for the structure of benzene?
3. What effects does benzene vapor have on the body?
4. Draw the structure of chlorobenzene.
5. Why is there only one monosubstitution product with any given substituent for benzene?
6. Why are there three and only three disubstitution products for benzene?
7. Draw the structures for the following compounds: (a) 1,3-dibromobenzene; (b) 1,4-diiodobenzene; (c) 1-chloro-2-iodobenzene.
8. Give IUPAC names for the following compounds.

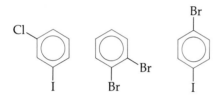

9. Name the compounds in question 8 using the prefix system.
10. Draw the structure of each of the following compounds: (a) *o*-dichlorobenzene; (b) *m*-bromochlorobenzene; (c) *p*-xylene.
11. Draw the structure of toluene and indicate one medical use for it.
12. What are xylenes used for?
13. Draw the structure of naphthalene; of α-naphthol.
14. Why does naphthalene have only two possible monosubstitution products?
15. Draw the structures of anthracene and phenanthrene. Why is the phenanthrene structure important biologically?
16. Draw the structure of phenol. For what is this compound used?
17. What are cresols? How many cresols are there? For what are they used?
18. Draw the structures of resorcinol and hexylresorcinol.
19. How can benzaldehyde be prepared? Write the equation.
20. What kind of compound is acetophenone?

21. How can benzoic acid be prepared? Write the equation.
22. Write the formula for sodium benzoate. What is it used for?
23. What is methyl salicylate used for? sodium salicylate?
24. Draw the structure of aspirin. What is its chemical name?
25. List several uses for aspirin.
26. Why is the use of aspirin contraindicated after surgery?
27. How can acetanilide be prepared?
28. Draw the structure of sulfanilic acid. Of what medical importance is this compound?
29. To what do sulfa drugs owe their activity?
30. What is the general structure for the following types of aromatic compounds: acids, aldehydes, amines, amides?
31. Write the reaction between benzoic acid and ammonia; between benzoic acid and methylamine.

B

32. Compare the structures of cyclohexane and cyclohexene; of cyclohexene and benzene.
33. How do the compounds in question 32 react with halogens?
34. Are anthracene and phenanthrene isomers? anthracene and naphthalene? Explain.
35. Benzpyrene contains how many rings? Where is it found? Why is it a dangerous compound?
36. Many germicides have a phenol coefficient of 500 or over. What does this mean?
37. Why are BHA and BHT added to foods?
38. Compare the use of R and Ar in "general formulas."
39. Name several aromatic ketones of biologic importance.
40. Why is niacinamide an important biologic compound? What does it do?

Practice Test

1. ![benzene ring with CH₃ group] is called ———.

   a. aniline
   b. toluene
   c. phenol
   d. phenanthrene

2. 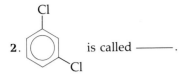 is called ———.

   a. *o*-dichlorobenzene
   b. *m*-dichlorobenzene
   c. *p*-dichlorobenzene
   d. 2,4-dichlorobenzene

3. Vitamin D and the sex hormones have the same basic structure as ———.
   a. anthracene   b. naphthalene
   c. phenanthrene   d. benzpyrene

4. The oxidation of toluene yields ———.
   a. benzpyrene   b. phenol
   c. acetophenone   d. benzoic acid

5. Aspirin is used as a(n) ———.
   a. antipyretic   b. analgesic
   c. both of these   d. neither of these

6. Which of the following structures represents phenol?

   a. 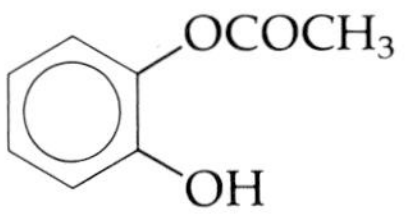   b.

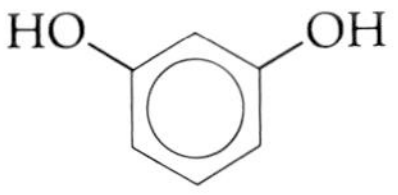

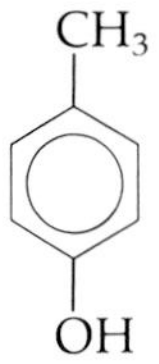

7. The chemical name for aspirin is ———.
   a. salicylic acid   b. acetylsalicylic acid
   c. benzpyretic acid   d. methyl salicylate

8. 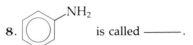 is called ———.

   a. benzene   b. nitrobenzene
   c. aniline   d. phenacetin

9. 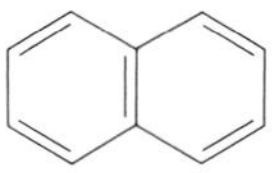 is called ———.

   a. dibenzene   b. anthracene
   c. naphthalene   d. benzpyrene

10. A substance used as a germicide is ———.
   a. benzene   b. acetophenone
   c. benzoic acid   d. phenol

20

Heterocyclic Compounds

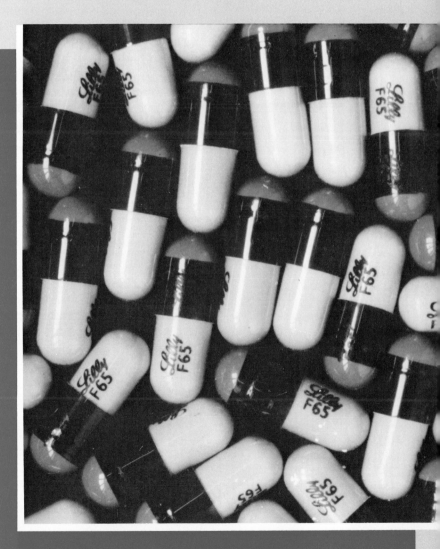

Capsules of Tuinal, a barbiturate. Each capsule contains
equal quantities of quinalbarbitone and amylobarbitone,
which act as sedative.

Heterocyclic compounds are ring compounds that contain some element other than carbon in the ring. The elements most commonly found in the ring other than carbon are nitrogen, sulfur, and oxygen. Heterocyclics are primarily five- and six-membered rings, sometimes joined to one another.

Some Common Heterocyclic Compounds

A few of the common heterocyclic rings in structural and abbreviated forms are shown in Figure 20-1. Also indicated there are the biologically and medically important compounds based on these structures. The compound after which the group is named is given in color.

The DNA (deoxyribonucleic acid) structure, which is the basis of genetic information in the cell, contains four heterocyclic rings; two of these have a purine base, and two have a pyrimidine base. The purines are adenine and guanine; the pyrimidines are cytosine and thymine. The cell messenger, ribonucleic acid (RNA) (see Chapter 34), contains the pyrimidine uracil in place of thymine. The structures of these important heterocyclics are

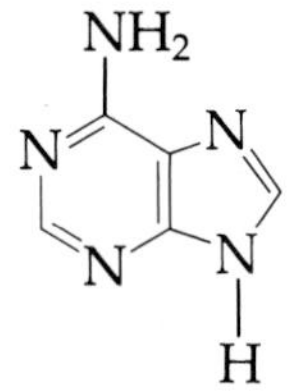

adenine 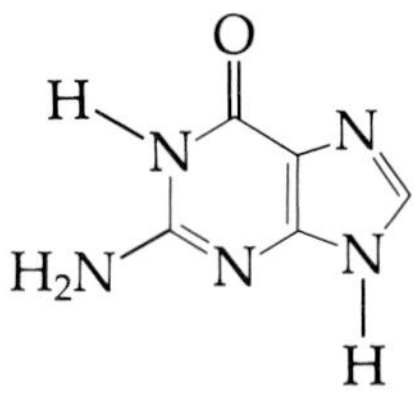　　　guanine 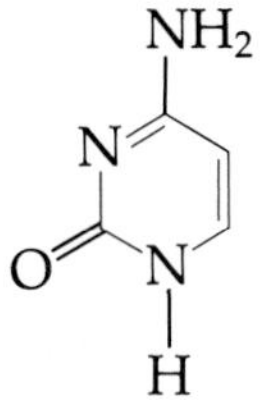　　　cytosine 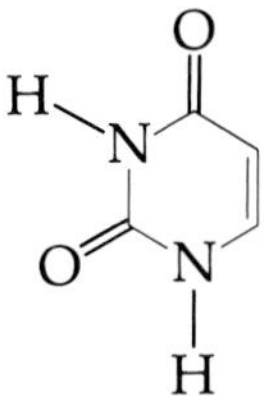　　　uracil 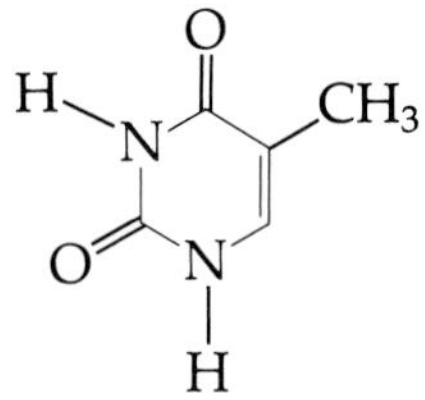　　　thymine

One important class of derivatives of pyrrole is called the porphyrins. Porphyrins are formed by joining four pyrrole rings by means of —CH— bridges. The simplest porphyrin is called prophin.

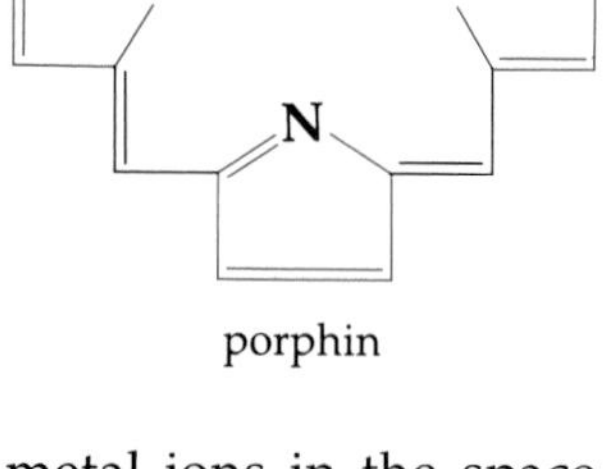

porphin

Porphyrins bind metal ions in the space between the four nitrogens. Examples of prophyrins of biologic importance are hemo-

Figure 20-1 Some common heterocyclic compounds. Below the name of each (in color) are listed several compounds of biological and medical importance that are based on the particular ring structure.

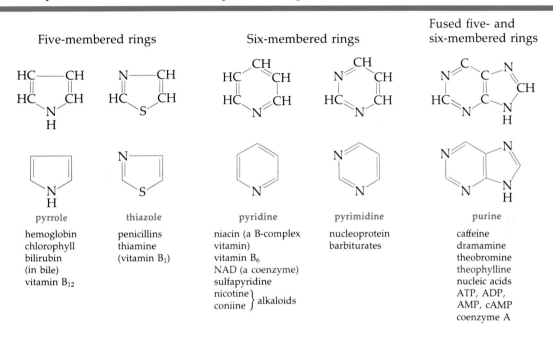

Five-membered rings		Six-membered rings		Fused five- and six-membered rings
pyrrole	thiazole	pyridine	pyrimidine	purine
hemoglobin	penicillins	niacin (a B-complex	nucleoprotein	caffeine
chlorophyll	thiamine	vitamin)	barbiturates	dramamine
bilirubin	(vitamin B_1)	vitamin B_6		theobromine
(in bile)		NAD (a coenzyme)		theophylline
vitamin B_{12}		sulfapyridine		nucleic acids
		nicotine ⎫ alkaloids		ATP, ADP,
		coniine ⎭		AMP, cAMP
				coenzyme A

globin, myoglobin, and the cytochromes (all of which bind iron), chlorophyll (which binds magnesium), vitamin B_{12} (which binds cobalt), and cytochrome oxidase (which binds copper).

Psychoactive Drugs

Psychoactive drugs affect the central nervous system. There are several categories of psychoactive drugs, depending upon their action and the responses they elicit. Among them are the narcotic analgesics, sedative *hypnotics, stimulants, and *psychedelics. An example of each category will be discussed in the following sections. Most of these drugs are heterocyclic compounds containing one or more nitrogen atoms in their rings.

Narcotic Analgesics — Opiates

Opiates are morphine narcotics that are obtained from the opium poppy. The dried unripened pod of this plant is the source of opium, which is a complex mixture of many narcotics. Among them are morphine, codeine, and papaverine. Heroin can be prepared from morphine by a process called acetylation. Synthetic opiates, compounds with morphinelike properties, include meperidine (Demerol) and methadone, but meperidine (Demerol) is itself addictive. The structures of morphine and meperidine are

morphine (R = H; R' = H)   meperidine

In codeine, the R' is replaced by a methyl group. In heroin the R and R' are replaced by acetyl groups.

The opiates depress the central nervous system. The principal effects are analgesia, sedation, drowsiness, lethargy, and *euphoria. The major reason for the abuse of opiates is the relief of apprehension and elevation of mood. However, as time passes, the euphoria wears off and the user becomes apathetic and suddenly falls asleep. Opiates depress the respiratory center, and death from an overdose results from respiratory failure.

Morphine is used as an analgesic for the relief of postoperative pain, for cardiac pain, and for conditions for which other narcotics are ineffective. The chief disadvantage of the continued use of morphine is that it leads to addiction.

Codeine (methylmorphine) depresses the cough center of the brain and so is used in cough medicines. It is also used in the management of pain.

Papaverine, another opiate, is an *antispasmodic; it is a nonspecific smooth muscle relaxant. The drug has been used to dilate blood vessels but is now practically obsolete.

One of the most commonly abused opiates is heroin, which is a bitter-tasting, water-soluble compound. Heroin is two and one-half times as powerful as morphine and has a correspondingly greater possibility of addiction. Its use in the United States has been outlawed except for research purposes. A user of heroin develops a dependence on the drug and also builds up a tolerance to it. As the tolerance develops, increased amounts of heroin are required for the same effect, so addicts often use amounts that would be fatal to a nonuser.

Rapid intravenous injection of heroin produces a warm flushing of the skin and a sensation in the lower abdomen described by addicts as a "thrill." Morphine behaves similarly. Actually, heroin is converted into morphine by the body.

If an addict is without the drug for 10 to 12 hr, the following symptoms appear: vomiting, diarrhea, tremors, pains, restlessness, and mental disturbances. A pregnant heroin addict can pass the dependency on to the child, who must then spend the first few days of life withdrawing from the drug. If not treated properly, the infant may die within a week.

Heroin and other opiates reduce aggression and sexual drive and therefore are unlikely to induce violent crime. However, many individuals commit crimes to pay for their habit.

The currently preferred treatment for heroin addiction is the use of methadone. Methadone is itself an addictive narcotic, but it does not induce euphoria. It eliminates the desire for heroin and also reduces the withdrawal symptoms that normally accompany abstention from heroin. The amount of methadone used is gradually reduced as withdrawal symptoms lessen.

Substitutes for the opiates have been developed for the treatment of intense pain, chronic coughing, or chronic diarrhea. Opiate antagonists counteract overdoses. Naloxone is the best known of these. When given intravenously, it will remove the effect of an opiate within 1 to 2 min.

Brompton's cocktail, a mixture of morphine and cocaine in alcohol, is given orally to relieve the pain of terminal cancer. The original Brompton cocktail, developed in England, contained heroin as a narcotic analgesic, cocaine as a central nervous system stimulant, and alcohol as a flavor enhancer. Syrup and chloroform water were added to improve the mixture's taste and texture.

Sedative Hypnotics — Barbiturates

The barbiturates are a class of synthetic drugs that have sedative and hypnotic (sleep-producing) properties when administered in therapeutic dosages. The structure of a barbiturate is

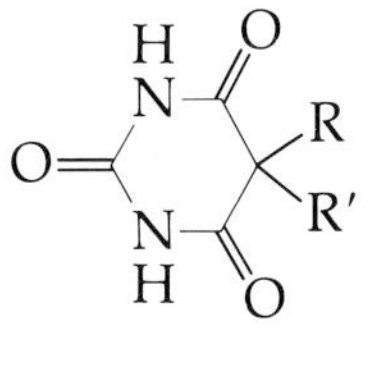

$$
\begin{array}{c}
\text{structure of a barbiturate ring with } O=, \text{ two } N-H \text{ groups, two } C=O \text{ groups, and substituents } R \text{ and } R'
\end{array}
$$

Table 20-1 indicates some of the barbiturates that can be prepared by varying the constituents R and R'.

Table 20-1 Composition and Names of Some Barbiturates

Name	R	R′	Trade Name
Phenobarbital	ethyl, C_2H_5—	phenyl, C_6H_5—	Luminal
Pentobarbital	ethyl, C_2H_5—	1-methylbutyl, $CH_3CH_2CH_2CH(CH_3)$—	Nembutal
Secobarbital	allyl, $CH_2{=}CHCH_2$—	1-methylbutyl, $CH_3CH_2CH_2CH(CH_3)$—	Seconal

In general, the more rapid the onset of action of a barbiturate, the shorter the duration of its physiologic effect. Phenobarbital is a long-acting barbiturate (10 to 12 hr), whereas secobarbital (Seconal) and pentobarbital (Nembutal) are short-acting drugs (3 to 4 hr).

Phenobarbital is used as an anticonvulsant agent in the treatment of epilepsy by potentiation of inhibitory pathways in the nervous system.

In general, barbiturates depress the central nervous system. A person who takes repeated doses of a barbiturate develops a tolerance for the drug and soon requires many times the original amount to produce the desired effect. Small amounts of barbiturates produce the symptoms of chronic alcoholism—the user is sociable, relaxed, and good humored, but his or her ability to react and alertness are decreased. Large amounts of the drug result in blurred speech, confusion, mental sluggishness, and loss of emotional control and may induce a coma. A barbiturate is especially dangerous when taken with alcohol, which is a depressant. The two combined depress the nervous system twenty times as much as barbiturate alone and can lead to death.

Barbiturates are a definite hazard to the user because, even though the tolerance builds up with increased use, the lethal dose remains approximately the same, thus reducing an addict's margin of safety.

In the United States the names of all barbiturates end in *-al*, even though the compounds are ketones rather than aldehydes. In the British Commonwealth, the ending *-one* is used.

Stimulants — Amphetamines

The amphetamines, known as "pep pills" or uppers, are one type of synthetic drugs that stimulate the central nervous system. These drugs are useful in the treatment of such conditions as narcolepsy (an overwhelming compulsion to sleep) and mild depression. Their use for obesity is no longer recommended since the reduction in appetite induced by amphetamines is temporary. Frequent misusers of amphetamines are overtired truckdrivers and businesspersons, students cramming for exams, and athletes who need a "pickup" before a game. Prolonged use or misuse leads to serious effects because the amphetamines tend to concentrate in the brain and cerebrospinal fluid. A tolerance to the drug develops rapidly, and increasing amounts lead to long periods of sleeplessness, loss of weight, and severe *paranoia.

The amphetamine Benzedrine was synthesized to simulate the action of epinephrine (adrenaline) (see page 629). Note the similarity of their structures.

Drug users refer to Benzedrine as "bennies" and to dextroamphetamine as "dexies." Amphetamine and methamphetamine, used intravenously, are known as "speed."

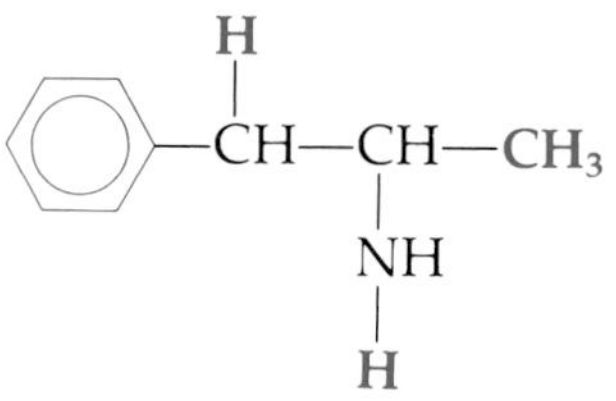

amphetamine
(Benzedrine)

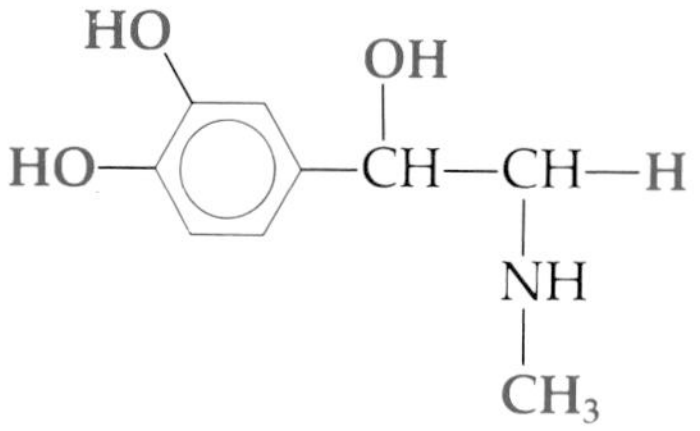

epinephrine
(adrenaline)

Psychedelics

The psychedelic drugs produce "visions." Among these drugs are the **hallucinogens**, drugs that stimulate sensory perceptions, and the **psychomimetics**, drugs that seem to mimic psychoses.

The "psychedelic state" includes several major effects. There is a heightened awareness of sensory input. The individual has an enhanced sense of clarity but reduced control over what is experienced. One part of the individual seems to be observing what the other part is seeing and doing. The environment may be perceived as harmonious and beautiful with interplays of light and color in minute detail. Colors may be heard and sounds may be seen. There is a loss of boundary between the real and the unreal.

The most abused of these drugs are the hallucinogens, such as LSD (lysergic acid diethylamide), mescaline (3,4,5-trimethoxy-phenylethylamine), and STP (2,5-dimethoxy-4-methylamphetamine) (the initials STP in the drug culture refer to serenity, tranquility, and peace).

The structures of LSD, mescaline, and STP are as follows.

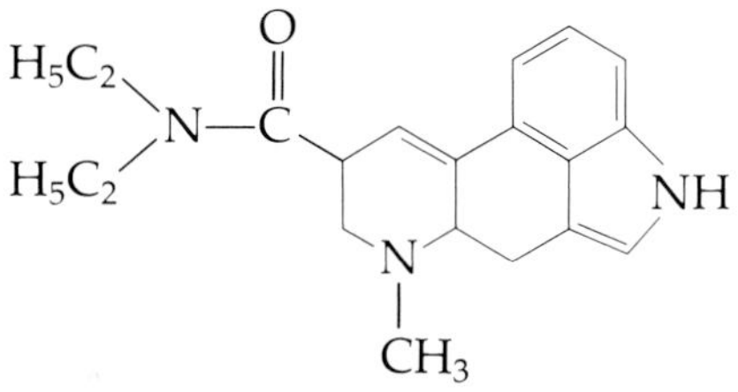

LSD

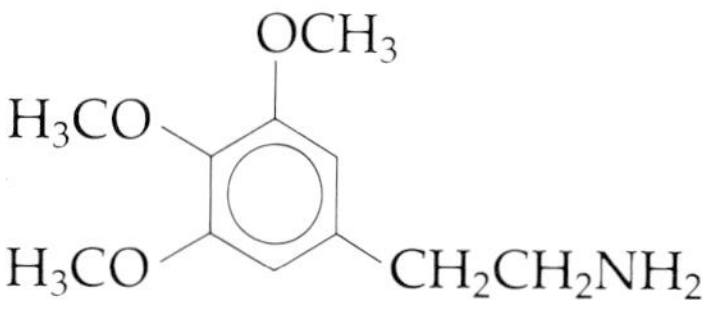

mescaline

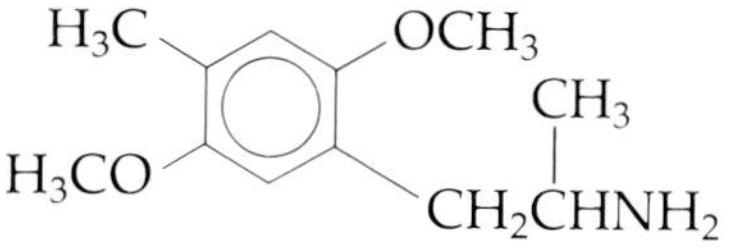

STP

Of all the hallucinogens, LSD ("acid") is the most powerful. An average "trip" dose of LSD is 0.1 mg. To obtain the same effect from STP requires 5 mg, and from mescaline 400 mg. Certain susceptible individuals feel the psychologic effects of LSD in doses as low as 20 to 25 μg (0.020 to 0.025 mg). The effects on the central nervous system begin within 1 hr and may last 8 to 12 hr.

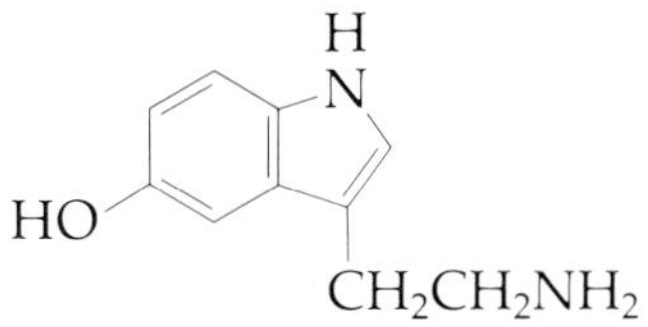

serotonin

In addition to the hallucinogenic effects, LSD causes dilation of the pupils of the eyes, increased blood pressure, *tachycardia, nausea, and increased body temperature.

It is believed that LSD exerts its effect by interfering with serotonin, a hormone found in the brain, which plays an important part in the thought process. Note the similarity in structure between LSD and serotonin.

LSD has been used by psychiatrists and psychologists in the treatment of *psychoneuroses and social delinquency. It has also been used for the treatment of terminal cancer patients because it lessens their pain and awareness.

Cannabis, also known as marijuana, hashish, "pot," "grass," or "Mary Jane," is obtained from the flowery tops of hemp plants. In the Middle East and North Africa the dried resinous extract of the tops is called hashish, and in the Far East charas. In the United States the term *marijuana* (or *marihuana*) is used to describe any part of the plant that induces psychic changes in humans.

Marijuana acts on both the central nervous system and the cardiovascular system. When it is smoked, the effects can be felt within a few minutes. If eaten with food, the effects are delayed about an hour. The physiologic effects include a voracious appetite, nausea, bloodshot eyes, increased pulse rate, dry mouth and throat, and dilation of the pupils. In addition, there is a sense of euphoria, intensified visual images, a keener sense of hearing, and a distortion of space and time.

Large doses of marijuana can induce hallucinations, delusions, and paranoic feelings. Anxiety may replace euphoria.

Marijuana is not a narcotic in that it is not a derivative of opium or synthetic substitutes, but it is so labeled by law.

Other Heterocyclic Compounds

Although many of the biologically active heterocyclic compounds contain nitrogen in the ring, other elements may also be present. For example, both thiamine and biotin, which are B vitamins, contain sulfur in the ring. Glucose has an oxygen in its ring structure.

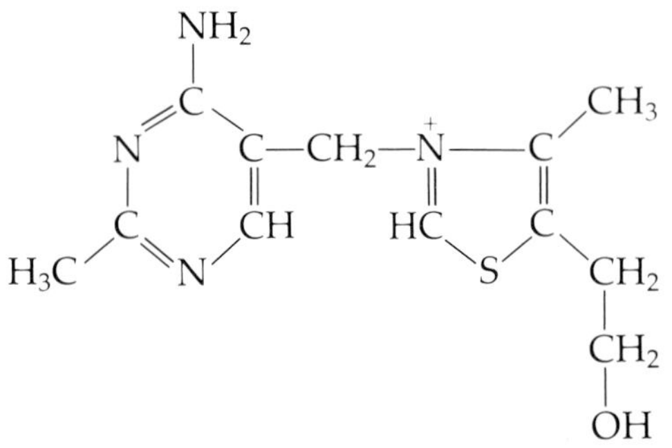

thiamine

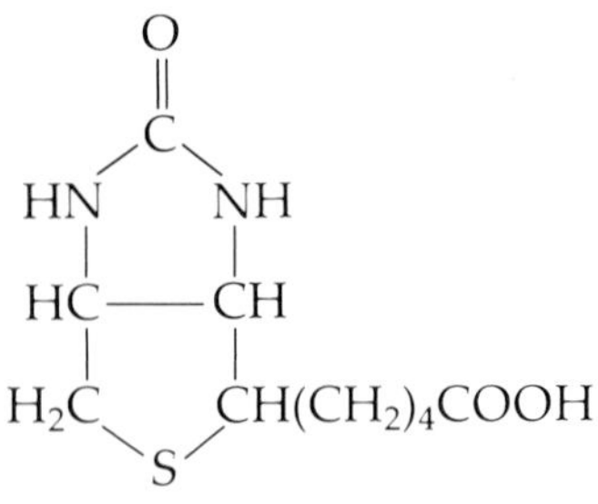

biotin

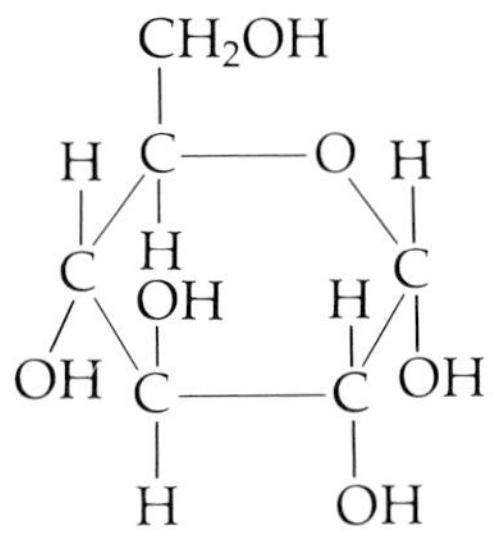

glucose

Alkaloids

One particular group of heterocyclic amines contains compounds called alkaloids. They are found chiefly in plants and usually have a marked physiologic effect on the body. Morphine, codeine, heroin, and LSD are alkaloids. Other examples follow.

Caffeine

Caffeine occurs in coffee and tea. It stimulates the central nervous system and also acts as a *diuretic. Caffeine is used in various headache remedies in conjunction with analgesic drugs.

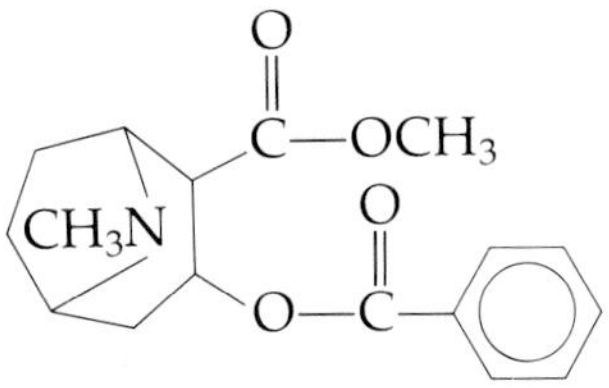

Cocaine

Obtained from the leaves of a coca shrub in South America, cocaine is administered medicinally in the form of a hydrochloride. Because it blocks nerve conduction, cocaine is widely used as a local anesthetic. However, it has largely been supplanted by synthetic drugs, which are less habit forming and less toxic.

Pharmaceutical chemists have produced derivatives of cocaine without habituation. Among the synthetic analogs are procaine (Novocaine), lidocaine (Xylocaine), and mepivacaine (Carbocaine).

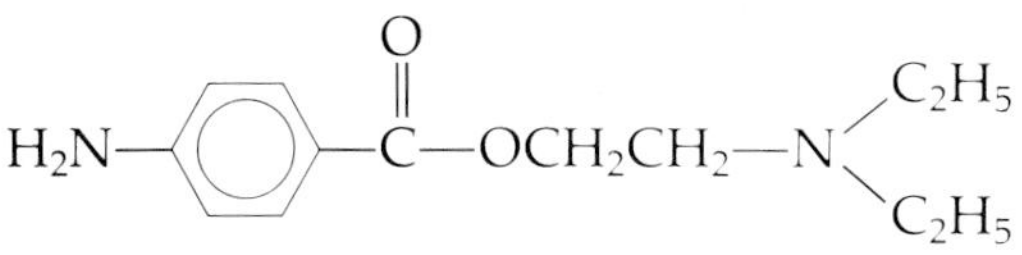

cocaine

procaine

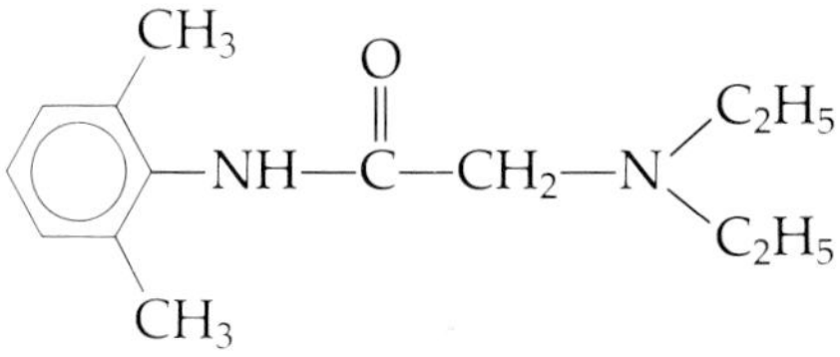

lidocaine

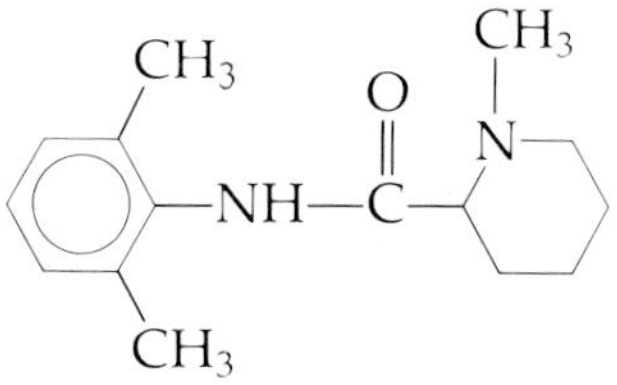

mepivacaine

Nicotine

Found in tobacco leaves, nicotine is one of the few water-soluble alkaloids. It has no therapeutic use but is used as an insecticide. Nicotine is found in the body after cigarette smoking. It increases the blood pressure and pulse rate and constricts the blood vessels.

Quinine

Obtained from the bark of the cinchona tree in South America, quinine was formerly used in the treatment of malaria. It has been replaced by less toxic and more effective synthetic antimalarial drugs. Quinine has also been used as an antipyretic and analgesic for muscular pain and headaches.

Quinine acts on skeletal muscle by three different mechanisms: (1) it increases the *refractory period by direct action on the muscle fiber; (2) it decreases the excitability of the motor end plate; and (3) it affects the distribution of calcium within the muscle fiber.

Ephedrine

Obtained from twigs of a Chinese plant, ma-huang, ephedrine is used to elevate blood pressure during spinal anesthesia. It is also used in the treatment of asthma, coughs, and colds and as a nasal decongestant.

Reserpine

Obtained from the shrub *Rauwolfia serpentina* in India and the Malay peninsula, reserpine has been used since 1952 for the treatment of hypertension but has some adverse *side effects. It is also used as a sedative, as a relaxant, and in the treatment of psychotic patients.

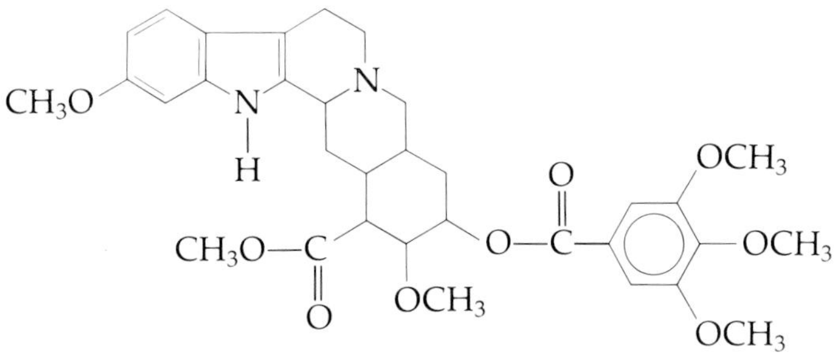

reserpine

Other Antipsychotic Drugs

Chlorpromazine (Thorazine) has been used against the symptoms of schizophrenia. Many compounds similar to chlorpromazine have been synthesized and act as tranquilizers.

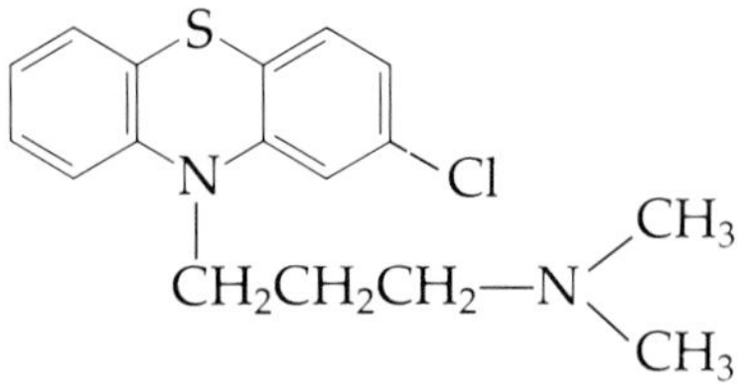

chlorpromazine

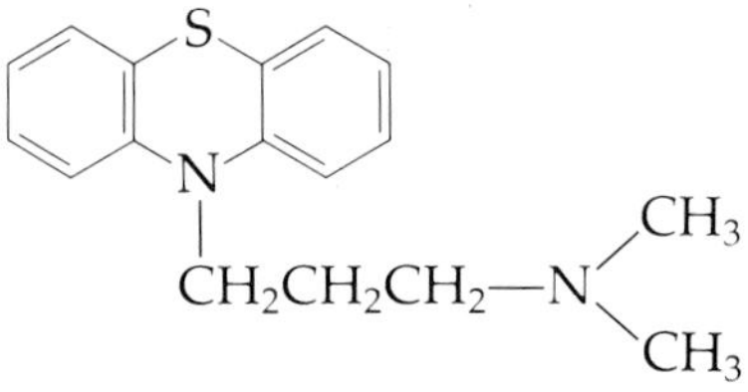

promazine

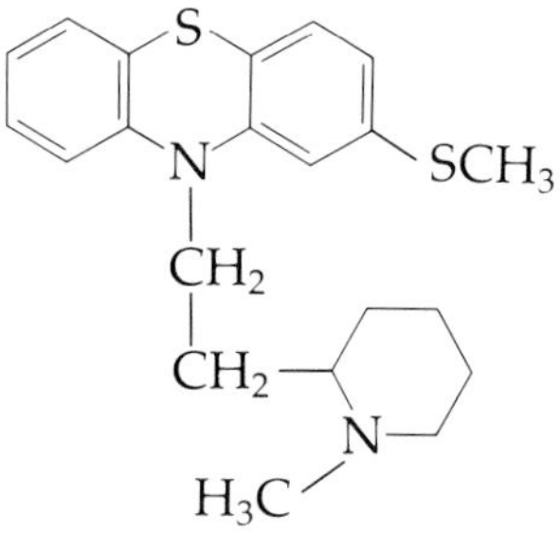

thioridazine

Heterocyclic Amines Found in the Brain

The compounds serotonin, norepinephrine, and dopamine function as neurotransmitters in the brain.

HO—[CH$_2$CH$_2$NH$_2$]   (structure with N, H)

serotonin

(structure: OH, H, HO, HO, N, H, H)

norepinephrine

(structure: HO, HO, N, H, H)

dopamine

Summary

Heterocyclics are compounds containing some element other than carbon in the ring, the most common being nitrogen, sulfur, and oxygen. Heterocyclics are usually five- or six-membered rings sometimes joined together.

The parent compounds of some of the more important heterocyclic compounds are pyrrole, a five-membered ring with a nitrogen in it; pyridine, a six-membered ring containing a nitrogen; pyrimidine, a six-membered ring containing two nitrogens; and purine, a six-membered ring containing two nitrogens joined to a five-membered ring containing two nitrogens.

The purines and the pyrimidines are the parent compounds for the base of the nucleic acids DNA and RNA.

Psychoactive drugs affect the central nervous system. Among the categories of psychoactive drugs are narcotic analgesics, sedative hypnotics, stimulants, and psychedelics.

The following heterocyclics are but a few of the many used medically: caffeine, as a stimulant; cocaine, as a local anesthetic; ephedrine, as a nasal decongestant; reserpine, for treatment of hypertension.

Questions and Problems

A

1. Define the term *heterocyclic*.
2. Which heterocyclic bases are used in the DNA and RNA molecules?
3. What are opiates?
4. Name two naturally occurring opiates; one semisynthetic opiate; two synthetic opiates.
5. What are the effects of opiates on the body?
6. For what purposes is morphine used? What is its chief disadvantage?
7. What is codeine used for? How is it related structurally to morphine?
8. What are the dangers of use of heroin? What are the symptoms of heroin withdrawal?
9. How is heroin addiction treated?
10. What properties do barbiturates have? How do they differ structurally?
11. What is one use for phenobarbital?
12. Why is there greater danger of a fatal overdose with barbiturates than with opiates?
13. What is narcolepsy? How is it treated?
14. Who are frequent abusers of amphetamines?
15. What are the effects of the use of psychedelic drugs?

16. What is the most powerful psychedelic drug?
17. What is believed to be the method of action of LSD?
18. Cannabis is known by what other names? What are the effects of small amounts of cannabis? large amounts?

B

19. Name one compound containing an oxygen in the ring; one with both a sulfur and a nitrogen.
20. Why are there warnings against the use of reserpine?
21. How does a hallucinogenic drug differ from a psychomimetic drug?
22. What effects does nicotine have on the human body?

Practice Test

1. Opiates are _____.
   a. sedative hypnotics   b. morphine narcotics
   c. stimulants   d. psychedelics
2. Amphetamines are known as _____.
   a. pep pills   b. downers
   c. hypnotics   d. depressants

3. LSD is an example of a (n) _____.
  a. sedative   b. hallucinogen
  c. depressant   d. analgesic

4. An example of a neurotransmitter is _____.
  a. reserpine   b. codeine
  c. promazine   d. serotonin

Heterocyclic compounds can contain which element in the ring in addition to carbon?
  a. N   b. O
  c. S   d. all of these

A heterocyclic compound found in the brain is _____.
  a. dopamine   b. reserpine
  c. amphetamine   d. morphine

7. A drug used in the treatment of hypertension is _____.
  a. codeine   b. dopamine
  c. reserpine   d. Thorazine

8. Caffeine acts as a(n) _____.
  a. analgesic   b. diuretic
  c. antipyretic   d. antihypertensive

9. One drug that depresses the central nervous system is _____.
  a. phenobarbital b. caffeine
  c. nicotine   d. reserpine

10. Which of the following is habit-forming?
  a. morphine   b. cocaine
  c. heroin   d. all of these

PART
III

Biochemistry

21

Carbohydrates

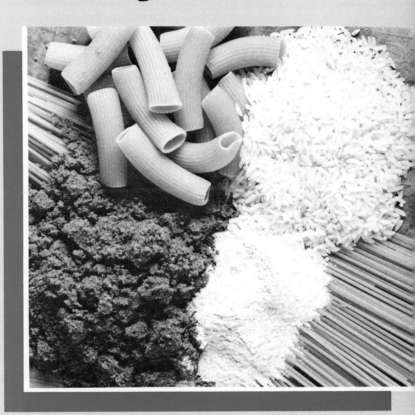

Carbohydrates include pasta, rice, flour, and brown sugar.

Carbohydrates (which contain the elements carbon, hydrogen, and oxygen) form a class of organic compounds that includes sugars, starches, and cellulose. Originally all known carbohydrates were considered to be hydrates of carbon because they contain hydrogen and oxygen in the ratio of 2 to 1 just as in water. The formula for glucose, $C_6H_{12}O_6$, was written as $C_6(H_2O)_6$. Likewise sucrose, $C_{12}H_{22}O_{11}$, was written as $C_{12}(H_2O)_{11}$. However, later investigation showed that rhamnose, another carbohydrate, has the formula $C_6H_{12}O_5$. It did not fit the general formula for a hydrate of carbon yet it was a carbohydrate. Also, such compounds as acetic acid ($C_2H_4O_2$) and pyrogallol ($C_6H_6O_3$) did fit such a system but were not carbohydrates.

Carbohydrates are now defined as polyhydroxyaldehydes or polyhydroxyketones or substances that yield these compounds on hydrolysis. **Polyhydroxy** means "containing several alcohol groups." Thus simple carbohydrates are alcohols and are also either aldehydes or ketones (they contain a carbonyl group, $>C=O$; see Chapter 18).

Classification

Carbohydrates are divided into three major categories: monosaccharides, disaccharides, and polysaccharides.

Monosaccharides (*mono-* means one) are simple sugars. They cannot be changed into simpler sugars upon hydrolysis (reaction with water).

Disaccharides (*di-* means two) are double sugars. On hydrolysis, they yield two simple sugars.

$$\text{disaccharides} \xrightarrow{\text{hydrolysis}} \text{2 monosaccharides}$$

Polysaccharides (*poly-* means many) are complex sugars. On hydrolysis, they yield many simple sugars.

$$\text{polysaccharides} \xrightarrow{\text{hydrolysis}} \text{many simple sugars}$$

Monosaccharides, or simple sugars, are called either **aldoses** or **ketoses**, depending upon whether they contain an aldehyde (—CHO) or a ketone ($>C=O$) group. Aldoses and ketoses are further classified according to the number of carbon atoms they contain. An aldopentose is a five-carbon simple sugar containing an aldehyde group. A ketohexose is a six-carbon simple sugar containing a ketone group.

Although there are simple sugars with three carbon atoms (trioses), four carbon atoms (tetroses), and five carbon atoms (pentoses), the hexoses (six-carbon simple sugars) are the most common in terms of the human body because they are the body's main energy-producing compounds.

Polysaccharides are sometimes called hexosans or pentosans, depending upon the type of monosaccharide they yield on hydrolysis. That is, hexosans on hydrolysis yield hexoses, and pentosans yield pentoses.

Origin

Plants pick up carbon dioxide from the air and water from the soil and combine them to form carbohydrates in a process called **photosynthesis.** Enzymes, chlorophyll, and sunlight are necessary. The overall reaction is represented in equation (21-1).

$$6\ CO_2 + 6\ H_2O \xrightarrow[\substack{chlorophyll, \\ enzymes}]{sunlight} \underset{sugar}{C_6H_{12}O_6} + 6\ O_2(g) \qquad (21\text{-}1)$$

However, it should be understood that even though reaction (21-1) appears simple, it is very complex, with many intermediate steps between the original reactants and the final products.

During photosynthesis oxygen is given off into the air, thus renewing our vital supply of this element.

The carbohydrate produced in reaction (21-1), $C_6H_{12}O_6$, is a monosaccharide. Plant cells also have the ability to combine two molecules of a monosaccharide into one of a disaccharide, as indicated in (21-2).

$$\underset{monosaccharide}{2\ C_6H_{12}O_6} \longrightarrow \underset{disaccharide}{C_{12}H_{22}O_{11}} + H_2O \qquad (21\text{-}2)$$

Reaction (21-2) is the reverse of hydrolysis—water is removed when two molecules of a monosaccharide combine.

Plant (and animal) cells can also combine many molecules of monosaccharide into large polysaccharide molecules. The n in equation (21-3) represents a number larger than 2.

$$\underset{monosaccharide}{n\ C_6H_{12}O_6} \longrightarrow \underset{polysaccharide}{(C_6H_{10}O_5)_n} + n\ H_2O \qquad (21\text{-}3)$$

This is an example of a polymerization reaction (see page 277).

Polysaccharides occur in plants as cellulose in the stalks and stems and as starches in the roots and seeds. Monosaccharides and disaccharides are generally found in plants in their fruits.

Plants as well as animals are able to convert carbohydrates into fats and proteins.

Oxygen—Carbon Dioxide Cycle in Nature

Although plants have the ability to pick up carbon dioxide from the air and water from the ground to form carbohydrates, animals are unable to do this and must rely on plants for their carbohydrates.

Animals oxidize carbohydrates in their bodies to yield carbon dioxide, water, and energy:

$$C_6H_{12}O_6 + 6\,O_2 \longrightarrow 6\,CO_2 + 6\,H_2O + \text{energy}$$

Again, this reaction is not as simple as it appears; many steps are involved between the reactants and the products, and many different enzymes are required.

It should be noted that this overall reaction during metabolism is the reverse of the one taking place during photosynthesis. Both reactions can be summarized by equation (21-4).

$$\text{energy} + 6\,CO_2 + 6\,H_2O \underset{\substack{\text{animal} \\ \text{metabolism}}}{\overset{\substack{\text{plant} \\ \text{photosynthesis}}}{\rightleftharpoons}} C_6H_{12}O_6 + 6\,O_2 \quad (21\text{-}4)$$

Thus, there is a cycle in nature. During photosynthesis, plants pick up carbon dioxide from the air and give off oxygen; both plants and animals pick up oxygen from the air and give off carbon dioxide.

During photosynthesis, the energy from the sun is needed for the reaction (that is, the reaction is *endothermic). During metabolism of these carbohydrates in animals, this same amount of energy is liberated (the reaction is *exothermic). Thus, all the energy from the burning of carbohydrates by animals comes originally from the sun. Plants store solar energy in carbohydrates, and this energy is utilized by all living organisms during the metabolic process. It has been estimated that only about 1 percent of the total solar energy falling on plants is converted into useful stored energy.

Isomerism

In Chapters 15 and 16 the topic of isomers was mentioned briefly. Isomers can be subdivided into four classifications: structural, functional, geometric, and optical.

Structural Isomers

Structural isomers are compounds with the same molecular formula, same functional groups, but different structural formulas. Examples are propyl alcohol and ispropyl alcohol, *o*-dichlorobenzene and *m*-dichlorobenzene, and 2-methylhexane and 3-methylhexane.

$$CH_3—CH_2—CH_2OH \qquad CH_3—\underset{\underset{OH}{|}}{CH}—CH_3$$

propyl alcohol isopropyl alcohol

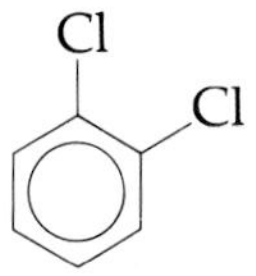

o-dichlorobenzene 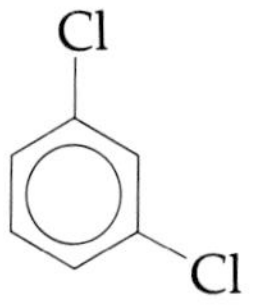 *m*-dichlorobenzene

$$CH_3\underset{\underset{CH_3}{|}}{CH}CH_2CH_2CH_2CH_3 \qquad CH_3CH_2\underset{\underset{CH_3}{|}}{CH}CH_2CH_2CH_3$$

2-methylhexane 3-methylhexane

Functional Isomers

Functional isomers are compounds having the same molecular formula but different functional groups. Examples are ethyl alcohol and dimethyl ether, and propanoic acid and 2-hydroxypropanal.

$$CH_3CH_2OH \qquad CH_3OCH_3$$

ethyl alcohol dimethyl ether

$$CH_3CH_2COOH \qquad CH_3\underset{\underset{OH}{|}}{CH}CHO$$

propanoic acid 2-hydroxypropanal

Geometric Isomers

Geometric isomers are compounds with the same molecular formula that have different structural formulas because either a double bond or a ring system prevents the rotation necessary to change one into the other.

If two carbon atoms are connected by a double bond, those two carbon atoms and the four groups attached to them all lie in a single plane. There are two possible isomeric structures, one having similar groups on the same "side" of the double bond and one having similar groups on opposite "sides" of the double bond. An example of geometric isomerism is

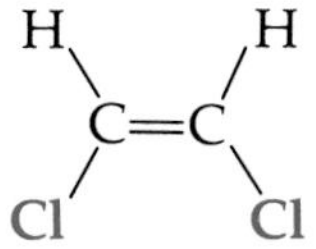

and

cis-1,2-dichloroethene *trans*-1,2-dichloroethene

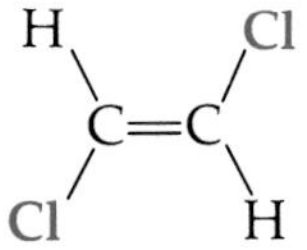

The prefix **cis-** means on the same side, and **trans-** means across or on opposite sides.

If the two groups or atoms attached to one of the doubly bonded carbon atoms are identical, only one structure is possible. The two groups or atoms attached to each carbon must be different for geometric isomerism to occur.

Cis–trans isomerism is also possible for ring structures. In cyclopropane, the three carbon atoms all lie in one plane. Substituents on adjacent carbons can be on the same side of the plane (cis) or on opposite sides of the plane (trans), such as

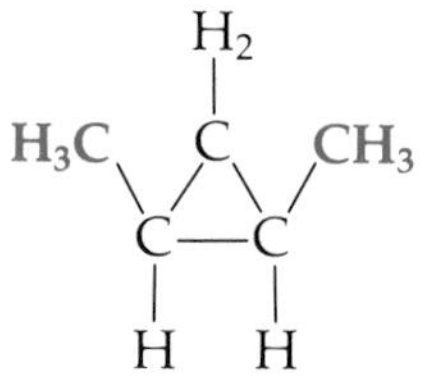

and

cis-1,2-dimethylcyclopropane   *trans*-1,2-dimethylcyclopropane

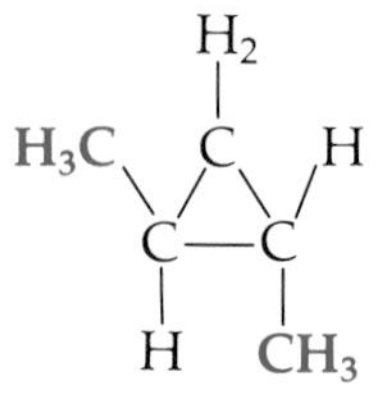

Cis–trans isomerism occurs in fatty acids (see Chapter 22) and is very important in the rhodopsin–vitamin A cycle (see page 595).

An example of a biologically active geometric isomer is *trans*-diethylstilbestrol (DES). This compound was formerly used during pregnancy, but its use has been discontinued because of its carcinogenic effects. The cis isomer is biologically inactive.

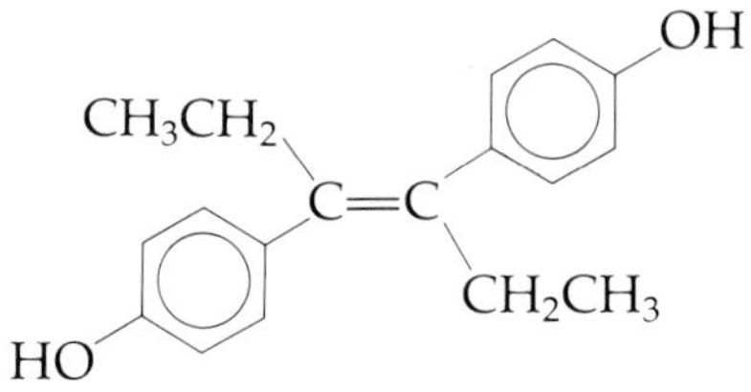

trans-diethylstilbestrol

Optical Isomers

Optical isomers are compounds with the same molecular formula but with structures that are the mirror image of one another. Such isomers rotate the plane of polarized light equally but in opposite directions. Let us see what this means and why.

Polarized light vibrates in one plane only, as opposed to ordinary light, which vibrates in all planes. When polarized light is passed through a solution of an optically active substance, the plane of polarized light is rotated (see Figure 21-1).

What causes such a rotation of the plane of polarized light? According to the van't Hoff, theory such an effect upon the plane of

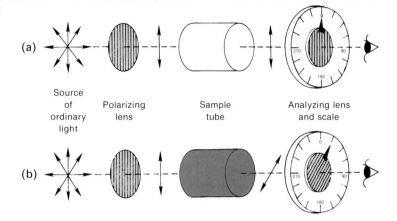

Figure 21-1 (a) Schematic diagram of polarimeter showing production of polarized light. (b) Rotation of polarized light by an optically active substance.

polarized light is due to the presence of one (or more) **asymmetric**[†] carbon atoms. An asymmetric carbon atom is one that has four different groups attached to it.

The simplest carbohydrate is glyceraldehyde. Glyceraldehyde has an aldehyde group at one end of the molecule, a primary alcohol at the other end, and a secondary alcohol in the middle. The central carbon atom in glyceraldehyde is asymmetric (chiral)—it has four different groups attached to it. They are

$$\text{—CHO} \quad \text{—OH} \quad \text{—CH}_2\text{OH} \quad \text{—H}$$

This compound can exist in two optically active forms. The two structures are mirror images of one another (see Figure 21-2); they are *not* superimposable. Your right and left hands are mirror images; they are not superimposable. You wear a right shoe and a left shoe; they too are mirror images and are not superimposable.

The Fischer projection formula is a two-dimensional representation of the mirror-image structures. In this system, the two isomers of glyceraldehyde are represented as

$$\begin{array}{ccc}
\text{CHO} & & \text{CHO} \\
| & & | \\
\text{H—C—OH} & \text{and} & \text{HO—C—H} \\
| & & | \\
\text{CH}_2\text{OH} & & \text{CH}_2\text{OH} \\
\text{D-glyceraldehyde} & & \text{L-glyceraldehyde}
\end{array}$$

The horizontal lines indicate bonds extending forward from the paper, and the vertical lines indicate bonds extending backward

The structure on the left of the page:

$$\begin{array}{c}
\text{CHO} \\
| \\
\text{H—C—OH} \\
| \\
\text{CH}_2\text{OH} \\
\text{glyceraldehyde}
\end{array}$$

† Some texts use the word **chiral** in place of asymmetric. An asymmetric carbon is also called a chiral center. Molecules that are not superimposable on their mirror images are said to be chiral and are optically active.

Figure 21-2   Ball-and-stick models of the two forms of glyceraldehyde.

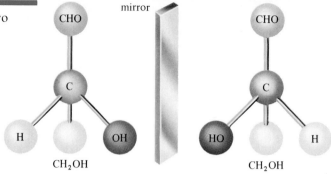

from the paper. Compare these two diagrams with those in Figure 21-2. The Fischer projection formulas are *always* written with the aldehyde (or ketone) group—the most highly oxidized—at the top. Thus, in the above Fischer formulas, the —H and —OH groups project forward from the paper and the —CHO and —CH$_2$OH groups project backward from the paper.

Fischer called the glyceraldehyde with the —OH group on the right side of the asymmetric carbon atom D-glyceraldehyde. Likewise, the one with the —OH group on the left side of the asymmetric carbon he called L-glyceraldehyde.

Glyceraldehyde is considered the parent compound from which more complex sugars can be derived. Those whose terminal structure (the primary alcohol and the asymmetric carbon) is similar to that of D-glyceraldehyde belong to the D series; those whose terminal end is similar to that of L-glyceraldehyde belong to the L series. Note in Figure 21-3 that D-erythrose and D-threose have structures similar to that of D-glyceraldehyde. The corresponding L structures are also similar. D and L refer to structure only, *not* to the direction of rotation of polarized light.

Figure 21-3 illustrates the D family of aldoses. The colored areas indicate the D conformation. A table of the mirror images of these aldoses would represent the L forms.

The number of optical isomers depends upon the number of asymmetric carbon atoms present in a compound and can be calculated by using the formula 2^n where n is the number of asymmetric carbons. Thus, glyceraldehyde, which has one asymmetric carbon, has 2^1 or 2 optical isomers, as was shown. Glucose, which will be discussed later in this chapter, has four asymmetric carbons and so has 2^4 or 16 optical isomers. Of these 16 isomers, 8 belong to the D series and 8 to the L series (one set of 8 is the mirror image of the other set).

Optical activity is of great importance in the body because many enzymes will interact with only one particular optical isomer. In the human body, the D series is the primary configuration for carbohydrates whereas the L series is the primary one for proteins.

Figure 21-3 The D family of aldoses.

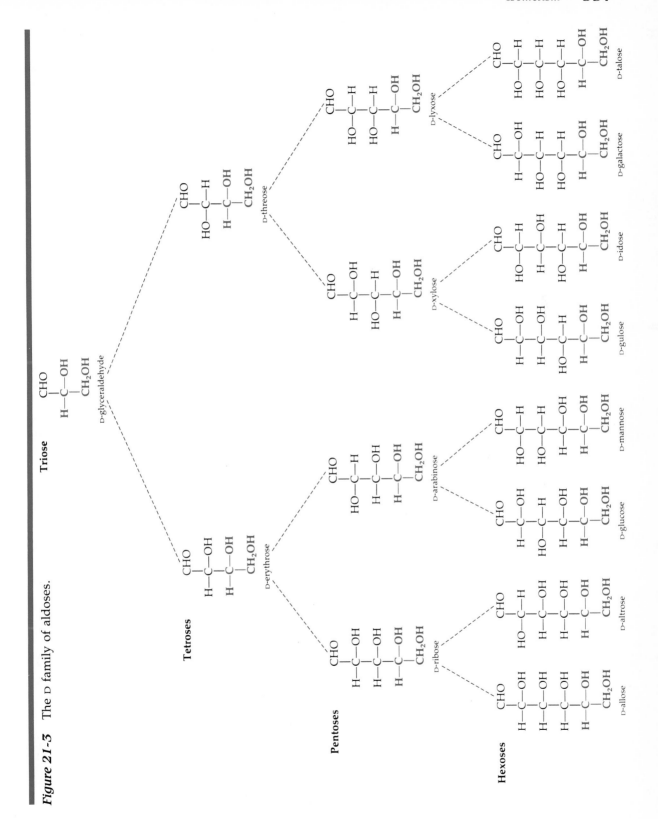

Monosaccharides

Monosaccharides are simple sugars. They cannot be broken down into other sugars. They are categorized according to the number of carbons they contain.

Trioses

A triose is a three-carbon simple sugar. Trioses are formed during the metabolic breakdown of hexoses in muscle metabolism. An example of a triose is glyceraldehyde (glycerose), whose optical isomers are shown on page 355.

Tetroses

Tetroses are four-carbon sugars. One tetrose, erythrose, is an intermediate in the hexose monophosphate shunt for the oxidation of glucose (see page 482).

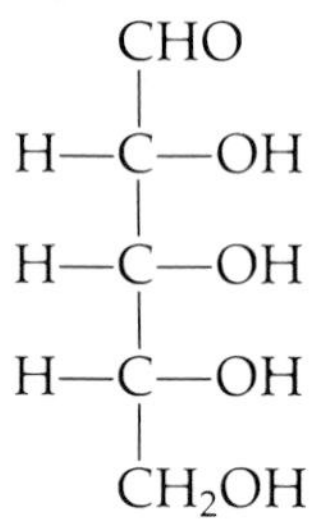

Pentoses

Pentoses are five-carbon sugar molecules. The most important of these are ribose and deoxyribose, which are found in nucleic acids. Ribose forms part of ribonucleic acid (RNA), and deoxyribose forms part of deoxyribonucleic acid (DNA). Both DNA and RNA are components of every cell nucleus and cytoplasm. The prefix *de-* means without, so **deoxy-** means without oxygen. Note that deoxyribose has one less oxygen atom than does ribose.

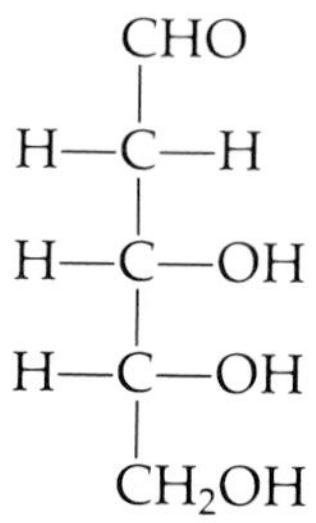

Structures (21-5), the Fischer projection representations for the pentoses, are called open-chain structures. However, the predominant form for pentoses is a ring structure. Recall that aldehydes react with alcohols to form hemiacetals (page 300). In the case of ribose, the aldehyde can react with the alcohol at carbon 4 to form two different compounds.

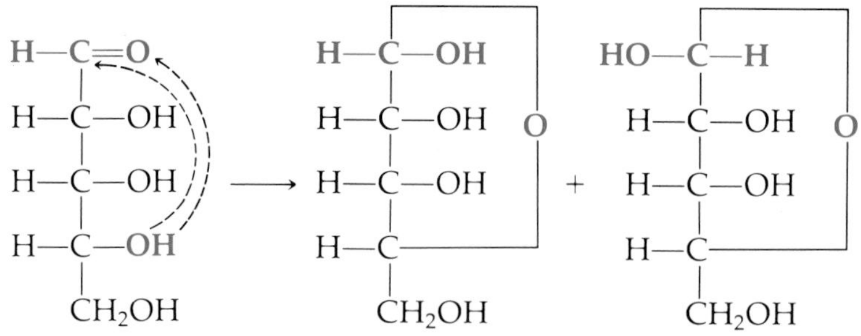

$$
\begin{array}{c}
\text{CHO} \\
| \\
\text{H—C—OH} \\
| \\
\text{H—C—OH} \\
| \\
\text{H—C—OH} \\
| \\
\text{CH}_2\text{OH} \\
\text{D-ribose}
\end{array}
\qquad (21\text{-}5)
$$

$$
\begin{array}{c}
\text{CHO} \\
| \\
\text{H—C—H} \\
| \\
\text{H—C—OH} \\
| \\
\text{H—C—OH} \\
| \\
\text{CH}_2\text{OH} \\
\text{D-deoxyribose}
\end{array}
$$

ribose $\longrightarrow$ α-ribose $+$ β-ribose

Another representation for the hemiacetal (ring) structure of ribose, and other monosaccharides, is called the Haworth projection (equation 21-6).

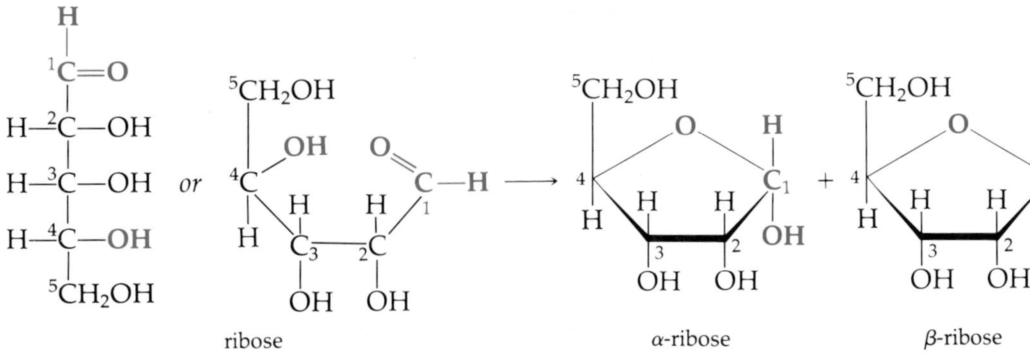

$$\text{(21-6)}$$

ribose α-ribose β-ribose

In this projection the ring, consisting of carbons at each corner and an oxygen where indicated, is considered to be in a plane perpendicular to the paper. The heavy line in the ring indicates the section closest to you; the lighter line, that part of the ring farther from you. The —H and —OH groups are above and below that plane. The Fischer and Haworth projections are related as follows.

1. The groups on the right side of the Fischer projection are written below the plane in the Haworth projection. Those on the left side are written above the plane.
2. One exception to rule 1 occurs at carbon 4 in pentoses and at carbon 5 in hexoses because of the nature of the reaction occurring there. At these carbons rule 1 is reversed.
3. At carbon 1 the α form is indicated by the —OH being written below the plane; the β form has the —OH above the plane.
4. In both projections the CH_2OH group, which has no asymmetric carbon, is written as a unit.

Other pentoses are ribulose, which is formed in the metabolic breakdown of glucose; lyxose, which is found in heart muscle; arabinose, which is found in gum arabic and the gum of the cherry tree; and xylose, which is obtained from the hydrolysis of wood, straw, corncobs, and similar plant materials.

Hexoses

The hexoses, the six-carbon sugars, are the most common of all the carbohydrates. Of the several hexoses, the most important as far as the human body is concerned are *glucose, galactose,* and *fructose.* All three of these hexoses have the same molecular formula, $C_6H_{12}O_6$, but different structural formulas; they are isomers.

Glucose

Glucose ($C_6H_{12}O_6$) is an aldohexose (page 350) and can be represented structurally as

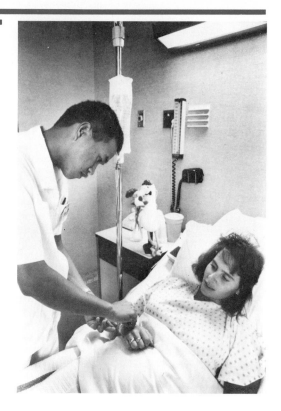

Figure 21-4 Intravenous infusion of glucose. [Saint Francis Hospital School of Nursing, Evanston, IL. © Jean Clough, 1988.]

D-glucose

Fischer projection

α-D-glucose†

β-D-glucose†

Haworth projection

Note that glucose contains four asymmetric carbon atoms (numbers 2, 3, 4, 5) and so has 2^4, or 16, optical isomers.

Medically, when the word glucose is mentioned, the D isomer is meant because that is the biologically active isomer. Likewise, for the

---

† The chemical names for these ring compounds are α-D-glucopyranose and β-D-glucopyranose, respectively, but this system of naming is of interest primarily to organic chemists.

other hexoses, the D isomer is commonly called by name only, without the prefix D.

Glucose is known commonly as dextrose or grape sugar. It is a white crystalline solid, soluble in water and insoluble in most organic liquids. It is found, along with fructose, in many fruit juices. It can be prepared by the hydrolysis of sucrose, a disaccharide, or by the hydrolysis of starch, a polysaccharide.

Glucose is the most important of all the monosaccharides. It is normally found in the bloodstream and in the tissue fluids. As will be discussed in Chapter 26, "Metabolism of Carbohydrates," glucose requires no digestion and can be given intravenously to patients who are unable to take food by mouth (see Figure 21-4).

Glucose is found in the urine of patients suffering from diabetes mellitus and is an indication of this disease. The presence of glucose in the urine is called glycosuria. Glucose may also show up in the urine during extreme excitement (emotional glycosuria), after ingestion of large amounts of sugar (alimentary glycosuria), or because of other factors that will be discussed in the chapter on the metabolism of carbohydrates.

Galactose

Galactose, an isomer of glucose, is also an aldohexose. The structures of galactose are

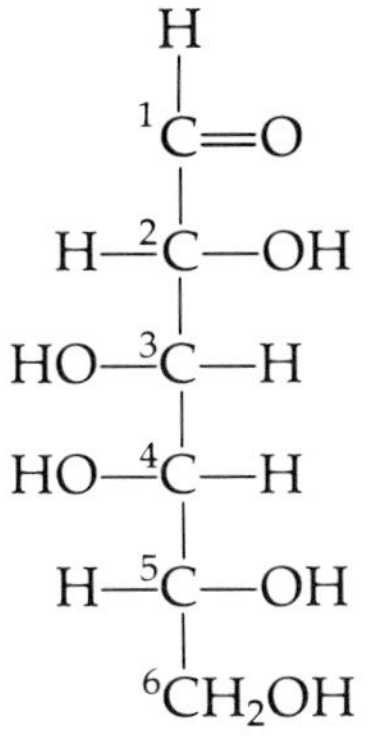

D-galactose 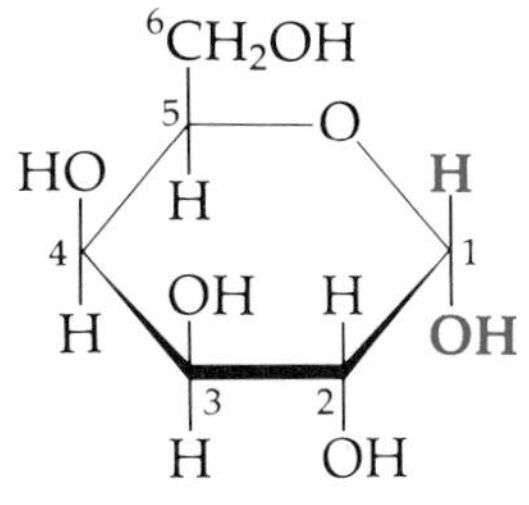   α-D-galactose 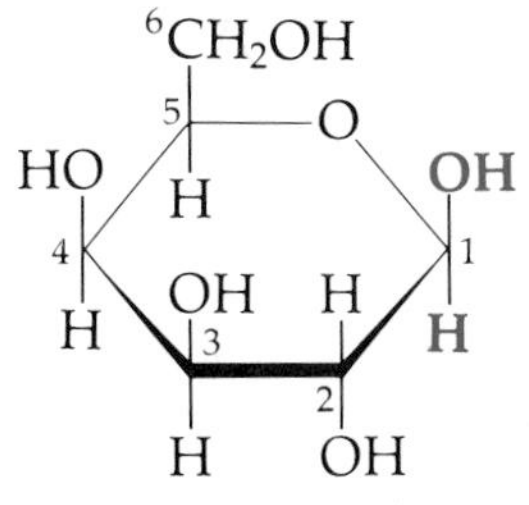   β-D-galactose

Glucose and galactose differ from each other only in the configuration of the H and OH about a single carbon atom. Two sugars that differ only in the configuration about a single carbon atom are called **epimers.** D-Galactose is converted to D-glucose in the liver by a specific enzyme called an epimerase. Galactose is present in some glycoproteins and glycolipids (see pages 397).

Galactosemia, a severe inherited disease, results in the inability of infants to metabolize galactose because of a deficiency of either the enzyme galactose 1-phosphate uridyl transferase or the enzyme galactokinase. The galactose concentration increases in the blood and urine (galactosuria).

Fructose

Fructose is a ketohexose. Its molecular formula, like that of glucose and galactose, is $C_6H_{12}O_6$. It, too can be represented as a straight-chain or as a ring compound (21-7). The ring structure is predominant. Note that the ring structure represents a hemiketal (see page 300).

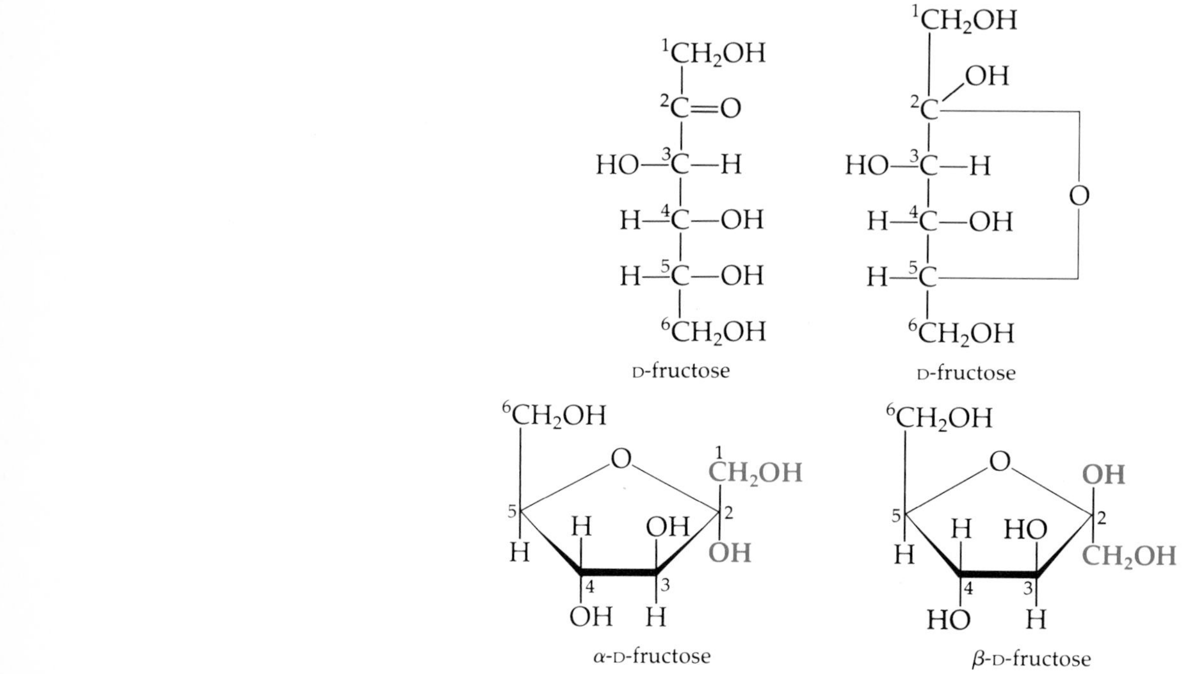

D-fructose

D-fructose

(21-7)

α-D-fructose

β-D-fructose

Fructose is often called levulose or fruit sugar. It occurs naturally in fruit juices and honey. It can be prepared by the hydrolysis of sucrose, a disaccharide, and also by the hydrolysis of inulin, a polysaccharide found in Jerusalem artichokes. Fructose is the most soluble sugar and also the sweetest of all sugars, being approximately twice as sweet as glucose.

Fructosemia, fructose intolerance, is an inherited disease due to a deficiency of the enzyme fructose 1-phosphate aldolase. An infant suffering from this disease experiences hypoglycemia, vomiting, and severe malnutrition. Such a condition is treated by placing the infant on a low fructose diet.

Reactions of the Hexoses

Hexoses, which are either aldoses or ketoses, show reducing properties. This reducing property is the basis of the test for sugar in the urine and in the blood. When a reducing agent is treated with an oxidizing agent such as Cu^{2+} complex ion,[†] a red-orange precipitate

---

[†] Fehling's solution is an alkaline solution of Cu^{2+} ions complexed with tartrate; Benedict's solution is an alkaline solution of Cu^{2+} ions complexed with citrate.

of copper(I) oxide (Cu_2O) is formed. The unbalanced equation for the reaction of an aldehyde with copper(II) complex ion can be written as follows.

$$\text{aldehyde} + \underset{\substack{\text{deep blue}\\\text{solution}}}{Cu^{2+}} \xrightarrow[\text{NaOH}]{\text{heat}} \text{acid} + \underset{\substack{\text{red-orange}\\\text{precipitate}}}{Cu_2O(s)} + \text{water}$$

In this reaction the aldehyde is oxidized to the corresponding acid. When glucose is treated with Cu^{2+} complex ion and the mixture is heated, the reaction is

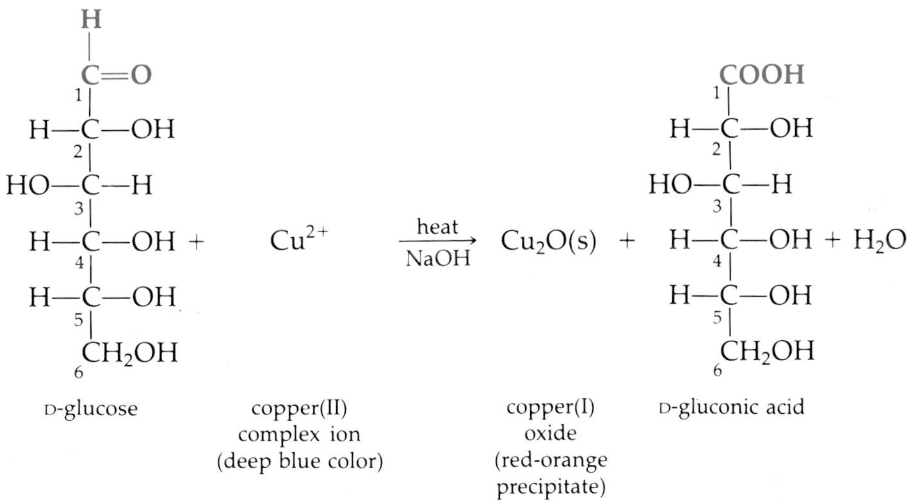

D-glucose copper(II) complex ion (deep blue color) copper(I) oxide (red-orange precipitate) D-gluconic acid

Laboratory tests for the presence of glucose in urine use Benedict's solution or Fehling's solution, both of which contain copper(II) complex ion. Clinitest tablets, which also contain a copper(II) complex, give a rapid quantitative measurement of the concentration of glucose present. If the blue liquid turns green, a trace of sugar (glucose) is present. This is frequently recorded as +. A yellow color, indicated by + +, indicates up to 0.5 percent sugar; an orange color, + + +, 0.5 to 1.5 percent; and a red color, + + + +, over 1.5 percent sugar.

Glucose does not normally appear in the urine for any extended period of time. Its persistent presence usually indicates that something is wrong with the metabolism of carbohydrates—such as diabetes mellitus.

Another laboratory test for the presence of a reducing sugar uses Tollens' reagent, which contains Ag^+ complex ion. In this reaction glucose is oxidized to gluconic acid as before and the silver complex ion is reduced to free silver, which appears as a bright shiny mirror on the inside of the test tube.

$$\text{glucose} + \text{Ag}^+ \xrightarrow[\text{NH}_4\text{OH}]{\text{heat}} \text{gluconic acid} + \text{Ag(s)} + \text{water}$$

Tollens'
reagent   silver mirror

Oxidation

An aldose contains an aldehyde group as well as several —OH groups. If the aldehyde end of the molecule is oxidized, the product is named an **-onic acid.** When the aldehyde end of glucose is oxidized, the product is called gluconic acid (see page 363). When the aldehyde end of galactose is oxidized, the product is called galactonic acid. If the alcohol at the end opposite the aldehyde (the other end of the molecule) is oxidized, the product is called a **-uronic acid.** The oxidation of the alcohol end of glucose yields glucuronic acid. Glucuronic acid is a minor product of glucose metabolism. It is also part of the heparin molecule (see page 374).

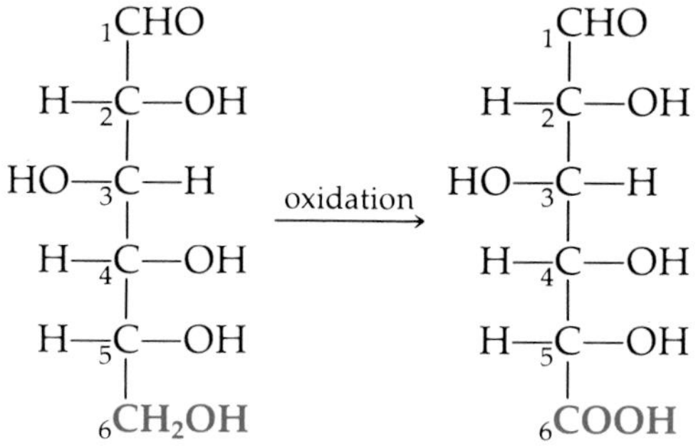

glucose   oxidation   glucuronic acid

Likewise, if the alcohol end of galactose is oxidized, galacturonic acid is formed.

If both ends of the glucose molecule are oxidized at the same time, the product is called saccharic acid.

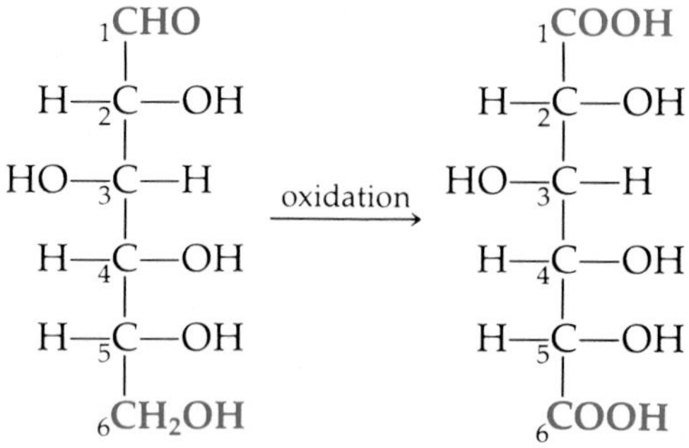

glucose   oxidation   saccharic acid

Reduction

The aldohexoses can be reduced to alcohols. When glucose is reduced, sorbitol is formed. Sorbitol accumulation in the eye is a major factor in the formation of cataracts due to diabetes.

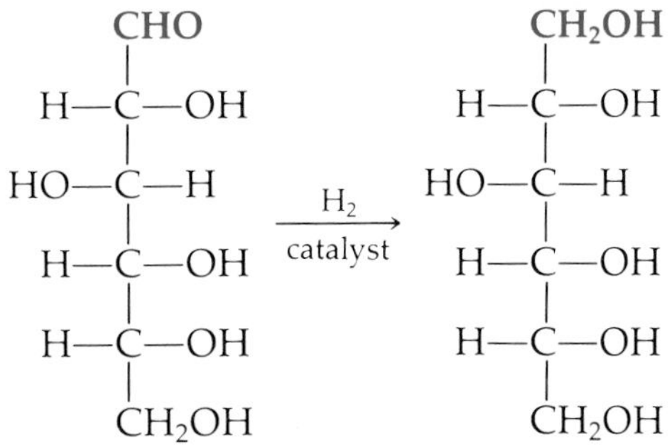

$$
\begin{array}{ccc}
\text{CHO} & & \text{CH}_2\text{OH} \\
\text{H—C—OH} & & \text{H—C—OH} \\
\text{HO—C—H} & \xrightarrow[\text{catalyst}]{\text{H}_2} & \text{HO—C—H} \\
\text{H—C—OH} & & \text{H—C—OH} \\
\text{H—C—OH} & & \text{H—C—OH} \\
\text{CH}_2\text{OH} & & \text{CH}_2\text{OH} \\
\text{glucose} & & \text{sorbitol}
\end{array}
$$

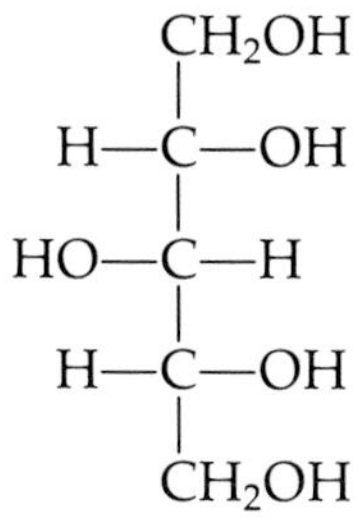

$$
\begin{array}{c}
\text{CH}_2\text{OH} \\
\text{H—C—OH} \\
\text{HO—C—H} \\
\text{H—C—OH} \\
\text{CH}_2\text{OH} \\
\text{xylitol}
\end{array}
$$

Reduction of galactose yields dulcitol, and reduction of fructose yields a mixture of mannitol and sorbitol.

A similar product, xylitol, was used as a "sweetening agent" in candy and gum but has been removed from the market because of potentially carcinogenic effects.

Fermentation

Glucose ferments in the presence of yeast, forming ethyl alcohol and carbon dioxide. This reaction will not readily occur in the absence of yeast. Yeast contains certain enzymes that catalyze this particular reaction.

$$
\underset{\text{glucose}}{\text{C}_6\text{H}_{12}\text{O}_6} \xrightarrow{\text{enzymes}} \underset{\text{ethyl alcohol}}{2\ \text{C}_2\text{H}_5\text{OH}} + 2\ \text{CO}_2
$$

Fructose will also ferment; galactose will not readily ferment. Pentoses do not ferment in the presence of yeast.

Formation of Phosphate Esters

Phosphate esters such as D-glyceraldehyde 3-phosphate and dihydroxyacetone phosphate are triose phosphate esters involved in glycolysis (see page 479). Glucose forms phosphate esters at carbons 1 and 6, as shown below,

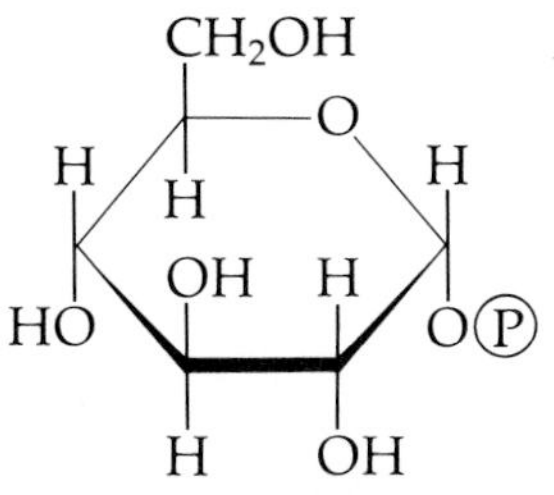

glucose 1-phosphate 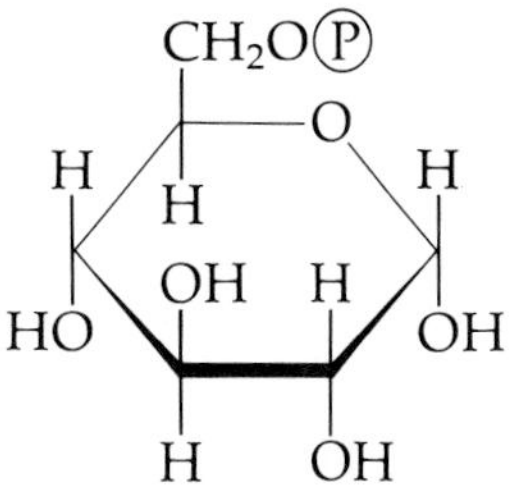   glucose 6-phosphate

and ribose forms a 1,5-diphosphate ester.

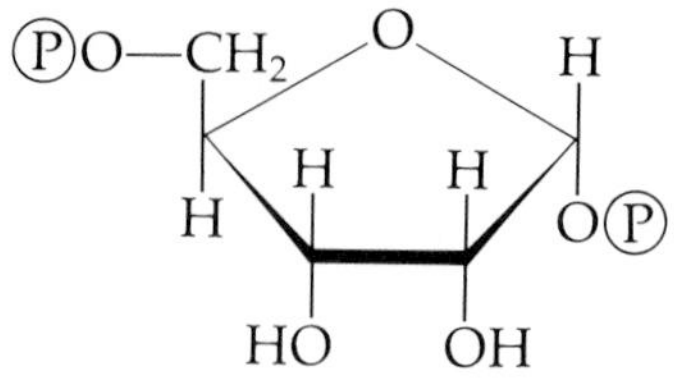

D-ribose 1,5-diphosphate

The phosphate group is abbreviated as $\circledP$, and the compound names are sometimes shortened to, for example, glucose 1-P.

Blood Group Substances

The outer layer of human plasma membranes contains many small carbohydrates. These membrane-bound carbohydrates act as biochemical markers. Among the many such markers are the blood group substances. They are primarily on the surface of the red blood cells (see page 538).

Amino Sugars

Amino sugars (hexosamines) contain an amino group in place of an —OH group. Three amino sugars have been found in nature. They are

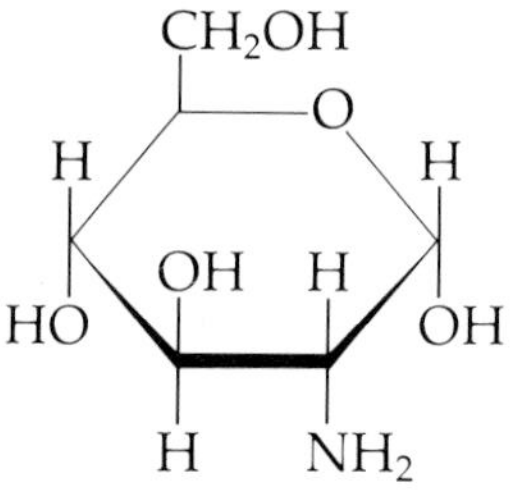

D-glucosamine 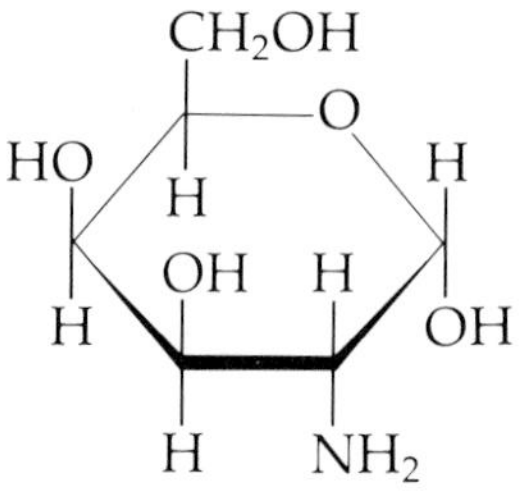　　　　　D-galactosamine 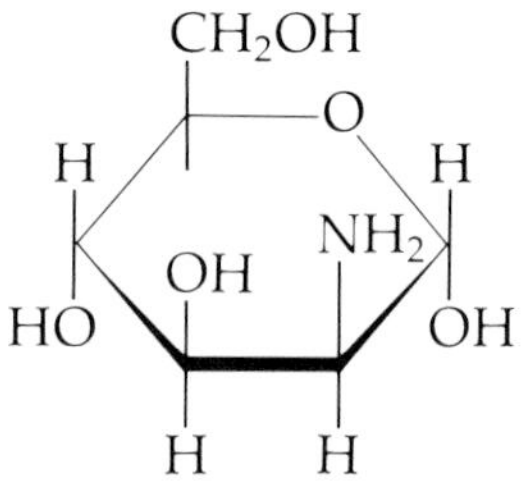　　　　　D-mannosamine

Erythromycin and carbomycin are examples of antibiotics that contain amino sugars.

Amido Sugars

When plasma glucose concentration is elevated over a period of time, normal hemoglobin covalently binds to glucose. The amount of glucosylated hemoglobin (HbA_{ic}) in the blood is used as a measure of the effectiveness of blood glucose control in a diabetic patient because the concentration of HbA_{ic} directly reflects the elevation of blood glucose over the preceding several days. (Blood glucose levels directly reflect the instantaneous glucose level.) About 7 to 11 percent of a diabetic patient's hemoglobin is HbA_{ic} compared to 4 to 6 percent for a nondiabetic person. The question arises whether glucose might combine with other proteins to produce some of the other complications associated with diabetes.

Disaccharides

There are three common disaccharides—sucrose, maltose, and lactose—all of which are isomers with the molecular formula $C_{12}H_{22}O_{11}$. On hydrolysis these disaccharides yield two monosaccharides. The general reaction can written as follows.

$$C_{12}H_{22}O_{11} \ + \ H_2O \ \xrightarrow{\text{hydrolysis}} \ C_6H_{12}O_6 \ + \ C_6H_{12}O_6$$

a disaccharide	a monosaccharide	a monosaccharide
sucrose	glucose	fructose
maltose	glucose	glucose
lactose	glucose	galactose

The disaccharides, just like the monosaccharides, are white, crystalline, sweet solids. Sucrose is very soluble in water; maltose is fairly soluble; and lactose is only slightly soluble. The disaccharides are also optically active; they rotate the plane of polarized light. However, even though they are soluble in water, they are too large to pass through cell membranes.

Structure

Disaccharides are formed by the combination of two monosaccharides. In the last section we noted that monosaccharides were either hemiacetals or hemiketals. If a hemiacetal or hemiketal (a monosaccharide) combines with an alcohol (another monosaccharide) an acetal or a ketal will be formed (see page 300). Such a bond between the two monosaccharides is called a *glycosidic linkage.

Consider the combination of a molecule of α-glucose with a molecule of β-glucose, as shown in equation (21-8). The products of such a reaction are β-maltose and water.

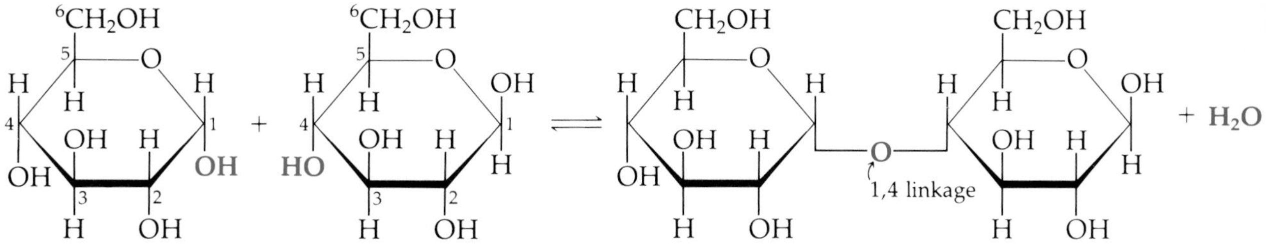

$$(21\text{-}8)$$

The linkage in β-maltose is between carbon 1 of one glucose and carbon 4 of the other glucose. Such a linkage is called an α-1,4 linkage; α-maltose has an —OH on carbon 1 of the second glucose molecule below the plane rather than above. Both α- and β-maltose exist, but the predominant form is the β.

If a molecule of β-glucose combines with another molecule of β-glucose, cellobiose, a compound with a β-1,4 glycosidic linkage, is formed.

Enzymes are specific in the type of glycosidic linkage (α or β) whose hydrolysis they can catalyze. For example, maltase (page 443) catalyzes the hydrolysis of maltose, which contains an α glycosidic linkage. It does not catalyze the hydrolysis of cellobiose, which contains a β glycosidic linkage.

Conversely, an enzyme that catalyzes the hydrolysis of β glycosidic linkages catalyzes the hydrolysis of cellobiose. Such β glycosidic linkage hydrolysis enzymes are not found in the human digestive system. That is why humans cannot digest cellulose and cellobiose, both of which have β glycosidic linkages.

Consider the reaction of a molecule of α-glucose with a molecule of β-fructose. The products of such a reaction are sucrose and water.

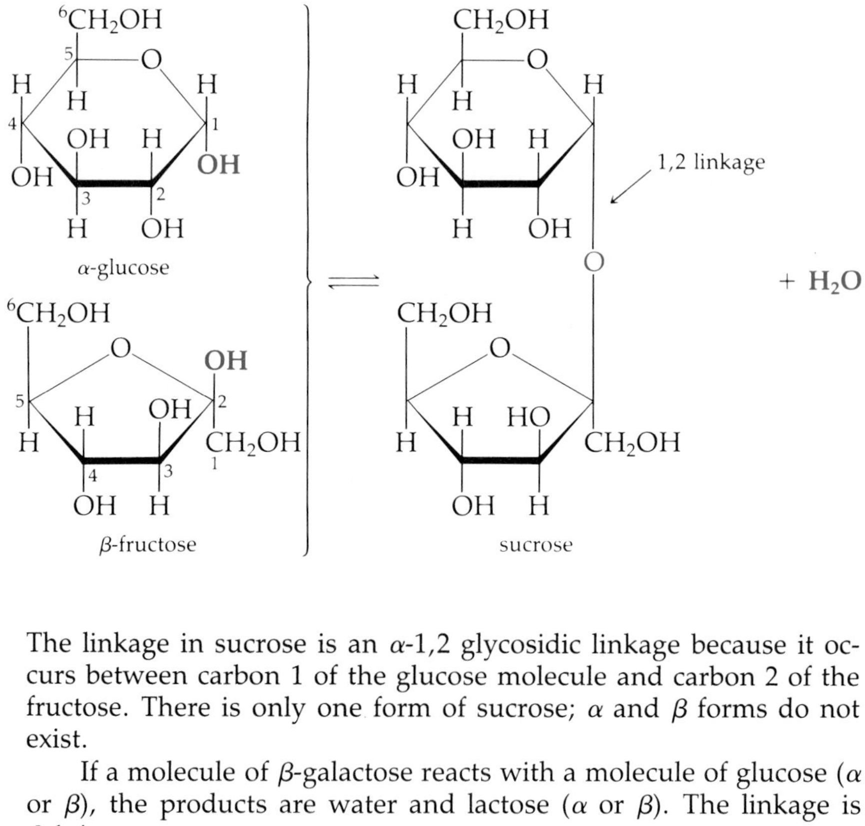

α-glucose

β-fructose

sucrose

1,2 linkage

+ H_2O

The linkage in sucrose is an α-1,2 glycosidic linkage because it occurs between carbon 1 of the glucose molecule and carbon 2 of the fructose. There is only one form of sucrose; α and β forms do not exist.

If a molecule of β-galactose reacts with a molecule of glucose (α or β), the products are water and lactose (α or β). The linkage is β-1,4.

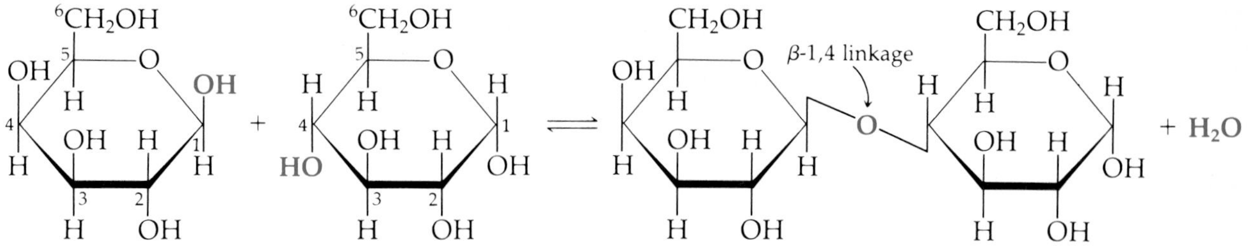

β-galactose $\quad$ α-glucose $\quad$ α-lactose

Reducing Properties

In maltose, the aldehyde groups are at carbon 1 in each of the original glucose molecules. Since the linkage is 1,4, one free aldehyde group remains. Therefore, maltose acts as a reducing sugar (see page 297).

In sucrose, the glucose part had the aldehyde at carbon 1 and the fructose part had the ketone group at carbon 2. Since the linkage is 1,2, neither group is free. Therefore, sucrose is not a reducing sugar.

Lactose, which has a 1,4 linkage, acts as a reducing sugar because both of the original aldehyde groups were on carbon 1 and one of them is free to react.

Fermentation

Sucrose and maltose will ferment when yeast is added because yeast contains the enzymes sucrase and maltase; lactose will not ferment when yeast is added because yeast does not contain lactase. The identity of a disaccharide can be deduced on the basis of its fermentation reaction and its reducing properties.

Suppose that a test tube contains a disaccharide, $C_{12}H_{22}O_{11}$. Is it sucrose, lactose, or maltose? The identity can be determined by the following method.

1. Mix the unknown disaccharide with alkaline Cu^{2+} complex and warm gently. If there is no reaction, the disaccharide must be sucrose. In this case, no further test is necessary to prove the identity of the disaccharide.
2. If the unknown disaccharide gives a positive test with alkaline Cu^{2+} complex, it must be either maltose or lactose. In this case, another sample of the disaccharide is mixed with yeast and allowed to stand to observe whether or not fermentation takes place. If the disaccharide does ferment, then it must be maltose. If it does not ferment, then it must be lactose.

The same two laboratory tests can be performed in reverse order with the same results.

1. Mix the unknown disaccharide with yeast, allow the mixture to stand, and observe whether fermentation takes place. If no fermentation is observed, the disaccharide is lactose, and no further test is necessary.
2. If fermentation does occur, the disaccharide is either sucrose or maltose. In this case take another sample of the unknown disaccharide, mix it with alkaline Cu^{2+} complex, and warm gently. If the color remains blue, the unknown disaccharide must be sucrose. If the unknown gives a positive test with alkaline Cu^{2+} complex, it must be maltose.

Sucrose

Sucrose is the sugar used ordinarily in the home. It is also known as cane sugar. Sucrose is produced commercially from sugar cane and sugar beets. It also occurs in sorghum, pineapple, and carrot roots.

When sucrose is hydrolyzed, it forms a mixture of glucose and fructose. This 50:50 mixture of glucose and fructose is called **invert sugar.** Honey contains a high percentage of invert sugar.

Maltose

Maltose, commonly known as malt sugar, is present in germinating grain. It is produced commercially by the hydrolysis of starch.

Lactose

Lactose, commonly known as milk sugar, is present in milk. It differs from the preceding sugars in that it has an animal origin. Certain bacteria cause lactose to ferment, forming lactic acid. When this reaction occurs, the milk is said to be sour. Lactose is used in high-calcium diets and in infant foods. Lactose can be used for increasing calorie intake without adding much sweetness. Lactose is found in the urine of pregnant women and, since it is a reducing sugar, it gives a positive test with Cu^{2+} complex ions.

Sweetness and Sugar Substitutes

When we speak of sugar, we think of a substance with a sweet taste. However, sweetness is not a specific property of carbohydrates. Many sugars have some degree of sweetness, but several synthetic compounds have much more sweetness without any calories. Table 21-1 shows the relative sweetness of a variety of sugars and synthetic compounds. Because of the *greater* sweetness of fructose, this substance is used in some low calorie foods because less is needed to provide the same level of sweetness.

calcium cyclamate   saccharin

aspartame

Table 21-1 Relative Sweetness of Sugars and Other Compounds

Fructose	173
Galactose	32
Glucose	74
Lactose	16
Maltose	32
Sucrose	100
Acesulfame K	20,000
Aspartame	20,000
Cyclamate	3,000
Saccharin	30,000

Cyclamates were first marketed in 1950, but after 1969 findings that large doses caused cancer in rats they were banned in the United States, although they are still in use in Canada. There has never been a connection between cyclamates and cancer or any other disease in humans.

When saccharin was first introduced, it was the only artificial sweetener marketed in the United States. In 1977 high doses of saccharin were reported to cause bladder cancer in rats. Saccharin was not banned because no other artificial sweetener was available as a replacement. Saccharin remains on the market along with a continuing controversy over its use.

In 1981 aspartame was approved for use as an artificial sweetener. The wide acceptance of the methyl ester of a dipeptide is partially due to the lack of the bitter aftertaste associated with saccharin. Aspartame is hydrolyzed to aspartic acid, phenylalanine, and methanol. Recent studies have indicated that aspartame is safe, and only those persons on a low phenylalanine diet (see page 660) need to avoid it.

In 1988 the FDA approved the use of acesulfame K, an artificial sweetener that is 200 times as sweet as sugar but has no calories. Acesulfame K can withstand temperatures of 300 to 400 °F, whereas aspartame can be used only to about 212 °F, so acesulfame K is expected to find extensive use in baked foods.

Polysaccharides

Polysaccharides are polymers of monosaccharides. Complete hydrolysis of polysaccharides produces many molecules of monosaccharides. The polysaccharides differ from monosaccharides and disaccharides in many ways, as is indicated in Table 21-2.

Polysaccharides can be formed from pentoses (five-carbon sugars) or from hexoses (six-carbon sugars). Polysaccharides formed from pentoses are called **pentosans.** Those formed from hexoses are called **hexosans** (or sometimes **glucosans**).

The hexosans (or glucosans) are the most important in terms of physiology. The hexosans have the general formula $(C_6H_{10}O_5)_x$,

Table 21-2 Comparison of Polysaccharides with Monosaccharides and Disaccharides

Property	Monosaccharides and Disaccharides	Polysaccharides
Molecular weight	Low	Very high
Taste	Sweet	Tasteless
Solubility in water	Soluble	Insoluble
Size of particles	Pass through a membrane[a]	Do not pass through a membrane
Test with Cu^{2+} complex ions (an oxidizing agent)	Positive (except for sucrose)	Negative

[a] Only monosaccharides.

where x is some large number. Some of the common hexosans are starch, cellulose, glycogen, and dextrin. All are made up of only glucose molecules, as shown in Figure 21-5.

Figure 21-5 Structures of some polysaccharides—glucose polymers in amylose, amylopectin, and cellulose.

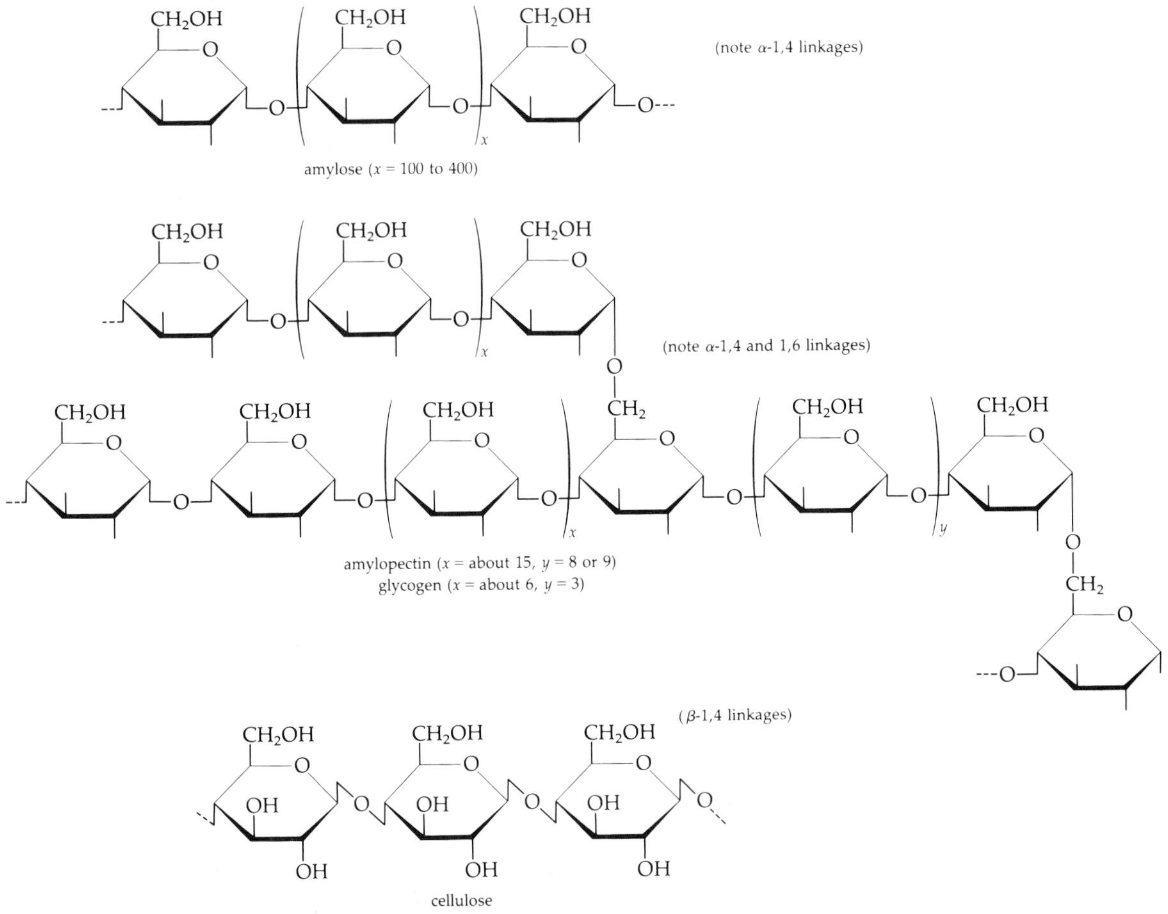

(note α-1,4 linkages)

amylose (x = 100 to 400)

(note α-1,4 and 1,6 linkages)

amylopectin (x = about 15, y = 8 or 9)
glycogen (x = about 6, y = 3)

(β-1,4 linkages)

cellulose

Starch

Plants store their foods in the form of starch granules. Starch is actually a mixture of the polysaccharides amylopectin and amylose (see Figure 21-5). Amylopectin is a branched polysaccharide present in starch to a large extent (80 to 85 percent). It is usually present in the covering of the starch granules. Amylose is a nonbranched polysaccharide present in starch to an extent of 15 to 20 percent.

Starch is insoluble in water. When starch is placed in boiling water, the granules rupture, forming a paste that gels on cooling. When a small amount of starch is added to a large amount of boiling water, a colloidal dispersion of starch in water is formed.

Starch gives a characteristic deep blue color with iodine. This test is used to detect the presence of starch because it is conclusive even when only a small amount of starch is present. That is, if iodine is added to an unknown and a blue color is produced, starch is present. This test can also be used to check for the presence of iodine. If starch is added to an unknown and a blue color is produced, iodine must be present.

When starch is hydrolyzed, it forms dextrins (amylodextrin, erythrodextrin, achroodextrin), then maltose, and finally glucose. Erythrodextrins turn red in the presence of iodine. Both maltose and glucose produce no color in the presence of iodine. Thus, it is possible to follow the hydrolysis of the starch by observing the changing colors when iodine is added.

$$\text{starch} \longrightarrow \text{erythrodextrins} \longrightarrow \text{maltose} \longrightarrow \text{glucose}$$
$$\text{blue} \qquad\qquad \text{red} \qquad\qquad \text{colorless} \qquad \text{colorless}$$

Cellulose

Wood, cotton, and paper are composed primarily of cellulose. Cellulose is the supporting and structural substance of plants. Like starch, cellulose is a polysaccharide composed of many glucose units. It is not affected by any of the enzymes present in the human digestive system and so cannot be digested. However, it does serve a purpose when eaten with other foods: it gives bulk to the feces and prevents constipation.

Cellulose does not dissolve in water or in most ordinary solvents. It gives no color test with iodine and gives a negative test with Cu^{2+} complex ions.

Cotton is nearly pure cellulose. When cotton fibers are treated with a strong solution of sodium hydroxide and then stretched and dried, the fibers take on a high luster. Such cotton is called *mercerized* cotton.

Cellulose is also used to make rayon. In this process, purified wood pulp (nearly pure cellulose) is converted into a viscous liquid called viscose by treatment with sodium hydroxide and carbon disulfide. When the viscose is forced through small openings in a block suspended in an acid solution, the cellulose is regenerated into fibers that can then be formed into threads.

Glycogen

Glycogen is present in the body and is stored in the liver and the muscles, where it serves as a reserve supply of carbohydrate. Glycogen has an animal origin, as opposed to the plant origin of starch.

Glycogen forms a colloidal dispersion in water and gives a red color with iodine. It gives no test with alkaline Cu^{2+} complex. Glycogen is formed in the body cells from molecules of glucose. This process is called **glycogenesis** (see page 475). When glycogen is hydrolyzed into glucose, the process is called **glycogenolysis.**

$$\text{glucose} \xrightleftharpoons[\text{glycogenolysis}]{\text{glycogenesis}} \text{glycogen}$$

Dextrin

Dextrin is produced during the hydrolysis of starch. Dextrin is an intermediate between starch and maltose. It forms sticky colloidal suspensions with water and is used in the preparation of adhesives. The glue on the back of postage stamps is a dextrin. Dextrin is also used when starch digestion may be a problem as with infants and elderly persons.

Heparin

Heparin is a polysaccharide used as a blood anticoagulant. It accelerates the inactivation of thrombin and other blood-clotting agents (see page 549). The structure of heparin consists of repeating units of glucuronic acid and glucosamine with some sulfate groups on the amino and hydroxyl groups. Heparin is the strongest organic acid present in the body.

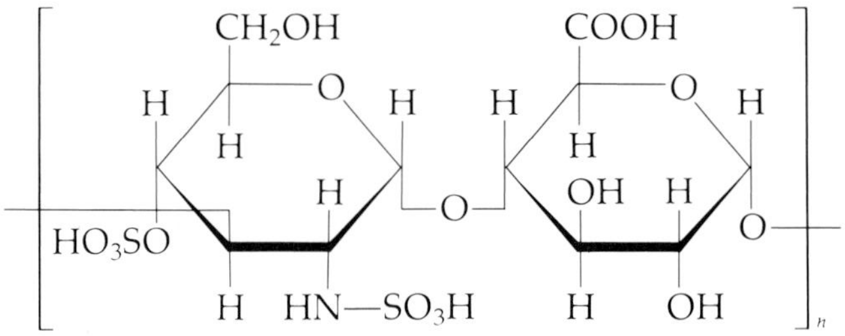

heparin

Dextran

Dextran (not the same as dextrin) is a polysaccharide produced by certain bacteria when they are grown on sucrose. There are various types of dextrans, differing in chain length and degree of branching. Medically, dextrans are used as blood extenders to hold water in the bloodstream and help prevent drops in blood volume and blood pressure. Dextrans growing on the surfaces of teeth are an important component of dental plaque.

Summary

Carbohydrates are polyhydroxyaldehydes or polyhydroxyketones or substances that yield these compounds on hydrolysis. Carbohydrates are divided into three categories: monosaccharides, disaccharides, and polysaccharides—based upon hydrolytic possibilities.

There are four types of isomers: structural, functional, geometric, and optical. Structural isomers have the same molecular formulas and the same functional groups but different structural formulas. Functional isomers have the same molecular formula but different functional groups. Geometric isomers have the same molecular formula but different structural formulas owing to a restricted rotation because of either a double bond or a ring system. Optical isomers have the same molecular formula but have structural formulas that are mirror images of each other.

Carbohydrates are formed in plants by a process called photosynthesis. Plant cells take carbon dioxide from the air and water from the ground and combine them in the presence of sunlight and chlorophyll to produce monosaccharides, at the same time giving off oxygen into the air. Plant cells also have the ability to convert the monosaccharides thus formed into disaccharides and polysaccharides. When carbohydrates are burned in the body, carbon dioxide and water are formed, thus returning these substances for reuse by plants.

Monosaccharides, or simple sugars, are either aldoses or ketoses, depending upon whether they contain an aldehyde or a ketone group. The six-carbon monosaccharides, the hexoses, are the most common in terms of the human body.

The most important hexoses in terms of the human body are glucose, fructose, and galactose. These compounds are isomers with the molecular formula $C_6H_{12}O_6$.

Glucose is an aldohexose whose structure may be represented as a linear or ring-shaped molecule. Glucose is commonly known as dextrose or grape sugar. It is the most important of all monosaccharides and is normally found in the bloodstream and in the tissue fluids.

Galactose is also an aldohexose. It occurs in nature as one of the constituents of lactose.

Fructose, a ketohexose, is commonly known as levulose or fruit sugar. Fructose is the sweetest of all sugars. It occurs in nature as one of the constituents of sucrose and is found free in fruit juices and in honey.

The hexoses are either aldehydes or ketones and can act as reducing agents. When a hexose is treated with alkaline Cu^{2+} complex (Fehling's solution, Benedict's solution, or Clinitest), a red-orange precipitate of Cu_2O is formed. This reaction is the basis for the test for sugar (hexoses) in the urine. Hexoses will also reduce Tollens' reagent (Ag^+ complex) to free silver.

Hexoses will ferment in the presence of enzymes found in yeast.

When the aldehyde end of a monosaccharide is oxidized, an *-onic* acid is formed.

When the alcohol end of a monosaccharide is oxidized, a -*uronic* acid is formed.

The three common disaccharides—sucrose, maltose, and lactose—are isomers with the molecular formula $C_{12}H_{22}O_{11}$. On hydrolysis a disaccharide yields two monosaccharides.

Of the three disaccharides only maltose and lactose show reducing properties with alkaline Cu^{2+} complex ions. Sucrose is not a reducing sugar.

Sucrose and maltose will ferment with yeast owing to the presence of the enzymes sucrase and maltase. Lactose will not ferment with yeast because of the absence of the enzyme lactase.

Polysaccharides are polymers of monosaccharides and yield monosaccharides upon hydrolysis.

Polysaccharides have a high molecular weight, are insoluble in water, are tasteless, and give negative tests for reducing sugars. These properties are the opposite of those for monosaccharides and disaccharides.

Three common polysaccharides are starch, cellulose, and glycogen. Plants store their food as starch; plants use cellulose as supporting and structural parts; animals use glycogen as a reserve supply of carbohydrate.

Questions and Problems

A

1. What is a carbohydrate?
2. What are aldoses? aldopentoses? ketohexoses?
3. What are structural isomers? Give an example.
4. What are geometric isomers? Give an example.
5. What are functional isomers? Give an example.
6. What are optical isomers? Give an example.
7. Draw the structures of (a) *cis*-1,2-dibromocyclopentane and (b) *trans*-1,2-dibromopropene.
8. What is meant by the term *asymmetric carbon atom?* Give an example.
9. How can the number of asymmetric carbon atoms be used to predict the number of optical isomers?
10. If a compound has three asymmetric carbons, how many optical isomers are possible?
11. What is the difference between ordinary light and polarized light?
12. What effect does a solution of an optical isomer have on the plane of polarized light?
13. What do the letters D and L refer to in terms of optical isomers?
14. What are epimers? Give an example.
15. Is the D or the L form the predominant one in terms of carbohydrates that the body uses?
16. What is the reference compound for the configurations of carbohydrates?
17. What is a hexosan? a pentosan?
18. What are the three types of carbohydrates?

19. How are carbohydrates formed in nature? Are they formed directly as polysaccharides?
20. Are animals able to synthesize carbohydrates from raw materials? Explain.
21. What are trioses? pentoses? hexoses?
22. Name the three important hexoses and draw their linear structural formulas.
23. How do the three important hexoses differ in molecular formula? in structure?
24. Why can glucose be given intravenously whereas sucrose cannot?
25. Where is glucose normally found in nature?
26. Where are fructose and galactose normally found in nature?
27. Where is glucose found in the body?
28. How do the hexoses affect alkaline Cu^{2+} complex ions? What use is made of this reaction?
29. What are the trade names given to alkaline Cu^{2+} complex ions used in testing for hexoses?
30. What is Tollens' reagent? What is it used for?
31. Do all hexoses ferment in the presence of yeast? Why or why not?
32. What product is formed when the aldehyde end of glucose is oxidized?
33. What product is formed when the alcohol end of glucose is oxidized?
34. What product is formed when both ends of glucose are oxidized?
35. What product is formed when the aldehyde end of glucose is reduced?
36. Do all three disaccharides act as reducing agents? Why?

37. Do all three disaccharides ferment in the presence of yeast? Why?
38. Explain how you could identify a disaccharide on the basis of its reducing action and its fermentation.
39. Compare the ring structures of the three disaccharides, and use these structures to explain their reaction to alkaline Cu^{2+} complex.
40. Where is sucrose found in nature?
41. Where is maltose found in nature?
42. Where is lactose found in nature?
43. Compare the properties of the polysaccharides with those of monosaccharides and disaccharides.
44. What is starch?
45. What is the test for the presence of starch? of iodine?
46. What products are formed when starch is slowly hydrolyzed? How can the presence of these products be detected?
47. For what purpose do plants use cellulose?
48. What is mercerized cotton?
49. For what purpose does the body use glycogen?
50. What is glycogenesis? glycogenolysis?
51. What are dextrins? How are they used commercially?
52. What type of liquid mixture does starch form in boiling water?

B
53. What is a hemiacetal? Draw the hemiacetal structure for α-D-glucose.
54. What is a hemiketal? Draw the hemiketal structure for α-D-fructose.
55. What is a Fischer projection? How does it indicate three-dimensional structure?
56. What is a Haworth projection? How does it indicate three-dimensional structure?
57. Draw the Fischer and Haworth projections for D-galactose.
58. What does the representation + + + + indicate in regard to the use of Clinitest tablets?

59. What is a glycosidic linkage?
60. What type of glycosidic linkage is present in maltose? sucrose? lactose?
61. Why are α and β forms of maltose possible?
62. Why can't the body's enzymes digest cellulose?

Practice Test
1. Animals store carbohydrate in the form of _____ .
   a. starch   b. cellulose
   c. dextran   d. glycogen
2. Which of the following contains glucose?
   a. starch   b. cellulose
   c. glycogen   d. all of these
3. Which carbohydrate is sweeter than sucrose?
   a. fructose   b. galactose
   c. lactose   d. maltose
4. An example of a five-carbon sugar is _____ .
   a. glyceraldehyde   b. ribose
   c. galactose   d. sucrose
5. An example of a disaccharide is _____ .
   a. glyceraldehyde   b. ribose
   c. galactose   d. sucrose
6. The difference between maltose and sucrose is that sucrose contains a _____ unit in place of a glucose unit.
   a. galactose   b. glyceraldehyde
   c. fructose   d. ribose
7. An asymmetric carbon atom is attached to how many different groups?
   a. 1   b. 2   c. 3   d. 4
8. If a compound contains two asymmetric carbons, how many optical isomers are possible?
   a. 2   b. 4   c. 6   d. 8
9. A carbohydrate that can be given intravenously is _____ .
   a. sucrose   b. lactose
   c. glucose   d. maltose
10. Which of the following is a reducing agent?
   a. sucrose   b. maltose
   c. starch   d. heparin

22

Lipids

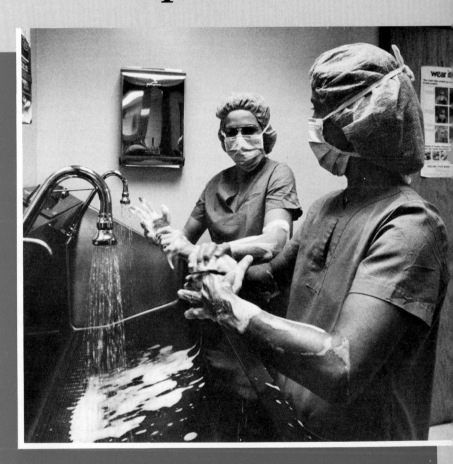

Nurses scrubbing before entering operating room.

General Properties

A second group of organic compounds that serve as food for the body is the lipids. In general, lipids

1. Are insoluble in water.
2. Are soluble in organic solvents such as alcohol, ether, acetone, and carbon tetrachloride.
3. Contain carbon, hydrogen, and oxygen; sometimes contain nitrogen and phosphorus.
4. Yield fatty acids on hydrolysis or combine with fatty acids to form esters.
5. Take part in plant and animal metabolism.

Fatty Acids

Both simple and compound lipids yield fatty acids on hydrolysis. Fatty acids are straight-chain organic acids. The fatty acids that are found in natural fats usually contain an even number of carbon atoms. Fatty acids can be either of two types: saturated or unsaturated. Saturated fatty acids contain only single bonds between carbon atoms. Unsaturated fatty acids contain one or more double bonds between carbon atoms.

Table 22-1 lists some of the common fatty acids and indicates where they are found in nature. Note that they all contain an even number of carbon atoms.

Unsaturated fatty acids have lower melting points than the corresponding saturated fatty acids, and the greater the degree of unsaturation, the lower the melting point. The 18-carbon saturated fatty acid, stearic acid, melts at 70 °C. The 18-carbon fatty acid with one double bond, oleic acid, melts at 13 °C. The 18-carbon fatty acid

Table 22-1 Common Fatty Acids

Name	Formula	Source
	Saturated Fatty Acids	
Butyric	C_3H_7COOH	Butter fat
Caproic	$C_5H_{11}COOH$	Butter fat
Caprylic	$C_7H_{15}COOH$	Coconut oil
Capric	$C_9H_{19}COOH$	Palm oil
Lauric	$C_{11}H_{23}COOH$	Laurel
Myristic	$C_{13}H_{27}COOH$	Nutmeg oil, coconut oil
Palmitic	$C_{15}H_{31}COOH$	Palm oil, lard, cottonseed oil
Stearic	$C_{17}H_{35}COOH$	Plant and animal fats such as lard, peanut oil
Arachidic	$C_{19}H_{39}COOH$	Peanut oil
	Unsaturated Fatty Acids	
Oleic	$C_{17}H_{33}COOH$ (contains 1 double bond)	Olive oil
Linoleic	$C_{17}H_{31}COOH$ (contains 2 double bonds)	Linseed oil
Linolenic	$C_{17}H_{29}COOH$ (contains 3 double bonds)	Linseed oil
Arachidonic	$C_{19}H_{31}COOH$ (contains 4 double bonds)	Animal tissues, corn oil, linseed oil

Table 22-2 Average Percentage of Fatty Acids in Fats and Oils

	Saturated				Unsaturated			
	Myristic Acid	Palmitic Acid	Stearic Acid	Other Acids	Oleic Acid	Linoleic Acid	Other Acids	Iodine Number
Vegetable oils								
Cottonseed oil	0–3	17–23	1–3	—	23–44	34–55	0–1	103–115
Corn oil	0–2	8–10	1–4	—	36–50	34–56	0–3	116–130
Animal fats								
Butter	8–13	25–32	8–13	4–11	22–29	3	3–9	26–45
Lard	1	25–30	12–16	—	41–51	3–8	5–8	46–66

with two double bonds, linoleic acid, melts at $-5\,°C$; and the 18-carbon fatty acid with three double bonds, linolenic acid, melts at $-10\,°C$.

Unsaturated fatty acids can be subdivided into the following categories.

1. Monounsaturated, those that contain only one double bond.
2. Polyunsaturated, those that contain many double bonds.
3. Eicosanoids, which include the prostaglandins, leukotrienes, prostacyclins, and thromboxanes (see page 397).

Linoleic acid is called the **essential fatty acid**—it is essential for the complete nutrition of the human body. It cannot be synthesized in the body and must be supplied from food we eat. Arachidonic and linolenic acids, which were formerly also designated as essential fatty acids, can be synthesized in the body from linoleic acid. Linoleic acid is found in large concentrations in corn, cottonseed, peanut, and soybean oils but *not* in coconut or olive oils. One of the functions of this essential fatty acid is in the synthesis of the prostaglandins (see page 398).

The absence of the essential fatty acid from the diet of an infant causes loss of weight and also *eczema. These conditions can be cured by administering corn oil or linseed oil. Commercial boiled linseed oil should never be used for this purpose because it can contain litharge, a lead compound that is poisonous to the body.

The percentages of fatty acids in corn oil, linseed oil, butter, and lard are listed in Table 22-2. The percentages are given as averages because the percent composition of a fat or oil can vary considerably because of weather conditions or the type of food eaten by the animal or both.

In addition to these straight-chain fatty acids, there are also cyclic fatty acids. An example is chaulmoogric acid, whose formula is

$$\begin{array}{c} HC \overset{\displaystyle CH}{\diagup} \quad CH{-}(CH_2)_{12}{-}COOH \\ | \qquad\quad | \\ H_2C{-\!\!-\!\!-}CH_2 \end{array}$$

chaulmoogric acid

Chaulmoogric acid occurs in chaulmoogra oil and has been used in the treatment of leprosy, although it has been supplanted in this use by new and more effective drugs.

Oleic acid ($C_{17}H_{33}COOH$) occurs in nature as the cis configuration (see page 353), as do most naturally occurring unsaturated fatty acids. The trans form is called elaidic acid.

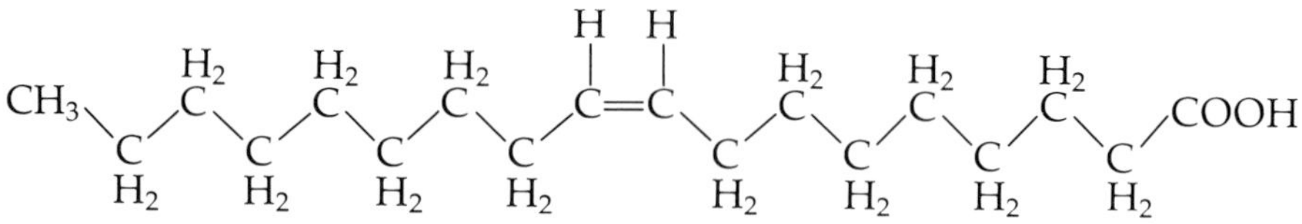

oleic acid (cis form)

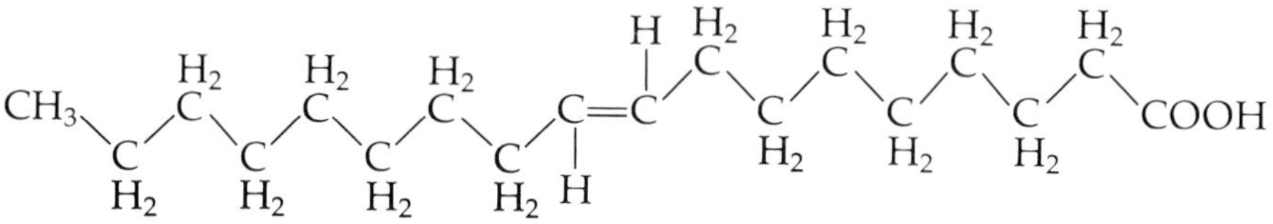

elaidic acid (trans form)

Classification of Lipids

Lipids are divided into three main categories: simple, compound, and derived.

Simple Lipids

Simple lipids are esters of fatty acids. The hydrolysis of a simple lipid may be expressed as

$$\text{simple lipid} + H_2O \xrightarrow{\text{hydrolysis}} \text{fatty acid(s)} + \text{alcohol}$$

If the hydrolysis of a simple lipid yields three fatty acids and *glycerol*, the simple lipid is called a *fat* or an *oil*. If the hydrolysis of a simple lipid yields a fatty acid and a high molecular weight monohydric alcohol, the simple lipid is called a *wax*.

Compound Lipids

Compound lipids on hydrolysis yield one or more fatty acids, an alcohol, and some other type of compound. In this category are phospholipids and glycolipids (also called cerebrosides because they are found in the cerebrum of the brain).

Phospholipids undergo hydrolysis as follows.

$$\text{phospholipid} + H_2O \xrightarrow{\text{hydrolysis}} \text{fatty acid} + \text{alcohol} + \text{phosphoric acid} + \text{a nitrogen compound}$$

Phospholipids are further subdivided into (1) phosphoglycerides, in which the alcohol is glycerol, and (2) phosphosphingosides, in which the alcohol is sphingosine.

Glycolipids (glycosphingolipids) undergo hydrolysis as follows.

$$\text{glycolipid} + H_2O \xrightarrow{\text{hydrolysis}} \text{fatty acid} + \text{a carbohydrate} + \underset{\substack{\text{a nitrogen-containing} \\ \text{alcohol}}}{\text{sphingosine}}$$

Other compound lipids include the sulfolipids and aminolipids as well as the lipoproteins.

Derived Lipids

Derived lipids are compounds produced when simple and compound lipids undergo hydrolysis. Derived lipids include such substances as fatty acids, glycerol, other alcohols, sterols (solid alcohols having a high molecular weight), fatty aldehydes, and ketone bodies (see page 499).

Fats and Oils

Structure

Fats are esters formed by the combination of a fatty acid with one particular alcohol, glycerol. If one molecule of glycerol reacts with one molecule of stearic acid (a fatty acid), glyceryl monostearate is formed.

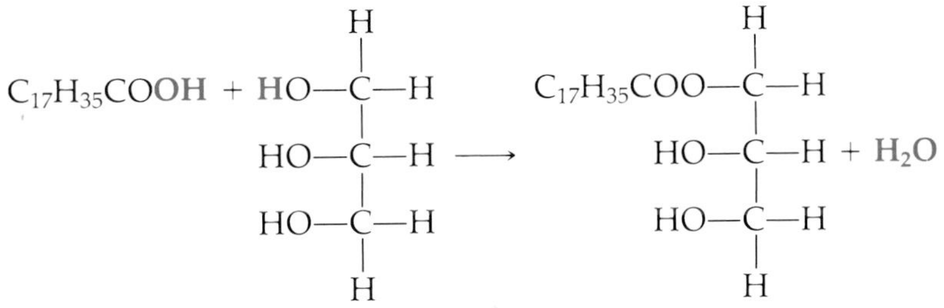

stearic acid　　　glycerol　　　glyceryl monostearate

The product of this reaction can react with a second molecule and then with a third molecule of stearic acid.

Glyceryl tristearate (also called tristearin) is formed by the reaction of one molecule of glycerol with three molecules of stearic acid. Since stearic acid is a saturated fatty acid, the product is a fat. As the degree of unsaturation of the fatty acids increases, the melting point decreases (see page 380). Fats with a melting point below room temperature are called oils.

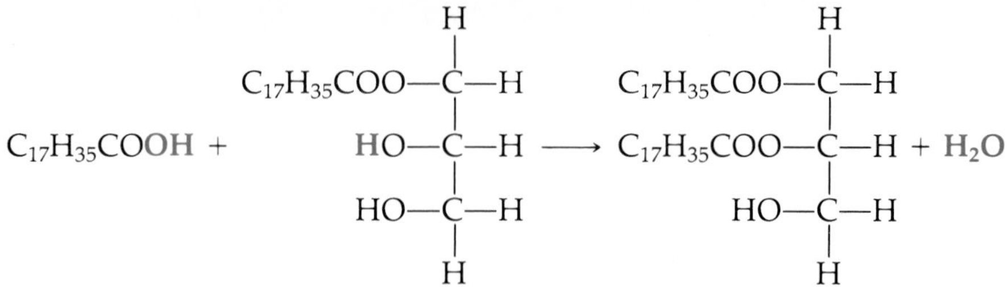

$$C_{17}H_{35}COOH \; + \quad \begin{array}{c} H \\ | \\ C_{17}H_{35}COO-C-H \\ | \\ HO-C-H \\ | \\ HO-C-H \\ | \\ H \end{array} \quad \longrightarrow \quad \begin{array}{c} H \\ | \\ C_{17}H_{35}COO-C-H \\ | \\ C_{17}H_{35}COO-C-H \\ | \\ HO-C-H \\ | \\ H \end{array} \; + \; H_2O$$

glyceryl distearate

$$C_{17}H_{35}COOH \; + \; \begin{array}{c} H \\ | \\ C_{17}H_{35}COO-C-H \\ | \\ C_{17}H_{35}COO-C-H \\ | \\ HO-C-H \\ | \\ H \end{array} \quad \longrightarrow \quad \begin{array}{c} H \\ | \\ C_{17}H_{35}COO-C-H \\ | \\ C_{17}H_{35}COO-C-H \\ | \\ C_{17}H_{35}COO-C-H \\ | \\ H \end{array} \; + \; H_2O$$

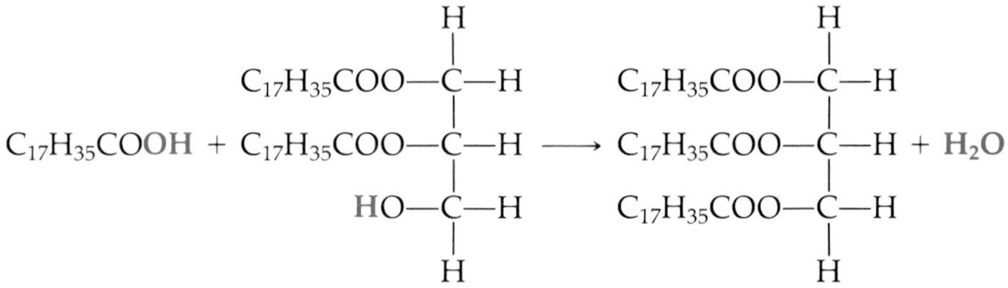

glyceryl tristearate,
a fat

The glycerol molecule contains three —OH groups and so combines with three fatty acid molecules. However, these fatty acid molecules do not have to be the same. Fats and oils can contain three different fatty acid molecules, which can be saturated, unsaturated, or some combination of these.

An example of a mixed triglyceride[†] formed from the reaction of glycerol with three different fatty acid molecules follows. The fatty acids are oleic, stearic, and linoleic.

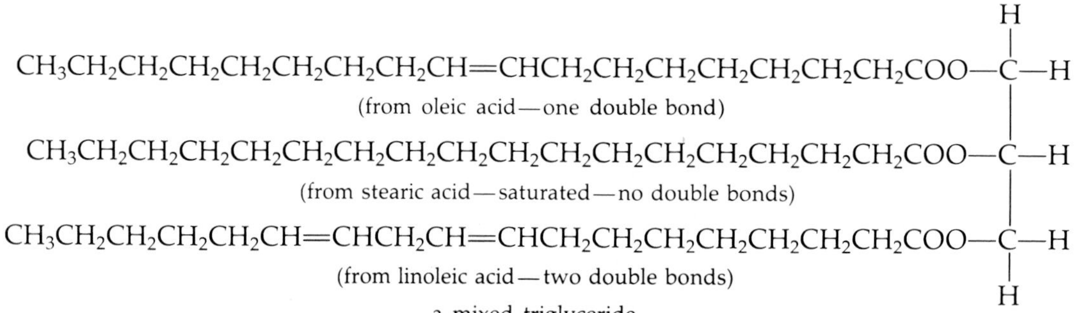

$$CH_3CH_2CH_2CH_2CH_2CH_2CH_2CH_2CH=CHCH_2CH_2CH_2CH_2CH_2CH_2CH_2COO-\overset{\displaystyle H}{\underset{\displaystyle |}{\overset{\displaystyle |}{C}}}-H$$

(from oleic acid—one double bond)

$$CH_3CH_2CH_2CH_2CH_2CH_2CH_2CH_2CH_2CH_2CH_2CH_2CH_2CH_2CH_2CH_2CH_2COO-\overset{\displaystyle |}{C}-H$$

(from stearic acid—saturated—no double bonds)

$$CH_3CH_2CH_2CH_2CH_2CH=CHCH_2CH=CHCH_2CH_2CH_2CH_2CH_2CH_2COO-\underset{\displaystyle H}{\overset{\displaystyle |}{\underset{\displaystyle |}{C}}}-H$$

(from linoleic acid—two double bonds)

a mixed triglyceride

Oleic acid has a cis configuration around its double bond; linoleic acid has a cis–cis configuration.

The preceding formula for a mixed triglyceride can be written in condensed form as shown in structure (22-1).

[†] According to the International Union of Pure and Applied Chemistry (IUPAC) and the International Union of Biochemistry (IUB), monoglycerides are to be designated as monoacylglycerols, diglycerides as diacylglycerols, and triglycerides as triacylglycerols. However, the glyceride name continues in general use.

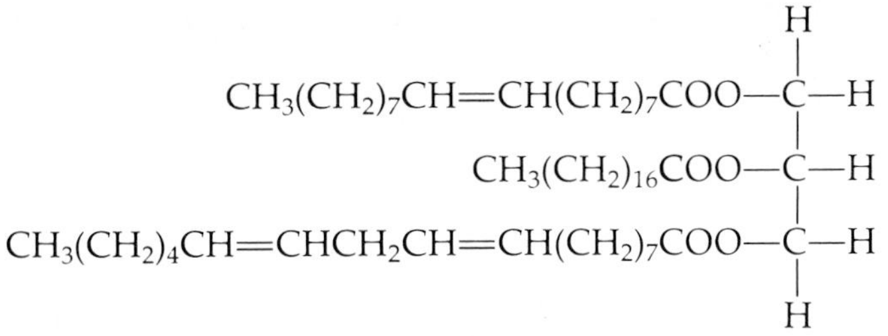

$$CH_3(CH_2)_7CH=CH(CH_2)_7COO-\overset{\displaystyle H}{\underset{\displaystyle |}{C}}-H$$

$$CH_3(CH_2)_{16}COO-\overset{|}{C}-H \quad (22\text{-}1)$$

$$CH_3(CH_2)_4CH=CHCH_2CH=CH(CH_2)_7COO-\overset{|}{\underset{\displaystyle |}{C}}-H$$

$$H$$

or simply as

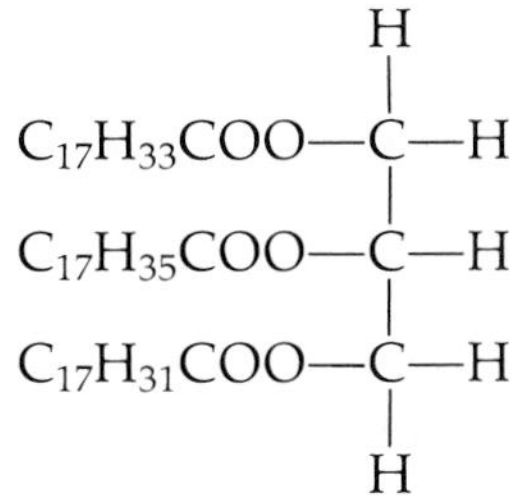

$$C_{17}H_{33}COO-\overset{\displaystyle H}{\underset{\displaystyle |}{C}}-H$$

$$C_{17}H_{35}COO-\overset{|}{C}-H \qquad (22\text{-}2)$$

$$C_{17}H_{31}COO-\overset{|}{\underset{\displaystyle |}{C}}-H$$

$$H$$

A more correct representation of the triglyceride structure (22-2) is

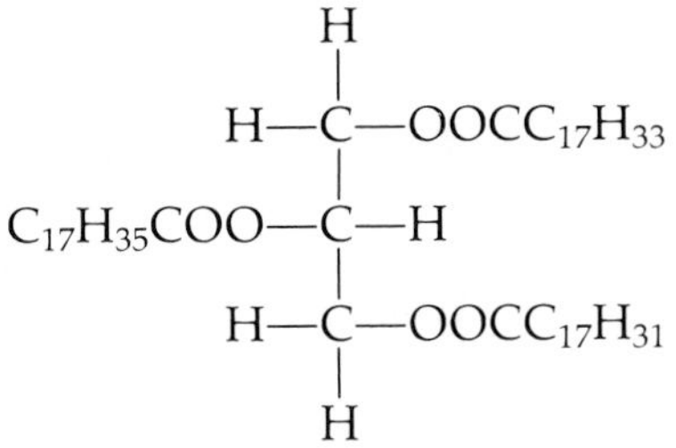

$$H-\overset{\displaystyle H}{\underset{\displaystyle |}{C}}-OOCC_{17}H_{33}$$

$$C_{17}H_{35}COO-\overset{|}{C}-H$$

$$H-\overset{|}{\underset{\displaystyle |}{C}}-OOCC_{17}H_{31}$$

$$H$$

indicating the L configuration of most naturally occurring trigly-cerides, but for simplicity we will use structure (22-2).

Iodine Number

Unsaturated fats and oils will readily combine with iodine, whereas saturated fats and oils will not do so very readily. The more unsaturated the fat or oil, the more iodine it will react with.

The iodine number of a fat or oil is the number of grams of iodine that will react with the double bonds present in 100 g of that fat or oil. The higher the iodine number, the greater the degree of unsaturation of the fat or oil. The iodine numbers of some fats and oils are listed in Table 22-2.

In general, animal fats have a lower iodine number than vegetable oils. This indicates that vegetable oils are more unsaturated. This increasing unsaturation is also accompanied by a change of state: animal fats are solid, and vegetable oils are liquid. Fats have iodine numbers below 70, oils above 70.

Animal and vegetable oils should not be confused with mineral oil, which is a mixture of saturated hydrocarbons, or with essential oils, which are volatile aromatic liquids used as flavors and perfumes.

Use of Fats in the Body

Fats serve as a fuel in the body, producing more energy per gram than either carbohydrate or protein. Metabolism of fat produces 9 kcal/g, whereas the metabolism of either carbohydrate or protein produces 4 kcal/g.

Fats also serve as a reserve supply of food and energy for the body. If a 70-kg person stored energy in the form of carbohydrate rather than fat, that person would weigh an additional 55 kg. Fat is stored in the adipose tissue and serves as a protector for the vital organs. That is, fats surround the vital organs to keep them in place and also act as shock absorbers. Fats in the outer layers of the body act as heat insulators, helping to keep the body warm in cold weather. Fats act as electrical insulators and allow rapid propagation of nerve impulses. The fat content of nerve tissue is particularly high. Fats are a constituent of lipoproteins, which are found in cell membranes and in the mitochondria and also serve as a means of transporting lipids in the bloodstream.

Physical Properties

Pure fats and oils are generally white or yellow solids and liquids, respectively. Pure fats and oils are also odorless and tasteless. However, over a period of time fats become rancid; they develop an unpleasant odor and taste.

Fats and oils are insoluble in water but are soluble in such organic liquids as benzene, acetone, and ether. Fats do not diffuse through a membrane. Fats are lighter than water and have a greasy feeling. Fats and oils form a temporary emulsion when shaken with water. The emulsion can be made permanent by the addition of an emulsifying agent such as soap. Fats and oils must be emulsified by bile in the body before they can be digested.

Chemical Reactions

Hydrolysis

When fats are treated with enzymes, acids, or bases, they hydrolyze to form fatty acids and glycerol. When tripalmitin (glyceryl tripalmitate) is hydrolyzed, it forms palmitic acid and glycerol and requires three molecules of water. Recall that in the formation of a fat, water is a product.

When fats are hydrolyzed to fatty acids and glycerol, the glycerol separates from the fatty acids and can be drawn off and purified. Glycerol is used both medicinally and industrially.

$$
\begin{array}{c}
\text{H} \\
| \\
\text{C}_{15}\text{H}_{31}\text{COO}-\text{C}-\text{H} \\
| \\
\text{C}_{15}\text{H}_{31}\text{COO}-\text{C}-\text{H} \\
| \\
\text{C}_{15}\text{H}_{31}\text{COO}-\text{C}-\text{H} \\
| \\
\text{H}
\end{array}
+ 3\ \text{H}_2\text{O} \xrightarrow[\text{enzyme}]{\text{heat}} 3\ \text{C}_{15}\text{H}_{31}\text{COOH} +
\begin{array}{c}
\text{H} \\
| \\
\text{HO}-\text{C}-\text{H} \\
| \\
\text{HO}-\text{C}-\text{H} \\
| \\
\text{HO}-\text{C}-\text{H} \\
| \\
\text{H}
\end{array}
$$

| tripalmitin | palmitic acid | glycerol |

Saponification

Saponification is the heating of a fat with a strong base such as sodium hydroxide to produce glycerol and the salt of a fatty acid.

$$
\begin{array}{c}
\text{H} \\
| \\
\text{C}_{17}\text{H}_{35}\text{COO}-\text{C}-\text{H} \\
| \\
\text{C}_{17}\text{H}_{35}\text{COO}-\text{C}-\text{H} \\
| \\
\text{C}_{17}\text{H}_{35}\text{COO}-\text{C}-\text{H} \\
| \\
\text{H}
\end{array}
+ 3\ \text{NaOH} \xrightarrow{\text{heat}} 3\ \text{C}_{17}\text{H}_{35}\text{COONa} +
\begin{array}{c}
\text{H} \\
| \\
\text{HO}-\text{C}-\text{H} \\
| \\
\text{HO}-\text{C}-\text{H} \\
| \\
\text{HO}-\text{C}-\text{H} \\
| \\
\text{H}
\end{array}
$$

| tristearin | sodium stearate, a soap | glycerol |

The sodium (or potassium) salt of a fatty acid is called a soap. Reactions and properties of soaps are discussed later in this chapter.

Hydrogenation

Fats and oils are similar compounds except that oils are more unsaturated; that is, oils contain many double bonds. These double bonds can change to single bonds upon the addition of hydrogen. Vegetable oils can be converted to fats by the addition of hydrogen in the presence of a catalyst. This process is called hydrogenation. Hydrogenation is used to produce the so-called vegetable shortenings used in the home. Oleomargarine is prepared by the hydrogenation of certain fats and oils with the addition of flavoring and coloring agents, plus vitamins A and D. Compounds that give butter its characteristic flavor are sometimes added.

$$
\begin{array}{c}
\text{H} \\
| \\
\text{C}_{17}\text{H}_{33}\text{COO}-\text{C}-\text{H} \\
| \\
\text{C}_{17}\text{H}_{33}\text{COO}-\text{C}-\text{H} \\
| \\
\text{C}_{17}\text{H}_{33}\text{COO}-\text{C}-\text{H} \\
| \\
\text{H}
\end{array}
+ 3\ \text{H}_2 \xrightarrow{\text{catalyst}}
\begin{array}{c}
\text{H} \\
| \\
\text{C}_{17}\text{H}_{35}\text{COO}-\text{C}-\text{H} \\
| \\
\text{C}_{17}\text{H}_{35}\text{COO}-\text{C}-\text{H} \\
| \\
\text{C}_{17}\text{H}_{35}\text{COO}-\text{C}-\text{H} \\
| \\
\text{H}
\end{array}
$$

| triolein, an oil (contains double bonds) | tristearin, a fat (contains single bonds) |

In actual practice, vegetable oils are not completely hydrogenated. Enough hydrogen is added to produce a solid at room temperature. If the oil were completely hydrogenated, the solid fat would be hard and brittle and unsuitable for cooking purposes.

As should be expected, hydrogenation lowers the iodine number to a value within the range of fats.

Acrolein Test

The acrolein test, which is a test for the presence of glycerol, is sometimes used as a test for fats and oils, since all fats and oils contain glycerol.

When glycerol is heated to a high temperature, especially in the presence of a dehydrating agent such as potassium bisulfate ($KHSO_4$), a product called acrolein results.

$$
\begin{array}{c}
\text{H} \\
| \\
\text{H}-\text{C}-\text{OH} \\
| \\
\text{H}-\text{C}-\text{OH} \\
| \\
\text{H}-\text{C}-\text{OH} \\
| \\
\text{H}
\end{array}
\quad \xrightarrow[\text{KHSO}_4]{\text{heat}} \quad
\begin{array}{c}
\text{H}-\text{C}=\text{O} \\
| \\
\text{H}-\text{C} \\
|| \\
\text{H}-\text{C} \\
| \\
\text{H}
\end{array}
\quad + \ 2 \ \text{H}_2\text{O}
$$

glycerol   acrolein

This substance is easily recognized by its strong, pungent odor. When fats or oils are heated to a high temperature or are burned, the disagreeable odor is that of acrolein.

Rancidity

Fats develop an unpleasant odor and taste when allowed to stand at room temperature for a short period of time. That is, they become rancid. Rancidity is due to two types of reactions—hydrolysis and oxidation.

Oxygen present in the air can oxidize some unsaturated parts of fats and oils. If this oxidation reaction produces short-chain acids or aldehydes, the fat turns rancid, as evidenced by a disagreeable odor and taste. Since oxidation, as well as hydrolysis, takes place more rapidly at higher temperatures, fats and foods containing a high percentage of fats should be stored in a cool place. Oxidation of fats, especially in hydrogenated vegetable compounds, can be inhibited by the addition of antioxidants, substances that prevent oxidation. Two naturally occurring antioxidants are vitamin C and vitamin E.

When butter is allowed to stand at room temperature, hydrolysis takes place between the fats and the water present in the butter. The products of this hydrolysis are fatty acids and glycerol. One of the fatty acids produced, butyric acid, has the disagreeable odor that causes one to say that the butter is rancid. The catalysts necessary for the hydrolysis reaction are produced by the action of microorganisms present in the air acting on the butter. At room temperature this reaction proceeds rapidly so that the butter soon turns rancid. This effect can be overcome by keeping the butter refrigerated and covered.

Soaps

Soaps are produced by the saponification of fats. Soaps are salts of fatty acids. When the saponifying agent used is sodium hydroxide, a sodium soap is produced. Sodium soaps are bar soaps. When the saponifying agent used is potassium hydroxide, a potassium soap is produced. Potassium soaps are soft or liquid soaps.

Soaps can also be produced by the reaction of a fatty acid with an inorganic base, although this method is much too expensive to be of commercial value.

$$C_{17}H_{35}COOH + NaOH \longrightarrow C_{17}H_{35}COONa + H_2O$$

stearic acid   sodium   sodium stearate,
hydroxide   a soap

Various substances can be added to soaps to give them a pleasant color and odor. Floating soaps contain air bubbles. Germicidal soaps contain a germicide. Scouring soaps contain some abrasive. Tincture of green soap is a solution of a potassium soap in alcohol.

Calcium and magnesium ions present in hard water react with soap to form insoluble calcium and magnesium soaps

$$2 \text{ Na soap} + Ca^{2+} \longrightarrow \text{Ca soap(s)} + 2 \text{ Na}^+$$

$$2 \text{ Na soap} + Mg^{2+} \longrightarrow \text{Mg soap(s)} + 2 \text{ Na}^+$$

The soap "precipitate" is mostly organic and floats to the top rather than sinking to the bottom as most precipitates do. This precipitated soap is seen as "the ring around the bathtub." More soap is required to produce a lather in hard water than in soft water (see page 177).

Zinc stearate is an insoluble soap used as a dusting powder for infants. It has antiseptic properties but is irritating to mucous membranes. Zinc undecylenate is used in the treatment of athlete's foot.

Children who suffer from celiac disease cannot absorb fatty acids from the small intestine. The unabsorbed fatty acids combine with calcium ions to form insoluble calcium compounds, or soaps. These calcium compounds are eliminated from the body, and the body will become deficient in calcium unless additional amounts of this necessary element are given to the child.

Cleansing Action

Soaps are cleansing agents. Consider a soap molecule such as sodium stearate.

Figure 22-1 Soap in an
oil–water mixture.

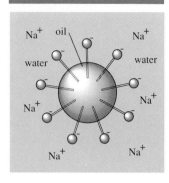

$$CH_3—(CH_2)_{16}—COONa$$

The long-chain aliphatic part is nonpolar, whereas the carboxylate part is polar. A simplified representation of a soap molecule is —○, the line representing the nonpolar part and the circle the polar part.

In general, nonpolar compounds dissolve in nonpolar liquids, and polar compounds dissolve in polar liquids (see page 186). If soap is added to a mixture of water and oil and then shaken rapidly, the nonpolar end of the soap molecule will dissolve in the oil, a nonpolar liquid. At the same time, the polar end of the soap molecule will dissolve in the water, a polar liquid. The nonpolar end of the soap molecule is said to be *hydrophobic (water-repelling). The polar end is *hydrophilic (water-loving). The carboxylate end of the soap molecule, which is in the water, yields sodium ions, which are free to move about. Structures such as this are called micelles (see Figure 22-1).

Note that the oil drop has a negative charge because of the negative ends of the soap molecules sticking out into the water. This negatively charged oil drop will repel all other oil drops, which will have acquired a like charge. That is, the oil will have become emulsified, with the soap acting as the emulsifying agent.

This is the manner in which soap cleanses, since most dirt is held on skin and clothing by a thin layer of grease or oil. Mechanical washing causes the oil or grease to break up into small drops. The soap then emulsifies that oil or grease, which can then be easily washed away. Soap also acts as a surfactant (page 195); it lowers the surface tension of the water, making emulsification easier.

Soap has little effect as an antibacterial agent. Nurses and surgeons in the operating room scrub for at least 10 min to remove most of the debris, such as keratin and natural fats, from the skin. A germicidal soap, one that contains a germ-killing compound, usually is used.

Detergents

Detergents (syndets) are synthetic compounds used as cleansing agents. They work like soaps but are free of several of the disadvantages that soaps have.

Detergents work as well in hard water as they do in soft water. That is, calcium and magnesium salts of detergents are soluble and do not precipitate out of solution (as do calcium and magnesium soaps). Recall that soaps do not work as well in hard water because insoluble calcium and magnesium salts precipitate out of solution. Detergents are generally neutral compounds compared to soaps, which are usually alkaline or basic substances. Therefore, detergents can be used on silks and woolens but soaps cannot. Detergents are used for washing clothes and also as cleansing agents in toothpastes and toothpowders.

Detergents are sodium salts of long-chain alcohol sulfates. For example, sodium lauryl sulfate can be prepared by treating lauryl alcohol, a 12-carbon alcohol, with sulfuric acid and then neutralizing with sodium hydroxide. The reactions are

$$C_{11}H_{23}CH_2OH + H_2SO_4 \longrightarrow C_{11}H_{23}CH_2OSO_3H + H_2O$$

$$\text{lauryl alcohol} \qquad\qquad\qquad \text{lauryl hydrogen sulfate}$$

$$C_{11}H_{23}CH_2OSO_3H + NaOH \longrightarrow C_{11}H_{23}CH_2OSO_3Na + H_2O$$

$$\begin{array}{c}\text{lauryl hydrogen}\\\text{sulfate}\end{array} \qquad\qquad \begin{array}{c}\text{sodium lauryl sulfate,}\\\text{a detergent}\end{array}$$

Note that the detergent, like a soap, has a nonpolar part and a polar part.

Detergents containing straight chains are **biodegradable** and do not cause water pollution, whereas those containing branched chains are nonbiodegradable and cause pollution.

Waxes

A wax is a compound produced by the reaction of a fatty acid with a high molecular weight monohydric alcohol such as myricyl alcohol ($C_{30}H_{61}OH$) or ceryl alcohol ($C_{26}H_{53}OH$). Carnauba wax is largely $C_{25}H_{51}COOC_{30}H_{61}$, an ester of myricyl alcohol. Beeswax is largely $C_{15}H_{31}COOC_{30}H_{61}$, also an ester of myricyl alcohol.

Note that waxes are primarily esters of long-chain fatty acids with an even number of carbon atoms and long-chain alcohols, also with an even number of carbon atoms. The number of carbon atoms is usually 26 to 34. The alcohol may also be a steroid such as lanosterol. The wax thus produced, lanolin, is widely used in cosmetics and ointments.

Waxes are insoluble in water, nonreactive, and flexible; hence, waxes make excellent protective coatings. Excessive loss of water through the feathers of birds, through the fur of animals, and through the leaves of plants is prevented by the presence of waxes.

Some of the common waxes are listed in Table 22-3.

Paraffin wax is different from these waxes because it is merely a mixture of hydrocarbons and is not an ester.

Table 22-3 Common Waxes

Name	Source	Use
Beeswax	Honeycomb of bee	Polishes and pharmaceutical products
Spermaceti	Sperm whale	Cosmetics and candles
Carnauba	Carnauba palm	Floor waxes and polishes
Lanolin	Wool	Skin ointments

Compound Lipids

Phospholipids

Phospholipids are phosphate esters and can be divided into two categories—phosphoglycerides and phosphosphingosides—depending on whether the alcohol is glycerol or sphingosine. As indicated on page 382, phospholipids also contain a nitrogen compound. Phospholipids are found in all tissues in the human body, particularly in brain, liver, and spinal tissue.

Phospholipids also occur in the membranes of all cells. Their peculiar properties are responsible for passage of various substances into and out of the cells.

Consider a phosphoglyceride whose structure can be represented as

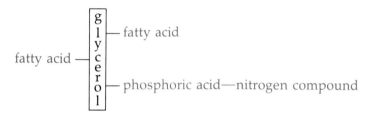

At carbons 1 and 2 of the glycerol there are esters of fatty acids. At carbon 3 there is a phosphate group, which in turn is bonded to a nitrogen compound. There are many different phosphoglycerides, depending upon the types of fatty acids bonded to the glycerol and also on the identity of the nitrogen compound bonded to the phosphate group. Most phosphoglycerides have a saturated fatty acid connected at carbon 1 and an unsaturated fatty acid at carbon 2.

The phosphate group and the nitrogen compound are polar substances, whereas the fatty acid molecules are nonpolar.

The fatty acid chains are *hydrophobic*—they point away from water. The other end of the molecule, the one containing the nitrogen compound and phosphoric acid, is *hydrophilic* and dissolves in water. Molecules of this type, with a hydrophobic (nonpolar) and a hydrophilic (polar) end, are said to be *amphipathic*.

Cell Membranes

Cell membranes serve two important functions. (1) They act as a mechanical support to separate the contents of a cell from its external environment, and (2) a structural support for certain proteins that serve to transport ions and polar molecules across the membrane. Some of these proteins act as "gates" and "pumps" to move certain materials through the membrane but exclude others. Other membrane proteins act as "receptor sites" by which molecules outside the cell can send messages inside the cell. An example of such a protein is the hormone insulin, which regulates the metabolism of glucose in certain cells but does not cross membranes itself. Instead, the insulin

Table 22-4 Percentages of Lipids in Certain Membranes

Myelin	82
Red blood cells	48
Mitochondria	
Outer membrane	48
Inner membrane	24

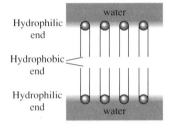

reacts with the specific receptor protein on the outer surface of the membrane, and that receptor protein in turn communicates the specific message to the inside of the cell.

Cell membranes are composed, on the average, of 40 to 50 percent lipids and 50 to 60 percent protein. However, these percentages vary considerably even among the cells of the same individual, as shown in Table 22-4. In addition, membranes also contain cholesterol and a small amount of carbohydrate. The cell membrane is relatively fluid, as it must be in order to account for the flexibility of the cell and its deformation without disruption of structure.

How do lipids form a membrane, and where do the proteins fit into that membrane? Recall that phospholipids have a polar end and a nonpolar end. In the marginal diagram, the long lines represent the nonpolar (hydrophobic) end, the fatty acid chains, whereas the circles represent the polar (hydrophilic) end, the end that dissolves in water. Such amphipathic molecules can form a bilayer, as shown at the left.

Such a simple model is not satisfactory for cell membranes because this type of bilayer would be highly impermeable to ions and most polar molecules, which cannot pass through the center of such a system. The current theory of cell membranes involves a phospholipid bilayer in which are embedded proteins, as shown in Figure 22-2.

In this type of membrane, called the fluid mosaic model, some proteins are embedded in the surface of the membrane, others are embedded in the center, and still others provide channels all the way through in order to transport polar molecules and ions through the membrane. The proteins are believed to be mobile and to move through the bilayer as well as across its surface.

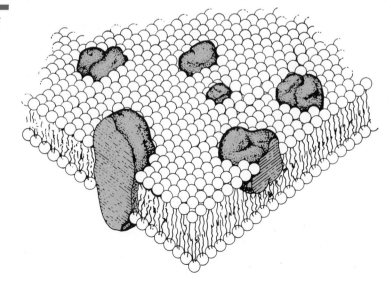

Figure 22-2 Fluid-mosaic model of plasma membrane structure.

Phosphoglycerides

One category of phospholipid is the phosphoglycerides, whose general structure is shown below. Phosphoglycerides can in turn be subdivided into several types, depending upon the nitrogen compound present. Among these are the lecithins and the cephalins.

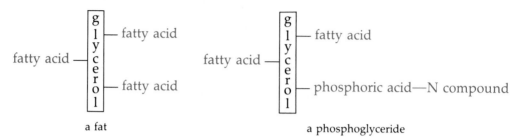

a fat   a phosphoglyceride

Lecithins Lecithins, now called phosphatidylcholines, are compounds that are particularly important in the metabolism of fats by the liver. In lecithins, the nitrogen compound is choline, an alcohol. Choline is an example of a quaternary ammonium compound (see page 311).

$$HOCH_2-CH_2-\overset{+}{N}-(CH_3)_3\ OH^-$$

choline

A typical formula for a lecithin (phosphatidylcholine) is shown as structure (22-3). In most lecithins, the fatty acid at carbon 1 of the glycerol is saturated, whereas the fatty acid at carbon 2 is unsaturated. The carbon marked with an asterisk is asymmetric, indicating optical activity. Naturally occurring lecithins have the L form (see page 356).

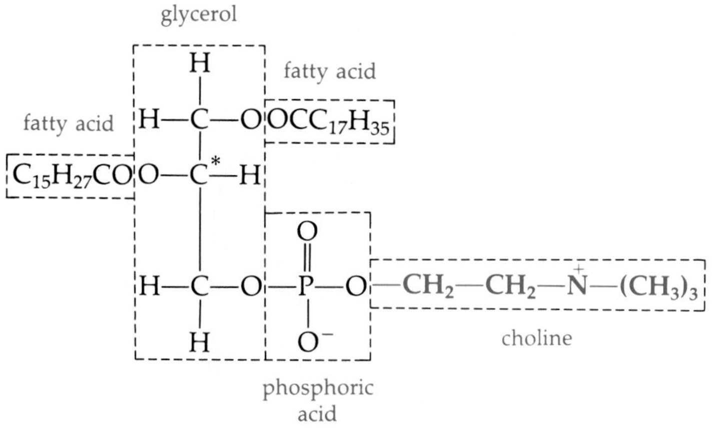

glycerol

$$
\begin{array}{c}
\text{H} \\
| \\
\text{H}-\text{C}-\text{O}-\text{OCC}_{17}\text{H}_{35} \quad \text{fatty acid}\\
| \\
\text{C}_{15}\text{H}_{27}\text{CO}-\text{O}-\overset{*}{\text{C}}-\text{H} \\
| \\
\text{H}-\text{C}-\text{O}-\overset{\overset{\text{O}}{\|}}{\text{P}}-\text{O}-\text{CH}_2-\text{CH}_2-\overset{+}{\text{N}}-(\text{CH}_3)_3 \\
| \qquad\quad | \\
\text{H} \qquad \text{O}^-
\end{array}
$$

(22-3)

fatty acid $\;C_{15}H_{27}CO$

choline

phosphoric acid

phosphatidylcholine, a lecithin

Lecithins are insoluble in water but are good emulsifying agents. They are also good sources of phosphoric acid, which is needed for the synthesis of new tissue. Lecithin is abundant in egg yolk and soybeans. It is used commercially as an emulsifying agent in dairy products and in the manufacture of mayonnaise.

Fats are partly converted to lecithins in the body and are transported as lecithins from one part of the body to another. Lecithins are widely distributed in all cells and have both metabolic and structural functions in membranes.

Dipalmityl lecithin (lecithin where the two fatty acids are palmitic acid) is a very good surface active agent (see page 195). It prevents adherence of the inner surfaces of the lungs.

Removal of one molecule of fatty acid from lecithin produces lysolecithin. The removal of this molecule of fatty acid is catalyzed by the enzyme lecithinase A, which is found in the venom of poisonous snakes. The venom is poisonous because it produces lysolecithin, which in turn causes hemolysis—the destruction of the red blood cells.

Cephalins Cephalins (22-4) are similar to lecithins except that another nitrogen compound, ethanolamine, is present instead of choline. The newer name for this compound is phosphatidylethanolamine.

$$\text{HO—CH}_2\text{—CH}_2\text{—NH}_2$$

ethanolamine

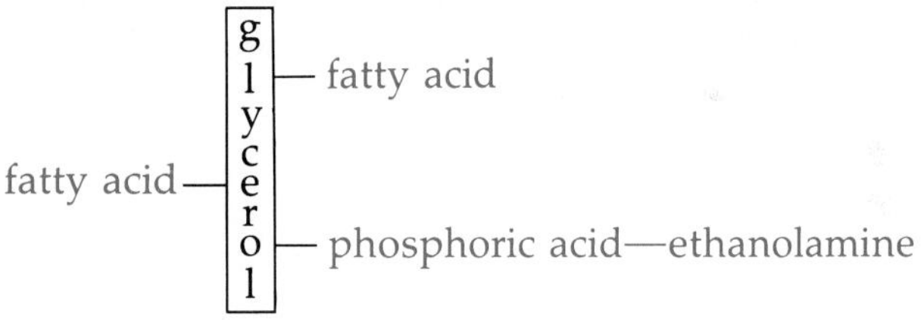

fatty acid —[**glycerol**]— fatty acid ; phosphoric acid—ethanolamine

(22-4)

a cephalin

Cephalins are important in the clotting of the blood and also are sources of phosphoric acid for the formation of new tissue.

Similar compounds, phosphatidylserine and phosphatidylinositol, are abundant in brain tissue. In these compounds, serine and inositol, respectively, occur in place of the choline or ethanolamine of (22-3) and (22-4).

$$\text{HO—CH}_2\text{—}\overset{\overset{\displaystyle NH_2}{|}}{\text{CH}}\text{—COOH}$$

serine

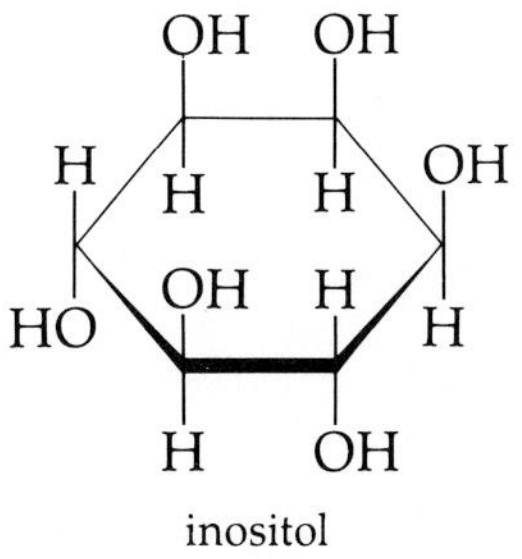

inositol

Plasmalogens Plasmalogens structurally resemble lecithin and cephalin but have an unsaturated ether at carbon 1 instead of an ester. The fatty acid at carbon 2 of the glycerol is usually unsaturated. Plasmalogens constitute up to 10 percent of the phospholipids found in membranes of brain and muscle cells. The structure of a plasmalogen is

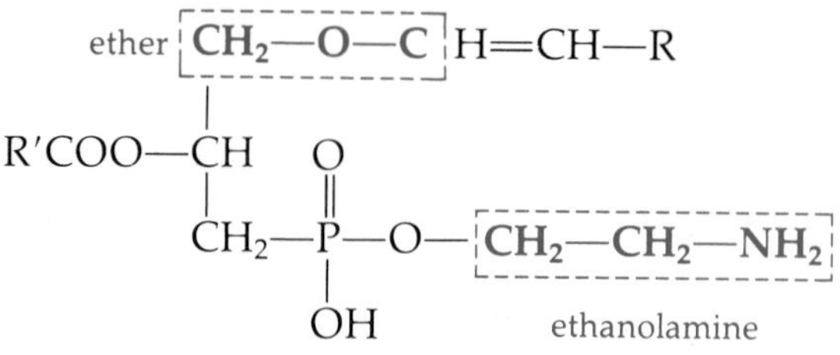

$$\text{ether}\ \boxed{CH_2-O-C}H=CH-R$$

$$R'COO-CH$$

$$CH_2-\overset{\overset{\displaystyle O}{\|}}{\underset{\underset{\displaystyle OH}{|}}{P}}-O-\boxed{CH_2-CH_2-NH_2}$$

$$\text{ethanolamine}$$

Phosphosphingosides

Phosphosphingosides, also called sphingolipids, differ from phosphoglycerides in that they contain the alcohol sphingosine in place of glycerol. One particular type of sphingolipid, called sphingomyelin, is present in large amounts in brain and nerve tissue. The general formula for a sphingolipid, is given below and the structural formula for sphingomyelin is structure (22-5).

In Niemann–Pick disease, a disease of infancy or early childhood, sphingomyelins accumulate in the brain, liver, and spleen. Accumulation of the sphingomyelins results in mental retardation and early death. It is caused by the lack of a specific enzyme, sphingomyelinase.

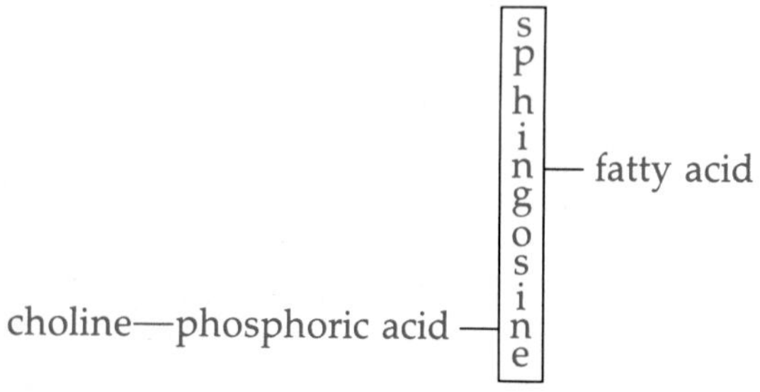

sphingosine

choline—phosphoric acid —$\begin{array}{c}s\\p\\h\\i\\n\\g\\o\\s\\i\\n\\e\end{array}$— fatty acid

a sphingolipid

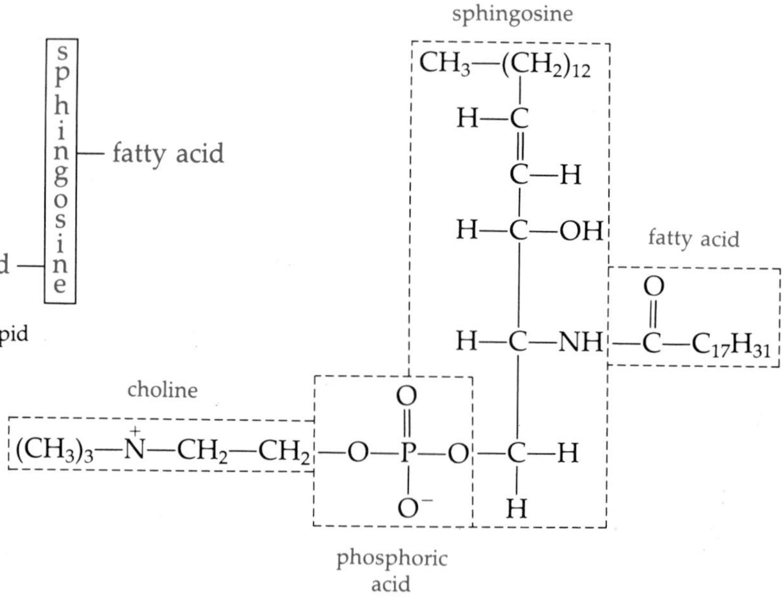

$$\boxed{CH_3-(CH_2)_{12}}$$
$$H-C$$
$$\|$$
$$C-H$$
$$H-C-OH$$
$$\text{fatty acid}$$
$$\boxed{\overset{\overset{\displaystyle O}{\|}}{H-C-NH-C-C_{17}H_{31}}}$$

choline

$$\boxed{(CH_3)_3-\overset{+}{N}-CH_2-CH_2}-O-\boxed{\overset{\overset{\displaystyle O}{\|}}{P}-O}-C-H$$
$$\overset{|}{O^-}\quad \overset{|}{H}$$

phosphoric
acid

sphingomyelin

(22-5)

Note that the fatty acid in sphingomyelin is bonded to an —NH$_2$ group rather than to an —OH group as in phosphoglycerides.

Glycolipids

Glycolipids are similar to sphingomyelins except that they contain a carbohydrate, often galactose, in place of the choline and phosphoric acid. The general structure for a glycolipid is

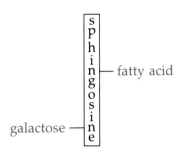

As in sphingomyelins, the fatty acid is bonded to an —NH$_2$ group on the sphingosine molecule.

Glycolipids produce no phosphoric acid on hydrolysis because they do not contain this compound. Glycolipids are also called cerebrosides because they are found in large amounts in the brain tissue.

Among the glycolipids are kerasin, cerebron, nervon, oxynervon, and the gangliosides. These compounds differ primarily in the identity of the fatty acid attached to the sphingosine.

In Gaucher's disease glycolipids accumulate in the brain and cause severe mental retardation and death by age 3. Juvenile and adult forms of this disease are characterized by enlarged spleen and kidneys, hemorrhaging, mild anemia, and fragile bones. This disease is caused by the lack of a specific enzyme, β-glucosidase.

In the absence of a particular enzyme, hexosaminidase A, glycolipids accumulate in the tissues of the brain and eyes. This effect, called Tay–Sachs disease, is usually fatal to infants before they reach age 2.

Derived Lipids

Eicosanoids

The eicosanoids are a biologically active group of compounds derived from arachidonic acid. They are extremely potent compounds with a variety of actions, as will be discussed in the following paragraphs. Among the eicosanoids are the prostaglandins, the thromboxanes, prostacyclin, and the leukotrienes.

The Prostaglandins

The prostaglandins consist of 20-carbon unsaturated fatty acids containing a five-membered ring and two side chains. One side chain has seven carbon atoms and ends with an acid group (COOH). The other chain contains eight carbon atoms with an —OH group on the third carbon from the ring (see structure given). The E series of prostaglandins has, in addition to four asymmetric carbon atoms, a trans configuration.

Prostaglandins are derived from arachidonic acid, which is formed from the essential fatty acid linoleic acid. The structures of arachidonic acid and prostaglandin E_1 (PGE_1) are

arachidonic acid   prostaglandin E_1 (PGE_1)

The abbreviation PGE_1 refers to prostaglandin E with one double bond. Likewise, PGE_2 refers to prostaglandin E with two double bonds.

Prostaglandins have been isolated from most mammalian tissues, including the male and female reproductive systems, liver, kidneys, pancreas, heart, lungs, brain, and intestines. The richest source of prostaglandins is human seminal fluid.

Prostaglandins have a wide range of physiologic effects. They seem to be involved in the body's natural defenses against all forms of change including those induced by chemical, mechanical, physiologic, and pathologic stimuli. Aspirin and other anti-inflammatory drugs appear to partially operate by inhibiting prostaglandin synthesis. Prostaglandins are involved at the cellular level in regulating many body functions, including gastric acid secretion, contraction and relaxation of smooth muscles, inflammation and vascular permeability, body temperature, and blood platelet aggregation. Prostaglandins stimulate steroid production by the adrenal glands and also stimulate the release of insulin from the pancreas. Prostaglandins markedly stimulate the movement of calcium ions from bone. Excessive production of prostaglandins by malignant tissue may provide a partial answer for the *hypercalcemia and *osteolysis observed in patients afflicted with such a condition.

Prostaglandins have also been used clinically to induce abortion or to induce labor in a term pregnancy, to treat hypertension, to relieve bronchial asthma, and to heal peptic ulcers.

Prostaglandins increase cyclic adenosine monophosphate (cAMP) (see page 618) in blood platelets, thyroid, corpus luteum, adenohypophysis, and lungs but decrease cAMP in adipose tissue.

Prostaglandin E_1 (PGE_1) is now used to strengthen babies born with cyanotic congenital heart disease, "blue babies," to prepare them for corrective surgery. Another use of prostaglandins is in animal husbandry, particularly in the management of breeding farm animals.

Prostacyclin and Thromboxanes

Prostacyclin is so called because it contains a second five-membered ring in addition to the ring found in prostaglandins. Thromboxanes have a cyclopentane ring interrupted by an oxygen atom. The structures of prostacyclin and thromboxane B_2 (TXB_2) are shown below.

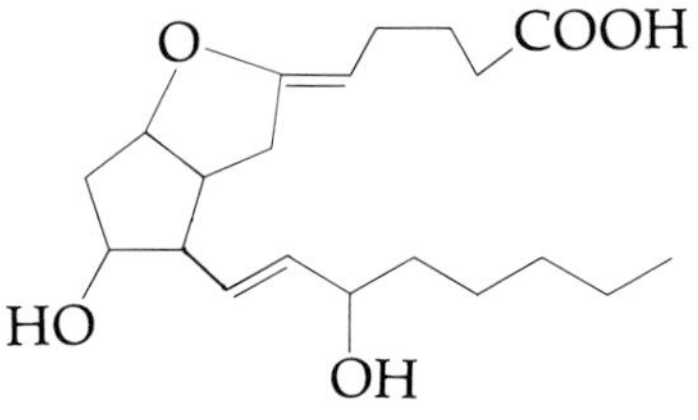

prostacyclin

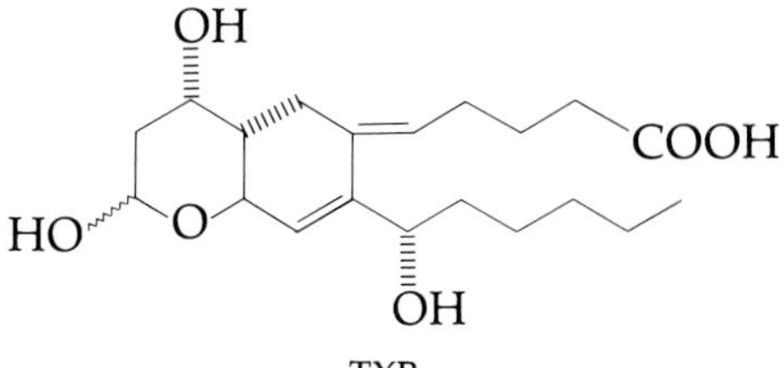

TXB_2

Prostacyclin is a potent inhibitor of platelet aggregation and is a powerful vasodilator.

Thromboxanes have an effect opposite to that of prostacyclin. They are potent aggregators of blood platelets and have a profound contractive effect on a variety of smooth muscles.

Thromboxanes function in conjunction with prostacyclin in maintaining a healthy vascular system. Both substances exert their influence by regulating the production of cAMP with the thromboxanes acting as inhibitors and prostacyclin as a stimulator of cAMP production.

Leukotrienes

Leukotrienes are another group of eicosanoids derived from arachidonic acid. The *tri* refers to three alternate sets of double bonds in the molecule. One of this group, leukotriene C, is involved in the body's allergic responses. It constricts air passages to the bronchi during an asthma attack. The structure of leukotriene C is

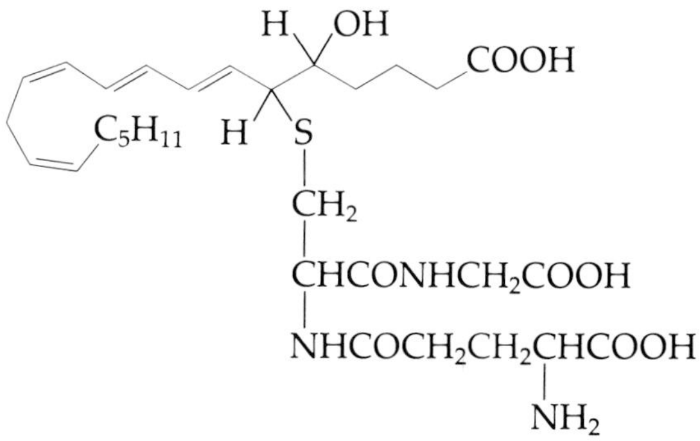

leukotriene C

Steroids

Steroids are high molecular weight tetracyclic (four-ring) compounds. Those containing one or more —OH groups and no C=O groups are called *sterols*. The most common sterol is cholesterol, which is found in animal fats but not in plant fats. Cholesterol is found in all animal tissues, particularly in brain and nervous tissue, in the bloodstream, and as gallstones. Cholesterol aids in the absorption of fatty acids from the small intestine. Several theories have been proposed relating cholesterol to coronary malfunction, but no definite evidence has yet been found linking the two.

Most of the body's cholesterol is derived or synthesized from other substances such as carbohydrates and proteins as well as from fats. The rest comes from the diet.

Atherosclerosis, a form of arteriosclerosis, results from the deposition of excess lipids, primarily triglycerides and cholesterol, from the bloodstream. Of these two, cholesterol poses a greater threat to the well-being of a person, although excess triglycerides also present a significant risk. One way of combating heart disease and atherosclerosis is to reduce the concentration of lipids in the bloodstream—either by reducing lipid intake or by the use of antihyperlipidemic drugs, those that tend to reduce blood lipid levels. It has been found that certain unsaturated fish and vegetable oils, when substituted for saturated fats, lower the serum cholesterol level.

Another substance, mevinolin, isolated from the fungus *Aspergillus terreus,* inhibits the synthesis of cholesterol and may be helpful in the prevention of atherosclerosis.

Ergosterol is a sterol similar to cholesterol. When ergosterol is irradiated (exposed to radiation) with ultraviolet light, one of the products formed is calciferol (vitamin D_2, see page 596).

Other steroids include bile salts, the sex hormones, and the hormones of the adrenal cortex. The similarities of the structure of some of these steroids are indicated in the following structures.

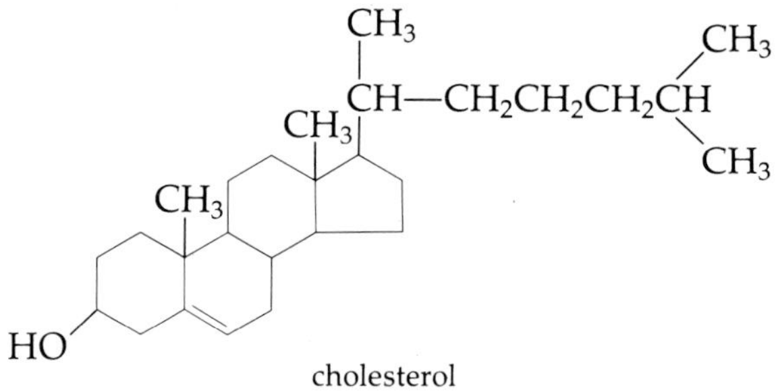

cholesterol

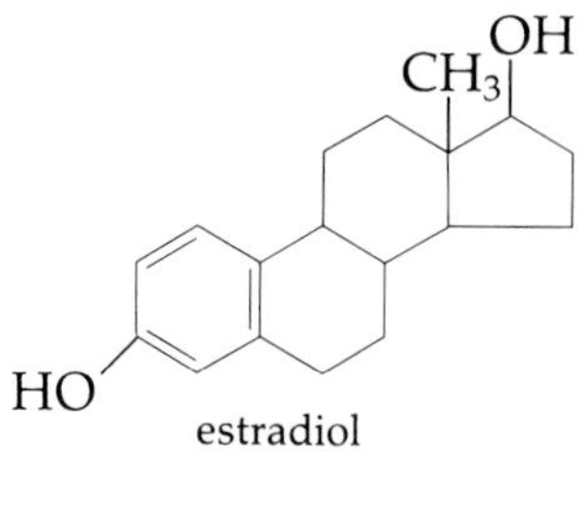

estradiol

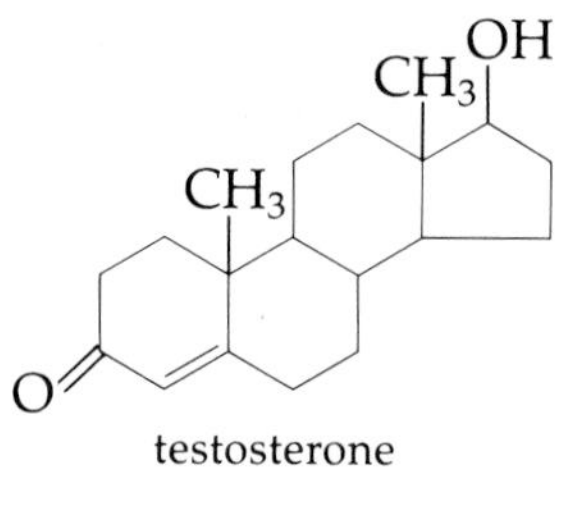

testosterone

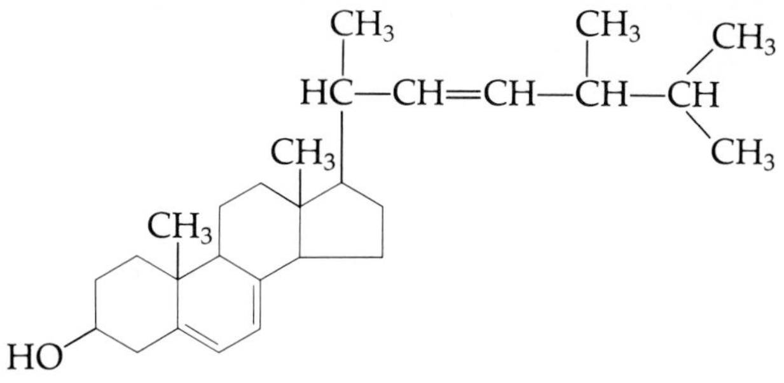

$$\underset{\text{ergosterol}}{\text{HO}}$$

The following structural formula appears:

CH₃ groups and chain: HC—CH=CH—CH—CH with CH₃ substituents, attached to the steroid ring system, HO at the bottom left.

ergosterol

Summary

Lipids yield fatty acids on hydrolysis or combine with fatty acids to form esters. Lipids are insoluble in water but are soluble in organic solvents such as ether, acetone, and carbon tetrachloride.

Lipids can be classified into three types: simple, compound, and derived. Simple lipids are esters of fatty acids. A simple lipid that yields fatty acids and glycerol upon hydrolysis is called a fat or oil. A simple lipid that, on hydrolysis, yields fatty acids and a high molecular weight alcohol is called a wax.

Compound lipids yield fatty acids, alcohol, and some other type of compound on hydrolysis.

Derived lipids are compounds produced when simple or compound lipids undergo hydrolysis. Fatty acids are straight-chain organic acids. Those found in nature usually contain an even number of carbon atoms. Saturated fatty acids have only single bonds between carbon atoms. Unsaturated fatty acids have one or more double bonds in the molecule and occur in nature in the cis form.

The essential fatty acid, so called because it is necessary in the diet, is linoleic acid.

Unsaturated fats and oils react with iodine, whereas saturated ones do not. The iodine number of a fat or oil is the number of grams of iodine that will react with (the double bonds present in) 100 g of that fat or oil.

Fats serve as fuel for the body—1 g of fat produces 9 kcal as compared to only 4 kcal/g of carbohydrate. Fats protect nerve endings and also act as insulators to keep the body warm in cold weather.

Fats and oils are odorless and tasteless when pure. They are insoluble in water but are soluble in organic solvents. Fats and oils must be emulsified before being digested.

When fats are hydrolyzed, they form fatty acids and glycerol. When fats are saponified, they form salts of fatty acids (soaps) and glycerol.

When oils are hydrogenated, the double bonds are changed to single bonds, and the (liquid) oil becomes a (solid) fat.

When fats or oils are heated to a high temperature, especially in the presence of a dehydrating agent, a product known as acrolein is produced. The odor of burning fat or oil is due to the presence of acrolein.

When fats are allowed to stand at room temperature, they become rancid because of hydrolysis and oxidation. Keeping fats cool prevents their becoming rancid.

Soaps are salts of fatty acids—sodium soaps are solid or bar soaps and potassium soaps are soft or liquid soaps.

Calcium and magnesium soaps are insoluble in water and are formed when sodium or potassium soaps are used in hard water. The precipitated calcium and magnesium soaps are seen as the "ring around the bathtub."

Detergents are similar to soaps in their cleansing properties. However, detergents do not precipitate in hard water because their calcium and magnesium compounds are soluble.

A wax is a compound produced by the reaction of a fatty acid with a high molecular weight monohydric alcohol.

Phospholipids contain fatty acids, an alcohol, a nitrogen compound, and phosphoric acid. Two types of phospholipids are phosphoglycerides and phosphosphingosides.

Phosphoglycerides include phosphatidylcholines (lecithins) and phosphatidylethanolamines (cephalins). Lecithins are important in the metabolism of fats by the liver and also are a source of phosphoric acid, which is needed for the formation of new tissue. Cephalins are important in clotting of the blood and also are a source of phosphoric acid for the formation of new tissue.

Sphingomyelins are an example of phosphosphingosides and are present in large amounts in brain and nerve tissue.

Glycolipids, also called cerebrosides, are found in large amounts in brain tissue.

Phospholipids are found in all cell membranes as bilayers and are responsible for the passage of various substances into and out of the cells.

The eicosanoids are a biologically active group of compounds derived from arachidonic acid. Among the eicosanoids are the prostaglandins, the thromboxanes, prostacyclin, and leukotrienes.

Steroids are high molecular weight, four-ring compounds. Those containing —OH groups are called sterols. The most common sterol in the body is cholesterol. Other steroids are vitamin D, bile salts, sex hormones, and hormones of the adrenal cortex.

Questions and Problems

A

1. State the general properties of lipids.
2. What are simple lipids?
3. What is the difference between a fat and a wax? a fat and an oil?
4. What is a compound lipid? Give examples of several.
5. What is a derived lipid? Give examples of several.
6. What is a fatty acid?
7. What is the difference between a saturated and an unsaturated fatty acid?
8. What is unusual about the number of carbon atoms in naturally occurring fatty acids?
9. What is the essential fatty acid? Why is it important?
10. Why should commercial boiled linseed oil never be used to overcome a deficiency of the essential fatty acid?
11. What fatty acid has been used in the treatment of leprosy?
12. What is the iodine number of a fat or oil? What determines this number?
13. Which have higher iodine numbers, animal or vegetable fats?
14. Draw the structure of glyceryl tripalmitate.
15. Draw the structure of the fat formed from the reaction of glycerol with palmitic, stearic, and oleic acids.
16. What is the function of fat in the body?
17. What are the general physical properties of fats and oils?
18. What products are formed when a fat is hydrolyzed?
19. What products are formed when a fat is saponified?
20. How can an unsaturated oil be changed to a saturated fat? Is the fat thus formed completely saturated? Why?
21. What is the test for the presence of a fat or oil?

22. What causes rancidity in a fat or oil?
23. Compare soaps with detergents on the basis of structure.
24. Compare soaps with detergents on the basis of their reaction with hard water.
25. What causes a soap to float?
26. What is a germicidal soap? How is it made?
27. Why do surgeons scrub so vigorously before surgery?
28. What causes celiac disease?
29. What is a phospholipid?
30. What are the types of phospholids?
31. What is a lysolecithin? How is it produced? What effect does it have on the body?
32. Compare the structures of the phosphoglycerides.
33. How do phospholipids help control the passage of materials into or out of cells?
34. What is meant by the term *amphipathic*?
35. What causes atherosclerosis? What can be done to prevent it?
36. What causes Niemann–Pick disease? Tay–Sachs disease? Gaucher's disease?
37. What is the most common sterol in the body?
38. Give the names of several steroids. Where is each found?
39. Why does the elimination of cholesterol-containing foods in the diet have little effect on the body's cholesterol? What can be done to reduce serum cholesterol?

B
40. Compare the structures of oleic and elaidic acids.
41. Explain the cleansing action of soap.
42. What do the terms *hydrophobic* and *hydrophilic* mean?
43. What are biodegradable detergents?
44. Why are waxes important biologically?
45. What effect do prostaglandins have on Ca^{2+} concentration?
46. What is prostacyclin? For what purpose is it useful?
47. How does the fluid-mosaic model account for the movement of polar substances through a membrane?

48. What does the abbreviation PGF_2 indicate?
49. How do prostacyclin and thromboxanes affect the production of cAMP?
50. From what compound are the eicosanoids derived? What are the main types of eicosanoids?

Practice Test

1. Which of the following is an unsaturated fatty acid?
   a. lauric   b. linoleic
   c. palmitic   d. stearic
2. Fats and oils contain which of the following?
   a. sodium   b. phosphate
   c. glycerol   d. carbohydrate
3. Oils have more _____ than fats.
   a. carbon   b. iodine
   c. double bonds   d. oxygen
4. Which compound would convert a fatty acid into a soap?
   a. NaOH   b. HCl
   c. CO_2 d. O_2
5. Cholesterol is a _____.
   a. prostaglandin   b. triglyceride
   c. phospholipid   d. sterol
6. Lipids _____.
   a. are insoluble in water
   b. yield fatty acids on hydrolysis
   c. are soluble in organic solvents
   d. all of these
7. The higher the iodine number of a lipid, the greater the degree of _____.
   a. aromaticity   b. acidity
   c. oxidation   d. unsaturation
8. Molecules with a polar and a nonpolar end are called _____.
   a. amphipathic   b. amphoteric
   c. isomers   d. polymers
9. A lipid necessary for blood clotting is _____.
   a. lecithin   b. cephalin
   c. prostaglandin   d. leukotriene
10. An example of a steroid is _____.
   a. cholesterol   b. sphingomyelin
   c. prostaglandin   d. arachidonic acid.

23

Proteins

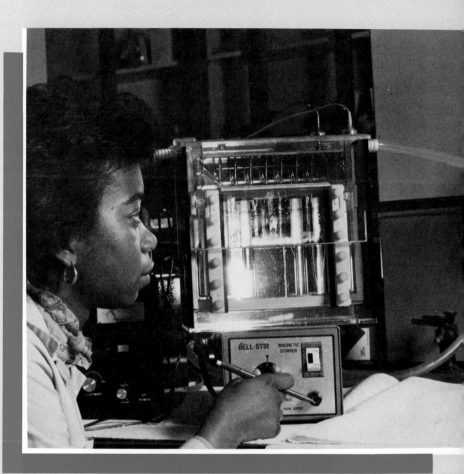

Scientist working on machine that separates proteins.

Other than water, proteins are the chief constituents of all cells of the body. Proteins are much more complex than either carbohydrates or fats. All proteins contain the elements carbon, hydrogen, oxygen, and nitrogen. Most proteins also contain sulfur, some contain phosphorus, and a few, such as hemoglobin, contain some other element.

Sources

Plants synthesize proteins from inorganic substances present in the air and in the soil. Animals cannot synthesize proteins from such materials. Animals must obtain proteins from plants or from other animals who in turn have obtained them from plants.

Animals excrete waste materials containing many nitrogen compounds. These nitrogen compounds along with decaying animal and plant matter are converted into soluble nitrogen compounds by soil bacteria. Plants in turn use these soluble nitrogen compounds to manufacture more protein, thus completing a cycle. A simplified version of the nitrogen cycle is shown in Figure 23-1. As the figure shows, another part of the cycle involves bacteria and gaseous nitrogen in the air.

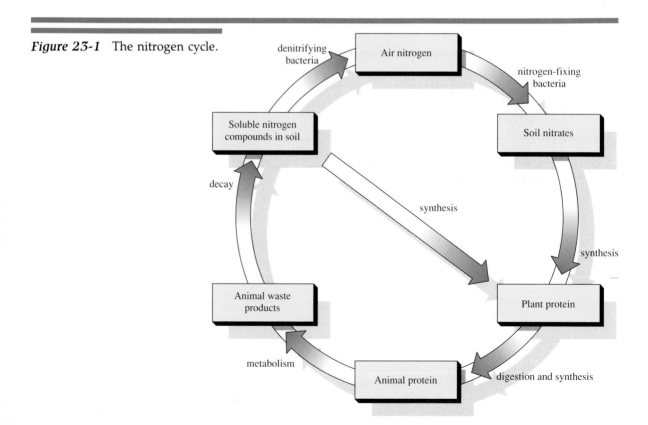

Figure 23-1 The nitrogen cycle.

Functions

The word **protein** is derived from the Greek word *proteios*, which means "of first importance." Proteins function in the body in the building of new cells, the maintenance of existing cells, and the replacement of old cells. Thus, proteins are the most important type of compound in the body. Proteins are also a valuable source of energy in the body. The oxidation of 1 g of protein yields 4 kcal—just as does the oxidation of 1 g of carbohydrate.

Proteins are involved in the regulation of metabolic processes (hormones), in the catalysis of biochemical reactions (enzymes), in the transportation of oxygen (hemoglobin), in the body's defense against infection (antibodies), in the transmission of impulses (nerves), and in muscular activity (contraction). Proteins are components of skin, hair, and nails as well as connecting and supporting tissue.

Molecular Weight

Proteins have very high molecular weights. A comparison of the molecular weights of proteins with those of carbohydrates and fats can be seen in Table 23-1.

Table 23-1 Molecular Weights of Various Proteins, Carbohydrates, and Fats

Type of Compound	Molecular Weight
Inorganic compounds	
Water	18
Sodium chloride	58.5
Plaster of paris	290
Organic compounds	
Benzene	78
Ethyl alcohol	46
Carbohydrates	
Glucose	180
Sucrose	342
Lipids	
Tristearin	891
Cholesterol	384
Proteins	
Insulin	12,000–48,000[a]
Lactalbumin	17,500
Hemoglobin	68,000
Serum globulin	180,000
Fibrinogen	450,000
Thyroglobin	630,000
Hemocyanin	9,000,000
Tobacco mosaic virus	59,000,000

[a] Even though the minimum calculated molecular weight of human insulin is 5734 (see structure, Figure 23-2), insulin exists in several different polymeric forms depending upon pH, temperature, and concentration.

Amino Acids

Proteins are polymers built up from simple units called amino acids. Hydrolysis of proteins yields amino acids. There are 20 known amino acids that can be produced by the hydrolysis of protein.[†] All these amino acids, except glycine, which has no asymmetric carbon, have the L configuration. Compare structures of L-amino acid and L-glyceraldehyde, in (23-1). Certain microorganisms can prepare D-amino acids, which are used as antibiotics.

Composition

An amino acid is an organic acids that has an amine (—NH$_2$) group attached to a chain containing an acid group. Although the amine group can be anywhere on the chain, amino acids found in nature usually have the amine group on the alpha (α) carbon—that is, the carbon atom next to the acid group. [The second carbon from the acid group is the beta (β) carbon; then come the gamma (γ) and delta (δ) carbons.] α-Amino acids can be represented by the general formula in (23-1), where R can be many different radicals.

$$\begin{array}{ccc}
\text{COOH} & & \text{CHO} \\
| & & | \\
\text{H}_2\text{N}-\text{C}-\text{H} & & \text{HO}-\text{C}-\text{H} \\
| & & | \\
\text{R} & & \text{CH}_2\text{OH} \\
\text{L-amino acid} & & \text{L-glyceraldehyde}
\end{array} \qquad (23\text{-}1)$$

Amino acids can be divided into two groups, polar and nonpolar, depending upon the polarity of the R group attached to the α carbon. If the R group is primarily hydrocarbon, which is nonpolar, then the amino acid will not dissolve in a polar liquid such as water. Thus, amino acids containing a nonpolar R group will be insoluble in water. Exceptions to this are glycine and alanine.

An R group that is polar, such as —OH, —SH, —NH$_2$, or —COOH, produces an amino acid that is polar. Such amino acids are soluble in water. Table 23-2 lists the polar and nonpolar amino acids.

The body can synthesize some, but not all, of the amino acids that it needs. Those that it cannot synthesize must be supplied from food consumed. These are called the nutritionally **essential amino acids** and are listed in Table 23-3 along with the daily requirements per kilogram of body weight.

[†] γ-Aminobutyric acid (GABA) is an amino acid that is not incorporated into protein. It is manufactured almost exclusively in the brain and spinal cord and functions as an inhibitory transmitter. A specific deficiency of GABA in the brain occurs in Huntington's *chorea, an inherited neurologic *syndrome characterized by uncontrollable body movements.

Table 23-2 Nonpolar and Polar Amino Acids

Structure	Name	IUPAC Abbreviation	Isoelectric Point
Nonpolar			
$\begin{array}{c} COOH \\ | \\ H_2N-C-H \\ | \\ CH_3 \end{array}$	Alanine	Ala	6.00
$\begin{array}{c} COOH \\ | \\ H_2N-C-H \\ | \\ CH \\ \diagup \quad \diagdown \\ CH_3 \quad CH_2-CH_3 \end{array}$	Isoleucine	Ile	6.02
$\begin{array}{c} COOH \\ | \\ H_2N-C-H \\ | \\ CH_2 \\ | \\ CH \\ \diagup \quad \diagdown \\ H_3C \quad CH_3 \end{array}$	Leucine	Leu	5.98
$\begin{array}{c} COOH \\ | \\ H_2N-C-H \\ | \\ CH_2-CH_2-S-CH_3 \end{array}$	Methionine	Met	5.74
$\begin{array}{c} COOH \\ | \\ H_2N-C-H \\ | \\ CH_2 \\ \end{array}$ (phenyl ring)	Phenylalanine	Phe	5.48
Proline ring structure: $\begin{array}{c} COOH \\ | \\ HN \quad C-H \\ | \qquad CH_2 \\ H_2C \quad C \\ \quad H_2 \end{array}$	Proline	Pro	6.30
$\begin{array}{c} COOH \\ | \\ H_2N-C-H \\ | \\ CH_2 \\ \end{array}$ (indole ring)	Tryptophan	Trp	5.89

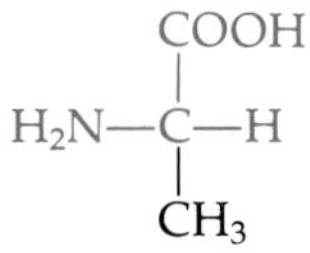

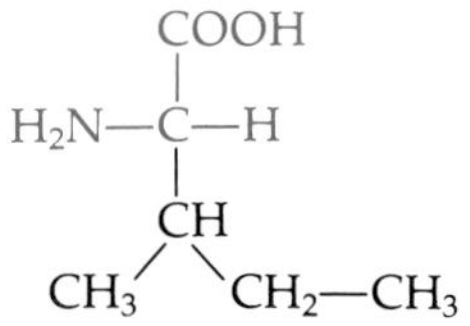

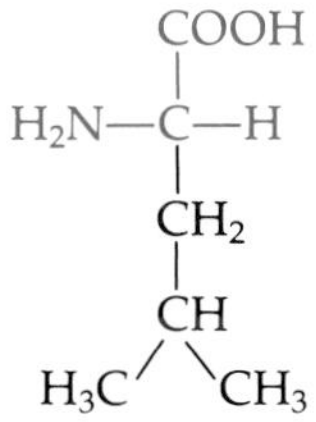

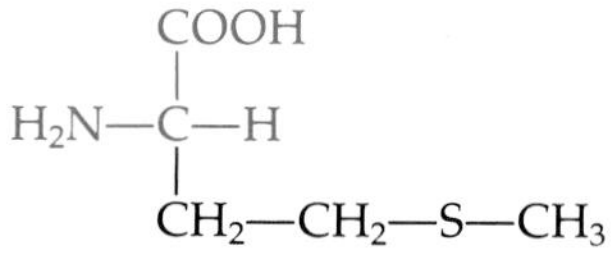

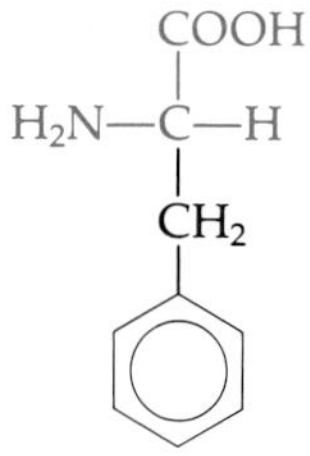

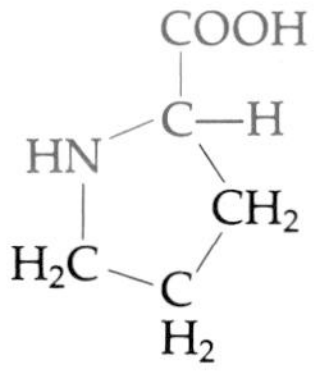

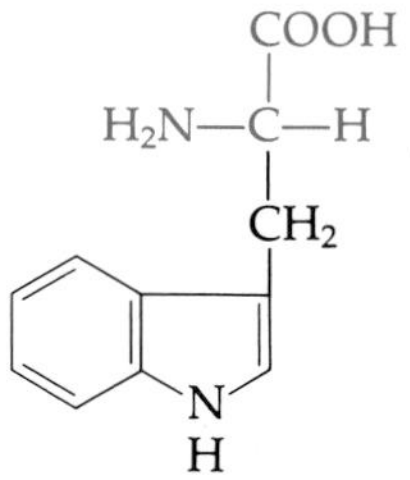

Table 23-2   (continued)

Structure	Name	IUPAC Abbreviation	Isoelectric Point
COOH | H_2N-C-H | CH H_3C CH_3	Valine	Val	5.96
Polar			
COOH | H_2N-C-H | $CH_2-CH_2-CH_2-NH-C$ NH NH_2	Arginine	Arg	10.76
COOH | H_2N-C-H | CH_2-COOH	Aspartic acid	Asp	2.97
COOH | H_2N-C-H | CH_2-C O NH_2	Asparagine	Asn	5.41
COOH | H_2N-C-H | CH_2 | SH	Cysteine	Cys	5.07
COOH | H_2N-C-H | CH_2-CH_2-COOH	Glutamic acid	Glu	3.22
COOH | H_2N-C-H | CH_2-CH_2-C O NH_2	Glutamine	Gln	5.65
COOH | H_2N-C-H | H	Glycine	Gly	5.97

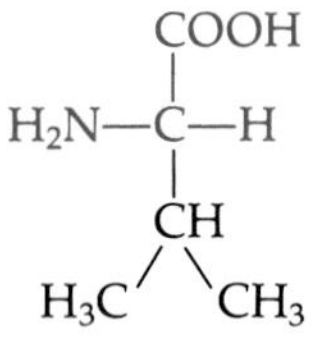

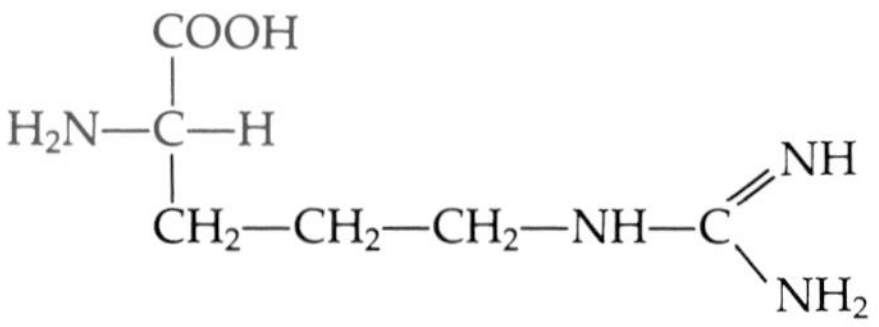

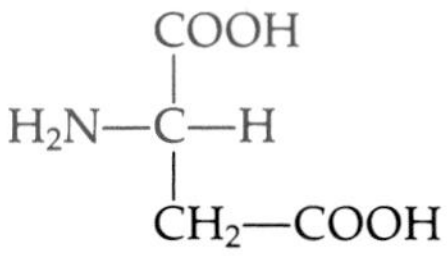

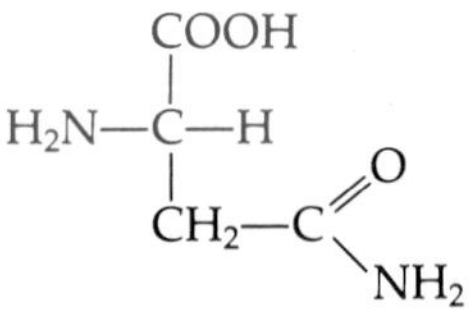

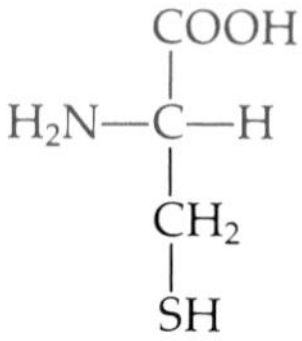

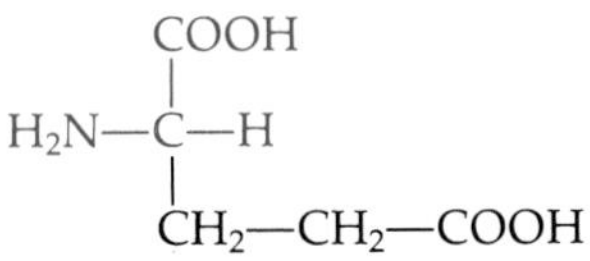

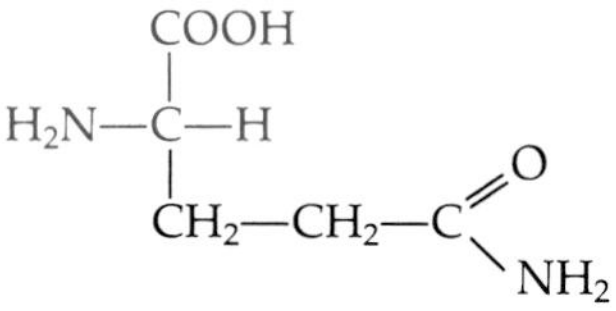

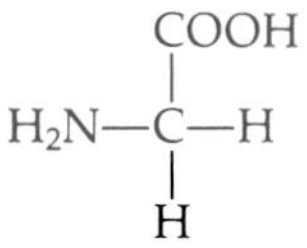

Table 23-2 (continued)

Structure	Name	IUPAC Abbreviation	Isoelectric Point
$\begin{array}{c} COOH \\ | \\ H_2N-C-H \\ | \\ CH_2 \\ | \\ C=CH \\ HN\diagdown\diagup N \\ C \\ H \end{array}$	Histidine	His	7.59
$\begin{array}{c} COOH \\ | \\ H_2N-C-H \\ | \\ CH_2-CH_2-CH_2-CH_2-NH_2 \end{array}$	Lysine	Lys	9.74
$\begin{array}{c} COOH \\ | \\ H_2N-C-H \\ | \\ CH_2OH \end{array}$	Serine	Ser	5.68
$\begin{array}{c} COOH \\ | \\ H_2N-C-H \\ | \\ H-C-OH \\ | \\ CH_3 \end{array}$	Threonine	Thr	5.60
$\begin{array}{c} COOH \\ | \\ H_2N-C-H \\ | \\ CH_2 \\ (\text{phenol ring}) \\ OH \end{array}$	Tyrosine	Tyr	5.66

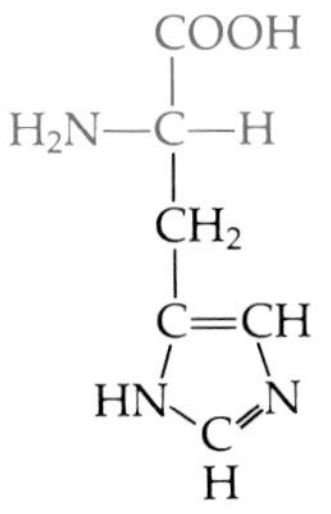

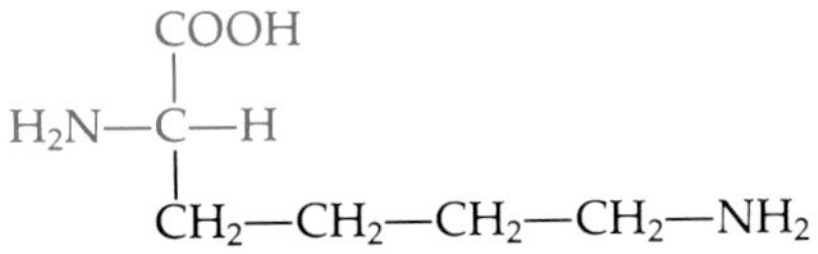

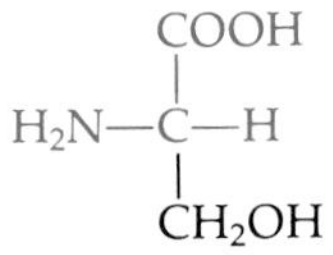

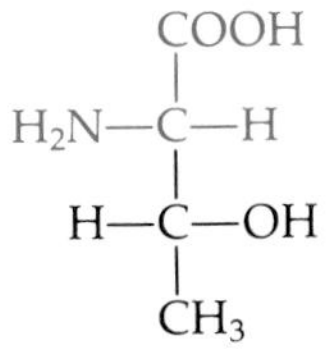

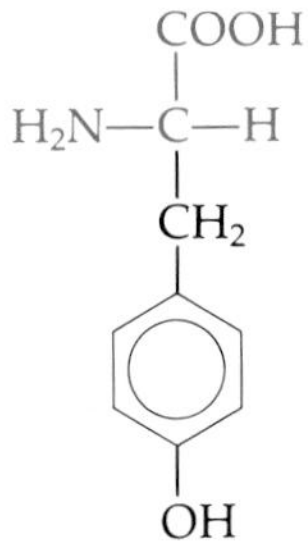

Table 23-3 Essential Amino Acids and Their Daily Requirements in Milligrams per Kilogram of Body Weight

	Adult	Infant		Adult	Infant
Isoleucine	14	83	Threonine	9	68
Leucine	17	135	Tryptophan	4	21
Lysine	14	99	Valine	13	92
Methionine	15	58	Arginine[a]	16	72
Phenylalanine	19	141	Histidine	—[b]	33

[a] Can be synthesized by the body but too slowly to be of practical use.
[b] Required primarily during infancy.

Table 23-4 Isoelectric Points of Some Proteins

Protein	Isoelectric Point (pH)
Egg albumin	4.7
Casein	4.6
Hemoglobin	6.7
Insulin	5.3
Serum globulin in blood	5.4
Fibrinogen in blood	5.6

Amphoteric Nature

Amino acids contain the —COOH group, which is acidic, and the —NH_2 group, which is basic (see page 311). In solution, the carboxyl group can donate a hydrogen ion to the amino group, forming a dipolar ion, called a **zwitterion**.

$$R\text{—}\underset{\underset{NH_2}{|}}{CH}\text{—}COOH \longrightarrow R\text{—}\underset{\underset{NH_3^+}{|}}{CH}\text{—}COO^-$$

amino acid　　　　　　　　zwitterion form of an amino acid

Amino acids are amphoteric compounds; that is, they can react with either acids or bases. When an amino acid is placed in a basic solution, it forms a negatively charged ion that will be attracted toward a positively charged electrode. In an acid solution, the amino acid forms a positively charged ion that will be attracted toward a negatively charged electrode.

$$R\text{—}\underset{\underset{NH_3^+}{|}}{CH}\text{—}COOH \underset{H^+}{\overset{}{\rightleftharpoons}} R\text{—}\underset{\underset{NH_3^+}{|}}{CH}\text{—}COO^- \overset{OH^-}{\rightleftharpoons} R\text{—}\underset{\underset{NH_2}{|}}{CH}\text{—}COO^-$$

positively charged ion　　　zwitterion　　　negatively charged ion
(in acid solution)　　　　　　　　　　　　　　(in basic solution)

Since amino acids are *amphoteric, proteins, which are made up of amino acids, are also amphoteric. This amphoteric nature of proteins accounts for their ability to act as buffers in the blood; they can react with either acids or bases to prevent an excess of either.

At a certain pH (that is, a certain hydrogen ion concentration) amino acids will not migrate toward either the positive or the negative electrode. At this pH, amino acids will be neutral; there will be an equal number of positive and negative ions. This point is called the *isoelectric point (see Table 23-2).

Proteins, which are composed of amino acids, also have an isoelectric point, which is different for each protein. At its isoelectric point, a protein has a minimum solubility, a minimum viscosity, and also a minimum osmotic pressure. At a pH above the isoelectric point, a protein has more negative than positive charges. At a pH below the isoelectric point, a protein has more positive than negative charges. The isoelectric points of a few proteins are listed in Table 23-4.

Dipeptides

Proteins consist of many amino acids joined together by what is called a **peptide linkage** or a **peptide bond** (see page 314). Suppose that a glycine molecule reacts with an alanine molecule. This reaction can occur in two different ways. The amine part of the glycine may react with the acid part of the alanine [equation (23-2)], or the acid part of the glycine may react with the amine part of the alanine [equation (23-3)].

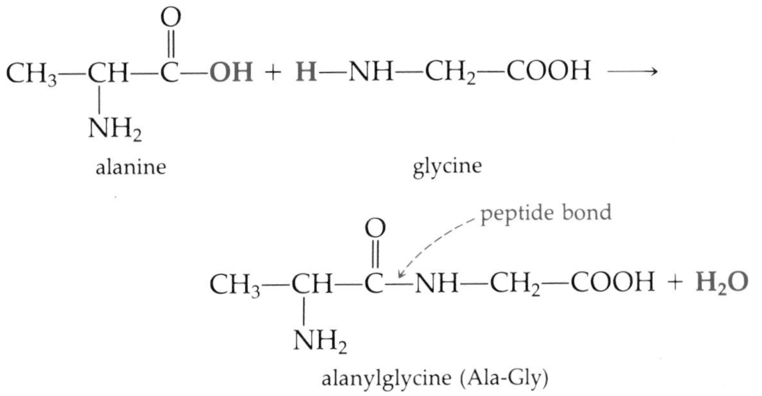

$$CH_3-CH-\overset{\overset{\displaystyle O}{\|}}{C}-OH + H-NH-CH_2-COOH \longrightarrow$$
$$\underset{NH_2}{|}$$

alanine   glycine

peptide bond

$$CH_3-CH-\overset{\overset{\displaystyle O}{\|}}{C}-NH-CH_2-COOH + H_2O \quad (23\text{-}2)$$
$$\underset{NH_2}{|}$$

alanylglycine (Ala-Gly)

$$NH_2-CH_2-\overset{\overset{\displaystyle O}{\|}}{C}-OH + HNH-CH-COOH \longrightarrow$$
$$\underset{CH_3}{|}$$

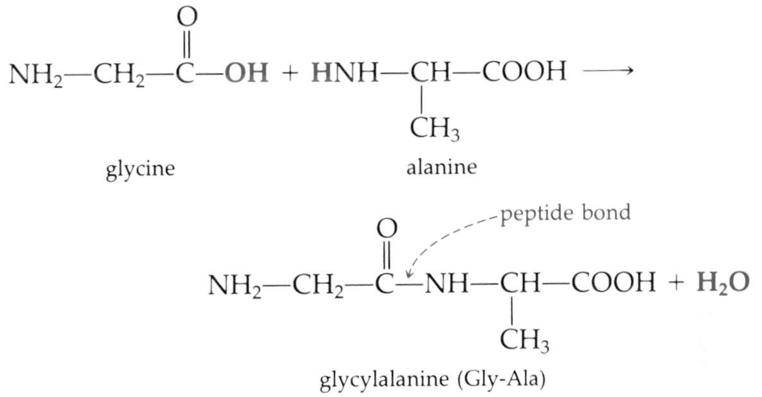

glycine   alanine

peptide bond

$$NH_2-CH_2-\overset{\overset{\displaystyle O}{\|}}{C}-NH-CH-COOH + H_2O \quad (23\text{-}3)$$
$$\underset{CH_3}{|}$$

glycylalanine (Gly-Ala)

When two amino acids combine, the product is called a **dipeptide.** When three amino acids combine, the product is called a **tripeptide.** When four or more amino acids join together, the product is called a **polypeptide.** For just two amino acids, glycine and alanine, two different combinations have already been indicated— glycylalanine and alanylglycine, where the first member of each group acts as the one furnishing the —OH from the acid group. For three different amino acids—such as glycine, alanine, and valine— there are six possible combinations (or tripeptide linkages).

1. Glycylalanylvaline (Gly-Ala-Val)
2. Glycylvalylalanine (Gly-Val-Ala)
3. Alanylglycylvaline (Ala-Gly-Val)
4. Alanylvalylglycine (Ala-Val-Gly)
5. Valylglycylalanine (Val-Gly-Ala)
6. Valylalanylglycine (Val-Ala-Gly)

By convention, peptides are written with the —NH$_2$ end (the N terminal) at the left and the —COOH (the C terminal) at the right.

Figure 23-2   Structure of human insulin. Note that insulin contains disulfide bridges between Cys groups. Breaking these disulfide bridges inactivates insulin.

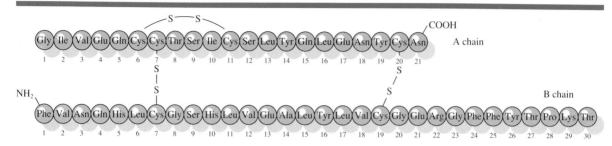

Proteins contain a large number of peptide linkages, and the number of possible combinations of the many amino acids in the formation of a protein is beyond all comprehension. Insulin (see Figure 23-2) illustrates the peptide linkages. It has an A chain, containing 21 amino acids, and a B chain, which contains 30 amino acids. Note that the two chains are connected by two disulfide bridges (see page 416).

Structure

Figure 23-3   The α-helical secondary structure of a protein.

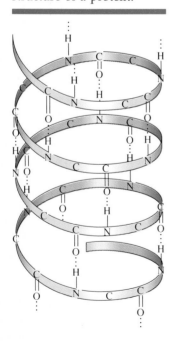

When a protein is hydrolyzed (by acids, bases, or certain enzymes), it breaks down into smaller and smaller units, eventually forming amino acids. Likewise, when amino acids combine (under the influence of certain enzymes), they first form dipeptides, then tripeptides, then polypeptides, and so on, until they eventually form a protein.

$$\text{protein} \underset{}{\overset{H_2O}{\rightleftharpoons}} \text{proteoses} \underset{}{\overset{H_2O}{\rightleftharpoons}} \text{peptones} \underset{}{\overset{H_2O}{\rightleftharpoons}} \text{polypeptides} \underset{}{\overset{H_2O}{\rightleftharpoons}}$$

$$\text{tripeptides} \underset{}{\overset{H_2O}{\rightleftharpoons}} \text{dipeptides} \underset{}{\overset{H_2O}{\rightleftharpoons}} \text{amino acids}$$

Proteins have a three-dimensional structure that can be considered as being composed of simpler structures.

The **primary structure** of a protein refers to the number and sequence of the amino acids in the protein. These amino acids are held together by peptide bonds. The primary structure of human insulin is indicated in Figure 23-2.

A slight change in the amino acid sequence can change the entire protein (see normal hemoglobin and sickle cell hemoglobin, page 659).

The **secondary structure** of a protein refers to the regular recurring arrangement of the amino acid chain. One such arrangement, called the α helix, occurs when the amino acids form a coil or spiral. The coil consists of loops of amino acids held together by hydrogen bonds [between the —H of the —NH$_2$ of one amino acid and the O of the C=O of the acid part of another amino acid (see Figure 23-3)].

Figure 23-4 A β-pleated
structure formed by hydrogen
bonds between two polypeptide
chains.

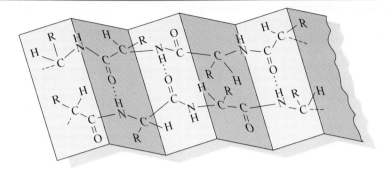

Each turn of the helix contains an average of 3.6 amino acids. Such a structure is both flexible and elastic. Hair and wool are examples of protein in such a helical structure.

When the amino acids are coiled, they can form either a right- or left-handed spiral. However, since α-amino acids in protein are all of the ʟ configuration, the coils always are right-handed.

A second type of secondary structure, the β pleated sheet (also called the pleated sheet) consists of parallel strands of polypeptides held together by hydrogen bonds. Such a structure (Figure 23-4) is flexible but not elastic. Silk has such a pleated sheet structure. It is strong, but resistant to stretching. This type of structure is less common than the α helix.

The **tertiary structure** of a protein refers to the specific folding and bending of the coils into specific layers or fibers (see Figure 23-5). It is the tertiary structure that gives proteins their specific biologic activity (see enzymes, page 436). Tertiary structures are stabilized by several types of bonds as indicated in Figure 23-6. Salt bridges (a) are formed between positively and negatively charged

Figure 23-5 Tertiary
structure of a protein. Note
that the coiled helix
represents the secondary
structure, which in turn is
made up of the various
amino acids in the
sequence specified by
mRNA (see page 654).

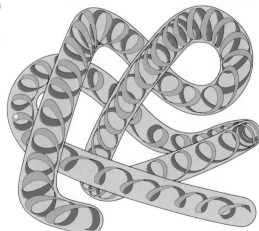

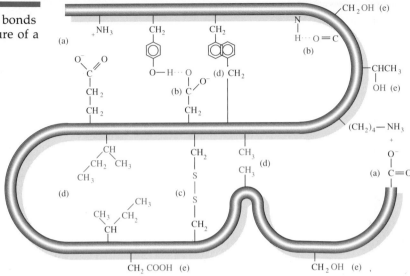

Figure 23-6 Various types of bonds that stabilize the tertiary structure of a protein.

groups within the protein molecule. Examples of such groups are the carboxyl and amino side chains found in glutamic acid, lysine, arginine, and aspartic acid. Hydrogen bonds (b) can form between different segments of the coil. Disulfide bonds (c) can form between cysteine groups in different parts of the coil. Hydrophobic bonds (d) can be formed. In general, nonpolar amino acids are folded on the "inside" of the protein, and polar amino acids are on the "outside" (e), where they can react with water molecules to form polar group interactions (also hydrogen bonds).

Some proteins have a **quaternary structure**, which occurs when

Figure 23-7 Quaternary structure of hemoglobin.

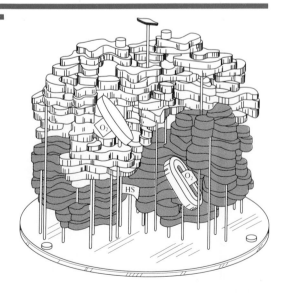

several protein units, each with its own primary, secondary, and tertiary structure, combine to form a more complex unit. An example of a protein with a quaternary structure is hemoglobin (Figure 23-7). It consists of two identical α chains (light-colored) and two identical β chains (dark-colored). Each chain enfolds a heme (iron-containing) group. The oxygen-binding site is marked O_2.

Percent Composition

The average percentage of nitrogen present in protein is 16 percent; that is, about one-sixth of protein is nitrogen. Because protein is the major food that contains nitrogen, the chemist can determine the amount of protein present in a food substance by determining the amount of nitrogen present. This amount is about one-sixth of the amount of protein present. Therefore, the amount of protein in the food can be calculated by multiplying the weight of nitrogen by 6 and converting this to a percentage of the total. For example, suppose that a 100-g sample of food yielded 4 g of nitrogen upon chemical analysis. Since the amount of nitrogen in protein is one-sixth of the total amount of protein present, the amount of protein present is 6 × 4 g, or 24 g. Then the percentage of protein present in the original 100-g is 24 percent.

Classification

Proteins are divided into three categories—simple, conjugated, and derived. On hydrolysis, simple proteins yield only amino acids or derivatives of amino acids. On hydrolysis, conjugated proteins yield amino acids plus some other type of compound. Conjugated protein consists of a simple protein combined with a nonprotein compound. Derived proteins are produced by the action of chemical, enzymatic, and physical forces on the other two classes of protein. Derived proteins include proteoses, peptones, polypeptides, tripeptides, and dipeptides. They also can be hydrolyzed to amino acids.

Proteins are classified according to their composition, according to their function, or according to their shape.

Classification According to Composition

Simple proteins are classified according to their solubility in various solvents and also as to whether they are coagulated by heat (see Table 23-5). Conjugated proteins are classified according to the nature of the nonprotein portion of the molecule (see Table 23-6).

Glycoproteins
Glycoproteins are proteins containing carbohydrates in varying amounts. Glycoproteins have molecular weights from 15,000 to over 1 million. The carbohydrates present in glycoproteins include the

Table 23-5 Properties of Simple Proteins

Type of Protein	Solubility	Coagulated by Heat	Examples
Albumins	Soluble in water, precipitated by saturated salt solution	Yes	Egg albumin; serum albumin; lactalbumin
Globulins	Soluble in dilute salt solution, insoluble in water and moderately concentrated salt solution	Yes	Serum globulin; lactoglobulin; vegetable globulin
Albuminoids	Insoluble in all neutral solvents and in dilute acid and alkali	No	Keratin in hair, nails, feathers; collagen
Histones	Soluble in water and very dilute acid; insoluble in very dilute NH_4OH	No	Nucleohistone in thymus gland; globin in hemoglobin

Table 23-6 Conjugated Proteins

Type	Prosthetic Group (Nonprotein Portion of the Combination)	Examples
Nucleoproteins	Nucleic acid	Chromosomes
Glycoproteins	Carbohydrates	Mucin in saliva
Phosphoproteins	Phosphate	Casein in milk
Chromoproteins	Chromophore group (color-producing group)	Hemoglobin, hemocyanin, flavoproteins, cytochrome
Lipoproteins	Lipids	Fibrin in blood
Metalloproteins	Metals	Ceruloplasmin (containing Cu) and siderophilin (containing Fe) in blood plasma

hexoses mannose and galactose, the pentoses arabinose and xylose, and sialic acids, which are derivatives of neuraminic acid.

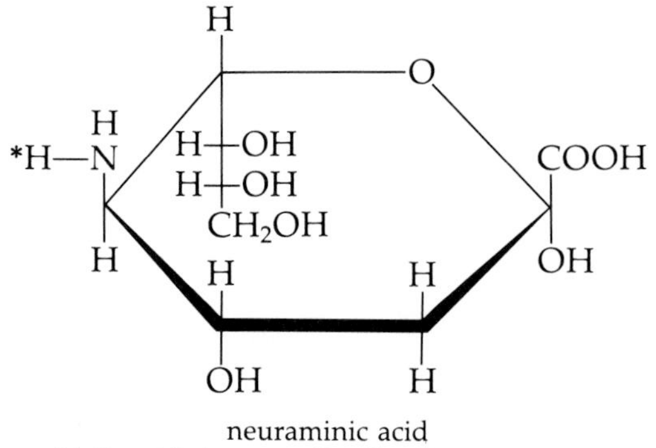

neuraminic acid
(sialic acids have an R—C=O group
in place of the H marked with the asterisk)

Glucose is not found in glycoproteins, except for collagen.

Glycoproteins are present in most organisms including animals, plants, bacteria, viruses, and fungi. Human cell membranes are about 5 percent carbohydrate present as glycoproteins and glyco-lipids. Glycophorin is a glycoprotein found in the membranes of human *erythrocytes.

In addition to their functions in membranes, glycoproteins serve in the following ways: as structural proteins (collagen); as lubricants (mucin and mucous secretions); as transportation molecules for vita-mins, lipids, minerals, and trace elements; as immunoglobulins such as interferon; as hormones such as thyrotropin (TSH); as enzymes such as the hydrolases and nucleases; as hormone receptor sites; and for the specification of human blood types.

Lipoproteins

Lipoproteins, which are proteins containing lipids, are part of cell membranes (see page 392).

Plasma Lipoproteins Lipids such as cholesterol and triglycerides are not soluble in water and thus need to be complexed to a water-soluble carrier protein (lipoprotein). Plasma lipoproteins consist of a neutral lipid core of triglyceride and cholesterol ester that is sur-rounded and stabilized by free cholesterol, protein, and phospho-lipid. The relative proportions of nonpolar lipid, protein, and polar lipid determine the density, size, and charge of the resulting lipoproteins. The density of lipoproteins has been used to classify them, as indicated in Table 23-7.

Chylomicrons are produced in the intestinal mucosa and are used to transport dietary lipids into the blood plasma via the thoracic lymph duct. They are removed from the plasma with a half-life of 5 to 15 minutes. They are responsible for the creamed-tomato-soup appearance of blood following a meal containing fats.

Table 23-7 Lipoproteins

Type	Density (g/mL)	Components				
				Cholesterol		
		Protein	Triglycerides	Free	Ester	Phospholipids
Chylomicrons	Less than 0.95	1	85–95	1–2	1–2	3–6
Very low density lipoproteins (VLDL)	0.95–1.006	10	50–60	4–8	10	15–20
Intermediate density lipoproteins (IDL)	1.006–1.019	16	35	7	25	17
Low density lipoproteins (LDL)	1.019–1.063	22	10	10	38	20
High density lipoproteins (HDL)	1.063–1.21	45–50	3	5	15–20	25–30

Very low density lipoproteins (VLDL) transport triglycerides synthesized by the liver to the other parts of the body. Their breakdown leads to the production of the transient intermediate-density lipoproteins (IDL) and the end product low-density lipoprotein (LDL). LDL provides cholesterol for cellular needs. LDL is thought to promote coronary heart disease by first penetrating the coronary artery wall and then depositing cholesterol to form atherosclerotic plaque.

High-density lipoproteins (HDL) are involved in the catabolism of other lipoproteins. They incorporate the cholesterol and phospholipid released by a lipoprotein. HDLs may also remove excess cholesterol from peripheral tissue.

Elevated LDL levels have been associated with increased risk of developing coronary artery disease, whereas elevated HDL levels appear to reduce the risk. Women have higher HDL levels than men (55 vs. 45 mg/100 mL), and this may account for women's lower rate of heart disease. Aerobic exercise increases HDL levels (marathon runners average 65 mg/100 mL). Moderate alcohol consumption has also been shown to increase HDL levels.

Classification According to Function

Proteins can also be classified according to their biologic function. Table 23-8 lists the various categories of proteins classified by this method.

Classification According to Shape

Proteins can also be classified according to their shape and dimensions. Globular proteins consist of polypeptides folded into the shape of a "ball." They have a length-to-width ratio of less than 10. Globular proteins are soluble in water or form colloidal dispersions and have an active function. Proteins classified as globular are hemoglobin, albumin, and the globulins.

Table 23-8 Proteins Classified According to Function

Type of Protein	Example	Use
Structural	Collagen	In structure of connective tissue
	Keratin	In structure of hair and nails
Contractile	Myosin, actin	In muscle contraction
Storage	Ferritin	In storage of iron needed to make hemoglobin
Transport	Hemoglobin	In carrying oxygen
	Serum albumin	In carrying fatty acids
Hormones	Insulin	In metabolism of carbohydrates
Enzymes	Pepsin	In digestion of protein
Protective	Gamma globulin	In antibody formation
	Fibrinogen	In blood clotting
Toxins	Venoms	Poisonous

Fibrous proteins consist of parallel polypeptide chains that are coiled and stretched out. They have a length-to-width ratio greater than 10. Fibrous proteins are insoluble in water. Examples include collagen, fibrin, and myosin.

Properties

Colloidal Nature

Proteins form colloidal dispersions in water. Being colloidal, protein will pass through a filter paper but not through a membrane. The inability of protein to pass through a membrane is of great importance in the body. Proteins present in the bloodstream cannot pass through the cell membranes and should remain in the bloodstream. Since proteins cannot pass through membranes, there should be no protein material present in the urine. The presence of protein in the urine indicates damage to the membranes in the kidneys—possibly nephritis.

Denaturation

Denaturation of a protein refers to the unfolding and rearrangement of the secondary and tertiary structures of a protein without breaking the peptide bonds (see Figure 23-8). A protein that is denatured loses its biologic activity. When the conditions for denaturation are mild, the protein can be restored to its original conformation by

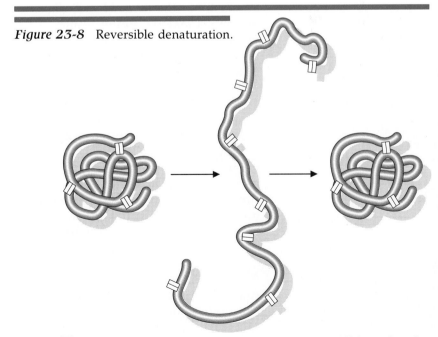

Figure 23-8 Reversible denaturation.

▭ = Areas of forces (—S—S—, hydrogen bonding, ionic, etc.) stabilizing conformation carefully reversing the conditions that caused the denaturation. This is called **reversible denaturation.** If the conditions that caused the denaturation are drastic, the process is irreversible; the protein will coagulate or precipitate from solution. Proteins can be denatured by a variety of agents, as indicated in the following section.

Reagents or Conditions that Cause Denaturation

Alcohol

Alcohol coagulates (precipitates) all types of protein except prolamines. Alcohol (70 percent) is used as a disinfectant because of its ability to coagulate the protein present in bacteria (see page 284). Alcohol denatures protein by forming hydrogen bonds that compete with the naturally occurring hydrogen bonds in the protein. Such a process is not reversible.

Salts of Heavy Metals

Heavy metal salts, such as mercuric chloride (bichloride of mercury) or silver nitrate (lunar caustic), precipitate protein. These denature protein irreversibly by disrupting the salt bridges and the disulfide bonds present in the protein. They are very poisonous if taken internally because they coagulate and destroy protein present in the body. The antidote for mercuric chloride or silver nitrate when these poisons are taken internally is egg white. The heavy metal salts react with the egg white and precipitate out. (The egg white colloid has a charge opposite to that of the heavy metal ion and so attracts it.) The precipitate thus formed must be removed from the stomach by an emetic or the stomach will digest the egg white and return the poisonous material to the system.

Dilute silver nitrate solution is used as a disinfectant in the eyes of newborn infants. Stronger solutions of silver nitrate are used to cauterize fissures and destroy excessive granulation tissues.

Heat

Gentle heating causes reversible denaturation of protein whereas vigorous heating denatures protein irreversibly by disrupting several types of bonds. Egg white, a substance containing a high percentage of protein, coagulates on heating. Heat coagulates and destroys protein present in bacteria. Hence sterilization of instruments and clothing for use in operating rooms requires the use of high temperatures. The presence of protein in the urine can be determined by heating a sample of urine, which will cause the coagulation of any protein material that is present.

Alkaloidal Reagents

Alkaloidal reagents, such as tannic acid and picric acid, form insoluble compounds with proteins. Alkaloidal reagents denature protein irreversibly by disrupting salt bridges and hydrogen bonds.

Tannic acid has been used extensively in the treatment of burns. When this substance is applied to a burn area, it causes the protein to precipitate as a tough covering, thus reducing the amount of water loss from the area. It also reduces exposure to air. Newer drugs have taken the place of tannic acid for burns, but an old-fashioned remedy still in use for emergencies involves the use of wet tea bags (which contain tannic acid).

Radiation

Ultraviolet or X-rays can cause protein to coagulate. The radiation denatures irreversibly by disrupting the hydrogen bonds and the hydrophobic bonds present in the protein. In the human body the skin absorbs and stops ultraviolet rays from the sun so they do not reach the inner cells. Proteins in cancer cells are more susceptible to radiation than those present in normal cells, so X irradiation is used to destroy cancerous tissue (see page 64).

pH

Changes in pH can disrupt hydrogen bonds and salt bridges, causing irreversible denaturation. Proteins are coagulated by such strong acids as concentrated hydrochloric, sulfuric, and nitric acids. Casein is precipitated from milk as a curd when it comes into contact with the hydrochloric acid of the stomach. Heller's ring test is used to detect the presence of albumin in urine. A layer of concentrated nitric acid is carefully placed under a sample of urine in a test tube. If albumin is present, it will precipitate out as a white ring at the interface of the two liquids. If acid or base remains in contact with protein for a long period of time, the peptide bonds will break.

Oxidizing and Reducing Agents

Oxidizing agents such as bleach and nitric acid and reducing agents such as sulfites and oxalates denature protein irreversibly by disrupting disulfide bonds (see page 416).

Salting Out

Most proteins are insoluble in saturated salt solutions and precipitate out unchanged. To separate a protein from a mixture of other substances, the mixture is placed in a saturated salt solution [such as $NaCl$, Na_2SO_4, or $(NH_4)_2SO_4$]. The protein precipitates out and is removed by filtration. The protein can then be purified from the remaining salt by the process of dialysis (see page 208).

Color Tests

Color tests for the presence of proteins depend on the presence of certain amino acids in that protein. It may be necessary to try several tests before deciding whether a substance is a protein or not.

Xanthoproteic Test

The word xanthoproteic means yellow protein. The test consists of adding concentrated nitric acid to a protein. The protein will then turn yellow and precipitate. Anyone who has spilled nitric acid on his or her hands will recall the yellow color produced by the reaction of the nitric acid with the protein of the skin. The xanthoproteic test works only for a protein that consists of amino acids containing a benzene ring, such as tyrosine, phenylalanine, or tryptophan.

Biuret Test

If a protein suspension is made alkaline with sodium hydroxide solution and copper(II) sulfate solution is added, a violet color is produced. This test is positive for substances that contain two or more peptide linkages—that is, such substances as proteins or polypeptides. It is negative for amino acids, which do not contain a peptide linkage, and for dipeptides, which contain only one peptide linkage.

Millon's Test

Millon's reagent consists of mercury dissolved in nitric acid (forming a mixture of mercuric and mercurous nitrates). When Millon's reagent is added to a protein, a white precipitate forms. This white precipitate on heating turns brick red. Millon's test is specific for the amino acid tyrosine.

Hopkins—Cole Test

In the Hopkins–Cole test, a protein is mixed with glyoxalic acid, CHOCOOH, and then carefully placed over a layer of concentrated sulfuric acid in a test tube. If the amino acid tryptophan is present, a purple color will appear at the area of contact of the two liquids.

Ninhydrin Test

When proteins are boiled with ninhydrin (a benzene-type compound), a blue purple color and CO_2 are produced. This test indicates the presence of α-amino acids or peptide groups.

Chromatography

Chromatography is a method used to separate very complex mixtures of amino acids, proteins, or lipids, It may be used for very small samples or for components present in extremely low concentrations. There are several types of chromatography. Among them are column, paper, thin-layer, and gas–liquid chromatography.

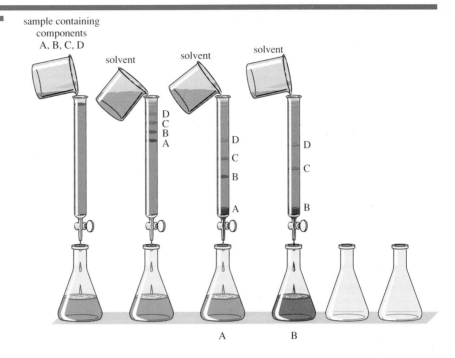

Figure 23-9 Column chromatography for a mixture of four components.

Column Chromatography

In column chromatography a glass tube is packed with a chemically inert material that tends to adsorb solids on its surface. A solution containing a sample is poured into the column, where it is held by packing material at the top (see Figure 23-9). A solvent in which the components of the mixture have different solubilities is poured into the column, gradually flushing the components of the mixture down the column. The component of the mixture that is most soluble (or least strongly adsorbed by the inert material) moves down the column most rapidly, whereas the component that is least soluble in the solvent (or most strongly adsorbed by the inert material) moves downward the most slowly. As more solvent is added, the components move downward with greater and greater separation, until they can be collected individually.

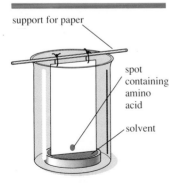

Figure 23-10 Paper chromatography.

Paper Chromatography

In paper chromatography, a piece of filter paper is used in place of the adsorbent. The paper is suspended vertically in a stoppered container. A drop of unknown is placed near the bottom of the paper, and solvent is added until its level is near the unknown (see Figure 23-10). As the solvent rises through the paper by capillary action, the components of the mixture are gradually separated. Different amino acids move at different rates and can be identified by

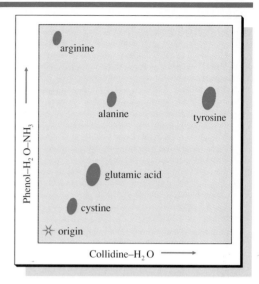

Figure 23-11 Two-dimensional paper chromatography.

calculations of ratios of distances moved. If, after the separation of a mixture into its components, the paper is dried, turned at right angles, and then placed in another solvent, a two-dimensional chromatogram is obtained. In this method, there is a further separation of the amino acids (see Figure 23-11).

Thin-Layer Chromatography

Thin-layer chromatography is very similar to paper chromatography, except that a glass plate covered with a thin layer of adsorbent is used in place of the filter paper. The advantages of this method over paper chromatography are the rapidity of separation and the choice of adsorbents, which permits separations not possible with filter paper.

A special adaptation of thin-layer chromatography is called *zone electrophoresis*. In this method a small amount of blood serum is applied to a cellulose acetate strip and an electric current is applied. The protein bands are separated and can be visualized after staining. The results of electrophoresis are shown on page 548.

Gas–Liquid Chromatography

Gas–liquid chromatography is probably the most widely used of all chromatographic methods. In this system, a column is packed with an inert material coated with a thin layer of a high-boiling, nonvolatile liquid. A very small sample (0.001 to 0.010 mL) is injected into the top of the column. The column is heated, and an inert carrier such as helium is passed through it. The mixture separates gradually as the more volatile components vaporize first. These more volatile components move faster than the others and are carried out of the column first. As each component of the mixture is eluted from the column, a detector produces an electric signal, which is proportional to the amount of the substance being detected. This electric signal is used to produce a tracing of concentration versus time. Such a trace is known as a chromatogram.

Summary

Proteins are high molecular weight compounds containing the elements carbon, hydrogen, oxygen, and nitrogen. Some proteins also contain other elements. Proteins are the chief constituents of all cells of the body.

Animals cannot synthesize protein from raw materials. They must obtain their protein from plants or from other animals, which in turn have obtained the protein from plants.

Protein serves to build new cells, to maintain existing cells, and to replace old cells in the body. Protein is necessary for the formation of the various enzymes and hormones in the body. The oxidation of protein yields 4 kcal/g.

Proteins are polymers of amino acids. Hydrolysis of protein yields amino acids with an amine group attached to the α carbon. The body can synthesize very few amino acids. Those that it needs and cannot synthesize are called essential amino acids. All the amino acids in the body, except glycine, have the L configuration. Amino acids in which the R group is primarily hydrocarbon are nonpolar and insoluble in water. Amino acids with polar R groups are polar compounds and soluble in water.

Amino acids are amphoteric—they react with either acids or bases because they contain an acid group (—COOH) and a basic group (—NH_2).

When an amino acid is placed in an acid solution, it forms a positive ion and migrates toward the negative electrode. When an amino acid is placed in a basic solution it forms a negative ion and migrates toward the positive electrode. At a certain pH, the isoelectric point, the amino acid will be neutral; it will not migrate toward either electrode.

When two amino acids combine, a dipeptide is formed. When four or more amino acids combine, a polypeptide is formed. Polypeptides in turn form peptones, then proteoses, and finally protein. The hydrolysis of protein proceeds through the same types of compounds in reverse order, forming proteoses, peptones, polypeptides, dipeptides, and amino acids.

The primary structure of a protein refers to the number and sequence of the amino acids in the protein chain. The secondary structure refers to the regular recurring arrangement of the amino acid chain into a coil or pleated sheet. The tertiary structure refers to the specific folding and bending of the coils into specific layers or fibers. The quaternary structure of a protein occurs when several protein units combine to form a more complex unit.

In general, the nitrogen content of protein is 16 percent.

Proteins can be classified according to composition, function, or shape.

Denaturation of a protein refers to the unfolding and rearrangement of the secondary and tertiary structures of a protein.

Proteins do not dissolve in water; rather they form colloidal dispersions. Proteins, being colloids, cannot pass through membranes and should not normally be present in the urine.

Proteins can be coagulated (precipitated) by means of alcohol, concentrated salt solutions, salts of heavy metals, heating, the use of alkaloidal reagents, concentrated inorganic acids, and X-rays.

Proteins give certain color tests based upon the presence of certain amino acids. Among these tests are the xanthoproteic test (for benzene ring proteins), the biuret test (for substances containing two or more peptide linkages), the Millon test (for the amino acid tyrosine), the Hopkins–Cole test (for the amino acid tryptophan), and the ninhydrin test (for α-amino acids).

Mixtures of amino acids, proteins, and lipids can be separated by chromatographic methods. Among the systems used are column, paper, thin layer, and gas–liquid chromatography.

Questions and Problems

A

1. All proteins contain which elements?
2. What additional elements are present in most proteins?
3. Where do plants obtain their protein?
4. Where do animals obtain their protein?
5. Describe the nitrogen cycle in nature.
6. Where did the word *protein* originate?
7. How does the energy value of protein compare with that of carbohydrate? with fat?
8. How does the molecular weight of a protein compare with that of other organic compounds?
9. What are amino acids? State the names and write the structure of three amino acids.
10. Name the essential amino acids.
11. Why are amino acids amphoteric? Why are they optically active? What configuration do they have?
12. What is meant by the term *isoelectric point* in relation to proteins?
13. What happens to the solubility of a protein at its isoelectric point?
14. What is a peptide linkage?
15. Write the reaction of glycine with valine in two different ways.
16. What is a polypeptide?
17. What amino acids are present in insulin?
18. Describe the types of structure of proteins.
19. When proteins are slowly hydrolyzed, what products are formed?
20. What percentage of protein is usually nitrogen? What use is made of this fact?
21. How are proteins classified?
22. List several types of simple protein.
23. List several types of conjugated protein.
24. Why should protein not normally be found in the urine?
25. Why is 70 percent alcohol a better disinfectant than 100 percent alcohol?
26. What is meant by the term *salting out* of a protein?
27. What is the antidote for $HgCl_2$ poisoning? Why must the stomach be pumped afterward?
28. What use is made of the fact that heat coagulates protein?
29. Name five color tests for protein. What reagent does each use? What is each test for?
30. Describe the separation of amino acids by column chromatography.
31. How does paper chromatography work? What is meant by the term *two-dimensional paper chromatography*?
32. What is thin-layer chromatography? What are its advantages over paper chromatography?
33. Describe gas–liquid chromatography.

B

34. Why are proteins important in the body?
35. Where are D-amino acids found? What function do they have?
36. Compare polar and nonpolar amino acids in terms of structure and solubility in water.
37. What is a zwitterion?
38. What is an α helix? What holds it together? Give two examples of such a structure.
39. Why are coils of proteins right-handed?
40. Compare an α helical structure with a β-pleated structure.
41. What types of bonding stabilize the tertiary structure of proteins?
42. What is a globular protein? fibrous protein? How do they differ?
43. Give an example of a protein that performs the indicated function: storage, transportation, enzyme, hormone.
44. Compare reversible and irreversible denaturation.

Practice Test

1. Amino acids can exist in several different forms. Which of the following represents the zwitterion form?

a. R—CH—COOH   b. R—CH—COO$^-$
      | |
      NH$_3^+$ NH$_3^+$

c. R—CH—COOH   d. R—CH—COO$^-$
      | |
      NH$_2$ NH$_2$

2. How many different peptide combinations are possible for three different amino acids?
   a. 1   b. 4   c. 6   d. 10

3. Which of the following is (are) involved in the tertiary structure of proteins?
   a. hydrogen bonds   b. disulfide bonds
   c. salt bridges   d. all of these

4. Which of the following will *not* denature a protein?
   a. alcohol   b. water
   c. acid   d. heat

5. High levels of _____ are associated with a decreased risk of coronary artery disease.
   a. VLDL   b. LDL
   c. chylomicrons   d. HDL

6. An example of an amino acid is _____.
   a. glycine
   b. benzene
   c. *p*-aminobenzoic acid
   d. toluene

7. Which structure refers to the number and sequence of amino acids?
   a. primary   b. secondary
   c. tertiary   d. quaternary

8. The average percent of nitrogen in protein is _____.
   a. 10   b. 16   c. 27   d. 38

9. Protein forms which type of liquid mixture in water?
   a. solution   b. suspension
   c. colloid   d. emulsion

10. A nonessential amino acid is _____.
    a. leucine   b. threonine
    c. valine   d. glutamine

24

Enzymes

Photomicrograph of mitochondria in heart muscle.

Enzymes are biologic catalysts. Catalysts are substances that increase the speed of a chemical reaction. Although a catalyst influences a chemical reaction, it is not itself permanently changed nor does it cause the reaction to occur; that is, a catalyst can increase the speed of a reaction but cannot cause that reaction if it would not occur in the absence of that catalyst. Since catalysts are not used up, they can be used over and over again.

Enzymes are organic catalysts produced by living organisms. Each enzyme will affect only specific substances called *substrates. Enzymes provide a chemical pathway that has a lower *activation energy than the same reaction uncatalyzed. How do enzymes differ from nonbiologic catalysts and why are there so many in the body? Enzymes are superior to other catalysts in several ways.

1. They have a much greater catalytic power. Consider the reaction

$$CO_2 + H_2O \xrightleftharpoons{\text{carbonic anhydrase}} H_2CO_3$$

which takes place in red blood cells (see page 552). The enzyme carbonic anhydrase increases the reaction rate over 10 million times that of the same reaction without the presence of the enzyme. In general, enzymes increase the rate of reaction from over 1 million to over 1 trillion times faster than a corresponding reaction without an enzyme. Few catalysts can cause such an increase in reaction rate.

2. Enzymes act on specific substrates and can distinguish between molecules that have very similar structures. Some enzymes are specific for one particular compound and will not catalyze the reaction of closely related compounds. Other enzymes will catalyze reactions of compounds that have a common structure; that is, they can catalyze a whole series of reactions. For example, an enzyme might catalyze the reaction of D-glyceraldehyde and have no effect on the reaction of L-glyceraldehyde. That is, the enzyme is specific for one compound only. An example of an enzyme with a more general action is trypsin, which catalyzes the hydrolysis of the peptide bonds in many different proteins.

3. The activity of enzymes is closely regulated whereas that of catalysts is difficult to control. There are substances in cells that can increase or decrease the activity of an enzyme and thus control the rate of the particular reaction that a cell requires.

Enzyme Reactions

Enzymes are proteins[†] and therefore undergo all the reactions that proteins do. That is, enzymes can be coagulated by heat, alcohol,

---

† Recent discoveries show that ribonucleic acid (RNA) (see page 646) can act as an enzyme, but such enzymatic reactions are beyond the scope of this book.

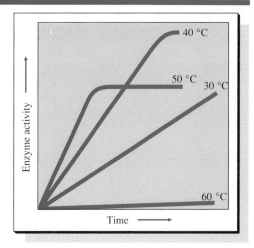

Figure 24-1 Effect of temperature on enzyme activity.

strong acids, and alkaloidal reagents. Many enzymes have now been prepared in crystalline form.

Temperature Requirement

The speed of all chemical reactions is affected by temperature: the higher the temperature, the faster the rate of the reaction (see Figure 24-1). This is also true for reactions involving enzymes. However, if the temperature is raised too much, the enzyme (protein) will be inactivated by denaturation and will be unable to function.

The best temperature for enzyme function—the temperature at which the rate of a reaction involving an enzyme is the greatest—is called the **optimum temperature** for that particular enzyme. At higher temperatures, the enzyme will coagulate and will be unable to function. At temperatures below the optimum temperature, the rate of reaction will be slower than the maximum rate.

Because many enzymes have an optimum temperature near 40 °C, or close to that of body temperature, they function at maximum efficiency in the body.

Role of pH

Each enzyme has a pH range within which it can best function (see Figure 24-2). This is called the **optimum pH range** for that particular enzyme. For example, the optimum pH of pepsin, an enzyme found in gastric juice, is approximately 2, whereas the optimum pH of trypsin, an enzyme found in pancreatic juice, is near 8.2. If the pH of a substrate is too far from the optimum pH required by the enzyme, that enzyme cannot function at all. However, since body fluids contain buffers, the pH usually does not vary too far from the optimum values.

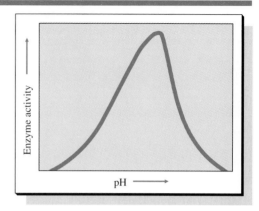

Figure 24-2 Effect of pH on enzyme activity.

Effects of Concentrations

As with all chemical reactions, the speed is increased with an increase in concentration of reactants. With an increased concentration of substrate, the rate of the reaction will increase until the available enzyme becomes saturated with substrate. Also with an increase in the amount of enzyme, the rate of reaction will increase, assuming an unlimited supply of substrate.

Activators and Inhibitors

Inorganic substances that tend to increase the activity of an enzyme are called **activators.** For example, the magnesium ion (Mg^{2+}) is an inorganic activator for the enzyme phosphatase, and the zinc ion (Zn^{2+}) is an activator for the enzyme carbonic anhydrase (see page 552).

An enzyme **inhibitor** is any substance that will make an enzyme less active or render it inactive. Enzyme inhibitors that bind to the active site and so block access by the substrate (see page 437) are called competitive inhibitors. Other inhibitors that bind to another site on the enzyme to render it less active or inactive are called noncompetitive inhibitors. They act by changing the conformation of the enzyme, thereby reducing or stopping its activity.

Heat, changes in pH, strong acids, alcohol, and alkaloidal reagents can all denature protein. These are examples of **nonspecific inhibitors;** they affect all enzymes in the same manner. **Specific inhibitors** affect one single enzyme or group of enzymes. In this category are most poisonous substances, such as cyanide ion (CN^-), which inhibits the activity of the enzyme cytochrome oxidase (see page 485).

Poisons

Many enzyme inhibitors are poisonous because of their effect on enzyme activity. Mercury and lead compounds are poisonous because they react with sulfhydryl (—SH) groups of an enzyme and so

Figure 24-3 Mercury and lead compounds are noncompetitive inhibitors, reacting with an enzyme to change its conformation.

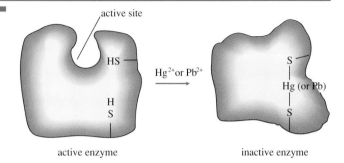

change its conformation (see Figure 24-3). The subsequent loss of enzyme activity leads to the various symptoms of lead and mercury poisoning, such as loss of equilibrium, hearing, sight, and touch, which are generally irreversible.

Organic phosphorus compounds are frequently poisonous because they act as competitive inhibitors for cholinesterase, a neurotransmitter found in the brain. Such organic phosphorus compounds are called nerve poisons. Among them are the insecticides malathion and parathion and various war gases.

Once a neurotransmitter is blocked, the receptor sites in the brain "fire" repeatedly. This overstimulates the muscles. The heart beats rapidly and irregularly; convulsions occur, and death ensues rapidly. However, there are antidotes for several organic phosphorus poisons. They function by displacing the poisonous substance from the enzyme.

Drugs

While some enzyme inhibitors are poisonous, others, such as the sulfa drugs (see page 330), are beneficial to life.

Penicillin acts as an enzyme inhibitor for transpeptidase, a substance that bacteria need to build their cell walls (see page 446). If the cell wall is lacking, osmotic pressure causes the bacterial cell to burst and die. However, new strains of bacteria have developed an enzyme, penicillinase, that inactivates penicillin. To destroy these new strains, synthetically modified penicillins have been prepared so that this antibiotic remains effective.

Cyclic AMP (cAMP) acts as a chemical messenger to regulate enzyme activity within the cells that store carbohydrate and fat. Without cAMP the activity of all the enzymes working at maximum speed within the cells would soon create chaos. It also appears that an inadequate supply of cAMP can lead to one type of the uncontrolled cell growth that we call cancer.

Mode of Enzyme Activity

How do enzymes act? Why are they so specific toward certain substrates?

Figure 24-4
(a) Representation of an active site in an enzyme.
(b) Denatured enzyme—parts of the active site are no longer in close proximity.

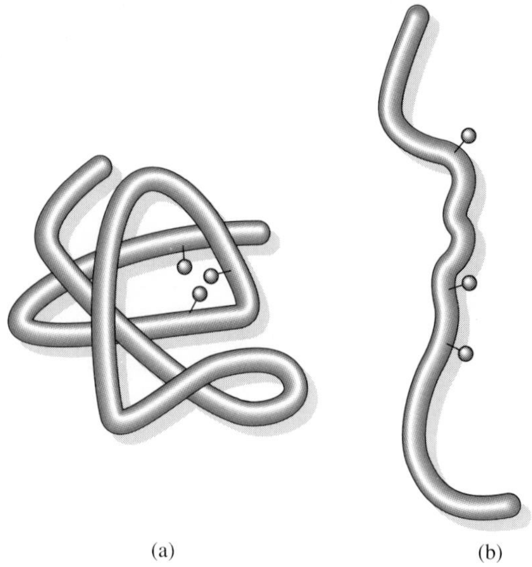

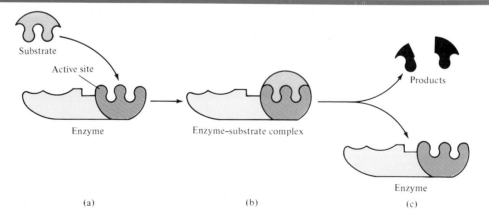

(a)   (b)

Each enzyme contains an "active site"—that section of the molecule at which combination with the substrate takes place. The active site consists of different parts of the protein chain (the enzyme). These parts are brought close together by the folding and bending of the protein chain (the secondary and tertiary structures), so that the active site occupies a relatively small area. The fact that enzymes (proteins) can be denatured by heat (changed in three-dimensional configuration) indicates the importance of structural arrangement (see Figure 24-4).

Figure 24-5 Schematic representation of the interaction of enzyme and substrate. (a) The active site on the enzyme and the substrate have complementary structures and hence fit together as a key fits a lock. (b) While they are bonded together in the enzyme–substrate complex, the catalytic reaction occurs. (c) The products of the reaction leave the surface of the enzyme, freeing the enzyme to combine with another molecule of substrate.

Substrate

Active site

Products

Enzyme

Enzyme–substrate complex

Enzyme

(a)   (b)   (c)

Figure 24-6 Enzyme–inhibitor complex.

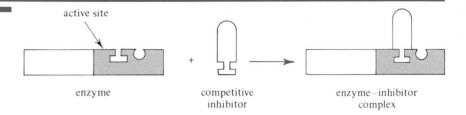

active site

enzyme   +   competitive   enzyme–inhibitor
inhibitor   complex

It is believed that enzyme activity occurs in two steps. First, the active site of the enzyme combines with the substrate to form an enzyme–substrate complex (see Figure 24-5b). This enzyme–substrate complex then breaks up to form the products and the free enzyme, which can react again (Figure 24-5c). According to this theory (the lock-and-key model), the substrate must "fit" into the active site of the enzyme—hence the specificity of that enzyme. For substrates that have an optically active site, it is believed that there must be three points of attachment between the substrate and the active site of the enzyme so that only one of the two optical isomers "fits." In the body, enzymes are specific for the L-amino acids and the D-carbohydrates.

A more recent version of the activity of an enzyme, the induced-fit model, suggests that the active site is not rigid, as in the lock-and-key model, but flexible. That is, the site changes in conformation upon binding to a substrate in order to yield an enzyme–substrate fit.

If some other substance should fit into the active site of the enzyme, it could prevent that enzyme from reacting with the substrate. Such a substance is called a **competitive inhibitor** (see Figure 24-6).

An example of a competitive inhibitor is sulfanilamide. Its structure is similar to that of *p*-aminobenzoic acid (PABA), which certain bacteria require for growth (see page 330). The sulfa inhibits the growth of the bacteria by competing for the active site on the enzyme.

Another type of inhibition is called **noncompetitive inhibition.** As the name implies, a noncompetitive inhibitor does not compete for the active site (see page 449).

Apoenzymes and Coenzymes

It was mentioned previously that most enzymes are proteins. However, some enzymes are simple proteins—they yield only amino acids on hydrolysis. Examples of such enzymes are pepsin and trypsin (page 460). Other enzymes are conjugated proteins—they contain a protein and a nonprotein part. Both parts must be present before the enzyme can function. The protein part is called the *apoenzyme and the nonprotein (organic) part is called the *coenzyme.

Coenzymes

Coenzymes are not proteins and so are not inactivated by heat. Examples of coenzymes are the vitamins or compounds derived from vitamins.

The reaction involving a coenzyme can be written as follows

$$\text{coenzyme} + \text{apoenzyme} \longrightarrow \text{enzyme}$$

Coenzyme A [CoA, structure (24-1)] is essential in the metabolism of carbohydrates, lipids, and proteins in the body. It also functions in certain acetylation reactions. Hydrolysis of CoA yields pantothenic acid (a B vitamin), adenine (a purine), ribose (a sugar), phosphoric acid, and mercaptoethanolamine (a sulfur compound).

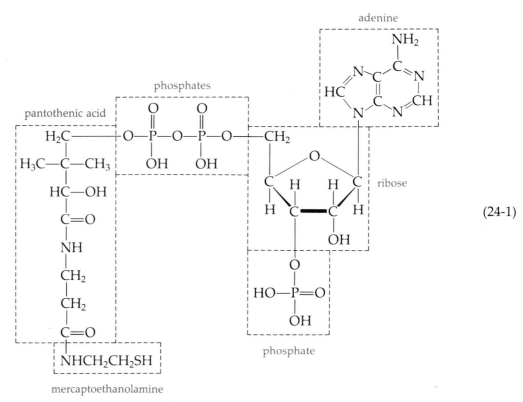

(24-1)

coenzyme A

Acetyl coenzyme A (acetyl CoA) has an acetyl group attached to the sulfur atom. Acetyl CoA functions in the oxidation of food in the Krebs (critic acid) cycle (see Chapter 26).

Nicotinamide is a very important constituent of two coenzymes—nicotinamide adenine dinucleotide (NAD^+) and nicotinamide adenine dinucleotide phosphate ($NADP^+$). These coenzymes are involved in most oxidation–reduction reactions in the mitochondria. They also take part in the Krebs cycle. The structure of NAD^+ is indicated in formula (24-2).

When NAD^+ is reduced to NADH, another hydrogen atom is attached to the carbon atom indicated by an asterik in structure (24-2). The charge on the nitrogen atom is eliminated, and the double bond between that nitrogen and the marked carbon is changed to a single bond. Recent evidence indicates that a defect in the enzyme responsible for the conversion of NADH to NAD^+, NADH dehydrogenase, may be involved in cystic fibrosis.

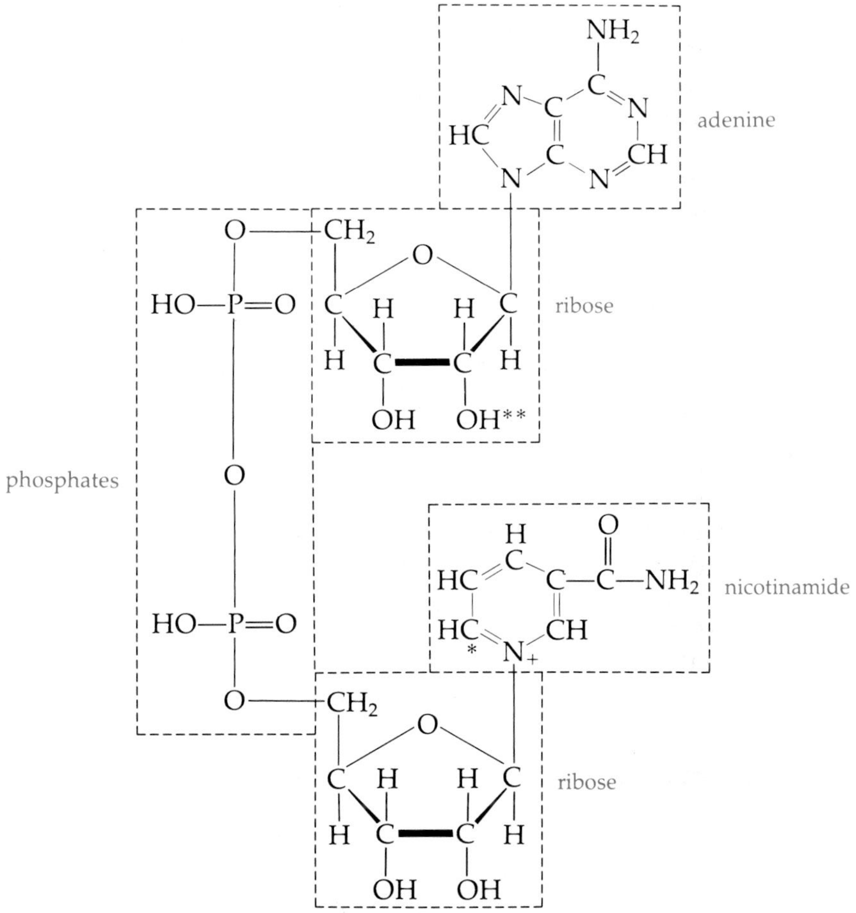

(24-2)

nicotinamide adenine dinucleotide (NAD^+)

In $NADP^+$, another phosphate group has been added in the position marked with a double asterisk in NAD^+ [structure (24-2)], so that $NADP^+$ contains three phosphate groups.

Coenzyme Q is found in the *mitochondria and has a structure similar to those of vitamin K and vitamin E (see pages 600 and 602). The structure of coenzyme Q is

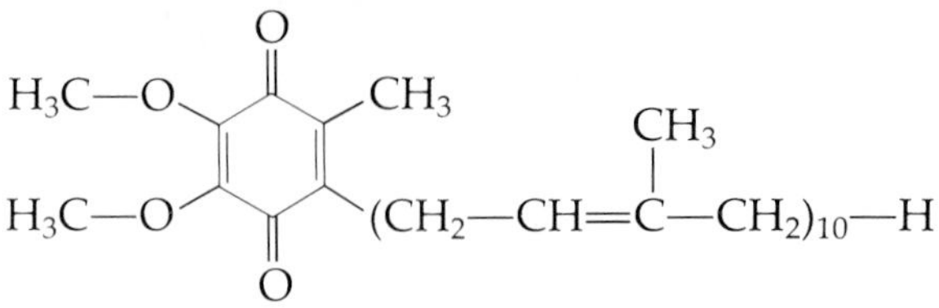

$$H_3C-O \quad CH_3$$

$$H_3C-O \quad (CH_2-CH=\overset{\overset{\displaystyle CH_3}{|}}{C}-CH_2)_{10}-H$$

coenzyme Q

Coenzyme Q functions in electron transport and in oxidative phosphorylation.

Coenzymes frequently contain B vitamins as part of their structure. Many coenzymes involved in the metabolism of amino acids contain pyridoxine, vitamin B_6. Other B vitamins, such as riboflavin, pantothenic acid, nicotinamide, thiamine, and lipoic acid, are found in coenzymes involved in oxidation–reduction reactions. B vitamins such as folic acid and cobalamin (vitamin B_{12}) are part of different coenzymes.

Reactions involving enzymes and coenzymes usually are written as a series of interconnected equations. An example of such a series of reactions is diagram (24-3). In step 1 a substrate, S, is oxidized by NAD^+, which in turn is reduced to NADH. However, the NAD^+ must be regenerated so that it can act again. Therefore, in step 2, NADH is oxidized to NAD^+ by a flavoprotein, which in turn is reduced to flavoprotein H_2. In step 3, the flavoprotein is regenerated by the cytochrome in which Fe^{2+} is oxidized to Fe^{3+}. Finally, in step 4, the cytochrome is regenerated by oxygen, with the final product being water. Such a series of reactions is called a *chain*. Series (24-3) illustrates the steps in the respiratory chain (see page 471).

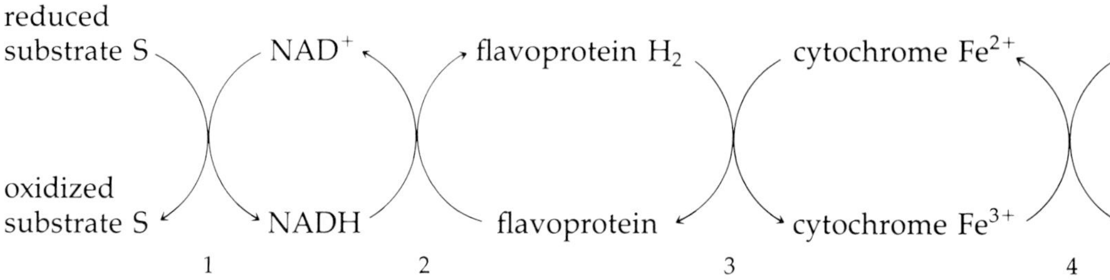

reduced
substrate S NAD^+ flavoprotein H_2 cytochrome Fe^{2+} H_2O

(24-3)

oxidized
substrate S   NADH   flavoprotein   cytochrome Fe^{3+} O_2

1   2   3   4

Nomenclature

Formerly enzymes were given names ending in *-in*, with no relation being indicated between the enzyme and the substance it affects— the substrate. Some of the enzymes named under this sytem are listed in Table 24-1.

The current system for naming enzymes uses the name of the substrate or the type of reaction involved, with the ending *-ase*. Table 24-2 lists some enzymes and substrates named under the preferred system.

Table 24-1 Enzymes Named Under the Older System

Enzyme	Substrate
Rennin	Casein
Pepsin	Protein
Trypsin	Protein
Ptyalin	Carbohydrate

Table 24-2 Enzymes and Substrates or Reaction Types

Enzyme	Substrate or Reaction Type
Maltase	Maltose
Urease	Urea
Proteases	Proteins
Carbohydrases	Carbohydrates
Lipases	Lipids
Hydrolases	Hydrolysis reactions
Deaminases	Removing amines
Dehydrogenases	Removing hydrogens

Classification

The Commission on Enzymes of the International Union of Biochemistry has classified enzymes into six divisions. Each of these divisions can be further subdivided into several classes. The following paragraphs indicate these divisions and some of the classes. The older names of the enzymes are still in common use and are given throughout this text merely for simplicity. It is much easier to write *sucrase* than *α-glucopyrano-β-fructofuranohydrolase*.

Oxidoreductases

Oxidoreductases are enzymes that catalyze oxidation–reduction reactions between two substrates. The enzymes that catalyze oxidation–reduction reactions in the body are important because these reactions are responsible for the production of heat and energy. Recall that oxidation–reduction requires a transfer of electrons. Many of these enzymes are present in the mitochondria, which are

Figure 24-7 A typical cell.

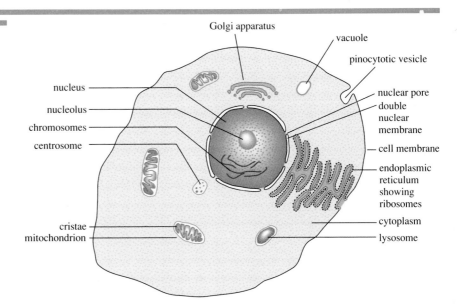

the most prominent structural and functional units in animal cells, except for the nucleus (see Figure 24-7). Mitochondria are sausage-shaped objects ranging in size from 0.2 to 5 μm (1 μm is one-millionth of a meter). The cytoplasm of a typical cell contains from 50 to 50,000 mitochondria.

Inside the mitochondrion is a double-layered membrane that is folded back and forth to form sacs called cristae. These cristae have a tremendous amount of surface area. On the surfaces of these membranes are enzyme-containing particles. Coenzymes are present in the fluid between the membranes. The mitochondria are highly organized and hold their enzymes in a definite spatial arrangement for the most efficient control of various cellular processes.

The mitochondria play an important part in oxidative phosphorylation in the Krebs cycle, which relates to carbohydrate metabolism (Chapter 26). Mitochondria contain enzymes needed for the oxidation of fatty acids. They also are involved in the metabolism of amino acids.

In addition to these functions, the mitochondria manufacture adenosine triphosphate (ATP), the main energy-supplying substance in the cell.

Dehydrogenases

Dehydrogenases catalyze the removal of hydrogen from a substrate.

Oxidases
Oxidases activate oxygen so that it will readily combine with a substrate.

Catalases
Catalases catalyze the decomposition of hydrogen peroxide to water and oxygen.

Peroxidases
Peroxidases catalyze the decomposition of organic peroxides to hydrogen peroxide and water.

Transferases

Transferases are enzymes that catalyze the transfer of a functional group between two substrates. Under this category are the kinases, which acid in the transfer of phosphate groups.

Hydrolases—The Hydrolytic Enzymes

The hydrolytic enzymes—hydrolases—catalyze the hydrolysis of carbohydrates, esters, and proteins. They are named for the substrate upon which they act. Some hydrolases are present in the cytoplasm in organelles called lysosomes.

Carbohydrases

Enzymes that catalyze the hydrolysis of carbohydrates into simple sugars are called carbohydrases. The various carbohydrases are

1. Ptyalin, or salivary amylase, for the hydrolysis of starch to dextrins and maltose.
2. Sucrase, for the hydrolysis of sucrose to glucose and fructose. Sucrase occurs in the intestinal juice.
3. Maltase, for the hydrolysis of maltose to glucose. Maltase occurs in the intestinal juice.
4. Lactase, for the hydrolysis of lactose to glucose and galactose. Lactase occurs in the intestinal juice.
5. Amylopsin, or pancreatic amylase, for the hydrolysis of starch to dextrin and maltose. Pancreatic amylase occurs in the pancreatic juice.

Esterases

Esterases are enzymes that catalyze the hydrolysis of esters into acids and alcohols. The various types of esterases are

1. Gastric lipase, for the hydrolysis of fats to fatty acids and glycerol.
2. Steapsin, or pancreatic lipase, for the hydrolysis of fats to fatty acids and glycerol.
3. Phosphatases, for the hydrolysis of phosphoric acid esters to phosphoric acid.

Proteases

Proteases are enzymes that catalyze the hydrolysis of protein to derived protein and amino acids. These are of two types: proteinases and peptidases.

The proteinases, for the hydrolysis of proteins to peptides, are as follows.

1. Pepsin, found in the gastric juice, for the hydrolysis of protein to polypeptides.
2. Trypsin, found in the pancreatic juice, for the hydrolysis of protein to polypeptides.
3. Chymotrypsin, found in the pancreatic juice, for the hydrolysis of protein to polypeptides.

The peptidases, for the hydrolysis of polypeptides to amino acids, are as follows.

1. Aminopeptidases, from the intestinal juices.
2. Carboxypeptidases, from the pancreatic juice.

Nucleases

Enzymes that catalyze the hydrolysis of nucleic acids are called nucleases. Examples are ribonuclease and deoxyribonuclease.

Lyases

Lyases are enzymes that catalyze the removal of groups from substrates by means other than hydrolysis, usually with the formation of double bonds. An example is fumarase, which catalyzes the change of fumaric acid to L-malic acid in the Krebs cycle (see page 486).

Isomerases

Isomerases are enzymes that catalyze the interconversion of optical, geometric, or structural isomers. One example is retinal isomerase, which catalyzes the conversion of 11-*trans*-retinal to 11-*cis*-retinal (see page 595). Another example is alanine racemase, which catalyzes the conversion of L-alanine to D-alanine, which is the form the body can use.

Ligases

Ligases are enzymes that catalyze the coupling of two compounds with the breaking of pyrophosphate bonds (see page 477). One example is the enzyme that catalyzes the formation of malonyl CoA during lipogenesis (see page 502).

Enzymes of the Kidneys

If the blood pressure drops, as in the case of hemorrhaging or in hypokalemia (page 562), the kidneys secrete the enzyme renin (sometimes considered a hormone) into the bloodstream. The following reactions occur.

$$\text{angiotensinogen} \xrightarrow{\text{renin}} \text{angiotensin I} \xrightarrow[\text{enzyme}]{\text{converting}} \text{angiotensin II}$$

Angiotensin II increases the force of the heartbeat and constricts the arterioles, thus causing an increase in blood pressure. Angiotensin II brings about the contraction of smooth muscle and also triggers the release of the hormone aldosterone (page 628), which aids in the retention of water. Actually, angiotensin II is the most powerful *vasoconstrictor known. It is an octapeptide; angiotensin I is a decapeptide.

Angiotensinogen is an α-globulin produced in the liver.

Various conditions can cause the kidneys to release renin. Among these are a decrease in blood pressure and increased amounts of prostaglandins and β-adrenergic agents. On the other hand, factors that tend to decrease the release of renin include vasopressin, angiotensin II, β-adrenergic antagonists, prostaglandin inhibitors, and an increase in blood pressure.

Chemotherapy

Chemotherapy is the use of chemicals to destroy infectious micro-organisms and cancerous cells without damaging the host's cells. These chemicals function by inhibiting certain cellular enzyme reactions. Among the chemotherapeutic agents are the antibiotics and the *antimetabolites.

Antibiotics

Antibiotics are compounds produced by one microorganism that are toxic to another microorganism. They function by inhibiting enzymes that are essential to bacterial growth. Among the most commonly used antibiotics are penicillin and tetracycline.

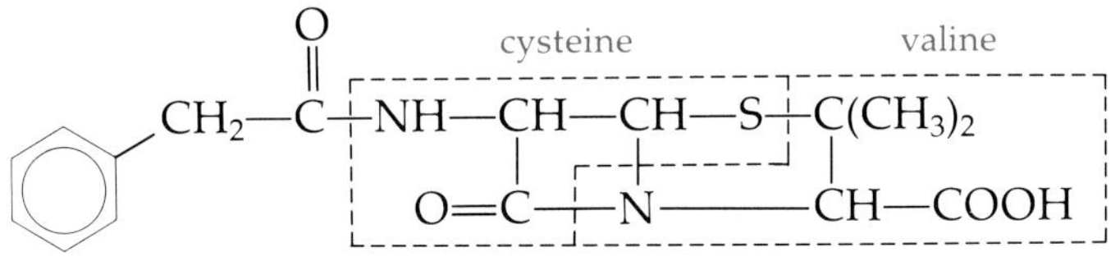

penicillin G

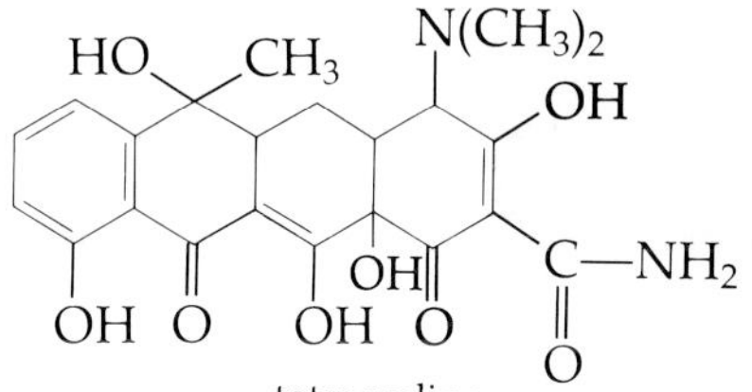

tetracycline

Various strains of penicillin contain different groups attached to the cysteine–valine combination. Recall that both of these substances are amino acids.

The antibacterial action of penicillins is a result of their inhibition of a final stage in bacterial cell wall synthesis. In these final stages long polysaccharide chains are cross-linked together by short peptide chains. The last step in this process is the formation of a

Figure 24-8 The final cross-linkage of polysaccharide chains in the bacterial cell wall.

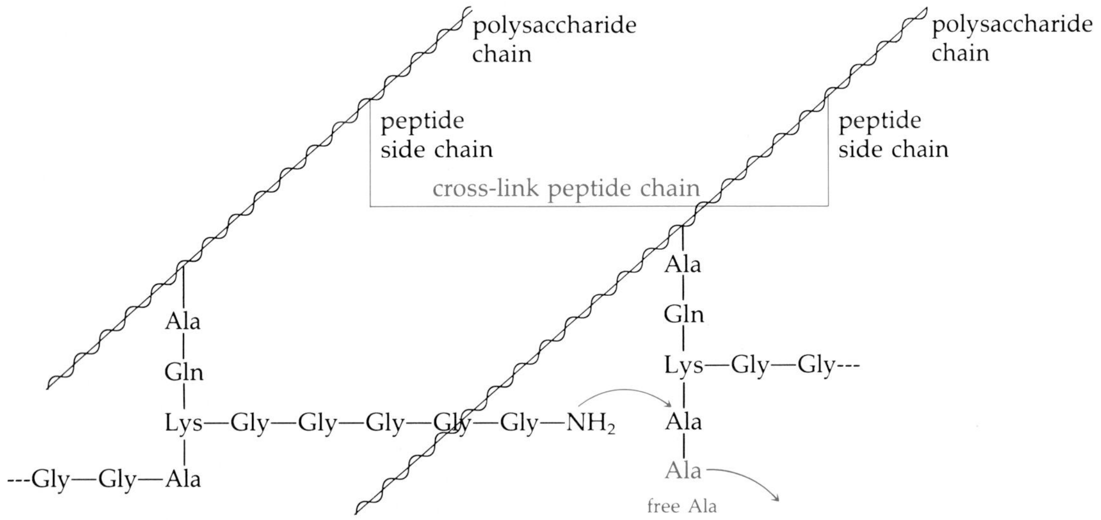

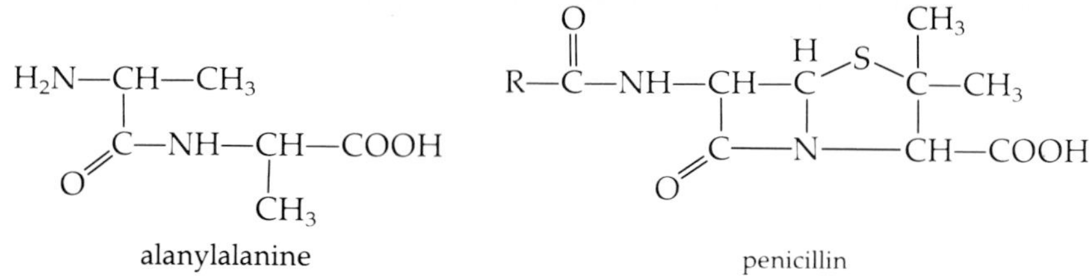

Figure 24-9 Comparison
of the structures of
alanylalanine
(D-alanyl-D-alanine and
penicillin.

alanylalanine   penicillin

Figure 24-9 Comparison of the structures of alanylalanine (D-alanyl-D-alanine and penicillin.

peptide bond between alanine on one peptide chain and glycine on another (see Figure 24-8). Penicillin inhibits this step, probably by irreversibly bonding to the active site of the enzyme. The similarity between alanylalanine (Ala-Ala) from the peptide chain and penicillin is shown in Figure 24-9. There is no human enzyme that uses alanylalanine (Ala-Ala) as a substrate, and penicillin therefore does not interfere with our own enzymatic machinery. Allergic sensitivity to penicillin is due to the breakdown products of penicillin, especially to 6-aminopenicillinic acid.

Bacterial resistance to penicillin is due to the secretion of an enzyme, penicillinase, which cleaves the ring amide bond.

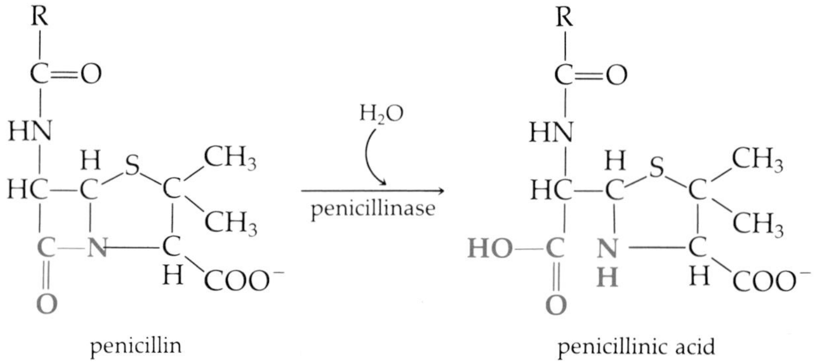

penicillin   penicillinic acid

This enzyme probably evolved as a detoxification mechanism, since its absence does not have any effect on the bacterial cell.

Antimetabolites

Antimetabolites are chemicals that have structures closely related to those of the substrates enzymes act on, thus inhibiting enzyme activity. One example of an antimetabolite is sulfanilamide (see page 330), whose structure is similar to that of *p*-aminobenzoic acid. Bacteria require folic acid as a coenzyme for their growth, and they synthesize it from *p*-aminobenzoic acid. Because their structure is similar to that of *p*-aminobenzoic acid, the sulfa drugs prevent the formation of folic acid and so inhibit the growth of the bacteria. This accounts for the use of sulfa drugs to fight bacterial infections.

Table 24-3 Enzymes of Diagnostic Value

Serum Enzyme	Major Diagnostic Use
Glutamic oxaloacetic transaminase (SGOT)	Myocardial infarction
Glutamic pyruvic transaminase (SGPT)	Infectious hepatitis
Trypsin	Acute pancreatic disease
Ceruloplasmin	Wilson's disease
Amylase	Liver and pancreatic disease
Acid phosphatase	Cancer of prostate
Alkaline phosphatase	Liver or bone disease
Creatine phosphokinase (CPK)	Myocardial infarction, muscle disorders
Lactate dehydrogenase (LDH)	Myocardial infarction, leukemia, anemia
Renin	Hypertension

Some *chemotherapeutic agents used in the treatment of cancer (*antineoplastic agents) are antimetabolites (mercaptopurine used in the treatment of leukemias); some are antibiotics (adriamycin used in the treatment of Hodgkin's disease); and others are alkylating agents, hormones, or natural products.

Clinical Significance of Plasma Enzyme Concentrations

The measurement of plasma enzyme levels can be of great diagnostic value. For example, the levels of glutamic pyruvic transaminase increase with infectious hepatitis; the levels of trypsin increase during acute disease of the pancreas; ceruloplasmin levels decrease during Wilson's disease (see page 566); and glutamic oxaloacetic transaminase levels rise rapidly after myocardial infarction (see page 514). Many other plasma enzymes are useful in the diagnosis of various diseases (see Table 24-3).

Isoenzymes

Isoenzymes, or isozymes, are enzymes with the same function but slightly different structural features. The reason for their existence is not known, but they are made use of clinically. Lactate dehydrogenase (LDH) (see page 479), creatine kinase, and alkaline phosphatase all occur in isoenzyme form and are of diagnostic value. LDH has five forms, and Table 24-4 lists some clinically useful diagnoses based on these isoenzymes. Different tissues will contain different relative amounts of each of these isoenzymes. This difference allows the determination of what tissues have been damaged.

Condition	Isoenzyme Pattern
Myocardial infarction	Moderate elevation of LDH_1; slight elevation of LDH_2
Acute hepatitis	Large elevation of LDH_5; moderate elevation of LDH_4
Muscular dystrophy	Elevation of LDH_1, LDH_2, LDH_3
Megaloblastic anemia	Large elevation of LDH_1
Sickle-cell anemia	Moderate elevation of LDH_1, LDH_2
Arthritis with joint effusions	Elevation of LDH_5

Table 24-4 Clinical Significance of Relative Amounts of LDH

Allosteric Regulation

Most of the body's biochemical processes take place in several steps, each one catalyzed by a particular enzyme, such as

$$A \xrightarrow{a_1} B \xrightarrow{b_1} C \xrightarrow{c_1} D \xrightarrow{d_1} E$$

where a_1, b_1, c_1, and d_1 are the enzymes needed to react with the given substrates. Thus B, the product of the first reaction, is in turn the substrate for the second enzyme reaction, and so on. The final product, E, may inhibit the activity of enzyme d_1. Such an inhibitor may be either competitive or noncompetitive (see page 437). Therefore, as the concentration of E increases, the activity of enzyme d_1 decreases and soon stops. Conversely, when the concentration of E is low, all the reactions proceed rapidly. Such a mechanism for regulating enzyme activity is called *feedback control*.

Enzymes whose activity can be changed by molecules other than those of the substrate are called **allosteric enzymes**. It is thought that the other molecules bind to the enzyme at sites other than the active site. This binding changes the three-dimensional structures of

Figure 24-10 An effector changes the shape of an enzyme so the substrate binds more readily.

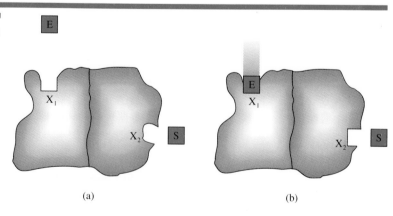

(a)   (b)

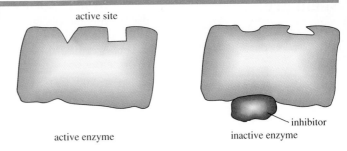

Figure 24-11 A noncompetitive inhibitor changes the shape of the enzyme so the substrate cannot react.

the enzymes. Some molecules increase the catalytic rate (effectors), and some decrease it (noncompetitive inhibitors). This control of key enzymes is of utmost importance to ensure that biologic processes remain coordinated at all times to meet the immediate metabolic needs of the cells. The key glycolytic enzyme phosphofructokinase (see page 479) is one such allosteric enzyme.

When an effector binds to a site on an enzyme, it changes the conformation of that enzyme not only at the active site but also at neighboring sections. Therefore, the substrate can bind more readily, and so the reaction rate is increased, as shown in Figure 24-10. In Figure 24-10a the active sites X_1 and X_2 are not readily available for substrate S. In Figure 24-11b an effector has changed the shape of the enzyme so that when one molecule of the substrate reacts, a second molecule can also react more easily.

A noncompetitive inhibitor binds to an enzyme, changing its shape so that the active site is less available to the substrate, thus decreasing enzyme activity. The inhibitor forms strong covalent bonds at another part of the enzyme, sometimes but not always near the active site, deforming the enzyme so that it cannot form an enzyme–substrate complex. Noncompetitive inhibitors do not need to have a structure similar to that of the substrate because they are not competing for the active site (see Figure 24-11).

Zymogens

Zymogens are inactive precursors of enzymes. Most digestive and blood-clotting enzymes exist in the zymogen form until activated. In the case of digestive enzymes, this is necessary to prevent digestion of pancreatic and gastric tissue. For blood clotting, it is to avoid premature formation of blood clots.

Zymogen	Active Form of Enzyme
pepsinogen $\rightarrow$	pepsin
trypsinogen $\rightarrow$	trypsin
prothrombin $\rightarrow$	thrombin

Summary

Enzymes are biologic catalysts that increase the speed of a chemical reaction but do not themselves change.

Most enzymes are proteins and will undergo all the reactions of proteins. The enzymes in the body function best at about 40 °C. Temperatures above or below body temperature will decrease the activity of enzymes.

Each enzyme has a certain pH at which it can function best.

An increase in the amount of enzyme will increase the rate of reaction. An increase in the amount of substrate will increase the rate of the reaction.

Inorganic compounds that increase the activity of an enzyme are called activators. Compounds that interfere with the activity of an enzyme are called inhibitors.

Enzymes contain an "active site" that binds to the substrate to form an enzyme–substrate complex. This complex yields the products and regenerates the enzyme.

Many enzymes contain two parts—a protein part and a nonprotein part. The protein part of an enzyme is called the apoenzyme.

Some enzymes require the presence of a substance called a coenzyme before they can act effectively. Coenzymes frequently contain the B vitamins or compounds derived from the B vitamins.

Under the older system of naming enzymes the substrate was not mentioned; the newer system indicates the substrate being acted upon. The names of enzymes under this system end in -*ase*.

Enzymes can be classified as oxidoreductases (enzymes that catalyze oxidation–reduction reactions between two substrates), transferases (which catalyze the transfer of a functional group between two substrates), hydrolases (which catalyze hydrolysis reactions), lyases (which catalyze the removal of groups from substrates by means other than hydrolysis), isomerases (which catalyze the interconversion of optical, geometric, or structural isomers), and ligases (which catalyze the coupling of two compounds with the breaking of pyrophosphate bonds).

Some hydrolytic enzymes are found in the lysosomes of the cytoplasm. The cytoplasm also contains mitochondria. These structural and functional units contain most of the oxidative enzymes and are deeply involved in the electron transport system of oxidation–reduction. The mitochondria also produce ATP, the cells' chief source of energy.

Chemotherapy is the use of chemicals to destroy infectious microorganisms and cancerous cells without damaging the host's cells. Among chemotherapeutic agents are antibiotics and antimetabolites.

Abnormal plasma enzyme concentrations are of clinical significance in the diagnosis of certain diseases.

Isoenzymes are enzymes with the same function but slightly different structural features.

Allosteric enzymes are key metabolic enzymes whose activity can be changed by molecules other than the substrate.

Zymogens are the precursors of enzymes.

Questions and Problems

A

1. What effect do enzymes have on the speed of a reaction?
2. Most enzymes are found to be what kind of compounds?
3. What is the effect of temperature upon an enzyme?
4. What is meant by the *optimum temperature* of an enzyme?
5. What is the effect of pH upon the activity of an enzyme?
6. What is the effect of concentration of enzyme upon the rate of the reaction?
7. What is the effect of concentration of substrate upon the rate of the reaction?
8. What is an apoenzyme? a coenzyme?
9. What is CoA used for? What products does it yield upon hydrolysis?
10. Compare the newer and older systems of naming enzymes.
11. What are the general classifications of enzymes?
12. What are carbohydrases? Name three. Indicate where they are found in the body and what substrate they act upon.
13. What are esterases? Name two. Where are they found? What substrate do they act upon?
14. Name two proteases. What substrate do they act upon? What products are formed?
15. What is a nuclease?
16. Name three types of oxidation–reduction enzymes.
17. What is a transferase? an isomerase?
18. Describe the appearance of the mitochondria. What functions do they serve?
19. What is coenzyme Q? What is its function?
20. How do enzymes function? Where do they get their specificity?
21. How do enzymes distinguish between optical isomers?
22. What is a competitive inhibitor? How does it work?
23. Distinguish between specific and nonspecific inhibitors.
24. What effect does denaturing have upon the "active site" of an enzyme? Why?
25. What is chemotherapy? Give an example of two chemotherapeutic agents.

B

26. How does an antimetabolite function?
27. How can plasma enzyme concentration be used to diagnose disease?
28. Discuss enzymes in terms of activation energy.
29. Compare competitive and noncompetitive inhibitors.
30. What is renin? Where is it found? Of what diagnostic value is it?
31. Compare the lock-and-key theory of enzymes with the induced-fit theory.
32. Of what clinical use is the determination of relative amounts of LDH?
33. What are allosteric enzymes? Why are they important?
34. What are zymogens? Why are they important?

Practice Test

1. Organic compounds of which element inhibit the enzyme cholinesterase?
   a. mercury   b. lead
   c. phosphorus   d. sulfur
2. Penicillin is used to treat _____.
   a. bacterial infection   b. virus disease
   c. cancer   d. all of these
3. Which enzyme would be most useful in determining whether a patient had a myocardial infarction?
   a. ceruloplasmin   b. SGOT
   c. renin   d. none of these
4. Most enzymes are _____.
   a. carbohydrates   d. lipids
   c. proteins   d. nucleic acids
5. Many _____ act as coenzymes.
   a. metals   b. phospholipids
   c. carbohydrates   d. vitamins
6. The optimum pH of the gastric juice is approximately _____.
   a. 2   b. 5   c. 7   d. 9
7. The protein part of an enzyme is called the _____.
   a. apoenzyme   b. coenzyme
   c. activator   d. inhibitor
8. Many oxidoreductases are found in the _____.
   a. cell nuclei   b. mitochondria
   c. cell membranes   d. cytoplasm
9. An example of a carbohydrase is _____.
   a. lipase   b. pepsin
   c. ligase   d. ptyalin
10. Inactive precursors of enzymes are called _____.
    a. allosteres   b. zymogens
    c. antimetabolites   d. antibiotics

25

Digestion

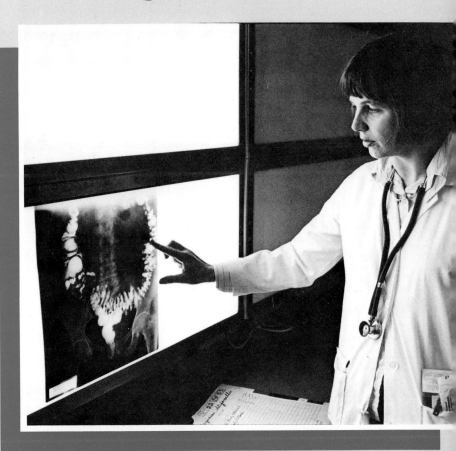

Radiologist pointing to an abnormality in the X-ray of the intestine.

Most foods (carbohydrates, fats, and proteins) are composed of large molecules that are usually not soluble in water. Before these foods can be absorbed through the alimentary canal, they must be broken down into smaller soluble molecules. Digestion is the process by which food molecules are broken down into simpler molecules that can be absorbed into the blood through the intestinal walls.

Digestion involves the use of hydrolases—the hydrolytic enzymes (Chapter 24). The hydrolases catalyze the hydrolysis of carbohydrates to monosaccharides, fats to fatty acids and glycerol, and proteins to amino acids.

However, not all foods require digestion. Monosaccharides are already in their simplest form and do not require digestion. Inorganic salts and vitamins also do not require digestion.

Digestion of food takes place in the mouth, the stomach, and the small intestine, each area having its own particular enzyme or enzymes that catalyze the hydrolytic reactions.

Salivary Digestion

During chewing, the food is mixed with saliva. The saliva moistens the food, so that swallowing is easier. Saliva is approximately 99.5 percent water. The remaining 0.5 percent consists of mucin, a glycoprotein that acts as a lubricant; several inorganic salts that act as buffers; and salivary amylase, an enzyme that catalyzes the hydrolysis of starches. Saliva also acts as an excretory fluid for certain drugs, such as morphine and alcohol, and for certain inorganic ions, such as K^+, Ca^{2+}, HCO_3^-, and SCN^-. Saliva has a pH range of 5.75 to 7.0, with an optimum pH of 6.6.

Gastric Digestion

When food is swallowed, it passes down the esophagus into the stomach, where it is mixed with the gastric juice. The gastric juice is secreted by glands in the walls of the stomach. When food enters the stomach, it causes the production of the hormone gastrin. Gastrin diffuses into the blood stream, which carries it back to the stomach, where it then stimulates the flow of gastric juice.

Approximately 2 to 3 L of gastric juice is secreted daily. Gastric juice is normally a clear, pale yellow liquid with a pH of 1 to 2. It is 97 to 99 percent water and up to 0.5 percent free hydrochloric acid. It is the presence of this hydrochloric acid that causes the gastric juice to have such a low pH. In certain pathologic conditions, the acidity of the stomach may be less than normal. Such a condition is known as *hypoacidity and is commonly associated with stomach cancer and pernicious anemia. Hyperacidity is a condition in which the stomach has too high an acid concentration. It is indicative of gastric ulcers, hypertension, or gastritis (inflammation of the stomach walls).

454

Hydrochloric acid in the stomach denatures protein so that the tertiary structure is lost. Thus the polypeptide chains unfold, making it easier for the proteases to act.

The low pH of the stomach also destroys most of the microorganisms that enter the body through the mouth.

The gastric juice contains the zymogen pepsinogen and the enzyme (gastric) lipase. In addition to these enzymes, a substance known as the intrinsic factor is secreted by the parietal cells in the walls of the stomach. Vitamin B_{12} must undergo a reaction with this intrinsic factor before it can be absorbed into the bloodstream. A lack of the intrinsic factor is associated with pernicious anemia.

Intestinal Digestion

The food in the stomach is very acidic. When this acid material enters the small intestine, it stimulates the mucosa to release the hormone secretin. Secretin, in turn, stimulates the pancreas to release the pancreatic juice into the small intestine.

Three different digestive juices enter the small intestine. These digestive juices are alkaline and neutralize the acid contents coming from the stomach. The three digestive juices entering the small intestine are (1) pancreatic juice, (2) intestinal juice, and (3) bile.

Pancreatic Juice

Pancreatic juice contains several enzymatic substances. Among these are trypsinogen, chymotrypsinogen, carboxypeptidase, pancreatic lipase, proelastase, and pancreatic amylase.

The pancreatic juice also contains cholesterol ester hydrolase, which acts on cholesterol esters and converts them into free cholesterol plus fatty acids; ribonuclease, which converts ribonucleic acids into nucleotides; deoxyribonuclease, which converts deoxyribonucleic acids into nucleotides; and phospholipase A_2, which acts on phospholipids and converts them into fatty acids and lysophospholipids.

Intestinal Juice

Intestinal juice contains several enzymes. Among these are aminopeptidase and dipeptidase. The intestinal juice also contains enzymes that catalyze the hydrolysis of phosphoglycerides, nucleoproteins, and organic phosphates. The intestinal mucosal cells contain the enzymes sucrase, maltase, and lactase.

Bile

Bile is produced in the liver and stored in the gallbladder. The gallbladder absorbs some of the water and other substances from the liver bile and so changes its composition slightly. When meat or fats enter the small intestine, they cause it to secrete a hormone, cholecystokinin (CCK). This hormone enters the bloodstream and is carried to the gallbladder, where it causes that organ to contract and empty into the duodenum through the bile duct.

Bile is a yellowish brown to green viscous liquid with a pH of 7.8 to 8.6. Because it is alkaline, it serves to neutralize the acid entering from the stomach. Primarily, bile contains bile salts, bile pigments, and cholesterol. Bile contains no digestive enzymes. Bile is the end product of cholesterol metabolism and is a primary determinant in cholesterol synthesis. Figure 25-1 illustrates the interrelation of cholesterol and bile.

Bile acts in the removal of many drugs and poisons from the body, in addition to removing such inorganic ions as Ca^{2+}, Zn^{2+}, and Hg^{2+}.

Bile Salts

Sodium glycocholate and sodium taurocholate are the two most important bile salts. They are both derived from cholic acid, a steroid similar to cholesterol in structure (see page 400).

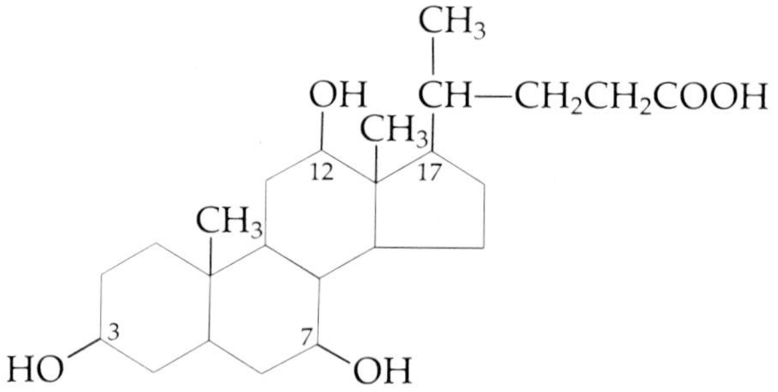

cholic acid

Bile salts have the ability to lower surface tension and increase surface area, thus aiding in the emulsification of fats. They also increase the effectiveness of pancreatic lipase (steapsin) in its digestive action on emulsified fats.

In addition, bile salts aid the absorption of fatty acids through the walls of the intestine. After absorption of these fatty acids, the bile salts are removed and carried back by portal circulation to the liver, where they are again returned to the bile. Bile salts also help stimulate intestinal motility.

Bile Pigments

The average red blood cells lasts about 120 days and is then destroyed. On the average, 100 million to 200 million red blood cells are destroyed per hour, a figure that roughly corresponds to $\frac{1}{5}$ ounce of hemoglobin daily. The hemoglobin is broken down into globin and heme. The body removes the iron from the heme and reuses it. The

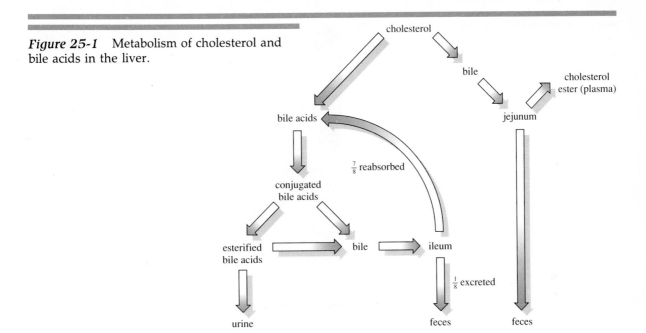

Figure 25-1 Metabolism of cholesterol and bile acids in the liver.

heme, with the iron removed, becomes biliverdin. Biliverdin is reduced in the reticuloendothelial cells of the liver, spleen, and bone marrow to form bilirubin, the main bile pigment excreted into the bile by the liver. In the intestines some bilirubin is converted to stercobilinogen and stercobilin, a pigment that gives the feces its characteristic yellow-brown color. Some bilirubin is absorbed into the bloodstream and comes to the liver where it is converted to urobilinogen and then to urobilin, which appears in the urine, giving that fluid its characteristic color. These reactions can be written as follows.

$$\text{hemoglobin} \longrightarrow \begin{array}{c} \text{heme} \\ + \\ \text{globin} \end{array} \longrightarrow \begin{array}{c} \text{biliverdin} \\ + \\ \text{iron} \end{array} \longrightarrow \text{bilirubin} \begin{array}{l} \nearrow \text{stercobilinogen} \longrightarrow \text{stercobilin} \\ \searrow \text{urobilinogen} \longrightarrow \text{urobilin} \end{array}$$

If the bile duct is blocked, the bile pigments remain in the bloodstream, producing jaundice. This disorder is recognizable by the yellow pigmentation of the skin. If the bile duct is blocked, no bile pigments can enter the intestine and the feces will appear clay-colored or nearly colorless.

Cholesterol

The body's excess cholesterol is excreted by the liver and carried to the small intestine in the bile. Sometimes the cholesterol precipitates in the gallbladder, producing gallstones. Figure 25-2 shows gallstones present in a gallbladder.

Figure 25-2 Granulated gallstones. [*Courtesy X-Ray Department, Michael Reese Hospital and Medical Center, Chicago, IL.*]

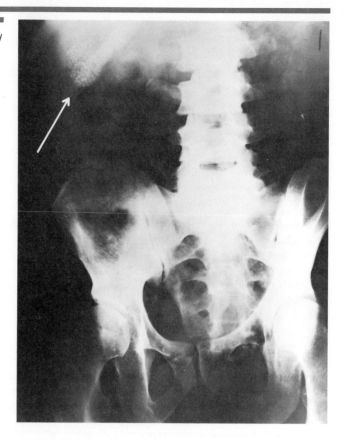

The body excretes about 1 g of cholesterol daily. Half is eliminated in the feces after conversion to bile acids. The other half is excreted as neutral steroids. However, much of the cholesterol secreted in the bile is reabsorbed.

Digestion of Carbohydrates

Digestion begins in the mouth, as chewing action reduces the size of the food particles. Thus they will have more surface area in contact with the digestive enzymes.

In the Mouth

Saliva contains salivary amylase (ptyalin), which catalyzes the hydrolysis of starch into maltose. However, this enzyme becomes inactive at a pH below 4 so that its activity ceases when it is mixed with the contents of the stomach, where the pH falls to about 1.5. Salivary amylase does not serve a very important function in digestion because the food does not remain in the mouth long enough for any appreciable hydrolysis to take place. Some hydrolysis of carbohydrates catalyzed by salivary amylase may take place in the stomach before the food is thoroughly mixed with the contents of the stomach, but this is of little importance because there are intestinal enzymes capable of hydrolyzing starch and maltose. The principal function of saliva is to lubricate and moisten the food so it can be easily swallowed.

In the Stomach

The stomach contains no carbohydrases, so no digestion of carbohydrates occurs there except for that catalyzed by salivary amylase. The activity of salivary amylase ceases as soon as it becomes mixed with the acid contents of the stomach.

In the Small Intestine

The major digestion of carbohydrates takes place in the small intestine through the action of enzymes in the pancreatic and intestinal juices.

The pancreatic juice contains the enzyme pancreatic amylase, which catalyzes the hydrolysis of starch and dextrins into maltose. The maltose thus produced is hydrolyzed to glucose through the activity of the enzyme maltase from the intestinal mucosal cells. The optimum pH of pancreatic amylase is 7.1. The intestinal mucosal cells also contain the enzymes sucrase and lactase, which catalyze the hydrolysis of sucrose and lactose, respectively.

$$(C_6H_{10}O_5)_n + H_2O \xrightarrow{\text{pancreatic amylase}} C_{12}H_{22}O_{11}$$
$$\text{starch} \qquad\qquad\qquad\qquad\qquad \text{maltose}$$

$$C_{12}H_{22}O_{11} + H_2O \xrightarrow{\text{maltase}} 2\ C_6H_{12}O_6$$
$$\text{maltose} \qquad\qquad\qquad\qquad \text{glucose}$$

If a monosaccharide such as glucose is eaten, digestion is not necessary because the monosaccharide is already in its simplest form and can easily undergo absorption into the bloodstream. Many adults cannot digest milk because they lack mucosal lactase. Such adults show milk intolerance with symptoms of abdominal cramps, bloating, and diarrhea.

Digestion of Fats

In the Mouth

A lingual lipase secreted by the dorsal surface of the tongue acts on triglycerides, particularly of the type found in milk. Lingual lipase has an optimum pH of 4.0 to 4.5 and a pH range of activity of 2.0 to 7.5, so it can continue its activity even at the low pH of the stomach.

In the Stomach

Although gastric lipase is present in the stomach, very little digestion of fats takes place because the pH of the stomach (1 to 2) is far below the optimum pH of that enzyme (7 to 8). Also, fats must be emulsified before they can be digested by lipase, and there is no mechanism for emulsification of fats in the stomach. However, if emulsified fats are eaten, a small amount of hydrolysis may take place in the stomach. In infants, whose stomach pH is higher, fat hydrolysis of milk may be of some importance.

In the Small Intestine

In the small intestine, the pancreatic lipase catalyzes the hydrolysis of fats into fatty acids and glycerol. This action is aided by the bile, which emulsifies the fats so that they can be acted upon readily by pancreatic lipase.

$$\text{fat + water} \xrightarrow{\text{pancreatic lipase}} \text{fatty acids + glycerol}$$

Digestion of Proteins

As the saliva contains no enzymes for the hydrolysis of protein, there is no digestion of protein in the mouth.

In the Stomach

The precursor enzyme pepsinogen is converted to pepsin when it is mixed with the hydrochloric acid of the stomach. Pepsin catalyzes the hydrolysis of protein to polypeptides.

$$\text{protein + water} \xrightarrow{\text{pepsin}} \text{polypeptides}$$

Rennin, an enzyme present in the gastric juice of calves, coagulates casein to form paracasein, which is precipitated by the calcium ions present in milk. Coagulated milk remains in the stomach longer than uncoagulated milk and so is more readily digested there. According to modern theories, rennin is not present in the gastric juice of humans.

In the Small Intestine

In the small intestine, the zymogen trypsinogen from the pancreatic juice is changed into trypsin by the intestinal enzyme enterokinase. Trypsin in turn changes chymotrypsinogen, another pancreatic zymogen, into chymotrypsin. Both trypsin and chymotrypsin catalyze the hydrolysis of protein, proteoses, and peptones to polypeptides. The optimum pH of trypsin and chymotrypsin is 8 or 9.

Table 25-1 Summary of Digestion

Type of Digestion	Location of Digestion	Digestive Juice and Enzymes	Substrate	Product
Salivary	Mouth	Saliva		
		Salivary amylase (ptyalin)	Starch	Dextrins
		Lingual lipase	Milk	Fatty acids + 1,2-diglycerides
Gastric	Stomach	Gastric juice		
		Hydrochloric acid	Pepsinogen	Pepsin
		Pepsin	Protein	Polypeptides
		Lipase	Fats	Fatty acids + glycerol
Intestinal	Small intestine	Intestinal juice		
		Enterokinase	Trypsinogen	Trypsin
		Aminopeptidase	Polypeptides	Amino acids
		Dipeptidase	Peptides	Amino acids
		Maltase	Maltose	Glucose
		Sucrase	Sucrose	Glucose + fructose
		Lactase	Lactose	Glucose + galactose
		Pancreatic juice		
		Trypsin	Protein	Polypeptides
		Chymotrypsin	Protein	Polypeptides
		Pancreatic amylase	Starch + dextrins	Maltose
		Pancreatic lipase	Fats	Fatty acids + glycerol
		Carboxypeptidase	Polypeptides	Amino acids
		Phosphatase	Organic phosphates	Free phosphates
		Polynucleotidases	Nucleic acids	Nucleotides
		Nucleosidases	Nucleosides	Purines, pyrimidines, phosphates, pentoses
		Elastase	Protein, polypeptides	Polypeptides, dipeptides
		Ribonuclease	Ribonucleic acid	Nucleotides
		Deoxyribonuclease	Deoxyribonucleic acid	Nucleotides
		Cholesterol ester hydrolases	Cholesterol esters	Cholesterol, fatty acids
		Phospholipase A_2	Phospholipids	Fatty acids, phosphates

The intestinal enzymes aminopeptidase and dipeptidase catalyze the hydrolysis of polypeptides and dipeptides into amino acids. Carboxypeptidase, an enzyme of the pancreatic juice, also catalyzes the hydrolysis of polypeptides to amino acids. Carboxypeptidase contains the element zinc.

Proelastase from the pancreatic juice is converted to elastase by trypsin. Elastase acts on protein and polypeptides to convert them into polypeptides and dipeptides, respectively.

Absorption of Carbohydrates

As we have seen, the principal digestion of carbohydrates takes place in the small intestine, where the polysaccharides and disaccharides are hydrolyzed into monosaccharides—glucose, fructose, and galactose. The monosaccharides are transported through the walls of the small intestine directly into the bloodstream by diffusion and by active transport of galactose and glucose. The blood carries the monosaccharides to the liver and then into the general circulation to all parts of the body. The monosaccharides can be oxidized to furnish heat and energy. Some of the monosaccharides are converted to glycogen, a polysaccharide, and stored in the liver or the muscles, and the rest are converted to fat and stored in the adipose tissue.

Absorption of Fats

The digestion of fats takes place primarily in the small intestine. The end products of digestion of fats—mono- and diglycerides, fatty acids, and glycerol—pass through the intestinal mucosa, where they are reconverted into triglycerides and phosphoglycerides, which then enter the lacteals, the lymph vessels in the villi in the walls of the small intestine. From the lacteals these products pass into the thoracic duct (a main lymph vessel) and then into the bloodstream.

Bile salts are necessary for this absorption process. After the absorption of the fatty acids, the bile salts are returned to the liver to be excreted again into the bile. Fatty acids of less than 10 to 12 carbon atoms are transported through the intestinal walls directly into the bloodstream, as is any free glycerol present.

Absorption of Proteins

The end products of the hydrolysis of proteins are amino acids. Absorption of amino acids occurs chiefly in the small intestine and is an active, enzyme-requiring process resembling the active transport of glucose. There are six or more specific transport systems for amino acids being carried into the bloodstream.

1. A system for small neutral amino acids such as glycine.
2. A system for large, neutral amino acids such as phenylalanine.
3. A system for basic amino acids such as lysine.
4. A system for acidic amino acids such as aspartic acid.
5. A system for proline.
6. A system for very small peptides.

Amino acids compete with one another for absorption via a particular pathway. Thus, high levels of leucine lower the absorption of isoleucine and valine.

Occasionally, proteins also escape digestion and are absorbed directly into the blood. This occurs more often in the very young since the permeability of their intestinal mucosa is greater, allowing the passage of antibodies of colostral milk. This passage of protein into the blood may be sufficient to cause immunologic sensitization and related food allergies.

The L-amino acids are absorbed more rapidly than the D isomers and pass through the capillaries of the villi directly into the bloodstream, which carries them to the tissues to be used to build or replace tissue. The amino acids can also be oxidized to furnish energy. Although the body can store carbohydrate and fat, it cannot store protein.

Absorption of Iron

Iron differs from practically all other inorganic materials in that the amount in the body is controlled by its absorption, not its excretion. Body stores of iron are conserved very efficiently except during menstrual flow.

Normally, only 5 to 10 percent of orally ingested iron is absorbed. Most of it is absorbed in the upper duodenum region of the small intestine. The stomach acidity along with reducing agents such as vitamin C (ascorbic acid) and proteins help reduce dietary iron to stable, absorbable, water-soluble Fe^{2+} complexes.

Several factors make absorption of iron difficult.

1. Most iron has to be reduced from Fe^{3+} to Fe^{2+}.
2. The relatively high pH of the small intestine increases the formation of insoluble iron compounds.
3. Iron forms insoluble compounds with bile salts.
4. Phosphates form insoluble iron compounds.
5. Iron absorption requires stomach acidity.

A lack of HCl in the stomach (achlorhydria) prevents the absorption of most of the iron, and stomach antacids may also reduce iron absorption.

Absorption of Vitamins

Most vitamins are absorbed in the upper small intestine. The fat-soluble vitamins (A, D, E, K) need fat and bile salts to be absorbed.

Table 25-2 Sites of Absorption of Body Nutrients	Small Intestine	Large Intestine
	Glucose, other monosaccharides	Bile salts
	Monoglycerides	Some electrolytes
	Glycerol	Water
	Cholesterol	
	Amino acids	
	Peptides	
	Vitamins	
	Some electrolytes, e.g., Fe^{2+}, Ca^{2+}, PO_4^{3-}	

Taking a multivitamin capsule with water does not provide the fat necessary for the absorption of the fat-soluble vitamins. Vitamin B_{12} absorption depends on its binding to an intrinsic factor produced in the stomach (see page 455). This complex along with calcium ions finds acceptor sites in the lower small intestine.

Formation of Feces

After the absorption of monosaccharides, glycerol, fatty acids, and amino acids, the remaining contents of the small intestine pass into the large intestine. The large intestine contains undigestible material (such as cellulose), undigested food particles, unused digestive juices, epithelial tissues from the walls of the digestive system, bile pigments, bile salts, and inorganic salts. The material passing into the large intestine is semifluid and contains much water. Most of this water and some salts are reabsorbed through the walls of the large intestine, leaving behind a residue called feces. Little or no digestion takes place in the large intestine.

The conditions in the large intestine are ideal for the growth of bacteria and usually one-fourth to one-half of the feces consist of bacteria. The bacteria cause the fermentation of carbohydrates to produce hydrogen, carbon dioxide, and methane gases, as well as acetic, butyric, and lactic acids. The gases can cause distention and swelling of the intestinal tract, producing a feeling of discomfort. The acids may be irritating to the intestinal mucosa and cause diarrhea. Particularly in infants, the acids can cause *excoriated buttocks. Infants who have this condition are usually given a diet that is high in protein and low in carbohydrate because the fermentation bacteria act on carbohydrates and not on protein. The bacteria also produce vitamin K.

Some of the amino acids undergo decarboxylation because of the action of intestinal bacteria to produce toxic amines called ptomaines (see page 518). For example, decarboxylation produces cadaverine from the amino acid lysine, putrescine from ornithine, and histamine from histidine. These toxic substances are reabsorbed from the large intestine, carried to the liver where they are detoxified, and then excreted in the urine.

The amino acid tryptophan undergoes a series of reactions to form the compounds indole and skatole, which are primarily responsible for the odor of the feces.

tryptophan   indole   skatole (methylindole)

Substance Malabsorbed	Symptoms
Lactose	Milk intolerance
Vitamin K	Bleeding, bruising
Iron, vitamin B_{12}, folates	Anemia
Protein products	Edema
Calcium, magnesium, vitamin D	Tetany
Calcium, protein products	Osteoporosis

Table 25-3 Effects of Malabsorption of Nutrients

Defects of Carbohydrate Digestion and Absorption

Some individuals have a deficiency of lactase that causes an intolerance to milk. Symptoms of such a deficiency include abdominal cramps, diarrhea, and *flatulence. These symptoms are caused by the accumulation of lactose, which, because of its osmotic activity, holds water. Another cause of these symptoms is the fermentation of lactose by intestinal bacteria, which produces both gases and irritation.

Many newborns also have a lactose intolerance after a bout of diarrhea has temporarily decreased the level of lactase enzymes in the intestinal tract. Treatment consists of the removal of lactose-containing milk for several days until the intestinal tract has repaired itself. Adults who lack lactase enzymes must avoid milk and milk products.

A few individuals have an inherited sucrase deficiency. Symptoms, which occur in early childhood, are similar to those of a lactase deficiency.

Some persons have a disaccharidase deficiency that causes disacchariduria, an increase in disaccharide excretion.

Malabsorption of glucose and galactose is a congenital condition due to a defect in a carrier mechanism. However, fructose absorption is normal because absorption of fructose does not depend upon that carrier system.

If absorption of nutrients is disturbed, various types of symptoms may occur, as illustrated in Table 25-3.

Summary

Digestion is the process by which foods are broken down (hydrolyzed) into simple molecules that can then be absorbed through the intestinal walls. Digestion takes place in the mouth, the stomach, and the small intestine.

Digestion begins in the mouth. Food is mixed with saliva, which moistens the food and makes swallowing easier. Saliva contains an enzyme that begins the hydrolysis of carbohydrates (starch). Saliva has a pH of 5.75 to 7.0 and is approximately 99.5 percent water.

In the stomach the food is mixed with the gastric juices. The gastric juices contain hydrochloric acid and have a pH of 1 to 2. The gastric juice contains the zymogen pepsinogen and the enzyme gastric lipase. Upon contact with hydrochloric acid, pepsinogen is converted into pepsin, which then catalyzes the hydrolysis of protein to polypeptides. Gastric lipase is not an important enzyme because it can act only on emulsified fats and there is practically no emulsified fat present in the stomach.

The food leaving the stomach enters the small intestine, where it is mixed with three digestive juices—those from the pancreas, those from the intestinal walls, and bile from the gallbladder.

Bile is produced in the liver and stored in the gallbladder. Bile is alkaline and neutralizes the acid entering the small intestine from the stomach. Bile salts lower surface tension and help emulsify fats; they aid in the absorption of fatty acids; they help stimulate intestinal motility.

Bile also contains pigments that come from the breakdown of hemoglobin. These bile pigments give the feces and the urine their characteristic color. If cholesterol crystallizes in the gallbladder, gallstones are produced.

Digestion of carbohydrates begins in the mouth. No digestion of carbohydrate takes place in the stomach except that catalyzed by salivary amylase from the saliva. Even this activity ceases when the food from the mouth reaches the low pH of the stomach. The major digestion of carbohydrates takes place in the small intestine with the aid of the enzymes pancreatic amylase, maltase, lactase, and sucrase.

No digestion of fats takes place in the mouth. No digestion of fats takes place in the stomach either unless the fats are already emulsified. In that case their hydrolysis is catalyzed by gastric lipase, although to a very limited extent. Fats are emulsified in the small intestine by the action of bile and then acted on by the enzyme pancreatic lipase.

Digestion of protein begins in the stomach with the aid of pepsin. The digestion of protein continues in the small intestine with the aid of the enzymes trypsin, chymotrypsin, carboxypeptidase, aminopeptidase, and dipeptidase.

The monosaccharides produced by the digestion of carbohydrates pass through the villi of the small intestine and enter the bloodstream.

Fatty acids and glycerol, products of the digestion of fats, are converted into glycerides in the intestinal mucosa by the action of bile and pass through the lacteals into the thoracic duct and then into the bloodstream. Fatty acids of less than 10 to 12 carbon atoms pass directly into the bloodstream through the villi.

Amino acids, from the digestion of protein, pass through the villi of the small intestine into the bloodstream.

Undigested food, undigestible foods, unused digestive juices, epithelial tissues from the walls of the digestive system, bile salts, inorganic salts, and water pass from the small intestine into the large intestine. Most of the water and some salts are reabsorbed. The remaining material is excreted as feces.

Questions and Problems

A

1. Why is digestion necessary?
2. Do all foods require digestion? Why?
3. Where does digestion take place?

4. What does saliva contain? What does it do? What is its pH?
5. What is the pH of gastric juice? What causes this pH?
6. What is hypoacidity? What may cause it?
7. What is hyperacidity? What may cause it?

8. Gastric juice contains what enzymes?
9. What is the difference between pepsinogen and pepsin?
10. What converts pepsinogen into pepsin?
11. What is the function of secretin?
12. Why is gastric lipase considered an unimportant enzyme?
13. What is the intrinsic factor? What disease is associated with a lack of this substance?
14. Three different digestive juices enter the small intestine. What are they?
15. What enzymes are present in the pancreatic juice? What does each do?
16. What enzymes are present in the intestinal juice? What does each do?
17. Where is bile produced? Where is it stored?
18. What is the function of cholecystokinin? Where is it produced?
19. What is the pH of bile? What does it contain?
20. What are the functions of bile salts?
21. How are bile pigments formed? Where are they excreted?
22. What is urobilin? Where is it found?
23. What happens when the bile duct is blocked?
24. What can cause gallstones?

B
25. Describe the digestion of carbohydrate. Where are the end products absorbed?
26. Describe the digestion of fats. Where are the end products absorbed?
27. Describe the digestion of protein. Where are the end products absorbed?
28. Describe the formation of feces.
29. What are the functions of the large intestine?
30. What may cause excoriated buttocks in infants?
31. What can be done to overcome this effect?
32. What causes the characteristic odor of feces?
33. Where are some of the toxic putrefactive products detoxified? Where are they excreted?
34. What is enterokinase? Where does it function?

Practice Test

1. Most of the digestion of food takes place in the _____.
   a. mouth   b. stomach
   c. small intestine d. large intestine
2. Most of the absorption of food takes place in the _____.
   a. mouth   b. stomach
   c. small intestine d. large intestine
3. The end product of starch digestion is _____.
   a. maltose   b. glucose
   c. fructose   d. galactose
4. The process of digestion involves _____.
   a. hydrolysis   b. hydration
   c. oxidation   d. hydrogenation
5. Adults who cannot tolerate milk or milk products probably lack the enzyme _____.
   a. pepsin   b. sucrase
   c. amylase   d. lactase
6. HCl in the stomach _____.
   a. destroys microorganisms
   b. denatures protein
   c. converts pepsinogen to pepsin
   d. all of these
7. The part of the digestive system that has the lowest pH is _____.
   a. the mouth   b. the stomach
   c. the small intestine d. the large intestine
8. Bile is produced in the _____.
   a. small intestine b. pancreas
   c. liver   d. stomach
9. The principal monosaccharide carried by the bloodstream is _____.
   a. glucose   b. sucrose
   c. fructose   d. lactose
10. Bile salts _____.
    a. lower surface tension
    b. aid in the absorption of fatty acids
    c. stimulate intestinal motility
    d. all of these

26

Metabolism of Carbohydrates

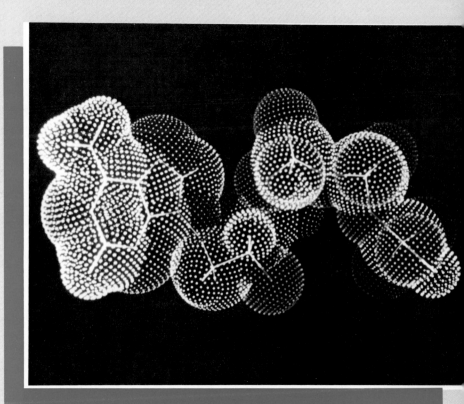

Computer-generated molecular model of adenosine triphosphate, the universal carrier of living systems.

Introduction

There are two major questions in biochemistry.

1. How do cells obtain energy from their environment?
2. How do cells synthesize the building blocks of their macro-molecules?

The answers lie with the chemical reactions collectively known as **metabolism.** Although there are hundreds of reactions in metabolism, the number of kinds of reactions is quite small. The metabolic pathways are regulated in simple common ways.

Pathways that produce ATP (adenosine triphosphate), the currency of energy in biologic systems, pertain to the first question. ATP is used for three major purposes: muscle contraction and movement, active transport of molecules and ions, and synthesis of biologic molecules.

Metabolism occurs in stages. In the first stage (biodegradation), large molecules in food are broken down into small units. This stage involves digestion and absorption. In the second stage (biosynthesis), these numerous small molecules are converted into a few very simple units that play the central role in metabolism. The third stage consists of the energy production mainly associated with the Krebs cycle and oxidative phosphorylation (Figure 26-1).

An important general principle of metabolism is that the biosynthetic and biodegradative pathways are almost always separated. This separation is necessary both for energetic reasons and for metabolic control. Frequently this separation is enhanced by compartmentalization of the pathways. For example, fatty acid oxidation occurs in the mitochondria, whereas fatty acid synthesis takes place in the cytoplasm.

An overview of the principal pathways to be studied in this and the following chapters is given in Figure 26-2.

Concentration of Sugar in the Blood (Blood Sugar Level)

The end products of carbohydrate digestion are the monosaccharides glucose, fructose, and galactose. Both fructose and galactose are converted to glucose in the liver so that the major monosaccharide remaining in the bloodstream is glucose.

The amount of glucose present in the blood will vary considerably, depending upon whether the measurements were taken $\frac{1}{2}$ hr after eating, 1 hr after eating, or during a period of fasting. The normal quantity of glucose present in 100 mL of blood taken after a period of fasting is 70 to 100 mg. This value is called the **normal fasting blood sugar.** Soon after a meal, the blood sugar level may

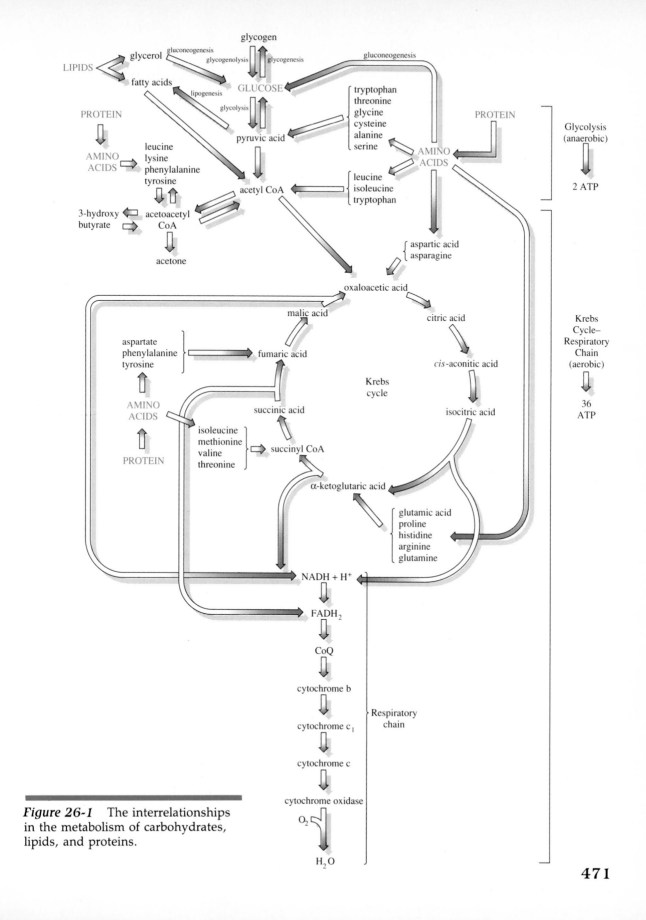

Figure 26-1 The interrelationships in the metabolism of carbohydrates, lipids, and proteins.

471

Figure 26-2 Three stages of metabolism.

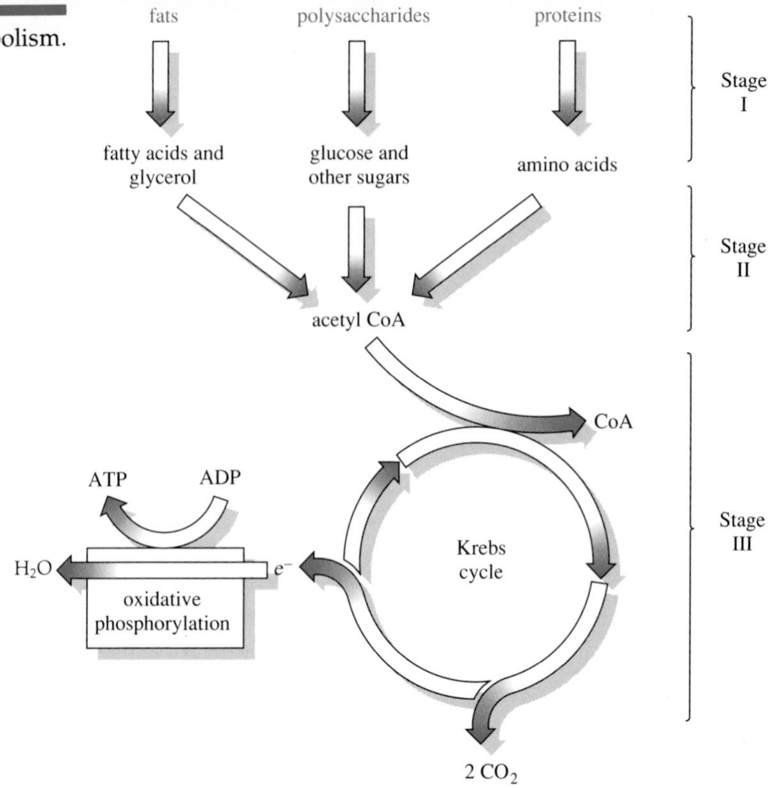

fats   polysaccharides   proteins

Stage I

fatty acids and glycerol   glucose and other sugars   amino acids

Stage II

acetyl CoA

CoA

ATP   ADP

Stage III

H_2O

oxidative phosphorylation

e^-

Krebs cycle

$2\,CO_2$

Figure 26-3 Blood sugar levels in milligrams per 100 mL of blood.

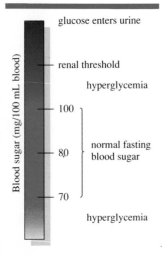

Blood sugar (mg/100 mL blood)

glucose enters urine

renal threshold

hyperglycemia

— 100

normal fasting blood sugar

— 80

— 70

hyperglycemia

rise to 120 to 130 mg per 100 mL of blood or even higher. However, the level soon drops, so that after $1\frac{1}{2}$ to 2 hr it again returns to its normal fasting value.

Blood sugar levels below 70 produce hypoglycemia, and levels over 100 produce hyperglycemia (Figure 26-3). The brain metabolizes approximately 120 g of glucose daily, and hypoglycemia can reduce the brain's energy supply, causing dizziness and loss of consciousness.

During the time of fasting, even though the body is continuously using glucose for the production of heat and energy, the amount of glucose present in the blood remains fairly constant, usually in the range of 70 to 100 mg per 100 mL. How does the body regulate the amount of glucose present in the blood, and what happens when these control mechanisms do not function properly?

When the glucose level in the blood rises, as happens after the digestion of a meal, the liver removes the excess glucose and converts it to glycogen, a polysaccharide. This process is called **glycogenesis.** This glycogen may be stored in the liver and in muscle tissue. However, only a certain amount of glycogen can be stored in the liver and muscle; the rest is changed to fat and stored as such.

Cellular uptake of glucose is greatly increased by the release of insulin (see page 490). Glucose is also removed from the blood by the normal oxidative reactions that take place continuously throughout the body.

Glucose does not normally appear in the urine except in amounts too small to be detected by Benedict's solution. However, if the blood sugar level rises above 170 to 180 mg per 100 mL of blood, the sugar "spills over" into the urine. The point at which the sugar spills over into the urine is called the renal threshold. The presence of glucose in the urine is called glycosuria.

Thus, the factors that remove excess glucose from the blood are (1) insulin secretion, (2) glycogenesis and storage as glycogen, (3) conversion to fat, (4) normal oxidation reactions in the body, and (5) excretion through the kidneys when the renal threshold is exceeded.

After a period of 2 to 3 hr, the liver converts glycogen back to glucose. This process is called glycogenolysis. Liver glycogen will be exhausted after 10 hr. The body then switches to making glucose from amino acids; this process is called gluconeogenesis (see Figure 26-14).

Source of Energy

Where does a muscle get its energy to contract? Where does the body get the energy necessary to synthesize protein, to send nerve impulses, to perform countless other functions?

The energy necessary for the body functions comes from certain high-energy compounds, compounds that yield a large amount of energy on hydrolysis. The key compound of this type is adenosine triphosphate (ATP). Hydrolysis of ATP to adenosine diphosphate (ADP) and inorganic phosphate liberates about 7600 cal/mol. This hydrolysis breaks one of the high-energy phosphate bonds, designated by $\sim$ in structures (26-1) and (26-2).

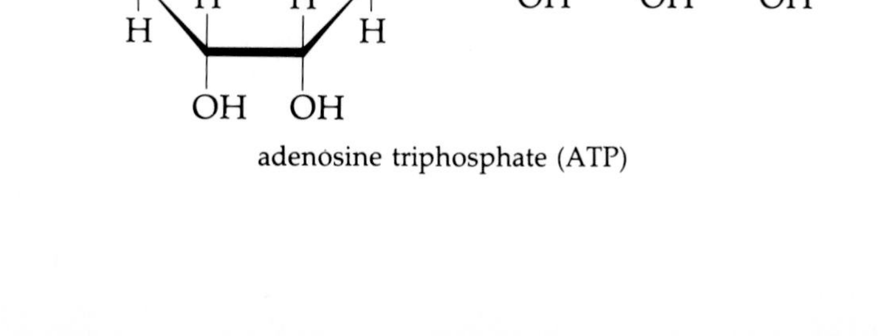

adenosine triphosphate (ATP)

This formula for ATP can be abbreviated as

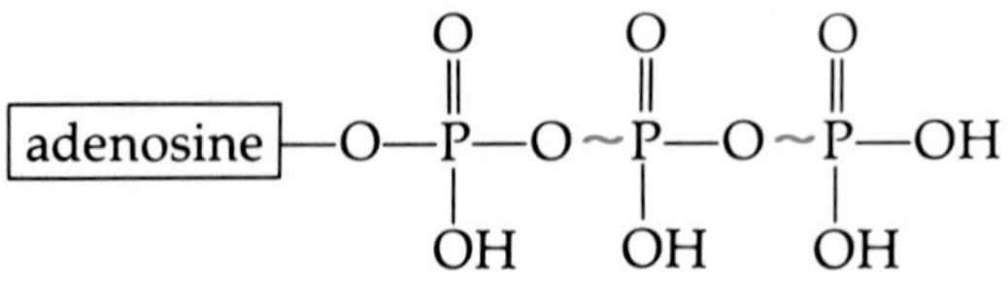

$$\boxed{\text{adenosine}}\!-\!O\!-\!\overset{\overset{\displaystyle O}{\|}}{\underset{\underset{\displaystyle OH}{|}}{P}}\!-\!O\!\sim\!\overset{\overset{\displaystyle O}{\|}}{\underset{\underset{\displaystyle OH}{|}}{P}}\!-\!O\!\sim\!\overset{\overset{\displaystyle O}{\|}}{\underset{\underset{\displaystyle OH}{|}}{P}}\!-\!OH \quad \text{or} \quad A\!-\!P\!\sim\!P\!\sim\!P \quad (26\text{-}1a)$$

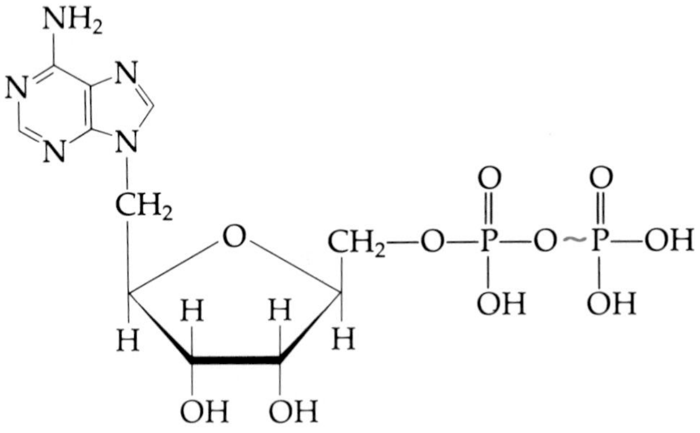

adenosine diphosphate (ADP)   (26-2)

This formula for ADP can be abbreviated as

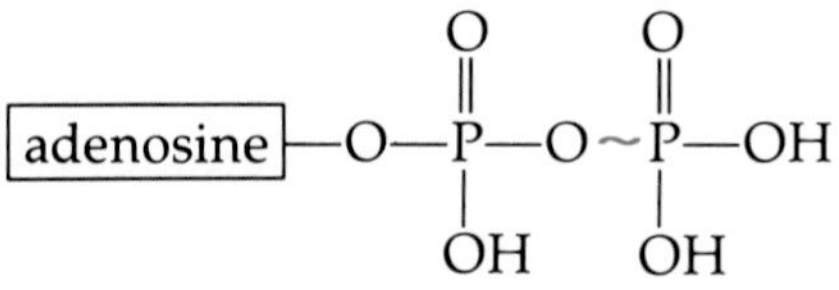

$$\boxed{\text{adenosine}}\!-\!O\!-\!\overset{\overset{\displaystyle O}{\|}}{\underset{\underset{\displaystyle OH}{|}}{P}}\!-\!O\!\sim\!\overset{\overset{\displaystyle O}{\|}}{\underset{\underset{\displaystyle OH}{|}}{P}}\!-\!OH \quad \text{or} \quad A\!-\!P\!\sim\!P \quad (26\text{-}2a)$$

However, the supply of ATP in the body is limited. There must be some mechanism for regenerating this high-energy compound so it will be available for continued use. How does the body change ADP back to ATP? To accomplish this, a high-energy phosphate group must be added to the ADP. This process is called phosphorylation and can be represented by the equation

$$\text{ADP} + \text{phosphate ion} + \text{fuel} \longrightarrow \text{ATP} + \text{fuel residue}$$

One of the fuels used in this reaction is glucose. The oxidation of glucose, the steps and enzymes required, the products produced, and the involvement of ADP and ATP in these processes will be discussed later in this chapter.

The metabolism of carbohydrates in humans is categorized as follows.

1. **Glycogenesis,** the synthesis of glycogen from glucose.
2. **Glycogenolysis,** the breakdown of glycogen to glucose.

3. **Glycolysis,** the oxidation of glucose or glycogen to pyruvic or lactic acid.
4. **Hexose monophosphate shunt,** an alternative oxidative path for glucose.
5. **Krebs (citric acid) cycle** and **electron transport chain,** the final oxidative paths to carbon dioxide and water.
6. **Gluconeogenesis,** the formation of glucose from noncarbohydrate sources.

Glycogenesis

Glycogenesis is the formation of glycogen from glucose. This process occurs primarily in the liver and the muscles. The liver may contain up to 5 percent glycogen after a high-carbohydrate meal but may contain almost no glycogen after 12 hr of fasting.

The overall conversion of glucose to glycogen and vice versa can be written as

$$n \; C_6H_{12}O_6 \; \underset{\text{glycogenolysis}}{\overset{\text{glycogenesis}}{\rightleftharpoons}} \; (C_6H_{10}O_5)_n + n \; H_2O$$

glucose   glycogen

However, the reaction is by no means as simple as this equation indicates. There are several steps involved, each one catalyzed by a particular enzyme.

The first step in glycogenesis involves the conversion of glucose to glucose 6-phosphate (abbreviated as glucose 6-P), a phosphate ester of glucose. ATP from the liver cells serves as a source of the phosphate group. After the loss of the phosphate group, ADP is left. The enzyme *glucokinase* is necessary to catalyze this reaction. Insulin is involved in the phosphorylation of glucose by glucokinase.

$$\text{glucose} + \text{ATP} \; \xrightarrow[\text{insulin}]{\text{glucokinase}} \; \text{glucose 6-P} + \text{ADP}$$

Glucose 6-phosphate is then rearranged so that the phosphate group moves from the number 6 position to the number 1 position, producing glucose 1-phosphate. The enzyme required for this reaction is *phosphoglucomutase.*

$$\text{glucose 6-P} \; \xrightarrow{\text{phosphoglucomutase}} \; \text{glucose 1-P}$$

Glucose 1-phosphate then reacts with uridine triphosphate (UTP) to form uridine diphosphate glucose (UDPG). (Uridine triphosphate is similar to adenosine triphosphate except that uracil takes the place of adenine.) The enzyme is *UDPG pyrophosphorylase.*

$$\text{glucose 1-P} + \text{UTP} \; \xrightarrow{\text{UDPG pyrophosphorylase}} \; \text{UDPG} + \text{pyrophosphate}$$

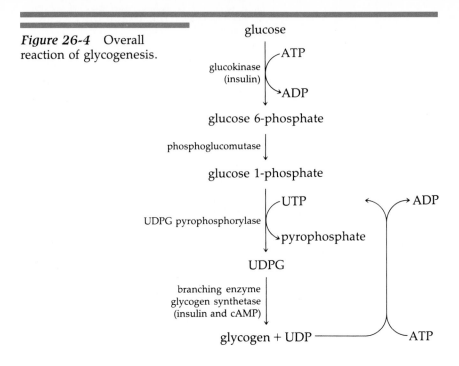

Figure 26-4   Overall reaction of glycogenesis.

Then the glucose molecules in UDPG (activated glucose molecules) are joined together to form glycogen. The enzymes necessary here are *glycogen synthetase* and a branching enzyme. The former enzyme is regulated by both insulin and cyclic adenosine monophosphate (cAMP). The latter enzyme aids in the formation of 1,4- and 1,6-glycosidic linkages.

$$\text{UDPG} \xrightarrow[\text{branching enzyme}]{\text{glycogen synthetase}} \text{glycogen} + \text{UDP}$$

The UDP formed in this reaction then reacts with ATP to regenerate UTP.

$$\text{UDP} + \text{ATP} \longrightarrow \text{UTP} + \text{ADP}$$

A diagram of the overall reaction is shown in Figure 26-4.

Glycogenolysis

In glycogenolysis we might expect the reverse of all the reactions of glycogenesis, but it should be noted that the first reaction, the one involving glucokinase, is *not* a reversible reaction.

In glycogenolysis, glycogen is converted to glucose 1-phosphate by the enzyme phosphorylase a and then to glucose 6-phosphate by the enzyme phosphoglucomutase.

Glucose 6-phosphate is then converted to glucose by the enzyme glucose 6-phosphatase, an enzyme found in the liver but not in the muscle. Therefore, muscle glycogen cannot serve as a source of blood glucose.

cAMP (see page 618) is involved in the conversion of glycogen into glucose 6-phosphate in both the liver and the muscles. When the body is under stress, it produces hormones that are carried by the bloodstream to the liver cells, where they activate the enzyme adenyl cyclase, which in turn causes the production of cAMP from ATP. cAMP then activates a protein kinase, which in turn activates a phosphorylase b kinase. This phosphorylase b kinase then activates phosphorylase a, which triggers the conversion of glycogen into glucose. The same type of reaction is involved in the conversion of muscle glycogen into glucose 6-phosphate. cAMP also deactivates the enzyme glycogen synthetase, thereby stopping glycogenesis.

Glycogenesis and glycogenolysis can be summarized as in Figure 26-5.

Glycogen Storage Diseases

Glycogen is normally stored in the liver and the muscles. In several inherited diseases, glycogen cannot be reconverted to glucose so it begins to accumulate.

Figure 26-5 Combined reactions of glycogenolysis and glycogenesis.

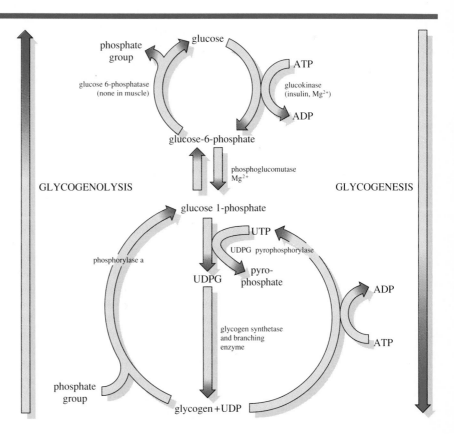

In *von Gierke's disease*, the enzyme glucose 6-phosphatase is lacking in the liver, and glycogen accumulates in that organ. As this occurs, hypoglycemia, ketosis, and hyperlipemia also occur, and the liver enlarges because of the increase in stored glycogen.

In *Pompe's disease* glycogen accumulates in the lysosomes because of a lack of a lysosomal enzyme that acts in the breakdown of glycogen.

In *Forbes's disease or Cori's limit dextrinosis* a debranching enzyme is absent so, as in von Gierke's disease, glycogen accumulates. However, the symptoms due to Forbes's disease are not as severe as those of von Gierke's disease.

Other glycogen storage diseases include *Andersen's disease* (in which death occurs in the first year of life because of liver and cardiac failure), *McArdle's syndrome* (in which individuals have a greatly diminished tolerance to exercise because of a lack of a muscle enzyme involved in glycogenolysis), and *Tarui's disease* (which is due to a phosphorylase deficiency in the liver).

Embden—Meyerhof Pathway: Glycolysis

The breakdown of glycogen resupplies the energy used up during muscle contraction. This breakdown involves a series of steps, each catalyzed by a particular enzyme.

The breakdown of glycogen to pyruvate (pyruvic acid) and lactate (lactic acid), called the Embden–Meyerhof pathway, is the first phase of muscle contraction. This process is also called *glycolysis* and is an *anaerobic process—that is, each step takes place without oxygen. Glycolysis supplies the ATP needed for muscle contraction. Over 90 percent of the energy in the red blood cells is produced by glycolysis.

The overall reaction of glycolysis can be summarized as

$$\text{glucose 6-P} + 2\,\text{ADP} \longrightarrow 2\,\text{pyruvic acid (or pyruvate)} + 2\,\text{ATP}$$
glucose 6-phosphate
(from glycogen or glucose)

The ATP formed is available for muscular work. As ATP is used, it is changed to ADP and must then be regenerated. This regeneration can be accomplished through the above pathway or by the use of another anaerobic sequence,

$$\text{creatine phosphate} + \text{ADP} \longrightarrow \text{creatine} + \text{ATP}$$

The steps in glycolysis are given below and, with the structural formulas of the intermediary products, in Figure 26-6.

Glycogen is changed to glucose 1-phosphate by the catalytic action of the enzyme phosphorylase a. Glucose 1-phosphate is then

Figure 26-6 Glycolysis—the Embden–Meyerhof pathway.

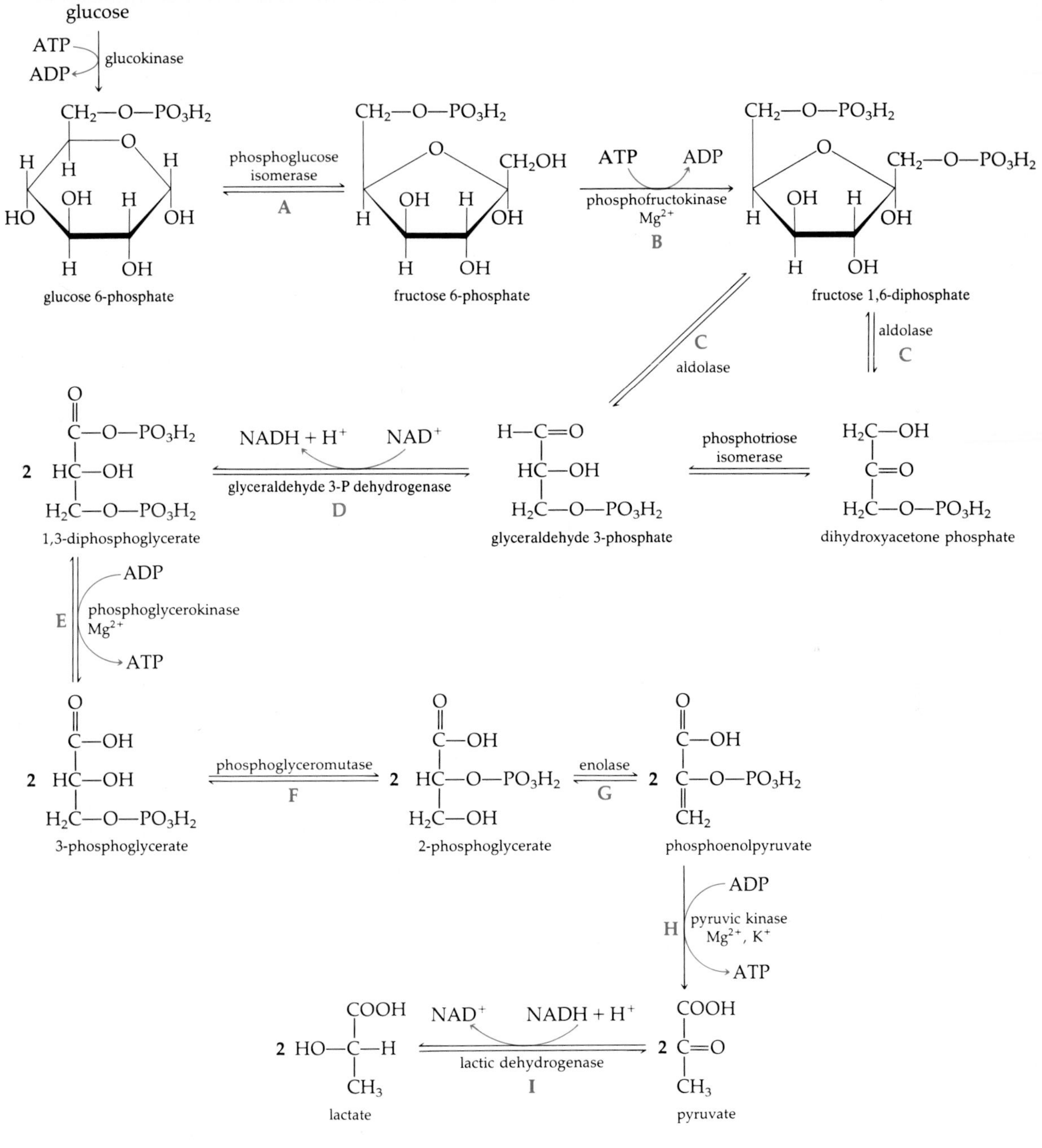

changed to glucose 6-phosphate by the enzyme phosphoglucomutase. Glucose 6-phosphate could also be formed directly from glucose by the action of ATP and the enzyme glucokinase.

Step A   Glucose 6-phosphate is changed to fructose 6-phosphate by the action of the enzyme phosphoglucose isomerase.

Step B   Fructose 6-phosphate is changed to fructose 1,6-diphosphate by the action of the enzyme phosphofructokinase. Note that during this reaction, ATP is converted to ADP.

Step C   Fructose 1,6-diphosphate is changed to the three-carbon compounds glyceraldehyde 3-phosphate and dihydroxyacetone phosphate. The enzyme involved is aldolase. Dihydroxyacetone phosphate is converted to glyceraldehyde 3-phosphate by the action of the enzyme phosphotriose isomerase.

Step D   Glyceraldehyde 3-phosphate is changed to 1,3-diphosphoglycerate by the action of the enzyme glyceraldehyde 3-phosphate dehydrogenase. During this reaction, nicotinamide adenine dinucleotide (NAD^+) is reduced to $NADH + H^+$.

Step E   1,3-Diphosphoglycerate is changed to 3-phosphoglycerate by the action of the enzyme phosphoglycerokinase. In this reaction, two ADPs (one for each three-carbon compound) are changed to two ATPs.

Step F   3-Phosphoglycerate is changed to 2-phosphoglycerate by the action of the enzyme phosphoglyceromutase.

Step G   2-Phosphoglycerate is changed to phosphoenolpyruvate by the action of the enzyme enolase.

Step H   Phosphoenolpyruvate is changed to pyruvate by the action of the enzyme pyruvic kinase. During this reaction, two ADPs are changed to two ATPs (one for each three-carbon compound).

Step I   Pyruvate is changed to lactate through the enzyme lactic dehydrogenase. At the same time, NADH and H^+ are changed to NAD^+

The sequence of reactions involved in glycolysis is summarized in Figure 26-7. Three of these reactions are irreversible—the conversions of glucose to glucose 6-phosphate, fructose 6-phosphate to fructose 1,6-diphosphate, and phosphoenolpyruvate to pyruvate. The rate of glycolysis is primarily controlled by the allosteric enzyme phosphofructokinase. This enzyme is inhibited by high levels of ATP and, conversely, is stimulated by low ATP levels. Glycolysis also provides compounds for biosynthesis, and thus phosphofructokinase is regulated by other sources for these compounds, such as fatty acids and citrate, as well.

In addition, red blood cells have two enzymes that allow for the formation of 2,3-diphosphoglycerate (2,3-DPG) from 1,3-diphosphoglycerate. 2,3-DPG is a regulator of oxygen transport in red blood cells. When the O_2 level in the blood drops, 2,3-DPG increases and more O_2 is delivered to the cells.

Figure 26-7 The sequence of events in glycolysis.

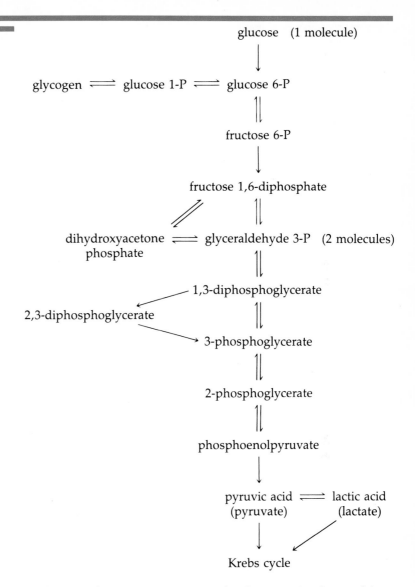

According to these reactions, muscle glycogen is changed into pyruvic acid and then to lactic acid. However, only about one-fifth of the lactic acid thus formed is oxidized to carbon dioxide and water, resupplying the energy used up during muscle contraction. The other four-fifths of the lactic acid is changed back to glycogen, reversing the above reactions. Part of the lactic acid is changed back to glycogen in the muscle. The rest of the lactic acid is carried to the liver by the bloodstream, where it is converted to liver glycogen. In addition, for lactic acid to be utilized for energy, it must first be converted to pyruvic acid. Lactic acid is a metabolic dead end.

The oxidation of some of the lactic acid, the aerobic sequence, produces a large amount of energy. The oxidation of one molecule of lactic acid converts 18 molecules of ADP to ATP.

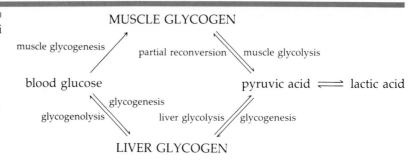

Figure 26-8 Lactic acid or Cori cycle.

After a muscle contracts and relaxes, the net change is a partial loss of glycogen. Glycogen can be replenished by the conversion of blood glucose to muscle glycogen (muscle glycogenesis).

This conversion of glycogen to lactic acid (lactate) and partial reconversion to glycogen, the lactic acid or Cori cycle, is shown in Figure 26-8.

Another Oxidative Pathway: Hexose Monophosphate Shunt

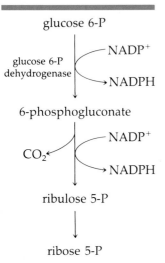

Figure 26-9 The hexose monophosphate or pentose shunt.

The oxidation of glucose to lactic acid can also proceed through a series of reactions called the **hexose monophosphate shunt**, or the **pentose shunt**. This sequence is important because it provides the five-carbon sugars needed for the synthesis of nucleic acids and nucleotides and because it makes available NADPH, the reduced form of nicotinamide adenine dinucleotide phosphate ($NADP^+$), a coenzyme necessary for the synthesis of fatty acids and steroids. This pathway is much more active in adipose tissue than in muscle. Note that ATP is not generated in this sequence. The ribose can be further metabolized to glyceraldehyde 3-phosphate.

A genetic deficiency of glucose 6-phosphate dehydrogenase (see Figure 26-9) is a major cause of hemolytic anemia. A lack of glucose 6-phosphate dehydrogenase also provides resistance to malaria. However, at the same time a lack of this enzyme reduces the level of NADPH in red blood cells. The major role of NADPH in red blood cells is the maintenance of proper levels of the antioxidase glutathione. Without adequate glutathione, red blood cells are easily oxidized by a variety of drugs.

The Aerobic Sequence: Krebs Cycle

The aerobic sequence converts lactic and pyruvic acids (from anaerobic glycolysis) through a series of steps to carbon dioxide and water. This series of reactions is called the **Krebs cycle** or **citric acid cycle** (see Figure 26-10). The Krebs cycle uses oxygen transported to the

Figure 26-10 The Krebs or citric acid cycle.

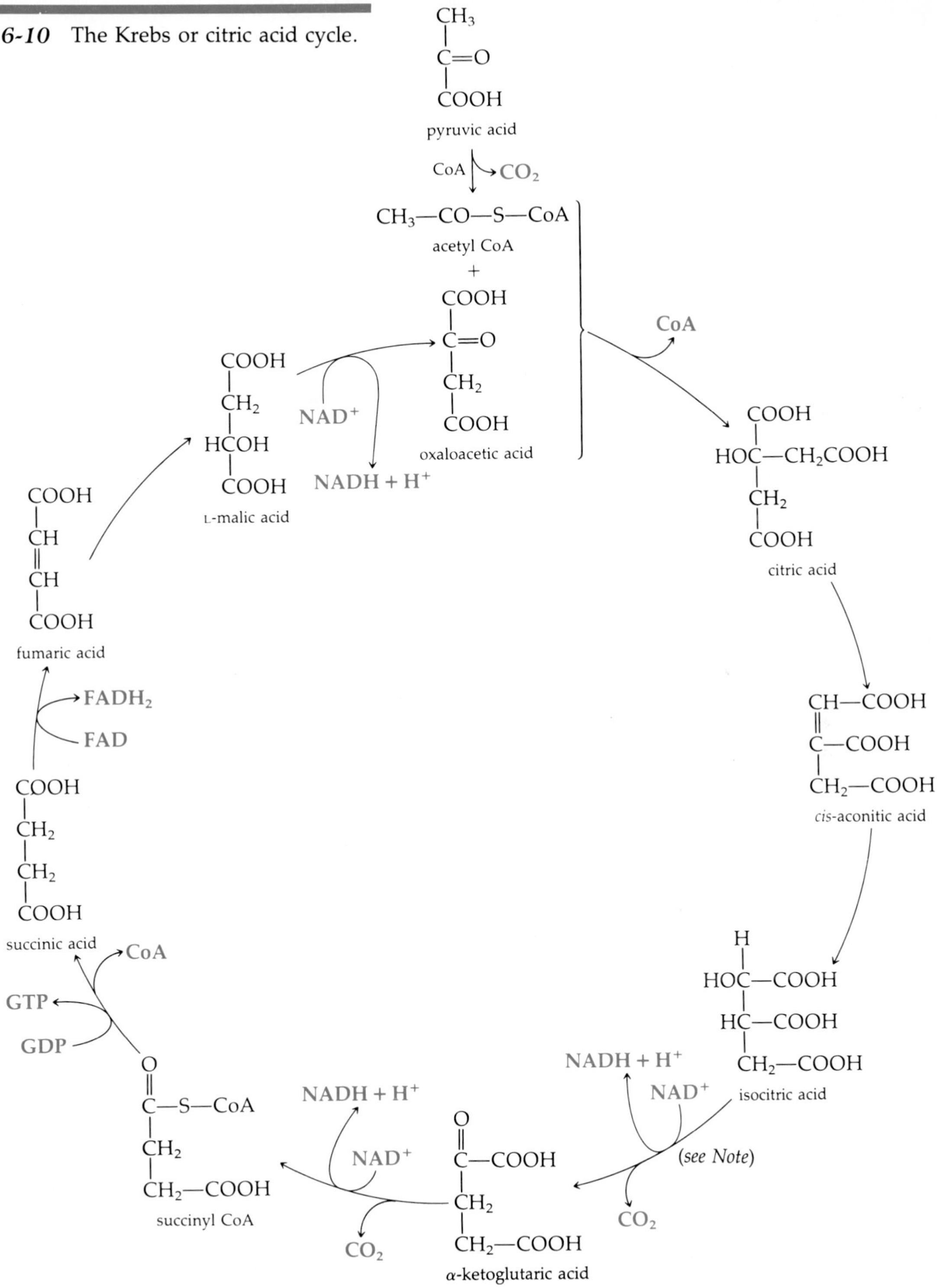

CH₃
|
C=O
|
COOH

pyruvic acid

CoA ↘ CO_2

$CH_3—CO—S—CoA$

acetyl CoA

+

COOH
|
C=O
|
CH₂
|
COOH

oxaloacetic acid

NAD^+

NADH + H⁺

COOH
|
CH₂
|
HCOH
|
COOH

L-malic acid

CoA

COOH
|
HOC—CH₂COOH
|
CH₂
|
COOH

citric acid

COOH
|
CH
‖
CH
|
COOH

fumaric acid

FADH₂

FAD

CH—COOH
‖
C—COOH
|
CH₂—COOH

cis-aconitic acid

COOH
|
CH₂
|
CH₂
|
COOH

succinic acid

CoA

GTP

GDP

O
‖
C—S—CoA
|
CH₂
|
CH₂—COOH

succinyl CoA

H
|
HOC—COOH
|
HC—COOH
|
CH₂—COOH

isocitric acid

NADH + H⁺

NAD^+

(see Note)

CO_2

NADH + H⁺

NAD^+

O
‖
C—COOH
|
CH₂
|
CH₂—COOH

α-ketoglutaric acid

CO_2

Note: An enzyme-bound complex of oxalosuccinic acid occurs here as an intermediary.

Table 26-1 Production of ATP from the Oxidation of One Molecule of Glucose

Source of ATP	Number of Molecules of ATP Formed from 1 Molecule of Glucose	Other Compounds Formed
Glycolysis	2	2 NADH
Pyruvic acid → acetyl CoA	0	2 NADH
Krebs cycle	2	2 GTP + 6 NADH + 2 FADH$_2$
Oxidative phosphorylation of 2 NADH from glycolysis	6	
Transportation across mitochondrial membrane	−2	
2 NADH from pyruvic acid → acetyl CoA reaction	6	
6 NADH from Krebs cycle	18	
2 FADH$_2$ from Krebs cycle	$\underline{4}$	
	36	

GTP = guanosine triphosphate.

cells by hemoglobin, hence the term *aerobic*. This cycle takes place in the mitochondria (see page 442 and Figure 26-13).

During the complete oxidation of one molecule of glucose, 36 molecules of ATP are produced, as indicated in Table 26-1. Note that most of the ATP formed comes from oxidative phosphorylation.

Assuming 7.6 kcal per high-energy phosphate bond, the overall sum is 36 × 7.6 kcal, or 274 kcal. Theoretically, 686 kcal should be produced from 1 mole of glucose; thus, the efficiency of conversion is approximately 40 percent.

The first step in the aerobic process is the formation of active acetate from pyruvic acid. This active acetate is the acetyl derivative of coenzyme A, or acetyl CoA. Acetyl CoA is the converting substance in the metabolism of carbohydrates, fats, and proteins. Acetyl CoA becomes the "fuel" for the Krebs cycle. As will be noted, acetyl CoA reacts with oxaloacetic acid and goes through the cycle. At the end of the cycle, oxaloacetic acid is regenerated and picks up another molecule of acetyl CoA to carry it through the sequence. During the cycle, acetyl CoA is oxidized to carbon dioxide, and at the same time NADH and FADH$_2$ are produced. These enter into the electron transport chain that functions on the inner membranes of the mitochondria. The overall reaction for the Krebs cycle can be summarized by the following equation.[†]

$$\text{acetyl CoA} + 3\ \text{NAD}^+ + \text{FAD} + \text{GDP} + \text{P}_i + 2\ \text{H}_2\text{O} \longrightarrow$$

$$2\ \text{CO}_2 + \text{CoA} + 3\ \text{NADH} + 2\ \text{H}^+ + \text{FADH}_2 + \text{GTP}$$

[†] FAD = flavin adenine dinucleotide; GDP = guanosine diphosphate.

Figure 26-11 The electron transport system.

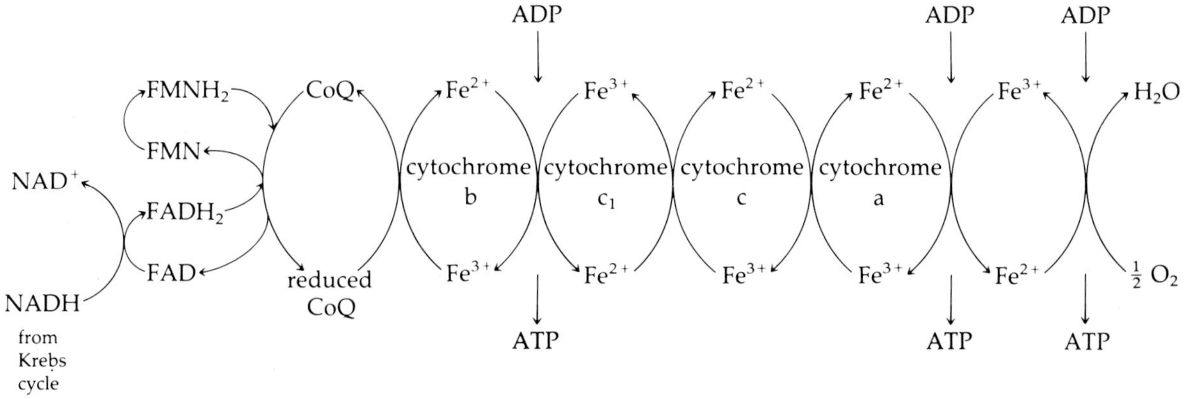

Oxidative Phosphorylation: Electron Transport System

In the electron transport chain, or electron transport system, electrons are transferred from NADH through a series of steps to oxygen, with the regeneration of NAD^+ and FAD, and the formation of H_2O. The energy yielded is used for the formation of ATP (see Figure 26-11). The overall reaction in the electron transport chain is also called oxidative phosphorylation and proceeds because large amounts of NADH and $FADH_2$ are produced from the Krebs cycle (see Figure 26-12) and also because there is a plentiful supply of oxygen in the tissues. The overall reaction is

$$NADH + H^+ \; 3 \; ADP + P_i + \tfrac{1}{2}O_2 \rightarrow NAD^+ + 3 \; ATP + H_2O$$

Involved in this electron transport chain are NADH, $FADH_2$, coenzyme Q, and several cytochromes, which are complexes containing heme (recall that heme is part of the hemoglobin molecule).

One of the cytochromes, cytochrome oxidase, is a complex that binds oxygen, reduces it with electrons received from other cytochromes in the electron transport system, and finally converts that oxygen to water. Note that the oxygen in this sequence (which is therefore an aerobic sequence) reacts only at the last step.

One unusual fact about cytochrome oxidase is that it contains two metals—iron and copper. This particular enzyme is mainly responsible for the introduction of oxygen into the metabolic processes of the Krebs cycle and so is considered absolutely vital to life. If the cytochrome oxidase function is blocked, as with cyanide poisoning, all cellular activity stops very quickly. It is believed that the cytochrome oxidase functions by transferring electrons from copper to iron to oxygen.

Note that another high-energy compound produced by the Krebs cycle is guanosine triphosphate (GTP). GTP is considered to be interchangeable with ATP.

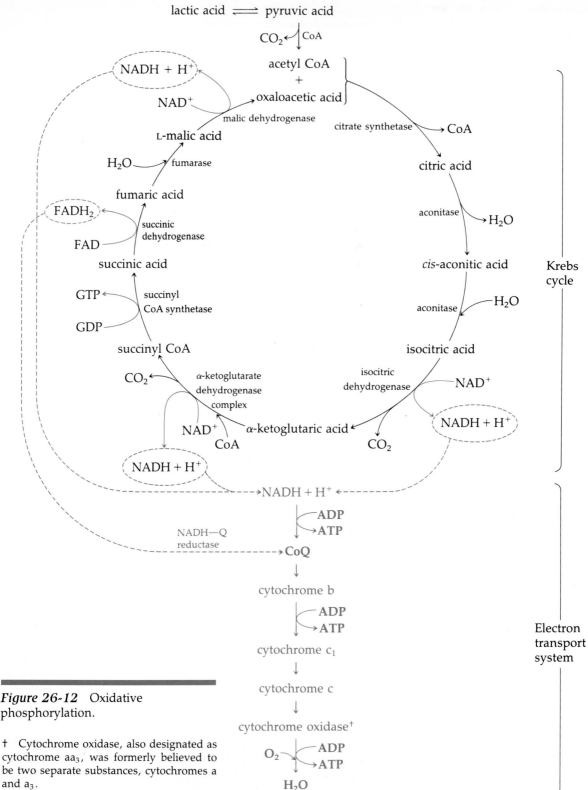

Figure 26-12 Oxidative phosphorylation.

† Cytochrome oxidase, also designated as cytochrome aa_3, was formerly believed to be two separate substances, cytochromes a and a_3.

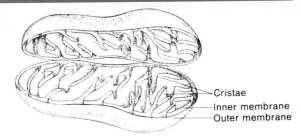

Figure 26-13 Three-dimensional representation of a mitochondrion.

Cristae
Inner membrane
Outer membrane

The Chemiosmotic Theory

The chemiosmotic theory, proposed by Peter Mitchell in England in 1961, endeavors to explain oxidative phosphorylation in terms of the movement of protons (H^+) through the mitochondrial membrane (see Figure 26-13). The main principles of this theory are

1. The mitochondrial membrane is impermeable to ions, particularly to H^+, which accumulates outside the membrane, causing an electrochemical potential difference across the membrane.
2. The synthesis of ATP occurs under the influence of an enzyme on the inside of the inner mitochondrial membrane.
3. ATP synthesis occurs because of the movement of protons (H^+) through special ports in the membrane (not through the membrane itself) from the outside to the inside of the inner mitochondrial membrane.
4. The potential difference drives membrane-located ATP synthase.
5. The respiratory chain is folded into three oxidation–reduction loops in the membrane, each loop corresponding to a part of the respiratory chain.

Gluconeogenesis

Gluconeogenesis is the formation of glucose from noncarbohydrate substances such as amino acids and glycerol. This process takes place primarily in the liver, although it also occurs to a small extent in the kidneys.

A continuous supply of glucose is necessary for normal body functions. If blood glucose levels fall too low (severe hypoglycemia), brain dysfunction may occur, which can lead to coma and eventually to death. Also, glucose is the only fuel that supplies energy to the skeletal muscles under anaerobic conditions.

Glucose is the precursor of lactose in the mammary glands and also is actively used by the fetus.

Thus gluconeogenesis meets the body's needs for glucose when sufficient carbohydrate is unavailable.

Gluconeogenesis is increased by high-protein diets and de-

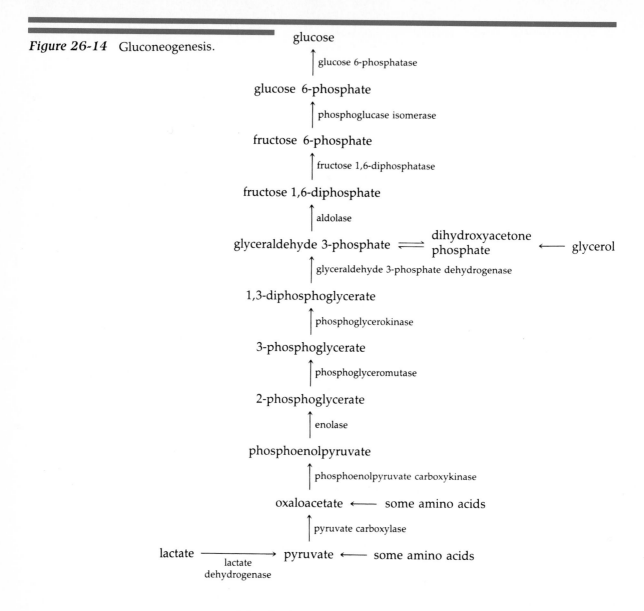

Figure 26-14 Gluconeogenesis.

creased by high-carbohydrate diets. During starvation, gluconeogenesis supplies glucose from the amino acids of the tissue protein.
In severe diabetes, gluconeogenesis not only from food protein but
also from tissue protein may lead to emaciation. Since several reactions in glycolysis are not reversible, gluconeogenesis includes
some additional reactions (see Figure 26-14).

Interconversion of Hexoses

A typical meal containing starch, sucrose, and lactose loads the liver
with galactose and fructose. These sugars must be converted into

Figure 26-15 Interconversion of hexoses.

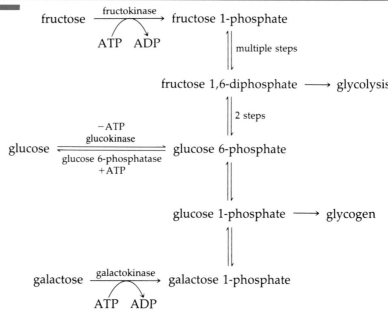

glucose. Figure 26-15 summarizes the reactions involved in these conversions. These reactions indicate that there is no essential carbohydrate. Adequate levels of any hexose can be used to produce the others.

Galactosemia is a disease associated with the inability to convert galactose into glucose. The reverse reaction, the conversion of glucose into galactose, is not affected, so galactose accumulates. High levels of galactose produce mental retardation and cataracts and stunt growth. The removal of all galactose-containing foods reverses these symptoms except mental retardation, which may or may not be reversible. Galactose-deprived patients are able to provide the galactose necessary for biosynthetic purposes by utilizing the glucose-to-galactose conversion.

Overall Scheme of Metabolism

The interrelationship of glycogenesis, glycogenolysis, glycolysis, gluconeogenesis, and the Krebs (critic acid) cycle, along with the metabolism of fats and proteins, is indicated in Figure 26-1.

Hormones Involved in Regulating Blood Sugar

The liver plays a vital function in controlling the normal blood sugar level by removing sugar from and adding sugar to the blood. The activity of the liver in maintaining the normal blood sugar level is in turn controlled by several different hormones. Among these are insulin, epinephrine, and glucagon. The hormones of the anterior pituitary, the adrenal cortex, and the thyroid also have a definite effect upon carbohydrate metabolism.

Insulin

Insulin is a hormone produced by the β cells of the islets of Langerhans in the pancreas. Insulin performs the following functions:

1. It aids in the transportation of glucose across cell membranes.
2. It accelerates the oxidation of glucose in the cells.
3. It increases the transformation of glucose to glycogen (glycogenesis) in the muscle and also in the liver. (Insulin controls the phosphorylation of glucose to glucose 6-phosphate by means of the enzyme glucokinase; see page 475.)
4. It depresses the production of glucose (glycogenolysis) in the liver.
5. It promotes the formation of fat from glucose.

Thus, the principal function of insulin may be said to be the removal of glucose from the bloodstream and a consequent lowering of the blood sugar level.

Diabetes Mellitus

If the amount of insulin is decreased or eliminated (either because of decreased activity of the islets of Langerhans or by the degeneration of these cells), the blood sugar level will rise. Increased blood sugar level (hyperglycemia) leads to glycosuria (glucose in the urine) because the renal threshold is exceeded.

Also, the lack of insulin in diabetes leads to an increased oxidation of fatty acids as a source of ATP. Increased oxidation of fatty acids leads to an accumulation of acetoacetic acid, β-hydroxybutyric acid, and acetone. These substances, commonly known as ketone bodies, form faster than they can be oxidized and removed and so accumulate in the blood (and urine). A higher-than-normal concentration of these substances in the blood is known as ketosis.

The presence of ketone bodies affects the pH of the blood, since two of the three compounds are acids. If the ketone bodies accumulate and lower the pH of the blood, a condition known as acidosis exists. A decreased pH reduces the ability of hemoglobin to carry oxygen; therefore acidosis can be very serious. Prolonged acidosis first causes nausea, then depression of the central nervous system, severe dehydration, deep coma (known as diabetic coma), and finally death.

Prompt injection of insulin will alleviate the symptoms accompanying high blood sugar. Persons suffering from diabetes mellitus can lead normal lives provided that they receive insulin as needed. Since insulin is a protein, it cannot be taken orally (it would be digested as are all proteins) and so must be administered by injection.

Epinephrine

Epinephrine is a hormone secreted by the *medulla of the adrenal glands. It stimulates the formation of glucose from glycogen in the liver (glycogenolysis) (see page 475) and so has an action opposite to that of insulin. Insulin removes glucose from the bloodstream, whereas epinephrine increases the amount of glucose present in the blood.

During periods of strong emotional stress, such as anger or fright, epinephrine is secreted into the bloodstream, where it promotes glycogenolysis in the liver. This increases the amount of glucose in the blood, making that glucose readily available as the body needs it to meet the emergency situation. The amount of glucose may then exceed the renal threshold (hyperglycemia) and sugar will appear in the urine. This is one example of how the presence of sugar in the urine may be due to a condition other than diabetes.

Glucagon

Glucagon is a hormone produced by the α cells of the pancreas. Its effects are opposite to those of insulin. Glucagon raises blood sugar levels by stimulating the activity of the enzyme phosphorylase in the liver, which changes liver glycogen to glucose. The activity of phosphorylase depends upon cAMP. Glucagon also increases gluconeogenesis from amino acids and from lactic acid. Glucagon has no effect on phosphorylation in the muscles.

Glucose Tolerance Test

A positive Benedict's test on a urine specimen indicates that the patient *may* have diabetes mellitus. However, this test is by no means conclusive proof because the presence of sugar in the urine may be due to other conditions, such as pregnancy, emotional disturbances, large intake of fruit or fruit juices, and genetic disorders (idiopathic pentosuria, fructosuria, or galactosuria).

A patient who is suspected of being diabetic is given a glucose tolerance test (see page 673). The glucose tolerance test is a valuable diagnostic tool because it indicates the ability of the body to utilize carbohydrate. A decreased utilization may indicate diabetes, whereas an increased utilization may indicate Addison's disease, hypopituitarism, or hyperinsulinism.

Summary

The end products of carbohydrate digestion are the monosaccharides. The major monosaccharide in the bloodstream is glucose. Some of the blood glucose is converted to glycogen in the liver and in the muscle. This process is called glycogenesis. Other glucose is constantly being oxidized to furnish energy for the body. If the blood sugar level rises too much, the excess spills over into the urine. The presence of glucose in the urine is known as glycosuria.

The oxidation of glucose produces energy, which is stored in high-energy compounds, especially adenosine triphosphate (ATP). Hydrolysis of ATP to adenosine diphosphate (ADP) and inorganic phosphate liberates about 7.6 kcal.

Glycogenesis—the formation of glycogen from glucose—takes place in the liver and in muscle tissue. There are many intermediate steps involved in this conversion, each being catalyzed by a specific enzyme.

Glycogenolysis—the conversion of glycogen to glucose—takes place primarily in the liver.

The breakdown of muscle glycogen to pyruvic acid (pyruvate) and lactic acid (lactate), a process that requires no oxygen (anaerobic), is termed glycolysis. Glycolysis supplies most of the ATP needed for muscle contraction. Glycolysis proceeds through a series of steps, each catalyzed by a specific enzyme. In this process three molecules of ADP are converted to ATP, which is then available for muscular work.

About one-fifth of the lactic acid formed in glycolysis is oxidized to carbon dioxide and water. The other four-fifths is converted back to liver glycogen. The cycle of glucose–glycogen–lactic acid–glycogen is known as the lactic acid cycle.

The oxidation of glucose to lactic acid can also proceed through a series of reactions called the hexose monophosphate shunt or the pentose shunt. This sequence is important because it provides five-carbon sugars needed for the synthesis of nucleic acids and nucleotides and also because it makes available the reduced form of $NADP^+$, a coenzyme necessary for the synthesis of fatty acids.

The aerobic sequence for the oxidation of lactic and pyruvic acids is called the Krebs or citric acid cycle. This cycle takes place in the mitochondria.

The first step in the Krebs cycle is the formation of acetyl CoA from pyruvic acid. This acetyl CoA, also called active acetate, is the fuel for the Krebs cycle. Acetyl CoA reacts with oxaloacetic acid and then goes through a series of steps, each catalyzed by a particular enzyme. At the completion of the cycle, oxaloacetic acid is regenerated and then picks up another molecule of acetyl CoA to carry through the same cycle again. The NADH and $FADH_2$ produced in the Krebs cycle enter the electron transport chain in the mitochondria. The overall reaction, called oxidative phosphorylation, involves oxygen and produces ATP. Several coenzymes and cytochromes are involved.

Glucose can also be formed from noncarbohydrate substances such as amino acids, fatty acids, and glycerol. Such a process is termed gluconeogenesis.

Thus carbohydrate metabolism interrelates the processes of glycogenesis, glycogenolysis, glycolysis, gluconeogenesis, and the Krebs cycle.

The liver controls the blood sugar level. This activity is governed by several hormones. Among these are insulin, epinephrine, and glucagon.

Insulin, a hormone secreted by the pancreas, accelerates oxidation of glucose in the cells, increases glycogenesis, decreases glycogenolysis, and promotes the formation of fat from glucose. Thus insulin removes glucose from the bloodstream.

Epinephrine, a hormone of the adrenal medulla, changes liver glycogen to glucose and muscle glycogen to lactic acid. Epinephrine is secreted into the bloodstream during periods of emotional stress.

Glucagon, a hormone secreted by the pancreas, has an effect opposite to that of insulin. It raises blood sugar levels.

The glucose tolerance test is given to a patient suspected of being a diabetic. He or she is fed glucose and the blood sugar level is checked for several hours.

Questions and Problems

A

1. Which monosaccharide is the principal one remaining in the bloodstream after passing through the liver?
2. What is meant by the term *normal fasting blood sugar*?
3. What is glycogenesis? Where does it occur?
4. What happens to excess carbohydrate that cannot be immediately utilized or converted to glycogen?
5. What is meant by the term *renal threshold*?
6. What is glycosuria?
7. How can glucose be removed from the bloodstream?
8. What types of compounds does the body use to store energy?
9. Indicate the hydrolysis reaction of ATP. How much energy is produced in this reaction?
10. Name two ways in which ADP can be converted to ATP.
11. Describe the process of glycogenesis, indicating the intermediary products and enzymes necessary.
12. What is glycogenolysis?
13. Are glycogenesis and glycogenolysis reversible in both muscle and liver? Why?
14. What is glycolysis? Is it an aerobic or an anaerobic sequence?
15. Does glycolysis supply most of the body's energy?
16. Indicate the steps in glycolysis and the enzymes required at each step.
17. Is all the lactic acid formed in glycolysis oxidized to carbon dioxide and water?
18. What happens to most of the lactic acid formed during glycolysis?
19. After a muscle contracts and then relaxes, how does it replenish its glycogen?
20. Diagram and label the lactic acid cycle.
21. What is the hexose monophosphate shunt? Why is it important?

22. Explain briefly what happens in the Krebs cycle (citric acid cycle).
23. What is the function of acetyl CoA in the Krebs cycle?
24. How much energy is produced by the oxidation of glucose? What is the efficiency of conversion?
25. How does the energy produced during the Krebs cycle compare with that produced during glycolysis?

B

26. Diagram and label the Krebs cycle.
27. What is gluconeogenesis? What might increase it?
28. Show the interrelationship of glycolysis, gluconeogenesis, the Krebs cycle, glycogenesis, and glycogenolysis.
29. Why is acetyl CoA not considered to be a source of glucose?
30. What are the functions of insulin?
31. When is epinephrine secreted? What does it do?
32. Where is glucagon formed and what is its function?
33. What is cytochrome oxidase? What does it do? What is unusual about it? Why is it considered essential to life?
34. Where does the electron transport system function? Describe the steps involved.

Practice Test

1. Glycogenesis occurs primarily in the _____.
   a. brain　　　　　　　b. liver
   c. stomach　　　　　　d. kidneys
2. The chief end product of glycolysis is _____.
   a. glucose　　　　　　b. pyruvic acid
   c. acetyl CoA　　　　　d. CO_2
3. 2,3-DPG, a regulator of oxygen transport in red blood cells, is produced during _____.
   a. glycolysis　　　　　b. Krebs cycle
   c. gluconeogenesis　　　d. glycogenesis

4. Which of the following is *not* produced directly in the Krebs cycle?

   a. NADH　　　　　　b. FADH$_2$
   c. GTP　　　　　　 d. ATP

5. The Krebs cycle converts _____ into CO$_2$.

   a. pyruvate　　　　b. lactate
   c. glucose　　　　 d. acetyl CoA

6. Which of the following functions are performed by insulin?

   a. increase glucose transport into cells
   b. increase cellular oxidation of glucose
   c. decrease the breakdown of glycogen into glucose
   d. all of these

7. The hexose monophosphate shunt is important because it produces _____.

   a. ATP　　　　　　b. NADH
   c. NADPH　　　　 d. GTP

8. Normal fasting blood sugar levels are approximately _____ mg/100 mL.

   a. 50 to 70　　　　b. 70 to 100
   c. 100 to 150　　　d. 150 to 180

9. ATP contains how many high-energy bonds?

   a. one　　b. two　　c. three　　d. four

10. Gluconeogenesis takes place primarily in the _____.

   a. stomach　　　　b. liver
   c. small intestine　d. pancreas

27

Metabolism of Fats

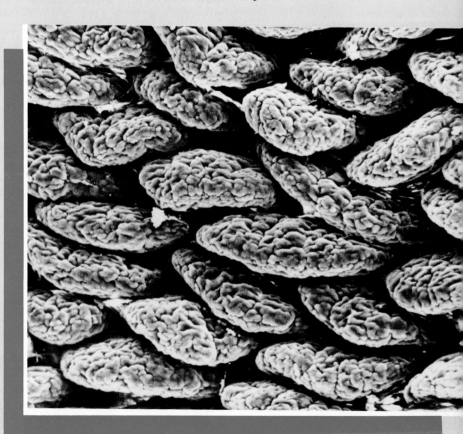

Electron microscope photo of intestinal villi where the products of fat digestion appear as resynthesized fats prior to their metabolism.

Plasma Lipid Levels

During digestion, fats and phospholipids are emulsified and then hydrolyzed into fatty acids and glycerol. The products are synthesized into triglycerides in the intestinal mucosa and flow into the thoracic duct and then into the bloodstream. However, such substances are insoluble in water and likewise in the blood. In order to be transported by the blood, fats and phospholipids form a complex with plasma (water-soluble) protein. Such complexes are called lipoproteins (see page 419). Table 30-2 (page 545) indicates lipid levels in the blood.

Abnormalities of lipid metabolism lead to various types of hypolipoproteinemia or hyperlipoproteinemia. The most common type of abnormality is diabetes, where a deficiency of insulin leads to hypertriacylglycerolemia.

Absorption of Fat

The digestion of fats takes place primarily in the small intestine, with hydrolysis yielding fatty acids and glycerol. Prior to their digestion, the fats have been emulsified by the bile salts. The products of fat digestion pass through the *lacteals of the *villi into the *lymphatics, where they appear as resynthesized fats. From the lymphatics, the fats flow through the thoracic duct into the bloodstream and then to the liver. After a meal the fat content of the blood rises and remains at a high level for several hours, then gradually decreases to the fasting level.

In the liver some of the fats are changed to phospholipids, so the blood leaving the liver contains both fats and phospholipids. These phospholipids, such as sphingomyelin and lecithin (see page 394), are necessary for the formation of nerve and brain tissue. Lecithins (phosphatidyl cholines) are also involved in the transportation of fat to the tissues. Cephalin, another phospholipid, is involved in the normal clotting of the blood. From the liver, some fat goes to the cells, where it is oxidized to furnish heat and energy. The fat in excess of what the cells need is stored as adipose tissue.

Lipolysis, the hydrolysis of triacylglycerols (triglycerides) to fatty acids and glycerol, is under the control of cAMP and various hormones. Glucagon and epinephrine stimulate the production of cAMP and so increase lipolysis. However, insulin and the prostaglandins (see page 397) depress the levels of cAMP and so decrease the rate of lipolysis.

Oxidation of Fat

The oxidation of fat (triglyceride) actually involves the oxidation of the two hydrolysis products—glycerol and fatty acids. General aspects of the controls on triglyceride breakdown into glycerol and

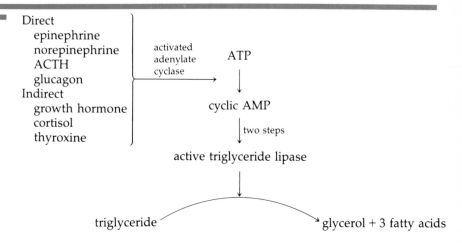

Figure 27-1 Breakdown of triglyceride in adipose tissue.

Direct
  epinephrine
  norepinephrine
  ACTH
  glucagon
Indirect
  growth hormone
  cortisol
  thyroxine

activated adenylate cyclase

ATP
↓
cyclic AMP
↓ two steps
active triglyceride lipase
↓

triglyceride → glycerol + 3 fatty acids

fatty acids are summarized in Figure 27-1. This sequence is blocked by insulin or high levels of glucose.

Oxidation of Glycerol

The glycerol part of a fat is oxidized to dihydroxyacetone phosphate, as is indicated in the following sequence. Recall that dihydroxyacetone phosphate is part of the glycolysis sequence (see page 478). This compound can be converted into glycogen in the liver or muscle tissue or into pyruvic acid, which enters the Krebs cycle. Thus, the glycerol part of a fat is metabolized through the carbohydrate sequence.

$$
\begin{array}{ccc}
CH_2OH & CH_2OH & CH_2OH \\
| & | & | \\
CHOH & HO-C-H & C=O \\
| & | & | \\
CH_2OH & CH_2-O-PO_3H_2 & CH_2-O-PO_3H_2 \\
glycerol & \alpha\text{-glycerophosphate} & dihydroxyacetone\ phosphate
\end{array}
$$

ATP ADP
glycero-kinase
phosphatase

NAD⁺ NADH + H⁺
glycerophosphate dehydrogenase

Oxidation of Fatty Acids

There are several theories about the oxidation of fatty acids. The original one, proposed by Knoop in 1905 and still preferred today, is called the β-oxidation theory. This theory involves the oxidation of the second carbon atom from the acid end of the fatty acid molecule—the β carbon atom. In this process β oxidation removes two carbon atoms at a time from the fatty acid chain. That is, an 18-carbon fatty acid is oxidized to a 16-carbon fatty acid, then to a 14-carbon fatty acid, and so on, until the oxidation process is complete. A simplified version of such an oxidation is shown in Figure 27-2.

Figure 27-2   Oxidation of a fatty acid.

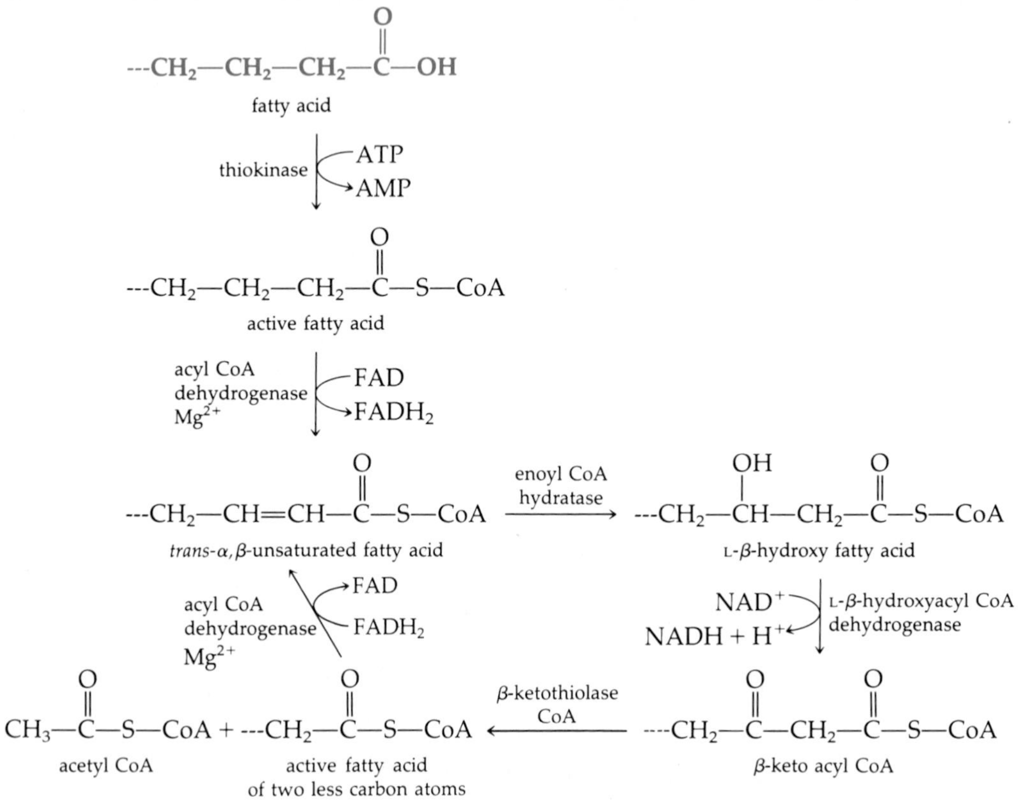

The acetyl CoA thus produced enters the Krebs cycle and the new molecule of active fatty acid goes through the same sequence again, each time losing two carbon atoms until the entire fatty acid molecule has been oxidized. This sequence presupposes the presence of fatty acids containing an even number of carbon atoms, a condition usually encountered in nature. The $FADH_2$ and the $NADH + H^+$ enter the respiratory chain.

If fatty acids containing an odd number of carbon atoms are oxidized, they follow the same steps except that the final products are acetyl CoA and propionyl CoA. The propionyl CoA is changed in a series of steps to succinyl CoA, which then enters the Krebs cycle, as does the acetyl CoA. These reactions require the presence of vitamin B_{12} and also biotin.

The unsaturated fatty acids are metabolized slowly. They must first be reduced by some of the dehydrogenases found in the cells. Then they can follow the fatty acid cycle for oxidation.

Energy Produced by Oxidation of Fatty Acids

The oxidation of 1 g of fat produces more than twice as much energy as the oxidation of 1 g of carbohydrate. Let us see why.

Table 27-1 ATP Formed from the Oxidation of a 16-Carbon Fatty Acid

Source	Number of ATP Molecules Formed per C_{16} Molecule
7 FADH$_2$	14
7 NADH	21
Initial activation of fatty acid	−2
8 acetyl CoA	96
	$\overline{129}$

The oxidation of acetyl CoA through the Krebs cycle yields 12 high-energy phosphate bonds (ATP) per molecule of acetyl CoA. If we consider the oxidation of palmitic acid, a 16-carbon fatty acid, eight 2-carbon units will be formed during the β-oxidation cycle. These eight 2-carbon units will yield $8 \times 12 = 96$ ATP. However, 2 ATP are used up in the initial activation of the fatty acid. In addition, it has been calculated that palmitic acid will produce 35 ATP as it goes through the fatty acid cycle (7 FADH$_2$, each equivalent to 2 ATP, and 7 NADH, each equivalent to 3 ATP). That is, the net number of ATP molecules produced will be $96 - 2 + 35 = 129$ (see Table 27-1).

Considering each mole of ATP as requiring 7.6 kcal for formation, 129×7.6 kcal, or 980 kcal is needed. The theoretic yield from 1 mole of palmitic acid is 2340 kcal, so that the efficiency of conversion is 980/2340, or 42 percent, with the remainder of the energy being produced as heat. (Other fatty acids and glycerol are also oxidized, so the net result is that fats produce much more energy than do carbohydrates.)

Ketone (Acetone) Bodies

In a diabetic patient, or any other situation in which carbohydrate metabolism is restricted, the body uses oxaloacetate to produce glucose for the brain and muscles. This reduces the amount of oxaloacetate available for the Krebs cycle, and acetyl CoA cannot be properly metabolized. When this occurs, the acetyl CoA is changed to acetoacetyl CoA, which is converted into acetoacetic acid in the liver by

Figure 27-3 Formation of ketone (acetone) bodies.

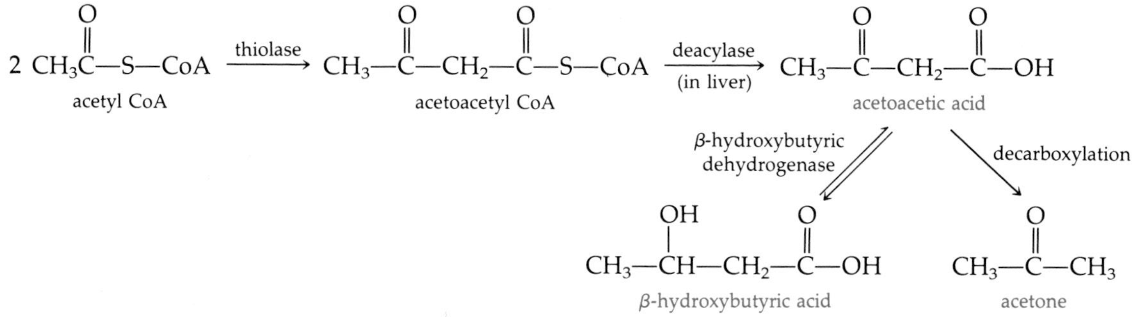

the enzyme deacylase. Acetoacetic acid may be changed into acetone and β-hydroxybutyric acid, as shown in Figure 27-3.

These three substances—acetoacetic acid, β-hydroxybutyric acid, and acetone—are commonly called acetone bodies, or ketone bodies. They are carried by the blood to the muscles and tissues, where they are converted back to acetoacetyl CoA and then oxidized normally. However, during diabetes, the production of these ketone bodies by the liver exceeds the ability of the muscles and tissues to oxidize them, so they accumulate in the blood.

Ketosis

The excess accumulation of ketone bodies in the blood is called **ketonemia.** The excess accumulation of ketone bodies in the urine is called **ketonuria.** The overall accumulation of ketone bodies in the blood and the urine is called ketosis. During ketosis, acetone can be detected on the patient's breath because it is a volatile compound and is easily excreted through the lungs. Ketosis may occur with diabetis mellitus, in starvation or severe liver damage, or on a diet high in fats and low in carbohydrates.

During diabetes mellitus the body is unable to oxidize carbohydrates and instead oxidizes fats, leading to an accumulation of ketone bodies in the blood and the urine. These ketone bodies are acidic and tend to decrease the pH of the blood. The lowering of the pH of the blood is termed acidosis and may lead to a fatal coma. During acidosis an increased amount of water intake is needed to eliminate the products of metabolism. Unless the water intake of a diabetic is increased, dehydration will occur. Dehydration of diabetics can also be caused by polyuria due to an increased amount of glucose in the urine.

Likewise, during prolonged starvation or on a high-fat, low-carbohydrate diet, the body tends to burn fat instead of carbohydrate, leading to ketosis and acidosis.

In severe liver damage, the liver cannot store glycogen in the required amounts. The resulting shortage of carbohydrates needed for the normal oxidation of fats leads to ketosis.

Storage of Fat

Fat in excess of that required for the normal oxidative processes of the body is stored as adipose tissue under the skin and around the internal organs. This stored fat serves several important purposes.

1. Reserve supply of food.
2. Support for the internal organs.
3. Shock absorber for the internal organs.
4. Insulation of interior of the body against sudden external changes in temperature.

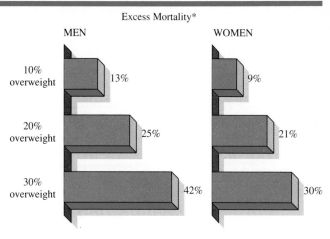

Figure 27-4 Mortality table. The percentages represent excess mortality compared with mortality of standard risks (mortality of standard risk = 100 percent). [*Courtesy Metropolitan Life Insurance Co., New York.*]

The fat stored in the body is in equilibrium with that in the bloodstream. That is, the fats stored in the adipose tissue do not merely remain there as inert compounds until they are needed. They are continuously being used and replaced, and there is always a dynamic transfer of fats between the bloodstream and the storage tissues.

Obesity (20 percent or more over normal weight) is a condition in which excess fat is deposited as adipose tissue. An obese person eats more food than his or her body can burn up, and the excess is converted to fat and stored as adipose tissue. For every 9 kcal of food eaten in excess of the body's requirements, 1 g of fat is deposited.

Most people have a tendency to become overweight as they grow older. This is because they require less food for the maintenance of their bodies and because they exercise less than younger people.

In general obesity leads to a shortened life expectancy, as indicated in Figure 27-4. An overweight person runs a higher than normal risk of developing cardiovascular disease, diabetes, or liver disease. A weight greater than 10 percent above that considered normal for a person's age and height can cause medical problems. The answer to obesity lies in proper dieting under the supervision of a doctor because the metabolism of the body is a highly intricate mechanism that can very easily be disturbed.

Excessive accumulation of triglycerides in the liver leads to cirrhosis and impaired liver function. Such an accumulation may be due to either increased levels of free plasma fatty acids or blockage in the production of plasma lipoproteins from the free fatty acids.

Chronic alcoholism also leads to hyperlipidemia and eventually to cirrhosis.

Lipogenesis

Lipogenesis—the conversion of glucose to fats—takes place in the liver and in the adipose tissue, with the latter place predominating. Insulin is necessary for lipogenesis both in the liver and in the adipose tissue. Lipogenesis is reduced during fasting or during a high-fat diet, whereas it is increased during a high-carbohydrate diet.

The synthesis of fatty acids occurs in the mitochondria and in the cell cytoplasm, especially the latter. The process in the mitochondria involves the lengthening of fatty acid chains of moderate length, whereas the cytoplasmic processes involve the synthesis of fatty acids from acetyl CoA.

Steps in the synthesis of fatty acids from acetyl CoA are

Step 1   Acetyl CoA is changed to malonyl CoA.

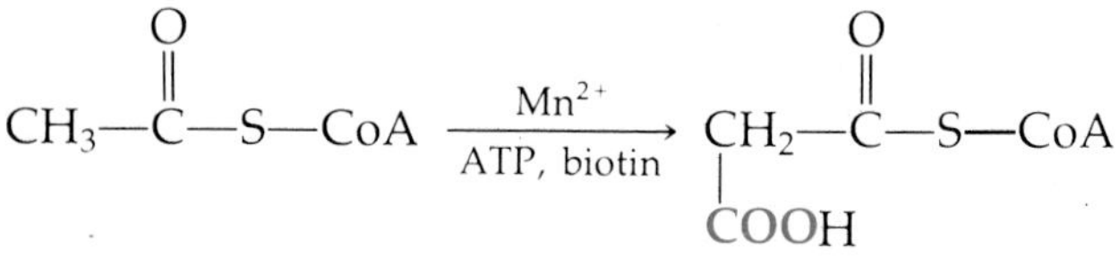

$$CH_3-\overset{\overset{O}{\|}}{C}-S-CoA \xrightarrow[\text{ATP, biotin}]{Mn^{2+}} \underset{\underset{COOH}{|}}{CH_2}-\overset{\overset{O}{\|}}{C}-S-CoA$$

acetyl CoA   malonyl CoA

Note that one carbon atom has been added to the chain.

Step 2   Malonyl CoA reacts with another molecule of acetyl CoA to form acetoacetyl complex.

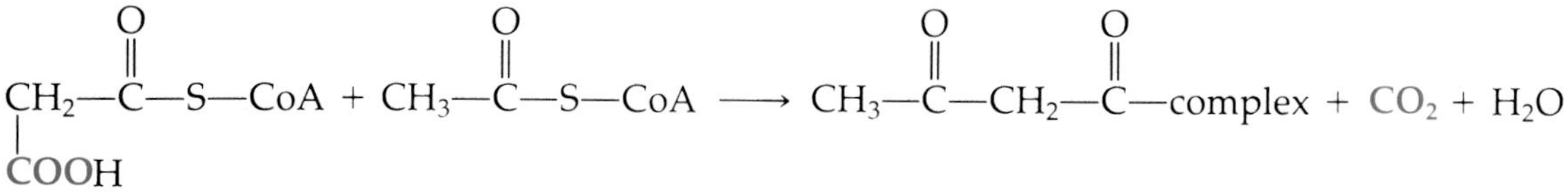

$$\underset{\underset{COOH}{|}}{CH_2}-\overset{\overset{O}{\|}}{C}-S-CoA + CH_3-\overset{\overset{O}{\|}}{C}-S-CoA \longrightarrow CH_3-\overset{\overset{O}{\|}}{C}-CH_2-\overset{\overset{O}{\|}}{C}-complex + CO_2 + H_2O$$

malonyl CoA   acetyl CoA   acetoacetyl complex

Note that the additional carbon atom from the previous equation has been removed. It was used mainly to activate the α carbon in the acetyl CoA so that the condensation reaction could take place. The enzyme complex required here is called fatty acid synthetase and consists of seven enzymes.

Step 3   The keto group of the acetoacetyl complex is reduced to the corresponding alcohol by NADPH.

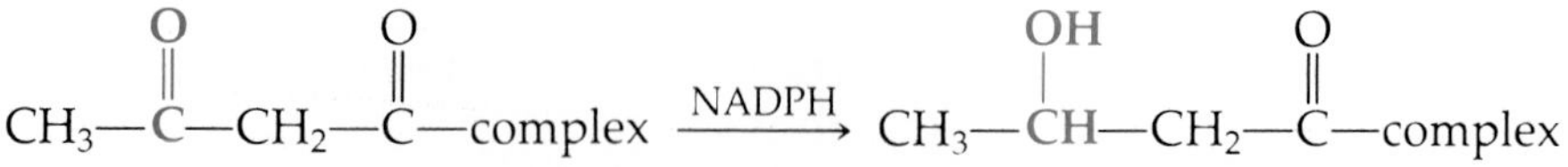

$$CH_3-\overset{\overset{O}{\|}}{C}-CH_2-\overset{\overset{O}{\|}}{C}-complex \xrightarrow{NADPH} CH_3-\overset{\overset{OH}{|}}{CH}-CH_2-\overset{\overset{O}{\|}}{C}-complex$$

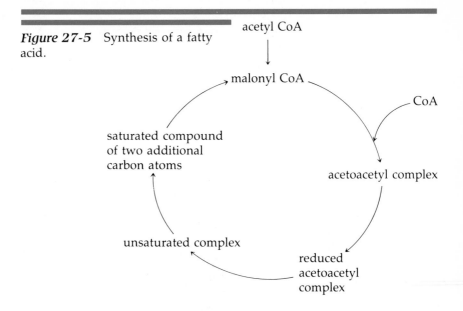

Figure 27-5   Synthesis of a fatty acid.

Step 4   The alcohol produced in step 3 is dehydrated to form an unsaturated compound.

$$CH_3—\underset{\underset{OH}{|}}{CH}—CH_2—\underset{\overset{O}{||}}{C}—complex \longrightarrow CH_3—CH=CH—\underset{\overset{O}{||}}{C}—complex + H_2O$$

Step 5   The unsaturated compound in step 4 is reduced by NADPH to the corresponding saturated compound.

$$CH_3—CH=CH—\underset{\overset{O}{||}}{C}—complex \xrightarrow{NADPH} CH_3—CH_2—CH_2—\underset{\overset{O}{||}}{C}—complex$$

The product of step 5 goes through the cycle again and again, each time adding two carbon atoms until the fatty acid the body requires has been synthesized. The process is diagrammed in Figure 27-5.

Synthesis of Phospholipids

The phospholipids are very important because they form (along with protein) the framework of most of the cell membrane system (see page 392). Figure 27-6 shows the formation of a diglyceride as an intermediate step in the formation of a fat (a triglyceride).

Figure 27-6 Synthesis of a triglyceride.

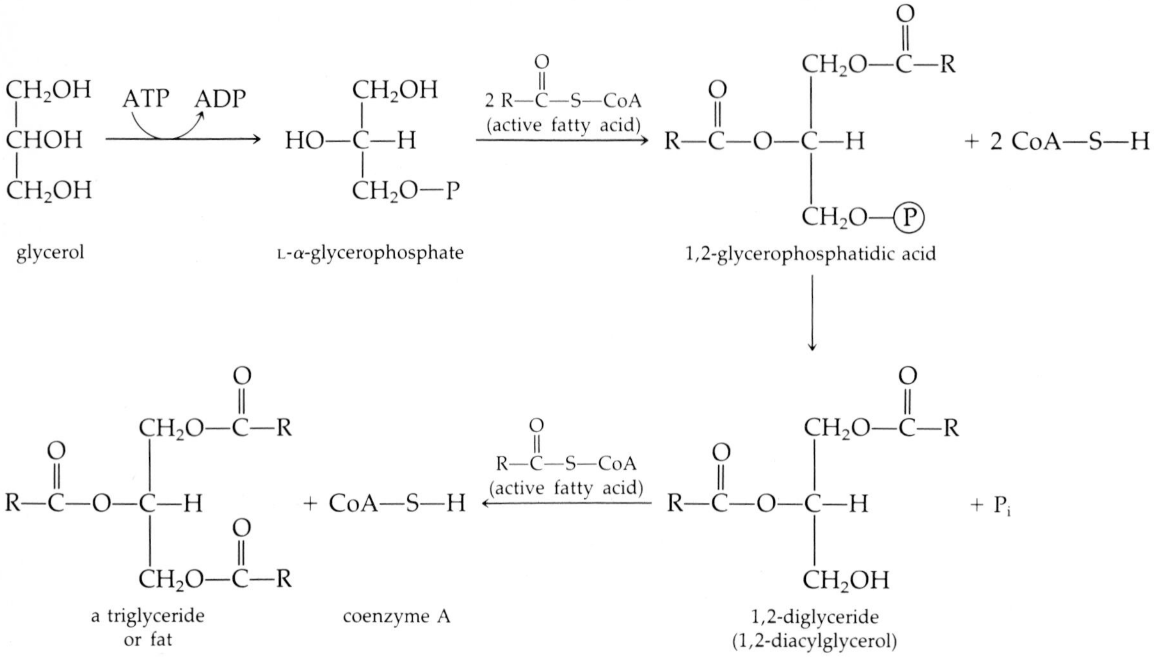

The 1,2-diglyceride may also be converted into a phospholipid such as lecithin as follows.

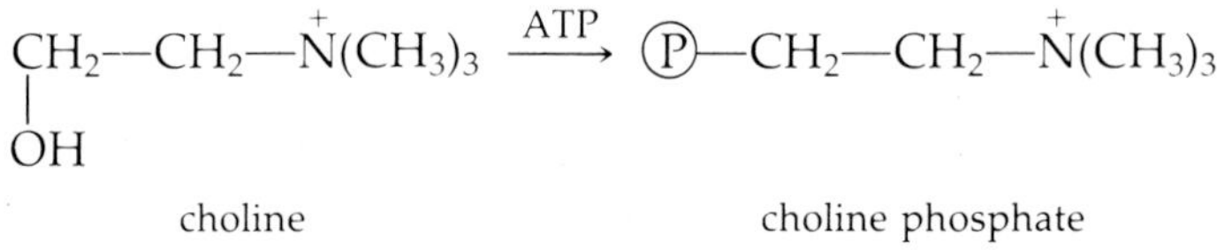

$$CH_2-CH_2-\overset{+}{N}(CH_3)_3 \xrightarrow{\text{ATP}} \text{P}-CH_2-CH_2-\overset{+}{N}(CH_3)_3$$

choline   choline phosphate

$$\text{choline}-\text{P} + \underset{\substack{\text{cytidine}\\\text{triphosphate}}}{\text{CTP}} \longrightarrow \text{CDP choline}$$

In this reaction the CTP acts as an activator for the choline phosphate.

The CDP choline reacts with a diglyceride to form phosphatidylcholine (lecithin).

$$\text{1,2-diglyceride} + \text{CDP choline} \longrightarrow$$

phosphatidylcholine
(lecithin)

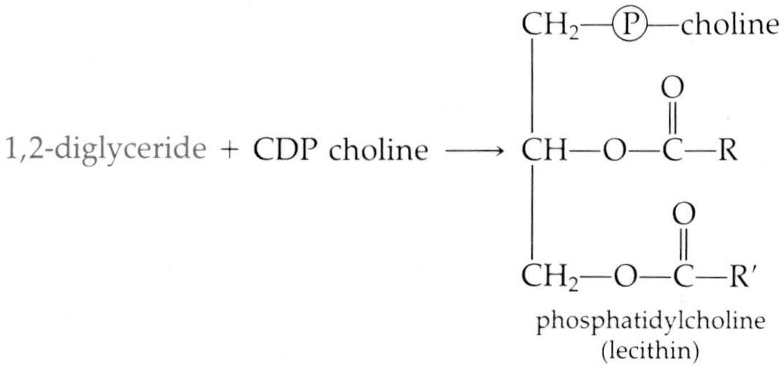

If ethanolamine, inositol, or serine is used in place of choline, the corresponding phospholipid—phosphatidylethanolamine, phosphatidylinositol, or phosphatidylserine—is formed.

Cholesterol

Cholesterol is found in all cells of the body but particularly in brain and nerve tissue. Cholesterol occurs in animal fat but not in plant fat. An adult normally ingests about 0.3 g cholesterol daily from such foods as egg yolk, meat fats, liver, and liver oils. In addition, it has been estimated that the body manufactures about 1 g of cholesterol daily.

Cholesterol normally is eliminated in the bile. However, sometimes it settles out in the gallbladder as gallstones. If cholesterol deposits in the walls of the larger arteries, the condition is known as atherosclerosis, a type of hardening of the arteries. When this occurs, there is a decrease in the usable diameter of the blood vessels. The elasticity of the arterial walls decreases. There is an interference with the rate of blood flow because there is greater friction due to the irregular lining of the blood vessels. This irregular lining may also cause clots as the blood flows over that type of surface and lead to myocardial infarction.

Cholesterol is important as a precursor of several important steroids such as vitamin D, the sex hormones, and the adrenocortical hormones (the hormones of the cortex of the adrenal glands).

Synthesis

Cholesterol is synthesized primarily in the liver, but the adrenal cortex, skin, testes, aorta, and intestines are also able to synthesize

Figure 27-7 Comparison of normal artery wall with those containing fatty deposits. [*Reprinted by permission, American Heart Association.*]

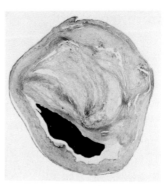

(a)
Normal artery

(b)
Fatty deposits in vessel wall

(c)
Plugged artery with fatty deposits and clot

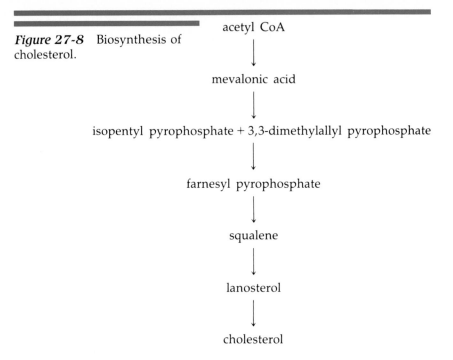

Figure 27-8 Biosynthesis of cholesterol.

acetyl CoA

↓

mevalonic acid

↓

isopentyl pyrophosphate + 3,3-dimethylallyl pyrophosphate

↓

farnesyl pyrophosphate

↓

squalene

↓

lanosterol

↓

cholesterol

it. This synthesis takes place in the microsomal and cytosomol fraction of the cell. Acetyl CoA is the starting material and is also the source of all the carbon atoms in cholesterol (see Figure 27-8).

Lipid Storage Diseases

Glycolipids (see page 397) are components of nerve and brain tissue. If these compounds accumulate because of a breakdown in the enzyme system, genetic abnormalities such as Tay–Sachs disease, Niemann–Pick disease, and Gaucher's disease occur (see page 661).

Summary

Emulsified fats pass through the lacteals of the villi into the lymphatics through the thoracic duct to the liver. In the liver some of the fats are changed to phospholipids, which are necessary for the formation of nerve and brain tissue. Some fat is stored in the adipose tissue; some is oxidized to furnish energy.

Glycerol from a fat is oxidized to dihydroxyacetone phosphate, which is part of the glycolysis sequence. That is, the glycerol part of a fat is metabolized through the carbohydrate sequence.

Fatty acids pass through a β-oxidation cycle in which two carbon atoms are removed at a time and converted to acetyl CoA, which then enters the Krebs cycle.

The oxidation of one molecule of a 16-carbon fatty acid yields a total of 129 ATP molecules, with an efficiency of about 42 percent.

Ketone (acetone) bodies are normally produced in the β-oxidation process. However, they are produced only in small amounts and do not normally accumulate. If excess ketone bodies accumulate in the blood, a condition known as ketonemia exists. Accumulation of ketone bodies in the urine is termed ketonuria. The overall accumulation of ketone bodies is called ketosis; it occurs in the abnormal metabolism of carbohydrates.

Fat in excess of the body's needs is stored as adipose tissue. Stored fat serves as a reserve supply of food, as a support for the internal organs, as a shock absorber for the internal organs, and as an insulator for the body.

When more than a normal amount of fat is deposited in the adipose tissue, the resulting condition is termed obesity. Obesity may be due to a glandular disorder or simply to overeating.

The conversion of glucose to fats, lipogenesis, takes place in the liver and in the adipose tissue.

The synthesis of fatty acids occurs both inside and outside the mitochondria, with the latter being the predominant site.

Phospholipids can be synthesized from glycerol. If the 1,2-diglyceride formed from glycerol reacts with cytidine diphosphate choline, then the phospholipid lecithin is formed. If instead the 1,2-diglyceride combines with cytidine diphosphate ethanolamine, then the phospholipid cephalin is formed.

Cholesterol is found in all cells of the body but particulary in the nerve tissue. Cholesterol is normally eliminated in the bile, but if it settles out in the gallbladder, gallstones are formed. If cholesterol deposits on the walls of the arteries, atherosclerosis occurs.

Cholesterol is an important precursor for vitamin D, for the sex hormones, and for the adrenocortical hormones.

Cholesterol is synthesized in the liver beginning with acetyl CoA and proceeding through a series of steps.

Questions and Problems

1. Where are fats absorbed? How do they get to the liver?
2. What happens to fats in the liver?
3. Describe the oxidation of the glycerol part of a fat.
4. What is the β-oxidation theory of fats? What are the end products of this cycle? What becomes of these products?
5. What types of products are formed during the oxidation of fatty acids with an even number of carbon atoms? an odd number of carbon atoms?
6. What happens to unsaturated fatty acids before they can be oxidized?
7. Compare the energy produced by the oxidation of 1 g carbohydrate and 1 g fat.
8. In the oxidation of a fat, in which part of the sequence is most of the energy produced?
9. The oxidation of one molecule of a 16-carbon fatty acid produces how many ATP molecules?
10. What are the ketone bodies? How are they interrelated structurally?
11. Where are ketone bodies formed?
12. What is ketonemia? ketonuria? ketosis? Under what conditions might ketosis occur?
13. Where is excess fat stored? What functions does this fat have?
14. What conditions lead to obesity?
15. What is lipogenesis? Where does it occur?
16. How does lipogenesis in the mitochondria compare with the oxidation of a fat?
17. Why are phospholipids important?
18. Describe the synthesis of lecithin.
19. How does the synthesis of lecithin compare with that of cephalin?
20. Where is cholesterol found in the body? Where is it produced?
21. Why is cholesterol important in the body?
22. If cholesterol is deposited in the walls of the large arteries, what is the condition termed?
23. If cholesterol is deposited in the gallbladder, what is formed?

Practice Test

1. Which of the following hormones does *not* cause triglyceride breakdown?
   a. glucagon
   b. insulin
   c. epinephrine
   d. cortisol
2. The glycerol portion of a triglyceride is metabolized through the _____ sequence.
   a. amino acid
   b. fatty acid
   c. carbohydrate
   d. nucleic acid
3. The production of ketone bodies occurs whenever the body has limited _____ to metabolize.
   a. amino acids
   b. fatty acids
   c. carbohydrates
   d. nucleic acids
4. The end product of fatty acid oxidation is _____
   a. lactate
   b. acetone
   c. pyruvate
   d. acetyl CoA
5. Cholesterol is involved in _____.
   a. gallstone formation
   b. atherosclerosis
   c. formation of sex hormones
   d. all of these

6. In the oxidation of a fatty acid, how many carbon atoms are removed at a time?
   a. one
   b. two
   c. three
   d. four
7. Restricted carbohydrate metabolism causes _____.
   a. acidosis
   b. accumulation of acetone
   c. ketosis
   d. all of these
8. Lipogenesis takes·place primarily in _____.
   a. the liver
   b. adipose tissue
   c. both of these
   d. neither of these
9. Fat serves as a(n) _____.
   a. reserve supply of food
   b. support for the internal organs
   c. insulation for the body
   d. all of these
10. Fat is transported through the bloodstream as _____.
    a. phospholipids
    b. lipoproteins
    c. glycoproteins
    d. all of these

28

Metabolism of Proteins

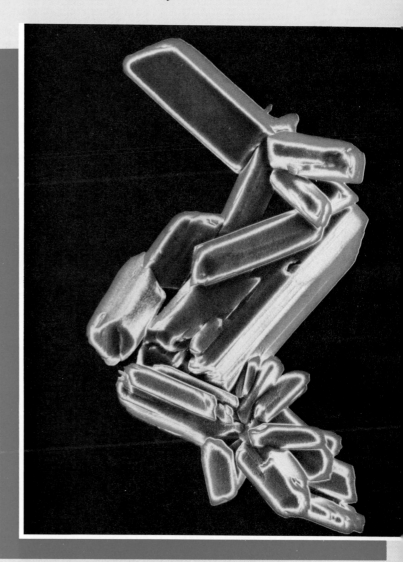

Scanning electron micrograph of glycine, the simplest amino acid.

Functions of Protein in the Body

During digestion, proteins are hydrolyzed into amino acids, which are then absorbed into the bloodstream through the villi of the small intestine. These amino acids enter the amino acid pool of the body (see Figure 28-1).

The amino acids from the amino acid pool serve many functions. For example, they

1. Convert to tissue protein to build new tissue.
2. Convert to tissue protein to replace old tissue.
3. Aid in the formation of hemoglobin.
4. Aid in the formation of some hormones.
5. Aid in the formation of enzymes.
6. Are used in synthesis of other amino acids that the body needs.
7. Serve as a source of energy when they are catabolized.
8. Are used to form nucleic acids, neurotransmitters, and other substances needed for body functions.

Nitrogen Balance

The body can store carbohydrates (as glycogen in the liver and the muscles) and fats (as adipose tissue and in tissues around the internal organs). However, the body cannot store protein. The amino acids that result from the digestion of protein are either used for the synthesis of new tissues, the replacement of old tissues, and the formation of various required body substances such as hormones and enzymes, or they are converted to fat or oxidized to furnish energy.

Because the body cannot store protein and because protein contains nitrogen, the amount of nitrogen taken in each food should

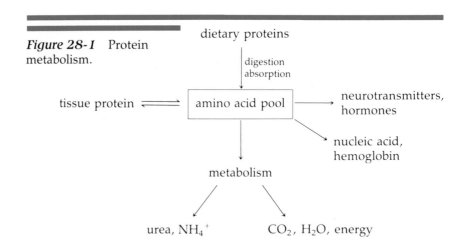

Figure 28-1 Protein metabolism.

510

usually equal the amount of nitrogen excreted per day (for a normal adult whose weight remains constant). This takes into consideration the normal replacement of worn-out tissue where the reaction is merely one of exchange between one amino acid and another.

A person whose body excretes as much nitrogen per day as is taken in with food is said to be in nitrogen balance. Children exhibit a positive nitrogen balance because they take in more nitrogen in their food than they excrete. This is because children need amino acids to build growing body tissues. Any body condition marked by the growth of new tissues will exhibit a positive nitrogen balance. An example of this is seen in persons recovering from a wasting illness where the body needs to rebuild tissues and so does not excrete as much nitrogen (amino acids) as it takes in.

Conversely, if the body excretes more nitrogen than it acquires from food, a negative nitrogen balance exists. Conditions that can produce a negative nitrogen balance are starvation, malnutrition, prolonged fever, and various wasting illnesses. However, a person can survive for a reasonable period of time in negative nitrogen balance, as when dieting, since body protein will be used for essential purposes.

However, when inadequate protein intake leads to a negative nitrogen balance, the individual will eventually develop a protein deficiency disease called kwashiorkor. This disease is characterized by a wasting away of fat and muscle and a degeneration of many of the internal organs (see page 547).

Protein malnutrition also develops in seriously ill patients who cannot eat by mouth. Such patients include those who have had major surgery or trauma. If protein is not provided along with other foods and minerals, the person will have a negative nitrogen balance and will have a much slower rate of recovery. In such cases, a procedure called *hyperalimentation* is used whereby glucose and hydrolyzed protein are administered intravenously.

A person consuming about 300 g of carbohydrate, 100 g of fat, and 100 g of protein (2500 kcal) per day will excrete about 16.5 g of nitrogen per day. About 95 percent of this nitrogen is eliminated through the kidneys and the remaining 5 percent in the stool. Most of the nitrogen is excreted in the form of urea (see page 515).

Synthesis of Protein

As was discussed in Chapter 23, proteins are synthesized from amino acids through the various intermediaries such as peptides and polypeptides. The body takes the amino acids produced by the digestion of protein and recombines them into the protein that it needs in the various parts of the body. (See Chapter 34 for the mechanism of protein synthesis.)

Some amino acids that the body needs can be synthesized from other amino acids. However, there are certain amino acids that the

Table 28-1 Amino Acids in the Body

Nutritionally Essential Amino Acids	Nutritionally Nonessential	
	Amino Acid	Precursor
Isoleucine	Alanine	Pyruvate
Leucine	Arginine	Glutamate
Lysine	Asparagine	Aspartate
Methionine	Aspartate	Oxaloacetate
Phenylalanine	Cysteine	Serine, homocysteine
Threonine	Glutamate	α-Ketoglutarate
Tryptophan	Glutamine	Glutamate
Valine	Glycine	Serine
For Infants:	Hydroxylysine	
Arginine[a]	Proline	Arginine
Histidine[a]	Serine	3-Phosphoglycerate
	Tyrosine	Phenylalanine

[a] Infants cannot manufacture arginine or histidine and must obtain them in their diet.

body needs but cannot synthesize. These amino acids must be supplied in the food if the body is to function normally.

The eight nutritionally essential (ten for infants) and the nonessential amino acids (with their precursors) are listed in Table 28-1. Recall that the body requires the essential amino acids in the diet but can manufacture the nonessential amino acids that it requires from other amino acids (see page 408).

Some proteins contain all of the nutritionally essential amino acids; those that do not are called incomplete proteins. Two common incomplete proteins are gelatin, which is lacking in tryptophan, and zein (from corn), which is lacking in both tryptophan and lysine.

Biosynthesis of Nonessential Amino Acids

Most of the time the mixture of amino acids provided by the protein in our diet is not in the proportions required by our bodies. Consequently, it is necessary to rearrange the amino acid pool metabolically. Table 28-1 shows the sources of nonessential amino acids that can be synthesized by the body provided that there are adequate supplies of amine nitrogen.

Many of these syntheses involve transamination (see following section). One particular biosynthetic pathway deserves a closer look. This pathway involves the methyl cycle, which is illustrated in Figure 28-2. Note that the methyl cycle consumes three high-energy phosphate bonds. Several important methylated compounds are produced as by-products. The methyl carrier is tetrahydrofolate (THF), which is the active form of the B vitamin folic acid (see page 610). Several anticancer drugs, such as methotrexate, block the synthesis of methylated DNA by preventing the transfer of methyl groups into the cycle by THF.

Figure 28-2   The methyl cycle.

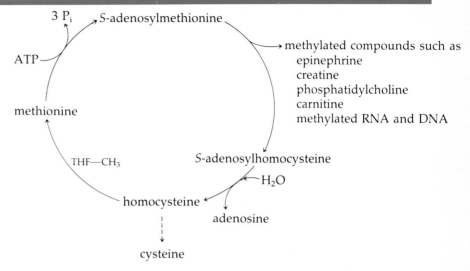

Transamination

Transamination is a reaction in which one or more amino acids are converted into other amino acids. When an α-amino acid and an α-keto acid react, they interconvert to form another α-amino acid and another α-keto acid. All other amino acids can be converted to glutamic acid by transamination. In this way the body can manufacture the amino acids it needs.

Transamination is catalyzed by enzymes called transaminases or aminotransferases. An essential part of the active site of a transaminase is pyridoxal phosphate, the coenzyme form of vitamin B_6. Transaminases are used in the diagnosis of a variety of disorders. For example, SGOT (serum glutamic oxaloacetic transaminase) levels are increased after myocardial infarction and with cirrhosis of the liver, and SGPT (serum glutamic pyruvic transaminase) levels are increased during infectious hepatitis. Decreased serum transaminase levels occur during pregnancy and with vitamin B_6 deficiency.

An example of transamination is the reaction of glutamic acid (an α-amino acid) and oxaloacetic acid (an α-keto acid) to form α-ketoglutaric acid (another α-keto acid) and aspartic acid (another α-amino acid).

Note that this transamination reaction is catalyzed by the enzyme GOT (glutamic oxaloacetic transaminase). This enzyme occurs in high concentration in heart muscle. Increased levels of GOT in the bloodstream (called serum GOT or SGOT) indicate myocardial infarction, which results from the reduction in blood flow to the heart muscle caused by a clot in the coronary artery (see page 668).

In addition to transamination, the body has other processes for the synthesis of several nonessential amino acids.

Some genetic diseases (see page 658) are caused by a deficit or lack of enzymes that catalyze the synthesis of certain amino acids. One such disease is PKU (phenylketonuria). Normally the body converts phenylalanine (an α-amino acid) to tyrosine (another α-amino acid). The required enzyme is phenylalanine hydroxylase.

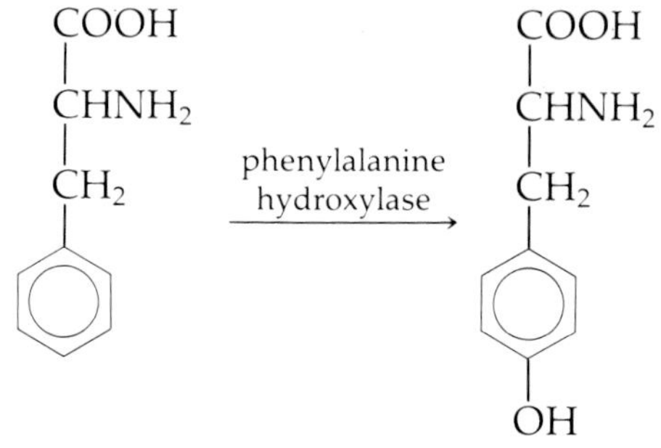

phenylalanine   tyrosine

If this enzyme is lacking (due to a genetic deficiency), then tyrosine cannot be produced. Instead, phenylalanine is converted into phenylpyruvic acid (a transamination reaction).

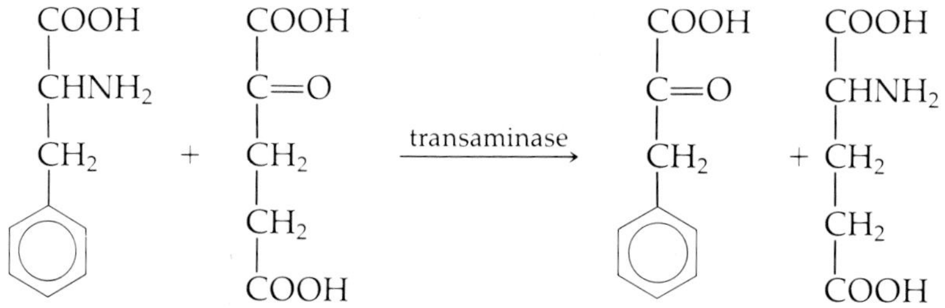

phenylalanine   α-ketoglutaric acid   phenylpyruvic acid   glutamic acid

Phenylpyruvic acid accumulates in the bloodstream and is eliminated in the urine. PKU refers to the presence of phenyl ketones in the urine. PKU is characterized by severe mental retardation.

Body's Requirements of Protein

A certain minimum daily amount of protein is required for the normal replacement of body tissues. This amount, however, may be greatly increased by increased metabolism such as during high fevers. However, for the normal adult keeping a constant weight, the recommended daily intake of protein is approximately 0.8 g per kilogram of body weight. This amounts to approximately 46 g of protein per day for the adult female and 56 g per day for the adult male.

Catabolism of Amino Acids

The amino acids that the body does not need for tissue building or that are not of the correct type for this purpose are broken down to ammonia, carbon dioxide, and water, at the same time producing heat and energy. Such a process is called *catabolism*, which is defined as the breakdown or oxidation of large molecules into smaller molecules with the release of energy. *Anabolism* is the buildup of large molecules necessary for life and is a process requiring energy. The total of all anabolic and catabolic reactions is termed *metabolism*.

Deamination

Deamination (also called oxidative deamination) is a catabolism reaction whereby the α-amino group of an amino acid is removed, forming an α-keto acid and ammonia. Deamination occurs primarily in the liver and the kidneys under the catalysis of the enzyme amino acid oxidase.

$$CH_3-\underset{\underset{NH_2}{|}}{CH}-COOH \xrightarrow{\text{amino acid oxidase}} CH_3-\underset{\underset{O}{\|}}{C}-COOH + NH_3$$

alanine   pyruvic acid   ammonia
(an α-amino acid)   (an α-keto acid)

The α-keto acid produced by this process can undergo several types of reactions.

1. It can be catabolized to carbon dioxide, water, and energy in the citric acid cycle.
2. It can be converted to carbohydrates (glycogen) or to fat.
3. It can be reconverted to a different amino acid by transamination.

Formation of Urea

The ammonia formed from the deamination of amino acids combines with carbon dioxide to form urea and water. This process takes place in the liver. The overall reaction can be written as

$$2\ NH_3 + CO_2 \xrightarrow{\text{enzymes}} NH_2CONH_2 + H_2O \quad (28\text{-}1)$$

ammonia   carbon dioxide   urea   water

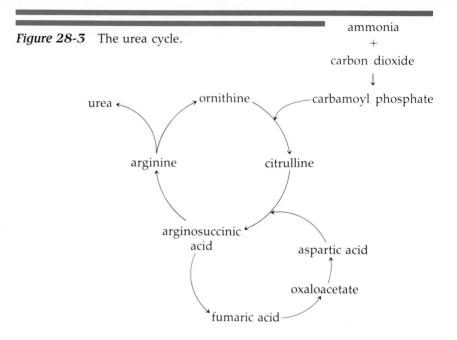

Figure 28-3   The urea cycle.

but the actual process is certainly not as simple as that. It consists of a series of steps, each catalyzed by an appropriate enzyme.

Ammonia is a toxic by-product of the deamination of amino acids and must be removed from the body, predominantly in the form of the compound urea. Three amino acids are involved in the conversion of ammonia to urea—arginine, citrulline, and ornithine. The pathway for the conversion of ammonia to urea is called the urea cycle (see Figure 28-3).

The first step in this cycle is the reaction of ammonia and carbon dioxide to form carbamoyl phosphate. In this reaction, ATP is converted to ADP. This reaction is catalyzed by N-acetylglutamic acid and the enzyme carbamoyl phosphate synthetase in the presence of magnesium ions. Lack of this enzyme produces the very serious disorder hyperammonemia. The body has only one principal way to dispose of excess nitrogen, and this is by the excretion of urea.

$$NH_3 + CO_2 + H_2O + 2\ ATP \xrightarrow[\substack{N\text{-acetylglutamic} \\ acid, \\ carbamoyl\ phosphate \\ synthetase}]{Mg^{2+}} \underset{\substack{carbamoyl \\ phosphate}}{H_2N\overset{\overset{\displaystyle O}{\|}}{-}C-O-\text{\textcircled{P}}} + 2\ ADP + P_i$$

In the second step, carbamoyl phosphate combines with ornithine to form citrulline. This reaction is catalyzed by the liver enzyme ornithine *trans*-carbamoyl transferase. Lack of this enzyme will produce a different hyperammonemia.

$$
\begin{array}{ccc}
\underset{\text{ornithine}}{
\begin{array}{l}
\text{CH}_2\text{—NH}_2 \\
| \\
(\text{CH}_2)_2 \\
| \\
\text{HC—NH}_2 \\
| \\
\text{COOH}
\end{array}}
+
\underset{\substack{\text{carbamoyl} \\ \text{phosphate}}}{
\text{H}_2\text{N}\overset{\overset{\text{O}}{\|}}{-}\text{C—O—}\textcircled{P}}
&
\xrightarrow[\text{transferase}]{\substack{\text{ornithine} \\ \textit{trans-}\text{carbamoyl}}}
&
\underset{\text{citrulline}}{
\begin{array}{l}
\text{CH}_2\text{—NH—}\overset{\overset{\text{O}}{\|}}{\text{C}}\text{—NH}_2 \\
| \\
(\text{CH}_2)_2 \\
| \\
\text{HC—NH}_2 \\
| \\
\text{COOH}
\end{array}}
\end{array}
$$

Next citrulline reacts with aspartic acid (derived from transamination of oxaloacetate) to form arginosuccinic acid. This reaction takes place in the presence of ATP, magnesium ions, and the enzyme arginosuccinate synthetase.

$$
\begin{array}{ccccc}
\underset{\text{citrulline}}{
\begin{array}{l}
\text{CH}_2\text{—NH—}\overset{\overset{\text{O}}{\|}}{\text{C}}\text{—NH}_2 \\
| \\
(\text{CH}_2)_2 \\
| \\
\text{HC—NH}_2 \\
| \\
\text{COOH}
\end{array}}
&+&
\underset{\text{aspartic acid}}{
\begin{array}{l}
\text{COOH} \\
| \\
\text{H}_2\text{N—CH} \\
| \\
\text{CH}_2 \\
| \\
\text{COOH}
\end{array}}
&
\xrightarrow[\substack{\text{arginosuccinate} \\ \text{synthetase}}]{\text{ATP} \quad \text{AMP} \atop \text{Mg}^{2+},}
&
\underset{\text{arginosuccinic acid}}{
\begin{array}{l}
\text{CH}_2\text{—NH—}\overset{\overset{\text{NH}}{\|}}{\text{C}}\text{—NH—CH—COOH} \\
| \qquad\qquad\qquad | \\
(\text{CH}_2)_2 \qquad\qquad \text{CH}_2 \\
| \qquad\qquad\qquad | \\
\text{HC—NH}_2 \qquad\quad \text{COOH} \\
| \\
\text{COOH}
\end{array}}
\end{array}
$$

Arginosuccinic acid is cleaved (split) hydrolytically into arginine and fumaric acid. Some fumaric acid may be converted back to aspartic acid, and some enters the Krebs cycle (Chapter 26).

$$
\begin{array}{ccc}
\underset{\text{arginosuccinic acid}}{
\begin{array}{l}
\text{CH}_2\text{—NH—}\overset{\overset{\text{NH}}{\|}}{\text{C}}\text{—NH—CH—COOH} \\
| \qquad\qquad\qquad | \\
(\text{CH}_2)_2 \qquad\qquad \text{CH}_2 \\
| \qquad\qquad\qquad | \\
\text{HC—NH}_2 \qquad\quad \text{COOH} \\
| \\
\text{COOH}
\end{array}}
&
\xrightarrow[\text{lyase}]{\text{arginosuccinate}}
&
\underset{\text{arginine} \qquad\qquad \text{fumaric acid}}{
\begin{array}{l}
\text{CH}_2\text{—NH—}\overset{\overset{\text{NH}}{\|}}{\text{C}}\text{—NH}_2 \qquad\qquad \text{COOH} \\
| \qquad\qquad\qquad\qquad\qquad\quad | \\
(\text{CH}_2)_2 \qquad\qquad\qquad\qquad\quad \text{CH} \\
| \qquad\qquad\qquad + \qquad\quad \| \\
\text{HC—NH}_2 \qquad\qquad\qquad\qquad \text{CH} \\
| \qquad\qquad\qquad\qquad\qquad\quad | \\
\text{COOH} \qquad\qquad\qquad\qquad\quad \text{COOH}
\end{array}}
\end{array}
$$

Finally, arginine is split hydrolytically by the liver enzyme arginase into ornithine and urea. The ornithine can then go through the cycle again, and the urea is excreted.

$$
\begin{array}{ccc}
\underset{\text{arginine}}{
\begin{array}{l}
\text{CH}_2\text{—NH—}\overset{\overset{\text{NH}}{\|}}{\text{C}}\text{—NH}_2 \\
| \\
(\text{CH}_2)_2 \\
| \\
\text{HC—NH}_2 \\
| \\
\text{COOH}
\end{array}}
&
\xrightarrow{\text{arginase}}
&
\underset{\text{ornithine} \qquad\qquad \text{urea}}{
\begin{array}{l}
\text{CH}_2\text{—NH}_2 \\
| \\
(\text{CH}_2)_2 \qquad\qquad\qquad\quad \overset{\text{O}}{\overset{\|}{}} \\
| \qquad\qquad\qquad\qquad\qquad\quad \\
\text{HC—NH}_2 \quad + \quad \text{H}_2\text{N—C—NH}_2 \\
| \\
\text{COOH}
\end{array}}
\end{array}
$$

To summarize, the urea cycle is shown diagrammatically in Figure 28-3.

The blood picks up the urea from the liver and carries it to the kidneys, where it is excreted in the urine. Urea is the principal nitrogen end product of protein metabolism and contains a large percentage of the total nitrogen excreted by the body. (This is one means the body has of removing ammonia.)

Decarboxylation

The decarboxylation (removal of a —COOH group) of an amino acid yields a primary amine. The carboxyl group that is removed is converted to carbon dioxide. The enzyme involved in a decarboxylation reaction requires pyridoxal phosphate as a coenzyme. The decarboxylation reaction can be summarized as follows.

$$\underset{\substack{\text{α-amino acid}}}{R\overset{\displaystyle H}{\underset{\displaystyle NH_2}{-\overset{|}{\underset{|}{C}}-COOH}}} \xrightarrow[\text{pyridoxal phosphate}]{\text{amino acid decarboxylase}} \underset{\text{primary amine}}{R-CH_2-NH_2} + CO_2$$

Several naturally occurring amines are formed by the decarboxylation of amino acids, for example,

$$\textit{amino acid} \xrightarrow{\text{decarboxylation}} \textit{primary amine}$$

amino acid		primary amine
histidine	$\longrightarrow$	histamine
lysine	$\longrightarrow$	cadaverine
ornithine	$\longrightarrow$	putrescine
tyrosine	$\longrightarrow$	tyramine

Some decarboxylation reactions are brought about by intestinal bacteria that attack amino acids, producing toxic amines called ptomaines (see page 464). This process is common in the spoilage of food protein.

The amino acid tryptophan undergoes a series of reactions to form the compounds indole and methylindole (skatole). These two compounds produce the characteristic odor of feces.

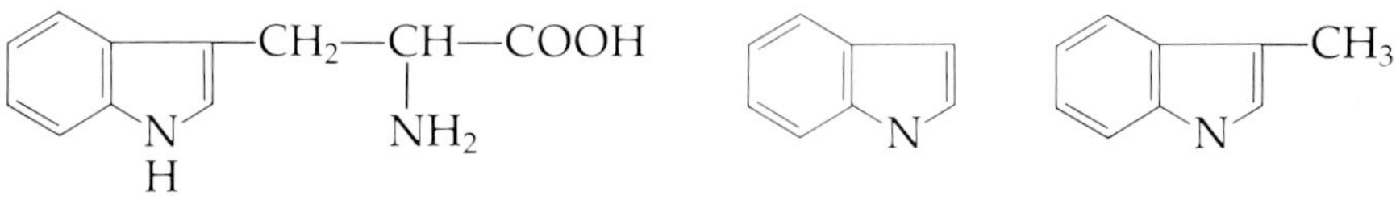

tryptophan   indole   skatole

Metabolism of the Carbon Portion of Amino Acids

Once the nitrogen has been removed from an amino acid, the carbon portion can be used as an energy source. The carbon skeletons can be converted into a variety of compounds (see Figure 28-6). The ones converted to acetyl CoA and acetoacetyl CoA are ketogenic, whereas those converted to pyruvate or Krebs cycle intermediates are glucogenic, since the body can synthesize glucose from them. Only leucine is purely ketogenic. The metabolism of lysine is poorly understood, and although Figure 28-6 shows lysine to be ketogenic, feeding experiments have shown that it is only slightly involved in producing ketone bodies. Amino acid catabolism thus provides energy and a wide variety of precursors and intermediates.

Metabolism of Hemoglobin

A red blood cell has a life span of about 120 days. After that period of time, the hemoglobin is catabolized. The globin (protein) part is metabolized as is any other protein. The heme is metabolized and excreted as waste products, but the iron is reused. The normal diet supplies about 12 to 15 mg of iron per day, but of this amount only about 1 mg per day may be absorbed. When hemoglobin is metabolized, 20 to 25 mg of iron is released per day. This amount must be reused or else the body will suffer a serious loss of iron. Stomach acidity is needed to aid in the absorption of iron.

The body synthesizes hemoglobin at the same rate as it metabolizes it. There are three important component parts of hemoglobin: Fe^{2+}, globin (a protein), and the porphyrin ring. Figure 28-4 summarizes the biosynthesis of hemoglobin.

Figure 28-4   Biosynthesis of hemoglobin.

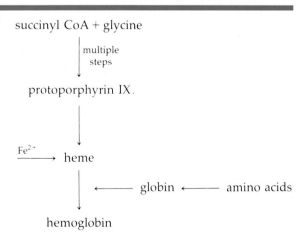

When an *erythrocyte ruptures, the hemoglobin ring is broken and the products formed are globin, ions, and biliverdin, a blue-green pigment. This process takes place in the reticuloendothelial cells of the liver, spleen, and bone marrow. Biliverdin is rapidly reduced to bilirubin, an orange-yellow pigment, by the enzyme bilirubin reductase, also in the reticuloendothelial cells. From there the bilirubin is transported to the liver as a bilirubin–albumin complex with the aid of serum albumin. In the liver, bilirubin is converted to bilirubin diglucuronide, which is then excreted into the bile. The bile flows into the small intestine. In the small intestine the bilirubin diglucuronide is changed to stercobilinogen and then to stercobilin for excretion into the stool and also into urobilinogen and then to urobilin for excretion into the urine. These reactions are shown in Figure 28-5. *Jaundice is the condition in which abnormal amounts of bilirubin accumulate in the blood. Patients with jaundice exhibit a characteristically yellow skin due to the presence of bilirubin.

If hemolysis takes place at an abnormally high rate, so that bilirubin accumulates in the blood, the condition is termed **hemolytic jaundice.** If the bile duct is obstructed so that bile cannot enter the intestinal tract, bilirubin again accumulates in the blood. This condition is termed **obstructive jaundice** and is characterized by white or clay-colored stools because decomposition products of bilirubin are not present. If the liver is damaged in such diseases as infectious hepatitis or cirrhosis, bilirubin cannot be removed and a jaundiced condition results.

There are two types of bilirubin. Direct bilirubin is bilirubin diglucuronide (a conjugated form), which is quite water soluble; indirect bilirubin (nonconjugated) is not very water soluble.

Figure 28-5 Metabolism of hemoglobin.

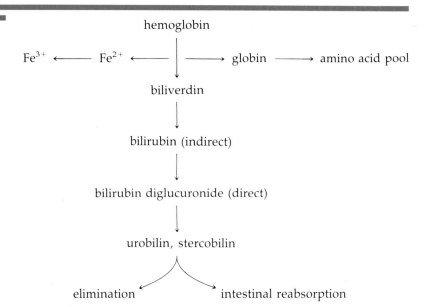

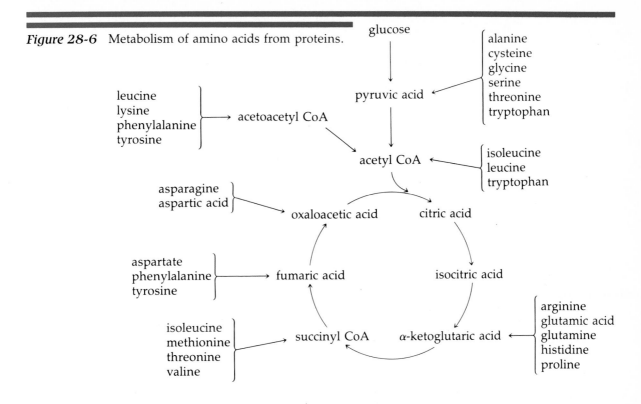

Figure 28-6 Metabolism of amino acids from proteins.

Overview of Protein Metabolism

The metabolism of protein produces compounds, such as pyruvic acid, that can enter the Krebs cycle, other products that can enter the glycogen-forming cycle, and still others that can enter the lipogenesis cycle, as indicated in Figure 28-6.

Summary

During digestion, proteins are hydrolyzed into amino acids. These amino acids may be used for the synthesis of new tissue, for the replacement of old tissue, and for the formation of enzymes and hormones, or they may be oxidized to furnish energy. The body cannot store protein (amino acids).

A person who excretes as much nitrogen daily as he or she takes in is said to be in nitrogen balance. Children have a positive nitrogen balance because they need extra proteins for growth and so excrete less. Malnutrition and prolonged fever may lead to a negative nitrogen balance.

Amino acids that the body cannot synthesize and that must be supplied in the food are called essential amino acids. A protein that does not contain all the essential amino acids is called an incomplete protein.

Normal adults require 46 to 56 g of protein per day.

Amino acids that the body does not need are catabolized into carbon dioxide and water. This catabolism may be oxidative deamination (deamination) whereby the α-amino group is removed, forming an α-keto acid and ammonia. The α-keto acid may be catabolized to carbon dioxide and water and energy through the Krebs cycle; it may be converted to glycogen or to fat; or it may undergo transamination, whereby a different amino acid is formed.

The ammonia formed from the deamination of an amino acid unites with carbon dioxide and water to form urea. This process, which takes place in the liver, proceeds through a cycle called the urea cycle.

Amino acids may also be decarboxylated to form primary amines.

When hemoglobin is metabolized, the protein part is metabolized as usual. The iron is used over again, and the remaining heme part goes through a series of steps, eventually ending up as urobilinogen and urobilin in the urine and stercobilinogen and stercobilin in the feces.

When excessive amounts of bilirubin, one of the intermediate products of the metabolism of hemoglobin, accumulate in the blood, the condition is known as jaundice.

Questions and Problems

1. What are the functions of protein in the body?
2. Can the body store protein? carbohydrate? fat?
3. What is meant by the term *nitrogen balance*?
4. What might cause a positive nitrogen balance?
5. What might cause a negative nitrogen balance?
6. What is an essential amino acid? List the essential amino acids.
7. What is an incomplete protein?
8. What is the normal daily requirement of protein?
9. What is deamination? Where does it occur in the body?
10. What products are produced by a deamination reaction?
11. What is transamination? What is its function in the body?
12. What various processes can an α-keto acid undergo?
13. Where is urea formed?
14. GOT is found in high concentration in which part of the body? What use is made of this fact?
15. What is decarboxylation? What type of product is produced by the decarboxylation of an amino acid?
16. What are ptomaines? How are they produced?
17. What is biliverdin? Where is it formed?
18. What are the end products of the metabolism of hemoglobin? Which are found in the urine? in the feces?
19. What is jaundice? What can cause it?
20. Indicate the relationship between the metabolism of carbohydrates, lipids, and proteins.
21. What is hyperalimentation, and where is it used?

Practice Test

1. The basic way the body has to eliminate excess nitrogen is to convert it to _____.
   a. uric acid   b. NH_3
   c. NH_4^+ d. urea
2. The cause of PKU is an inability to convert phenylalanine to _____.
   a. acetyl CoA   b. tyrosine
   c. glutamate   d. alanine
3. The metabolism of hemoglobin produces _____.
   a. cadaverine   b. protoporphyrin
   c. uric acid   d. bilirubin
4. Transamination of glutamic acid will produce the amino acid _____.
   a. histamine   b. glutamate
   c. alanine   d. aspartic acid
5. An example of an essential amino acid is _____.
   a. aspartate   b. lysine
   c. proline   d. serine
6. Protein is used in the formation of _____.
   a. hemoglobin   b. enzymes
   c. some hormones   d. all of these
7. A positive nitrogen balance occurs _____.
   a. during malnutrition
   b. in young children
   c. during a prolonged fever
   d. all of these
8. Deamination takes place in the _____.
   a. liver   b. stomach
   c. small intestine   d. pancreas
9. The decarboxylation of an amino acid yields _____.
   a. a keto acid   b. a primary amine
   c. an aldehyde   d. uric acid
10. The principal end product of protein metabolism is _____.
    a. creatine   b. phosphate
    c. urea   d. amino acids

29

Body Fluids: Urine

Medical laboratory technician conducting urinalysis test.

Excretion of Waste Material

The waste products of the body are excreted through the lungs, the skin, the intestines, and the kidneys. The liver also excretes waste products—the bile pigments and cholesterol.

The lungs eliminate water and carbon dioxide through the expired air. The skin eliminates water in the form of perspiration. Included in the perspiration are small amounts of inorganic and organic salts. The feces, excreted from the large intestine, contain undigested and undigestible material plus the excretory products from the liver—the bile pigments and cholesterol—some water, and some organic and inorganic salts. The primary excretory organs of the body, however, are the kidneys, which excrete water and water-soluble compounds including nitrogen compounds from the catabolism of amino acids.

The kidneys are important not only for their excretory function but also because of their role in the control and regulation of water, electrolyte, and acid–base balances in the body.

Formation of Urine

Blood flows to the kidneys through the renal arteries. From the renal arteries the blood passes into the arterioles and then into the capillaries of the kidneys. These capillaries coil up to form a glomerulus, a rounded ball of capillaries. Around the glomerulus is a structure called Bowman's capsule (see Figure 29-1). Each Bowman's capsule is connected by a tubule to a larger tube, which in turn carries the urine to the bladder where it is stored until it is excreted.

As blood flows into the kidney, the various soluble components diffuse into the glomeruli (there are over a million glomeruli in each kidney). The protein material in the blood cannot pass through the

Figure 29-1 Glomerulus and Bowman's capsule in the kidney.

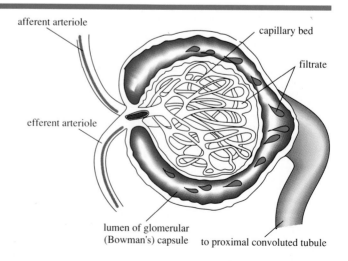

afferent arteriole

capillary bed

filtrate

efferent arteriole

lumen of glomerular (Bowman's) capsule

to proximal convoluted tubule

Table 29-1 Comparison of Composition of Blood Plasma and Urine

Constituent	Percentage in Blood Plasma	Percentage in Urine
Water	90–93	95
Glucose	97	0
Protein	7	0
Sodium	0.3	0.35
Ammonia	0.004	0.05
Phosphate	0.009	0.5
Urea	0.03	2.0
Sulfate	0.002	0.18

membranes (recall that proteins are colloids and colloids do not pass through membranes). The driving force for this diffusion of fluid through the walls of the glomerulus is the blood pressure. The liquid in the glomerulus thus has approximately the same composition as blood plasma except for the protein material.

As the fluid in the glomerulus passes down the tubule, a large proportion of the water is reabsorbed into the bloodstream. Also reabsorbed are the glucose, amino acids, and most inorganic ions. The remaining liquid, containing urea and other waste products, flows to the collecting tubules and then to the bladder.

Thus, the kidneys act as a very efficient filter, removing the waste materials but not the needed nutrients from the blood. Approximately 1 L of blood is filtered through the kidneys every minute. Of this amount, most of the water is reabsorbed, so the amount of urine excreted per day is less than 1 percent of the total amount of liquid filtered. If the kidneys are not functioning normally, an artificial kidney machine may be used (see page 209).

Table 29-1 compares the composition of blood plasma and urine.

General Properties of Urine

Volume

A normal adult excretes 600 to 2500 mL of urine per day. The amount depends on the liquid intake and also on the weather conditions. In hot weather, more water is lost through perspiration; therefore, the amount of urine formed is less. Conversely, a greater amount of urine is formed during cold weather or when the humidity is high, when little evaporation of perspiration takes place. Drugs such as caffeine (in coffee or tea) and also alcoholic beverages have a *diuretic effect—that is, they increase the flow of urine.

A decreased flow of urine is called **oliguria.** Such a condition may occur during a high fever when most of the water lost by the body is in the form of perspiration. Certain kidney diseases may also cause oliguria.

Anuria means a total lack of urine excretion. Anuria indicates extensive kidney damage such as may be caused by a blood transfusion of the wrong type. In this condition the blood cells disintegrate, releasing hemoglobin, which clogs the glomeruli and does not allow any excretion of urine. Bichloride of mercury also affects the kidneys and may cause oliguria or anuria.

Polyuria is a condition in which the amount of urine excreted is much greater than normal. It may be due to excessive intake of water or to certain pathologic conditions. Polyuria may be caused by such diuretics as alcohol or caffeine. Urea, a normal constituent of urine, is also a diuretic. A person on a high-protein diet will excrete more urea, which in turn causes the formation of more urine.

During diabetes insipidus the hormone vasopressin, which controls the reabsorption of water in the kidneys, is lacking or deficient. In this case the amount of urine is greatly increased, sometimes as high as 30 L per day.

Patients with diabetes mellitus show a definite polyuria because glucose is a diuretic. Excessive water loss during diabetes mellitus can lead to dehydration.

Specific Gravity

The specific gravity of the urine depends upon the concentration of the solutes. The greater the concentration of the solutes, the greater the specific gravity. A normal range of the specific gravity of the urine is 1.003 to 1.030. In cases of diabetes mellitus, the specific gravity will be higher because of a high concentration of sugar in the urine. In cases of diabetes insipidus, the specific gravity of the urine will be very low (close to 1.000) because of the large amounts of water being excreted.

pH

Urine is normally slightly acidic, with a pH range of 4.6 to 8.0 and an average value of about 6.3. However, the pH of urine varies with the diet.

Protein foods, such as meats, increase the acidity of the urine (lower the pH) because of the formation of phosphates and sulfates. The acidity of the urine is also increased during acidosis and with fever.

Conversely, the urine may tend to become alkaline on a diet high in vegetables and fruits or because of alkalosis, a condition that may be produced by excess vomiting.

Color

Normal urine is pale yellow or amber. The color, however, varies with the amount of urine produced and also with the concentration of the solutes in the urine. The larger the volume of urine excreted, the lighter the color. The greater the concentration of solutes, the darker the color. The color of the urine is due to urobilin and urobilinogen (see page 520). Various other components of the urine may cause it to have different colors. The presence of blood in the urine gives it a reddish color. A reddish color may also be due to eating beets or rhubarb, or taking cascara, a cathartic. Homogentisic acid (an intermediary in the metabolism of phenylalanine and tyrosine) colors the urine brown. The drug methylene blue colors the urine green.

Freshly voided urine is clear and usually contains no sediment. However, on standing it may become cloudy and develop sediment because of the precipitation of calcium phosphate.

Odor

Fresh urine has a distinctive odor, but this odor may be modified by the presence of other substances. In patients with ketosis the odor of acetone may be detected. Diet can also modify the odor of urine. For example, when asparagus is eaten, the urine may have a sulfurlike odor.

Normal Constituents

Approximately 50 to 60 g of dissolved solid material is excreted daily in the urine of the average person. This solid material has both inorganic and organic constituents (see Table 29-2). The inorganic constituents of urine make up approximately 45 percent of the total solids; the organic constituents comprise the other 55 percent.

Organic Constituents

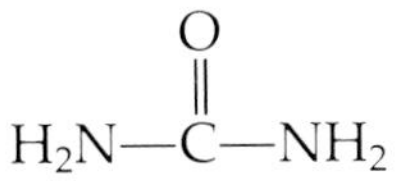

Urea

The principal end product of the metabolism of protein, urea, comprises about one half of the total solids in the urine.

$$H_2N-\overset{\overset{\textstyle O}{\|}}{C}-NH_2$$

urea

Table 29-2 Constituents of Urine (Amount Excreted per Day)

Constituent	Amount (g)	Approximate Percent
Organic		
Urea	25–30	40–50
Uric acid	0.7	1
Creatinine	1.4	2.5
Creatine	0.06–0.15	0.1–0.25
Others	0.1–1	0.1–1
Inorganic		
Chloride ion	9–16	15–25
Sodium ion	4	6
Phosphates	2	3
Sulfates	2.5	4
Ammonium ions	0.7	1
Other ions and inorganic constituents	2.5	4

Figure 29-2 The ravaging effects of gout are manifest in the deformed hands of a patient afflicted with this crippling disease. [*Courtesy National Institute of Arthritis and Metabolic Diseases, National Institutes of Health, Bethesda, MD.*]

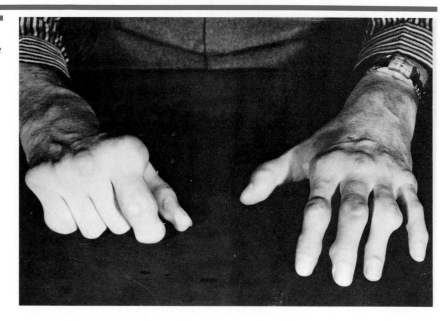

Uric Acid

Uric acid, a product of the metabolism of purines (see page 647) from nucleoprotein, is only slightly soluble in water and is excreted primarily as urate salts. The structure of uric acid is

uric acid

When urine is allowed to stand, uric acid may crystallize and settle out, since it is only slightly soluble in water or acid solution.

The average daily excretion of uric acid is about 0.7 g, but an increase in nucleoproteins in the diet will cause an increased excretion of uric acid. The output of uric acid is also increased in leukemia, in severe liver disease, and in various stages of gout. Deposits of urates and uric acid in the joints and tissues are also characteristic of gout, so that this disease appears to be a form of arthritis (see Figure 29-2).

Under certain conditions uric acid or urates crystallize in the kidneys and are called kidney stones, or *calculi (Figure 29-3).

Creatinine

Creatinine is a product of the breakdown of creatine. The amount of creatinine excreted per day is fairly constant regardless of the protein

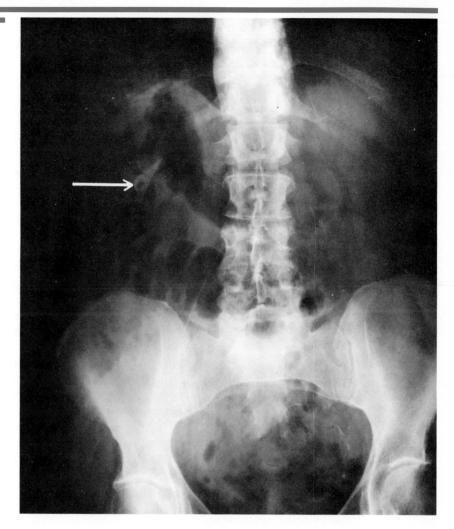

Figure 29-3 Stone in right kidney. [*Courtesy Dr. Leonard Berlin, Director of Radiology, Skokie Valley Community Hospital, Skokie, IL.*]

intake. The number of milligrams of creatinine excreted in the urine within a 24-hr period per kilogram of body weight is the **creatinine coefficient** of that individual. The creatinine coefficient of the normal male is 20 to 26; that of females is 14 to 22.

The average adult excretes 1.4 g of creatinine daily and about 0.06 to 0.15 g of creatine in the same period of time. Formation of creatinine appears necessary for the excretion of most of the creatine.

$$
\begin{array}{c}
\underset{\displaystyle\|}{NH} \\
C\!-\!NH_2 \\
| \\
N\!-\!CH_2\!-\!COOH \\
| \\
CH_3
\end{array}
\longrightarrow
\begin{array}{c}
\underset{\displaystyle\|}{NH} \\
C\!-\!\!-\!\!-\!\!-\!NH \\
| \qquad\quad | \\
N\!-\!CH_2\!-\!C\!=\!O \;+\; H_2O \\
| \\
CH_3
\end{array}
$$

    creatine   creatinine

Creatine

Creatine is produced in the body from three amino acids: arginine, methionine, and glycine. Creatine is normally present in muscle, brain, and blood, both as free creatine and as creatine phosphate. Recall the reaction

$$\text{creatine phosphate} + \text{ADP} \rightarrow \text{creatine} + \text{ATP}$$

Creatinuria is a condition in which abnormal amounts of creatine occur in the urine. It may occur during starvation, diabetes mellitus, prolonged fevers, wasting diseases, and hyperthyroidism. Creatinuria may also occur in pregnancy.

Other Organic Constituents

Also present in small amounts are amino acids, allantoin (from partial oxidation of uric acid), hippuric acid, urobilin, urobilinogen, and biliverdin. The presence of cAMP helps diagnose parathyroid function. Patients with hyperparathyroidism excrete significantly more cAMP and those with hypoparathyroidism excrete significantly less cAMP than persons with normal parathyroid glands (see page 627).

Urobilinogen excretion is increased in hemolytic anemias and in liver disease. Biliverdin excretion is increased in certain liver and biliary diseases.

Tests for pregnancy are based on the fact that the implantation of a fertilized ovum in the placental tissue produces the hormone chorionic gonadotropin, which is excreted in the urine. The presence of this hormone is the basis of the "rabbit test" for pregnancy. Other tests for the presence of chorionic gonadotropin (for pregnancy) are based on the fact that this hormone will react with specific *antibodies and can give a result in a doctor's office or at home in a very short time.

In addition to these constituents, the urine also normally contains very small amounts of vitamins, other hormones, and enzymes. Urinalysis for these substances is of diagnostic value.

Inorganic Constituents

The inorganic constituents of urine are the various positive and negative ions that make up the inorganic compounds being excreted. Among these are the following.

Chloride Ions

Between 9 and 16 g of chloride ion is excreted daily, mostly as sodium chloride. The amount of chloride ion varies with the intake, which is primarily sodium chloride. The excretion of sodium chloride is decreased in fevers and in some stages of *nephritis.

Sodium Ions

The amount of sodium ion excreted varies with the intake and the body's requirement. However, it is usually about 4 g per day.

Phosphates

The amount of phosphates present in the urine also depends upon the diet; the amount is higher when the diet contains foods high in phosphorus (nucleoproteins and phospholipids). An increase in excreted phosphates is found in certain bone diseases and in hyperparathyroidism. A decrease in phosphates is found in hypoparathyroidism, in renal diseases, and during pregnancy.

Sulfates

The sulfates in the urine are derived from the metabolism of sulfur-containing proteins, so the amount of sulfur is influenced by the diet. Sulfates are found as both organic and inorganic salts.

Ammonium Ions

The hydrolysis of urea produces such ammonium compounds as chlorides, sulfates, and phosphates in the urine.

Other Ions

In addition to the sodium and ammonium ions, other positive ions present in the urine are calcium, potassium, and magnesium.

The amount of calcium ions in the urine is increased in hyperthyroidism, hyperparathyroidism, and osteoporosis and decreased in hypoparathyroidism and vitamin D deficiency.

Nitrite ions in the urine indicate the presence of reducing bacteria in urinary tract.

The normal ratio of sodium to potassium in the urine is 2 parts sodium to 1 part potassium. The ratio of sodium to potassium is increased in Addison's diseasee.

Magnesium concentration in the urine is decreased in chronic alcoholism.

Abnormal Constituents

Protein

Because proteins are colloids and because colloids cannot pass through membranes, urine should not normally contain protein. **Proteinuria** denotes the presence of protein in the urine. Sometimes it is termed **albuminuria** because albumin is the smallest plasma protein and is the protein most frequently found in the urine. In cases of kidney disease, such as nephritis and nephrosis, and in severe heart disease, protein appears in the urine. The presence of protein due to such disorders is frequently called renal proteinuria or renal albuminuria to distinguish it from false albuminuria, which is a temporary harmless condition. False albuminuria, often called orthostatic albuminuria, is found in certain patients who stand for a long period of time. It is due to the constriction of the kidneys' blood vessels and disappears when the patient lies down. Small amounts of protein may also be found in the urine after severe muscular exercise, but they soon disappear.

The tests for the presence of protein in the urine are based on the fact that protein coagulates when heated. When a sample of urine is heated, any protein (albumin) present will precipitate out as a white cloud. However, phosphates may also precipitate when the urine is heated. To prove that the cloudy substance is albumin, the urine, after heating, is acidified with dilute acetic acid. The acid will dissolve the phosphates but not the protein, so a cloudy precipitate in the urine after heating and acidification is a verification of the presence of protein.

Glucose

The presence of glucose in the urine is called **glycosuria.** Normally there is always a very small amount of glucose present in the urine, but this amount is too small to give a positive test with Benedict's solution.

Glucose may be found in the urine after severe muscular exercise, but this condition clears up when the body returns to normal. Glucose may also be found in the urine after a meal high in carbohydrates.

Glycosuria may be due to such diseases as diabetes mellitus or renal diabetes or to liver damage.

Other Sugars

Lactose and galactose may occur in the urine during pregnancy and lactation. Both of these sugars give a positive Benedict's test.

Pentoses may occur in the urine after consumption of foods such as plums, grapes, and cherries, which contain large amounts of these carbohydrates.

Ketone Bodies

Ketone (acetone) bodies are present in the urine during diabetes mellitus, in starvation, or in other circumstances with inadequate carbohydrate intake. Such a condition is termed ketonuria. The excretion of the ketone bodies, which are acidic compounds, requires alkaline compounds. This results in a depletion of the alkaline reserve of the blood and leads to acidosis. The kidneys produce more ammonia to neutralize these ketone bodies.

The test for the presence of ketone bodies in the urine is performed by adding sodium nitroprusside to a sample of urine and then making the mixture alkaline with ammonium hydroxide. The presence of ketone bodies is indicated by a pink-red color. Normal urine gives no color with this test.

Blood

The presence of blood in the urine is called hematuria. It may result from lesions or stones in the kidneys or urinary tract. The presence of free hemoglobin in the urine, hemoglobinuria, results from hemolysis of the red blood cells caused by an injection of hypotonic solution, severe burns, or blackwater fever.

Large amounts of blood in the urine can be detected by the reddish color. Small amounts do not color the urine enough to show any color change.

Bile

Normally, bile is excreted by the liver into the small intestine and eventually ends up in the feces. The presence of bile in the urine indicates obstruction to the flow of bile to the intestines. Bile in the urine is indicated by a greenish brown color. Bile in the urine is also indicated by the presence of a yellow foam when the urine is shaken.

Phenylpyruvic Acid

Phenylpyruvic acid, an intermediate product in the metabolism of the essential amino acid phenylalanine, is not normally present in urine. However, its presence can be detected in the urine of a person with phenylketonuria (PKU) (see page 000). If this disease is not detected and treated early, mental retardation results. For this reason, some states require screening of newborns for this disease.

Diuretics

Diuretics are drugs that promote loss of water and salts through the urine. Examples of diuretics are ethyl alcohol, caffeine, mannitol, and thiazides.

Phenolsulfonphthalein Test

Phenolsulfonphthalein (PSP) is a red dye used to test how well the kidneys are functioning. The dye is administered by intravenous or intramuscular injection, usually the former. Urine specimens are collected at frequent intervals—after 15, 30, 60, and 120 min. If the 15-min specimen contains 25 percent or more of the injected PSP, then the kidneys are functioning normally; 40 to 60 percent of the dye should be excreted within 1 hr; 20 to 25 percent more in the second hour.

Other kidney-function tests have taken the place of the PSP test. They are based upon the amount of urea eliminated in the urine compared to the amount present in the blood (urea clearance test) or the change in the specific gravity of the urine after the patient's fluid intake is restricted (concentration test).

Reagent Tablets, Papers, and Dipsticks

To help nonlaboratory personnel perform a variety of urine (or blood) tests easily, quickly, and economically, many tablets, papers, and dipsticks have been developed. In all cases, it is essential that the manufacturer's directions be followed.

Reagent tablets and some papers and dipsticks are used to test for one substance only. That is, a tablet or paper or dipstick can be used to test for glucose, but another kind of tablet, paper, or dipstick must be used to test for ketone bodies or any other substance.

Some papers and dipsticks contain several reagent strips. These can be used to test for several substances at the same time.

In all cases, whether using tablets, papers, or dipsticks, the color produced is compared to a color chart prepared by the manufacturer in order to read the results.

Summary

The principal excretory organs of the body are the kidneys, which also control and regulate the water balance, electrolyte balance, and pH of body fluids.

The waste materials in the blood are picked up by the kidneys and are excreted in the urine.

Approximately 600 to 2500 mL of urine is excreted daily, the amount depending on fluid intake, weather conditions, humidity, and certain diuretic substances.

A decreased flow of urine is called oliguria; anuria is a total lack of urine; polyuria is excess urine formation.

The specific gravity of urine varies between 1.003 and 1.030, and the pH ranges from 4.6 to 8.0, with an average value of 6.3. Urine is normally pale yellow or amber, but certain components may cause another color.

Approximately 50 to 60 g of solid material, both organic and inorganic, is excreted daily. The principal organic constituent of urine is urea, the end product of protein metabolism. Another important constituent is uric acid, a product of the metabolism of purines. Gout is characterized by an increase of uric acid in the urine and blood and the deposition of uric acid or urate salts in the joints and tissues. Uric acid and urates may also crystallize in the kidneys as kidney stones.

The average adult excretes 0.06 to 0.15 g of creatine and 1.4 g of creatinine daily. Creatine is produced from three amino acids: arginine, methionine, and glycine. Creatinine is produced from creatine by a dehydration reaction. Creatinuria is the condition in which abnormal amounts of creatine occur in the urine.

Inorganic constituents of urine are the following ions: chloride, sodium, phosphate, sulfate, ammonium, and some small amounts of calcium, potassium, and magnesium.

16. A-DNA computer model. *[Richard Pas-ro/National Institutes of Health-FDA]*

17. Photomicrographs: Vitamins. *[Tore Johnson/ANA from Woodfin Camp, Inc.]*

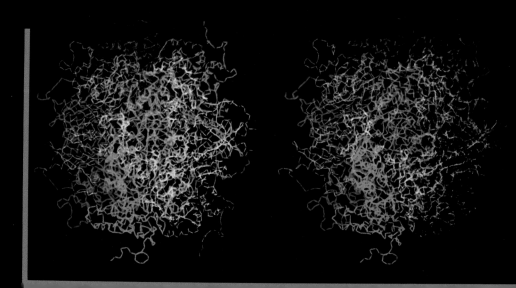

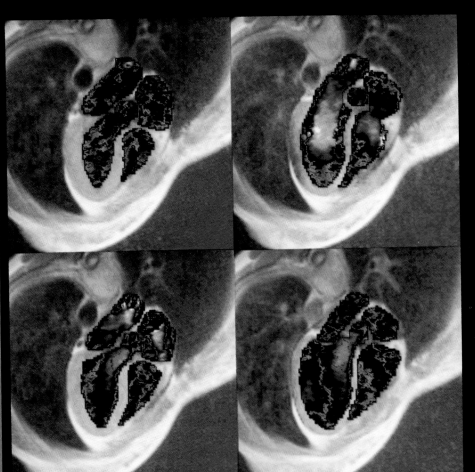

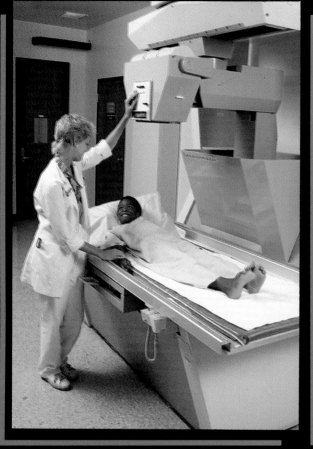

21. Medical imaging: X-raying a patient. *[Robert Frerck/Woodfin Camp, Inc.]*

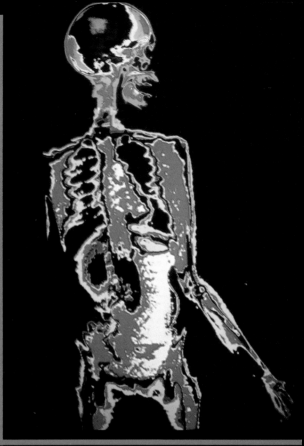

22. Medical imaging: A density scan. *[Howard Sochurek/Woodfin Camp, Inc.]*

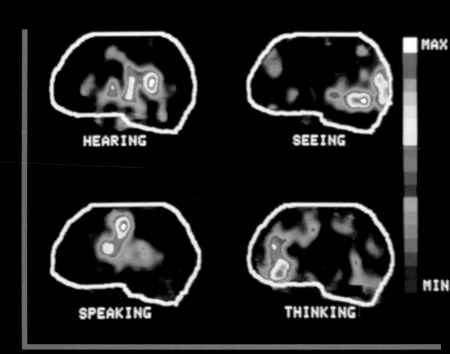

23. Medical imaging: PET scans. *[Mallinckrodt Institute of Radiology]*

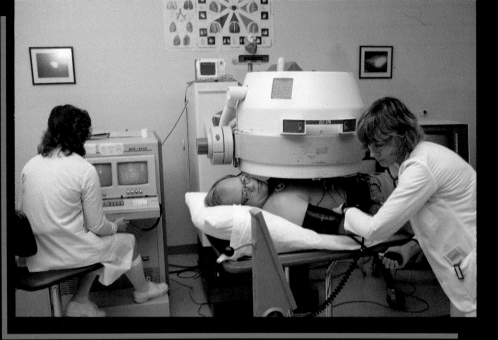

24. Medical imaging: Heart stress test. *[Ulrike Welsch]*

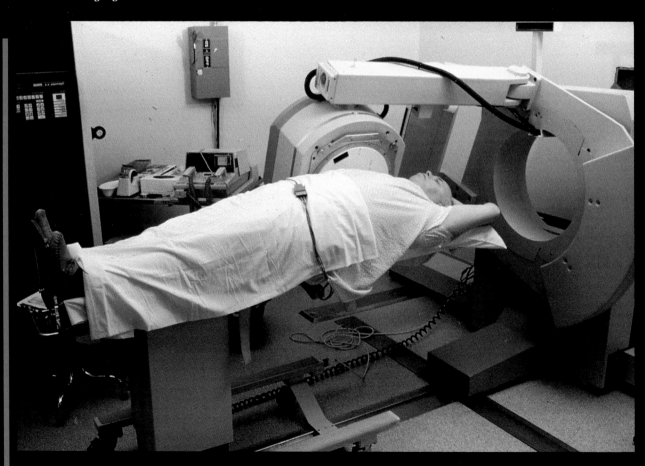

25. Medical imaging: CT scan. *[National Institutes of Health]*

Abnormal constituents in the urine are protein (proteinuria, or albuminuria) due to kidney disease; glucose (glycosuria) due to diabetes mellitus or liver damage; ketone bodies due to diabetes mellitus or starvation (ketonuria); blood (hematuria) due to lesions in the kidneys or urinary tract; and bile due to an obstruction of the flow of bile to the intestines.

Diuretics increase the output of water and salts in the urine.

Various tests may be performed to see if the kidneys are functioning normally. Among these is the phenolsulfonphthalein test (PSP) in which a red dye is administered intravenously and the amount of color measured in the urine at specified intervals of time.

Questions and Problems

1. How does the body excrete waste products?
2. What are the principal excretory organs of the body? What additional functions do they have?
3. What is a glomerulus? Bowman's capsule?
4. Describe the formation of urine. Where is the urine stored?
5. What forces the fluids through the membrane in the kidneys?
6. What happens to the nutrients and the water that also filter through the membranes in the kidneys?
7. What volume of urine does an adult excrete daily? What might affect this amount?
8. What is a diuretic?
9. What is oliguria? What might cause this condition?
10. What is anuria? What might cause this condition?
11. What might cause polyuria?
12. What does the hormone vasopressin do? If this hormone is lacking, what will be the effect on the body?
13. What is the pH range of urine? What might affect the pH of the urine?
14. What might affect the color of the urine?
15. What might cause urine to become cloudy upon standing?
16. How much solid material is excreted daily in the urine?
17. What is the principal organic constituent of the urine? Where does it come from? What is its structural formula?
18. Where does the uric acid in the urine come from?
19. Draw the structure of uric acid.
20. How much uric acid is excreted daily? What might affect this amount?
21. What is gout? What other conditions may cause an increased amount of uric acid in the urine?

22. What might cause kidney stones?
23. What is the creatinine coefficient?
24. How does the structure of creatine compare with that of creatinine?
25. Where does the body obtain its creatine?
26. Where is creatine normally present in the body?
27. What is creatinuria and what causes it?
28. Name several inorganic ions normally found in the urine.
29. How is the amount of phosphates in the urine affected by various diseases?
30. What is proteinuria?
31. What might cause albuminuria?
32. What is false albuminuria?
33. Describe the test for the presence of protein in the urine. Is it necessary to acidify the urine during this test? Why?
34. What is glycosuria? What causes this condition?
35. Could other sugars besides glucose be present in the urine? under what conditions?
36. What might cause the presence of (ketone) acetone bodies in the urine?
37. Describe the test for the presence of ketone bodies in the urine.
38. What is hematuria? What might cause this condition?
39. The presence of bile in the urine might indicate what condition?
40. Describe the PSP test for kidney function.

Practice Test

1. Urine is normally _____.
   a. slightly acidic
   b. neutral
   c. slightly alkaline
   d. variable over quite a pH range

2. The main organic compound in urine is _____.
   a. creatine   b. creatinine
   c. uric acid   d. urea
3. The main anion in urine is _____.
   a. HCO_3^- b. PO_4^{3-} c. SO_4^{2-} d. Cl^-
4. All of the following are normally absent from urine except _____.
   a. glucose   b. protein
   c. uric acid   d. ketone bodies
5. A decreased output of urine is called _____.
   a. anuria   b. oliguria
   c. polyuria   d. hematuria
6. Polyuria occurs in _____.

   a. diabetes   b. kidney damage
   c. high fever   d. all of these
7. The color of urine is due to the presence of
   a. hemoglobin   b. blood
   c. urobilin   d. stercobilin
8. Kidney stones may contain _____.
   a. sodium chloride   b. creatine
   c. urea   d. uric acid
9. Glycosuria indicates the presence of _____.
   a. glucose   b. glycogen
   c. amino acid   d. glycose
10. An example of a diuretic is _____.
   a. glucose   b. caffeine
   c. morphine   d. creatinine

30

Body Fluids: The Blood

Various types of blood in a blood ''bank.''

Functions

Blood has been called a circulating tissue. It carries oxygen, minerals, and food to the cells and carries carbon dioxide and other waste products away from the cells. It also transports hormones, enzymes, and blood cells. Blood regulates body temperature by transferring heat from the interior to the surface capillaries. The blood buffers maintain the pH of the body at its optimum value. Blood contains a clotting system that protects the body against hemorrhage and defense mechanisms against infection.

Composition

Blood consists of two parts—the suspended particles and the suspending liquid, the plasma. The suspended particles in the blood are the red blood cells, white blood cells, and platelets.

Red Blood Cells (Erythrocytes)

Normally there are 4.5 million to 5.0 million red blood cells per cubic millimeter of blood. Since there is about 6 qt of blood in the human body, the total number of red blood cells is approximately 30 trillion (30,000,000,000,000). An excess of red blood cells is called **polycythemia**; a shortage is called **anemia.** Each day 200 billion new red blood cells are formed in the bone marrow.

Erythropoietin, a hormone that stimulates red blood cell forma-

Figure 30-1 Scanning electron micrograph of a human red blood cell showing the typical biconcave disc shape. [*Dr. Jeremy Burgess/Science Photo Library, Photo Researchers, Inc.*]

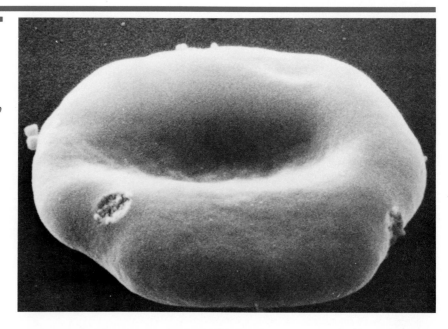

tion, is a glycoprotein with a molecular weight of approximately 35,000. Erythropoietin is formed by the action of a substance produced by the kidneys (renal erythropoietic factor) on a globulin in the blood plasma. The production of this hormone is increased by hypoxia in the kidneys, by cobalt salts, and by androgens. If the kidneys do not function properly, the patient may become anemic (see page 543).

White Blood Cells (Leukocytes)

The number of white blood cells normally present in the blood ranges from 5000 to 10,000 per cubic millimeter. White blood cells are larger than red blood cells. White blood cells have a nucleus; red blood cells do not. There are several types of white blood cells; among these are the basophils, eosinophils, lymphocytes, monocytes, and neutrophils.

The white blood cells attack and destroy harmful microorganisms and thus serve as one of the body's defenses against infection. A white blood cell count above normal usually indicates an infection. For diagnostic purposes, a **differential count** is sometimes ordered. This count gives the percentages of each of the various types of leukocytes present.

Platelets (Thrombocytes)

The number of thrombocytes or platelets ranges from 250,000 to 400,000 per cubic millimeter of blood. Platelets are smaller than red blood cells and do not have a nucleus. Platelets contain cephalin (phosphatidylethanolamine), a phospholipid that is involved in the clotting of the blood.

Blood Plasma

Approximately 92 percent of the plasma is water. The solids dissolved or colloidally dispersed in the blood make up the other 8 percent. The most important of all plasma solids are the blood proteins. These include serum albumin, the globulins, and fibrinogen.

The plasma proteins (primarily albumin) maintain the osmotic pressure of the blood, thus regulating the water and acid–base balance in the body.

The globulins are a mixture of protein molecules labeled α, β, and γ. The γ-globulins or immunoglobulins (antibodies) (see page 579) function against infection and disease. The α- and β-globulins include glycoproteins and lipoproteins, which function in the transportation of oligosaccharides, lipids, steroids, and hormones. Two particular globulins, transferrin and ceruloplasmin, transport iron and copper, respectively, in the plasma. Fibrinogen and prothrombin, two other globulins, function in the clotting of the blood.

Blood plasma also contains small amounts of lipids and carbohydrates (glucose), inorganic salts, waste products (such as urea, uric acid, carbon dioxide, creatinine, and ammonia), enzymes, vitamins, hormones, and antibodies. The inorganic ions present in the blood plasma serve to regulate the acid–base balance of the body.

On standing, freshly drawn blood soon forms a clot. When the clot settles, a yellowish liquid remains. This liquid is called **blood serum.** Blood serum is blood without the blood cells and without the fibrinogen necessary for the clotting of the blood. Blood plasma can be separated from the solid parts of the blood by centrifuging, during which the cells settle at the bottom of the test tube and the plasma remains above them. To keep the blood from clotting, an anticoagulant must be added before centrifuging.

Plasma Substitutes

Plasma volume can be greatly reduced in cases of severe burns, when water intake is too low, or following severe diarrhea, excessive vomiting, or polyuria. In such cases, plasma volume can be increased by intravenous infusion of an isotonic saline solution (0.9 percent NaCl).

However, if there has been a loss of blood volume due to hemorrhaging, intravenous saline solutions are not adequate because they lack plasma protein, salts, and blood cells. In such cases, transfusions of whole blood, plasma, or plasma substitutes may be used. If whole blood is to be used for a transfusion, careful matching of blood types is a necessity.

Blood volume extenders, such as dextrans, have been used. Although such substances are able to maintain the osmotic pressure of the blood, they have an adverse effect upon the blood clotting process.

One group of plasma substitutes are the perfluorocarbons, which are clear, colorless emulsions fortified with inorganic salts and other ingredients. These compounds have the ability to carry oxygen to the tissues and carbon dioxide from the tissues. However, they cannot assume the functions of white blood cells, platelets, clotting factors, and other blood components.

Most transfusions are given to replace lost blood volume and red blood cells, and plasma substitutes will work well in such cases. Also, plasma substitutes can be given to all patients regardless of blood type. Plasma substitutes are particularly useful in transplantation surgery, in which large volumes of blood are required for perfusion of organs outside the body. Use of plasma substitutes in such cases saves a large volume of whole blood for other needs.

Plasma substitutes are also useful in certain types of anemias in which the ability of the blood to carry oxygen is greatly reduced.

Patients whose religious beliefs prevent them from receiving blood transfusions can be given blood substitutes, such as perfluorocarbons which have a shelf life measured in years as compared to just a few weeks for whole blood. But the emulsion must be kept frozen until use or it begins to break down.

General Properties of Blood

Oxygenated blood has a characteristic bright red color; deoxygenated blood has a dark purplish color.

The specific gravity of whole blood ranges from 1.054 to 1.060, whereas that of blood plasma ranges from 1.024 to 1.028.

Blood is normally slightly alkaline, with a pH range of 7.35 to 7.45. If the pH of the blood falls slightly below 7.35, the condition is termed **acidosis.** If the pH of the blood rises slightly above 7.45, the condition is termed **alkalosis.** If the pH of the blood changes more than a few tenths from the normal values, the results are usually fatal.

The viscosity of the blood is approximately 4.5 times that of water and varies according to the number of cells, the quantity of protein, the temperature, and the amount of water present in the body (see page 196).

Blood Analysis

For most laboratory tests, 5 mL of blood is collected from a vein in the arm before the patient is given breakfast. If blood plasma is to be used for the test, an anticlotting agent such as potassium oxalate is added to the blood sample. If blood serum is to be tested, the blood is allowed to clot and the serum is poured off.

The usual blood chemistry tests, the normal ranges of the results, and the clinical significance of these tests are indicated in Table 30-1. Other tests include blood gas analysis and lactate/pyruvate determination.

Blood Volume

Approximately 8 to 9 percent of the total body weight is blood. The volume of the blood in the body amounts to 5 to 7 L in the adult. The volume increases in fever and pregnancy and decreases during diarrhea and hemorrhaging. Since blood volume can be rapidly replaced, a small loss of blood due to bleeding or to donating blood has no serious effect on the body.

Blood volume can be determined by injecting a suitable dye into the bloodstream, waiting about 10 min, and then withdrawing a sample to determine the concentration of the dye. A radioactive tracer can also be injected into the bloodstream and the blood volume determined by the amount of dilution of the radioactivity in a sample taken a short time later.

Table 30-1 Normal Composition of Blood

Determination[a]	Normal Range[b] (mg/100 mL)	Clinical Significance	
		Increased in	Decreased in
Calcium (s)	8.5–10.3 (4.5–5.3 mEq/L)[c]	Hyperparathyroidism, Addison's disease, malignant bone tumor, hypervitaminosis D	Hypoparathyroidism, rickets, malnutrition, diarrhea, chronic kidney disease, celiac disease
Cholesterol, total (s, p)	150–265	Diabetes mellitus, obstructive jaundice, hypothyroidism, pregnancy	Pernicious anemia, hemolytic jaundice, hyperthyroidism, tuberculosis
Uric acid (s, p)	Male, 3–9 Female, 2.5–7.5	Gout, leukemia, pneumonia, liver and kidney disease	
Urea nitrogen (s, p)	8–25	Mercury poisoning, acute glomerulonephritis, kidney disease	Pregnancy, low-protein diet, severe hepatic failure
Nonprotein nitrogen (b, s)	15–35	Kidney disease, pregnancy, intestinal obstruction, congestive heart failure	Low-protein diet
Creatine (s)	3–7	Nephritis, renal destruction, biliary obstruction, pregnancy	
Creatinine (b, s)	0.7–1.5	Nephritis, chronic renal disease	
Glucose (s, p)	70–100	Diabetes mellitus, hyperthyroidism, infections, pregnancy, emotional stress, after meals	Starvation, hyperinsulinism, Addison's disease, hypothyroidism, extensive hepatic damage
Chlorides (s, p)	96–106 mEq/L	Nephritis, anemia, urinary obstruction	Diabetes, diarrhea, pneumonia, vomiting, burns
Phosphate, inorganic (p)	3–4.5	Hypoparathyroidism, Addison's disease, chronic nephritis	Hyperparathyroidism, diabetes mellitus
Sodium (s)	136–145 mEq/L	Kidney disease, heart disease, pyloric obstruction	Vomiting, diarrhea, Addison's disease, myxedema, pneumonia, diabetes mellitus
Potassium (s)	3.5–5 mEq/L	Addison's disease, oliguria, anuria, tissue breakdown	Vomiting, diarrhea
Carbon dioxide (s, p)	Adults, 24–29 mEq/L Infants, 20–26 mEq/L	Tetany, vomiting, intestinal obstruction, respiratory disease	Acidosis, diarrhea, anesthesia, nephritis
Hemoglobin (b)	Male, 14–18 g/100 mL Female, 12–16 g/100 mL	Polycythemia	Anemia

[a] Key to abbreviations: b = blood; s = serum; p = plasma.   [b] Milligrams per 100 mL is also called milligram percent.
[c] mEq/L = milliequivalents per liter.

Hemoglobin

Hemoglobin is a conjugated protein made up of a protein part, globin, and an iron-containing part, heme. Heme contains four pyrrole groups joined together with an iron ion in the center. The structure of pyrrole is

$$
\begin{array}{ccc}
HC & \text{———} & CH \\
\parallel & & \parallel \\
HC & & CH \\
& \diagdown\,N\,\diagup & \\
& H &
\end{array}
$$

pyrrole

The structure of heme is indicated in Figure 30-2. Various hydrocarbon side chains are attached to the pyrrole rings in this compound. Four heme molecules combine with one globin molecule to form one molecule of hemoglobin.

The structures of cytochrome c, part of the oxidative phosphorylation sequence of the Krebs cycle (see page 482), and of chlorophyll, a plant pigment, are shown in Figure 30-2 for comparison with that of heme. Note that whereas both heme and cytochrome c have an iron ion (Fe^{2+}) in the center of four pyrrole rings, chlorophyll has a magnesium ion (Mg^{2+}).

Fetal Hemoglobin

Fetal hemoglobin binds oxygen more tightly than hemoglobin so the fetus can draw oxygen from its mother's blood. The structure of fetal hemoglobin is similar to that of hemoglobin, and soon after birth fetal hemoglobin transforms to normal hemoglobin.

Anemia

If the hemoglobin content of the blood falls below normal, the condition is called anemia. Anemia can result from a decreased rate of production of red blood cells, from an increased destruction of red blood cells, or from an increased loss of red blood cells.

A decreased rate of production of red blood cells may be due to various diseases that destroy or suppress the activity of the blood-forming tissues. Among these diseases are leukemia, *multiple myeloma, and Hodgkin's disease. Radiation and certain drugs such as benzene and gold salts also decrease the activity of the blood-forming tissues. Another frequent cause of decreased red blood cell production is a diet lacking in iron and protein, particularly in infancy and childhood and during pregnancy. Anemia may also be related to a genetic defect that affects the production of hemoglobin (see Chapter 34). Pernicious anemia, a failure of red blood cell production, is due to a lack of vitamin B_{12} or of the intrinsic factor (see page 455).

Figure 30-2   Structural formulas of heme, cytochrome c, and chlorophyll a.

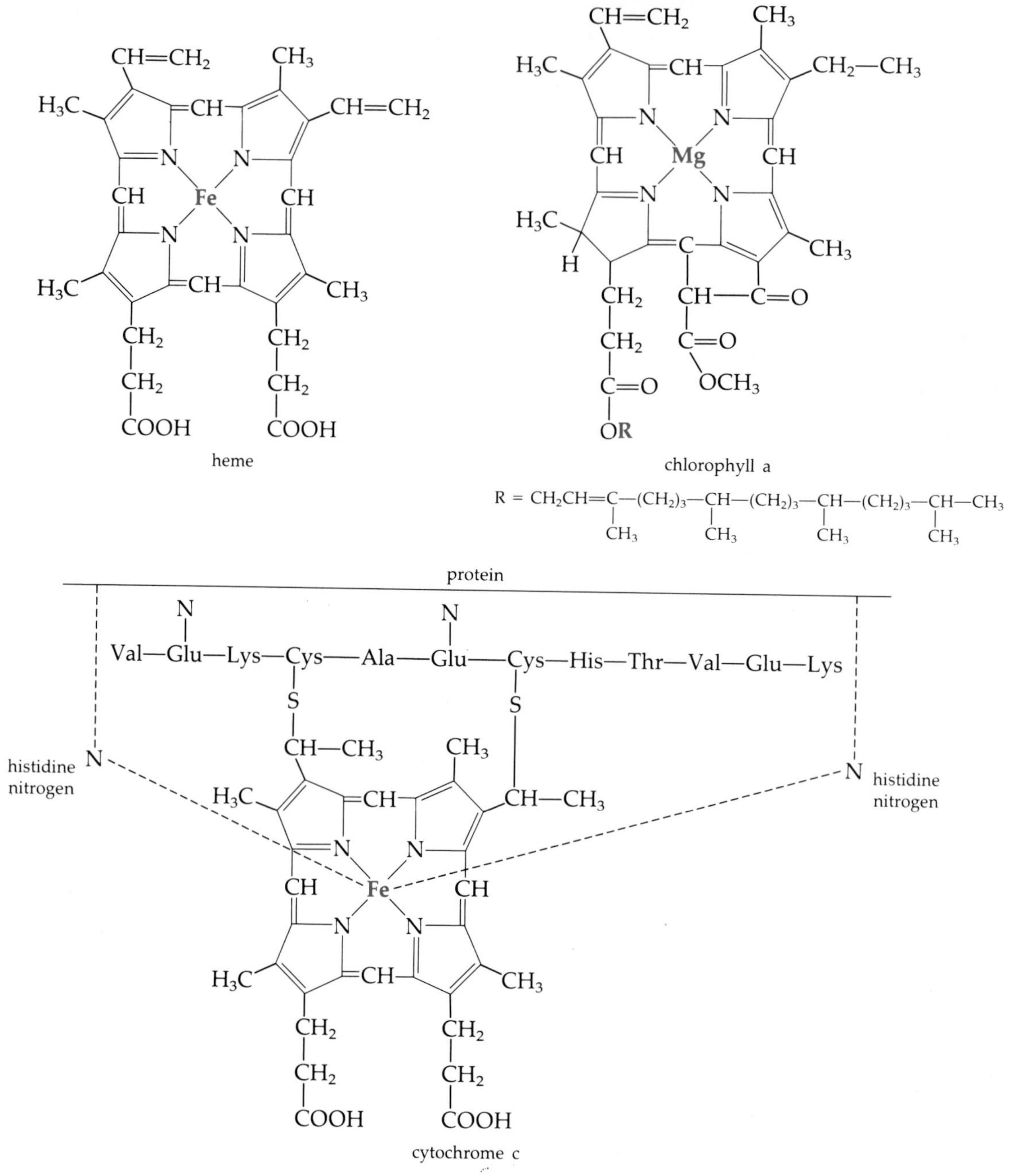

heme

chlorophyll a

$R = CH_2CH=C-(CH_2)_3-CH-(CH_2)_3-CH-(CH_2)_3-CH-CH_3$

cytochrome c

A decrease in red blood cells may be caused by several poisons or infections that can cause hemolysis. Carbon monoxide (CO) is a poisonous gas because hemoglobin combines with it approximately 210 times as fast as it does with oxygen. The compound formed between hemoglobin and carbon monoxide—carboxyhemoglobin—

is very stable, so only a small amount of hemoglobin is left to carry oxygen. If the CO content of the air is 0.02 percent, nausea and headache occur; if the CO content of the air rises to 0.1 percent, unconsciousness will occur within 1 hr and death within 4 hr. Other poisonous gases such as hydrogen sulfide (H_2S) and hydrocyanic acid (HCN) have similar effects on hemoglobin.

An excessive loss of hemoglobin may be due to hemorrhaging.

Myoglobin

Myoglobin is a globular protein similar to hemoglobin in that it also contains a heme group. However, myoglobin differs in that it has a single chain of 153 amino acids attached to the heme group.

Hemoglobin is involved in the transportation of oxygen to the tissues. Myoglobin stores oxygen in red muscle tissue such as heart muscle. Actually, the red color of such muscle tissues is due to the presence of myoglobin. Like fetal hemoglobin, myoglobin binds oxygen to itself more strongly than normal hemoglobin does and so stores it for the needs of the heart.

Under conditions of oxygen deprivation (such as severe exercise), myoglobin releases oxygen to the heart muscle for the synthesis of ATP. The oxygen must then be replaced from the hemoglobin.

Plasma Lipids

Plasma lipids can be divided into several classes, as shown in Table 30-2.

Metabolism of lipids produces much of the body's energy. In order to be metabolized, lipids must first be transported by the bloodstream. But lipids are hydrophobic and are not soluble in blood (mostly water). The transportation of lipids in the bloodstream is accomplished by associating insoluble lipids with phospholipids (which are polar) and then combining them with cholesterol and protein to form a *hydrophilic lipoprotein complex*. In this manner triglycerides, which are derived from intestinal absorption of fat or from the liver, are transported as chylomicrons and very low density

Table 30-2 Lipids in Human Blood Plasma

Lipid	Range (mg/100 mL)
Triglycerides	80–120
Phospholipids	123–390
Lecithin	50–200
Cephalin	50–130
Sphingomyelins	15–35
Cholesterol (total)	107–320
Free fatty acids	6–16
Total lipids	360–820

lipoproteins (VLDL) (see page 421). Free fatty acids, which are released from the adipose tissue, are carried in the bloodstream in an unesterified form as an albumin–fatty acid complex. That is, many types of lipids are transported in the bloodstream as plasma lipoproteins.

Plasma Proteins

The plasma proteins constitute about 7 percent of the plasma and are usually divided into three groups: albumin, globulins, and fibrinogen. Approximately 55 percent of the plasma protein is albumin, 38.5 percent globulins and 6.5 percent fibrinogen.

Albumin

Albumin in the blood functions in the regulation of osmotic pressure. The control of osmotic pressure in turn affects the water balance in the body.

Albumin, like other plasma proteins, cannot pass through the walls of the blood vessels (because they are colloids, and colloids cannot pass through membranes). Since albumin is the principal plasma protein and the smallest plasma protein both in size and weight (it consists of a single chain of 610 amino acids), it accounts for most of the colloid osmotic pressure of the blood.

The effect of albumin (and other plasma proteins) on water balance (both filtration and reabsorption) has been hypothesized by Starling, as indicated in the following paragraphs and also in Figure 30-3.

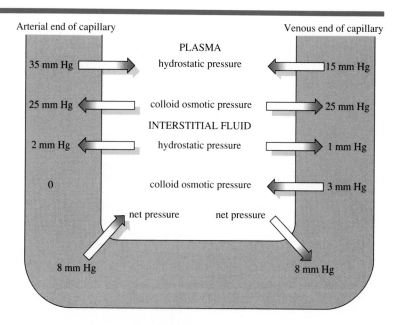

Figure 30-3   Effect of albumin (and other plasma proteins) on water balance.

When blood enters the arterial end of a capillary, it exerts a *hydrostatic (blood) pressure of 35 mm Hg, forcing fluid outward from the blood vessel. At the same time, the colloid osmotic pressure of the plasma, 25 mm Hg, pulls fluid back into the blood vessel. The *interstitial fluid exerts a hydrostatic pressure of 2 mm Hg, which forces fluid out of the tissues back into the blood. Since there is almost no protein in the tissue fluids at the end of the capillary, the colloid osmotic pressure of the interstitial fluid is 0 mm Hg. The net result of the pressure acting outward from the blood (35 and 0 mm Hg) and the pressure acting inward toward the blood (25 and 2 mm Hg) causes a net pressure of 8 mm Hg, forcing fluid out of the blood at the arterial end of the capillary. Thus, there is a net outward filtration from the capillary.

At the venous end of the capillary, the following conditions exist: The hydrostatic pressure of the blood is 15 mm Hg, forcing fluid outward. (Note that the hydrostatic pressure of the blood at the venous end is less than at the arterial end of the capillary.) The colloid osmotic pressure of the plasma remains at 25 mm Hg, pulling fluid inward. The interstitial fluid hydrostatic pressure of 1 mm Hg forces fluid out of the tissues back into the blood, whereas the interstitial fluid colloid osmotic pressure of 3 mm Hg causes fluid to flow back to the tissues.[†] The net result of two pressures acting outward at the venous end of the capillary (15 and 3 mm Hg) and two pressures acting inward (25 and 1 mm Hg) gives a net pressure of 8 mm Hg inward, causing reabsorption of materials. This reabsorption is aided by the *lymphatics.

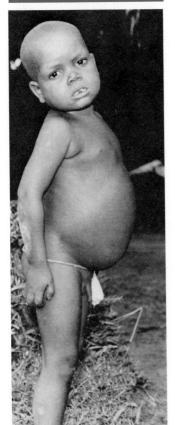

Figure 30-4 The potbelly of this undernourished child is a sign of malnutrition. [AFIP]

If the plasma proteins (primarily albumin) are present in decreased amounts (as in nephritis or during a low-protein diet), the osmotic pressure of the plasma decreases. This decreased osmotic pressure of the blood causes a greater net pressure outward at the arterial end of the capillary and a lower net inward venous pressure at the venous end of the capillary. When this occurs, water (fluid) accumulates in the tissues. Such a condition is known as *edema.

Kwashiorkor (see Figure 30-4), a protein deficiency disease, is characterized by edema of the abdomen and extremities. In children, a swollen belly is characteristic. Kwashiorkor is caused by a drop in plasma protein, particularly albumin. Under these circumstances, water moves from the bloodstream into the tissues, causing swelling.

Edema can also occur because of heart disease, whereby there is an increase in venous hydrostatic pressure. Many terminal illnesses are accompanied by edema. This becomes a serious problem, and tapping and draining may be necessary. Concentrated albumin infusions (25 g in 100 mL diluent) are helpful in the treatment of shock, to increase blood volume, and to remove fluid from the tissues.

[†] This colloid osmotic pressure is caused by small amounts of plasma protein that pass through the capillary membranes and tend to accumulate at the venous end of the capillaries.

The amount of albumin present in the blood is lowered in liver disease because albumin is formed in the liver.

Another function of albumin in the blood is to act as a carrier for fatty acids, trace elements, and many drugs.

Globulins

The globulins present in the plasma can be separated into different groups by a process known as *electrophoresis (see page 206), whereby the charged protein particles migrate at varying rates to electrodes of opposite charge, with albumin migrating the fastest. The distribution of the plasma proteins is shown in Figure 30-5. As can be seen in the illustration, the globulins are subdivided into alpha (α), beta (β), and gamma (γ). The globulins form complexes (loose combinations) with such substances as carbohydrates (mucoprotein and glycoprotein), lipids (lipoprotein), and metal ions (transferrin for iron and ceruloplasmin for copper). The amount of transferrin is decreased in such diseases as pernicious anemia and liver disease. The amount of ceruloplasmin is decreased in Wilson's disease (see page 566). These complexes can be transported to all parts of the body.

The γ-globulins (immunoglobulins) include the antibodies with which the body fights infectious diseases. γ-Globulin has been found to contain as many as 20 different antibodies for immunity against such diseases as measles, infectious hepatitis, poliomyelitis, mumps, and influenza. The most important use of serum electrophoresis is as an aid in the diagnosis of diseases in which abnormal proteins appear in the blood (*multiple myeloma and *macroglobulinemia), or when a protein component is either present in decreased amounts or lacking altogether, as in agammaglobulinemia.

Some people lack the ability to make γ-globulin. These people are quite susceptible to infections because they have no antibodies to counteract such diseases. The lack of γ-globulin is called **agamma-**

Figure 30-5 Distribution of plasma proteins during electrophoresis.

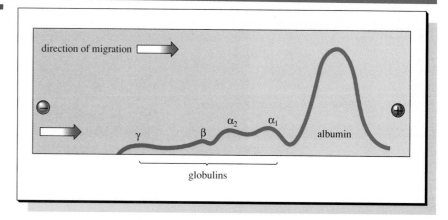

globulinemia and can be counteracted by the administration of γ-globulin.

Fibrinogen

Fibrinogen is the plasma protein involved in the clotting of the blood. It is manufactured in the liver, so any disease that destroys liver tissue causes a decrease in the amount of fibrinogen in the plasma. Fibrinogen is a soluble plasma glycoprotein with a molecular weight of 340,000.

Blood Clotting

When the skin is ruptured, blood flows out and soon forms a clot. When blood is taken from a vein and placed in a test tube, it soon forms a clot. Why does blood clot when it is removed from its normal place in the circulatory system? Why doesn't it clot in the blood vessels themselves?

When blood clots, a series of reactions occurs in which the soluble plasma protein fibrinogen is converted into insoluble fibrin. Fibrin precipitates in the form of long threads that cling together to form a spongy mass, which entraps and holds the blood cells, forming a clot.

When a blood vessel is cut, the blood comes into contact with the exposed tissue. This contact activates two separate systems of coagulation—the intrinsic system, in which all components necessary to form a clot are found in the blood, and the extrinsic system, which needs a component released by the tissues. Figure 30-6 illustrates the three mechanisms of *hemostasis—vasoconstriction, platelet aggregation, and fibrin clot. A vasoconstriction occurs promptly at the site of the injury, thus reducing blood flow. Also,

Figure 30-6   Reactions involved in hemostasis.

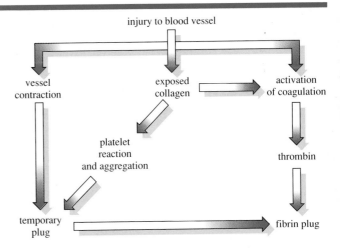

the platelets form a temporary plug in the injured capillaries. Prostaglandins and thromboxanes are required for this latter action. Since aspirin interferes with the action of prostaglandins, aspirin interferes with the clotting process.

Calcium ions are necessary for the clotting of blood. The coagulation of freshly drawn blood samples can be prevented by adding a substance (such as potassium oxalate) that removes calcium ions from solution. However, since oxalates are poisonous, this method can be used only if the blood is to be analyzed in a laboratory.

To prevent clotting in blood used for transfusions, sodium citrate is added. This substance removes the calcium ions from the blood by forming calcium citrate, a compound that is almost completely nondissociated. A deficiency of vitamin K reduces the production of prothrombin, without which the blood cannot clot.

In one form of hemophilia, factor X_a is missing, leading to an increase in blood clotting time.

Anticoagulant drugs such as bishydroxycoumarin (Dicumarol) reduce the conversion of prothrombin to thrombin and so keep blood from clotting rapidly. Heparin speeds up the removal of thrombin from several minutes to a few seconds. Anticoagulant drugs may be used after surgery to prevent clots from forming in the cut blood vessels. A clot formed in a blood vessel is called a **thrombus.** A thrombus in a blood vessel does no harm if it remains where it was formed because it is slowly reabsorbed. However, if the clot breaks loose and travels through the blood vessels, it may lodge in and obstruct a blood vessel leading to the heart or brain, causing paralysis or death.

Breakdown of fibrin in blood clots (fibrinolysis) occurs within a few days after formation and is the result of the action of the enzyme plasmin. Plasmin is activated by a variety of circumstances including stress, exercise, snake venom, and streptokinase from streptococci. Two drugs for dissolving and preventing blood clots, streptokinase

Figure 30-7　Fibrinolytic system.

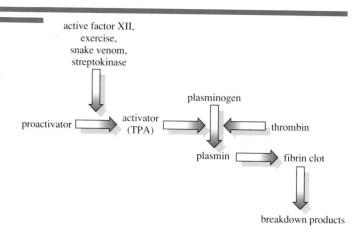

and the genetically engineered TPA (tissue plasmin activator), both activate plasmin. Figure 30-7 shows a condensed version of the fibrinolytic system.

Respiration

The tissues require oxygen for their normal metabolic processes; they must also eliminate carbon dioxide. Oxygen is carried from the lungs to the tissues by the hemoglobin of the blood. In the tissues the oxgyen is given up by the hemoglobin and the waste carbon dioxide from the tissues is picked up and carried to the lungs.

The inspired air has a higher concentration of oxygen than does the blood in the alveoli of the lungs. Gases always diffuse from an area of high concentration to one of lower concentration, so the oxygen diffuses from the lungs (high concentration) into the blood (lower concentration). In the blood the oxygen combines with the hemoglobin. Very little oxygen is actually dissolved (uncombined) in the blood. When the oxygen-rich blood reaches the tissues, it gives up its oxygen to the cells because those cells are using up oxygen and have a lower concentration of that gas than the blood. At the same time, the cells have a higher concentration of carbon dioxide than the blood; therefore, that gas diffuses from the cells into the bloodstream. The blood carries the carbon dioxide to the lungs. There it is in contact with air, which has a lower carbon dioxide concentration, and so it passes from the blood into the lungs where it is exhaled.

The transportation of gases in respiration includes the nine steps described in the next section.

Transportation of O_2 and CO_2

The transportation of oxygen and carbon dioxide is summarized in Figure 30-8, where the interrelationships of the following steps are shown.

1. As oxygen passes from the alveoli of the lungs into the bloodstream, some of it dissolves in the blood plasma. However, most of it reacts with hemoglobin (here represented by the formula HHb) to form oxyhemoglobin, HbO_2^-.

$$HHb + O_2 \longrightarrow HbO_2^- + H^+$$

Oxygen is carried to the cells in this form.
2. At this same time, carbon dioxide, which is produced during metabolic processes in the cells, diffuses from the tissues into the blood. A small amount dissolves directly in the plasma, but most of the carbon dioxide diffuses into the red blood cells, where it reacts with water in the cells to form carbonic acid. This reaction

Figure 30-8 Transportation of oxygen and carbon dioxide.

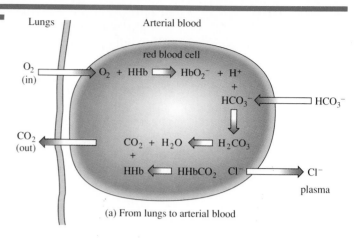

(a) From lungs to arterial blood

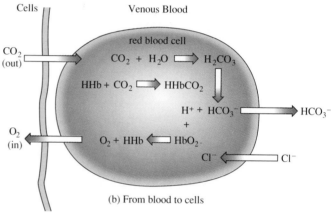

(b) From blood to cells

takes place rapidly under the influence of the enzyme carbonic anhydrase.

$$CO_2 + H_2O \xrightarrow{\text{carbonic anhydrase}} H_2CO_3$$

3. The carbonic acid thus formed ionizes to yield hydrogen and bicarbonate ions.

$$H_2CO_3 \longrightarrow H^+ + HCO_3^-$$

4. In the tissues, oxyhemoglobin reacts with the hydrogen ions to yield oxygen and hemoglobin (HHb).

$$HbO_2^- + H^+ \longrightarrow HHb + O_2$$

The release of oxygen to the tissues is enhanced by a decrease in the pH and by the presence of 2,3-diphosphoglyceric acid. Most of the hemoglobin travels back to the lungs to pick up more oxygen. But some of the hemoglobin reacts with carbon dioxide to form carbaminohemoglobin (here represented as $HHbCO_2$).

$$HHb + CO_2 \longrightarrow HHbCO_2$$

5. The bicarbonate ions from step 3 do not remain in the red blood cells because those cells can hold only a small amount of that ion. Therefore, the *excess* bicarbonate ions diffuse outward into the blood plasma. Red blood cells cannot stand a loss of negative ions, so to counteract this outflow of bicarbonate ions, chloride ions from the blood plasma flow into the red blood cells. This process is called the **chloride shift**.

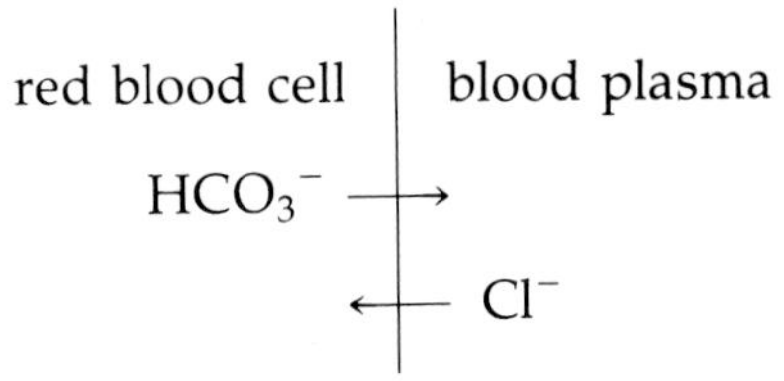

$$\text{red blood cell} \quad | \quad \text{blood plasma}$$

$$HCO_3^- \longrightarrow$$

$$\longleftarrow Cl^-$$

The bicarbonate ions can then act as buffers in the plasma.
6. In the lungs, the bicarbonate ions react with the hydrogen ions produced in step 1 to form carbonic acid.

$$HCO_3^- + H^+ \longrightarrow H_2CO_3$$

7. The carbonic acid thus formed rapidly decomposes into carbon dioxide and water under the influence of the enzyme carbonic anhydrase.

$$H_2CO_3 \xrightarrow{\text{carbonic anhydrase}} H_2O + CO_2$$

8. As the bicarbonate ions are used up in steps 6 and 7, more bicarbonate ions from the plasma flow into the red blood cells. At the same time, chloride ions migrate outward from the red blood cells. This is the **reverse chloride shift**.

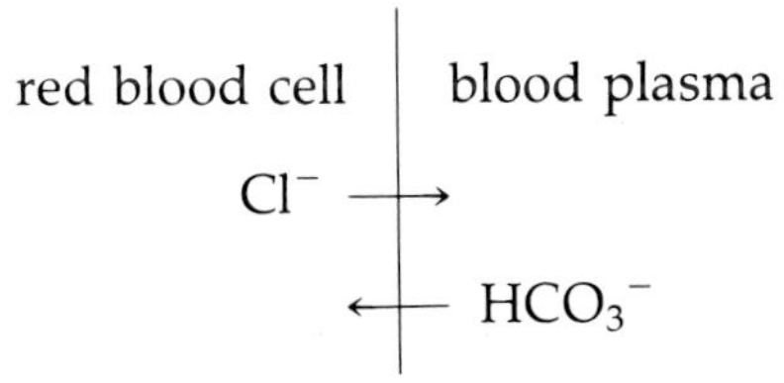

$$\text{red blood cell} \quad | \quad \text{blood plasma}$$

$$Cl^- \longrightarrow$$

$$\longleftarrow HCO_3^-$$

9. At the same time, the carbaminohemoglobin formed in step 4 decomposes in the lungs to yield hemoglobin and carbon dioxide.

$$HHbCO_2 \longrightarrow HHb + CO_2$$

Acid–Base Balance

The normal pH range of the blood is 7.35 to 7.45. When the pH falls below this range, the condition is termed *acidosis. *Alkalosis occurs when the pH rises above its normal value. Acidosis is more common than alkalosis because many of the metabolic products produced during digestion are acidic. The ability of the blood buffers to neutralize acid is called the **alkaline reserve** of the blood. In acidosis, the alkaline reserve decreases; during alkalosis, it increases.

How does the blood maintain the pH when acidic or basic substances are continuously being added to it?

The blood retains its fairly constant pH because of the presence of buffers. These buffers are present both in the blood plasma and in the red blood cells. Those in the plasma are primarily sodium buffers; those in the blood cells are mainly potassium buffers. Recall that buffers are substances (usually a weak acid and a salt of a weak acid) that resist change in pH (see page 242). The blood buffers consist of

1. Bicarbonate buffers.
2. Phosphate buffers.
3. Protein buffers (including hemoglobin and oxyhemoglobin).

Bicarbonate Buffers

The bicarbonate buffer system in the red blood cells consists of carbonic acid (H_2CO_3) and potassium bicarbonate ($KHCO_3$). The bicarbonate buffer system in the blood plasma consists of carbonic acid and sodium bicarbonate ($NaHCO_3$). If a strong acid (such as HCl) is added to a sample of blood, it will react with the salt part of the buffer and undergo the following reactions:

$$HCl + KHCO_3 \longrightarrow H_2CO_3 + KCl \qquad \text{(in blood cells)}$$

$$HCl + NaHCO_3 \longrightarrow H_2CO_3 + NaCl \qquad \text{(in blood plasma)}$$

The carbonic acid (H_2CO_3) produced is part of the original buffer. Note that the strong acid, HCl, has been replaced by a very weak one, H_2CO_3. The other products, KCl and NaCl, are neutral salts and will not affect the pH of the system.

If a strong base such as KOH or NaOH is added to a sample of blood, the following reactions will occur with the bicarbonate buffer systems:

$$KOH + H_2CO_3 \longrightarrow KHCO_3 + H_2O \qquad \text{(in blood cells)}$$

$$NaOH + H_2CO_3 \longrightarrow NaHCO_3 + H_2O \qquad \text{(in blood plasma)}$$

The salts, $KHCO_3$ and $NaHCO_3$, are part of the original buffer systems and the water produced is neutral, so the pH again is unaffected.

In both cases (reaction with a strong acid or a strong base), more of the buffer is produced plus a neutral compound.

The bicarbonate buffers and the blood protein buffers play a major part in the control of the pH; the phosphate buffers have an important role inside the cell and in the urine.

Phosphate Buffers

The phosphate buffers consist of mixtures of K_2HPO_4 and KH_2PO_4 (also Na_2HPO_4 and NaH_2PO_4), which function similarly to the bicarbonate buffers in neutralizing excess acid and base.

$$HCl + K_2HPO_4 \longrightarrow KH_2PO_4 + KCl$$

$$KOH + KH_2PO_4 \longrightarrow \underset{\text{more buffer}}{K_2HPO_4} + \underset{\text{neutral compound}}{H_2O}$$

Hemoglobin Buffer
HHb
KHb

Oxyhemoglobin Buffer
$HHbO_2$
$KHbO_2$

Hemoglobin Buffers

The hemoglobin buffers account for more than half of the total buffering action in the blood. There are hemoglobin buffers and oxyhemoglobin buffers.

These buffers, as well as other proteins that act as buffers in the bloodstream, pick up excess acid or base to help keep the pH of the blood within its normal range.

Function of the Kidneys

The kidneys help maintain the acid–base balance of the blood by excreting or absorbing phosphates and also by forming ammonia. The acid substances in the blood combine with ammonia and are excreted as ammonium salts through the kidneys, saving sodium and potassium ions for the buffer systems.

Metabolic Acidosis

Metabolic acidosis is characterized by a drop in plasma pH due to a decrease in HCO_3^- concentration and a compensatory drop in the partial pressure of carbon dioxide, pCO_2. Metabolic acidosis may be caused by either the addition of H^+ or the loss of HCO_3^-.

The most common causes of metabolic acidosis are uncontrolled diabetes mellitus with ketosis (see page 499), renal failure, poisoning with acid substances, severe diarrhea with loss of HCO_3^-, lactic acidosis, and severe dehydration.

As the pH decreases, more H_2CO_3 is produced from the HCO_3^-. The H_2CO_3 in turn decomposes to CO_2 and H_2O.

$$HCO_3^- + H^+ \rightleftharpoons H_2CO_3 \rightleftharpoons H_2O + CO_2$$

If the blood buffers cannot control the pH, then the excess carbon dioxide produced stimulates the respiratory center of the brain, making the person breathe faster, thus removing more carbon dioxide from the blood. The increased rate of respiration (hyperventilation) continues until the amount of carbon dioxide is too small to further stimulate the respiratory center of the brain, and breathing returns to its normal rate. In addition to hyperventilation, the kidneys respond by excreting H^+, primarily as NH_4^+.

Treatment, if the kidneys are functioning normally, consists of intravenous administration of HCO_3^-. This lowers the H^+ concentration and so increases pH to normal. Fluids are administered to replace lost water. A diabetic is given insulin therapy. Hemodialysis may be required if the kidneys are not functioning normally.

Metabolic Alkalosis

Metabolic alkalosis is characterized by an increase in plasma pH, an increase in HCO_3^- concentration, and a compensatory increase in pCO_2. Metabolic alkalosis is caused by a loss of H^+ or a retention of HCO_3^-.

Common causes of metabolic alkalosis are prolonged vomiting, gastric suction (both of which remove acid contents from the stomach), overdose of HCO_3^- (in the treatment of a gastric ulcer), loss of K^+ along with Cl^- in severe exercise, renal disease, massive blood transfusions, and diuretic therapy.

Normally, the kidneys are able to correct this situation by excreting large amounts of HCO_3^-. Therefore, persistence of metabolic alkalosis indicates impaired renal excretion.

Symptoms include slow respiration (hypoventilation) as the lungs try to conserve CO_2 (which comes from the H_2CO_3 needed to neutralize the excess alkalinity), numbness, convulsions, weakness, and muscle cramps.

The usual treatment involves the administration of NaCl, with KCl if the K^+ concentration is low. The increased Na^+ will enhance the excretion of HCO_3^-, thereby correcting the situation. Patients with metabolic alkalosis due to diuretic therapy can be given acetazolamide (Diamox) to increase HCO_3^- excretion.

Respiratory Acidosis

Respiratory acidosis is characterized by a decrease in plasma pH, an increase in pCO_2, and a variable compensatory increase in HCO_3^- concentration. Respiratory acidosis is due to impaired ventilation, either from inhibition of the respiratory center in the brain (sedatives, anesthetics, opiates, cardiac arrest) or impaired gas exchange across the lungs to the blood capillaries (poliomyelitis, emphysema, severe asthma, acute pulmonary edema, pneumonia).

The response to respiratory acidosis takes time. Buffers are ineffective, because HCO_3^- cannot buffer H_2CO_3. The kidneys respond by increasing H^+ excretion, but this occurs over hours to days. Once renal response has developed, it is usually capable of handling excessive acidity. However, the kidneys will not be able to correct the problem. Chronic respiratory acidosis is a relatively common problem most often seen in smokers with pulmonary disease.

Symptoms of respiratory acidosis include headache, blurred vision, fatigue, and weakness. Treatment of acute respiratory acidosis includes returning the breathing to normal. This may include mechanical assistance, use of bronchodilators, and small infusions of $NaHCO_3$. Chronic respiratory acidosis is usually not treated. Care must be taken with chronic patients since opiates and anesthetics will act as further respiratory depressants and induce further hypoventilation.

Respiratory Alkalosis

Respiratory alkalosis is characterized by an increase in plasma pH, a drop in pCO_2, and a variable decrease in HCO_3^- concentration. Such a condition may be caused by hyperventilation (hysteria), residence at high altitudes, congestive heart failure, pulmonary disease, aspirin overdose, excessive exercise, and cirrhosis. The problem originates with excessive excretion of CO_2. For instance, overstimulation of the respiratory control center in the brain by chemicals (aspirin) or the need for more O_2 (congestive heart failure, residence at high altitudes) leads to excessive loss of CO_2. The kidneys will attempt to compensate by increasing HCO_3^- excretion.

Symptoms of respiratory alkalosis are hyperventilation that cannot be controlled, possible convulsions, and lightheadedness. Treatment involves removal of the cause of the problem. Breathing into and out of a paper bag (to inhale CO_2 which has been exhaled) is also helpful.

Table 30-3 compares the changes in pH, pCO_2, and HCO_3^- concentration for respiratory and metabolic acidosis and alkalosis. The normal pH of the blood is 7.35 to 7.45; normal pCO_2 is 35 to 40 mm Hg; normal HCO_3^- is 25 to 30 mEq/L.

Table 30-3 Typical Changes in Plasma pH, pCO_2, and HCO_3^- Concentration for Various Acid–Base Disorders.

	pH	$[HCO_3^-]$	pCO_2
Metabolic acidosis	↓	↓	↓
Metabolic alkalosis	↑	↑	↑
Respiratory acidosis	↓	↑	↑
Respiratory alkalosis	↑	↓	↓

Fluid–Electrolyte Balance

Normally, in humans, water intake is balanced by water output. If the intake of water exceeds the output, **edema** results. If the output of water exceeds the input, **dehydration** may occur (see Figure 30-9).

Water Balance

The body replenishes its water supply in three ways.

1. By the ingestion of liquids (which are primarily, if not wholly, water).
2. By the ingestion of foods such as meats, vegetables, and fruits, all of which contain a very high percentage of water.
3. By metabolic processes taking place normally in the body. When carbohydrates, fats, and proteins are metabolized, water is produced. Approximately 14 mL of water is formed for every 100 kcal of energy released by the oxidation of foods. The oxidation of 100 g of carbohydrate produces 55 g of H_2O; the oxidation of 100 g of fat produces 107 g of H_2O; the oxidation of 100 g of protein produces 41 g of H_2O.

The total normal input of water in the body is approximately 2500 mL per day.
 The body loses water in several ways.

1. Through the kidneys, as urine.
2. Through the skin, as perspiration.
3. Through the lungs, as exhaled moisture.
4. Through the feces.

The total of these water losses should approximate that of the water intake, 2500 mL per day. However, the amounts of water lost by these methods may vary considerably. For example, the amount of moisture lost through the skin (by evaporation of perspiration) and through the lungs increases in the following circumstances.

1. During vigorous muscular exercise.
2. With an increased respiratory rate.
3. In a hot, dry environment.
4. During a fever.
5. When the skin receives severe burns.

Conversely, the amount of water lost through the kidneys increases following the ingestion of large amounts of water within a short period of time. It also increases because of the presence of large amounts of waste products in the bloodstream. However, when the body loses excess water through vomiting and/or diarrhea, the output of water through the kidneys is immediately lessened.

Figure 30-9 Water balance in the body.

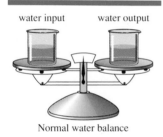

water input   water output

Normal water balance

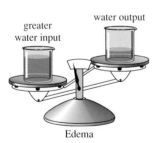

greater water input

water output

Edema

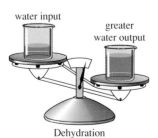

water input

greater water output

Dehydration

Distribution of Water in the Body

The water of the body is considered to be distributed in two major areas—intracellular (within the cells), 55 percent, and extracellular (outside the cells), 45 percent. The extracellular water, in turn, can be further divided into four areas:

1. Intravascular (plasma): fluid within the heart and blood vessels (7.5 percent).
2. Interstitial and lymph: fluids outside the cells (20 percent).
3. Dense connective tissue, cartilage, and bone (15 percent).
4. Transcellular fluids: extracellular fluid collections, including the salivary glands, thyroid gland, gonads, mucous membranes of the respiratory and gastrointestinal tracts, kidneys, liver, pancreas, cerebrospinal fluid, and fluid in the spaces within the eyes (2.5 percent).

Approximately 55 percent of the weight of an adult male and 50 percent of the weight of an adult female is water.

The principal difference between the blood plasma (the intravascular fluid) and the interstitial fluid is in the protein content. Proteins cannot pass through membranes and so they remain in the blood vessels. Thus, the protein content of the interstitial fluid is very low compared to that of the intravascular fluid, although the soluble electrolytes in each are approximately the same.

There are distinct differences in salt concentration in the intracellular and extracellular fluids.

The extracellular water in the dense connective tissue, cartilage, and bone and the extracellular water in the transcellular fluids do not readily interchange fluids and electrolytes with the rest of the body water. The rest of the body water—intracellular, intravascular, interstitial, and lymph—freely moves from one area to another within the body.

Electrolyte Balance

Table 30-4 indicates the electrolyte concentrations in the intravascular, interstitial, and intracellular fluids. The ions are divided into two groups—cations (positively charged ions) and anions (negatively charged ions). Concentrations of anions and cations are expressed in units of milliequivalents per liter (mEq/L). The unit milliequivalent measures the chemical and physiologic activity of an ionized substance, the electrolyte. Since milliequivalents are based upon ions, the term milliequivalents represents the number of charged particles or the number of both positive and negative charges present in a solution of an electrolyte. Recall that the number of positive charges must always equal the number of negative charges.

Table 30-4 shows that the electrolyte concentrations of the interstitial fluid are similar to those of the intravascular fluid, except

Table 30-4 Electrolyte Concentrations of Body Fluids (mEq/L)

	Intravascular	Interstitial	Intracellular
Cations			
Na^+	142	145	10
K^+	4	4	158
Mg^{2+}	3	2	35
Ca^{2+}	5	3	2
	154	154	205
Anions			
Cl^-	103	115	2
HCO_3^-	27	30	8
HPO_4^{2-}	2	2	140
SO_4^{2-}	1	1	—
$Protein^-$	16	1	55
$Organic\ acids^-$	5	5	—
	154	154	205

that there is more chloride ion and less protein concentration. Note also that in both the intravascular and the interstitial fluid, sodium is the principal cation and chloride the principal anion.

The intracellular fluid differs in concentrations from the intravascular and interstitial fluids in that potassium is the principal cation and phosphate the principal anion. Electrolyte concentrations in intracellular fluids are only approximate because they vary slightly from tissue to tissue.

Concentrations of electrolytes are frequently measured and expressed in the units milligrams per 100 mL. This can be changed to milliequivalents per liter by means of the following formula:

$$\frac{mg/100\ mL \times 10 \times ionic\ charge}{atomic\ weight} = mEq/L$$

For example, suppose a laboratory reports the serum potassium concentration as 15.6 mg/100 mL. The concentration of potassium in milliequivalents per liter can be calculated by using the preceding formula, the atomic weight (39), and the ionic charge (1), or

$$\frac{15.6 \times 10 \times 1}{39} = 4\ mEq/L$$

Osmotic activity depends upon the number of particles present in a solution regardless of whether they carry a charge or not. Thus, sodium ions, potassium ions, and chloride ions cause osmotic pressure, but so does glucose, which is a nonelectrolyte (that is, it carries no charge).

Osmotic activity is expressed by the unit **milliosmol**, which is a measure of the amount of work that dissolved particles can do in drawing a fluid through a semipermeable membrane. Osmotic activity is measured by means of an instrument called an osmometer.

Clinical Importance of Mineral Cations and Anions

Minerals required for the body can be divided into two main groups—*macrominerals*, which are required in amounts greater than 100 mg per day, and *microminerals* (trace elements), which are needed in amounts less than 100 mg per day.

Macrominerals

Sodium Ions

Sodium ions are the primary cations of the extracellular fluids. The principal functions of sodium ions are

1. To maintain the osmotic pressure of the extracellular fluid.
2. To control water retention in tissue spaces.
3. To help maintain blood pressure.
4. To maintain the body's acid–base balance by means of the bicarbonate buffer system.
5. To regulate the irritability of the nerve and muscle tissue and of the heart.

The average daily adult intake of sodium, as NaCl, is 5 to 15 g. About 95 percent of the sodium lost by the body passes through the kidneys. The body's sodium ion concentration is influenced by aldosterone, a hormone of the adrenal cortex. This hormone promotes the reabsorption of sodium ions in the kidney tubules. The antidiuretic hormone (ADH) promotes water absorption in the kidneys and so has a definite effect on extracellular sodium ion concentration.

Hyponatremia, a lower-than-normal serum sodium ion concentration, may be due to such causes as vomiting, diarrhea, hormone disorders, starvation, extensive skin burns, loss of sodium ions because of kidney damage, or use of diuretics. The clinical symptoms of hyponatremia are cold, clammy extremities, lowered blood pressure, weak and rapid pulse, oliguria, muscular weakness, and cyanosis (a dark purplish discoloration of the skin and mucous membrances due to decreased oxygenation of the blood). In addition, because of an increased plasticity of the tissues, hyponatremia frequently shows as fingerprinting over the sternum. In hyponatremia the specific gravity of the urine is less than 1.010.

Hypernatremia, a higher-than-normal serum sodium ion concentration, may be due to such causes as deficient water intake, excessive sweating, excessive water output (such as caused by diabetes insipidus), poor kidney excretion, rapid administration of sodium salts, hyperactivity of the adrenal cortex (as in Cushing's disease), and some cases of cerebral disease. The clinical symptoms of hypernatremia are dry, itchy mucous membranes, intense thirst, oliguria or anuria, rough dry tongue, and elevation of temperature.

The specific gravity of the urine rises above 1.030. In an extreme case of hypernatremia the symptoms include tachycardia (rapidly beating heart), edema, and cerebral disturbances.

Potassium Ions

Potassium ions are the principal cations of the intracellular fluid. Since the kidneys do not conserve potassium ions as well as they preserve sodium ions and since the body cannot store potassium ions, a depletion of this substance occurs readily in patients whose diets are low in potassium or who are excreting more potassium than they take in.

The principal functions of potassium ions in the body are

1. To maintain the osmotic pressure of the cells.
2. To maintain the electric potential of the cells.
3. To maintain the size of the cells.
4. To maintain proper contraction of the heart.
5. To maintain proper transmission of nerve impulses.

Potassium ions move into the cells during anabolic activity and move out of the cells during catabolic activity. The concentration of potassium ion is usually measured in terms of serum potassium because this is a much easier quantity to measure than cellular potassium concentration.

Hypokalemia, a lower-than-normal serum potassium ion concentration, can occur under the following conditions:

1. Too low an intake of potassium ions
   (a) During starvation or malnutrition.
   (b) In a diet deficient in potassium.
   (c) During intravenous infusions of fluids low or lacking in potassium ions.
2. Too great an output of potassium ions because of
   (a) The use of diuretics.
   (b) The use of corticosteroids (these hormones promote retention of sodium ions at the expense of potassium ions).
   (c) Prolonged vomiting.
   (d) Gastric suction and intestinal drainage.
   (e) Diarrhea.
   (f) Polyuria.

In addition, hypokalemia may be caused by a sudden shift of potassium ions from the extracellular fluid to the intracellular fluid. This could occur, for example, in the treatment of diabetic acidosis with insulin and glucose.

In general, hypokalemia occurs most frequently in conjunction with some other pathologic condition. The general symptoms of hypokalemia are a general feeling of being ill, lack of energy, muscular weakness, numbness of fingers and toes, apathy, dizziness on rising, and cramps, particularly in the calf muscles.

As hypokalemia develops to a greater extent, symptoms relating to the heart become evident. Among these are weak pulse, falling blood pressure, faint heart sounds, and changes in the ECG—first a flattening of the T wave, then inverted T waves with a sagging ST segment and AV block, and finally cardiac arrest.

Hypokalemia can be treated or prevented by giving the patient potassium intravenously (in the form of a potassium salt) or orally by the use of high-potassium foods such as veal, chicken, beef, pork, bananas, orange and pineapple juices, broccoli, and potatoes. Note that although there are many other foods high in potassium, they are usually also high in sodium. A patient on a high-potassium diet usually also has a low-sodium requirement; therefore, these other types of foods are not recommended.

Hyperkalemia, an increased serum potassium ion level, occurs

1. If the intake of potassium ions is too great because of too rapid an infusion of potassium ions or the administration of excess potassium ions intravenously.
2. If the output of potassium ions is too low because of renal failure or acute dehydration.
3. If there is a sudden shift of potassium ions from the intracellular fluid to the extracellular fluid because of severe burns, crush injuries (both of these could release potassium ions from the cells into the blood stream), or during acidosis.

The symptoms of hyperkalemia are a general feeling of ill-being, muscular weakness, listlessness, mental confusion, slower heart beat, poor heart sounds, *bradycardia, and eventually cardiac arrest. Characteristic changes in the ECG are elevated T waves, widening of the QRS complex, gradual lengthening of the PR interval, and final disappearance of the P wave.

Excess potassium ions can be removed either by dialysis or by the administration of glucose and insulin.

Calcium Ions

Most of the body's calcium is found in the bones and the teeth in the form of calcium carbonate and calcium phosphate. If the blood calcium ion concentration falls, it can readily be replenished from the bone. Conversely, if the blood calcium ion concentration rises, the amount replenished from the bones decreases (see pages 625 and 627).

The daily intake (adult) for calcium varies from 200 to 1500 mg and comes primarily from milk and milk products. Ionized calcium is present in body fluids and is important in blood coagulation (see page 549), in the regulation of membrane permeability, and in the normal functioning of nerve, heart, and muscle tissue.

Because of the great amount of calcium present in the bones, calcium is not required during intravenous therapy. In addition, calcium-containing solutions are not suitable for infusions because, if mixed with citrated blood, they may cause a clot in the drip tube.

Calmodulin is a calcium-modulating protein found in all cells. It has a molecular weight of 16,700 and consists of a chain of 148 amino acids. Calmodulin affects the synthesis and action of the prostaglandins and enhances blood platelet aggregation.

Phenothiazines, which are major tranquilizers used to treat psychoses, and phenytoin, which is used in the treatment of leprosy, may inactivate calmodulin, indicating a link between these conditions and calcium metabolism.

Hypocalcemia, a low serum calcium concentration, may be due to a hypoactive parathyroid gland (see page 627), the surgical removal of the parathyroid glands, or a large infusion of citrated blood. The symptoms of hypocalcemia include tingling of the fingertips, abdominal and muscle cramps, and tetany.

Hypercalcemia, an increased serum calcium concentration, may be caused by an overactive parathyroid or by a tumor of that gland. It may also be caused by the administration of excess vitamin D. The symptoms of hypercalcemia include hypotonicity of muscles, kidney stones, deep bone pain, and bone cavitation.

The serum calcium ion concentration decreases during hypoparathyroidism and rises during hyperparathyroidism (see page 627).

Magnesium Ions

Magnesium ion, like potassium ion, is found primarily in the intracellular fluid. Magnesium ions are essential for the proper functioning of the neuromuscular system.

Magnesium acts as an activator for more enzymes than any other metal ion in the body; it is necessary for over 100 metabolic reactions. The normal intake of magnesium is 15 to 30 mEq per day. Another unusual fact about magnesium is that it is the only positively charged ion that has a higher concentration in the cerebrospinal fluid than in the blood serum.

A deficiency of serum magnesium ions, **hypomagnesemia,** because of dietary intake is unusual because magnesium is a necessary element for chlorophyll, which is found in all green plant foods. A lower-than-normal magnesium ion concentration may be caused by such factors as

1. Chronic alcoholism.
2. Diabetic acidosis.
3. Prolonged intravenous infusion without magnesium ions.
4. Hypoparathyroidism.
5. Prolonged nasogastric suction.
6. Acute pancreatitis.
7. Severe malabsorption.

The clinical symptoms of a deficiency of magnesium ions are

1. Muscular tremors.
2. Convulsions.

3. Delirium.
4. Delusions.
5. Disorientation.
6. Hyperirritability.
7. Elevated blood pressure.

An excess of serum magnesium ions can be caused by severe dehydration or renal insufficiency. Excess magnesium ions act as a sedative. Extreme excesses may cause coma, respiratory paralysis, or cardiac arrest.

Chloride Ions

The chloride ion is the primary anion of the extracellular fluid. The body's intake of chloride ion is closely related to that of the sodium ion (see page 561).

One of the principal functions of the chloride ion is as a component of gastric hydrochloric acid. The chloride ion also serves an important function in the transportation of oxygen and carbon dioxide in the blood (see page 551).

Hypochloremia, a lower-than-normal serum chloride ion concentration, can be brought about by prolonged vomiting, profuse sweating, and diarrhea. This condition causes an alkalosis because of an increased concentration of bicarbonate ions. Hypochloremia can also occur when there is a marked loss of potassium ions.

Phosphate Ions

The phosphate ion is the primary anion of the intracellular fluid. Diets that are adequate in calcium usually contain more than enough phosphorus for the body's needs.

Most of the body's phosphate is present in the bones as calcium phosphate, a substance that gives the bones their rigidity, but phosphate is found in every cell of the body. Phosphate ions are important in the acid–base balance of the body. They constitute one of the body's buffer systems. Phosphates are also of great importance in the production of ATP, the body's principal energy compound. Serum phosphate levels are low in hyperparathyroidism and high in hypoparathyroidism and celiac disease.

Microminerals

Iron Ions

Iron ions are involved almost exclusively in cellular respiration. Iron is part of hemoglobin, myoglobin, and cytochromes as well as several oxidative enzymes. The formation of hemoglobin requires the presence of traces of copper. The best dietary sources of iron are the "organ meats," such as liver, heart, and kidneys. Other sources are egg yolk, fish, beans, and spinach.

Most of the iron present in the food we eat is in the iron(III) ion

(Fe^{3+}) form. In the digestive system, iron(III) ions are reduced to iron(II) ions (Fe^{2+}), which are then absorbed into the bloodstream from the stomach and duodenum. In the blood plasma, iron(II) ions are oxidized to iron(III) ions, which then become part of a specific protein—transferrin, a β-globulin. The conversion of iron(II) ions to iron(III) ions in the blood plasma is catalyzed by the enzyme ceruloplasmin, a copper-containing compound (see following paragraphs).

The liver, spleen, and bone marrow are able to extract the iron from transferrin and to store that iron in the form of two proteins—ferritin and hemosiderin. The bone marrow is also able to extract the iron from transferrin and to use that iron for the production of hemoglobin.

A deficiency of iron **(iron-deficiency anemia)** can result from a low intake of iron because of a diet high in cereal and low in meat, because of poor absorption of iron due to gastrointestinal disturbances or diarrhea, or because of excessive loss of blood. This type of anemia can be treated with a daily dose of ferrous sulfate in the diet, if absorption is normal.

Copper Ions

In addition to being required for the synthesis of hemoglobin, copper ions are necessary for certain oxidative enzymes, such as cytochrome oxidase (part of the oxidative phosphorylation sequence; see page 485) and uricase (which catalyzes the oxidation of uric acid to allantoin).

Copper is found in the brain in the form of cerebrocuprein, in the blood cells as erythrocuprein, and in the blood plasma as ceruloplasmin, an α-globulin. In **Wilson's disease,** there is a decreased concentration of ceruloplasmin in the blood. This disease is characterized by the presence of large amounts of copper in the brain along with an excessive urinary output of copper.

Copper aspirinate, a complex of copper and aspirin, has been found to be 20 times as effective as aspirin itself in the treatment of arthritis in animals. Its effect on humans is still questionable.

The average daily diet contains about 2.5 to 5 mg of copper, an amount that is considered to be adequate for the normal adult. The richest sources of copper are liver, nuts, kidney, raisins, and dried legumes.

Zinc Ions

Zinc is essential for normal growth and reproduction. It has a beneficial effect on wound healing and tissue repair. Zinc complexes with insulin are present in the β cells of the pancreas. Zinc is an essential component of several enzymes, such as alcohol dehydrogenase, alkaline phosphatase, carbonic anhydrase, and retinene reductase (found in the retina). Acetazolamide (Diamox) inhibits carbonic anhydrase activity by binding to the zinc atom present in the enzyme molecule. The absorption of zinc by the intestines involves pyridoxine, a B vitamin.

A deficiency of zinc can stunt growth and cause impaired wound healing, decreased sense of smell and taste, and hypogonadism.

An excess of zinc can cause gastrointestinal disturbances and vomiting.

Food sources of zinc include seafood (particularly oysters and clams), meat, liver, eggs, milk, and whole grain cereals. The recommended daily allowance for zinc is 15 mg per day for adults with an additional 15 mg per day during pregnancy and 10 mg per day during lactation. For children, the recommended allowance is 6 to 10 mg per day.

Other Ions

Zirconium chlorhydrate has been used in antiperspirant sprays, but its use has been discontinued. It was found that this compound could reach the user's lungs through inhalation. In the lungs, zirconium chlorhydrate could induce granulomas, which in turn cause growth of tumors. Aluminum chlorhydrate is still in use in antiperspirant sprays and appears to have no adverse effect upon the body.

Lithium salts are being used to treat manic-depressive psychoses and as antidepressants for some psychiatric patients. At one time, lithium chloride was employed as a sodium substitute, with dangerous and frequently fatal results. Patients on a therapeutic dosage of lithium salts may complain of fatigue, muscular weakness, nausea, and diarrhea. Slurred speech and hand tremors are noticeable. In larger doses, the central nervous system is affected and the patient may become unconscious or even go into a coma. Abnormalities in the *electroencephalogram are also common.

Cobalt is a constituent of vitamin B_{12}, which is necessary for the formation of red blood cells. Cobalt is present in almost all foods, so cobalt deficiency is quite rare.

Manganese is essential for normal bone structure, reproduction, and normal functioning of the central nervous system. Manganese is a constituent of several mitochondrial enzyme systems. A deficiency of manganese is not known in humans, indicating that the average daily diet supplies a sufficient amount of this substance. Good sources of manganese are nuts, whole grain cereals, vegetables, and fruits. Poor sources are meats, poultry, and fish.

Chromium is involved in the metabolism of glucose and in the proper activity of insulin. A deficiency of chromium leads to impaired glucose tolerance.

Iodine is necessary for the production of thyroid hormones (see page 624). A deficiency in children causes cretinism; in adults it causes goiter and myxedema. An excess of iodine produces psychotic symptoms and parkinsonism.

Molybdenum is involved in the metabolism of nucleic acids.

Summary

Blood carries oxygen, minerals, and food to the cells and carries carbon dioxide and other waste products from the cells. Blood also carries hormones, enzymes, antibodies, and blood cells. Blood regulates body temperature and maintains the pH of the body fluids.

Blood consists of two parts: the suspended particles, red blood cells, white blood cells, and platelets; and the suspending liquid, the plasma. When freshly drawn blood is allowed to clot and settle, the yellow liquid remaining is called blood serum.

Oxygenated blood has a bright red color, whereas deoxygenated blood has a dark purplish color. Blood has a pH range of 7.35 to 7.45. If the pH falls below 7.35, the condition is called acidosis. If the pH rises above 7.45, the condition is called alkalosis. The volume of blood in the body is 5 to 7 L.

The chemical analysis of blood samples is of great clinical significance. Increased or decreased amounts of some substances may indicate certain diseases.

Blood contains hemoglobin, a conjugated protein containing iron. Hemoglobin is composed of heme and globin. Cytochrome c, part of the oxidative phosphorylation sequence in the Krebs cycle, has a structure similar to that of hemoglobin. Chlorophyll a has a structure similar to heme but with a magnesium ion at the center instead of the iron.

If the hemoglobin content of the blood falls below normal, the condition is called anemia. Anemia may result from a decreased rate of production of red blood cells, an increased destruction of red blood cells, or an increased loss of red blood cells.

Plasma proteins are divided into three groups: albumin, globulins, and fibrinogen. Albumin regulates the osmotic pressure of the blood and controls the water balance of the body. α-Globulins and β-globulins form loose combinations with carbohydrates, metal ions, and lipids so that these substances can be transported to all parts of the body. γ-Globulins contain the antibodies with which the body fights infectious diseases. Fibrinogen is the plasma protein involved in the clotting of the blood.

When a blood vessel is cut, blood comes into contact with the exposed tissues. This activates both the extrinsic and intrinsic coagulation systems, producing a clot.

In respiration, the hemoglobin picks up oxygen to form oxyhemoglobin in the lungs. The oxyhemoglobin is carried to the tissues, where it gives up its oxygen. At the same time the blood picks up carbon dioxide and other waste products from the cells. Bicarbonate ions are involved in the hemoglobin–oxyhemoglobin cycle and also in the carbon dioxide removal cycle. Excess bicarbonate ion in the blood cells is shifted to the plasma in a process called the chloride shift whereby chloride ions take the place of the bicarbonate ions. In the lungs, as carbon dioxide is removed from the blood, a reverse chloride shift takes place.

The blood maintains a constant pH by the use of buffers, which react with acid (or basic) substances entering the blood and form new substances that are either neutral compounds or more of buffer salts.

Buffer systems in the blood are bicarbonate buffers, phosphate buffers, hemoglobin buffers, and protein.

Changes in CO_2 level in the blood and corresponding changes in HCO_3^- and pH may be caused by metabolic acidosis, metabolic alkalosis, respiratory acidosis, or respiratory alkalosis. Each condition involves different changes in these three factors.

Water intake must be balanced by water output. If water intake is greater than water output, edema occurs. If water output is greater than water intake, dehydration may occur.

Water in the body is considered to be distributed in two major areas—intracellular and extracellular. Extracellular water is further subdivided into interstitial water; intravascular and lymph water; water in dense connective tissues, cartilage, and bone; and water in transcellular fluids.

Each of the body's water compartments has its own concentration of electrolytes, all concentrations being expressed as milliequivalents per liter. An increase or decrease in the concentration of any one of the ions will have some effect on the body.

Serum osmotic pressure is affected primarily by the concentrations of sodium ions, bicarbonate ions, and chloride ions. Osmotic activity is measured in the unit milliosmol.

Questions and Problems

1. What are the functions of the blood?
2. What is the normal concentration of red blood cells in the body?
3. What is the normal concentration of white blood cells in the body?
4. What is polycythemia? anemia?
5. What is the function of the leukocytes?
6. Name several types of white blood cells.
7. What is a differential count? What is the function of the thrombocytes?
8. What are the functions of the inorganic ions present in the plasma?
9. What is the difference between blood plasma and blood serum?
10. What is the normal pH range of the blood? the specific gravity range? the viscosity?
11. Under what conditions might the blood volume increase? decrease?
12. How can blood volume be measured?
13. What type of compound is hemoglobin?
14. What is heme? Diagram its structure.
15. How does the structure of heme compare with that of chlorophyll a? with cytochrome c?
16. What conditions might cause a decreased production of red blood cells?
17. What causes pernicious anemia?
18. What conditions might cause the destruction of red blood cells?
19. What might cause an increased loss of hemoglobin?
20. Name the three groups of plasma proteins. What is the function of each?
21. Describe the effects of albumin on osmotic pressure in terms of Starling's hypothesis.
22. What is edema? What might cause such a condition?
23. What are the functions of the different types of globulins?
24. What is agammaglobulinemia?
25. Describe the clotting mechanism of the blood.
26. How can blood clotting be prevented in samples taken for laboratory analysis?
27. Why should oxalates never be used to prevent blood from coagulating when that blood is to be used in a transfusion?
28. What is the effect of a deficiency of vitamin K on blood clotting?
29. What is a thrombus? What might result from the presence of a thrombus in the brain?
30. Describe the process whereby the blood carries oxygen to the tissues.
31. How does the blood carry carbon dioxide from the tissues?
32. What is the function of carbonic anhydrase?
33. What is the chloride shift? the reverse chloride shift?
34. What is the alkaline reserve of the blood?
35. What types of buffers are present in the blood?
36. Where are potassium buffers located? sodium buffers?
37. Describe the reaction (in equation form) of an acid and a base with a bicarbonate buffer.
38. How does the body replenish its supply of water?
39. How does the body normally lose water?
40. Into what areas can the body's water be considered as being subdivided?
41. What are the principal functions of sodium ions in the extracellular fluid?
42. What may cause hyponatremia? What are its symptoms?
43. What may cause hypernatremia?
44. What are the principal functions of potassium ions in the body?

45. What may cause hypokalemia? What are its symptoms?
46. What may cause hyperkalemia? What are its symptoms?
47. Under what conditions may the serum calcium level change?
48. What are the symptoms of magnesium ion deficiency? What might cause this deficiency?
49. What are the principal functions of the chloride ion in the body?
50. What are the principal functions of the phosphate ion in the body? iron ions? copper ions?
51. A lab reports the magnesium ion concentration as 3.6 mg/100 mL. Express this as milliequivalents per liter.
52. If the colloid osmotic pressure of the plasma drops to 20 mm Hg, all other pressures remaining the same, what will be the effect on fluid flow? What will this condition be called?
53. What clinical tests would distinguish between metabolic acidosis and respiratory acidosis? between metabolic alkalosis and respiratory alkalosis?
54. Name two causes of metabolic acidosis. How does the body try to compensate for this condition?
55. Why is NaCl used to treat metabolic alkalosis?
56. What acid–base disturbance would each of the following represent? Use normal values of: pH, 7.4; pCO_2, 40 mm Hg; HCO_3^-, 24 mEq/L.
    (a) A 30-year-old woman has been vomiting persistently for 4 days. The following readings were obtained: blood pH, 7.55; HCO_3^-, 40 mEq/L; pCO_2, 56 mm Hg. What is the problem? How would you treat it?
    (b) A confused, disoriented man was brought to the emergency room. No history was obtained. The following lab results were obtained:
    blood pH, 7.10; HCO_3^-, 9 mEq/L; pCO_2, 23 mm Hg. What is the problem? What treatment might be used?
    (c) A 10-year-old boy with an acute asthma attack has the following lab results: blood pH, 7.25; HCO_3^-, 30 mEq/L; pCO_2, 58 mm Hg.
57. What causes kwashiorkor? What are its symptoms?

58. What are the advantages and disadvantages of the use of plasma substitutes?

Practice Test

1. A patient on an anticoagulant drug should avoid taking vitamin _____.
   a. A   b. B   c. E   d. K
2. The main buffers system in the blood uses _____.
   a. bicarbonate   b. phosphate
   c. protein   d. chloride
3. Albumin is most important for maintaining _____ in the blood.
   a. oxygen transport
   b. osmotic pressure
   c. acid–base balance
   d. carbon dioxide transport
4. The ion necessary for blood clotting is _____.
   a. Na^+ b. K^+ c. Ca^{2+} d. Fe^{2+}
5. The acid–base imbalance characterized by a decrease in pH, pCO_2, and HCO_3^- is _____.
   a. respiratory alkalosis
   b. respiratory acidosis
   c. metabolic acidosis
   d. metabolic alkalosis
6. The ion involved in contraction of the heart is _____.
   a. Na^+ b. K^+ c. Fe^{2+} d. Mg^{2+}
7. The globulins that include the antibodies are the _____.
   a. alpha   b. beta
   c. gamma   d. delta
8. The plasma protein involved in clotting is _____.
   a. albumin   b. fibrinogen
   c. heparin   d. lipase
9. The chloride shift occurs from the _____.
   a. blood cells to the plasma
   b. the plasma to the blood cells
   c. the tissues to the blood cells
   d. the blood cells to the tissues
10. Both pH and HCO_3^- increase during _____.
    a. metabolic acidosis
    b. metabolic alkalosis
    c. respiratory acidosis
    d. respiratory alkalosis

31

Immunology

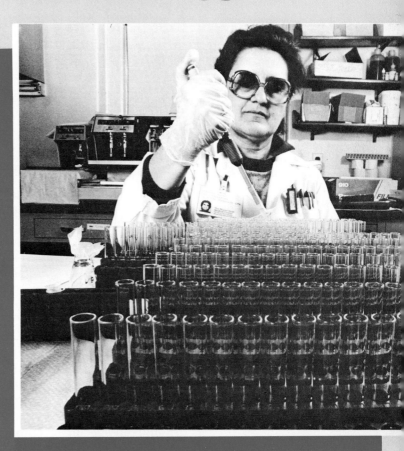

Technician preparing radioimmunoassays for different classes of immunoglobulins in plasma.

Introduction

Immunology is basically the study of adaptive responses of cells derived from the *hematopoietic system* to *macromolecules*. Any macromolecule that can induce a detectable immune response in the body is known as an *antigen*. Antigenic material may contain proteins and/or polysaccharides. Examples of antigenic materials include bacteria, viruses, red blood cells, and other tissue cells. The immune response to antigens includes a humoral (*immunoglobulin) and a cellular (T cell) response. The *immune system* is designed to protect the body from foreign *pathogens* while not attacking self-components. Thus the immune system must be able to distinguish between "self" and "nonself."

Cellular Components

The principal cells that respond to antigen are *macrophages, *granulocytes, *T cells, and *B cells*. Macrophages serve many roles. They are the most active of the body's *phagocytic cells and are important in antigen processing. Macrophages help B cells focus on the appropriate antigen. Macrophages are also important secretory cells. They produce components of the complement system (see page 582), hydrolytic enzymes, and *monokines*. Granulocytes are a group of less effective phagocytic cells. T cells are thymus-derived *lymphocytes with two general immunologic functions. Effector functions include some allergic reactions, transplant rejection, and tumor immunity. Regulatory functions are represented by the ability of some T cells to kill other cells (cytotoxicity) and to cause B cells to produce immunoglobulins. The role of B cells is to produce antibodies (immunoglobulins) that will bind with the particular antigen that stimulated their production. Both T cells and B cells are lymphocytes that are produced in the bone marrow and transformed elsewhere in the body. The immune response of T cells and B cells is specific for a particular antigen. This specificity is mediated through a system of antigen-specific receptors on the surface of T and B cells.

Immune Response

Primary Response

The immunologic response to an antigen changes the immune system. The initial exposure to an antigen produces a detectable level of immunoglobulin in the blood within 5 to 14 days of exposure (primary immune response). This is the time needed for antibody (immunoglobulin) levels to rise to detectable amounts. After this time, the antibody levels slowly increase for a few weeks before gradually declining (see Figure 31-1).

572

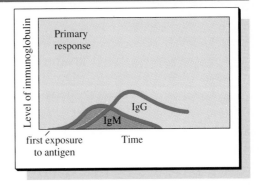

Figure 31-1 Primary and secondary immunologic responses.

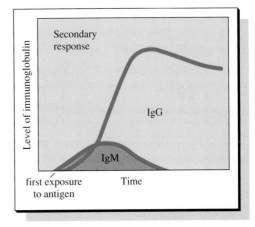

Secondary Response

Following a second exposure to antigen (booster dose) there is a secondary immune response. At first there is a slight sudden drop in antibody levels due to antigen-antibody combination. Within 2 or 3 days there is a 100-fold or more increase in antibody levels. The short time required for this increase compared to the primary response is due to the induction of memory cells by the primary response. This memory induction is important for four reasons.

1. It provides almost immediate protection.
2. It allows for repeated boost responses.
3. It provides for the development of a hyperimmune serum.
4. It will occur even if there is a lapse of several years between the first and second exposures.

Specificity and memory are the criteria used in deciding what constitutes an immunologic response. For example, adverse reactions to aspirin or sulfites seem neither to be specific nor to have memory. Therefore they are not considered to be immunologic responses.

Natural and Acquired Immunity

Natural Immunity

Natural immunity is an inherited nonspecific barrier that is effective against many kinds of pathogens. This statement is supported by many observations: humans get mumps, but dogs do not; AIDS occurs in humans but not in other animals; mammals get anthrax, but birds do not. Several factors are involved in natural immunity.

1. *Race or strain.* In addition to the above examples, it appears that blacks are more susceptible than whites to tuberculosis but are more resistant than whites to diphtheria, influenza, and gonorrhea.
2. *Sex.* There are apparent differences in susceptibility between males and females. These are currently being investigated and are probably due to hormonal differences.
3. *Nutrition.* Low protein-calorie diets produce a significant decrease in resistance to infection. Vitamin deficiencies cause a host of problems. Vitamin A deficiency allows for more skin infections. Folic acid deficiency lowers the number of T cells. Thiamine and riboflavin deficiency lower B cell activity. Vitamin C deficiency is related to increased bacterial infection.

    Overnutrition may also be harmful. It may increase susceptibility to viruses and bacteria, perhaps because when viral invasion occurs, more viruses are produced in healthy than in malnourished cells. Also many bacteria need iron for growth. When the body has an excess of iron, some of it becomes available for bacterial growth.
4. *Hormone-related resistance.* Evidence points to a link between hormonal imbalance and immunologic response. Staphylococcal, streptococcal, and fungal diseases occur more readily in people with diabetes. Also pregnancy is associated with not only marked hormonal changes but also an increase in the incidence of urinary tract infections and poliomyelitis.
5. *Age.* The age of an individual has a marked effect on immunity. This is an area of increased research interest. Some investigators believe that decrease in the ability of the immune system to function is a major cause of problems related to aging.

Acquired Immunity

Acquired immunity refers to the immunity that a person develops during his or her lifetime. It can be either active or passive. In *active immunity* the body produces its own antibodies and T cells, whereas in *passive immunity* the antibodies are acquired from an outside source.

Active immunity occurs whenever a person recovers from an infection. During the illness the body receives an antigenic stimulus that causes it to produce antibodies against that specific pathogen.

	Vaccine	Time Given
Table 31-1 Recommended Active Immunizations	Diphtheria, pertussis, tetanus	At 2, 4, and 6 months with boosters at $1\frac{1}{2}$ yr and at entrance to school; tetanus booster at 16
	Oral polio	At 2, 4, and 18 months
	Measles	1 yr
	Rubella	After 1 yr; before puberty in females
	Mumps	After 1 yr

Subsequent exposure allows the antigens to assist in the body's defenses. Immunizations function in this manner. Table 31-1 lists recommended immunizations.

Passive immunity can be conferred when antibodies pass from mother to fetus via the placenta during the later stages of pregnancy. These antibodies are almost entirely immunoglobulin G (IgG; see page 580). *Colostrum* secretions, on the other hand, are high in IgA and IgM (see page 580).

When children and adults acquire passive immunity, antibodies are injected directly into their bodies. Injections of hyperimmune serum and α-globulins can provide immunity to particular antigens. Table 31-2 compares active and passive immunity.

Regulation of Immune Response

Virtually all immune responses are automatically regulated in the body. This means that a balance exists between the mechanisms that tend to increase the response and those that tend to decrease it. There is currently great interest in this area because a number of immunologic problems are caused by imbalances in this regulatory system. Autoimmunity, immunologic tolerance, *immunosuppression, aging, and *AIDS are all related to regulation of the immune response.

		Active	Passive
Table 31-2 Comparison of Active and Passive Immunity	Source	Self	Outside
	Effectiveness	High	Moderate to low
	Method of acquisition	Disease itself, immunization	Maternal transfer, injection
	Time to develop	5–14 days	Immediate
	Ease of reactivation	Easy (booster)	Dangerous (because of *anaphylaxis)
	Use	Preventive	Preventive, therapeutic

Table 31-3 Known or Suspected Primary Immunologic Disorders Associated with Certain Diseases	1. Susceptibility to certain bacteria or viruses. 2. Predisposition to certain disease states such as multiple sclerosis, myasthenia gravis, insulin-dependent diabetes, ankylosing spondylitis, Addison's disease, rheumatoid arthritis, systemic lupus, aspermatogenesis. 3. Immune deficiencies of aging. 4. Allergies (ragweed allergy, hay fever). 5. Leukemias.

Major Histocompatibility Complex

The genetic regulation of immunity is linked to the **major histocompatibility complex* (MHC), a set of genes associated with the immune response and the body's response to transplanted organs. Table 31-4 summarizes some of the influences of MHC discovered so far.

Defects in this genetic and regulating control center of the immune response can result in disease. Table 31-3 lists some of the immunologic abnormalities associated with disease.

Phagocytic Cells

Phagocytosis is the recognition, engulfment, and digestion of some particles by cells. To initiate this process, agents that can attract the phagocytic cells to the particle must be present in the tissue. Substances that can act as chemical attractants include soluble bacteria products, histamine, compounds derived from the complement system, and T cells and their products. Contact between the phagocytic cell and a potential victim is, however, not always enough to bring about phagocytosis. Molecules called opsonins may be needed to promote the attachment of a phagocytic cell to its target particle. The most effective opsonin is the antibody to the foreign particle. When

Table 31-4 Lymphokines and Their Functions

Lymphokine	Function
Macrophage chemoactive factor	Attracts macrophages
Migration inhibition factor	Inhibits macrophage migration
Leukocyte inhibition factor	Inhibits neutrophil migration
Macrophage activation factor (MAF)	Enhances macrophage cytolytic activity
γ-Interferon	Same function as MAF
Fibroblast activation factor	Stimulates cell growth
Colony-stimulating factor (CSF)	Stimulates monocyte growth
Interleukin-2	Stimulates growth of activated T cells
Interleukin-3	Same function as CSF
B cell growth factor	Stimulates growth of B cells

the invading particle is coated with this antibody, phagocytosis is greatly increased.

Engulfment occurs when the phagocytic cell's plasma membrane flows completely around the particle, followed by ingestion of the particle. The digestion and destruction of the ingested particle are facilitated by acid and a variety of hydrolytic enzymes stored within the phagocyte. The more important macrophage components of destruction are several potent oxidizing agents. A resting phagocyte uses glycolysis to obtain energy. During *phagocytosis it shifts to using the hexose monophosphate shunt as its energy source. This shift allows for the production of the powerful oxidizing agents hydrogen peroxide (H_2O_2), the hydroxyl radical ($OH\cdot$), and, most important, the superoxide anion radical ($O_2^-\cdot$).

There are two types of phagocytic cells derived from bone marrow: the granulocytes, which circulate only in the blood and migrate to the site of the inflammation, and the macrophages, which are found in both blood and tissue. Macrophages are much more diverse in function and response than granulocytes.

Granulocytes

There are three types of granulocytes: neutrophils, eosinophils, and basophils. *Neutrophils* represent about 60 percent of the circulating white blood cells. They are about twice as large as red blood cells. Neutrophils enter the bloodstream each day from the bone marrow. They also leave the body each day via the urine, oral and pulmonary secretions, the gastrointestinal tract, and exudations from inflamed tissues. Neutrophils usually die after phagocytosis. They, along with the digested foreign matter and foreign residues, are transformed into the pus associated with acute inflammation. Their levels vary considerably from day to day because of many factors.

Eosinophils represent about 1 to 3 percent of the circulating white blood cells. Their major role is to defend the body against parasites. They are also involved in allergic reactions and are known to accumulate in the skin when an allergen is applied to the skin of an allergic individual.

Basophils constitute less than 1 percent of the circulating white blood cells. They are primarily involved in allergic reactions and contain significant amounts of heparin (a powerful anticoagulant) and histamine (a smooth muscle contractant). They have little or no phagocytic activity.

The relative levels of phagocytic activity are macrophages > neutrophils > eosinophils > basophils.

Macrophages

The precursors of macrophages are the monocytes found in the blood, which represent about 2 to 5 percent of the circulating white blood cells. After circulating for 8 to 10 hr, they enter the tissue and become macrophages. Tissue macrophages are not identical even though they all arise from a common precursor. When the monocytes enter a tissue, they undergo a metamorphosis that involves a rapid increase in size, in protein synthesis, and in enzyme content. The nature of these changes depends on the tissue in which they occur. The principal tissues in which these changes occur are the spleen, liver, lung, pancreas, neural system, and connective tissue.

Macrophages have lifetimes of months to years except when they become involved in the inflammatory response. In addition to taking part in phagocytosis, macrophages help regulate the immune response of T cells, fix complement, help bind antibodies, and secrete several compounds involved in the immune response.

Lymphocytes

T Cells

T cells have several functions. They regulate immunoglobulin production, mediate delayed hypersensitivity, and lyse virally infected cells. They originate in the bone marrow but mature and are transformed in the thymus gland (hence T, for thymus, cells). Activation of T cells is a dual process. Both macrophage-processed antigen and MHC gene products are necessary for activation. Interleukin-1 (IL-1), produced by macrophages, is also necessary. Without interleukin-1, T cell activation is minimal. IL-1 also has a number of other functions that will be discussed later.

Once T cells are activated, they begin to proliferate. In order for that to take place, interleukin-2 (IL-2; sometimes called T cell growth factor) is necessary. IL-2 is secreted by activated T cells and is responsible for expanding the role of T cells in the immune response and for helping the T cell–B cell collaborative process (see Figure 31-2).

Activated T cells release compounds called lymphokines. These compounds act in antigen-specific ways on other cells such as T and B cells. Some lymphokines and their functions are listed in Table 31-4. Several are currently being used experimentally in treating a variety of diseases.

The functions of T cells are many and varied. There are four major subsets of T cells.

1. *Helper T cells* (Th) represent about one-half of the total T cell population. Their role is to help B cells develop into immunoglobulin-producing cells. Helper T cells and macrophages are attacked by the AIDS virus.
2. *Suppressor T cells* (Ts) provide immune tolerance. Their role is opposite to that of the Th cells in some unknown way. They represent 10 percent of the total T cell population. Thus, in a

Figure 31-2 Interactions among T cell, B cell, and macrophage.

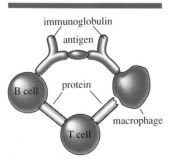

immunoglobulin
antigen
B cell
protein
macrophage
T cell

	T Cell	B Cell
Where tissue modified	Thymus	Bone marrow
Complement reception	No	Yes
Tissue distribution	High in thoracic duct, lymph, blood	High in spleen, low in blood
Life span	Long	Short
Cellular products	Lymphokines	Immunoglobulins
Sensitivity to immunosuppression	Lower	Higher

Table 31-5 Comparison of T and B Cells

normal person, there would be about five times as many Th as Ts cells. In AIDS there is an imbalance in this 5:1 Th/Ts ratio, with many more Ts cells than Th cells.

3. *Cytotoxic T lymphocytes* (CTL) are T cells that destroy target cells such as transplanted tissues, tumor cells, and virally infected host cells.

4. *Delayed hypersensitivity T cells* (T_{dh}) are very important in the defense against viruses, fungi, and other organisms that replicate intracellularly. They are antigen specific and release several compounds that participate in allergic hypersensitivity reactions.

Natural killer (NK) cells are similar to T cells, but they are active without the need of prior exposure to an antigen. They probably form the first line of defense against neoplastic and virally infected cells.

B Cells

The cellular source of antibodies is the B cells. Table 31-5 shows the similarities and differences between T and B cells. Pre-B cells are made in the bone marrow and mature into B cells in the bone marrow. Their activation and proliferation require both antigen and Th cells. Activation converts a B cell into a plasma cell that is able to produce and secrete antibodies. Over 40 percent of the total production of protein by a plasma cell may be immunoglobulin. Thus a single plasma cell can release thousands of antibody molecules per second. Each type of B cell usually produces only one type of immunoglobulin. Figure 31-2 summarizes the possible interactions among macrophages, T cells, and B cells.

Immunoglobulins

*Immunoglobulins are protein molecules that carry antibody activity. They have the ability to combine with the substance responsible for their formation (the antigen). There are five classes of immunoglobulins: IgG, IgA, IgM, IgD, and IgE. Each immunoglobulin contains at

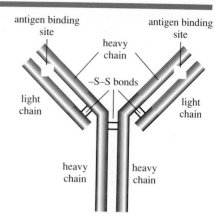

Figure 31-3   Diagram of basic immunoglobulin molecule.

least one basic unit composed of four polypeptide chains. One pair of identical polypeptide chains is approximately twice as large as the other pair of identical polypeptide chains, so they are referred to as heavy (H) chains and light (L) chains, respectively. The chains are held together by disulfide bonds. Figure 31-3 illustrates an immunoglobulin molecule.

The structure of the H chains determines the immunoglobulin class. Within each polypeptide chain there is a variable region, which allows each antigen-determined immunoglobulin molecule to be unique, and a constant region that is relatively the same in all immunoglobulins. The constant region determines the biologic function of the molecule.

IgG

Immunoglobulin G (IgG) represents about 75 percent of the total immunoglobulins in the body. It is the only immunoglobulin that can cross the placenta and is responsible for the protection of the newborn for the first few months following birth. It also takes part in complement fixation (see page 582).

IgA

Immunoglobulin A (IgA) is the predominant immunoglobulin in body secretions. It provides direct protection against local infections through its abundance in saliva, tears, bronchial secretions, nasal mucosa, prostate fluid, vaginal secretions, and intestinal secretions.

IgM

Immunoglobulin M (IgM) is prominent in early immune responses to most antigens and is the predominant antibody involved in ABO blood group mismatches. It is also the most efficient fixer of complement.

IgD

Immunoglobulin D (IgD) is present only in trace amounts. Its role is currently unknown.

IgE

Immunoglobulin E (IgE) has a high affinity for mast cells and is involved in allergic reactions. Mast cells are very populous in connective tissue, in the lungs and the uterus, and around blood vessels. Upon combination with certain specific antigens called allergens, IgE antibodies trigger the release from mast cells of large amounts of serotonin, heparin, and histamine, all of which are vasoactive and produce the symptoms associated with allergic reactions.

Interleukins

Interleukin-1

Interleukin-1 (IL-1) is a polypeptide product of many cells. Its principal source is macrophages. Macrophages containing antigen produce interleukin-1, which then promotes growth and differentiation of activated T cells. This process includes increased cytotoxicity for T cells, enhanced natural killer activity, increased helper T cell activity, and decreased suppressor T cell activity. Interleukin-1 also influences non-T cells—it increases antibody production of B cells, elevates acute-phase protein levels, increases levels of several metals in the plasma, and causes fever, anorexia, bone resorption, and somnolence. A number of agents decrease interleukin-1 activity; these include steroidal drugs, aspirin, and cyclosporine.

Interleukin-2

Activated helper T cells produce interleukin-2 (IL-2) for T cell growth. Interleukin-2 also helps T cells produce several lymphokines such as interferons. Overproduction of interleukin-2 almost never occurs; however, impaired production is associated with several diseases such as lupus, AIDS and other immunodeficiency diseases, and advanced metastatic cancers. Glucocorticosteroids, cyclosporine, and suppressor T cells inhibit production of interleukin-2. Several difficult-to-treat cancers have been treated with interleukin-2 to enhance host antitumor activities.

Interferons

Interferons are peptides that exert nonspecific antiviral activity. Originally identified on the basis of their antiviral activities, they are now known to have potent immunoregulating functions as well.

Table 31-6 Uses of Monoclonal Antibodies

Diagnostic	Therapeutic (Currently Experimental)
Leukocyte identification	Antitumor therapy (can be coupled with a drug or radioisotope to enhance effect)
HLA antigen detection	
Viral detection and identification	
Parasite identification	Immunosuppression
Hormone detection	
Carcinoprotein detection	Fertility control
Typing of leukemias	Drug toxicity reversal (e.g., digitalis intoxication)
Tumor-related antigen detection	

Three types of interferon have been identified. Interferon-α is produced primarily by leukocytes; interferon-β is produced by macrophages and fibroblasts; and interferon-γ is a lymphokine produced by activated T cells. The major viral effect of interferons is the inhibition of growth and development of viruses. The effects of interferons on the immune system are quite complex. In general, such effects are immunoenhancing either by increasing Th function or by decreasing Ts function. Many effects of interferon, however, are dependent on the type of interferon, the dose, and the time of treatment with respect to the course of the disease.

Monoclonal Antibodies

The production of *monoclonal antibodies has allowed immunologists to prepare virtually unlimited amounts of antibodies that are chemically, physically, and immunologically pure. Monoclonal antibodies are identical copies of an antibody that consists of only one kind of H chain and one kind of L chain. Table 31-6 lists some current and potential uses for monoclonal antibodies, and Table 31-7 lists some advantages and disadvantages of their use.

Complement System

When an antigen combines with circulating antibodies, cells are lysed, bacteria are consumed, leukocytes are attracted to the antigen, and histamine is released. All of these processes involve both path-

Table 31-7 Advantages and Disadvantages of Using Monoclonal Antibodies

Advantages	Disadvantages
Very high sensitivity	Too specific
Unlimited supply	Decreased affinity for antigen
Well-identified immunologic properties	Decreased complement function, high cost

ways of the complement system. These are the antibody-activating pathway and the alternate pathway triggered by bacterial cell membrane structures. The complement pathways are a very complicated series of plasma enzyme reactions somewhat similar to blood clotting. The major consequence of the complement pathway is the punching of holes through the cell membranes of antigens coated by IgM or IgA. These holes allow leakage of ions followed by cell lysis. Leukocytes, thrombocytes, erythrocytes, and gram-negative bacteria are the most susceptible to the complement system.

Organ and Tissue Transplants

Transplantation of tissue and organs is an artificial biologic event that has provided a way to treat organ failure. Skin grafts are regularly used to repair injuries in burn patients. Corneal transplants have restored sight to thousands. Kidney transplants have extended the lives of many people with renal failure. Bone marrow transplants are used to treat aplastic anemia, leukemia, and acute radiation poisoning. Heart and liver transplants have become common in recent years, and progress is being made in lung, pancreas, and brain tissue transplants.

Survival of a transplant depends on how the host reacts to it. The less antigenic the graft, the less the host reacts. The antigenicity is determined by the proteins associated with the human leukocyte antigen (HLA).

An *autograft* is a tissue transplant within the same individual. An example is removing skin from a healthy site and transplanting it to a damaged site on a burn patient. Similarly, in leukemia patients, bone marrow is withdrawn and then reinjected at a later date after the patient has been irradiated and the marrow has been purged of leukemia cells in an attempt to induce remission. Bone marrow injections have also been used experimentally to overcome the effects of high doses of chemotherapy.

A *syngraft* is a graft of tissue obtained from a genetically identical individual (an identical twin), so there are no antigenic differences. *Allografts* are tissue exchanges between genetically nonidentical members of the same species. In clinical transplantations nearly all transplants are allografts. Grafts from HLA-identical siblings give the best results, but the use of grafts from parents, offspring, and nonidentical siblings increases the likelihood of a successful transplant and decreases the amount of *immunosuppression needed. *All* allografts, however, will require immunosuppression because of HLA differences between the individuals.

Xenografts are grafts between different species. While very rarely used, animal hearts and kidneys have been transplanted into humans.

In addition to HLA typing, blood ABO compatibility is necessary because of preformed antibodies in the recipient.

Table 31-8 Types of Rejections

Type	Time After Transplantation	Probable Mechanism	Prognosis/ Treatment
1. Hyperacute			
a. Immediate	Minutes	Preformed antibodies	No treatment; rare if donor and recipient are ABO compatible
b. Accelerated	1–5 days	T cells	Very poor; does not usually respond to antirejection therapy
2. Acute			
a. Cellular	After 14 days	Delayed hypersensitivity	Generally responds to antirejection therapy
b. Humoral	After 7 days	Antibodies	Much less responsive to antirejection therapy; high risk of rejection within event year
3. Chronic	Months to years	T cells and antibodies	Mixed results; antirejection therapy ineffective; some patients deteriorate rapidly, others maintain suboptimal function for years

The major unwanted event following a transplant is the rejection of the tissue by the recipient. This can occur very rapidly or rather slowly depending on the antigenic differences between donor and patient, the possibility that the recipient has been previously sensitized to the donor's antigens, and the degree of immune function impairment. The three types of rejection episodes are summarized in Table 31-8.

Prevention of Rejection

Besides antigen and blood type matching, a variety of immunosuppressive agents are used. The four most common will be briefly described.

Azathioprine is a purine compound that inhibits the immunoglobulin and T cell response by interfering with purine metabolism. It is used with prednisone (see below) to prevent rejection. It is ineffective in treating a rejection episode.

Corticosteroids such as prednisone are used to prevent rejection and to reverse acute rejection episodes. Current evidence suggests that they function by inhibiting interleukin-1 and indirectly suppressing interleukin-2. In addition, the corticosteroids may be involved in the lysis of activated T cells.

Antilymphocyte globulin (ALG) is a potent immunosuppressant used to prevent and reverse acute rejection. It is usually used in conjunction with azathioprine and prednisone. ALG causes the lysis of circulating lymphocytes and the depletion of T cells. ALG consists of an IgG fraction that has antibodies relatively specific to human lymphocytes.

The use of cyclosporine (a fungally derived cyclic peptide) is the most significant recent advance in transplantation. Cyclosporine is best used in the prevention of rejection episodes. Its multiple effects include inhibition of lymphocyte activation and also inhibition of interleukin-1 and interleukin-2 production. The benefits of cyclosporine treatment include earlier release from the hospital, less bone marrow suppression, and fewer viral and bacterial infections. Its disadvantages include high cost, unpredictable absorption, and very high renal toxicity.

The adverse effects of all the above chemical immunosuppressive agents include unusually high rates of infection and a greatly increased risk of developing cancer.

Serology

Although blood transfusions have become routine, they still carry immunologic risks as well as the risk of infections. Most of these risks can be eliminated by careful testing of both blood donor and recipient before transfusion. Hepatitis remains the most serious risk, despite the emergence of AIDS.

The surfaces of red blood cells contain large numbers of antigens. These antigens are classified into blood groups. While there are many (over 40) different blood groups, their clinical significance depends on two factors: the frequency of occurrance of antibodies on the antigen and the potency of the antibody.

The blood group antigens are probably inherited according to simple Mendelian genetics. Table 31-9 summarizes the ABO blood group. When a patient with antibodies against a red blood cell antigen is transfused with incompatible blood, intravascular hemolysis occurs. Often this triggers intravascular coagulation. The severity

Table 31-9 Blood Group System

Blood type	Antigen on Red Blood Cells	Antibody in Serum	Genotype	Frequency (percent of population)		
				Caucasian	Black	Oriental
A	A	Anti-B	AA, AO	41	28	38
B	B	Anti-A	BB, BO	11	17	22
O	Neither A nor B	Both anti-A and anti-B	OO	45	51	30
AB	Both A and B	Neither anti-A nor anti-B	AB	3	4	10

of the reaction depends upon the strength of the antibody and the number of transfused cells.

The immunology and genetics of the *Rh antigen* are very complicated. Its precise chemical nature is unknown. It is common to 85 percent of the population. An Rh-negative person does not automatically have antibodies to Rh antigens. Immunization to the Rh antigen comes from either an improper transfusion of Rh-positive blood into an Rh-negative person or fetal immunization of an Rh-negative mother by an Rh-positive fetus.

Erythroblastosis fetalis is a severe hemolytic disease that develops when an Rh-negative mother carries an Rh-positive child and the mother produces anti-Rh antibodies. These antibodies cross the placenta and cause the destruction of fetal red blood cells. Depending on the amount of antibodies produced, the fetus may abort or be born with a hemolytic disease. This disease is rarely associated with the first Rh-incompatible pregnancy, but subsequent pregnancies increase the risk. The disease can be prevented by injecting the mother with Rh immunoglobulin within 72 hr after delivery of an Rh-incompatible baby.

Allergies

The immune response in allergies is the same as immunologic response in disease. With allergies, the resulting inflammatory reactions are undesirable and produce the disease. Table 31-10 lists the major classes of allergic reactions. Both B and T cells are involved.

Autoimmunity

Autoimmunity is a state in which the body's natural tolerance to self has been compromised. As a result, antibodies react with normal body constituents, causing an autoimmune disease. It is becoming

Table 31-10 Types of Allergic Reactions

Type	Immunoglobulin Involved	Cellular Involvement[a]	Chemicals Released	Examples
*Anaphylactic	IgE	Mast cells, basophils	Heparin, histamine	Hay fever, food allergies, injectable antibiotic (penicillin)
Cytotoxic	IgG or IgM	RBCs, WBCs, platelets	Complement system	Drug allergy, transfusion reaction
Immune complex	IgG or IgM	Host tissue	Immune complex	Serum sickness
T cell	None	T cells	Lymphokines	Contact dermatitis

[a]RBCs, red blood cells; WBCs, white blood cells.

Table 31-11 Some Diseases with Autoimmune Component

Myasthenia gravis	Rheumatoid arthritis
Diabetes	Pernicious anemia
Autoimmune hemolytic anemia	Rheumatic fever
Systemic lupus erythematosus	Multiple sclerosis

apparent that autoimmune responses are not as rare as was once thought and that not all autoimmune responses are harmful. An abnormal autoimmune response is sometimes a primary cause and at other times a secondary cause of many human diseases. Currently there is no unifying concept to explain the origin or pathogenesis of the various autoimmune disorders. Studies support the idea that autoimmune diseases result from a wide variety of genetic and immunologic abnormalities, which differ from individual to individual. These abnormalities may express themselves early or late in life depending on the presence or absence of a variety of superimposed accelerating factors such as viruses, hormones, and abnormal genes. Table 31-11 lists some diseases thought to involve autoimmunity.

Immunodeficiencies

Until the emergence of acquired immune deficiency syndrome (*AIDS) a few years ago, known defects in immunity were rare. Immunodeficiency diseases may involve any component of the immune system including the lymphocytes, phagocytes, and all the complement system. Only AIDS will be discussed here.

AIDS is a highly lethal epidemic immunodeficiency disease that is caused by two viruses called HIV-1 and HIV-2 (both viruses are associated with similar but slightly different diseases that infect the T helper cells). The immunologic abnormalities found in AIDS are those of a severe and profound immune deficiency. They include an absence of delayed hypersensitivity, lymphopenia, a greatly decreased level of T helper cells, and impaired natural killer cell function.

Although antibodies to a wide variety of antigens remain at high levels, there is evidence of impaired B cell function as well. It appears that the incubation period for the virus can be quite long, up to 8 years (or more).

Patients with this syndrome can develop life-threatening opportunistic infections, Kaposi's sarcoma, or both. The infectious agents include most of the bacterial, fungal, and parasitic agents associated with cellular immunodeficiency.

In mid-1987 the incidence of cases of AIDS by patient groups broke down as follows: homosexual/bisexual men, 66 percent; homosexual intravenous (IV) drug users, 8 percent; IV drug users, 17 percent; heterosexuals, 4 percent; hemophiliacs and other blood

Table 31-12 Distribution of AIDS Cases in United States Population as of December 1988	Percent of Cases	
	Male	Female
Adult		
Homosexual/bisexual	73	0
IV drug users	20	52
Hemophiliac	1	0
Heterosexual contact	2	29
Blood transfusion	2	11
Undetermined	3	8
Children		
Hemophiliac	10	1
Babies born to AIDS mothers	71	84
Blood transfusion	15	11
Undetermined	4	4

recipients, 2 percent; and undetermined, 3 percent. Table 31-12 gives 1988 data for adults and children. At the present time there is no known effective cure. Some drug therapies such as AZT appear to prolong and improve a patient's life.

Summary

The immune system is made up of a variety of cells designed to protect the body against antigens. The principal cells of the immune system are macrophages, B lymphocytes (B cells), and T lymphocytes (T cells). Macrophages engulf and digest foreign particles. B cells make and secrete specific immunoglobulins. T cells produce lymphokines that activate and amplify immune responses.

The first exposure to an antigen produces a primary immune response in which up to 2 weeks is needed to produce detectable antibodies. Upon second exposure to the same antigen, antibodies are evident within only 2 or 3 days.

Immunity may be either natural or acquired. Acquired immunity, the immunity a person develops during his or her lifetime, is classified as active if the person's body produces its own antibodies and passive if the antibodies are transferred from another source.

The genetic regulation of the immune response is controlled by the major histocompatibility complex (MHC). Among the conditions influenced by the MHC are immune response, susceptibility to allergies, autoimmune disorders, and suitability of organ transplants.

T lymphocytes have several functions: regulating immunoglobulin production, producing allergic reactions, and lysing virally infected cells. Among the chemicals produced by T cells are interleukins, which have a variety of roles in the immune response and the resulting inflammation, and interferons, which have antiviral activity. There are four types of T cells: (1) T helper cells, which activate B cells into immunoglobulin production; (2) T suppressor cells, which may participate in immune tolerance; (3) cytotoxic T cells, which destroy target cells; and (4) delayed hypersensitivity T cells, which are involved in many allergic reactions.

B lymphocytes produce immunoglobulins when activated. Immuno-globulins are proteins that have antibody abilities. There are five types of immunoglobulins: IgG, IgA, IgM, IgD, and IgE. Each is highly specific for a particular antigen, with which it combines to form an immune complex. Phagocytosis usually follows this attachment.

Monoclonal antibodies are identical copies of an antibody. They are currently being studied for a variety of applications.

The complement system is composed of several plasma enzymes involved in the lysis of certain antigenic cells.

Successful organ and tissue transplants depend on several factors. The most important of these is the similarity of the human leukocyte antigens. Rejection can sometimes be controlled through the application of immuno-suppression therapy. Azathioprine, corticosteroids, antilymphocyte globulin, and cyclosporine are all used for this. Blood transfusions require the matching of red blood cells to prevent adverse immune reactions.

Erythroblastosis fetalis is a severe hemolytic disease that develops when an Rh-negative mother carries an Rh-positive child. This problem is prevented by giving the mother an injection of Rh antibody.

Allergies are unwanted immune reactions due to either T cells or immunoglobulins.

Autoimmunity is a lack of self-tolerance. Many diseases such as lupus, multiple sclerosis, and rheumatoid arthritis involve autoimmunity. Others probably will be discovered.

The most common immunodeficiency disease is AIDS. It is caused by viruses that infect T helper cells and macrophages. The result is a loss of immunity that allows several life-threatening infections to occur.

Questions and Problems

1. What system in the body produces the cells for the immune system?
2. The immune system is designed to protect us against foreign pathogens. What are these pathogens called?
3. What must these pathogens contain to activate the immune system?
4. Give three examples of cells that will activate the immune system.
5. Name the three principal cells that respond to antigens.
6. What are the principal products of B cells?
7. What are the principal products of T cells?
8. What two factors are used in deciding whether a response is immunologic?
9. Regulation of the immunologic response is very important. Give three current areas of investigation of imbalance in this response.
10. List four factors that influence the regulation of the immunologic response.
11. What is the major histocompatibility complex? Why is it important?
12. What is phagocytosis?
13. What are opsonins? What acts as the most effective opsonin?
14. What are the most important components of the digestion and destruction process carried out during phagocytosis?
15. What are the two types of phagocytic cells?
16. Discuss the origin and role of macrophages.
17. What two substances are necessary for T cell activation?
18. List five lymphokines and their functions.
19. List the four types of T cells and their functions.
20. List the five classes of immunoglobulins and their roles.
21. What determines the class to which an immunoglobulin belongs?
22. What are interferons? What are their functions?
23. What are monoclonal antibodies? What possible therapeutic use do they have?
24. What transplants have the highest success rates? Why?
25. What does survival of a transplant depend on?
26. Nearly all clinical transplants are of what type?
27. Name three types of rejection episodes and discuss them in terms of time, mechanism, treatment, and prognosis.

28. The following have been used in the prevention of organ rejection: azathioprine, corticosteroids, antilymphocyte globulin, and cyclosporine. Discuss the uses, advantages, and disadvantages of each.

29. What is the complement system? What is its function?

30. What blood type is the most common in the United States? Which is the rarest?

31. What is erythroblastosis fetalis? How can it be prevented?

32. List the four types of allergic reactions and give an example of each.

33. Name four diseases thought to have some autoimmune component.

Practice Test

1. The immunoglobulin that provides protection to newborns because it can cross the placenta is _____.

   a. IgA   b. IgM   c. IgG   d. IgE

2. The AIDS virus infects which T cells?

   a. Th   b. Ts   c. T_{dh} d. CTL

3. Interleukin-1 is necessary to activate _____.

   a. B cells   b. T cells
   c. interferon   d. macrophages

4. Autoimmunity is usually thought *not* to contribute to _____.

   a. diabetes
   b. rheumatoid arthritis
   c. lupus
   d. tuberculosis

5. Contact dermatitis such as that caused by poison ivy is due to _____.

   a. T cells   b. IgE
   c. IgM   d. macrophages

6. The most significant recent drug advance in transplantation is _____.

   a. prednisone   b. cyclosporine
   c. ALG   d. azathioprine

7. The most destructive oxidative species in phagocytosis is _____.

   a. H_2O b. O_2 c. H_2O_2 d. $O_2^- \cdot$

32

Vitamins

Wet beriberi with dependent edema predominantly involving the trunk and lower extremities.

Animals fed on a diet consisting only of purified carbohydrates, fats, proteins, minerals, and water will lose weight and develop certain deficiency diseases. Something else must be administered to sustain normal life. The additional substances required are called vitamins.

The name *vitamin* was originally *vitamine* because the first one found was an amine, hence the name *vital amine*, or *vitamine*. Subsequent studies of other such substances showed that they were not all amines so the "e" was dropped.

Vitamins are similar to hormones in many ways. Vitamins and hormones are carried by the bloodstream to the various parts of the body where they are needed. Vitamins and hormones are required by the body only in extremely small amounts. Neither vitamins nor hormones furnish energy by themselves, although vitamins function with certain enzymes to control energy changes in the body. One important difference between vitamins and hormones is that most vitamins must be supplied in the diet whereas hormones are synthesized by the body.

Vitamins are divided into two major groups—fat-soluble vitamins and water-soluble vitamins. The fat-soluble vitamins include vitamins A, D, E, and K and are usually found associated with lipids in natural foods. They are transported in the bloodstream by the lipoproteins. Fat-soluble vitamins are not excreted in the urine but do appear in the feces.

The water-soluble vitamins include vitamin C and the B-complex vitamins. Although the vitamins have a letter designation and sometimes a subscript in addition, such as vitamin B_1, the chemical names are being more and more widely used. For example, the chemical name for vitamin B_1 is thiamine.

More and more people eat a balanced diet, take vitamins, or are under medical care. This has helped to produce a healthier population. One piece of evidence that points in this direction is the continual overthrowing of previous athletic records.

Fat-Soluble Vitamins

Vitamin A

Source

Vitamin A is found in fish liver oils, butter, milk, and to a small extent kidneys, fat, and muscle meats. The precursor of vitamin A (the substance from which vitamin A can be made) is called provitamin A and is found in yellow fruits and vegetables such as peaches, apricots, sweet potatoes, carrots, and tomatoes and in leafy green vegetables.

Structure

Vitamin A is a high molecular weight alcohol known as retinol. Vitamin A has an all-trans structure (see page 353).

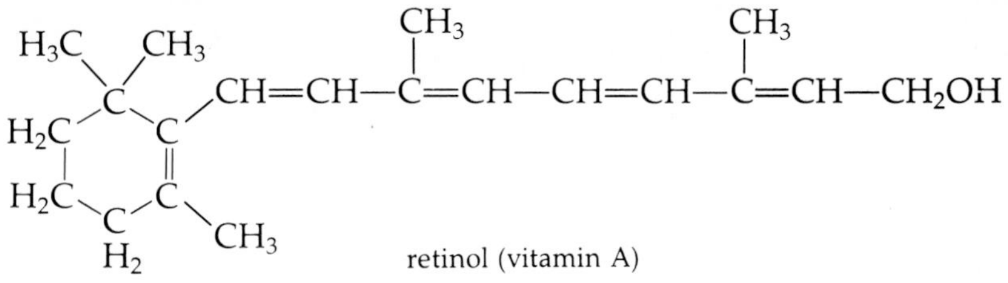

retinol (vitamin A)

Provitamin A is a compound that can be converted into vitamin A. One such provitamin A is β-carotene. The conversion of β-carotene into retinal (vitamin A aldehyde) and then to retinol (vitamin A) is shown in Figure 32-1.

It should be noted that one molecule of β-carotene produces two molecules of vitamin A. Because β-carotene must be metabolized to produce vitamin A, β-carotene is only one-sixth as effective as oral vitamin A itself.

Provitamin A is transformed into vitamin A in the intestinal walls of some animals, including rats and pigs, but in the liver in humans.

Properties

Vitamin A is soluble in fats but not in water. It is stable to heat, acid, and alkali but is destroyed by oxidation. (Recall that double bonds

Figure 32-1 Conversion of β-carotene to vitamin A.

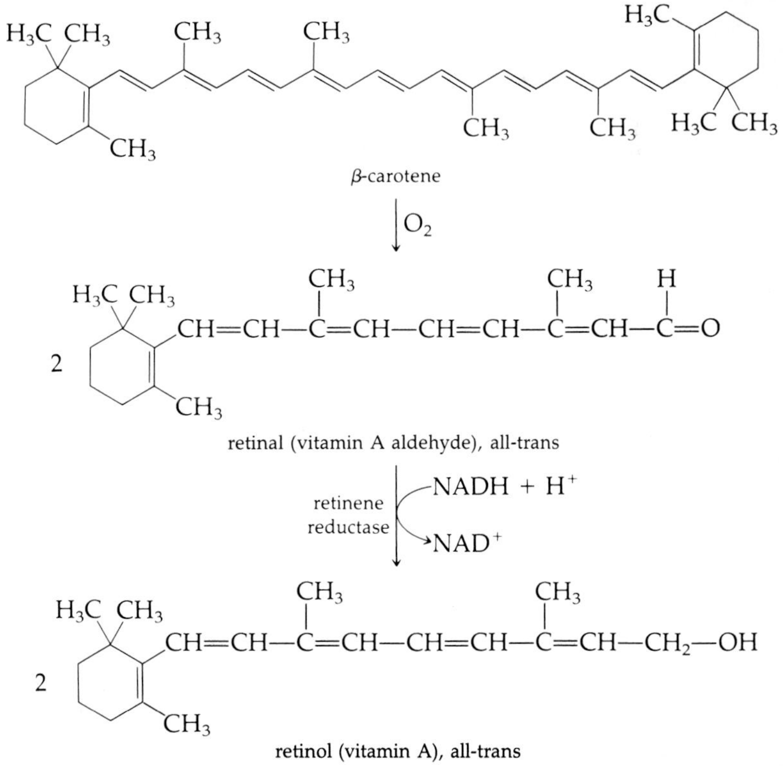

are usually quite susceptible to oxidation.) Ordinary cooking does not destroy vitamin A. The vitamin A present in butter is destroyed when the butter turns rancid (becomes oxidized).

Daily Requirement

The recommended daily dosages of vitamin A for the normal adult male and female are 1000 and 800 retinol equivalents (5000 and 4000 IU), respectively. One retinol equivalent is 1 μg of retinol or 6 μg β-carotene. One international unit (IU) of vitamin A is equivalent to 0.3 μg of retinol or 0.6 μg of β-carotene. The daily requirements of vitamin A are increased to 1000 retinol equivalents (5000 IU) during pregnancy and to 1200 retinol equivalents (6000 IU) during lactation. A child requires 400 to 700 retinol equivalents (2000 to 3300 IU) daily.

Vitamin A is necessary for normal growth and development, reproduction, mucus secretion, and lactation. It is necessary for the synthesis of the membranes around the lysosomes and the mitochondria and acts to regulate membrane permeability. Vitamin A plays an important role in the functioning of the retina and in the maintenance of the integrity of epithelial tissues. Recent evidence suggests that vitamin A can protect against a variety of cancers.

Effect of Deficiency

Vitamin A was discovered by the observation that certain animals did not grow on a diet low in some animal fats. However, this effect on growth is characterized by a lack of other vitamins as well. A lack of vitamin A causes a shrinking and hardening of the epithelial tissues of the membranes in the eyes, digestive tract, respiratory tract, and genitourinary tract. Such a hardening is called keratinization.

When keratinization occurs in the lining of the respiratory tract, the patient is more likely to suffer from colds, pneumonia, and other respiratory infections because of the drying of the membranes.

When keratinization occurs in the eyes, the tear ducts become keratinized and are no longer able to secrete tears to wash the eyes. When this occurs, bacteria are able to attack the corneal tissue of the eyes, producing an infection called xerophthalmia. In this disease, the cornea becomes cloudy and does not allow light to pass through, so sight is lost permanently.

An early symptom of the lack of vitamin A is nyctalopia, or night blindness. A person with night blindness cannot see very well in dim light because of lack of visual purple (rhodopsin) in the retina of the eyes. This pigment is acted on and changed by light and is then regenerated in the presence of retinal, an oxidized form of vitamin A. If there is a lack or deficiency of vitamin A, the rhodopsin is regenerated very slowly; thus, there is an impairment in vision at night.

The role of vitamin A in the visual process and the part it plays in the regeneration of rhodopsin are illustrated in the following simplified cycle.

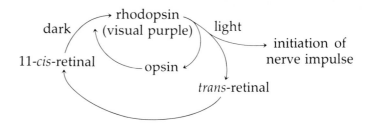

The role of calcium ions in the visual cycle has recently been elucidated. In the dark, calcium and sodium ions carry a current across the membranes of the retinal rod cells. Ion channels in these membranes are held open by cyclic guanosine monophosphate (cGMP). When a photon of light strikes the eye, cGMP is destroyed, the ion channels close, and the current stops.

Calcium ions control how rapidly the cGMP is replaced by binding to the enzyme that synthesizes cGMP. Therefore, calcium ions control how rapidly the visual system recovers after exposure to light.

A lack of vitamin A can also produce keratinization of the epithelial cells of the genital system and might cause sterility. A lack of vitamin A causes deformities in the teeth of young animals.

All of these symptoms can be cured by the administration of vitamin A.

Retinoic acid, produced by the oxidation of retinal, serves as a carrier for polysaccharides in the synthesis of glycoproteins. A derivative of retinoic acid, 13-*cis*-retinoic acid, has been used in the treatment of severe acne.

Effect of Excess

Hypervitaminosis A, an intake of vitamin A far in excess of normal daily requirements, has such early symptoms as irritability, loss of appetite, fatigue, and itching. These symptoms usually disappear within a week after the withdrawal of the vitamin.

Acute cases of hypervitaminosis A have been observed following ingestion of polar bear liver, which contains up to 35,000 IU of vitamin A per gram. Symptoms are drowsiness, sluggishness, vomiting, severe headache, and generalized peeling of the skin after 24 hr. Membranes have increased permeability and decreased stability, leading to mitochondrial swelling, lysosomal rupture, and eventually death.

Eating a great many carrots on a daily basis may result in *carotenosis*, a benign condition characterized by a yellowing of the skin.

Storage and Absorption

Vitamin A is stored in the liver until it is needed. Mother's milk contains 10 to 100 times as much vitamin A as ordinary milk. Zinc is necessary for vitamin A to be mobilized from the liver, indicating that vitamin A metabolism may be adversely affected by a zinc deficiency.

Mineral oil should never be taken immediately before or after a meal as it coats the mucosa and interferes with the absorption of carotenes.

Since vitamin A is fat-soluble, it cannot be absorbed from the intestinal tract without the presence of bile. Any interference with the flow of bile to the intestines will also cause a deficiency of this fat-soluble vitamin.

Vitamin D

Source
Vitamin D is sometimes called "the sunshine vitamin." The richest sources of vitamin D are oils from such fish as cod or halibut and the flesh of such oily fish as sardines, salmon, and mackerel. Milk is not a very good source of vitamin D, although its vitamin D content can be increased by irradiation with ultraviolet light. The vitamin D content of the body can be increased by exposure of the skin to ultraviolet rays from the sun. Care must be taken to avoid overexposure and consequent sunburn.

Structure
The D vitamins are a group of sterols with considerable differences in their potency. The two most important vitamins in the D group are vitamin D_2 (ergocalciferol, or activated ergosterol) and vitamin D_3 (activated 7-dehydrocholesterol or cholecalciferol). Vitamin D_1, originally called vitamin D, was found to be a mixture of vitamins D_2 and D_3.

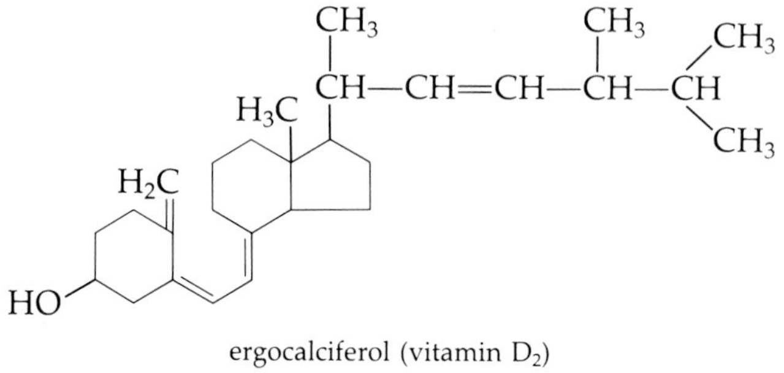

ergocalciferol (vitamin D_2)

Vitamin D_3 is similar to vitamin D_2, the difference being in the structure of the side chain. Vitamin D_3 is made by the irradiation of 7-dehydrocholesterol (equation 32-1), so it is frequently called "activated 7-dehydrocholesterol."

Ergosterol, the precursor of vitamin D_2, occurs in plants, whereas 7-dehydrocholesterol, the precursor of vitamin D_3, occurs in the skin of humans and animals. 7-Dehydrocholesterol in the skin is isomerized to vitamin D_3 under the influence of ultraviolet light.

Vitamin D must acquire two hydroxyl groups in order to be biologically active. The first hydroxyl group is added in the liver and the second in the kidneys, as indicated in Figure 32-2 for vitamin D_3.

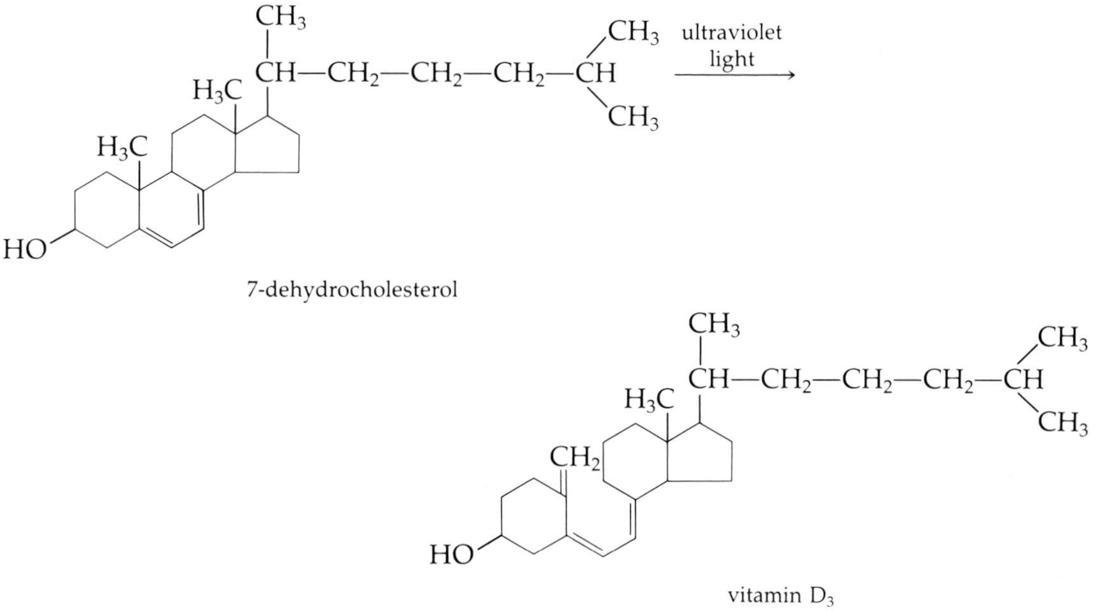
ultraviolet light
7-dehydrocholesterol
vitamin D$_3$

(32-1)

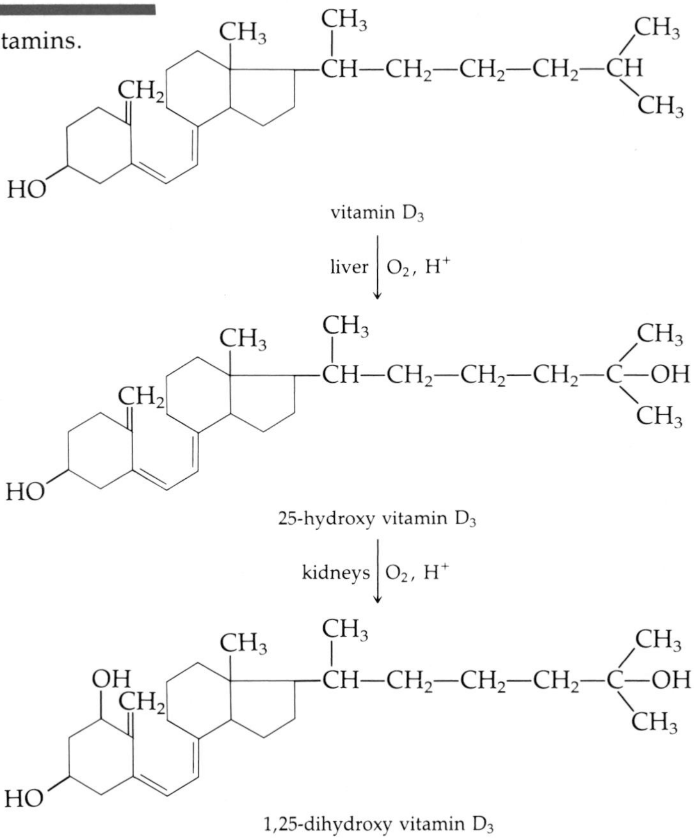
vitamin D$_3$
liver
O_2, H^+
25-hydroxy vitamin D$_3$
kidneys
O_2, H^+
1,25-dihydroxy vitamin D$_3$

Figure 32-2 Vitamin D$_3$, one of the D vitamins.

The same reactions occur with vitamin D_2. Patients who cannot convert vitamin D to its active form, because of liver or kidney disease, often develop *osteomalacia, *osteoporosis, or other bone diseases. Such patients can now be treated with synthetic active-form vitamin D.

Properties
The D vitamins are soluble in fats and insoluble in water. They are stable to heat, resistant to oxidation, and unaffected by cooking. Vitamin D_2 has a greater potency in humans than vitamin D_3, whereas the reverse is true in chickens.

Daily Requirement
The daily requirement for children is 400 IU. Daily requirements for adults are not stated because, in most instances, exposure to sunlight is sufficient to supply the body's need. However, 400 IU is recommended daily for women during pregnancy and lactation. One international unit of vitamin D is defined as the biologic activity of 0.025 μg of ergocalciferol.

Physiologic Action
The principal action of vitamin D is to increase the absorption of calcium and phosphorus from the small intestine. It also increases the release of calcium and phosphate in the bones and is necessary for normal growth and development. Vitamin D is required for the proper activity of the parathyroid hormone (see page 626) and so is used therapeutically in the treatment of hypoparathyroidism.

A lack of vitamin D may cause hypocalcemia and hypophosphatemia.

Effect of Deficiency
Rickets, a disease primarily of infancy and childhood, was previously thought to be due to a deficiency of vitamin D. This disease is now believed to be due to a lack of sunshine on the skin. Why? Because sunshine is necessary for the synthesis of calciferol in the skin.

Rickets is characterized by an inability to deposit calcium phosphate in the bones. The bones become soft and pliable; they bend and become deformed. The joints become enlarged, and the ribs become beaded. The knobby or beaded appearance of the ribs is called **rachitic rosary.** A child who has rickets does not grow. He or she develops such symptoms as nervousness, irritability, a bulging abdomen, loss of weight, loss of appetite, anemia, and a delayed development of the teeth. Injection of small amounts of calciferol, or one of its derivatives, or adequate exposure to sunlight can prevent or cure rickets (see Figure 32-3).

A diet low in phosphorus and vitamin D may produce *osteomalacia, or adult rickets. Since adult rickets is a rare condition,

Figure 32-3 A child with marasmus PCM and vitamin D deficiency (rickets) combined with marked deformity of his chest. [*AFIP 73-10185*]

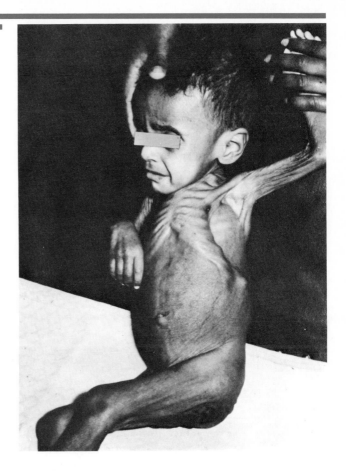

this suggests that adults need less vitamin D than children do. When an adult develops rickets, there is no bulging of the joints since the growth of the bones in an adult is already complete. There is, however, some softening of the bones with accompanying deformities. This disease occurs most often in women after repeated pregnancies during which there was a deficiency of vitamin D.

A lack of calcium and vitamin D in the diet can cause *osteoporosis in the adult. This disease, like osteomalacia, is characterized by decalcification and softening of the bones but to a much greater extent.

Effect of Excess

Hypervitaminosis D may show such early symptoms as weakness, lassitude, fatigue, nausea, vomiting, and diarrhea, all of which are associated with hypercalcemia. Later symptoms include calcification of soft tissues, including the kidneys and the lungs. Treatment of hypervitaminosis D consists of the immediate withdrawal of the vitamin, increased fluid intake, a diet low in calcium, and the administration of glucocorticoids.

Vitamin E

Source

Vitamin E is found in milk, eggs, fish, muscle meats, cereals, leafy vegetables such as lettuce, spinach, and parsley, and plant oils such as cottonseed oil, corn oil, and peanut oil. Wheat germ oil is particularly rich in vitamin E.

Structure

There are several vitamins E. The most important of these is called α-tocopherol; β- and δ-tocopherols are less active, and γ-tocopherol is inactive.

As indicated by the ending of the name, *-ol*, vitamin E is an alcohol. The structure of α-tocopherol is

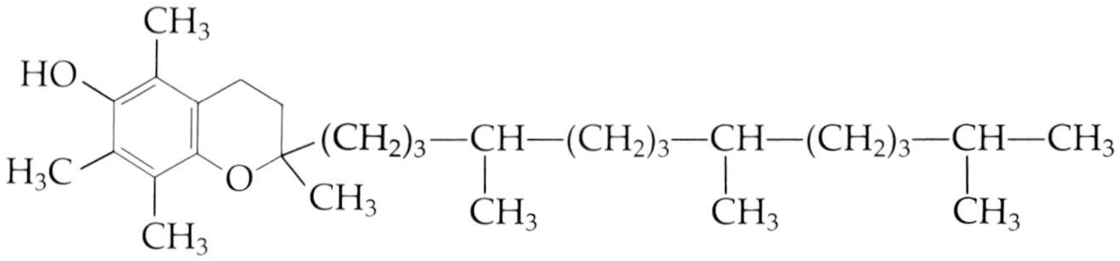

α-tocopherol

Properties

α-Tocopherol, the most important and most active of the E vitamins, is a colorless to pale yellow oil. It is soluble in fats and fat solvents but insoluble in water. Vitamin E is stable to heat but is destroyed by ultraviolet light and by oxidizing agents.

The activity of the tocopherols appears to be due to their antioxidant properties. They are very effective in preventing the oxidation of vitamin A and unsaturated fatty acids. There is also evidence that vitamin E functions as a cofactor in oxidative phosphorylation reactions. Vitamin E is believed to protect the lung tissues from damage by oxidants present in polluted air.

Vitamin E also plays an important role in selenium metabolism. Selenium is required for pancreatic function. This includes absorption of lipids along with vitamin E. Selenium is present in the enzyme glutathione peroxidase, which helps destroy peroxides and reduces the requirement of vitamin E for maintenance of membranes. Selenium also aids in the retention of vitamin E in plasma lipoprotein.

Conversely, vitamin E reduces selenium requirements in the body by preventing the loss of that element.

In 1989 the daily requirements for selenium were set at 70 μg for a male and 55 μg for a female.

Premature infants suffering from hemolytic anemia show very low levels of α-tocopherol in their blood. When given vitamin E supplements, people suffering from hemolytic anemia improve greatly. It is now believed that vitamin E is essential in infant metabolism. The U.S. Food and Drug Administration requires that commercial milk substitutes sold as infant foods contain adequate amounts of vitamin E. Recent evidence has shown that vitamin E is helpful in the diets of pregnant and lactating women. Vitamin E also is useful in treating older persons suffering from circulatory problems.

Vitamin E is required in higher animals such as cattle and poultry for fertility. The name *tocopherol* is derived from the Greek *tokos* meaning "childbirth" and *pherein* meaning "to bear."

Daily Requirement
The international unit of vitamin E is defined as the activity of 1 mg of *dl*-α-tocopherol acetate. The recommended daily requirement of vitamin E is 15 IU for the adult male and 12 IU for the adult female.

Effect of Deficiency
Vitamin K is found in the green leafy tissues of such plants as spinach, cabbage, and alfalfa. Vitamin K is also found in putrefied fish meal, liver, eggs, and cheese. Fruits contain very little vitamin K. produce embryos that develop normally for a few weeks but then die and are reabsorbed.) Vitamin E has not proven useful in treating sterility in humans.

Some animals on a vitamin E–deficient diet develop muscular dystrophy, resulting in paralysis. Administration of vitamin E helps these animals overcome such effects. Vitamin E has not been found effective in treating muscular dystrophy in humans, and there is no evidence that vitamin E has any therapeutic use at all.

Severely impaired absorption of fat in the intestines can cause vitamin E deficiency in humans. The symptoms of such a disease are muscular weakness, creatinuria, and fragile red blood cells.

There appears to be a relationship between vitamins A and E. Vitamin E may help in the absorption, storage, and utilization of vitamin A. Vitamin E also appears to offer protection against large overdoses of vitamin A. No definite side effects of large doses of vitamin E are known.

Vitamin K

Source
Vitamin K is found in the green leafy tissues of such plants as spinach, cabbage, and alfalfa. Vitamin K is also found in putrefied fish meal, liver, eggs, and cheese. Fruits contain very little vitamin K.

Structure
There are two naturally occurring K vitamins, K_1 and K_2. Vitamin K_1 is a yellow oil whose structure is

vitamin K₁

menadione

Vitamin K_2 differs from K_1 only in the nature of the side chain. Vitamin K_1 is produced in plants and K_2 by intestinal bacteria. A synthetic compound, menadione, exhibits vitamin K activity. It is converted in the body to vitamin K.

Compare the structure of the K vitamins with that of coenzyme Q (page 440), which functions in oxidative phosphorylation.

Properties

The K vitamins are soluble in fats and insoluble in water. They are stable to heat but are destroyed in acidic and alkaline solutions. They are also unstable to light and oxidizing agents.

Daily Requirement

In 1989 the daily requirement of vitamin K was set at 80 μg for a male and 65 μg for a female. The average diet supplies sufficient vitamin K; in addition, intestinal bacteria are able to synthesize this vitamin for their host.

A deficiency of vitamin K will occur only as a result of prolonged use of a broad-spectrum antibiotic coupled with a diet lacking or low in vitamin K. Vitamin K deficiency does occur in breast-fed infants because they do not have intestinal bacteria and because breast milk is a poor source of vitamin K. Vitamin K is usually given as a supplement during the first five months after birth. Some infants are born with a deficiency of prothrombin and are subject to bleeding. Other infants have no ability to manufacture vitamin K in the intestine. Without proper care such infants may die of brain hemorrhage. This condition can be alleviated by administering vitamin K to the mother before delivery or to the infant shortly after birth.

Effect of Deficiency

A deficiency of vitamin K may also occur when fat absorption is impaired as in biliary or pancreatic disease or in the atrophy of intestinal mucosa.

Vitamin K is known as the antihemorrhage vitamin. It is necessary for the production of prothrombin in the liver. When there is a deficiency of vitamin K, there is a lack of prothrombin and thus a prolonged clotting time for the blood. Vitamin K is also necessary as a cofactor for oxidative phosphorylation reactions.

Vitamin K is absorbed from the small intestine with the help of bile. In conditions in which bile does not enter the small intestine, such as in obstructive jaundice, vitamin K is not absorbed. This condition leads to a tendency to bleed for a long period of time after an injury or when undergoing surgery. This effect can be overcome by administering both bile and vitamin K to the patient.

Vitamin K is also used therapeutically as an antidote for anti-coagulant drugs such as Dicumarol.

Effect of Excess
Large doses of vitamin K can cause hemolysis in infants and aggravate hyperbilirubinemia.

Water-Soluble Vitamins

The water-soluble vitamins include the B-complex vitamins and vitamin C. The B complex represents a whole series of vitamins, many of which act as cofactors or coenzymes in various oxidative reactions. Each B-complex vitamin has a different physiologic activity. The B complex, also called the vitamin B family, comprises the following vitamins:

1. Vitamin B_1 (thiamine)
2. Vitamin B_2 (riboflavin)
3. Niacin
4. Pyridoxine
5. Pantothenic acid
6. Biotin
7. Folic acid
8. Vitamin B_{12} (cobalamin)

Since these vitamins are water-soluble, they can be excreted in the urine and thus they rarely accumulate in toxic amounts.

Because they are water-soluble, the B vitamins must be provided continually in the diet. The only exception is vitamin B_{12}, cobalamin, because the human liver can store several years' supply of this vitamin.

Vitamin B_1 (Thiamine)

Source
Thiamine occurs in yeast, milk, eggs, meat, nuts, and whole grains. Vegetables and fruits contain very little vitamin B_1. Synthetic vitamin B_1 is now being added to flour and bread to enrich their vitamin content.

Structure
Thiamine has been crystallized as a hydrochloride with the following structure:

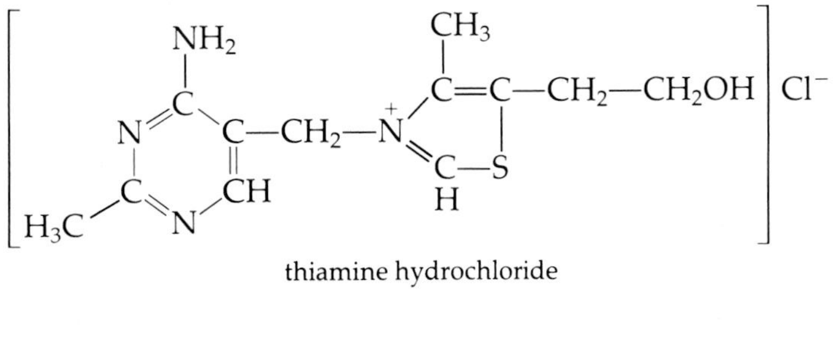

thiamine hydrochloride

Properties

Thiamine is soluble in water and also in alcohol up to 70 percent. It is insoluble in fats and fat solvents. Thiamine is stable in acid solution but is destroyed in alkaline and neutral solutions. Thiamine is quite stable to heat; it can be sterilized for 30 min at 120 °C without appreciable loss in activity. Thiamine hydrochloride, a salt produced by treating thiamine with hydrochloric acid, is more soluble than thiamine itself and so is generally used whenever this vitamin is required.

Daily Requirement

It is difficult to determine the amount of thiamine required daily. It depends upon several factors. The body's requirement of thiamine increases during a fever, increased muscular activity, hyperthyroidism, pregnancy, and lactation. The thiamine requirement of the body also increases during a diet high in carbohydrates, whereas it decreases with a diet high in fat and protein. Thiamine cannot be stored in the body to any significant degree.

The recommended thiamine daily intake is 0.3 to 0.5 mg for infants, 0.7 to 1.2 mg for children, 1.0 to 1.1 mg for adult females, and 1.2 to 1.4 mg for adult males.

Effect of Deficiency

A deficiency of vitamin B_1 (thiamine) causes a lack of appetite, arrested growth, and loss in weight.

Alcoholism is the most common cause of thiamine deficiency in the United States. This is due to low vitamin intake plus the high calorie content of alcohol.

As the lack of thiamine continues, a disease called beriberi develops in humans (in animals this disease is called polyneuritis). Beriberi occurs mainly in the Orient where fish and polished rice (both lacking in vitamin B_1) are the chief diet. In beriberi there is a degeneration of certain nerves leading to the mucsles. When pressure is applied along the nerves, severe pain is felt. The muscles served by these nerves become stiff and atrophy from disuse. Cardiovascular symptoms also occur. These are palpitation, tachycardia, an enlarged heart, and an abnormal electrocardiogram. Finally, death may occur because of heart failure. Beriberi can be prevented or treated by a diet containing thiamine.

Thiamine is necessary for the normal metabolism of carbohydrates. The vitamin is changed in the liver to thiamine pyrophosphate (TPP), which acts as a coenzyme (cocarboxylase) for the decarboxylation of pyruvic and α-keto acids and also acts in transketolase reactions. In the Krebs cycle, cocarboxylase is necessary for the conversion of pyruvic acid to acetyl CoA and also for the conversion of α-ketoglutaric acid to succinyl CoA (see page 483). In thiamine pyrophosphate, the —OH group is replaced by two phosphate groups. During a deficiency of thiamine, pyruvic acid accumulates in the blood and carbohydrates are not properly metabolized.

Thiamine also functions in the utilization of pentoses in the hexose monophosphate shunt and in some amino acid syntheses.

Vitamin B₂ (Riboflavin)

Source
Riboflavin occurs in many of the same sources as thiamine. It is found in yeast, milk, liver, kidney, heart, and leafy vegetables. Cereals contain very little riboflavin unless it is added artificially.

Structure
Riboflavin or vitamin B_2 has been found to consist of a five-carbon sugar alcohol (ribitol) and a pigment (flavin). Its structure is

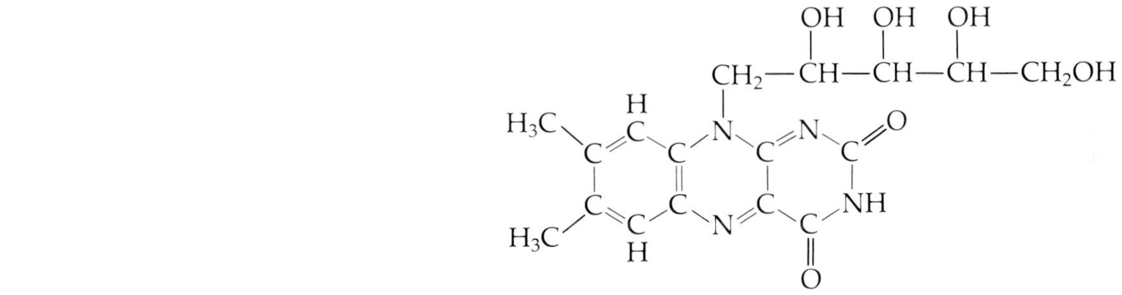

riboflavin

Properties
Riboflavin is an orange-red crystalline solid, slightly soluble in water and alcohol but insoluble in fats and fat solvents. In water solution riboflavin forms a greenish yellow fluorescent liquid. Riboflavin is destroyed by light and alkaline solutions but is fairly stable to heat and so is not destroyed by cooking.

Riboflavin acts as a coenzyme in two different forms—flavin adenine dinucleotide (FAD) and flavin mononucleotide (FMN). These riboflavin coenzymes act as acceptors for the transfer of protons between NAD^+ and $NADP^+$ and the cytochromes, which transport electrons in the mitochondria (see page 485).

Daily Requirement
The daily requirement of riboflavin is 0.8 to 1.4 mg for children, 1.4 to 1.7 mg for adult males and 1.2 to 1.3 mg for adult females, with increased amount needed during pregnancy and lactation.

Effect of Deficiency
A deficiency of riboflavin in humans produces lesions in the corners of the mouth (cheilitis), inflammation of the tongue (glossitis), and lesions on the lips and around the eyes and nose. There is also an inflammation of the skin (dermatitis) and a clouding of the cornea of the eye.

Because riboflavin is so sensitive to light, newborn infants with hyperbilirubinemia when treated with phototherapy show signs of riboflavin deficiency. This condition persists even when riboflavin supplements are given.

In rats a deficiency of riboflavin produces dermatitis, clouding of the corneas, and loss of hair.

Niacin

Source
Niacin (formerly known as nicotinic acid or vitamin B₅) is widely distributed in plants and animals. It is found in liver, kidney, and heart as well as in yeast, peanuts, and wheat germ. Milk, eggs, and fruit contain some niacin but are generally classified as poor sources of that vitamin.

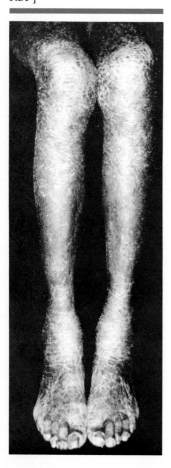

Figure 32-4 The roughened skin, dark in patches, that is characteristics of the niacin deficiency disease pellagra. [*Custom Medical Stock Photo* © 1983 *Carroll H. Weiss, RBP*]

Structure and Properties
Niacin and niacinamide (nicotinamide), which it readily forms in the body, have the following structures.

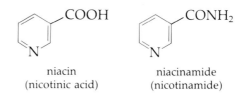

niacin
(nicotinic acid)

niacinamide
(nicotinamide)

Niacin is slightly soluble in water but quite soluble in alkali. It is insoluble in fats. Niacin is stable to alkalis and acid and to heat and light, and it is not destroyed by cooking.

Niacinamide, along with thiamine and riboflavin, serves as a coenzyme in tissue oxidation. It functions in the mitochondria in the form of NAD⁺ (nicotinamide adenine dinucleotide) and NADP⁺ (nicotinamide adenine dinucleotide phosphate) (see page 439).

Daily Requirement
The recommended daily intake of niacin is 16 to 19 mg for males and 13 to 15 mg for females with a slight increase in requirements for adolescents and during pregnancy and lactation. However, these requirements can be greatly affected by the protein of the diet because the amino acid tryptophan can supply much of the body's needed niacin (60 mg of tryptophan equals 1 mg of niacin). Some niacin may also be synthesized by intestinal bacterial action and thus become available for use in the body.

Effect of Deficiency
Niacin was originally called nicotinic acid or the antipellagra factor. The word **pellagra** comes from the Italian words *pelle agra* meaning rough skin. A deficiency of niacin in humans produces serious con-

sequences, and a deficiency in this vitamin is usually accompanied by a deficiency in other substances also. In pellagra there is a dermatitis (skin rash or lesions) and an inflammation of the mouth and tongue (glossitis). These symptoms are accompanied by diarrhea and then dementia (see Figure 32-4).

A lack of niacin in dogs produces tongue lesions called black tongue. A diet containing niacin is effective in curing pellagra in humans and also niacin deficiency diseases in animals.

Pellagra was once quite common in the southern states where the diet consisted chiefly of corn and fat pork. Corn has a low tryptophan content and can give very little niacin. Fat pork also has very little niacin. Thus there was a deficiency of this vitamin, leading to pellagra. With an improvement in the diet, especially with the addition of foods containing niacin (or tryptophan), pellagra is not as common in the United States as it was previously.

High doses of niacin (nicotinic acid) but not niacinamide can produce such symptoms as skin flushing, gastrointestinal distress, and pruritus as well as lower serum cholesterol levels.

Pyridoxine

Source
Pyridoxine is found in yeast, liver, egg yolk, and the germ of various grains and seeds. It is also found to a limited extent in milk and leafy vegetables.

Structure
Pyridoxine was originally called vitamin B_6, or the rat antidermatitis factor. Subsequent work showed that vitamin B_6 is a mixture of pyridoxine, pyridoxal, and pyridoxamine. The generally accepted term for these compounds is pyridoxine because these compounds are readily interconvertible.

Daily Requirement
The recommended daily allowance of pyridoxine for the adult male is 2.0 mg, and for the adult female 1.6 mg, per day.

Effect of Deficiency
A deficiency of pyridoxine in rats produces dermatitis in the paws, nose, and ears. A deficiency in dogs and pigs produces anemia. If the deficiency of pyridoxine is continued for a long period of time, these animals suffer from epileptiform fits. A deficiency of pyridoxine in infants produces convulsions. A deficiency of this vitamin in adults produces such symptoms as dermatitis, sore tongue, irritability, and apathy. A diet containing pyridoxine (or vitamin B_6) will alleviate these symptoms.

The generation of niacin from the amino acid tryptophan requires the presence of the coenzyme pyridoxal phosphate. Therefore, pellagra frequently accompanies pyridoxine deficiency.

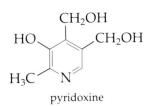

A widely used antituberculosis drug, isonicotinic acid hydrazide (isoniazid) can induce isolated pyridoxine deficiency by forming a compound with pyridoxal. This compound is excreted in the urine, thus producing a deficiency of pyridoxal.

The antihypertensive drug hydralazine and the chelating agent penicillinamine will also cause pyridoxine deficiencies. Pyridoxine should not be taken by patients receiving levodopa treatment for Parkinson's disease because pyridoxine enhances the drug's deactivation.

Pyridoxal phosphate and pyridoxamine phosphate serve as coenzymes for the decarboxylation of amino acids, taking part in the reactions occurring primarily in the gray matter of the central nervous system. It is believed that a deficiency of these coenzymes interferes with decarboxylation reactions in the central nervous system and so leads to epileptiform seizures. Pyridoxal phosphate and pyridoxamine phosphate also serve as coenzymes in amino acid metabolism. Further, pyridoxine is involved in the absorption of zinc by the intestines.

Pantothenic Acid

Source

Pantothenic acid has a widespread distribution in nature. Its name comes from the Greek word meaning "from everywhere." Good sources of pantothenic acid are egg yolk, yeast, kidney, and lean meats. Other fairly good sources are skimmed milk, broccoli, sweet potatoes, and molasses.

Structure and Properties

Pantothenic acid is a viscous yellow oil, soluble in water but insoluble in fat solvents such as chloroform. It is stable in acid and alkaline solution. Pantothenic acid is one of the constituents of coenzyme A (CoA) (see page 438), which is involved in the metabolism of carbohydrates, fats, and proteins and in the synthesis of cholesterol (see page 506).

The structure of pantothenic acid is

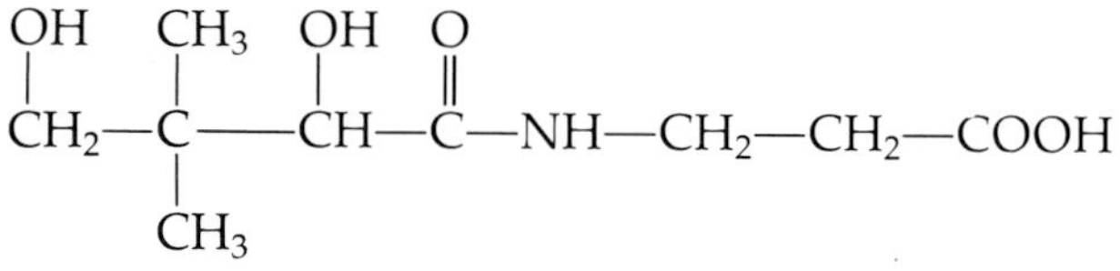

$$\underset{CH_3}{\underset{|}{CH_2}}-\overset{\overset{OH}{|}}{C}-\overset{\overset{CH_3}{|}}{\underset{|}{C}}-\overset{\overset{OH}{|}}{CH}-\overset{\overset{O}{\|}}{C}-NH-CH_2-CH_2-COOH$$

pantothenic acid

The daily human requirement of pantothenic acid is not known, but estimates of 4 to 7 mg are usually considered adequate and are easily met with an ordinary diet.

Effect of Deficiency

Pantothenic acid was originally known as chick antidermatitis factor because it was a substance that prevented dermatitis in chicks.

Rats and dogs who were given a diet deficient in pantothenic acid showed a loss of pigmentation from their hair. The black hair of such animals turned gray but returned to its original black color upon the addition of pantothenic acid to the diet. There is no evidence that this vitamin is of significant value in restoring hair color in humans. There is little evidence of pantothenic acid deficiency in humans.

A deficiency of pantothenic acid in animals causes degeneration in the adrenal cortex and a failure in reproduction.

Biotin

Biotin, another member of the vitamin B complex, is widely distributed in nature. Rich sources of this vitamin are liver, egg yolk, kidney, yeast, and milk. Biotin was formerly known as the anti-egg-white injury factor. This name was given to it because rats fed raw egg white failed to grow and also developed dermatitis. Raw egg white contains a protein, avidin, that combines with biotin and renders it unavailable to the animal.

The structure of biotin is

$$
\begin{array}{c}
O \\
\parallel \\
C \\
HN \diagup \ \diagdown NH \\
HC \!-\!\!-\!\!-\! CH \\
H_2C \diagdown_S \diagup CH \!-\! CH_2 \!-\! CH_2 \!-\! CH_2 \!-\! CH_2 \!-\! COOH
\end{array}
$$

biotin

An artificially produced deficiency of biotin in humans causes scaly dermatitis, nausea, muscle pains, and depression. These symptoms are rapidly relieved by the administration of a diet containing biotin.

Biotin is supplied by the action of intestinal bacteria, in humans as well as in animals, so that a deficiency of this vitamin is unlikely in most cases, except on a severely restricted diet. Biotin functions as a coenzyme for carboxylation reactions in the formation of fatty acids (see page 502).

Folic Acid

Folic acid (folacin) occurs in green leaves, yeast, liver, kidney, and cauliflower. The structure of folic acid (which contains a pteridine nucleus, *p*-aminobenzoic acid, and glutamic acid) is

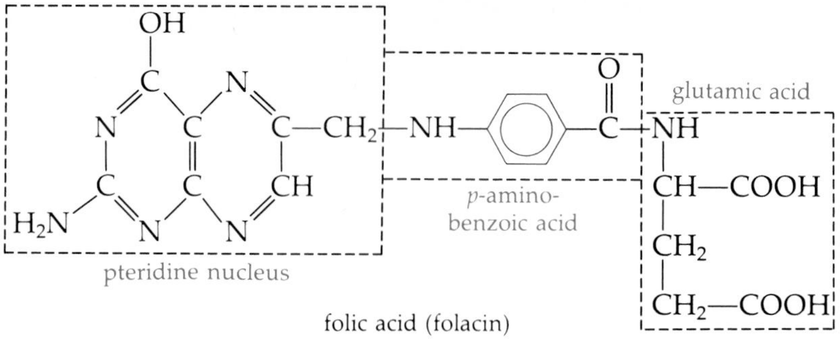

folic acid (folacin)

Actually, this compound, which is also called pteroylglutamic acid (PGA), is only one of several related compounds in the folic acid group. Others are pteroic acid, pteroyltriglutamic acid, and pteroyl-heptaglutamic acid.

In 1989 the daily requirements for folic acid were set at 200 μg for a male and 180 μg for a female.

A deficiency of folic acid causes *megaloblastic anemia and gastrointestinal dusturbances.

Folic acid in its reduced form, tetrahydrofolic acid, acts as a coenzyme for the transfer of methyl groups in the formation of such compounds as choline and methionine.

Folic acid is primarily involved in reactions involving transfer of methyl groups. This type of reaction plays an essential role in the synthesis of hemoglobin, nucleic acids, and methionine.

Treatment with antileukemic drugs such as methotrexate involves inhibiting the conversion of folic acid into its active form tetrahydrofolate. Without this folate, cancer cells (as well as other cells) cannot grow.

Vitamin B$_{12}$ (Cobalamin)

Source
Plants do not contain vitamin B$_{12}$ (cobalamin). Microorganisms are able to synthesize it. The best sources of cobalamin are liver, kidney, fish, eggs, milk, oysters, and clams.

Structure and Properties
Cobalamin is an odorless, tasteless, reddish crystalline compound, soluble in water and alcohol and insoluble in fat solvents such as ether and acetone. One unusual property of this vitamin is that it contains the element cobalt (4.35 percent).

A cyano group (CN$^-$) is usually attached to the central cobalt atom during isolation of the vitamin. This compound is known as cyanocobalamin. The cyano group must be traded for a methyl group in the body before cobalamin can be activated. The structure of cyanocobalamin is shown on the next page.

Daily Requirement
The recommended daily amount of vitamin B$_{12}$ (cobalamin) is 2.0 μg.

Effect of Deficiency

Vitamin B_{12} is important in the transfer of methyl groups. Thus, many symptoms of vitamin B_{12} and folate deficiency are similar. Vitamin B_{12} is involved in the maintenance of the myelin sheath, in the synthesis of nucleic acids and hemoglobin, and in the metabolism of lipids and carbohydrates. Pernicious anemia is an anemia that resembles megaloblastic anemia. Pernicious anemia has several neurologic complications. It results from a dietary lack of vitamin B_{12} (strict vegetarianism) or of the intrinsic factor.

Vitamin B_{12} (cobalamin) is also called the anti-pernicious anemia factor. It is absorbed from the small intestine in the presence of hydrochloric acid and the intrinsic factor (see page 455). In the absence of the intrinsic factor, vitamin B_{12} cannot be absorbed, and this leads to pernicious anemia. Injection of a small amount of cobalamin will produce remarkable improvement in that disease.

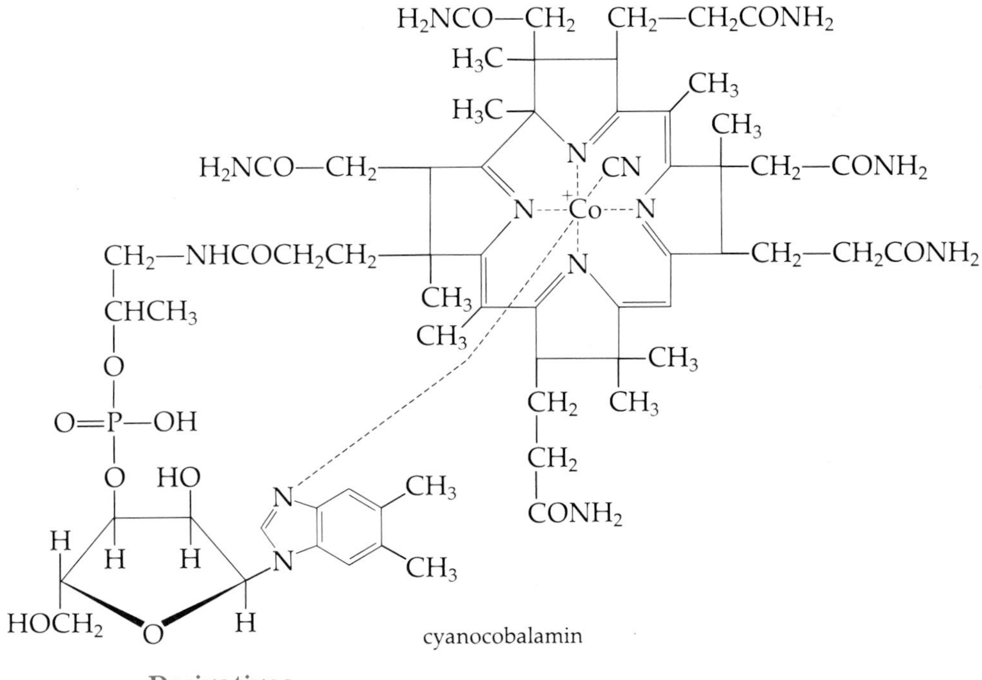

cyanocobalamin

Derivatives

Coenzyme B_{12} is one of several coenzymes derived from vitamin B_{12}. These coenzymes are required for hydrogen transfer and isomerization in the conversion of methyl malonate to succinate, thus involving both carbohydrate and fat metabolism.

Coenzyme B_{12} differs from vitamin B_{12} in that a 5-deoxyadenosine group replaces the CN group attached to the central cobalt.

Related Substances

There are several substances frequently classified along with the B vitamins although they are not truly vitamins. Among these are choline, inositol, *p*-aminobenzoic acid, and lipoic acid.

Choline

Choline is an essential substance as far as the body is concerned, although it is generally not classified as a vitamin because it can be synthesized in the required amounts in the body. It is required in much larger amounts than vitamins.

Choline is a viscous, colorless liquid, soluble in water and alcohol but insoluble in ether. Its structure is

$$\left[\begin{array}{c} CH_3 \\ CH_3 - \overset{+}{N} - CH_2 - CH_2 - OH \\ CH_3 \end{array} \right] Cl^-$$

choline

A deficiency in choline leads to such symptoms as fatty liver. Young rats on a diet deficient in choline also had hemorrhagic degeneration of the kidneys. Older rats who survived these symptoms developed cirrhosis. Chicks and young turkeys develop perosis or slipped tendon disease.

Choline is a constituent of lecithin (phosphatidylcholine) and sphingomyelin and is important in brain and nervous tissue. A deficiency of choline will not develop in a person on a high-protein diet because proteins supply the amino acids from which the body can synthesize this compound.

Choline is a constituent of acetylcholine, which is present in nerve cells and aids in the transmission of nerve impulses by the following means: When a nerve cell is stimulated, it releases acetylcholine. This acetylcholine in turn stimulates the adjacent nerve cell to release acetylcholine. This process continues as a chain reaction until the impulse reaches the brain. Once a nerve cell has passed its impulse on to the next cell, the enzyme acetylcholinesterase hydrolyzes acetylcholine into acetic acid and choline. The nerve cells then use these two compounds to regenerate the acetylcholine for the next impulse.

Inositol

It was found that mice on a synthetic diet containing all of the known vitamins still failed to grow. Also there was an effect on their hair and impaired lactation. Pantothenic acid, which affects hair in mice, did not help these animals. However, the symptoms were overcome by the addition of a compound obtained from cereal grain. This compound was inositol. Inositol is required for the growth of yeasts, mice, rats, guinea pigs, chickens, and turkeys.

Inositol is found in liver, milk, vegetables, yeast, whole grains, and fruits. The molecular formula for inositol is $C_6H_{12}O_6$, so it is an isomer of glucose.

Inositol, along with folic acid and *p*-aminobenzoic acid, is not a true vitamin. However, it is included with the B group even though no specific role in human nutrition has been established.

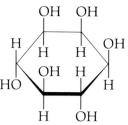

inositol
(hexahydroxycyclohexane)

NH$_2$

COOH

p-aminobenzoic acid
(PABA)

NH$_2$

SO$_2$NH$_2$

sulfanilamide

p-Aminobenzoic Acid

p-Aminobenzoic acid (PABA) is a growth factor for certain microorganisms. It forms part of the folic acid molecule and is believed to be necessary for the formation of that vitamin. However, humans are incapable of using *p*-aminobenzoic acid to produce folic acid. *p*-Aminobenzoic acid is formed upon the hydrolysis of folic acid.

Compare the structure of sulfanilamide with that of *p*-aminobenzoic acid. Sulfonamides exert their antibacterial action by acting as antimetabolites to *p*-aminobenzoic acid. One theory is that the sulfa drugs are similar in structure to PABA and attach themselves to an enzyme required in the metabolism of PABA, thus blocking the use of that substance (see page 446). Persons taking sulfa drugs should be cautioned against taking vitamins containing PABA.

p-Aminobenzoic acid is not regarded as a vitamin for humans, but it is included here because some microorganisms need it for the synthesis of folic acid.

Lipoic Acid

Lipoic acid was first detected in the studies of the growth of lactic acid bacteria. It is fat-soluble and so was called lipoic acid. Its structure is

$$CH_2-CH_2-CH-(CH_2)_4-COOH$$
$$\underset{S\rule{2cm}{0.4pt}S}{|\hspace{3cm}|}$$

lipoic acid

As far as is known, lipoic acid is not required in the diet of higher animals and no deficiency effects have been noted.

Actually, lipoic acid is not a true vitamin; however, because its coenzyme function in carbohydrate metabolism is closely related to that of thiamine, it is classified along with the B-vitamin group. Lipoic acid functions, along with thiamine, in the initial decarboxylation of α-keto acids to form acetyl CoA for the Krebs cycle.

Vitamin C (Ascorbic Acid)

Source

Fresh fruits and vegetables, including oranges, lemons, grapefruit, berries, melons, tomatoes, and raw cabbage, are excellent sources of vitamin C. Dry cereals, legumes, milk, meats, and eggs contain very little of this vitamin.

Structure and Properties

Vitamin C, ascorbic acid, is a white crystalline substance soluble in water and alcohol but insoluble in most fat solvents. It is a strong reducing agent and is easily oxidized in air, especially in the presence of such metallic ions as Fe^{3+} or Cu^{2+}. Ascorbic acid is rapidly destroyed by heating. For this reason, cooking of foods in copper pots should be avoided because this will destroy the vitamin C content of the foods being cooked.

Vitamin C can easily be oxidized to dehydroascorbic acid. Both ascorbic acid and dehydroascorbic acid are biologically active, and both have been synthesized in the laboratory.

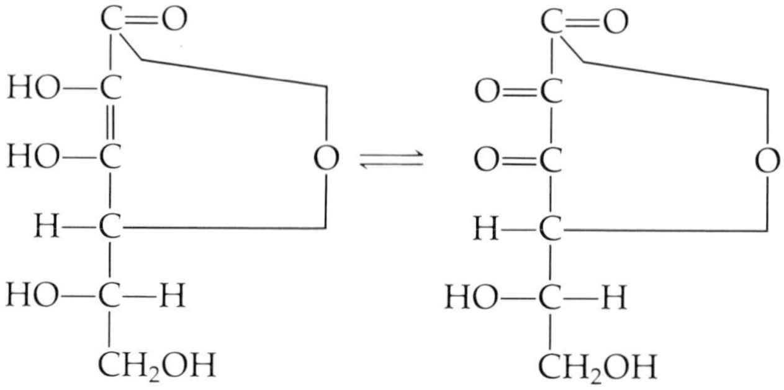

ascorbic acid   dehydroascorbic acid

Daily Requirement
The recommended daily allowance of ascorbic acid for an adult is 60 mg; the amount is increased to 80 mg during pregnancy and 100 mg during lactation.

Effect of Deficiency
Plants and all animals except guinea pigs, humans, and other primates are able to synthesize ascorbic acid and are resistant to diseases caused by a lack of this vitamin. A deficiency of vitamin C produces a disease known as scurvy. The symptoms in humans are swollen, bleeding gums, pain in the joints, decalcification of the bones, loss of weight, and anemia.

A deficiency of vitamin C prevents the body from forming and maintaining the intercellular substance that cements the tissues together. A lack of this intercellular substance in the capillaries leads to rupturing and subsequent hemorrhaging in these vessels and to the formation of weak bones and atrophy of the bone marrow, accompanied by anemia, and also accounts for loosening of the teeth and spongy gums. All of these symptoms are relieved by the addition of ascorbic acid to the diet.

Vitamin C is also believed to function in oxidation–reduction reactions and in cellular respiration.

Summary

The body requires vitamins in addition to carbohydrates, fats, and proteins but in much smaller amounts. Vitamins must be supplied in the diet, whereas hormones are synthesized by the body.

Vitamins are divided into two types: fat-soluble and water-soluble.

Vitamin A is found in fish liver oils, in butter, and in milk. Vitamin A is an alcohol. The recommended daily adult dose of vitamin A is 1000 retinol equivalents (5000 IU) for the adult male and 800 retinol equivalents (4000 IU) for the adult female. Vitamin A has an all-trans structure.

A lack of vitamin A produces keratinization in the membranes of the eyes, digestive tract, respiratory tract, and genitourinary tract. Nyctalopia is also due to a deficiency of vitamin A. Vitamin A is stored in the liver.

Vitamin D is found in fish liver oils such as cod or halibut. The vitamin D content of milk is increased by irradiation. The D vitamins are sterols with a structure similar to that of cholesterol. Vitamin D functions to increase the absorption of calcium and phosphorus from the small intestine. Vitamin D also functions in the deposition of calcium phosphate in the teeth and bones. A lack of vitamin D produces rickets.

Vitamin E is found in many foods; wheat germ is particularly rich in this substance. The E vitamins have antioxidant properties and act as a cofactor in oxidative phosphorylation reactions. Vitamin E prevents sterility in animals, but its necessity in humans has not yet been determined.

Vitamin K is known as the antihemorrhage vitamin. It is necessary for the production of prothrombin in the liver.

Vitamin B_1 (thiamine) occurs naturally in yeast, milk, and whole grains and also may be made synthetically. A deficiency of this vitamin causes a lack of appetite, arrested growth, and weight loss. A prolonged deficiency of vitamin B_1 leads to the disease known as beriberi (or polyneuritis in animals). Thiamine is also necessary for the normal metabolism of carbohydrates. This vitamin acts as a coenzyme for the decarboxylation of pyruvic acid. The coenzyme, known as cocarboxylase, is necessary in the Krebs cycle for the conversion of α-ketoglutaric acid to succinyl CoA.

Vitamin B_2 (riboflavin) is found in the same sources as thiamine. Riboflavin acts as a coenzyme in two different forms: flavin adenine dinucleotide (FAD) and flavin mononucleotide (FMN). These coenzymes act as acceptors for the transfer of protons between NAD^+ and $NADP^+$ and the cytochromes.

Niacin (or nicotinic acid) is widely distributed in nature. A deficiency of this vitamin produces a condition known as pellagra. Niacin is an important constituent of two coenzymes: nicotinamide adenine dinucleotide (NAD^+) and nicotinamide adenine dinucleotide phosphate ($NADP^+$). These coenzymes are involved in most oxidation–reduction reactions in the mitochondria.

Pyridoxine is involved in the decarboxylation of amino acids and is also required for certain transaminase reactions.

Pantothenic acid is one of the constituents of CoA, which is involved in the metabolism of carbohydrates, fats, and proteins as well as in the synthesis of cholesterol.

Biotin functions in the activation of carbon dioxide for carboxylation reactions in the formation of fatty acids.

Folic acid is concerned with the transfer of methyl groups in the formation of such compounds as choline and methionine.

Vitamin B_{12}, cobalamin, is called the anti-pernicious anemia factor. It is also involved in the synthesis of certain amino acids and of choline and is necessary for the formation of coenzyme B_{12}.

Choline is a constituent of lecithin and so is important in brain and nerve tissue. Choline is also a constituent of acetylcholine, which aids in the transmission of nerve impulses.

p-Aminobenzoic acid (PABA) is formed upon the hydrolysis of folic acid. This substance is a growth factor for certain microorganisms.

Inositol is required for the growth of yeasts, mice, and rats.

Lipoic acid is closely related to thiamine in the initial oxidation of α-keto acids.

Vitamin C, ascorbic acid, can be synthesized by plants and most animals but not by humans. A lack of this vitamin produces a disease known as scurvy. A lack of ascorbic acid also causes a lack of the intercellular substance that cements the tissues together.

Questions and Problems

1. Compare vitamins with hormones
2. Which vitamins are water-soluble? fat-soluble?
3. What are the sources of vitamin A?
4. Why should mineral oil never be taken immediately before or after a meal?
5. What is β-carotene? How is it used in the body?
6. What are the effects of a deficiency of vitamin A?
7. Where is vitamin A stored in the body? How is it absorbed?
8. Why does the body need vitamin A? What causes hypervitaminosis A? What are the symptoms?
9. Describe the role of vitamin A in the visual process.
10. What are the sources of vitamin D?
11. Why is vitamin D called the sunshine vitamin?
12. What type of compound is vitamin D? Are all the D vitamins equally potent?
13. What are the daily adult requirements of vitamin A? vitamin D?
14. What are the functions of vitamin D in the body?
15. What are the effects of a lack of vitamin D? of an excess?
16. What are the sources of vitamin E? What are its properties? its functions?
17. What are the sources of vitamin K? What are its properties? its functions?
18. What are the sources of vitamin B_1? What are its properties? its functions?
19. What are the effects of a lack of thiamine?
20. What are the sources of riboflavin?
21. What are the sources of niacin? of pyridoxine?
22. What are the effects of a deficiency of niacin? of pyridoxine? How does the body use these substances?
23. What are the sources of pantothenic acid? its properties? the effects of a deficiency?
24. What are the functions of lipoic acid, biotin, folic acid, *p*-aminobenzoic acid, inositol, and cobalamin?
25. What are the sources of vitamin C? its properties? its functions?
26. What are the effects of a deficiency of ascorbic acid?
27. How do sulfonamides exert their antibacterial effect?
28. What is carotenosis?
29. How are nerve impulses transmitted?
30. How does vitamin E affect selenium metabolism?

Practice Test

1. A vitamin that is not fat-soluble is vitamin _____.
   a. A   b. K   c. C   d. D
2. The only vitamin that contains a metal is B_{12}. It contains _____.
   a. Cu   b. Mg   c. Zn   d. Co
3. Niacin is important in the prevention of _____.
   a. beriberi   b. scurvy   c. pellagra   d. rickets
4. Which of the following vitamins is toxic if taken in excess?
   a. A   b. B_{12} c. C   d. biotin
5. The vitamin used by the body to make CoA is _____.
   a. biotin   b. riboflavin   c. pyridoxine   d. pantothenic acid
6. A vitamin involved in the visual cycle is vitamin _____.
   a. A   b. B   c. C   d. D
7. Rickets is caused by a lack of vitamin _____.
   a. A   b. B_1 c. C   d. D
8. Pellagra is caused by a lack of _____.
   a. niacin   b. pyridoxine   c. riboflavin   d. inositol
9. A lack of which vitamin can lead to hemorrhaging?
   a. B_1 b. C   c. D   d. K
10. Beriberi is caused by a lack of vitamin _____.
   a. B_1 b. B_2 c. B_6 d. B_{12}

33

Hormones

Teaching a child how to inject insulin.

Hormones exert a very important influence on the regulation of body processes. Hormones are produced in the endocrine glands and are secreted directly into the bloodstream. Hormones may be proteins, polypeptides, amino acids, or steroids.

There is an internal balance and interaction among the various endocrine glands.

Hormones act in several ways. Some stimulate RNA production in target cell nuclei and thus increase the production of enzymes. Some hormones stimulate enzyme synthesis in the ribosomes through the translation of information carried by messenger RNA. Others are involved in the transportation of various substances across membranes. Hormones also affect the levels of cAMP, which in turn activates many kinase enzymes (see following section). The action of most protein hormones is inhibited by the absence or severely reduced concentration of calcium ions.

Both vitamins and hormones are necessary in only very small amounts; however, unlike vitamins, hormones are produced in the body. Hormones are produced in a particular organ and are carried by the bloodstream to some other body part, where they cause a specific physiologic effect. The level of hormones in the bloodstream ranges from 10^{-6} M to 10^{-12} M.

Cyclic Adenosine Monophosphate

Cyclic adenosine monophosphate (cAMP) is present in almost every type of body cell. It is produced through the activity of the enzyme adenylate cyclase (found in the membranes of the cell walls) on ATP. cAMP is not found in plants.

When a hormone, a chemical messenger, reaches its target cell, it interacts with the cell membrane receptors that are specific for that hormone. The adenylate cyclase in the cell then triggers the production of cAMP from ATP. The cAMP thus released activates the enzyme systems that in turn catalyze protein synthesis by the target cell, which produces the characteristic effects of that cell.

The scheme of the reactions is shown in Figure 33-1.

The concentration of cAMP in the cells is controlled primarily by two methods: (1) by regulating its rate of synthesis and (2) by changing it into an inactive form, AMP, through the action of certain enzymes.

Abnormalities in the metabolism of cAMP may explain the effects of certain diseases. For example, the bacteria that cause cholera produce a toxin that stimulates the intestinal cells to accumulate cAMP. The excess cAMP instructs the cells to secrete a salty fluid. The accumulation and subsequent loss of large amounts of this salty fluid and the resulting dehydration, if unchecked, can cause cholera to be fatal.

cAMP activates a protein kinase that catalyzes phosphorylation

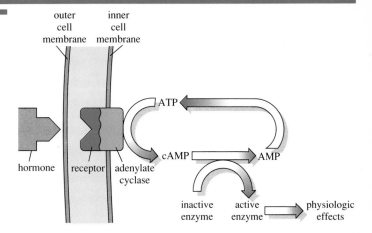

Figure 33-1 Mechanism of hormone action.

and thus activates RNA polymerase. RNA polymerase, in turn, stimulates RNA production, which then functions as a messenger for protein synthesis (see page 653). In this way, cAMP influences the synthesis of protein.

The role of cAMP in the breakdown of liver and muscle glycogen to glucose is indicated on page 477.

The structure of cAMP is as follows.

cyclic adenosine monophosphate (cAMP)

A similar compound, cyclic guanosine monophosphate (cGMP), is involved in the actions of the F prostaglandins, whose effects run counter to those of the E prostaglandins. It has been suggested that cAMP and cGMP function reciprocally in regulating cellular activity, but much more research needs to be done in this area.

Cardiac Hormones

The atrium of the heart produces atriopeptin, a cardiac hormone involved in fluid, electrolyte, and blood pressure homeostasis. Atriopeptin is stored in the *cardiocytes as the prohormone atriopeptinogen, which consists of 126 amino acids. The hormone atriopeptin, which has a chain of 28 amino acids, is continually released in small amounts and so is always present in the blood.

Elevated vascular volume triggers a series of events. First, more atriopeptin is released, which causes the kidneys to increase the rate of renal blood flow and thereby the rate of glomerular filtration. Thus urine volume is increased, as is sodium excretion. Renin activity (see page 444) is decreased, and so is aldosterone (see page 629) production. At the same time the amount of vasopressin is increased (see page 634). The resulting decrease in vascular volume causes a negative feedback that decreases the levels of atriopeptin.

Hormone of the Pineal Gland

The pineal gland is a cone-shaped body deep in the back of the brain. For years it was thought to be a *vestigial organ with no function. However, in 1958 a hormone called melatonin was isolated from this gland. Melatonin is an unusual hormone in that it is more prevalent in the body during the hours of darkness than during daylight hours. That is, melatonin production rises as darkness falls. The brain interprets this increase in melatonin as a signal to go to sleep for the night. The light of dawn shuts down melatonin production.

It has been found that patients suffering from melancholia have unusual melatonin rhythms. In these patients, maximum amounts of the hormone occur at dawn or midnight instead of at about 2 A.M. as in healthy individuals.

Travelers given melatonin before long flights do not suffer from jet lag, indicating that somehow melatonin keeps the mind in synchronization with the outside world.

In children melatonin inhibits sex hormones and so restrains sexuality. As the melatonin levels fall during puberty, the activity of the sex hormones rises.

Hormones of the Blood

Bradykinin is a nonapeptide produced in the bloodstream by the cleavage of larger protein molecules whenever body tissue is injured. The structure of this hormone is

Arg-Pro-Pro-Gly-Phe-Ser-Pro-Phe-Arg

Bradykinin is the most potent pain-producing chemical known and is one of the body's messengers telling the brain that an injury has occurred. One theory states that when the body is injured the damaged nerve sends an impulse to the brain and the ruptured cells release an enzyme that produces bradykinin, which in turn ruptures the nearby capillaries, allowing the red blood cells to leak out. The excess fluid and enlarged blood vessels cause the skin to swell and redden.

Hormones of the Gastrointestinal System

The pyloric mucosa produces a hormone called **gastrin**. Gastrin is absorbed into the bloodstream and is carried back to the stomach, where it stimulates the secretion of hydrochloric acid in the gastric juice and also produces greater motility of the stomach. Gastrin also stimulates the secretion of pepsin and the intrinsic factor (see page 455). Gastrin is a polypeptide containing 17 amino acids.

Secretin is formed in the mucosa of the small intestine when the acidic chyme enters the duodenum. Secretin stimulates the pancreas to release the pancreatic juices into the small intestine. Secretin is a polypeptide containing 27 amino acids, 14 of which are identical to those found in glucagon (page 623).

Cholecystokinin (CCK) is secreted when fat enters the duodenum. CCK stimulates the contraction and emptying of the gallbladder into the small intestine. It also stimulates pancreatic enzyme secretion. CCK is a polypeptide consisting of 33 amino acids.

In addition to these gastrointestinal hormones, several others are involved in the gastrointestinal system. Among these are the following.

1. Enterogastrone, which inhibits gastric acid secretion.
2. Motilin, which stimulates gastric motility.
3. Enteroglucagon, which is involved in glycogenolysis.
4. Gastric inhibitory polypeptide (GIP), which inhibits gastric acid secretion and gastric motility and stimulates release of insulin by the pancreas.
5. Chymodenin, which stimulates secretion of chymotrypsin.

Hormones of the Pancreas

The pancreas secretes digestive enzymes (see Chapter 25) and four hormones—insulin, glucagon, somatostatin, and pancreatic polypeptide. Insulin and glucagon are involved in carbohydrate metabolism. Somatostatin is involved in local regulation of insulin and glucagon secretion. Pancreatic polypeptide affects gastrointestinal secretion.

Insulin

Insulin is a protein secreted by the β cells of the islets of Langerhans in the pancreas. Insulin has been isolated and crystallized. Crystalline insulin contains a small amount of zinc, which it obtains from the zinc-rich tissues in the pancreas. Small amounts of chromium are also needed for the synthesis of insulin. The amount of chromium in the body decreases with age. The amino acid sequence in insulin is indicated on page 414.

Insulin increases the rate of oxidation of glucose, facilitates the conversion of glucose to glycogen in the liver and the muscles, and increases the synthesis of fatty acids, protein, and RNA. Insulin functions to decrease blood sugar to its normal fasting level after it has been increased by digestion of carbohydrates by facilitating the transportation of glucose through the membranes into the cells.

Normally, the blood sugar level rises to 120 to 130 mg per 100 mL of blood after a meal containing carbohydrates. This level is returned to its normal fasting value by the action of insulin. If the islets of Langerhans are underactive or degenerated, little or no insulin is produced, so the blood sugar level remains high. This condition is termed hyperglycemia and is associated with diabetes mellitus. In diabetes mellitus, there is an increased blood sugar level; glucose appears in the urine, and there is formation of ketone bodies accompanied by acidosis. Injection of insulin will produce a rapid recovery from these symptoms. It will not, however, cure diabetes because these same symptoms will reappear after a short time. Thus insulin must be taken by diabetics for the rest of their lives.

Since insulin is a protein, it cannot be taken orally because it would be digested. Therefore, insulin is given by subcutaneous injection. Injections of insulin must be given two or three times a day to a patient with diabetes mellitus. However, when insulin is combined with protamine (a protein), the product, called protamine zinc insulin, is utilized much more slowly and is effective for more than 24 hr. Thus only one injection is needed daily.

If too large a dose of insulin is administered, the blood sugar level falls far below its normal fasting value. This condition is called hypoglycemia and is characterized by such symptoms as dizziness, nervousness, blurring of vision, and then unconsciousness. Such a state is called insulin shock and may be relieved by the administration of sugar, either orally or by injection. Hypoglycemia also occurs when there is a tumor on the islets of Langerhans in the pancreas.

Insulin is degraded primarily in the liver and kidneys by the enzyme glutathione insulin transhydroxylase.

Insulin is available in sterile solutions in which 1 mL contains 100 units (U-100); 1 unit of insulin is the amount required to reduce the blood sugar level of a normal 2-kg rabbit after a 24-hr fast from 120 to 45 mg per 100 mL.

Although insulin cannot be taken orally, there are certain *hypoglycemic substances that will lower the blood sugar level. These substances are effective in treating adult diabetes but not juvenile diabetes. They function by stimulating the β cells of the islets of Langerhans to produce insulin. Examples of such hypoglycemic agents are tolbutamide (Orinase) and chlorpropamide (Diabinese). The structural formula of tolbutamide is

$$CH_3$$

$$SO_2NH—C—NH—CH_2—CH_2—CH_2—CH_3$$

tolbutamide

Recent evidence indicates that these so-called hypoglycemic agents have many toxic side effects. Adult patients with a mild type of diabetes frequently can control it just as well by means of a proper diet.

Glucagon

When crude insulin was first used to lower blood sugar levels, it was noted that a temporary hyperglycemia occurred first and then, soon afterward, came the hypoglycemic effects expected of insulin. The unknown substance present in the crude insulin that gave the hyperglycemic effects was called the hyperglycemic glycogenolytic factor, or HGF. This factor is now known as **glucagon.** Glucagon has been isolated and crystallized. It is a polypeptide containing 29 amino acids in a single chain.

Glucagon has a different arrangement of amino acids than insulin. Unlike insulin, it contains no disulfide bridges. Glucagon contains methionine and tryptophan, which insulin does not. However, insulin contains cysteine, proline, and isoleucine, whereas glucagon does not.

Glucagon is formed in the α cells of the islets of Langerhans in the pancreas. Significant amounts of glucagon also come from A cells in the stomach and other parts of the gastrointestinal tract.

Glucagon causes an increase in the sugar content of the blood by stimulating phosphorylase activity in the liver. The reactions involved are as follows:

$$\text{glycogen + phosphate} \xrightarrow{\text{phosphorylase}} \text{glucose 1-phosphate}$$

$$\text{glucose 1-phosphate} \xrightarrow{\text{phosphoglucomutase}} \text{glucose 6-phosphate}$$

$$\text{glucose 6-phosphate} \xrightarrow{\text{phosphatase}} \text{glucose + phosphate}$$

Glucagon increases the formation of cAMP, which activates phosphorylase, and so increases the glucose content of the blood, causing hyperglycemia.

Glucagon stimulates the formation of glucose from amino acids (gluconeogenesis). It also increases the release of potassium ions from the liver. In the adipose tissue glucagon increases the breakdown of lipids to fatty acids and glycerol.

Glucagon has been used to treat hypoglycemic effects due to an overdose of insulin and insulin shock induced in the treatment of psychiatric patients.

Glucagon is also used in a diagnostic test for glycogen storage disease.

Hormones of the Thyroid Gland

The thyroid gland is an H-shaped gland consisting of one lobe on each side of the trachea with a piece of tissue connecting the two lobes. In the adult the thyroid gland weighs approximately 25 to 30 g.

The hormones of the thyroid gland regulate the metabolism of the body. They also affect the growth and development of the body.

The thyroid is filled with many small follicles that contain *colloid*, which contains the thyroid gland's stored hormones. Actually, the thyroid is the only endocrine gland in the body that is capable of storing appreciable amounts of hormone.

The thyroid gland contains the element iodine—one of the elements necessary for the proper functioning of the body. Iodine exists in the body in two different forms: as iodide ions and in the thyroid hormones.

Older, indirect methods for evaluating activity such as basal metabolic rate (BMR), protein-bound iodine (PBI), and butanol-extractable iodine (BEI) have been replaced by radioimmunoassay methods that are direct and more reliable.

The colloid of the thyroid contains the protein **thyroglobin**, a glycoprotein. This protein liberates **triiodothyronine** (T_3) and **thyroxine** (T_4), the principal thyroid hormones. T_3 is the major active form of the thyroid hormone and is three to five times as biologically active as T_4. Some mono- and diidothyronine are also formed, but these compounds are quickly deiodinated in the bloodstream and the freed iodine is used to form more thyroglobin. The structures of these hormones are

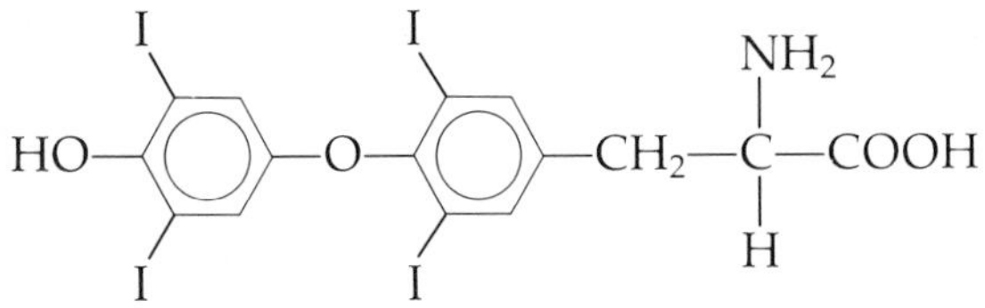

thyroxine (T_4) 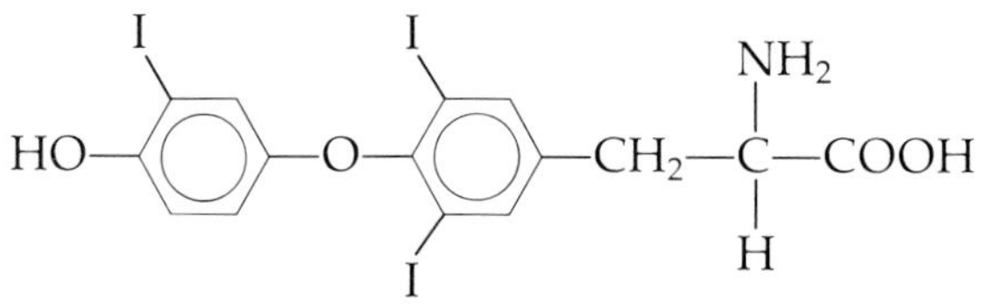   triiodothyronine (T_3)

The C cells in the thyroid gland produce the hormone **calcitonin** (thyrocalcitonin), which, along with the parathyroid hormone, regulates the calcium ions in the blood. It has been shown that the parathyroid hormone sustains the blood supply of calcium ions whereas calcitonin prevents the blood calcium ion concentration from rising above the required level.

Pure calcitonin was first isolated in 1968, and its structure was determined shortly thereafter. This hormone is a polypeptide containing a single chain of 32 amino acids. Calcitonin has now been synthesized in the laboratory, and its mode of action in the body has been investigated. It produces its effect by inhibiting the release of calcium ions from the bone to the blood. The release of calcitonin is stimulated by high levels of Ca^{2+} in the blood.

Medullary thyroid carcinoma, a disorder of the C cells of the thyroid, causes an abnormally high production of calcitonin.

Hypothyroidism

Hypothyroidism is a condition in which the thyroid gland does not manufacture sufficient thyroxine for the body's needs. It is usually due to a lack of iodine in the diet, particularly in parts of the country where the water and foods contain little iodine. Hypothyroidism may also be due to a disease of the thyroid gland or to its congenital absence.

The symptoms of hypothyroidism are sluggishness, weight gain, slower heartbeat, reduced metabolic rate, and loss of appetite. Hypothyroidism is easily remedied by the use of iodized salt as part of the normal diet.

Cretinism

If the thyroid gland is absent or fails to develop in an infant, the effects produced are called cretinism and the individual is called a cretin. Cretins have a greatly retarded growth, both physically and mentally. They are usually abnormal dwarfs with coarse hair and thick dry skin, obese with protruding abdomens. They are also underdeveloped mentally and sexually.

A cretin may develop normally if given thyroid hormones before he or she reaches adulthood (see Figure 33-2).

If the thyroid gland should atrophy after an individual reaches adulthood, the same symptoms as in cretinism appear, except that the individual remains adult in size. One very noticeable symptom is in the development of thick, coarse, dry skin. Such a condition is known as *myxedema. Persons with myxedema are also sluggish, have a lower pulse and metabolic rate and lower body temperature and are frequently anemic. They are also very sensitive to cold.

Myxedema can usually be cured by the administration of thyroxine.

Simple Goiter

Simple *goiter, also called colloid goiter or endemic goiter, is a condition in which the thyroid gland enlarges, usually because of a lack

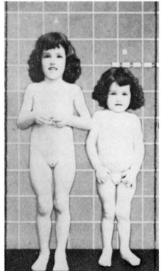
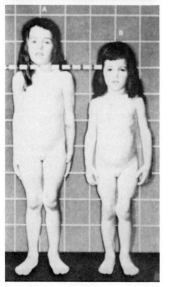

Figure 33-2 Cretin B is much shorter than her twin, A. After therapy with thyroid hormones the difference, although present, is greatly reduced. [*Courtesy Warner-Chilcott Laboratories, Morris Plains, NJ.*]

of iodine in the diet. The decreased production of thyroid hormones causes an increased production of TSH (thyroid-stimulating hormone), which in turn overstimulates the thyroid gland. Simple goiter is accompanied by a definite increase in the amount of colloid material in the thyroid gland and also an increase in the size of the neck itself.

Simple goiter occurs in areas where there is a deficiency of iodine in the food and drinking water. The condition can be successfully prevented or cured by the addition of iodine compounds (usually iodized salt) to the diet.

Hashimoto's Disease

Hashimoto's disease is a type of hypothyroidism in which all aspects of thyroid function may be impaired. This disease is caused by an attack on the thyroid gland by the body's own immune system.

Hyperthyroidism

Hyperthyroidism occurs when the thyroid gland produces excess thyroxine. The symptoms are an increased metabolic rate, bulging of the eyes (exophthalmos), nervousness, loss of weight, a rapid, irregular heartbeat, and an elevated body temperature. Such a condition is also called Graves' disease, Basedow's disease, or exophthalmic goiter. Hyperthyroidism may also be due to a tumor in the thyroid gland (toxic adenoma or Plummer's disease.)

Hyperthyroidism can be controlled or cured by surgical removal of part of the thyroid gland, by the oral administration of radioactive iodine, or by the use of antithyroid drugs. Hypertrophy of the endocrine glands can lead to toxic adenomas with a malignant potential if not treated promptly.

Radioactive iodine, used in the treatment of hyperthyroidism, is usually administered in the form of sodium iodide, NaI. The body converts the inorganic radioactive iodide into thyroglobin in the thyroid gland, thus subjecting that gland to radiation that will cut down its activity.

Hormones of the Parathyroid Glands

There are four small parathyroid glands attached to the thyroid gland. In humans these glands are reddish brown and together weigh 0.05 to 0.3 g. In early experimental thyroidectomies (removal of the thyroid glands) in animals, the parathyroid glands were also inadvertently removed. This caused the death of the animals.

The parathyroid glands produce parathyroid hormone, a hormone that influences the metabolism of calcium and phosphorus in the body. This hormone is a protein with a molecular weight of approximately 9500 and consists of a single polypeptide chain of 84 amino acids. The parathyroid gland cannot store this hormone, so the hormone is synthesized and secreted continuously. Recall that calcitonin (pages 624–25) is also involved in the regulation of calcium. Administration of vitamin A decreases parathyroid hormone, possibly by increasing calcium uptake into the parathyroid gland.

Surgical removal of the parathyroid glands causes hypoparathyroidism, characterized by such symptoms as muscular weakness, irritability, and tetany owing to a decrease in the calcium content of the blood plasma. Death occurs because of convulsions caused by the lack of calcium. At the same time as the calcium content of the plasma is decreasing, the calcium content of the urine is also decreasing and the phosphate content of the plasma is increasing.

The symptoms of hypoparathyroidism can be relieved by treatment with vitamin D or calcium salts, or both.

Hyperparathyroidism is an increase in the production of hormones by the parathyroid glands. It is usually due to a tumor of those glands (parathyroid adenoma) and produces such symptoms as decalcification of the bones followed by deformation and fractures of the bones, nausea, and polyuria. Deposits of calcium occur in soft tissues, and renal stones frequently occur.

In hyperparathyroidism, the calcium content of the blood plasma is high and the phosphate content low. The extra calcium in the blood is obtained from the bones, thus causing those bones to become decalcified. Urine calcium, phosphate, and cAMP are increased.

Hyperparathyroidism is usually treated by the surgical removal of the tumor of the parathyroid glands.

Hormones of the Adrenal Glands

The adrenal (suprarenal) glands are located close to the upper pole of the kidneys and weigh about 3 to 6 g each. The adrenal glands are divided into two distinct portions: the cortex, which is the outer portion, and the medulla, which is the inner portion (see Figure 33-3). Each of these portions is distinct both structurally and physiologically, and each produces its own hormones.

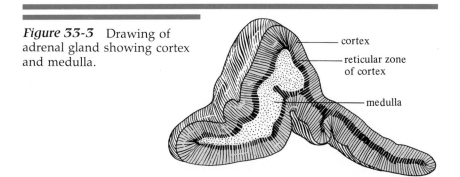

Figure 33-3 Drawing of adrenal gland showing cortex and medulla.

cortex

reticular zone of cortex

medulla

Hormones of the Adrenal Cortex

The hormones of the adrenal cortex are steroidal and fall into three categories:

1. **Glucocorticoids,** which primarly affect the metabolism of carbohydrates, lipids, and protein. Examples are corticosterone, cortisone (11-dehydroxycorticosterone), and cortisol (hydrocortisone).
2. **Mineralocorticoids,** which primarily affect the transportation of electrolytes and the distribution of water in the tissues. The most potent of this group is aldosterone.
3. **Androgens** or **estrogens,** which primarily affect secondary sex characteristics. The principal androgen is dehydroepiandrosterone.

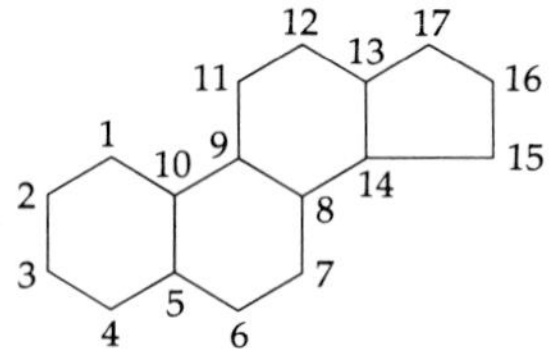

The steroids of the adrenal cortex function in the cell nucleus for the synthesis of RNA and protein. The steroid nucleus contains four fused carbon rings, numbered as shown at the left.

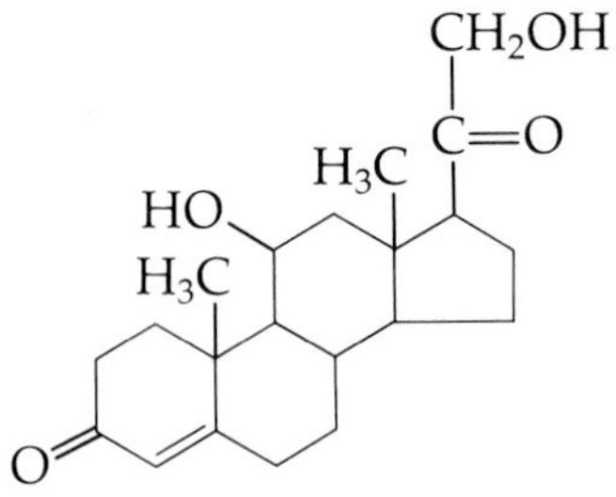

steroid nucleus

corticosterone

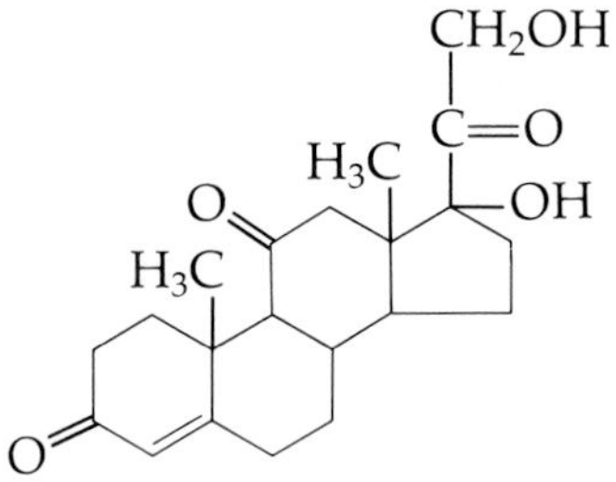

cortisone

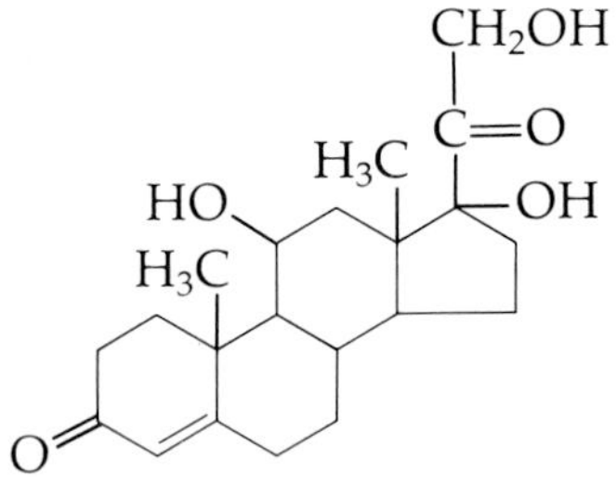

cortisol

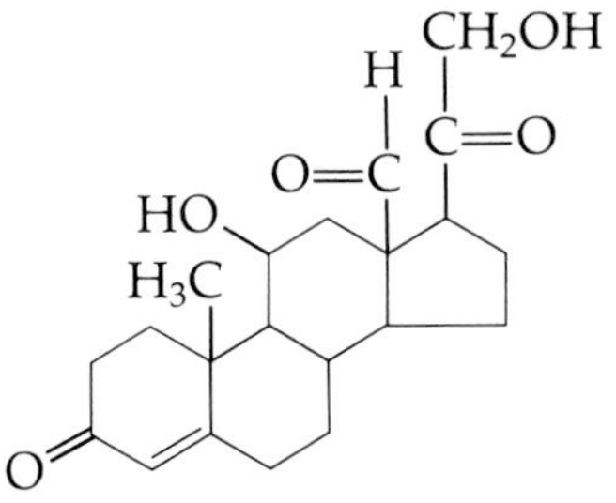

aldosterone

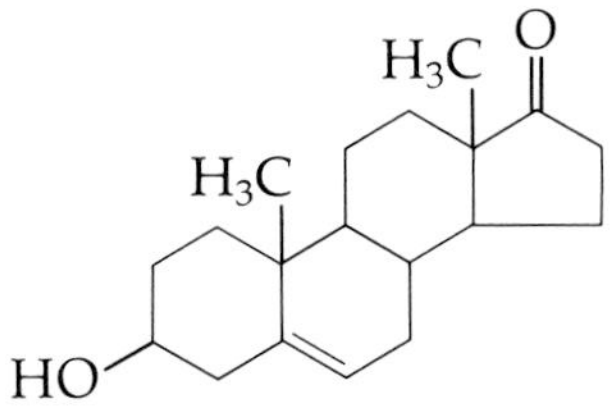

dehydroepiandrosterone

Effect of a Deficiency of Adrenal Cortex Hormones

If the adrenal glands are removed from an animal, it will soon die because of a lack of the hormones produced by the adrenal cortex. In humans, a hypofunctioning of the adrenal glands, because of a tuberculosis of those glands or in association with pernicious anemia, diabetes, or hypothyroidism, results in Addison's disease. It is characterized by an excessive loss of sodium chloride in the urine, low blood pressure, low body temperature, hypoglycemia, elevation of serum potassium, muscular weakness, a progressive brownish pigmentation of the skin, nausea, and loss of appetite.

The preceding symptoms, except for the pigmentation, are due to a lack of the salt- and water-regulating hormones—the mineralo-corticoids, primarily aldosterone. These hormones stimulate the kidney tubules to reabsorb sodium ions. If there is a lack of these hormones, sodium ions are not reabsorbed and are eliminated in the urine. Along with the sodium ions, chloride ions are also eliminated so that the urine is abnormally high in sodium chloride. Accompanying the sodium chloride loss by the body is a loss of water through osmosis. This loss in water in turn decreases both blood volume and blood pressure. The blood becomes more concentrated, and excretion of urea, uric acid, and creatinine is decreased, thereby increasing the concentration of these substances in the blood. Dehydration occurs rapidly and death occurs from circulatory collapse.

Another function of the hormones of the adrenal cortex is to stimulate the process of gluconeogenesis, the formation of glucose from amino acids. This process takes place in the liver. In the absence of the adrenocortical hormones, the glucocorticoids, the blood sugar falls and the glycogen stored in the liver and the muscles decreases considerably.

All of these symptoms, except the skin discoloration, can be relieved by the use of cortisonelike compounds—either naturally occurring extracts or synthetically prepared ones.

Effects of Hyperactivity of the Adrenal Cortex

Hyperactivity of the adrenal cortex (hyperadrenocorticism) is caused by a tumor on the adrenal cortex, by an overdosage of cortisone or ACTH, or by an increased production of ACTH (see page 632).

In children, hyperactivity of the adrenal cortex is manifested by early sexual development. Hyperactivity of the adrenal cortex in the adult female, which occurs more often than in the adult male, causes a decrease in feminine characteristics. The voice deepens, the breasts decrease in size, the uterus atrophies, and hair appears on the face. In the adult male, hyperactivity of the adrenal cortex is manifested by an increase in male characteristics—an increase in amount of body hair, increased size of sex organs, deeper voice.

Other effects of hyperadrenocorticism are hyperglycemia and glycosuria, retention of sodium ions and water, increased blood volume, edema, depletion of potassium ions (hypokalemia), and excessive gluconeogenesis.

Hormones of the Adrenal Medulla

Even though the medulla of the adrenal glands secretes three hormones—epinephrine (adrenaline), norepinephrine (noradrenaline), and dopamine—it is not essential to human life. That is, it can be removed without causing death.

OH

HO—C—H

CH$_2$—NH—CH$_3$

epinephrine

OH

HO—C—H

CH$_2$—NH$_2$

norepinephrine

OH

H—C—H

H—C—H

NH$_2$

dopamine

Dopamine and norepinephrine are precursors for epinephrine and account for 80 percent of the adrenal medulla hormones.

The hormones of the adrenal medulla (particularly epinephrine) are necessary for the body's adaptation to acute and chronic stress.

Under conditions of stress, epinephrine does several things. It rapidly provides fatty acids as the primary fuel for muscle action. It increases glycogenolysis and gluconeogenesis in the liver and decreases glucose uptake in muscle and other organs. It also depresses insulin release to preserve glucose for the central nervous system. Other effects are increased blood flow to the brain, stimulation of heart activity, increased rate of breathing, constriction of arterioles of the skin, and simultaneous dilation of the arterioles of the skeletal muscles.

Epinephrine also has the following effects.

1. It relaxes the smooth muscles of the stomach, intestines, bronchioles, and bladder. This relaxing effect on the muscles of the bronchioles makes epinephrine especially useful in the treatment of asthma and hay fever.
2. Epinephrine is sometimes used during minor surgery when it is administered along with a local anesthetic. The constriction of the arterioles by epinephrine prevents the anesthetic from spreading too rapidly from the site of the injection. Epinephrine can be injected directly into the heart muscle when that organ stops beating or when it does not start to beat in a newborn baby.
3. Epinephrine production is increased during anxiety, fear, or other stress. This extra epinephrine in turn causes a rise in blood sugar, frequently exceeding the renal threshold. In this case, glucose appears in the urine. Such a condition is called emotional glycosuria and disappears as soon as the stress is relieved.

Norepinephrine raises blood pressure by constricting the arterioles. It does not affect the heart itself and does not relax the muscles of the bronchioles as does epinephrine. Norepinephrine is found in the sympathetic nerves where it acts as a neurotransmitter.

The medulla of the adrenal glands rarely becomes diseased. To date, no deficiency effects of its hormones are known. However, certain tumors of the medulla of the adrenal glands stimulate these glands to produce excess hormones. The symptoms caused are intermittent hypertension leading to permanent hypertension and eventually to death from such complications as coronary insufficiency, ventricular fibrillation, and pulmonary edema.

Hormones of the Pituitary Gland

The pituitary gland (**hypophysis**) is located at the base of the brain. It consists of two parts: the anterior and intermediate lobes (the **adenohypophysis**) and the posterior lobe (the **neurohypophysis**). Each

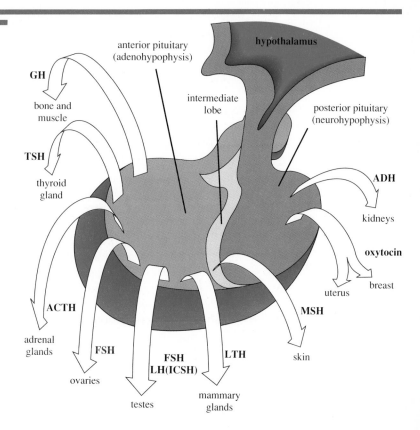

Figure 33-4 Hormones of the pituitary.

of these parts secretes or releases its own hormones. The pituitary gland has often been called the "master gland" of the body because it seems to exert a direct influence upon most of the other endocrine glands. It is now known that almost all of the secretory activity of the pituitary gland is controlled by a small area of the brain known as the **hypothalamus** (see Figure 33-4).

Hormones of the Anterior Lobe

Six hormones have been isolated from the anterior lobe of the pituitary gland.

1. Growth hormone (GH)
2. Thyrotropic hormone (TSH)
3. Adrenocorticotropic hormone (ACTH)
4. Prolactin, or lactogenic hormone (LTH)
5. Luteinizing hormone (LH), also known as the interstitial cell-stimulating hormone (ICSH)
6. Follicle-stimulating hormone (FSH)

Growth Hormone
Growth hormone (GH), also called somatotropin, is a protein with a molecular weight of about 21,500 and contains 191 amino acids. It is

present in much greater quantities than other pituitary hormones. Growth hormone stimulates the growth of the long bones at the epiphyses, stimulates the growth of soft tissue, and increases the retention of calcium ions. The growth hormone also increases protein synthesis, leading to a positive nitrogen balance. In the muscle, growth hormone increases amino acid transportation across membranes, also leading to increased protein, DNA, and RNA synthesis. Growth hormone causes the mobilization of fatty acids from fat deposits, providing cellular fuel. In the muscles, growth hormone antagonizes the effects of insulin; it inhibits glucose metabolism by muscle tissue.

Growth hormone also stimulates the mammary glands and increases the production of somatomedins from the liver. Somatomedins, in turn, foster sulfate incorporation into the cartilage.

Underactivity of the anterior lobe of the pituitary gland in children leads to a deficiency of the growth hormone, causing dwarfism. These children develop normally but do not grow in size. Unlike cretins, they are not mentally retarded, but they may be sexually underdeveloped.

An overactivity of the anterior lobe of the pituitary gland, possibly due to a tumor, results in an overproduction of growth hormone. When this occurs in a child, growth is stimulated and gigantism results. If the overactivity of this gland occurs during adulthood, the individual does not grow in size, but the bones in the hands, feet, and face grow, producing a condition known as *acromegaly.

Secretion of growth hormone is regulated by the growth-hormone-releasing factor (GHRF) from the hypothalamus.

Thyrotropic Hormone

The thyrotropic hormone is also called the thyroid-stimulating hormone (TSH). This hormone is a glycoprotein with a molecular weight of approximately 30,000. The release of TSH from the pituitary gland is regulated by the thyrotropic-releasing factor (TRF) from the hypothalamus.

A deficiency of TSH causes the thyroid gland to atrophy. As a result, thyroxine production ceases and the metabolic rate drops.

If this hormone is injected into an animal, the symptoms of hyperthyroidism appear—increased metabolic rate, increased heart rate, and *exophthalmos.

Adrenocorticotropic Hormone

The adrenocorticotropic hormone (ACTH) is a polypeptide with a molecular weight of 4500. ACTH contains 39 amino acids, but it has been shown that only the first 23 are required for activity. The remaining 16 vary according to the animal source. The release of ACTH by the pituitary gland appears to be regulated by the corticotropin-releasing factor (CRF) of the hypothalamus in response to various biologic stresses. ACTH has been prepared synthetically and is used medically.

ACTH stimulates the synthesis and release of the hormones by the adrenal cortex. ACTH, like many other hormones, controls its target tissue through cAMP. Administration of ACTH to a normal person causes retention of sodium ions, chloride ions, and water, elevation of blood sugar, and increased excretion of potassium ions, nitrogen, phosphorus, and uric acid.

Prolactin

Prolactin [lactogenic hormone or luteotropin (LTH)], was first identified by its property of stimulating the formation of "crop milk" in the crop glands of pigeons. The arrangement of the amino acids in prolactin has been determined. It is a protein with a molecular weight of about 23,000.

Prolactin initiates lactation. In mammals, a hormone produced by the placenta stimulates the growth of the mammary glands and at the same time inhibits the secretion of prolactin. At parturition the inhibiting effect of the placenta is not present, so prolactin is secreted and thus initiates lactation.

Luteinizing Hormone

Luteinizing hormone (LH) is also known as the interstitial cell-stimulating hormone (ICSH). This hormone is a glycoprotein with a molecular weight of about 40,000. It was the first pituitary gonadotropin whose sequence of amino acids was precisely determined.

LH stimulates the development of the testes in males and also causes an increased production of testosterone. In male animals who have had the pituitary gland surgically removed (hypophysectomy), administration of LH increases the weight of the seminal vesicles and also of the ventral lobe of the prostate gland.

In females, LH plays an important role in ovulation. It not only causes the production of the corpus luteum but sustains it and stimulates the production of progesterone. In hypophysectomized females, an injection of LH stimulates repair of interstitial ovarian tissues and also increases ovarian weight.

Follicle-Stimulating Hormone

Follicle-stimulating hormone (FSH) is a glycoprotein with a molecular weight of 25,000. It stimulates and initiates the development of the follicles of the ovary and prepares those follicles for the action of LH. It also stimulates the secretion of estrogen. In males it causes the growth of the testes and stimulates the production of spermatozoa.

Gonadotropic Hormones

The gonadotropic hormones LH and FSH (and also prolactin) are secreted by the anterior lobe of the pituitary gland. Removal of the anterior lobe in a male causes atrophy of the testes, prostate gland, and seminal vesicles. Removal in a female causes atrophy of the ovaries, uterus, and fallopian tubes.

LH and FSH as well as TSH consist of an α and a β chain of amino acids. The α chain is the same in all three, so biologic specificity must reside in the β chain. Both LH and FSH stimulate cAMP synthesis in appropriate target organs.

Hormone of the Intermediate Lobe

The pars intermedia or intermediate lobe of the pituitary gland secretes a hormone called **intermedin** or the **melanocyte-stimulating hormone** (MSH). It is a polypeptide with α and β parts.

MSH increases the deposition of melanin in the human skin, thus producing darker pigmentation. When the adrenal cortex is underactive, as in Addison's disease, more MSH is produced. This leads to an increased synthesis of melanin with the accompanying brown pigmentation of the skin. Epinephrine and, even more strongly, norepinephrine inhibit the action of the melanocyte-stimulating hormone.

Hormones of the Posterior Lobe

The posterior lobe of the pituitary gland contains two hormones—**vasopressin** and **oxytocin**—which are produced in the hypothalamus and stored in this lobe of the pituitary gland.

Vasopressin
Vasopressin, also called the antidiuretic hormone (ADH), stimulates the kidneys to reabsorb water. When the water content of the body is high, very little of this hormone is secreted and more water is eliminated. Conversely, when the water content of the body is low, more of this hormone is secreted, causing the kidney tubules to reabsorb more water. Thus, this hormone serves to regulate the water balance in the body.

Diabetes insipidus (see page 526) is caused by the absence of the antidiuretic hormone, which results in excess daily elimination of water (up to 30 L). Diabetes insipidus can be controlled by the administration of ADH.

Vasopressin, or ADH, stimulates the peripheral blood vessels to constrict and cause an increase in blood pressure. Because of this effect, it has been used to overcome low blood pressure caused by shock following surgery.

Vasopressin is a polypeptide whose arrangement of amino acids is indicated at the left.

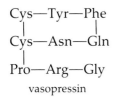

Cys—Tyr—Phe
|　　　　|
Cys—Asn—Gln
|
Pro—Arg—Gly

vasopressin

Oxytocin
Oxytocin is also a polypeptide whose structure is similar to that of vasopressin except that isoleucine occurs in place of phenylalanine, and leucine in place of arginine.

Oxytocin contracts the muscles of the uterus and also stimulates the ejection of milk from the mammary glands. Oxytocin is used in obstetrics when uterine contraction is desired.

Hormones of the Hypothalamus

The hypothalamus secretes certain neurohormones, some of which act as stimulators and others as inhibitors for the secretion of hormones by the anterior pituitary. These hormones are also called **releasing hormones** or **factors.** They have a relatively short life span in the bloodstream. Their half-life is about 2 to 3 min compared to 10 to 15 min for the growth hormone. Therefore, levels of releasing factors are 100 to 1000 times lower in the bloodstream than those of the growth hormone.

The first of the hypothalamic hormones to be isolated (and later synthesized) was the thyrotropin releasing hormone (TRH), also known as the thyrotropin releasing factor (TRF), which controls the release of the thyroid-stimulating hormone (TSH). TRH is a tripeptide containing three amino acids—glutamic acid, histidine, and proline. The structure of TRH is

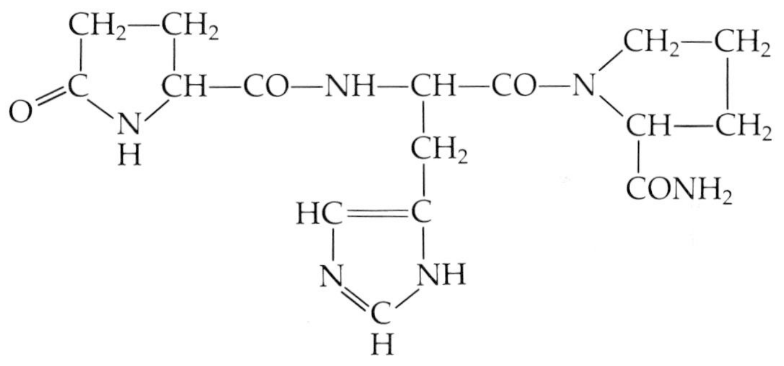

TRH, thyrotropin releasing hormone
(pyroglutamylhistidylproline amide)

TRH is highly specific and causes an increase of TSH within 1 min. It can be administered orally. TRH is used to distinguish between lesions in the pituitary and the hypothalamus. TRH also affects the central nervous system. Synthetic analogues of TRH have been prepared that enhance the effect on the central nervous system while not affecting the thyrotropin-releasing influence.

The second hypothalamic releasing factor that was isolated was the luteinizing hormone releasing hormone (factor), LHRH, which controls the release of the luteinizing hormone from the anterior pituitary. It is a decapeptide (10 amino acid chain), whose amino acid sequence is

Glu-His-Trp-Ser-Tyr-Gly-Leu-Arg-Pro-Gly

Figure 33-5 Structures of some hypothalamic releasing and release-inhibiting hormones.

CRH   Ser-Gln-Glu-Pro-Pro-Ile-Ser-Leu-Asp-Leu-Thr-Phe-His-Leu-Leu-Arg-
       Glu-Val-Leu-Glu-Met-Thr-Lys-Ala-Asp-Gln-Leu-Ala-Gln-Gln-Ala-His-
       Ser-Asn-Arg-Lys-Leu-Leu-Asp-Ile-Ala

GHRH   Tyr-Ala-Asp-Ala-Ile-Phe-Thr-Asn-Ser-Tyr-Arg-Lys-Val-Leu-Gly-Gln-Leu-
       Ser-Ala-Arg-Lys-Leu-Leu-Gln-Asp-Ile-Met-Ser-Arg-Gln-Gln-Gly-Glu-Ser-
       Asn-Gln-Glu-Arg-Gly-Ala-Arg-Ala-Arg-Leu

GHRIH
(somatostatin)   Ala-Gly-Cys-Lys-Asn-Phe-Phe-Trp-Lys-Thr-Thr-Phe-Ser-Cys

PRIH   HO—⟨◯⟩—CH₂—CH₂—NH₂

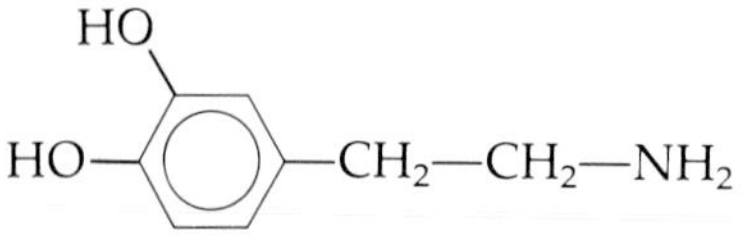

An injection of LHRH causes an increase in circulating LH in 1 to 2 min. LHRH also increases the amount of FSH and so may be identical with the follicle-stimulating hormone releasing hormone (factor), FSHRH.

Another releasing factor is corticotropin (ACTH) releasing hormone CRH, which has recently been synthesized. It contains 41 amino acids, as shown in Figure 33-5. Other releasing factors are growth hormone releasing hormone (factor), GHRH (see Figure 33-5); prolactin releasing hormone, PRH; melanocyte-stimulating hormone releasing hormone (factor), MSHRH.

In addition to the releasing factors, the hypothalamus also contains **release-inhibiting factors**—hormones that inhibit the release of the "releasing factors."

Among the release-inhibiting factors are the growth hormone release inhibiting factor, GHRIH or GIF. GIF, also called somatostatin, is a tetradecapeptide (14 amino acid chain; see Figure 33-5). This factor inhibits the release of the growth hormone. GIF also inhibits the release of insulin (see page 622), glucagon, gastrin, TSH, and FSH.

Other release-inhibiting factors are prolactin release inhibiting factor PRIH or PIF (see Figure 33-5); and follicle-stimulating hormone release inhibiting factor, FSHRIH or FSHRIF. Additional release-inhibiting factors have been postulated but have not been isolated or identified.

The hypothalamus also produces two hormones, which are stored in and secreted by the posterior lobe of the pituitary gland. They are vasopressin, the antidiuretic hormone (ADH), and oxytocin (see page 634).

In addition, the hypothalamus contains polypeptides called **endorphins.** One of these compounds, β-endorphin (see Figure 33-6),

Figure 33-6 Structures of some neurotransmitters.

NH$_2$CH$_2$COOH

glycine

NH$_2$CH$_2$CH$_2$CH$_2$COOH

γ-aminobutyric acid

CH$_2$CH$_2$COOH
|
NH$_2$CHCOOH

glutamic acid

β-endorphin   Tyr-Gly-Gly-Phe-Met-Thr-Ser-Glu-Lys-Ser-Asn-Lys-Phe-Leu-Thr-Val-
Leu-Pro-Thr-Gln-Ala-Ile-Val-Lys-Asn-Ala-His-Lys-Gly-Gln

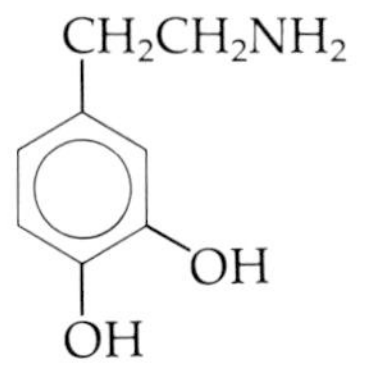

dopamine

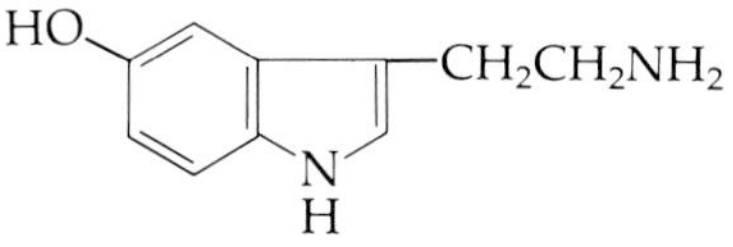

serotonin

exhibits morphinelike activity in the brain and has been used in the treatment of psychiatric patients. The α- and γ-endorphins have not yet been tried on humans, but in rats α-endorphin produces analgesia whereas γ-endorphin induces violent behavior.

A subclass of the endorphins is the **enkephalins,** which are pentapeptides. However, the enkephalins act separately in the body instead of in conjunction with the endorphins.

Two predominant enkephalins in the brain differ only in the end amino acid:

Tyr-Gly-Gly-Phe-Leu   leucine enkephalin
Tyr-Gly-Gly-Phe-Met   methionine enkephalin

Morphine and other opiates (see page 337) bind to opiate receptors in the brain. Endorphins and enkephalins also bind to these same receptors.

The enkephalins are involved with the sensation of pain. It is believed that enkephalins bind to the opiate receptors and prevent the transmission of pain impulses. Morphine also binds to opiate receptors and increases the pain-killing effect of enkephalins.

Opiates such as morphine are addictive. Why? In the cells, the synthesis of cAMP (see page 618) is catalyzed by the enzyme adenylate cyclase. Opiates and enkephalins inhibit the production of this enzyme so that the amount of cAMP in the cells decreases. Therefore, the cells try to compensate by synthesizing more of the enzyme. When this occurs, more opiate must be used in order to have the same pain-killing effect. Eventually, the body will adjust to the new amount of opiate and again cAMP will begin to increase. Thus, the addict requires more and more of the drug.

Conversely, if the opiate is withdrawn, the synthesis of adenylate cyclase is no longer inhibited and large amounts of cAMP are produced. It is the high concentration of cAMP that produces the symptoms of withdrawal.

The increased production of endorphins at about 20 to 30 min into a run may be responsible for a "runner's high" and for the addiction to running experienced by many runners.

It is believed that acupuncture works by triggering nerve impulses that stimulate the release of endorphins and enkephalins.

The Female Sex Hormones

The ovary secretes two different types of hormones. The follicles of the ovary secrete the follicular or estrogenic hormones. The corpus luteum that forms in the ovary from the ruptured follicle secretes the progestational hormones.

Estrogenic or Follicular Hormones

The maturing follicles of the overies produce the estrogenic hormones, which are also called **estrogens**. These hormones are **estradiol, estrone,** and **estriol.** Of these three hormones, estradiol is the parent compound and also the most active. The other two hormones are derived from estradiol. Estriol is the main estrogen found in the urine of pregnant women and also in the placenta. Estrone is in metabolic equilibrium with estradiol. Note the similarities of structures of these three hormones. One international unit (1 IU) of estrogen activity is equal to 0.1 mg of estrone.

Estradiol, estrone, and estriol are concerned with the maturation of the eggs (ova) and the maintenance of the secondary sex characteristics. In lower animals, the estrogens produce estrus, the urge for mating. The estrogens also suppress production of FSH, the follicle-stimulating hormone, by the pituitary gland. FSH initially starts the development of the follicle. Once the follicle begins to develop, further production of FSH is not needed and is inhibited by the estrogenic hormones. However, the estrogenic hormones stimulate the production of LH (luteinizing hormone).

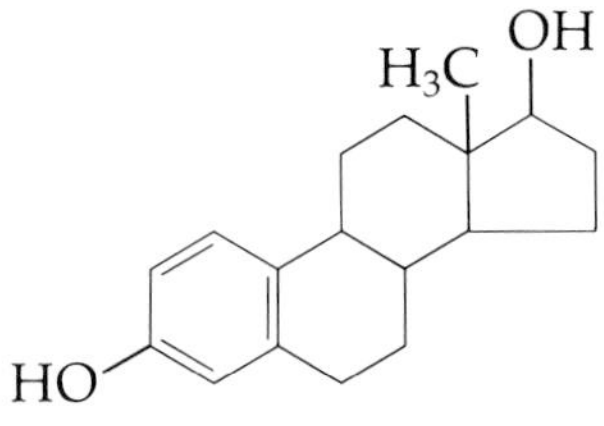

estradiol

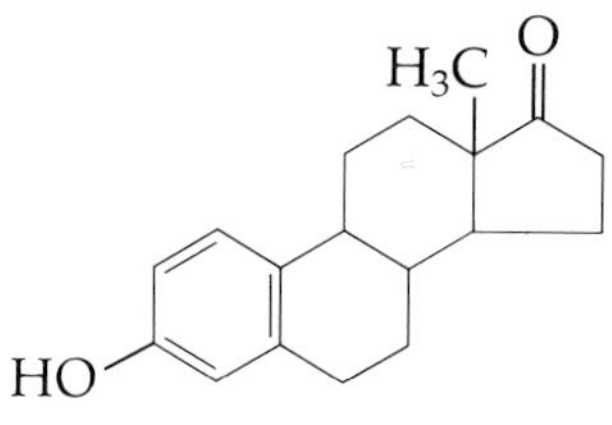

estrone

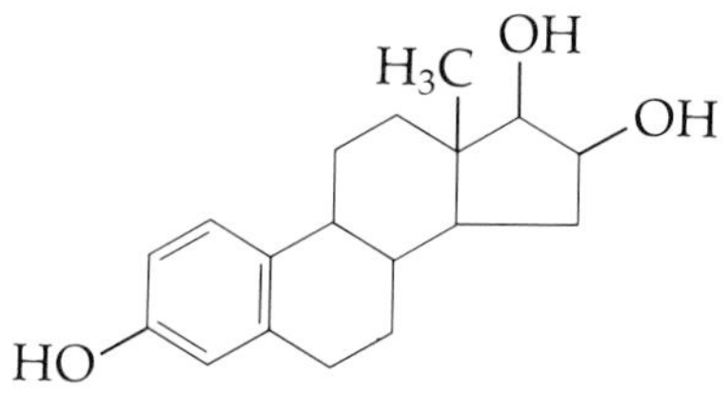

estriol

Synthetic estrogens such as diethylstilbestrol and ethinylestradiol have been developed. These synthetics can be given orally, but naturally occurring estrogens are destroyed in the digestive tract. Estrogens can be used in treatment of underdeveloped female characteristics.

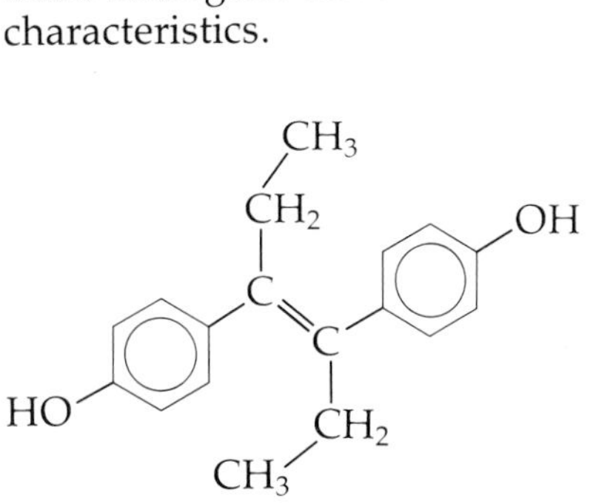

diethylstilbestrol 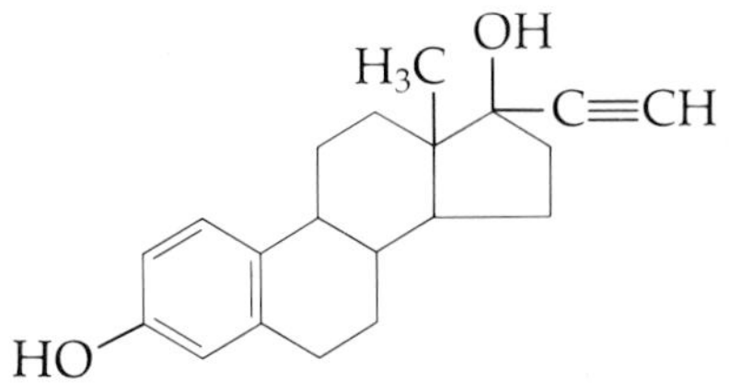   ethinylestradiol

Progestational Hormones

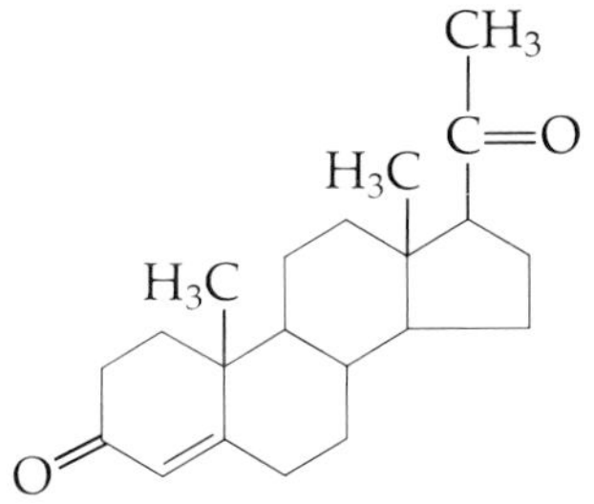

progesterone

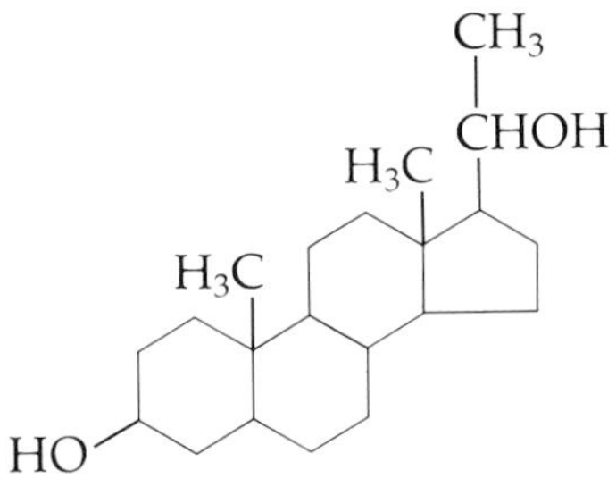

pregnanediol

The corpus luteum is produced in the follicle after the matured ovum is discharged into the uterus. The corpus luteum produces a hormone called **progesterone.** This hormone causes development of the endometrium of the uterus, preparing the uterus to receive and maintain the ovum, and it stimulates the mammary gland. Progesterone inhibits estrus, ovulation, and the production of LH, the hormone that initially stimulated ovulation, which led to the formation and maintenance of the corpus luteum. If the ovum is not fertilized, the corpus luteum breaks down and menstruation follows. If the ovum is fertilized, progesterone from the corpus luteum aids in the development of the placenta.

Progesterone is excreted as **pregnanediol** and is found in the urine.

In addition to progesterone, the corpus luteum also produces another hormone, **relaxin,** which is a polypeptide with a molecular weight of 5521. Relaxin also occurs in the placenta. It causes ligaments of the symphysis pubis to distend, increases dilation of the cervix in pregnant women at parturition, and also helps, along with estrogen and progesterone, to maintain gestation.

The Male Sex Hormones

Male hormones are produced primarily in the testes, although small amounts are also produced in the adrenal glands.

The male hormones are called **androgens.** They consist of testosterone and dihydrotestosterone, with the former being the major steroid secreted by the adult testes. The principal metabolic product of testosterone is dihydrotestosterone (DHT), the active form of the hormone in many tissues.

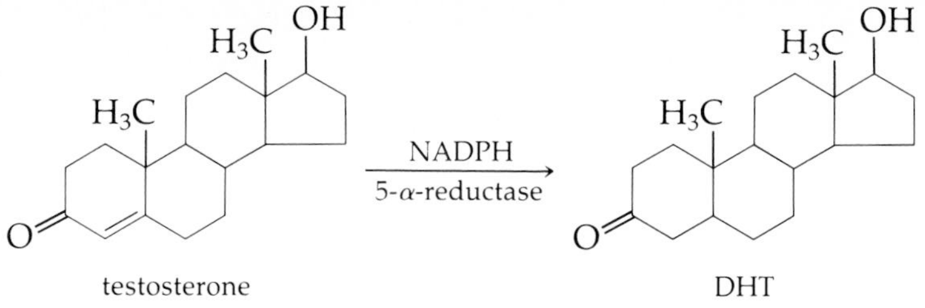

testosterone   DHT

The male hormones, mainly testosterone and dihydrotestosterone, are involved in (1) spermatogenesis, (2) development of the male sex organs, (3) development of secondary male characteristics, (4) male-pattern behavior, and (5) gene regulation. Note that the structures of the male sex hormones are similar to those of the female sex hormones. Both are derived from a common substance, cholesterol.

A deficiency in testosterone synthesis is called hypogonadism. If this occurs before puberty, secondary male characteristics fail to develop. If it occurs in the adult, many of the male characteristics regress.

Hormones of the Kidney

In addition to their excretory function, the kidneys act as endocrine glands. Among the hormones produced by the kidneys are **renin** (see page 444), which increases the force of the heartbeat and constricts the arterioles, and **erythropoietin** and **erythrogenin**, which cause the bone marrow to stimulate the production of red blood cells. The kidneys play an important role in the activation of vitamin D (see page 597). The kidneys also produce prostaglandins PGE_1, PGE_2, PGE_{20}, and possibly others. PGE_1 antagonizes the effects of vasopressin; PGE_2 increases excretion of sodium; and PGE_{20} decreases venous tone.

Hormones That Regulate Calcium Metabolism

Calcium ions regulate a number of body processes. Among these are coagulation of blood, enzyme reactions, secretory processes, neuromuscular excitability, plasma membrane transport, bone and teeth mineralization, recovery of the visual cycle, and release of hormones and neurotransmitters. To ensure that these processes function normally, calcium ions must be controlled within very narrow limits.

Plasma calcium exists in three forms—complexes with organic acids (6 percent); protein bound, mostly with albumin (47 percent); and ionized calcium ion (47 percent).

Three hormones control the body's calcium. They are the parathyroid hormone (see page 626), calcitriol, and calcitonin (see page 625).

Calcitriol stimulates intestinal absorption of calcium and phosphates. It is made from vitamin D_3 in the liver and kidneys. Thus a lack of calcitriol is due to a lack of vitamin D_3 and is another aspect of deficiency of that vitamin (see page 598).

vitamin D_3 calcitriol

Neurotransmitters

Hormones carry chemical messages from the endocrine glands to the specific body part that those hormones affect. Neurotransmitters carry chemical messages from one nerve cell to another. While hormones can travel a great distance in the body, neurotransmitters travel very short distances from one neuron to the neighboring one. While hormones are produced in the endocrine glands, neurotransmitters are produced in the neurons.

Some neurotransmitters are simple amino acids, some are polypeptides, and others are in a group called catecholamines (see Figure 33-6). When a nerve impulse travels along a neuron, it comes to a synapse, a space between that neuron and the adjacent one. The

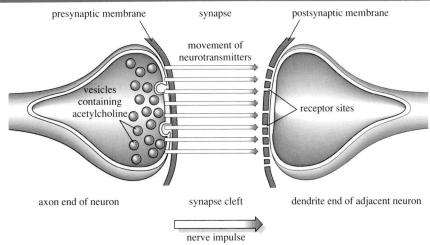

Figure 33-7 Transmission of nerve impulse between neurons.

presynaptic membrane   synapse   postsynaptic membrane

movement of neurotransmitters

vesicles containing acetylcholine

receptor sites

axon end of neuron   synapse cleft   dendrite end of adjacent neuron

nerve impulse

impulse cannot cross that gap. A neurotransmitter is needed to carry the impulse across (see Figure 33-7).

When the impulse reaches the presynaptic membrane, that membrane becomes depolarized and allows calcium ions in rapidly. The calcium ions cause the fusion of the vesicle membranes with the presynaptic membrane, thereby releasing the contents of those vesicles, the neurotransmitters. After the release of the neurotransmitters, the vesicles reform rapidly.

The neurotransmitters travel across the synaptic cleft and bind to receptor sites on the postsynaptic membrane, where they alter the permeability of that membrane to sodium and potassium ions and so produce a new impulse that can travel along that neuron.

The neurotransmitter must be removed or deactivated so further transmission of nerve impulses can occur. For example, acetylcholine, a neurotransmitter, is deactivated by the enzyme choline acetyltransferase and forms choline and acetic acid. Monoamine neurotransmitters such as serotonin, epinephrine, norepinephrine, and dopamine are deactivated by the oxidoreductase monoamine oxidase (MAO).

MAO inhibitors have been used in the treatment of hypertension and depression, but great care must be taken with their use.

Summary

Hormones are produced in the body's endocrine glands and are secreted directly into the bloodstream, which carries them to the body parts on which they produce their effects.

Cyclic adenosine monophosphate (cAMP) acts as a messenger that instructs the cells to respond to a particular hormone.

The pyloric mucosa produces gastrin, which stimulates the secretion of hydrochloric acid and also produces greater motility of the stomach.

The mucosa of the small intestine produces secretin, which stimulates the pancreas to release the pancreatic juices, and cholecystokinin, which stimulates the contraction and emptying of the gallbladder.

The pancreas secretes insulin, which increases the rate of oxidation of glucose and also facilitates the conversion of glucose to glycogen in the liver and the muscles.

The pancreas also secretes glucagon, which causes an increase in blood sugar content by stimulating phosphorylase activity in the liver.

The hormones of the thyroid gland regulate the metabolism of the body and also affect the growth and development of the body. The thyroid hormones contain the element iodine. The hormonal iodine carried by the blood is termed protein-bound iodine, or PBI. Hypothyroidism occurs when the thyroid gland does not manufacture sufficient thyroxine. The symptoms of hypothyroidism are sluggishness, weight gain, slower heartbeat, reduced metabolic rate, and loss of appetite.

If the thyroid gland is absent or fails to develop in an infant, the effect produced is called cretinism. If the thyroid gland in an adult should atrophy, the effect produced is called myxedema. If the thyroid gland enlarges, the condition is called simple goiter.

If the thyroid gland produces excess thyroxine, hyperthyroidism results; the symptoms are increased metabolic rate, bulging of the eye, nervousness, weight loss, and a rapid, irregular heartbeat. Certain drugs may be given to counteract the effects of an overactive thyroid gland.

Another hormone of the thyroid gland, calcitonin, keeps the calcium concentration of the blood from becoming too high by inhibiting the release of calcium ions from the bone.

The parathyroid glands produce a hormone, parathyroid hormone, which influences the metabolism of calcium and phosphorus in the body.

The adrenal glands are divided into two portions: the cortex and the medulla.

The hormones of the adrenal cortex are divided into three categories: the glucocorticoids, which primarily affect the metabolism of carbohydrate, fat, and protein; the mineralocorticoids, which primarily affect the transportation of electrolytes and the distribution of water in the body; and the androgens or estrogens, which primarily affect secondary sex characteristics. A hypofunctioning of the adrenal glands produces Addison's disease, which is characterized by an excessive loss of NaCl in the urine, low blood pressure, elevation of serum potassium, muscular weakness, brownish pigmentation of the skin, nausea, and loss of appetite.

The medulla of the adrenal glands secretes epinephrine and norepinephrine. Norepinephrine is a precursor of epinephrine. Epinephrine relaxes the smooth muscles, elevates blood pressure, and causes glycogenolysis in the liver.

The pituitary gland has three lobes: anterior, intermediate, and posterior. The flow of hormones from the pituitary gland is under the control of the hypothalamus.

The hormones of the anterior lobe of the pituitary gland are the growth hormone (GH), the thyrotropic hormone (TSH), the adrenocorticotropic hormone (ACTH), lactogenic hormone (LTH), luteinizing hormone (LH), and follicle-stimulating hormone (FSH).

The posterior lobe of the pituitary secretes the hormones vasopressin and oxytocin. Vasopressin stimulates the kidneys to reabsorb water. Oxytocin contracts the muscles of the uterus and stimulates the ejection of milk from the mammary glands.

The pars intermedia (intermediate lobe) of the pituitary gland secretes the hormone intermedin, which increases the deposition of melanin in the skin.

The hypothalamus secretes neurohormones, some of which stimulate and others of which inhibit the secretion of hormones by the anterior pituitary.

The ovaries secrete two different types of hormones. The follicles secrete the follicular or estrogenic hormones. The corpus luteum that forms in the ovary from the ruptured follicle secretes the progestational hormones.

The male sex hormones are produced in the testes. The principal male hormone is testosterone. Both male and female sex hormones are derived from cholesterol.

The hypothalamus also contains endorphins and enkephalins, which have morphinelike activity.

The kidneys, in addition to their excretory function, also act as endocrine glands.

Calcium ions regulate many body processes. Three hormones that regulate calcium are the parathyroid hormone, calcitriol, and calcitonin.

Neurotransmitters carry chemical messages from one nerve cell to another.

Questions and Problems

1. Compare hormones with vitamins.
2. In what different ways can hormones act?
3. Where is gastrin produced? What is its function?
4. Where is secretin formed? What is the function of secretin? of CCK?
5. What hormones are produced by the pancreas?
6. Insulin contains what metallic element?
7. Why can insulin not be taken orally?
8. What is protamine zinc insulin?
9. What are the functions of insulin in carbohydrate metabolism?
10. What is hyperglycemia?
11. Will insulin cure diabetes mellitus? Why?
12. What might cause hypoglycemia? What will relieve such a symptom?
13. What is the function of glucagon?
14. What unusual element is found in the thyroid gland?
15. List the symptoms of diabetes mellitus.
16. List the symptoms of hyperthyroidism.
17. What is a cretin? How may this defect be overcome?
18. What causes myxedema?
19. What are the symptoms of a goiter?
20. What are the symptoms of hypothyroidism?
21. What function does the parathyroid hormone perform?
22. What are the symptoms of hypoparathyroidism? of hyperparathyroidism?
23. What three types of hormones are produced by the adrenal cortex? What is their function? Give an example of each.
24. What causes Addison's disease?
25. Indicate the symptoms of a lack of the adrenocortical hormones.
26. What are the effects of hyperadrenocorticism?
27. What are the hormones of the medulla of the adrenal glands? What is the function of each?
28. List the hormones of the anterior lobe of the pituitary gland. What is the function of each?
29. Compare dwarfism with cretinism.
30. What are the symptoms of an overactive anterior lobe of the pituitary gland?
31. What are the hormones of the posterior lobe of the pituitary gland? What is the function of each?
32. What is MSH? Where is it formed? What is its function?
33. What types of hormones are produced by the hypothalamus? What do they do?
34. What type of structure do the hypothalamic hormones have?
35. Where are the estrogenic hormones produced? the progestational hormones?
36. Where are the male sex hormones produced?
37. Why is the pituitary gland no longer considered to be "the master gland" of the body?
38. Where are vasopressin and oxytocin produced? stored?
39. How do enkephalins prevent transmission of pain?
40. Why are opiates addictive?
41. Why does acupuncture work?
42. What hormones are produced by the kidneys? What are their functions?
43. What role does cAMP play in the cells? cGMP?

Practice Test

1. A hormone involved in carbohydrate metabolism is _____.
   a. calcitriol   b. insulin
   c. epinephrine   d. serotonin
2. An example of a neurotransmitter is _____.
   a. glucagon   b. secretin
   c. progesterone   d. epinephrine
3. A hormone that contains the element iodine is _____.
   a. somatostatin   b. calcitriol
   c. insulin   d. T_3
4. A symptom of hyperthyroidism is _____.
   a. exophthalmos   b. myxedema
   c. Hashimoto's   d. cancer
5. The hypothalamus produces _____.
   a. oxytocin   b. TSH
   c. MSH   d. calcitonin
6. Male hormones are called _____.
   a. melatonins   b. androgens
   c. estrogens   d. adrenergic
7. A hormone involved in fluid electrolyte balance is _____.
   a. atriopeptin   b. insulin
   c. oxytocin   d. thyroxin
8. The pineal gland produces which hormone?
   a. somatostatin   b. androgen
   c. melatonin   d. ACTH
9. Early sexual development may be due to an overproduction of _____.
   a. ACTH   b. oxytocin
   c. adrenaline   d. intermedin
10. Which of the following is *not* a hormone of the anterior lobe of the pituitary gland?
    a. GH   b. MSH   c. TSH   d. LTH

34

Heredity

A person's genetic makeup is determined by the genes of the parents.

A human being grows from one fertilized egg cell. As this organism divides and grows, it develops gradually into recognizable form. How do the cells "know" where the head should be, where the arms should be, that there should be two arms and not one or three? Why not claws instead of fingers, scales instead of skin?

Originally it was believed that the chromosomes somehow transmitted information from one generation to the next. The next step forward in the study of heredity was the theory that the chromosomes contained genes—each gene giving a particular characteristic to that individual; that is, there were genes for tallness or shortness, for blue eyes or brown eyes, for light hair or dark hair.

Recent studies have shown that genes are composed of deoxyribonucleic acid (DNA). In 1953 J. D. Watson (see Figure 34-1) and F. H. C. Crick of Cambridge University proposed a double helix structure for the DNA molecule. Before we discuss this structure and explain what a double helix is, let us review a little about nucleic acids.

Nucleic Acids

Figure 34-1 Dr. J. D. Watson, Harvard University, who was awarded the Nobel Prize for Medicine and Physiology in 1962, along with Dr. F. H. C. Crick and Dr. M. H. F. Wilkins, for the discovery of the molecular structure of deoxyribonucleic acid (DNA).

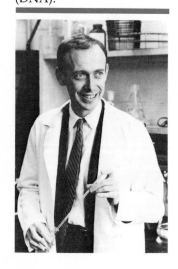

Nucleic acids were first isolated from the cellular nucleus, hence the name. Nucleic acids are macromolecules, huge polymers with molecular weights of over 100 million.

There are two main types of nucleic acids, **deoxyribonucleic acid (DNA)** and **ribonucleic acid (RNA)**. DNA is primarily responsible for the transfer of genetic information whereas RNA is primarily concerned with the synthesis of protein.

Hydrolysis of nucleic acids gives nucleotides, which can be considered the units that make up the polymer. Nucleotides can be further hydrolyzed to nucleosides and phosphoric acid. Hydrolysis of a nucleoside yields a sugar and a heterocyclic base.

$$\text{nucleic acid} \rightleftharpoons \text{nucleotide} \rightleftharpoons \text{nucleoside} + \text{phosphoric acid} \rightleftharpoons \text{heterocyclic base} + \text{sugar}$$

The sugar in both types of nucleic acids is a pentose. In RNA the sugar is ribose, whereas in DNA it is deoxyribose. The prefix *deoxy-* means "without oxygen." Note in the following structural diagrams that deoxyribose contains one less oxygen atom than ribose.

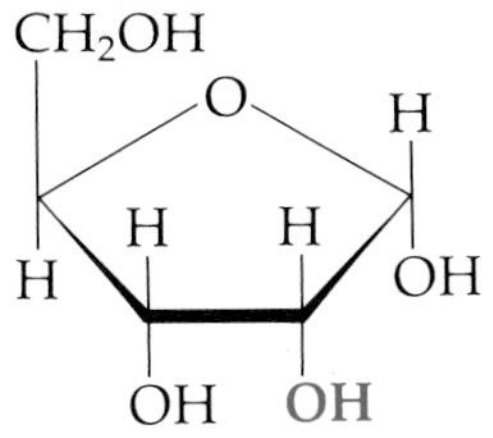

[Structural diagrams of ribose and deoxyribose sugars]

ribose 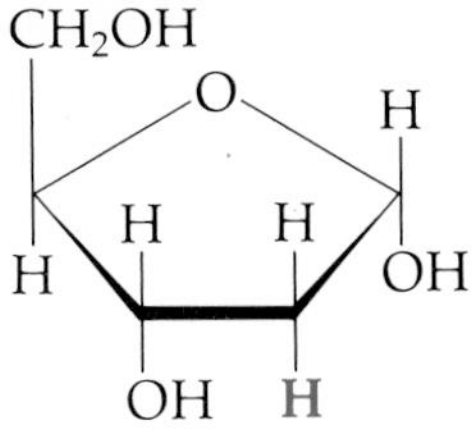   deoxyribose

Heterocyclic Bases

The heterocyclic bases present in nucleic acids are divided into two types—purines and pyrimidines. The two purines present in both DNA and RNA are **adenine** and **guanine**. The pyrimidine **cytosine** is present in both DNA and RNA, whereas **thymine** is found in DNA only and **uracil** is present in RNA only. The structures of these heterocyclic bases are indicated in the following sections.

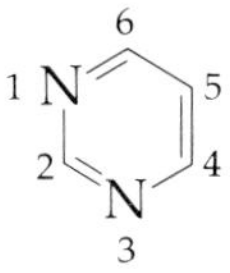

The Pyrimidines

pyrimidine

Pyrimidine is a six-membered heterocyclic ring containing two nitrogen atoms. Three important derivatives of pyrimidine found in nucleic acids are thymine (2,4-dioxy-5-methylpyrimidine), cytosine (2-oxy-4-aminopyrimidine), and uracil (2,4-dioxypyrimidine). Their structures are as follows:

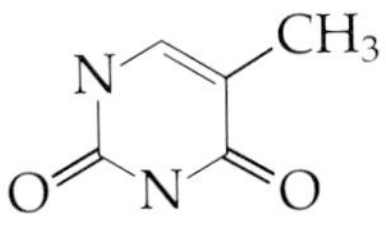

thymine 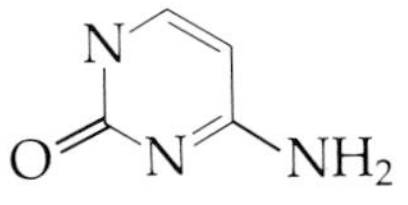   cytosine 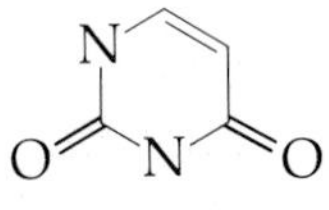   uracil

Other important compounds containing pyrimidines are thiamin (vitamin B_1) and the barbiturates.

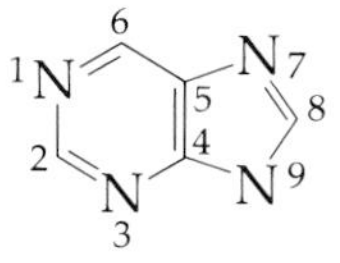

The Purines

purine

The purines that are found in nucleic acids are derivatives of a substance, purine, that does not occur naturally. As indicated by their structures, adenine is 6-aminopurine and guanine is 2-amino-6-oxypurine.

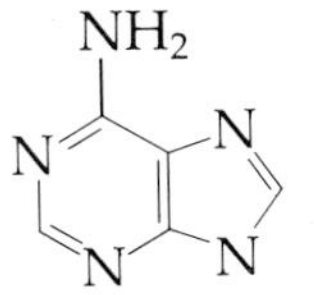

adenine 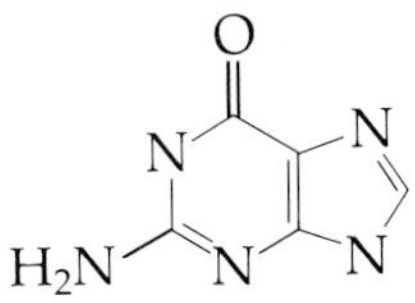   guanine

Other purines include caffeine and theophylline. Caffeine is a stimulant for the central nervous system and also a diuretic. Caffeine is found in coffee and tea. Its chemical name is 1,3,7-trimethyl-2,6-dioxypurine. Theophylline, 1,3-dimethyl-2,6-dioxypurine, is found in tea and is used medicinally as a diuretic and for bronchial asthma.

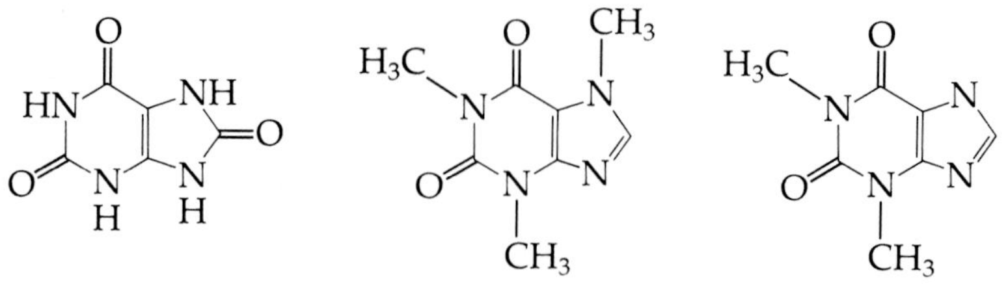

uric acid	caffeine	theophylline

The Adenosine Phosphates

One important nucleotide is adenosine monophosphate (AMP). It is formed by the reaction of adenosine (a nucleoside) with one molecule of phosphoric acid. If two phosphate groups react with adenosine, adenosine disphosphate (ADP) is formed; ATP, adenosine triphosphate, is formed when three phosphate groups react.

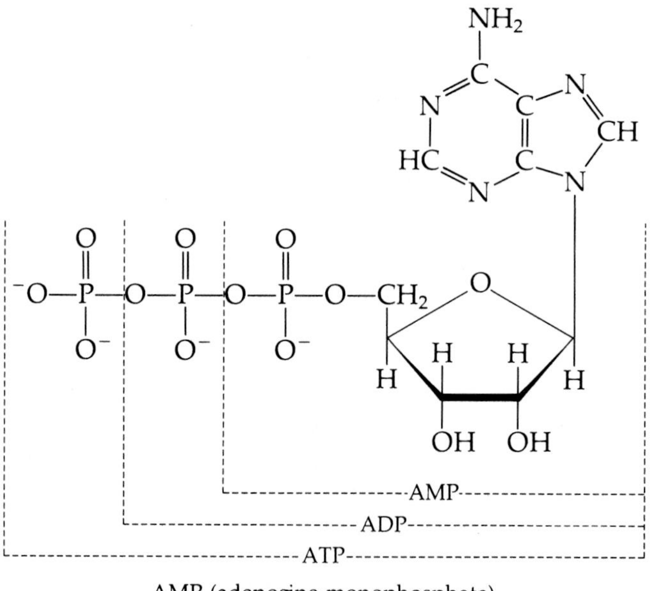

AMP (adenosine monophosphate)
ADP (adenosine diphosphate)
ATP (adenosine triphosphate)

ATP, ADP, and AMP are involved in various metabolic processes involving the storage and release of energy from their phosphate bonds.

Figure 34-2 Metabolism of nucleoproteins.

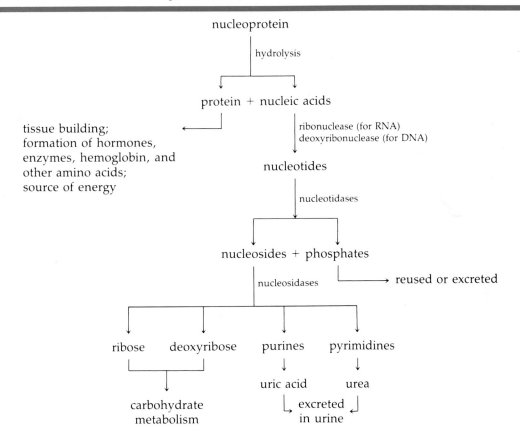

Metabolism of Nucleoproteins

Nucleoproteins are composed of proteins conjugated with nucleic acids. DNA and RNA, two nucleic acids, are essential constituents of chromosomes, viruses, and cell nuclei. Figure 34-2 illustrates the hydrolysis products of nucleoprotein.

Ribose and deoxyribose enter the normal carbohydrate metabolic pathway. The phosphates either are used to prepare new phosphate compounds or are excreted in the urine. The purines are converted through a series of reactions to uric acid, which is eliminated in the urine. The reactions are as follows.

The uric acid concentration in the blood is increased during the disease called gout (see pages 528 and 542). In this disease, uric acid crystals are deposited in and around the joints and in the cartilage, chiefly in the large toe and in the ear cartilage. Uric acid crystals may also be deposited in the form of gallstones or kidney stones.

The pyrimidines are eliminated in the urine in the form of urea:

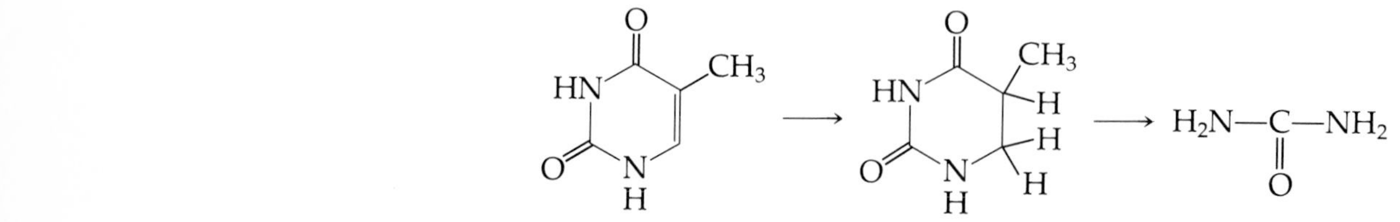

thymine (a pyrimidine) → dihydrothymine → urea

$H_2N—C—NH_2$

The Watson–Crick Model of DNA

The Watson–Crick model of DNA proposed a double-coiled chain consisting of two strands intertwined around one another. The sugar and phosphate groups form the backbone of the strands, and the heterocyclic groups form the connecting links. Watson and Crick also proposed that an adenine of one chain was bonded to a thymine of the opposite chain and a guanine of one chain was bound to a cytosine of the other chain.

The Watson–Crick model of DNA can also be shown as indicated in Figure 34-3, as a spiral ladder with the two sides held together by rungs consisting of heterocyclic nitrogen compounds (adenine, thymine, guanine, and cytosine).

If the chain is untwisted and straightened out, it can be represented as follows:

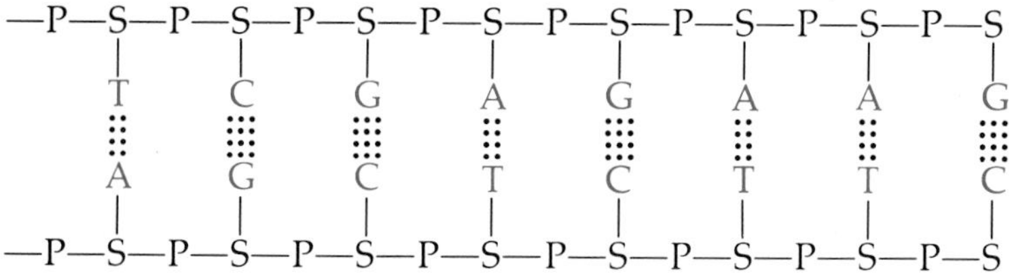

—P—S—P—S—P—S—P—S—P—S—P—S—P—S—P—S

T   C   G   A   G   A   A   G

A   G   C   T   C   T   T   C

—P—S—P—S—P—S—P—S—P—S—P—S—P—S—P—S

The solid lines indicate ordinary chemical bonding, and the dotted lines indicate hydrogen bonding. Recall that hydrogen bonds are much weaker than ordinary chemical bonds. Note that there are two hydrogen bonds between adenine and thymine and three hydrogen bonds between guanine and cytosine.

Figure 34-3 Structure of DNA. P is phosphate; S, sugar; G, guanine; C, cytosine; A, adenine; and T, thymine.

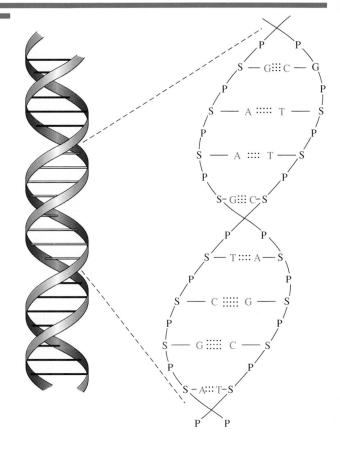

The DNA Code

All DNA molecules have the same sequence of deoxyribose and phosphates in the ladder part of the chain. The difference lies in the order of the adenine, thymine, cytosine, and guanine parts of the chain. This difference in sequence of the heterocyclic compounds constitutes the genetic code.

This code consists of only four letters—A, T, G, and C—representing adenine, thymine, guanine, and cytosine, respectively. In the DNA molecule these letters of the code are grouped in threes (which we shall discuss later). Thus the code might be ATC, TAC, AAA, CCT, and so on.

Simple bacterial viruses contain about 5500 nucleotides in their DNA molecules; 5500 nucleotides grouped in threes can produce 5500/3 or approximately 1800 coded pieces of information. These 1800 coded pieces of information are sufficient to describe that particular virus. If each piece of information corresponded to one letter of our alphabet, it would take 1800 letters to describe, genetically, that virus; 1800 letters corresponds roughly to the number of letters on one page of this book.

One DNA molecule in a human contains approximately 5,000,000,000 nucleotides, or enough to form 1,700,000,000 coded pieces of information (of three nucleotides to a group). If each of these coded pieces corresponded to a letter of our alphabet, it would take approximately 2500 volumes, each the size of this book, to describe a human being genetically.

Replication of DNA

When a cell divides, it produces two new cells with identical characteristics. This means that the DNA originally present must duplicate (or replicate) itself. How does the DNA molecule direct the synthesis of another identical DNA molecule?

Watson and Crick theorized that the DNA molecule unwinds, and each half acts as a template for nucleotide units to collect on and form a new chain. Recall that if adenine (A) is present on one chain, it can attract and hold only thymine (T), and cytosine (C) can attract and hold only guanine (G). Thus each half of the chain is highly specific in what it attracts. It forms complementary chains that coil up again and form two new DNA molecules.

This can be represented diagrammatically, as shown in Figure 34-4. This representation is necessarily quite general and approximate. The actual mechanism of DNA unwinding and untwisting for replication is quite complex, involving several enzymes such as DNA polymerase and DNA ligase, and is beyond the scope of this book.

Transfer of Information

The preceding paragraph indicates how the DNA molecule duplicates itself, but how does the DNA molecule transfer its coded information to the cell's ribosomes where the actual production of the protein called for by the code takes place?

The information contained in the DNA molecule is carried by another molecule called messenger RNA (mRNA). During replica-

Figure 34-4   Unwinding of DNA.

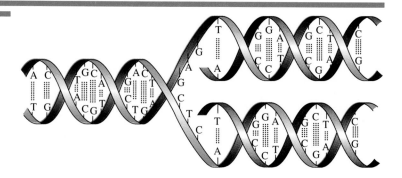

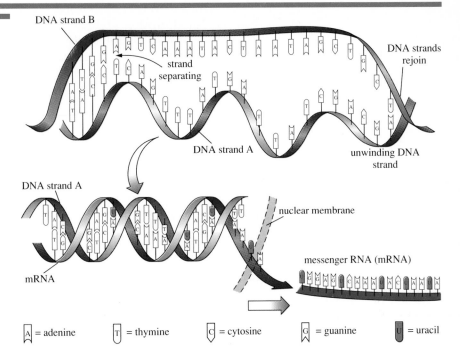

Figure 34-5 Formation of messenger RNA from DNA.

$\boxed{A}$ = adenine $\boxed{T}$ = thymine $\boxed{C}$ = cytosine $\boxed{G}$ = guanine $\boxed{U}$ = uracil

tion, DNA "unzips" and synthesizes two new chains to produce two identical molecules. When the DNA molecule synthesizes mRNA, the DNA again "unzips"; however, only one part of one strand of DNA acts as a template for the information of mRNA (see Figure 34-5). The enzyme required is called RNA polymerase. The process is known as transcription.

When mRNA is produced, there is one major difference in the attraction of nucleotides between it and DNA. In RNA, the partner of adenine (A) is uracil (U) and not thymine (T). Another difference is that RNA contains the pentose ribose, whereas DNA contains the pentose deoxyribose.

There are at least three types of RNA: (1) messenger RNA, (2) ribosomal RNA, and (3) transfer RNA. **Messenger RNA (mRNA)** is concerned with the transmission of genetic information from DNA to the site of protein synthesis, the ribosomes; it is found in the nucleus and the cytoplasm of the cell. **Ribosomal RNA (rRNA)**, which is the major fraction of the total RNA, combines with protein to form the ribosomes; therefore, ribosomes are an example of nucleoprotein. It is in the ribosomes that the "translation" of the DNA message takes place. **Transfer RNA (tRNA)** holds a specific amino acid for incorporation into a protein molecule, as indicated in the following paragraphs. A fourth type of RNA, heterogeneous RNA (hRNA), may be a precursor for mRNA. hRNA is found in the nucleus of the cell.

Let us see how mRNA and tRNA function in the ribosomes in the synthesis of a protein. The mRNA moves from the nucleus to the

Figure 34-6 Diagrammatic forms of transfer RNA (tRNA).

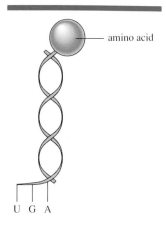

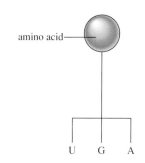

cytoplasm and then to the ribosomes, where it acts as a template for the formation of protein. In the cytoplasm is tRNA, which is a spiral form of RNA containing relatively few nucleotides. At one end of the tRNA are three nucleotides that are specific for a certain code on the mRNA. Amino acids in the cytoplasm are activated by ATP and are coupled to the other end of tRNA. This can be illustrated as shown in Figure 34-6.

If mRNA contains the coded units ACU, the tRNA that would attach itself must have the code UGA. [Recall that in RNA the adenine (A) is bonded to uracil (U) and cytosine (C) to guanine (G).] If the coded message in mRNA is UGC, the corresponding code in tRNA must be ACG:

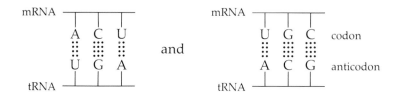

As the mRNA travels across the ribosome, the first coded group (of three letters) picks up and holds a corresponding tRNA. As the second coded group of the mRNA passes, it too picks up a corresponding tRNA. Attached to the opposite end of each tRNA is an amino acid. The amino acid attached to the end of the second tRNA becomes bonded to the amino acid at the end of the first tRNA. The amino acid at the end of the third tRNA in turn becomes bonded to the second amino acid.

Thus, as each tRNA attaches itself to the mRNA, the tRNA gives up its amino acid to form a chain of amino acids, or a protein. After giving up its amino acid, the tRNA leaves the mRNA and goes in search of another amino acid that it can pick up and use to repeat the sequence.

Thus, as the mRNA passes through the ribosome, it directs the gathering, in specified sequence, of the tRNAs, which in turn bond together their amino acids to form a protein.

This effect can be illustrated diagrammatically as shown in Figure 34-7.

In these diagrams, the first coded group in the mRNA—group CAG—attracts tRNA coded GUC. Attached to tRNA coded GUC is an amino acid that we have simply labeled 1. As the mRNA passes further along into the ribosome, the next coded group, code UUA, attracts tRNA coded AAU. The amino acid attached to the end of this tRNA is labeled 2. Amino acid 2 bonds itself to amino acid 1 as illustrated. At the same time, the tRNA coded GUC that held amino acid 1 goes off in search of another amino acid labeled 1 so it can be used again whenever the code calls for it. The third coded group in the mRNA, code UGA, attracts tRNA, coded ACU, with its attached amino acid 3. Then amino acid 3 bonds to amino acid 2, which is

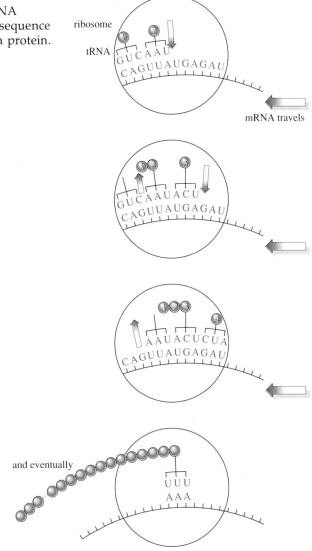

Figure 34-7   mRNA directs the proper sequence of tRNAs to form a protein.

already bonded to amino acid 1. So the chain of amino acids begins and continues as illustrated in the fourth figure. This process continues until the end of the coded groups of the mRNA. At this time, the protein is complete and moves out of the ribosome into the cytoplasm.

The Triplet Code

The DNA code consists of four letters, A, T, C, and G, grouped in threes. It is believed that the code in the DNA molecule specifies individual amino acids. Since there are 20 primary amino acids that occur in nature, there must be at least 20 different coded groups in the DNA molecule.

The 20 primary amino acids and their abbreviations are as follows:

alanine	Ala	glycine	Gly	proline	Pro
arginine	Arg	histidine	His	serine	Ser
asparagine	Asn	isoleucine	Ile	threonine	Thr
aspartic acid	Asp	leucine	Leu	tryptophan	Trp
cysteine	Cys	lysine	Lys	tyrosine	Tyr
glutamic acid	Glu	methionine	Met	valine	Val
glutamine	Gln	phenylalanine	Phe		

Decoding the Code

How can the DNA code be decoded? That is, how can we tell which amino acid is specified by a certain coded group?

In 1961 it was found that if a synthetic RNA composed only of uracil nucleotides was substituted for mRNA in a protein-synthesis system, a polypeptide was formed that contained only the amino acid phenylalanine. Since the synthetic mRNA contained only uracil, it must have the code group UUU. The corresponding group in the DNA molecule must be AAA. Thus we can say that the coded group AAA in the DNA molecule corresponds to the code UUU in mRNA, which in turn specifies the amino acid phenylalanine. Another

Table 34-1 Genetic Code in mRNA

First Letter	Second Letter				Third Letter
	A	C	G	U	
A	Lys	Thr	Arg	Ile	A
	Asn	Thr	Ser	Ile	C
	Lys	Thr	Arg	Met[b]	G
	Asn	Thr	Ser	Ile	U
C	Gln	Pro	Arg	Leu	A
	His	Pro	Arg	Leu	C
	Gln	Pro	Arg	Leu	G
	His	Pro	Arg	Leu	U
G	Glu	Ala	Gly	Val	A
	Asp	Ala	Gly	Val	C
	Glu	Ala	Gly	Val	G
	Asp	Ala	Gly	Val	U
U	[a]	Ser	[a]	Leu	A
	Tyr	Ser	Cys	Phe	C
	[a]	Ser	Trp	Leu	G
	Tyr	Ser	Cys	Phe	U

[a] Chain terminator.
[b] Chain initiator.

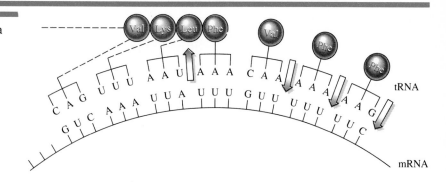

Figure 34-8 Synthesis of a polypeptide from mRNA.

synthetic mRNA, consisting solely of adenine nucleotides, produced a polypeptide containing only lysine. Thus the mRNA code for lysine is AAA and the corresponding DNA code must be TTT.

Table 34-1 indicates the amino acids and their coded groups.

Some experiments indicated that there may be more than one code group for a specific amino acid. That is, alanine is indicated by the mRNA codes GCU, GCC, GCA, and also GCG. Note that all of these code groups begin with the letters "GC." In general, the first two "letters" of the code are more important than the third.

The genetic code is universal. That is, all known organisms use this same code for the synthesis of their proteins.

Figure 34-8 indicates the synthesis of a polypeptide chain from mRNA, indicating the various amino acids specified by that mRNA.

Regulation of Protein Synthesis

When mRNA acts as a template for the synthesis of protein, only a small amount of the total information in DNA is used at one time to produce a certain type of protein. What determines whether mRNA is formed from a certain segment of the DNA? What "turns on" the DNA and what "turns it off"?

Since protein is not synthesized continuously but only as needed, DNA must normally be in a "repressed state." A repressor, which is a polypeptide, binds to a small segment of the DNA. This segment is called the operator site. As long as the repressor is bonded to the DNA, no mRNA is produced and so production of protein is inhibited. When a particular protein is needed, an inducer is formed. The inducer combines with the repressor, changing its shape so that it can no longer bind to the DNA. Once the repressor is removed from the DNA, synthesis of mRNA and hence protein can begin. When sufficient protein has been produced, the inducer is removed and the repressor once again binds to the DNA, stopping protein synthesis.

Certain drugs, such as tetracyclines and streptomycin, bind to the ribosomes of bacteria and prevent synthesis of protein. Hence these drugs are effective antibiotics.

Mutations

Suppose that one of the letters (nucleotides) in a DNA code group was substituted by another letter. Then the message would be miscopied and a mutation would occur. Such a change in the sequence of nucleotides may (1) give no detectable effect, particularly if the change is in the third letter of the code (see page 656); (2) cause a different amino acid to be incorporated into the chain (the results may be acceptable, partially acceptable, or totally unacceptable to the function of that protein); or (3) cause the termination of the chain prematurely so that the protein cannot function normally.

A mutation will also occur if one of the code letters is omitted or if one is added or if the order of the code letters is rearranged. Mutations occur because of exposure to radiation (in industry, medically, because of naturally occuring radiation and cosmic rays, or because of fallout) and possibly because of certain chemicals.

Genetic Diseases

Most genetic diseases are caused by a defective gene, which results in a loss of activity of some enzyme. Even though the body has thousands of enzymes, the loss of only one may be disastrous.

Consequences of Inherited Enzyme Deficiency

Consider a simple enzymatic reaction in which compound X is changed to compound Y under the influence of enzyme x_1, and then compound Y is changed to compound Z under the influence of enzyme y_1.

$$X \xrightarrow{x_1} Y \xrightarrow{y_1} Z$$

If enzyme y_1 is deficient, one of the following consequences might occur:

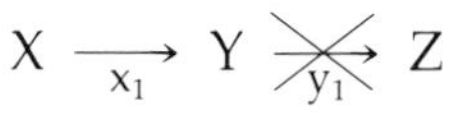

$$X \xrightarrow{x_1} Y \not\xrightarrow{y_1} Z$$

a. The body would be lacking in compound Z, with the resulting effect of that lack.

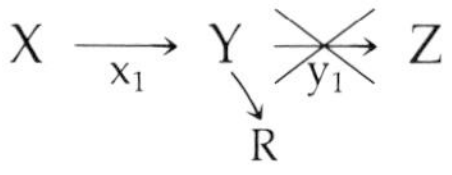

$$X \xrightarrow{x_1} Y \not\xrightarrow{y_1} Z$$
with Y → R

b. Compound Y might be changed to a harmful by-product R.

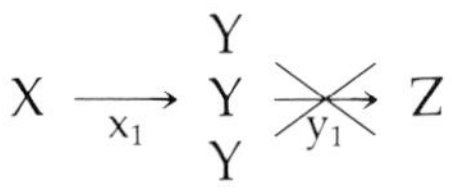

$$X \xrightarrow{x_1} Y \not\xrightarrow{y_1} Z$$

c. Compound Y might accumulate and could be harmful if present in large amounts.

In these cases, the genetic disease is caused by a lack of enzyme y_1, which in turn is due to a defective gene—one that might have been copied incorrectly, or changed, or even be lacking altogether. We shall see in the following paragraphs how a single simple change in the sequence of a code group in mRNA can lead to a genetic defect.

Common Genetic Diseases

The most common genetic disease in the United States is cystic fibrosis, which affects 1 individual in 2000. Another genetic disease, phenylketonuria, affects 1 in 10,000. Most states now require routine screening of newborn infants for this latter disease. Tay–Sachs disease affects 1 in 1000 among persons of Eastern European Jewish origin. One out of every 10 black persons in the United States carries a sickle cell gene.

Paget's disease, a disease of the bones, occurs frequently among people of English descent. Polynesians are prone to have clubfoot, and Finns tend to have kidney disease. There is a higher than average frequency of deafness in Eskimos and of glaucoma in Icelanders. Native Americans and Mexicans are very susceptible to diabetes and gallbladder disease. Impacted wisdom teeth occur frequently among Europeans and Asians but rarely among Africans. Today there are at least 1200 distinct inherited genetic diseases that have been identified.

Sickle Cell Anemia

Notice the arrangement for the 146 amino acids in the β chain of normal hemoglobin, shown in Figure 34-9. Each of these 146 amino acids was designated by a certain arrangement of three nucleotides in mRNA. Altogether these must have been 438 (3×146) nucleo-

Figure 34-9 Sequence of amino acids in normal hemoglobin β chain.

```
 1           5                    10                   15
Val-His-Leu-Thr-Pro-Glu-Glu-Lys-Ser-Ala-Val-Thr-Ala-Leu-Trp-Gly-Lys-Val-
   20                   25                   30                   35
Asn-Val-Asp-Glu-Val-Gly-Gly-Glu-Ala-Leu-Gly-Arg-Leu-Leu-Val-Val-Tyr-Pro-
         40                   45                   50
Trp-Thr-Gln-Arg-Phe-Phe-Glu-Ser-Phe-Gly-Asp-Leu-Ser-Thr-Pro-Asp-Ala-Val-
   55                   60                   65                   70
Met-Gly-Asn-Pro-Lys-Val-Lys-Ala-His-Gly-Lys-Lys-Val-Leu-Gly-Ala-Phe-Ser-
         75                   80                   85                   90
Asp-Gly-Leu-Ala-His-Leu-Asp-Asn-Leu-Lys-Gly-Thr-Phe-Ala-Thr-Leu-Ser-Glu-
              95                   100                  105
Leu-His-Cys-Asp-Lys-Leu-His-Val-Asp-Pro-Glu-Asn-Phe-Arg-Leu-Leu-Gly-Asn-
   110                  115                  120                  125
Val-Leu-Val-Cys-Val-Leu-Ala-His-His-Phe-Gly-Lys-Glu-Phe-Thr-Pro-Pro-Val-Gln-
         130                  135                  140                  146
Ala-Ala-Tyr-Gln-Lys-Val-Val-Ala-Gly-Val-Ala-Asn-Ala-Leu-Ala-His-Lys-Tyr-His
```

Figure 34-10 (a) Normal red blood cells. (b) A sickled cell. [*Courtesy Richard F. Baker, University of Southern California Medical School, Los Angeles.*]

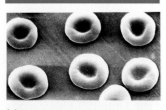

(a)

(b)

tides, arranged in the proper sequence in order to form this molecule of the β chain of hemoglobin.

Look at amino acid 6, the one in color. This amino acid, Glu, is glutamic acid. The mRNA code group for glutamic acid is either GAA or GAG. If the middle codon of this group is changed from A to U, the sequence becomes either GUA or GUG, both of which designate the amino acid valine (Val). That is, if there is a change in only one of the nucleotides, from A to U, on the sixth codon of the 146 amino acid chain of the β chain of hemoglobin, a different type of molecule is produced. This type of hemoglobin is called hemoglobin S and causes the genetic disease sickle cell anemia.

The red blood cells normally have a concave shape when deoxygenated (see Figure 34-10a). Red blood cells containing hemoglobin S look like a sickle (see Figure 34-10b). These sickle cells are more fragile than normal red blood cells, leading to anemia. They can also occlude capillaries, leading to thrombosis. The points and abnormal shapes of the sickle cells cause slowing and sludging of the red blood cells in the capillaries, with resulting hypoxia of the tissues. This produces such symptoms as fever, swelling, and pain in various parts of the body. Eventually the spleen is affected. Many victims of severe sickle cell anemia die in childhood.

Sickle cell anemia is a hereditary condition found primarily among Hispanics and blacks. Many of these people have the sickle cell trait but are relatively unaffected by it until there is a sharp drop in blood oxygen level, such as might be caused by strenuous exercise at high altitudes, underwater swimming, and drunkenness.

Phenylketonuria

Phenylketonuria (PKU) results when the enzyme phenylalanine hydroxylase is absent. A person with PKU cannot convert phenylalanine to tyrosine, and so the phenylalanine accumulates in the body, resulting in injury to the nervous system. In infants and in children up to age 6, an accumulation of phenylalanine leads to retarded mental development.

This disease can be readily diagnosed from a sample of blood or urine. Treatment consists of giving the patient a diet low in phenylalanine.

Galactosemia

Galactosemia results from the lack of the enzyme uridyl transferase, which catalyzes the formation of glucose from galactose. This disease may result in an increased concentration of galactose in the blood. Galactose in the blood is reduced in the eye to galacticol, which accumulates and causes a cataract. Ultimately, if galactose continues to accumulate, liver failure and mental retardation will occur. This disease can be controlled by the administration of a diet free of galactose.

Wilson's Disease

Wilson's disease is caused by the body's failure to eliminate excess Cu^{2+} ions because of a lack of ceruloplasmin (see page 566) or a failure in the bonding of copper ions to the copper-bonding globulin, or both factors. In this disease copper accumulates in the liver, kidneys, and brain. There is also an excess of copper in the urine. If deposition of copper in the liver becomes excessive, cirrhosis may develop. In addition, accumulation of copper in the kidneys may lead to damage of the renal tubules, leading to increased urinary output of amino acids and peptides.

Albinism

Albinism is caused by the lack of the enzyme tyrosinase, which is necessary for the formation of melanin, the pigment of the hair, skin, and eyes. Consequently, albinos have very white skin and hair. Although this disease is not serious, persons affected by it are very sensitive to sunburn.

Hemophilia

Hemophilia is caused by a missing protein, an antihemophilic globulin, which is important in the normal clotting process of the blood. Consequently, any cut may be life threatening to hemophiliacs, but the primary damage is the crippling effect of repeated episodes of internal bleeding into body joints.

Muscular Dystrophy

One form of muscular dystrophy, Duchesne muscular dystrophy, is caused by the lack of a protein called dystrophin. This disease primarily affects boys and causes progressive weakness and wasting of muscles. Victims are usually confined to a wheelchair by age 12 and die before age 30 because of respiratory failure. Currently, there is no treatment for this disease.

Dystrophin is a large protein with a molecular weight of 400,000. It is normally found in the muscle cell membrane, on the side facing the cell interior.

Other Genetic Diseases

Niemann–Pick disease is caused by a lack of the enzyme sphingomyelinase, which causes an accumulation of sphingomyelin in the liver, spleen, bone marrow, and lymph nodes. This disease affects the brain and causes mental retardation and early death.

Gaucher's disease is caused by a lack of the enzyme glucocerebrosidase, which is necessary for the cleavage of glucocerebrosides into glucose and ceramide. This disease is characterized by the accumulation of glycolipids in the spleen and liver. In children, Gaucher's disease causes severe mental retardation and early death (see page 397). In adults, the spleen and liver enlarge progressively but the disease is compatible with long life.

Tay–Sachs disease is due to a lack of the enzyme hexosaminidase A, leading to the accumulation of glycolipids in the brain and the eyes. Red spots show up in the retina, and there is also muscular weakness. This disease is fatal to infants before the age of 4.

Recombinant DNA (Gene Splicing)

The term recombinant DNA refers to DNA molecules that have been artificially created by splicing segments of DNA from one organism into the DNA of a completely different organism. Almost any organism (animal, plant, bacteria, virus) can serve as a DNA donor. Theoretically, any organism also can serve as a DNA acceptor, but to date the bacterium *Escherichia coli* (*E. coli*) has been used almost exclusively as a DNA acceptor. *E. coli* was selected because its genetic makeup has been thoroughly studied. Most of *E. coli*'s 3000 to 4000 genes are contained within a large, single, ringed chromosome. In addition to this large chromosome, there are plasmids, smaller closed loops of DNA consisting of only a few genes. It is these plasmids that act as DNA acceptors for *E. coli*, since the plasmids are known to multiply independently within the host cell as the cell itself replicates.

How does gene splicing work? First, the *E. coli* is placed in a detergent solution to break open the cells. The plasmids are extracted with *restriction enzymes*, which cut the plasmid DNA at specific points. DNA from another organism (also cleaved by a similar method) is inserted into the *E. coli* plasmid. The plasmid ring is then closed and the plasmid is put back into *E. coli*. The new cells are isolated and allowed to replicate (*E. coli* duplicates itself every 20 to 30 min). As the cells replicate, they synthesize the proteins coded for by their DNA, including those coded for by the inserted DNA (see Figure 34-11).

Recombinant DNA research has produced human insulin, interferon, growth hormone, somatostatin, and some vaccines. Theoretically, gene splicing could be used to find cures for genetic diseases and for cancer, and to manufacture various enzymes, hormones, antibodies, and other substances that are difficult to isolate and produce pharmaceutically.

Opponents of recombinant DNA research point out the considerable risks involved. They fear that a man-made strain of *E. coli* might be accidentally released and cause a worldwide epidemic and the death of large populations. They also believe that this type of research might be used politically to control human behavior.

Figure 34-11 Recombinant DNA in *Escherichia coli*.

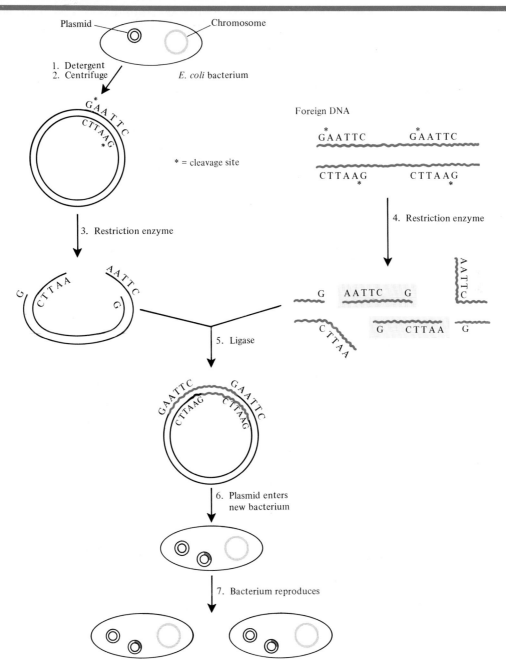

Genetic Markers

One of the newer topics of cancer research is genetic markers, defects in one part of a chromosome that it is believed can cause a certain disease. A genetic marker for Alzheimer's disease has been found on chromosome 21, the same chromosome that is associated with Down's syndrome. Since a chromosome can include many, many genes, much work is yet to be done in this area.

Oncogenes

Oncogenes are genes that appear to trigger uncontrolled or cancerous growth. Cancerous cells exhibit three general characteristics:

1. Uncontrolled growth.
2. Invasion of body tissues.
3. Spread to other body parts (metastasis).

In cancer, the genes controlling growth are abnormal, but little is known about how the cell growth is controlled.

Cancer is the second most common cause of death in the United States, after cardiovascular disease.

The incidence of cancer increases with age, and a wide variety of body organs can be affected. The abnormal production of enzymes, hormones, and proteins is frequently associated with cancer. These products are called *tumor markers*. Measurement of such markers is helpful in diagnosing and treating cancer. For example, the detection of the enzyme prostatic acid phosphatase may be associated with cancer of the prostate gland, and abnormal amounts of calcitonin may indicate carcinoma of the thyroid gland.

What causes cancer? It is believed that all cancer-causing agents can be grouped into three main categories: radiant energy, chemical compounds, and viruses.

Radiant energy such as X-rays and γ-rays are carcinogenic because of the formation of *free radicals in the tissues. Radiation can also damage DNA and so is *mutagenic.

Many *chemical compounds* in common use are known to be carcinogenic. As much as 75 percent of human cancers are caused by chemicals in the environment. Among these carcinogenic chemicals are benzene, asbestos, and fused-ring compounds such as the benzpyrene found in cigarette smoke (see page 323).

Oncogenic viruses contain either DNA or RNA. Under certain circumstances, infection of appropriate cells with polyoma virus or

Table 34-2 Some Oncogenic Viruses

Class	Members
DNA virus	
Papovavirus	Polyomavirus, SV 40 virus
Herpesvirus	Epstein–Barr virus, herpes simplex type 2 virus
Hepadnavirus	Hepatitis B virus
RNA virus	
Retrovirus type C	Leukemia virus

SV 40 virus can result in a malignant transformation. Table 34-2 lists some oncogenic viruses.

Summary

Originally it was believed that the cell's chromosomes transmitted information from one generation to the next. Then it was believed that the chromosomes contained genes, with each gene giving a particular characteristic to that individual. However, recent studies have shown that genes are composed of deoxyribonucleic acid (DNA), which stores and transfers hereditary characteristics.

The two types of nucleic acids are deoxyribonucleic acid (DNA), which is primarily responsible for the transfer of genetic information, and ribonucleic acid (RNA), which is primarily responsible for protein synthesis.

The Watson–Crick model of DNA proposes a double-coiled chain consisting of two strands interwined around one another. The chains are composed of alternating sugar (deoxyribose) and phosphates with the sugar being bonded to one of four different heterocyclic compounds—adenine, guanine, cytosine, and thymine (A, G, C, T). The model suggests that the adenine of one chain is always bonded to a thymine of the opposite chain and that cytosine is always bonded to guanine.

All DNA molecules have the same sequence of deoxyribose and phosphates in the ladder part of the chain. The difference lies in the order of adenine, thymine, guanine, and cytosine.

When a cell divides, the DNA molecule uncoils and each half acts as a template for the formation of a new chain.

During replication, the DNA molecule uncoils and both halves produce molecules identical to the original. However, when the DNA molecule synthesizes RNA, although the coil again unwinds, only one half acts as the template to produce RNA. There are three kinds of RNA: ribosomal RNA (rRNA), messenger RNA (mRNA), and transfer RNA (tRNA). A possible fourth is heterogeneous RNA (hRNA).

When mRNA is formed from DNA, guanine is still bonded to cytosine but adenine is bonded to uracil (U) rather than to thymine.

Messenger RNA moves through the cytoplasm to the ribosomes, where it acts as a template for the formation of protein. The ribosomes contain tRNA; tRNA carries a specific amino acid to the mRNA. The identity of this amino acid is determined by the code on the end of the tRNA, and this code is specific for another code on the mRNA.

As the mRNA moves through the ribosome, the first of its coded groups picks up and holds a corresponding coded group on tRNA. The second, third, and succeeding groups on mRNA do likewise. The specific amino acids carried by the tRNA become bonded to one another to form the designated protein, and the tRNAs are released to find more of the specific amino acids and begin the procedure again.

The DNA code consists of four letters (A, T, C, and G) arranged in groups of three. The three-letter coded groups specify the 20 primary amino acids occurring in nature.

If one of the code letters is changed or is missing, the information copied will be incorrect and a mutation will occur. Mutations occur because of exposure to radiation, both naturally and in industry and medicine, and possibly because of certain chemicals.

DNA directs the synthesis of enzymes (which are also proteins). If the message is transferred incorrectly, the proper enzyme will not be synthesized and so will be lacking in the body. This lack of a specific enzyme may lead to a genetic disease such as phenylketonuria, galactosemia, albinism, pentosuria, or hemophilia.

Oncogenes are genes that appear to trigger cancerous growth. Cancerous cells exhibit uncontrolled growth, invasion of body tissues, and metastasis.

Questions and Problems

1. What is the relationship between chromosomes, genes, and DNA?
2. What are nucleic acids?
3. Diagram the Watson–Crick model of DNA.
4. What types of bonds are present in the DNA molecules? Are they all equal in strength?
5. How does DNA replicate?
6. What are the three types of RNA? What is the function of each?
7. What is the difference in nucleotides in mRNA and DNA?
8. How do mRNA and tRNA function in the formation of protein?
9. What is the most common genetic disease in the United States?
10. How many primary amino acids occur in nature?
11. Name the primary amino acids and indicate their abbreviations.
12. How many units are present in one unit of the DNA code?
13. Describe briefly how the DNA code was decoded.
14. What amino acid does each of the following coded groups of mRNA represent?
    (a) ACG (b) GCA (c) UUU (d) GUU (e) CCC (f) CGA
15. What is a mutation in terms of the DNA molecule?
16. What is sickle cell disease? What causes it?
17. How do normal blood cells differ from sickle cells in (a) shape and (b) arrangement of amino acids?
18. What causes the following genetic diseases and what are the symptoms of each?
    (a) Phenylketonuria
    (b) Hemophilia
    (c) Galactosemia
    (d) Albinism
    (e) Wilson's disease
    (f) Tay–Sachs disease

19. Explain how a genetic disease might occur if the body is deficient in an enzyme.

Practice Test

1. The end product of purine metabolism is _____.
   a. uric acid   b. urea
   c. creatinine   d. ammonia
2. Which of the following is part of a nucleotide?
   a. purine or pyrimidine base
   b. five-carbon sugar
   c. phosphoric acid
   d. all of these
3. The RNA concerned with the transcription of genetic information is _____.
   a. ribosomal   b. messenger
   c. transfer   d. homogeneous
4. The number of nucleotides necessary to code for an amino acid is _____.
   a. one   b. two   c. three   d. four
5. The insertion of the wrong nucleotide into a DNA code group could result in _____.
   a. cloning   b. splicing
   c. recombination   d. mutation
6. Guanosine pairs with _____.
   a. adenine   b. thymine
   c. cytosine   d. uracil
7. How many different amino acids are involved in the genetic code?
   a. 3   b. 5   c. 10   d. 20
8. Which of the following is *not* a genetic disease?
   a. sickle cell anemia   b. phenylketonuria
   c. albinism   d. goiter
9. The two chains of DNA are held together by which type of bond?
   a. ionic   b. covalent
   c. hydrogen   d. disulfide
10. The most common genetic disease in the United States is _____.
    a. cystic fibrosis
    b. hemophilia
    c. galactosemia
    d. muscular dystrophy

35

Clinical Chemistry

A Red Cross laboratory technician studying the results of a cross-matching test prior to a blood transfusion.

Introduction

Clinical laboratory tests measure chemical changes in the body in the interest of diagnosis, therapy, and prognosis. This work consists of assaying the chemical constituents in the blood, urine, and other body fluids or tissues. The levels of most of these constituents remain relatively constant but often become altered during the course of a disease. This chapter will introduce a few of the more common laboratory tests.

Much of the actual chemistry is beyond the scope of this text. Also many of the laboratory tests are done by machine. Whenever possible a brief discussion of the chemistry involved in a test will be included.

Myocardial Infarction

A *myocardial infarction (MI) occurs when a deficient flow of blood to the heart damages the cardiac tissue. The decrease in blood flow may be due to a clot in a coronary artery or a narrowing of the coronary artery caused by an accumulation of plaque on the arterial wall.

As the cardiac cells die, their intracellular enzymes leak out into the plasma. Testing for the presence of these enzymes is helpful in the diagnosis of a myocardial infarction. The enzyme levels that are useful in MI detection are listed in Table 35-1.

Some enzyme activities increase after a myocardial infarction (for example, creatine kinase), and some increase much later and remain elevated (for example, lactic dehydrogenase). Each enzyme follows a different time schedule, as shown in Figure 35-1. Since the physician has no control over how long a patient will delay before coming for examination, tests for the different enzymes have been developed to help diagnose an MI over a period from 4 hr to 10 days (see Figure 35-1).

The concentration of an enzyme in the plasma or serum cannot

Table 35-1 Enzymes Used to Detect a Myocardial Infarction

Enzyme		Adult Upper Limit of Normal (ULN) (units/L)	Enzyme Activity Increase		
Name	Abbreviation		Usual Rise ($\times$ ULN)	Time for Maximum Rise (hr)	Time for Return to Normal (days)
Creatine kinase	CK	100	5–8	24	2–3
Isoenzyme MB	CK–MB	6	5–15	24	2–3
Aspartate aminotransferase	AST	25	3–5	24–48	4–6
Lactate dehydrogenase	LDH	290	2–4	48–72	7–12
Isoenzyme #1	LDH_1	100, or 34%	2–4	48–72	6–12
Isoenzyme #2	LDH_2	115, or 40%	1.1–3	48–72	6–10

Table 35-1   Enzyme levels following myocardial infarction.

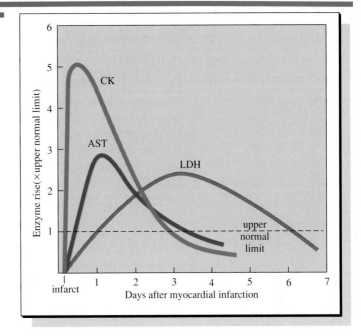

be measured readily because of the difficulty of directly determining the mass of a specific enzyme per unit volume of fluid. Instead the rate at which an enzyme catalyzes a reaction with a substrate is measured under specific conditions. The measurement may involve the decrease in substrate concentration with time, the increase of a product, or the increase or decrease of a cofactor such as NADH (see page 486).

Pancreatitis

It is often difficult to diagnose acute *pancreatitis without the help of laboratory tests. The patient usually complains of an intense pain in the upper abdomen that could be caused by several disorders. The two tests most commonly performed for diagnostic purposes are tests for serum amylase (for the digestion of starch; see page 459) and for serum lipase (for the digestion of triglycerides; see page 676).

Test for serum amylase involves measuring the disappearance of a starch substrate or the production of the reducing sugars glucose and maltose. Serum lipase activity is determined by measuring the amount of fatty acid produced from a known quantity of olive oil.

Acute pancreatitis is caused by a blockage of the pancreatic ducts or by injury to the pancreatic tissue by poisons, inflammation, or trauma. In any of these cases there is usually a decreased flow of pancreatic juice to the small intestine. Pancreatic juice and the accompanying amylase and lipase enter the bloodstream instead and show greatly increased levels there.

Urinalysis

Examination of the urine for signs of illness has been carried out for centuries. The traditional methods of examination include noting the appearance and odor of the urine. The main advances in urinalysis have been in the use of dipsticks or strips for semiqualitative information about the urine's contents. A routine urinalysis usually consists of an examination of a morning sample (taken upon arising) for color, odor, and specific gravity or osmolality; some qualitative and semiqualitative tests for pH, proteins, glucose and other reducing sugars, ketones, blood, and bilirubin; and a microscopic look at the urinary sediment.

Volume

Knowledge of the daily urinary output is important, but it requires a timed (24 hr) specimen. The daily output depends upon fluid intake and such factors as degree of excretion, temperature, salt intake, and hormone levels. The average adult excretes about 1 mL/min or 1400 ± 800 mL over a 24-hr period. A decreased urinary output is called *oliguria and may be caused by a drop in blood pressure, shock, hemorrhage, renal disease, calculi (stones), or tumors. Excessive urinary output is called *polyuria and may be caused by the excretion of large amounts of solutes such as salts after excessive salt intake or glucose when the patient has diabetes mellitus. Polyuria can also be caused by a deficiency of the antidiuretic hormone (ADH), by diuretics, or by excessive ingestion of fluids (see pages 525–26).

Color

Fresh blood imparts a reddish color to urine, while old blood makes the urine look smoky. Both indicate bleeding in the genitourinary tract. Bile pigments produce a green, brown, or deep yellow color that indicates liver disease.

Odor

The odor of urine is affected by some foods such as asparagus. With diabetes there may be the fruity odor of acetone. In maple syrup disease, the urine has the odor of maple syrup. Old urine samples or urine with certain infections can have a strong odor of ammonia.

Specific Gravity

The specific gravity (page 526) of urine is directly related to the amount of solutes excreted. It provides information about the kidneys' ability to concentrate the urine. In renal tubular disease, the concentrating ability is among the first functions to be lost.

pH

The pH of urine can vary from 4.6 to 8.0. Children usually have higher urine pH values than adults. Starvation and ketosis will lower the pH of the urine. Urine is seldom alkaline except after a large meal, in metabolic or respiratory alkalosis, after ingestion of large amounts of stomach antacids, or when ammonia-generating bacteria are present in the urine.

Protein

Small amounts of protein in the urine are normal, but usually the levels are below detection. Large, detectable amounts of protein (*proteinuria) usually indicate an injury to the kidneys' filtration process. However, temporary proteinuria can occur after a high fever or during excessive muscular exercise.

Glucose

A test for glucose is usually done to check for diabetes or the efficiency of insulin therapy. The dipstick is impregnated with glucose oxidase, peroxidase, and a *chromogen. The strip is dipped into urine and at the appropriate time the glucose level is determined by comparison of the strip's color to a color chart. False positives rarely occur, but false negatives can happen when a patient has a high vitamin C intake. Ascorbic acid inhibits the strip reaction.

Ketones

Ketones (acetone) in the urine indicate problems with carbohydrate metabolism (diabetes, low carbohydrate diet, starvation, anorexia, pregnancy). To test for ketones, the strip is impregnated with sodium nitroprusside and an alkaline buffer. In the presence of either acetone or acetoacetate, a lavender color is produced that is compared with a color chart.

Nonprotein Constituents

Creatinine, a waste product formed in muscle from creatine phosphate, appears in the urine and is not reabsorbed by the kidneys. Any condition that reduces the glomerular filtration rate (GFR) will lower the amount of creatinine excreted. Serum creatinine levels are elevated whenever there is a large reduction in the GFR or when urine elimination is obstructed. The concentration of serum creatinine is a much better indicator of renal function than that of either urea or uric acid because it is not affected by diet, exercise, or hormones as the other two are.

Serum Urea

Serum urea is the end product of nitrogen metabolism in the body (see page 428). Serum urea levels are most strongly influenced by the degree of protein metabolism. Increased serum urea levels may be due to a high protein diet, corticosteroids, and/or stress. Decreased levels occur late in pregnancy during rapid fetal growth, and in starvation or low protein diets.

Serum urea is measured by enzymatic conversion of the urea into ammonia. The liberated ammonia is converted into a blue dye that is measured with a *spectrophotometer.

Serum Urate

Serum urate is the end product of purine metabolism (see page 649). Its primary use is in the diagnosis and treatment of gout or diseases accompanied by the breakdown of large amounts of nucleic acid such as toxemia during pregnancy and radiation and chemotherapy treatments. Serum urate levels can be determined by the enzymatic oxidation of urate by hydrogen peroxide.

Serum Glucose

Disorders associated with glucose metabolism are raised plasma glucose levels (*hyperglycemia) or decreased plasma glucose levels (*hypoglycemia).

Hyperglycemia is harmful when glucose levels increase enough to cause cellular dehydration (coma can be produced by severe dehydration of brain cells). Other associated factors such as acidosis and electrolyte imbalances can also cause cellular damage. Chronic hyperglycemia can produce many pathological changes.

Normal hemoglobin becomes glycosylated (a reaction in which glucose covalently bonds to hemoglobin; see page 366) over a period of time when there are high levels of glucose in the blood. The determination of plasma levels of glycosylated hemoglobin is clinically useful in determining the effectiveness of blood glucose control because the amount of glycosylated hemoglobin varies directly with increases in plasma glucose.

Diabetes mellitus is by far the most important cause of hyperglycemia. Other causes include excess growth hormone, excess cortisol, and excess ACTH.

Whenever plasma glucose falls below 60 mg/100 mL, the condition is termed hypoglycemia. Since the brain is dependent on glucose for its energy, the clinical symptoms of hypoglycemia resemble cerebral anoxia: faintness, weakness, dizziness, tremors, anxiety, hunger, "cold sweat," mental confusion and motor incoordination. When plasma glucose levels drop below 40 mg/100 mL, consciousness is usually lost, although newborn infants may be able to drop to 25 to 30 mg/100 mL before this happens. If hypoglycemic episodes continue, brain damage and death result. Hypoglycemia can be caused by excess use of insulin; deficiency of cortisol, ACTH, or growth hormone; glycogen storage diseases; starvation; and severe liver damage.

Testing for Hyperglycemia and Hypoglycemia (Glucose Tolerance Test)

A patient with a suspected problem with glucose metabolism is given a glucose tolerance test. For 3 days prior to the test, the patient must be placed on a diet containing adequate protein, calories, and 150 g of carbohydrate per day. This stimulates the production of the liver enzymes necessary for the conversion of glucose into glycogen.

Then, after 12 hr of fasting, the patient is given approximately 1 g of glucose for each kilogram of body weight. The patient's blood sugar level is checked by withdrawing blood samples at regular intervals over several hours. The samples are chemically analyzed, and the concentration of sugar in the blood is plotted against time. In a normal person the blood sugar level rises from about 80 mg per 100 mL of blood to 130 mg per 100 mL of blood in about 1 hr. Then the blood sugar level gradually returns to normal after about $2\frac{1}{2}$ hr. In a diabetic patient, because no insulin is being secreted, the blood sugar level rises to an even higher level than in the normal individual and remains there for a much longer period of time. Then it slowly begins to return toward its normal value.

These results are illustrated in the graphs in Figure 35-2.

Figure 35-2 Criteria used for interpretation of glucose tolerance tests.

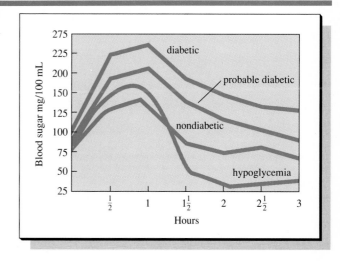

Time (hr after glucose)	Plasma Glucose Concentration (mg/dL)	Points[a]
0	130	1
1	195	0.5
2	140	0.5
3	140	1

Table 35-2 Wilkerson Point System for Interpretation of Oral Glucose Tolerance Test

From H. L. C. Wilkerson, "Diagnosis: Oral Glucose Tolerance Tests," in T. S. Danowski (Ed.), *Diabetes Mellitus* (New York: American Diabetes Association, 1964), p. 31.
[a] A total of 2 points establishes diabetes.

Also in the diabetic patient, large amounts of glucose appear in the urine because of the spillover when the renal threshold is exceeded.

The glucose tolerance test is a valuable diagnostic tool because it indicates the ability of the body to utilize carbohydrate. A decreased utilization may indicate diabetes, whereas an increased utilization may indicate Addison's disease, hypopituitarism, or hyperinsulinism.

The American Diabetes Association recommends the criteria given in Table 35-2 for establishing a diagnosis of diabetes mellitus.

Lipid Metabolism

All lipids, whether derived from dietary sources or synthesized in the body, must be transported by carriers because of their insolubility in water. This is accomplished by transforming lipids into *lipoproteins.

There are four classes of lipoproteins: chylomicrons, very low density lipoproteins (VLDL), low density lipoproteins (LDL), and high density lipoproteins (HDL). All lipoproteins contain definite amounts of phospholipids, cholesterol, and triglycerides (see page 419).

Classification of lipoproteins arises from two sets of properties: the density of the lipoproteins following ultracentrifugation and the charge carried by the lipoproteins, which aids in separation under *electrophoresis (see page 206).

The lipoproteins have different functions (see page 419). Several

Table 35-3 Classification of Lipoproteins

Classified by Density	Classified by Electrophoresis
Chylomicrons	Chylomicrons
VLDL	Prebeta
LDL	Beta
HDL	Alpha

Table 35-4 Serum HDL and LDL (Cholesterol) for Various Age Groups

Age Group	Average HDL (Cholesterol) (mg/100 mL)		Average LDL (Cholesterol) (mg/100 mL)
	Male	Female	
0–19	48	50	110
20–29	48	57	115
30–39	45	55	130
40–49	43	61	135
50–75	50	65	145

*hyperlipidemias associated with abnormal levels of lipoproteins have been identified.

The serum lipid tests most commonly requested are total cholesterol, HDL cholesterol, and triglycerides. An elevated concentration of serum cholesterol is considered to increase the risk of *atherosclerosis. If serum triglycerides are also elevated, the risk is increased. Other risk factors include smoking, obesity, stress, hypertension, and lack of exercise.

Serum Total Cholesterol

Normally, about 60 percent of total cholesterol is LDL, 22 percent is HDL, 13 percent is VLDL, and 5 percent is chylomicrons. Mean serum cholesterol concentrations for various age groups are shown in Table 35-4. Before menopause the serum cholesterol level for women does not differ from that of men, but after menopause it often becomes higher. Increases in total serum cholesterol may occur in hypothyroidism, in uncontrolled diabetes mellitus, in bile duct obstruction, and in several types of hyperlipidemias. Serum cholesterol levels also increase in late pregnancy but generally return to normal within a month after delivery. A decrease in serum cholesterol level is seen in hyperthyroidism, liver disease, anemias, starvation, and certain genetic diseases.

HDL and LDL Serum Cholesterol

Since cholesterol is primarily transported by two lipoproteins with different functions, total serum cholesterol has limited value in deciding the risk of coronary artery disease. High density lipoprotein (HDL) transports cholesterol from tissues to the liver while low density lipoprotein (LDL) transports cholesterol to the tissues, including blood vessels. HDL competes with LDL for cholesterol, and this may reduce cholesterol accumulation in blood vessel walls. High levels of HDL appear to reduce the risk of coronary artery disease. HDL levels can be increased by moderate to vigorous exercise, estrogens, and moderate alcohol consumption. Low levels of HDL increase the risk of cardiovascular disease. Factors that decrease

HDL levels include cigarette smoking, uncontrolled diabetes mellitus, male sex hormones, and very high carbohydrate diets.

Serum Triglycerides

Serum triglyceride concentrations are moderately elevated after a meal containing fat and peak about 4 to 5 hr after eating. Elevated triglyceride levels in the blood are not unusual in patients with atherosclerosis or a history of myocardial infarction. Very high triglyceride levels may cause pancreatitis. Elevated triglyceride levels are frequently seen in hypothyroidism, acute alcoholism, obstructive liver disease, and uncontrolled diabetes—and sometimes in insulin-treated diabetes.

Blood Gases

The body's acid–base status is determined by measuring the blood gases. Measurements are made simultaneously for blood pH, pCO_2, and pO_2 upon arterial blood drawn anaerobically into a capped syringe. After determining the pH and the pCO_2, the $[HCO_3^-]$ can be calculated from a graph. The following precautions are used for this test: heparinized arterial blood is used whenever possible; the sample is drawn and stored anaerobically without air bubbles; the sample is immediately placed in an ice bath and transported to the laboratory to minimize metabolic changes. The blood gas instrument uses three separate electrodes to determine values for pH, pCO_2, and pO_2. Table 35-5 lists the normal values of blood gases.

Potassium

Close control of extracellular potassium is essential because levels greater than 7.0 mEq/L may result in cardiac arrest. Low potassium levels are also dangerous because they may cause cessation of the heartbeat during contraction. Therefore, it is essential to notify the physician whenever a patient has a very high or very low potassium level. Normal serum potassium levels are about 4 mEq/L. Increased potassium levels may occur during acidosis, decreased urine output, shock, chronic renal failure, and decreased aldosterone production.

Table 35-5　Values of Blood Gases

	Arterial	Venous
pH	7.37–7.44	7.35–7.45
pCO_2		
Males	34–45 mm Hg	36–50 mm Hg
Females	31–42 mm Hg	34–43 mm Hg
pO_2	85–95 mm Hg	32–50 mm Hg

Decreased levels may be caused by prolonged diarrhea or vomiting, some diuretics, oversecretion of aldosterone, and low potassium intake.

Serum potassium can be measured using a potassium sensitive electrode similar to a pH electrode.

Summary

Clinical laboratory tests measure chemical changes in the blood, urine, other body fluids, and tissues for diagnosis, therapy, and prognosis.

Myocardial infarction (MI) can be diagnosed by measuring levels of various cardiac enzymes leaked into the blood following an MI. Creatine kinase, lactate dehydrogenase, and aspartate aminotransferase are used to determine the extent and approximate times of the MI.

Pancreatic function is diagnosed by tests for serum lipase and amylase activity.

Urinalysis includes examining the urine's daily volume, color, odor, pH, specific gravity, and contents. Contents frequently checked are protein, glucose, ketones, blood, bilirubin, creatinine, and sediment.

Serum urea reflects the level of protein metabolism, while serum urates are formed by the breakdown of purine nucleic acids.

Hyperglycemia and hypoglycemia are diagnosed from serum glucose levels. Both are potentially dangerous conditions that can result in coma or death. The primary cause of hyperglycemia is diabetes mellitus. Hypoglycemia can be due to either an insulin excess or a deficiency of several hormones. The most common test for either condition is the oral glucose tolerance test.

Lipoproteins are used to transport lipids in the blood. They are classified on the basis of density or electrophoresis separation. The most common tests are for serum triglycerides, total serum cholesterol, and HDL cholesterol. Elevated levels of either serum triglycerides or total serum cholesterol increases the risk of atherosclerosis. Increased levels of HDL cholesterol are associated with a decreased risk of atherosclerosis.

Blood gases are measured to determine the body's acid–base status. Normal values for arterial blood gases are pH = 7.4, pCO_2 = 40 mm Hg, and pO_2 = 90 mm Hg.

Serum potassium has an important influence on cardiac function. Either low or high levels of serum potassium are potentially life threatening.

Questions and Problems

1. How do cardiac enzymes enter the blood plasma?
2. What enzymes are used to diagnose a myocardial infarction?
3. How does the level of cardiac enzymes vary with time following a myocardial infarction?
4. What has been the major advance in urinalysis?
5. What does a routine urinalysis include?
6. What color does blood give to urine?
7. What factors can cause urine to be alkaline?
8. What disorders can produce ketones in the urine?
9. What is urea the end product of?
10. What is uric acid the end product of?
11. Plasma levels of glycosylated hemoglobin are useful in the treatment of what disorder?
12. What level of blood glucose produces weakness, confusion, lack of coordination, and unconsciousness?
13. What are some causes for hypoglycemia?
14. What is the basis for the classification of lipoproteins?

15. List the different lipoproteins.
16. What factors may cause an increase in serum cholesterol levels?
17. What factors may cause a decrease in serum cholesterol levels?
18. In which age group do men and women have the highest HDL level? highest LDL level?
19. What are the normal arterial values for pH, pCO_2, and pO_2?
20. Why does venous blood have a lower pH than arterial blood?
21. What organ is most affected by large changes in plasma K^+ levels?
22. What can cause a low potassium level? Name four.

Practice Test

1. Which of the following is *not* useful in diagnosing an MI?

   a. CK
   b. LD
   c. HLA
   d. AST

2. Which of the following would remain above normal 4 days after an MI?

   a. CK
   b. CK–MB
   c. AST
   d. LD

3. Levels of which compound are the best indicators of renal function?

   a. H_2O
   b. creatinine
   c. urea
   d. uric acid

4. Elevated levels of which lipoprotein are considered beneficial?

   a. LDL
   b. VLDL
   c. chylomicrons
   d. HDL

5. Normal arterial plasma pH is about _____.

   a. 7.4
   b. 7.6
   c. 7.0
   d. 7.2

6. A plasma potassium level of 6.5 mEq/L is considered _____.

   a. normal
   b. high
   c. low
   d. ineffective

APPENDIX

Exponential Numbers

Changing Exponential Numbers to Common Numbers

Mathematicians have developed a shorthand method for expressing very large or very small numbers. This system involves the use of a base number, 10, raised to some power. A power, or exponent, indicates how many times the base number, 10, is repeated as a factor. Thus,

1×10^2 (10 repeated as a factor 2 times) $= 1 \times 10 \times 10 = 100$
1×10^3 (10 repeated as a factor 3 times) $= 1 \times 10 \times 10 \times 10 = 1000$
1×10^5 (10 repeated as a factor 5 times) $= 1 \times 10 \times 10 \times 10 \times 10 \times 10$
$= 100,000$

Negative exponents are used to indicate numbers less than 1. A negative exponent indicates the reciprocal of the same number with a positive exponent:

$$1 \times 10^{-1} = \frac{1}{10^1} = 0.1$$

$$1 \times 10^{-2} = 1 \times \frac{1}{10^2} = 0.01$$

$$1 \times 10^{-4} = 1 \times \frac{1}{10^4} = 0.0001$$

Note: The positive exponent indicates how many places the decimal point must be moved to the right (from the number 1). Also, note that the negative exponent indicates how many places the decimal point must be moved to the left (including the number 1).

679

Any number that is not an exact power of 10 can be expressed as a product of two numbers, one of which is a power of 10. The other number is always written with just one figure to the left of the decimal point.

Example A-1   Express 6.2×10^3 as a common number.

A positive three (+3) exponent indicates that the decimal point should be moved three places to the right. Thus

$$6.2 \times 10^3 = (6.200) = 6200$$

Example A-2   Change 8.45×10^5 to a common number.

A +5 exponent indicates that the decimal point should be moved five places to the right.

$$8.45 \times 10^5 = (8.45000) = 845,000$$

Example A-3   Change 1.27×10^{-2} to a common number.

A −2 exponent indicates that the decimal point should be moved two places to the left.

$$1.27 \times 10^{-2} = (001.27) = 0.0127$$

Example A-4   Change 3.5×10^{-9} to a common number.

$$3.5 \times 10^{-9} = (0000000003.5) = 0.0000000035$$

Changing Common Numbers to Exponential Numbers

When a common number is changed to an exponential number, the decimal point is moved so that there is just one digit to the left of it. The exponent corresponds to the number of places the decimal point must be moved. If the decimal point is moved to the left, the exponent is positive; if it is moved to the right, the exponent is negative.

Example A-5   Change 4000 to an exponential number.

The decimal point must be moved to the left three places in order for just one digit to remain before that decimal point. Three places to the left indicates an exponent of 3 (+3).

$$4000 = (4000) = 4 \times 10^3$$

Example A-6   Change 604 to an exponential number.

$$604 = (604) = 6.04 \times 10^2$$

Example A-7   Change 0.00037 to an exponential number.

$$0.00037 = (0.00037) = 3.7 \times 10^{-4}$$

where the negative exponent indicates that the decimal point has been moved to the right.

Glossary

acetal Compound formed by the reaction of one molecule of an aldehyde with two molecules of alcohol. The general formula is

$$\begin{array}{c} OR' \\ | \\ R-C-H \\ | \\ OR'' \end{array}$$

acidosis Condition in which the pH of the blood drops from 7.35.

acromegaly Pathologic enlargement of the bones of the face, hands, and feet resulting from an overactive anterior lobe of the pituitary gland.

activation energy Minimum amount of energy molecules must possess in order for a reaction to occur.

adhesion The attraction or joining of two dissimilar substances.

aerobic Requiring oxygen.

aerosol Gaseous suspension of fine solid or liquid particles.

AIDS Acquired immune deficiency syndrome is a T cell immunodeficiency in previously healthy adults in association with opportunistic infection or Kaposi's sarcoma.

alkalosis Condition in which the pH of the blood rises from 7.45.

allograft Tissue exchange between genetically nonidentical members of the same species.

alloy Solid solution of two or more metals.

alveoli Air sacs in the lungs.

amoebicide Substance that destroys amoeba.

amphipathic Molecule having a hydrophobic and a hydrophilic end.

amphoteric Capable of reacting either as an acid or a base.

anaerobic Not requiring the presence of oxygen.

analgesic A pain killer.

anaphylaxis A reaction of immediate hypersensitivity that results from sensitization of mast cells by antibodies following exposure to antigen.

anion A negatively charged ion.

antibody A protein molecule that is released by a plasma cell and binds to an antigen.

antigen A substance that can induce a detectable immune response when introduced into an animal; stimulates production of antibodies.

antimetabolite Substance that inhibits utilization of metabolites.

antineoplastic agents Agents that inhibit the growth of malignant cells.

antipyretic Anything that reduces a fever.

antiseptic Substance capable of destroying disease-causing microorganisms.

antispasmodic Capable of preventing spasms or convulsions.

antitoxin An antibody capable of neutralizing a poison of biologic origin.

apoenzyme Protein part of an enzyme.

asphyxiation Unconsciousness or death caused by lack of oxygen.

astringent Substance that draws together or contracts tissue.

atherosclerosis Thickening of arterial wall, characterized by deposition of fatty substances.

autograft Tissue transplant within the same individual.

autoradiograph Self-picture taken by a radioactive substance.

bactericidal   Capable of killing bacteria.

bacteriostatic   Capable of inhibiting growth of bacteria without destroying them.

B lymphocyte (B cell)   A white blood cell, derived from the lymphoid tissue, that is not thymus dependent and is responsible for the production of antibodies.

bradycardia   Slower-than-normal heart rate.

bronchioles   Thin-walled extensions of the bronchial tubes in the lungs.

calculi   Stones, such as those found in the gallbladder or kidney

capillary action   Rise of fluid through a small opening.

carcinogenic   Cancer-causing.

cardiocyte   A heart cell.

cathartic   Laxative.

catheter   Slender, flexible tube for insertion into a body channel.

cation   Positively charged ion.

chain reaction   Self-sustaining nuclear reaction yielding products that cause further reactions of the same kind.

chemotherapeutic agent   Chemical used in treatment of disease, particularly cancer.

chorea   A nervous disorder marked by uncontrollable and irregular body movements.

chromogen   A substance without color that can be transformed into a colored dye by chemical reaction.

coenzyme   Small molecule that combines with the apoenzyme to form an active enzyme.

cohesion   Property of a substance sticking to itself.

colostrum   The first secretion of the mammary glands at the termination of pregnancy.

crenation   Shrinking of red blood cells in a hypertonic solution.

curie   Unit of radiation; 1 curie (Ci) equals 37 billion nuclear disintegrations per second.

diabetes mellitus   Disorder of carbohydrate metabolism characterized by inadequate secretion or utilization of insulin.

diuretic   Substance that increases output of urine.

eczema   Noncontagious inflammation of the skin, marked by the outbreak of lesions that become encrusted and scaly.

edema   Accumulation of fluid in the tissues.

electroencephalogram   Graphic record of the electrical activity of the brain.

electrophoresis   Separation of charged particles under the influence of an electric field.

emphysema   Pathologic enlargement of the alveoli in the lungs.

endothermic   Absorbing energy.

equilibrium   Dynamic state in which the rates of opposing reactions are equal.

erythroblastosis fetalis   A hemolytic disease in newborns that results from the development in the mother of anti-Rh antibody in response to Rh-positive fetal blood.

erythrocyte   Red blood cell.

euphoria   Feeling of well-being.

excoriated   Chafed.

exogenous   Developed from external causes.

exophthalmos   Abnormal protrusion of the eyeball.

exothermic   Releasing energy.

fission   Splitting of the nucleus of an atom.

flatulence   Excessive gas in the stomach and intestine.

free radical   A particle with an unpaired electron.

fusion   Combining of small nuclei to make a larger one.

glycosidic linkage   Bond formed between two monosaccharides.

goiter   Enlargement of the thyroid gland visible as a swelling in the front of the neck.

granulocyte   A white blood cell (neutrophil, basophil, or eosinophil) that contains cytoplasmic granules.

halogen   Any one of group VIIA nonmetals.

hematopoietic system   System responsible for the formation of blood cells.

hemiacetal   Compound formed by the reaction of one molecule of an aldehyde with one molecule of alcohol; general formula is

$$\begin{array}{c} OR' \\ | \\ R-C-H \\ | \\ OH \end{array}$$

hemiketal   Compound formed by the reaction of one molecule of a ketone with one molecule of alcohol; general formula is

$$\begin{array}{c} OR'' \\ | \\ R-C-R' \\ | \\ OH \end{array}$$

hemolysis Destruction of red blood cells caused by a hypotonic solution.

hemostasis Stoppage of bleeding or blood flow.

hertz Unit of frequency equal to 1 cycle per second.

hydrophilic Having an affinity for water.

hydrophobic Antagonistic to water.

hydrostatic Referring to the pressures liquids exert.

hyperbaric High pressure oxygen.

hypercalcemia Increased serum calcium concentration.

hyperglycemia Abnormally high blood sugar level.

hyperlipidemia An increased level of lipo-proteins in the blood.

hypertension High blood pressure.

hypertonic Salt concentration higher than that of the blood.

hypnotic Sleep inducer.

hypoacidity Lower-than-normal acidity.

hypoglycemia Abnormally low blood sugar level.

hypotonic Salt concentration less than that of the blood.

hypoxia Deficiency in the amount of oxygen reaching body tissues.

immune system Body system that recognizes foreign molecules (antigens) and acts to immobilize, neutralize, or destroy them.

immunoglobulin A glycoprotein that functions as an antibody. All antibodies are immunoglobulins, but it is not certain that all immunoglobulins are antibodies.

immunosuppression Suppression of the immunologic response by means of chemical, pharmacologic, physical, or immunologic agents.

interstitial fluid Fluid between tissues.

isoelectric point pH at which amino acid is neutral.

isotonic Having the same salt concentration as the blood.

IUPAC International Union of Pure and Applied Chemistry.

jaundice Yellowish pigmentation of the skin caused by deposition of bile pigments.

ketal Compound formed by the reaction of one molecule of a ketone with two molecules of alcohol; general formula is

$$R-\underset{\underset{OR'''}{|}}{\overset{\overset{OR''}{|}}{C}}-R'$$

lacrimator Substance that causes production of tears.

lacteals Lymphatic vessels arising from the villi of the small intestine.

laking Bursting of red blood cells so that hemoglobin is released into the plasma.

lymphatics Vessels that carry lymph.

lipoprotein A compound that contains both lipid and protein.

lymphokine Secretion products of lymphocytes (usually T cells) that are responsible for the multiple effects of cellular immune response.

macrocytic Referring to abnormally large red blood cells.

macroglobulinemia Disease characterized by an increase in blood serum viscosity and presence of highly polymerized globulins.

macromolecule A very large molecule containing hundreds or thousands of atoms.

macrophage Protective cell type common to lymphocytic tissue, connective tissue, and some body organs that phagocytizes tissue cells and bacteria.

major histocompatibility complex (MHC) A collection of genes associated with the immune response and transplantation antigens.

mammography X-ray examination of the breasts.

mast cell A tissue cell where serotonin, heparin, and histamine are stored.

medulla Inner part of a biologic structure, such as the adrenal medulla.

megaloblastic Referring to very large immature cells.

metastases Spreading of disease from original sites.

mitochondria Microscopic bodies, found in all cells, that play an important part in metabolic reactions and energy production.

monoclonal antibodies Identical copies of an antibody that contain only one kind of H chain and one kind of L chain.

monokine Substance produced by macrophages that affects the function of other cells.

multiple myeloma Disease of the bone marrow.

mutagenic Causing biologic mutation.

myocardial infarction Formation of dead tissue resulting from an obstruction of blood vessels supplying the myocardium.

myxedema Disease caused by decreased activity of the thyroid gland in adults.

neoplasms Abnormal new growth of tissues; tumors.

nephritis Inflammation of the kidneys.

occlusion   Substance containing trapped liquid and gaseous material.

oliguria   Very small amount of urine formation.

opsonin   Molecule that promotes attachment of a phagocytic cell and its object.

osmolarity   Measure of the concentration of particles in solution.

osmosis   Flow of solvent through a semipermeable membrane until concentrations on both sides are equal.

osteolysis   Bone destruction.

osteomalacia   Softening of the bones because of a deficiency of vitamin D, calcium, or phosphorus.

osteomyelitis   Inflammation of the bone marrow.

osteoporosis   Reduction in the quantity of bone.

pancreatitis   Inflammation of the pancreas.

paranoia   Chronic psychosis characterized by delusion of persecution or of grandeur.

pathogen   Disease-producing organism.

phagocyte   Cell that ingests and destroys foreign particles.

phagocytosis   Engulfment of particles by leukocytes.

plasmolysis   Shrinking of red blood cells because of addition of a hypertonic solution.

pneumonitis   Inflammation of the lungs.

polyuria   Large amount of urine formation.

positron   Positive electron.

proteinuria   Presence of more than 1 g/L of protein in urine.

pruritus   Itching.

psychedelic   Substance that causes hallucinations.

psychoneurosis   Neurosis based on emotional conflict.

rad   Unit of radiation equal to 100 ergs per gram of irradiated tissue.

rem   Unit of radiation equivalent to the absorption of 1 roentgen by a human.

refractory   Something that is not easily treated.

refractory period   A time interval, such as the interval following the excitation of a neuron or the contraction of a muscle, during which repolarization occurs.

Rh antigen   A system of human blood group antigens shared with the rhesis monkey.

resonance   Property exhibited by a compound that is represented by two or more structures differing only in the position of the valence electrons.

roentgen   Unit of radiation involving X-rays that produce 1 electrostatic unit of positive or negative charge in 1 cm^3 of air.

semipermeable membrane   A selectively permeable membrane.

side effect   Usually associated with pharmacologic results of therapy unrelated to the objective.

specific heat   Amount of heat required to raise the temperature of 1 g of a substance 1 °C.

spectrophotometer   An instrument for measuring the intensity of the various wavelengths of light transmitted by a substance or solution.

substrate   Molecule on which an enzyme acts.

surfactant   Surface-active agent.

synapse   Region between the axon of one neuron and the dendrite of an adjacent neuron.

syndrome   A group of signs and symptoms that collectively indicate a disease.

syngraft   Tissue exchange between two genetically identical individuals.

tachycardia   Excessively rapid heart beat.

tetrahedral   Having four sides with equal triangular faces.

therapeutic   Having healing powers.

thromboses   Blood clots.

tincture   Alcoholic solution.

T lymphocyte (T cell)   Thymus-processed white blood cell that participates in a variety of cell-mediated immune reactions and enhances the antibody response to most antigens.

ultrasonic   Very high frequency sound waves.

vasoconstrictor   Substance that constricts blood vessels.

vestigial   A degenerate or imperfectly developed organ or structure having little or no utility but which in an earlier stage of the individual or in preceding organisms performed a useful function.

villi   Projections arising from a mucous membrane.

virucide   Substance that destroys viruses.

viscosity   A measure of resistance to flow.

volatile   Easily vaporized.

xenograft   Tissue exchange between members of different species.

Answers to Odd-Numbered Questions and Problems

Chapter 1
(page 13)

1. 37 °C
3. 185 °F, 358 K
5. They are identical.
7. Kilogram, kelvin.
9. Gram, centimeter, millimeter, kiloliter, meter.
11. (a) 50.8 cm   (b) 10.0 kg   (c) 42.7 lb
    (d) 25.0 in.   (e) 3.9 in.   (f) 110 lb
    (g) 162.6 cm   (h) 80.0 kg
13. 60 in., 149.6 lb, 99 °F
15. 255 g
17. 14.7 mL
19. Thousands of watts, thousandths of a second, millions of curies, millionths of an ampere.
21. 3750 cm^3, 3.75 L, 0.00375 kL
23. Weight depends on gravity and varies; mass depends on the amount of matter and is constant.
25. 36 mg

Chapter 2
(page 29)

1. (a) Solids have definite shape; liquids and gases do not.
    (b) Solids and liquids have definite volume; gases do not.
    (c) Solids have high or low density; liquids usually have low density; gases have very low density.
    (d) Solids have particles that are closely adhering and tightly packed; liquids have particles that are mobile and relatively close to one another; gases have particles that are independent and relatively far apart.
    (e) Solids and liquids are incompressible; gases are highly compressible.
    (f) Solids and liquids expand slightly when heated; gases expand greatly when heated.
3. Heat of vaporization is the number of calories required to change 1 g of liquid to gas at the boiling point; heat of fusion is the number of calories required to change 1 g of solid to liquid at the melting point.
5. Potential energy is stored energy, energy of position; kinetic energy is active energy, energy of motion.
7. 6000 cal, 1.5 g protein.
9. Elements and compounds must be homogeneous. Mixtures may be.
11. (a) Mg   (b) Ca   (c) I
    (d) C   (e) P   (f) Zn
    (g) Cu   (h) Br   (i) Hg
    (j) B   (k) Ag   (l) Mn
13. No; some substances will burn or decompose when heated (paper, wood, etc).
15. (a) Zn, Cr   (b) Ca, P   (c) Fe, Cu

17. The particles no longer retain their orderly arrangement.
19. Because the particles are free to move in liquids and gases but are in fixed positions in a solid.
21. calorie
23. Metals conduct heat and electricity; nonmetals do not. Metals are lustrous, ductile, and malleable; nonmetals are not. Metals are usually solid at room temperature; nonmetals may be solid, liquid, or gas at room temperature.

Chapter 3

(page 45)

1.

N $\quad 1s^2 2s^2 2p^3$	Be $\quad 1s^2 2s^2$
F $\quad 1s^1 2s^2 2p^5$	S $\quad 1s^2 2s^2 2p^6 3s^2 3p^4$
Ne $\quad 1s^2 2s^2 2p^6$	Si $\quad 1s^2 2s^2 2p^6 3s^2 3p^2$

3.

	Electrons in Outer Energy Level	Electron Energy Levels
(a) Be	2	2
(b) P	5	3
(c) Br	7	4
(d) Ca	2	4
(e) Rb	1	5
(f) As	5	4
(g) Ra	2	7
(h) Zr	2	5
(i) Mn	2	4
(j) Ac	2	7
(k) Rn	8	6

5. An atom is the smallest portion of an element that retains all of the properties of that element.
7. In the nucleus. 9. Periods.
11. No, yes.
13. Left and center, right side, extreme right column.
15. Electrons in the outermost or highest energy level; elements with eight electrons in their outermost or highest energy level (except for helium).
17. They have the same number of electrons in their highest energy level.
19. It has six electrons in its highest energy level; it has four energy levels of electrons; it is a nonmetal.

Chapter 4

(page 78)

1. Helium nuclei, high-speed electrons, electromagnetic radiation.
3. (a) $^{12}_{6}C$ \quad (b) $^{24}_{11}Na$ \quad (c) $^{30}_{16}S$
5. Isotopes produced artificially by bombardment with various types of particles.
   (a) Tracing the path of carbon as it passes through the body in various types of compounds.
   (b) For thyroid diagnosis and treatment.
   (c) For measurement of absorption of iron from the digestive tract.
7. 64 mg
9. Used to take photos of various body parts and also for the treatment of superficial skin conditions.
11. Radiation causes ionization within the cell and may also produce substances that impair cellular metabolism; radiation can also cause sterility, leukemia, and cataracts.
13. From the natural background, from medical diagnosis and treatment, and from fallout.
15. ^{99m}Tc is injected into the bloodstream. A short while later a sample of blood is withdrawn and its radioactivity is measured. From the amount of radiation originally present and that in the sample withdrawn, the blood volume can be calculated.
17. Ultrasonography uses high-frequency sound waves in place of X-rays.
19. The CT scanner rotates in a circle around a patient, taking sharp, detailed X-rays of narrow cross sections of a body part. This information is fed into a computer, which produces a visible image of the section scanned.
21. When an alpha particle is emitted, atomic number decreases by 2 and mass number decreases by 4; when a beta particle is emitted, atomic number increases by 1 while mass number does not change; when a gamma ray is emitted, there is no change in either the atomic number or the mass number.

23. Nuclear fission involves breaking apart the nucleus, while fusion involves building up a nucleus. Both produce large amounts of energy.
25. The energy comes from the "loss" in mass; the amount of energy can be calculated by using Einstein's equation, $E = mc^2$.
27. Positron emission transverse tomography, a method of using positron-emitting substances to trace biologically active chemicals through the brain.
29. A method of measuring minute concentrations of substances present in the blood; it is based upon the body's antigen–antibody system.
31. By causing the ionization of water or by knocking an electron out of the water molecule; it alters the chemical balance within the cells, causing damage to the cell and its systems.
33. 2 mrem

Chapter 5

(page 101)

1. A molecule is a combination of two or more atoms.
3. A representation whereby the symbol indicates all of the atom except the electrons in the outer energy level. Each electron in the outer energy level is indicated by a dot.
5. Eight electrons in the outer energy level.
7. ionic.
9. (a) potassium iodide   (b) calcium sulfate
   (c) zinc sulfide   (d) sodium nitrate
   (e) aluminum chloride
11. H:H   H:Cl: H:S:H   H:P:H (with H above P)

    H:C:H (with H above and below C)   :N::N:
13. Single covalent bond—one in which electrons are shared equally as in Cl_2; double covalent bond—two pairs of electrons shared, as in CO_2; triple covalent bond—three pairs of electrons shared, as in N_2.
15. (a) phosphorus trichloride
    (b) sulfur dioxide
    (c) dinitrogen pentoxide
    (d) carbon disulfide
    (e) iodine chloride
17. (a) Ag_2SO_4 (b) $KHCO_3$

(c) $(NH_4)_2S$ (d) HNO_3
19. H_2, one molecule of hydrogen; 2 H, two atoms of hydrogen; CO_2, one molecule of carbon dioxide; CO, one molecule of carbon monoxide.
21. An empirical formula represents the relative number of each type of atom present in each molecule of a given compound; a molecular formula represents the actual number of atoms present in each molecule of a given compound. Yes; for some molecules they are the same.
23. A metallic atom, because the ion has the same nuclear charge but fewer electrons, and thus these electrons are pulled in closer; a nonmetal ion, because the ion has a greater charge in its electron energy levels than in its nucleus, so the energy levels are not held as closely as in the neutral atom.
25. A nonpolar molecule may contain symmetrically placed polar bonds.
27. Na^+, K^+
29. If the two elements have equal electronegativities, the bond will be nonpolar; if the electronegativities are unequal, the bond will be polar.
31. It has one electron that it can lose from its outer energy level.
33. No; one must have a positive oxidation number and the other a negative oxidation number so that the sum of the two equals zero.
35. No; hydrogen forms covalent bonds with only two electrons around it.

Chapter 6

(page 117)

1. (a) $2 Mg + O_2 \rightarrow 2 MgO$
   (b) $Zn + 2 HCl \rightarrow ZnCl_2 + H_2$
   (c) $C + O_2 \rightarrow CO_2$
   (d) $NaCl + AgNO_3 \rightarrow AgCl + NaNO_3$
   (e) $ZnSO_4 + 2 NaOH \rightarrow Zn(OH)_2 + Na_2SO_4$
   (f) $3 Fe + 2 O_2 \rightarrow Fe_3O_4$
   (g) $Mg + 2 AgNO_3 \rightarrow Mg(NO_3)_2 + 2 Ag$
   (h) $2 Al(OH)_3 + 3 H_2SO_4 \rightarrow$
   $$Al_2(SO_4)_3 + 6 H_2O$$
3. Nature of the reacting subtances, temperature, concentration of reacting substances, presence of catalyst, surface area.
5. One mole is the number of atoms in 12.000 g of ^{12}C. 342 g   7. 2.5
9. A dynamic state in which the rates of forward and reverse reaction are equal; Le Châtelier's principle states that when a stress is applied to a

reaction at equilibrium, the equilibrium will be displaced in such a direction as to relieve that stress.

11. 20 g   13. 3.01×10^{24}, 9.03×10^{24}

15. 80 g

17. By adding N_2 or H_2, by removing NH_3, by decreasing temperature; reverse of previous steps.

19. It increases the chances of collision between reactants.

21. Dynamic; both forward and reverse reaction continue, but at the same rate, so there is no net change.

Chapter 7

(page 133)

1. (a) Molecules are far apart, so there is a small mass in a large volume.
   (b) Molecules are far apart and can be forced closer together.
   (c) Collisions between molecules are perfectly elastic, so no energy is lost.
   (d) There is enough space between molecules of one gas for another gas to enter.
   (e) Molecules are in rapid motion in all directions and continue moving unless confined by the walls of a container.
   (f) At constant pressure, as the temperature is increased, the molecules move faster, and in order to maintain a constant pressure they must occupy a larger volume.
   (g) At constant temperature, when the volume of a gas is decreased, the same number of molecules is confined in a smaller volume; thus, they strike the walls of the container more often, increasing the pressure.

3. 579 mm Hg   5. 2.23 ft^3

7. When the pressure in a chest respirator is increased, the volume of the lungs is decreased, thus forcing air out; when the pressure in the respirator is decreased, the air in the lungs expands.

9. Because the partial pressure of CO_2 is higher in the blood than in the lungs, and gases always diffuse from a higher pressure area to one of lower pressure.

11. CH_4 13. Automobiles, tobacco smoke, the burning of fossil fuel.

15. (a) First, eye irritation and irritation of respiratory passages, followed later by increased airway resistance and acute bronchitis and pneumonitis.
    (b) Eye irritation and danger to persons with respiratory and cardiac problems.
    (c) Impairment of vision and judgment, and possibly strokes and hypertension.
    (d) Airway spasms, particularly dangerous to persons having lung problems.
    (e) Causes spasms in lung tissue and also destruction of cells in the lungs.

17. Since molecules are moving in all directions, they should exert a pressure equally in all directions.

19. When the bulb is squeezed, the pressure of air inside the bulb is increased. This increased pressure is transmitted to the mercury in the cuff. As the blood exerts a pressure against the cuff, this change in pressure is indicated on the dial attached to the sphygmomanometer.

21. Each molecule in a gas acts independently of the others. If the concentration of one type of molecule is higher, then more of that type will strike the walls of the container in a given period of time, causing a greater (partial) pressure.

23. Divers work under high pressure, causing more air to dissolve in the blood. If the pressure is reduced too quickly, the dissolved gases bubble out too rapidly, causing bubbles in the blood and blockage of some blood vessels.

25. A barometer consists of a glass tube filled with mercury, the open end of which was inverted into a dish of mercury. The mercury in the tube falls until the pressure of the air just balances the pressure of the column of mercury in the tube. As air pressure increases or decreases, the mercury level in the barometer rises or falls correspondingly.

27. 0.125 mol

29. Acid rain is formed by the reaction of rain with oxides of nitrogen and sulfur in the air. Acid rain affects marble and also mortar between bricks; it affects trees, fruits, and vegetables.

31. It prevents the sun's high-energy ultraviolet radiation from reaching the Earth's surface; chlorofluorocarbons and oxides of nitrogen.

Chapter 8

(page 149)

1. Oxygen is heavier than air, so a bottle is kept mouth upward to prevent the oxygen from escaping.

3. An oxide.

5. Care should be taken in handling and storing flammable fluids such as ether and alcohol. Smoking should not be allowed in a room where oxygen is in use, nor should body rubs with alcohol or powder be given in the presence of oxygen equipment in use.
7. Laboratory—heating $KClO_3$, electrolysis of water; commercial—liquefaction of air.
9. Oxygen is used for patients with pulmonary diseases, for patients with hypoxia, for cases of asphyxia, or for those who have been nearly drowned. Care must be taken not to allow smoking or active electrical equipment in rooms where oxygen is in use.
11. To stimulate breathing.
13. Caused by bubbles of nitrogen forming in the blood vessels when a diver has the air pressure reduced too suddenly. It can be overcome by building up the pressure around the diver and then slowly releasing it.
15. Colorless gas with almost no odor or taste, heavier than air, and a low solubility in blood; it is nonflammable but does support combustion; care must be taken with its use to avoid a fire, as with oxygen.
17. Because it forms a strong bond with hemoglobin so the blood is unable to carry sufficient oxygen to the tissues.
19. In acidic or neutral solutions.
21. When a metal reacts with oxygen, an oxide is formed; when an active metal reacts with oxygen, a peroxide is formed; when a very active metal reacts with oxygen, a superoxide is formed.
23. $0, -2, -1, -\frac{1}{2}$
25. It is removed through action of the enzyme catalase.
27. No; oxidation can involve other elements such as in the reaction of sodium with chlorine.
29. The presence of oxyhemoglobin.

Chapter 9

(page 163)

	Oxidizing Agent	Reducing Agent
1. (a)	Cl_2	KI
(b)	H_2SO_4	Al
(c)	$AgNO_3$	Cu
(d)	O_2	H_2

3. Electrons can never be lost by one substance unless another substance gains those electrons.

5. (a) Reaction of sodium with chlorine (Na is oxidized).
   (b) Reaction of carbon with oxygen ($C + O_2 \rightarrow CO_2$).
   (c) CH_3CH_2OH being oxidized to CH_3CHO.
7. See page 160.

9.	Oxidized	Reduced	Oxidizing Agent	Reducing Agent
(a)	Zn	HCl	HCl	Zn
(b)	P	HNO_3	HNO_3	P
(c)	$MnCl_2$	Br_2	Br_2	$MnCl_2$
(d)	$Mn(NO_3)_2$	$NaBiO_3$	$NaBiO_3$	$Mn(NO_3)_2$

Chapter 10

(page 181)

1. Because of a lack of dissolved gases and minerals.
3. Water expands upon freezing and so becomes less dense.
5. Increasing the pressure increases the boiling point, so food will cook properly.
7. It takes energy for a solution to evaporate. The energy comes from the liquid or from the skin, thus lowering the temperature.
9. $H:\overset{..}{\underset{..}{O}}:$ It is polar because it has a + and a
       H   − side.
11. No.   13. Acid.
15. Because ordinary water may contain bacteria and other microorganisms that might be harmful.
17. Boiling kills bacteria; boiling does not remove impurities unless they are volatile.
19. A salt combined with a definite amount of water of crystallization.
21. The property of a hydrate giving off water.
23. Bicarbonates of calcium and magnesium; removed by boiling.
25. Hard water is passed through a tank containing a zeolite. The calcium and magnesium compounds are changed to sodium compounds, which do not cause hardness. The calcium and magnesium compounds remain behind in the tank.
27. Any water that is not pure; oil and dead fish floating on the surface of water; bad taste to drinking water, foul odor along a waterfront, tainted fish that cannot be eaten, unchecked growth of aquatic weeds along the shore.

29. Bacteria, sewage, and wastes from papermills, food-processing plants, and other industrial wastes; thermal pollution.

31. Decreases the amount of oxygen dissolved in water and increases the rate of chemical reaction; may also be fatal to certain types of marine life. Caused by water used as a coolant in an industrial plant.

33. Molecules escaping from the surface of a liquid into the air above; directly.

35. The higher the vapor pressure of a liquid, the lower the boiling point.

37. Digestion involves the hydrolysis of carbohydrates, fats and proteins.

39. No; it should not be used by persons on restricted sodium intake.

Chapter 11

(page 212)

1. (a) 5 g boric acid dissolved in sufficient water to make 100 mL.
   (b) 1 g $KMnO_4$ dissolved in sufficient water to make 300 mL.
   (c) 117 g NaCl dissolved in sufficient water to make 2 L.
   (d) 4.5 g NaCl dissolved in sufficient water to make 500 mL.

3. Because blood cells are affected by other types of solutions.

5. Clear, homogeneous, do not settle, have variable composition, can be separated by physical means, pass through filter paper.

7. Not clear, homogeneous, do not settle, pass through filter paper but not membranes, exhibit Brownian movement and Tyndall effect.

9. The scattering of light by colloidal particles when a beam of light is passed through a colloidal dispersion.

11. The net downward force exerted on surface molecules of a liquid.

13. Raises boiling point and lowers freezing point.

15. If concentrations in blood and dialysate are the same, no waste products will be removed.

17. By adding an emulsifying agent.

19. They are smaller than openings in filter paper but larger than openings in membranes.

21. Alcoholic solution.

23. Add another crystal of solute. If it dissolves, solution was unsaturated; if it doesn't, solution was saturated.

25. pH and mineral content of dialysate will affect the body; also temperature control is important to avoid chilling the blood.

27. By adding excess solute to a saturated solution, warming, filtering off excess solute, and then allowing the solution to cool to room temperature; no; if shaken or if one crystal of the solute is added, excess solute will crystallize immediately and a saturated solution will remain.

29. 2.92 g   31. Take 101 g of KOH, dissolve in water, and dilute to 1.5 L.

33. KNO_3

35. A solid solution of two metals; an alloy containing mercury; alloys are used in artificial joints, amalgams in dental fillings.

37. (a) Take 100 mL of 10 percent stock solution and dilute to 500 mL.
    (b) Take 20 mL of 1:10 stock solution and dilute to 200 mL.

39. Attraction of unlike molecules; attraction of like molecules; flow of a liquid through a small tube or small openings; resistance of a liquid to flow.

41. In reverse osmosis, water flows from a dilute to a concentrated solution. This process is caused by the application of hydrostatic pressure to the system.

43. (a) 500 mosmol   (b) 400 mosmol
    (c) 550 mosmol

Chapter 12

(page 221)

1. Use a conductivity apparatus. If it conducts electricity, it is an electrolyte.

3. Decreases freezing point more than a nonionized solute does; each ion of an electrolyte has an effect upon the freezing point of the liquid.

5. (a) When electrolytes are placed in water, the molecules break up into ions.
   (b) Some ions have a positive charge, others a negative charge.
   (c) The sum of the positive charges equals the sum of the negative charges.
   (d) The conductance of electricity by solutions of electrolytes is due to the presence of ions.
   (e) Nonelectrolytes do not conduct electricity because of the absence of ions.
   (f) The greater effect of electrolytes on boiling point and on freezing point is due to the increased number of particles (ions) present.

7. Positively charged ions are attracted toward the negative electrodes; negatively charged ions are attracted toward the positive electrodes.
9. One that is highly or completely ionized; HNO_3 or H_2SO_4.
11. See table in text.
13. Sodium chloride solution has twice the osmolarity of an equimolar glucose solution and so should have twice the effect on an increase in the boiling point.
15. Strong electrolytes are highly ionized and in general undergo irreversible reactions; weak electrolytes behave oppositely.

Chapter 13

(page 236)

1. (a) A substance that yields hydrogen ions in solution.
   (b) A substance that yields hydroxide ions in solution.
   (c) Reaction of an acid with a base to produce a salt and water.
   (d) A measurement indicating the strength of an acid or a base.
3. Yield hydroxide ions in solution, have a bitter taste and a slippery feeling, react with indicators, neutralize acids.
5. Turns it from red to blue.
7. Otherwise they would react with the metal of the container.
9. Wash thoroughly with water and neutralize with sodium bicarbonate.
11. Calcium hydroxide, an antacid and antidote used for oxalic acid poisoning; magnesium hydroxide, an antacid and laxative; ammonium hydroxide, a heart and respiratory stimulant.
13. Blood, 7.35 to 7.45; urine, 5.5 to 7.0; saliva, 6.2 to 7.4.
15. For an acid, both definitions are essentially the same since a proton is a hydrogen ion; for bases, the Brønsted definition is broader because substances other than hydroxide ions can also accept protons.
17. Strong acids and bases are highly or completely ionized; weak acids and bases are slightly ionized.
19. KOH yields OH^- ions in solution; C_2H_5OH does not.
21. Because of the presence of dissolved CO_2, which forms the weak acid H_2CO_3.
23. 0.325 N KNO_3 25. 19.7 mL

Chapter 14

(page 244)

1. (a) Product, other than water, formed by reaction of an acid and a base.
   (b) Chemical reaction of a substance with water resulting in a breakdown of the substance and the water molecule.
   (c) Solution that maintains a constant pH upon addition of acid or base.
3. Any carbonates and phosphates, except those of sodium, potassium, and ammonium.
5. $Zn + CuSO_4 \rightarrow ZnSO_4 + Cu$
7. $2\,HCl + Na_2CO_3 \rightarrow 2\,NaCl + H_2O + CO_2$
9. Any ten from Table 14-5.
11. Acids or bases react with a buffer to produce a neutral compound plus more buffer so that the pH does not change.
13. They maintain a constant pH for the blood cells so enzymes can function best.
15. (a) 7   (b) above 7   (c) below 7

Chapter 15

(page 256)

1. He showed that an organic compound could be made from inorganic reagents.
3. A bond refers to a pair of shared electrons.
5. Because one molecular formula can represent two or more structural formulas. Each different structural formula represents a different compound with different properties.
7.

9. 3

11. The angle from the center of the tetrahedron where the carbon is, to any two adjacent hydrogens of the four hydrogens around it— 109.5 °.

Chapter 16

(page 279)

1. (a) pentane, ethane
   (b) butyne, propyne
   (c) octene, ethene

3. An alkane from which one hydrogen has been removed. Ethyl,

$$
C_2H_5\text{—} \quad \text{or} \quad H\text{—}\overset{\displaystyle H}{\underset{\displaystyle H}{C}}\text{—}\overset{\displaystyle H}{\underset{\displaystyle H}{C}}\text{—}
$$

5. Single bonds between carbon atoms.

7. By substitution,

$$CH_4 + Cl_2 \rightarrow CH_3Cl + HCl$$

9. Petroleum and natural gas.

11. 2,2,4-trimethylpentane

$$2\ C_8H_{18} + 25\ O_2 \rightarrow$$
$$16\ CO_2 + 18\ H_2O + \text{energy}$$

13. (a) CH_3—CH—CH_3
          |
          Cl
    (b) CH_3—CH_2—CH_2—CH_2—CH_3
            |
            CH_3
    (c) CH_3—C—CH_2—CH_3
            |
            I
    (d) CH_2—CH_2—CH_2
          | |
          Cl   Cl

15. (a) 2,3-dibromobutane
    (b) 3-bromo-1-butene
    (c) 4-ethylcyclopentene
    (d) 5-chloro-2-pentene

17. Polymers are formed when many molecules of a substance are joined together to make a very large molecule; polymers are used for synthetic heart valves, blood vessels, syringes, and drug containers.

Chapter 17

(page 291)

1. ROH

3. (a) methanol (b) ethanol
   (c) 2-propanol
   (d) propanol
   (e) 1,2,3-propanetriol

5. Seventy percent alcohol coagulates protein slowly enough that it can penetrate all the way through a cell before coagulation takes place.

7. It evaporates more slowly, but it is cheaper and nonflammable.

9. Alcohol rendered unfit for human consumption; used for industrial purposes.

11. By fermentation of sugar.

13. Because it is a constituent of the body's fat.

15. Primary, methanol or methyl alcohol; secondary, 2-propanol or isopropyl alcohol; tertiary, 2-methyl-2-propanol.

17.

$$H\text{—}\overset{H}{\underset{H}{C}}\text{—}\overset{H}{\underset{H}{C}}\text{—OH} + HO\text{—}\overset{H}{\underset{H}{C}}\text{—}\overset{H}{\underset{H}{C}}\text{—H} \xrightarrow{H_2SO_4}$$

$$H\text{—}\overset{H}{\underset{H}{C}}\text{—}\overset{H}{\underset{H}{C}}\text{—O—}\overset{H}{\underset{H}{C}}\text{—}\overset{H}{\underset{H}{C}}\text{—H} + H_2O$$

19. May produce nausea; irritating to membranes of respiratory tract; flammable.

21. By reaction of carbon monoxide and hydrogen; from fermentation of molasses or from the reaction of ethene and water.

23. $CH_3CH_2CH_2CH_2-O-CH_2CH_2CH_2CH_3 + H_2O$

$CH_3CH_2CH=CH_2 + H_2O$

Chapter 18

(page 314)

1.

$$\underset{\underset{H}{|}}{\overset{\overset{H}{|}}{H-C-OH}} + [O] \rightarrow \underset{\text{methanal}\ \text{(formaldehyde)}}{\overset{\overset{H}{|}}{H-C=O}}$$

3. RCHO   5. A ketone

7. Polymer of acetaldehyde; used as a hypnotic.

9.

$$\underset{\underset{H\ OH H}{|\ \ |\ \ |}}{\overset{\overset{H\ H\ H}{|\ \ |\ \ |}}{H-C-C-C-H}} + [O] \rightarrow \underset{\underset{H\ O\ H}{|\ \ \|\ \ |}}{\overset{\overset{H\ \ \ \ H}{|\ \ \ \ \ \ |}}{H-C-C-C-H}}$$

11. A ketone

13.

$$\underset{}{\overset{\overset{H}{|}}{H-C=O}} + [O] \rightarrow \underset{\text{methanoic acid}\ \text{(formic acid)}}{\overset{\overset{OH}{|}}{H-C=O}}$$

15. RCOOH

17. Citric acid in citrus fruits; acetic acid in vinegar; lactic acid in sour milk and muscles.

19. $HCOOH + C_2H_5OH \rightarrow HCOOC_2H_5 + H_2O$

ethyl formate

21. Organic compound derived from ammonia. Primary amine, RNH_2; secondary amine, R_2NH; tertiary amine, R_3N.

23. Organic acid that has an amine attached to them; acts as both acid and base.

25. They have an acid part (COOH) and a basic part (NH_2).

27.

$$\underset{}{\overset{\overset{O}{\|}}{H-C-OH}} + \underset{}{\overset{\overset{H}{|}}{H-N-H}} \xrightarrow{\text{heat}}$$

$$\underset{\text{formamide}}{\overset{\overset{O}{\|}}{H-C-NH_2}} + H_2O$$

29. Because it can be considered as being formed from carbonic acid and two molecules of ammonia.

31. (a) 3-chlorobutanal
    (b) 2,2-dichloropropanoic acid
    (c) 3-methylbutanal
    (d) 3-methylbutanone
    (e) 2-aminopropanoic acid
    (f) 4-methyl-3-hexanone
    (g) *N*-ethylmethanamide

33. Aldehydes are easily oxidized by $Cu(OH)_2$, whereas ketones are not.

35. A compound formed by the reaction of one molecule of a ketone with one (two) molecule(s) of an alcohol; important in the structure of monosaccharides.

Chapter 19

(page 334)

1.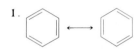

3. Decreases red and white blood cell counts; it is also carcinogenic.

5. Because all six positions on the benzene ring are identical.

7. (a) Br   (b) I   (c) Cl

9. *m*-Chloroiodobenzene, *o*-dibromobenzene, *p*-bromoiodobenzene.

11. CH_3

urine sample preservative.

13. OH

15.

Phenanthrene has the basic structure from which are derived the sex hormones, vitamin D, and cholesterol.

17. Methyl derivatives of phenol, three, as antiseptics.
19. By mild oxidation of toluene,

21. By oxidation of toluene,

23. To relieve pain in muscles and joints; an antipyretic and to relieve pain of arthritis and headache.
25. Antipyretic, analgesic, to treat rheumatic fever and arthritis.
27.

29. The *p*-aminobenzenesulfonamide group.
31.

33. By substitution; by addition; by substitution.
35. Five; product of burned coal or wood; it is carcinogenic.
37. They are food preservatives.
39. Estrone, testosterone, androsterone, progesterone.

Chapter 20

(page 345)

1. A ring compound that contains some element other than carbon in the ring.
3. Morphine narcotics obtained from the poppy plant.

5. Depress the central nervous system, producing analgesia, sedation, drowsiness, lethargy, and euphoria.
7. Used in cough medicines to depress the cough center of the brain; it is methylmorphine.
9. With methadone, which eliminates the desire for heroin and also reduces withdrawal symptoms.
11. As an anticonvulsant in the treatment of epilepsy.
13. An overwhelming compulsion to sleep; treated with amphetamines.
15. Produce visions or hallucinations.
17. It is believed to interfere with the action of serotonin, a hormone that plays an important part in the thought process.
19. Glucose, promazine
21. Hallucinogens stimulate sensory perception; psychomimetics mimic psychoses.

Chapter 21

(page 376)

1. A polyhydroxy aldehyde or ketone or substances that yield these compounds on hydrolysis.
3. Structural isomers have the same molecular formula and same functional groups, but different structural formulas.

   Example: $C-C-C-C-C$ and

5. Functional isomers have the same molecular formulas but different functional groups.

   Example: CH_3OCH_3 and CH_3CH_2OH

7. (a)

(b)

9. By using the formula 2^n, where n is the number of asymmetric carbon atoms.
11. Ordinary light vibrates in all planes, polarized light in only one plane.
13. They refer to the configuration of the OH group on the carbon atom next to the end of the chain containing the primary alcohol group; D on the right, L on the left.

15. D
17. A polysaccharide that on hydrolysis yields hexoses; a polysaccharide that on hydrolysis yields pentoses.
19. They are formed as monosaccharides from CO_2 in the air and H_2O from the ground under the action of sunlight and chlorophyll.
21. Three-carbon simple sugars; five-carbon simple sugars; six-carbon simple sugars.
23. All have the same molecular formula; glucose and galactose are both aldehydes with a different arrangement about carbon atom number 4; fructose is a ketone.
25. In fruit juices, particularly grape.
27. In the bloodstream and the tissues.
29. Clinitest, Fehling's solution, Benedict's solution.
31. No; glucose and fructose will, galactose will not because of a lack of a particular enzyme in yeast.
33. Glucuronic acid.   35. Sorbitol.
37. No; yeast contains sucrase and maltase but not lactase.
39. Sucrose has no free carbonyl group so is not a reducing sugar; lactose and maltose have a free carbonyl group, so they are reducing sugars.
41. In sprouting grain.
43. Polysaccharides have a high molecular weight, mono- and disaccharides have a low molecular weight, polysaccharides are tasteless, mono- and disaccharides are sweet; polysaccharides are insoluble in water, mono- and disaccharides are soluble; polysaccharides give a negative test with $Cu(OH)_2$, mono- and disaccharides give a positive test except fo sucrose; polysaccharides and disaccharides do not pass through membranes, monosaccharides pass through membranes.
45. Add I_2 and see if blue color develops; add starch and see if blue color develops.
47. As a supporting and structural substance.
49. As a reserve supply of carbohydrate.
51. Hydrolysis products of starch; used as glues.
53. A compound formed by the reaction of one molecule of an aldehyde with one molecule of an alcohol.

55. A three-dimensional representation of structure in which the horizontal lines indicate bonds extending forward from the plane and the vertical lines indicate bonds extending backward from the plane.
57.
59. A bond between two monosaccharides is such as in a ketal or an acetal.
61. The —OH on carbon number 1 can be either above or below the plane.

Chapter 22

(page 402)

1. Insoluble in water, soluble in organic solvents, yield fatty acids on hydrolysis or combine with fatty acids to form an ester, take part in metabolism, contain C, H, and O.
3. In a fat the alcohol is glycerol; in a wax the alcohol has a high molecular weight. A fat is a highly saturated compound; an oil is unsaturated.
5. A compound derived from simple and compound lipids on hydrolysis. Examples are fatty acids, glycerol, other alcohols, and sterols.
7. Saturated has single bonds between carbon atoms; unsaturated has one or more double bonds.
9. Linoleic; without it infants lose weight and develop eczema.
11. Chaulmoogric acid.   13. Vegetable fats.
15.
17. Odorless and tasteless when pure, insoluble in water but soluble in organic solvents; fats are solid and oils are liquids.

19. Salt of a fatty acid (soap) and glycerol.
21. Acrolein test; heat fat or oil with $KHSO_4$ and see if odor of acrolein is present.
23. Soaps are salts of fatty acids; detergents are salts of long-chain alcohol sulfates.
25. Presence of air bubbles.
27. To remove debris and bacteria from the skin.
29. A compound containing fatty acids, glycerol, a nitrogen compound, and phosphoric acid.
31. A compound produced when one molecule of fatty acid is removed from lecithin under the influence of the enzyme lecithinase A. It is very poisonous.
33. They form a bilayer, which helps control passage into and out of the cell.
35. Deposits of excess lipids from bloodstream; reduce intake of lipids or use hyperlipidemic drugs.
37. Cholesterol.
39. Because the body manufactures its own cholesterol; a diet low in cholesterol, particularly one containing unsaturated fish and vegetable oils.
41. Soap has a fat-soluble and a water-soluble end. The fat-soluble end dissolves in oil or grease and the water-soluble end sticks out into the water. The water-soluble end ionizes, leaving the oil drop with a negative charge that repels all other oil drops that have a like charge. Thus, the oil is emulsified and easily washed away.
43. Detergents that are degraded by natural conditions and do not cause pollution.
45. They stimulate the movement of calcium ions from bone.
47. In the fluid-mosaic model, proteins are embedded throughout the membrane and transport polar molecules and ions through the membrane.
49. Prostacyclin stimulates production of cAMP; thromboxanes inhibit production of cAMP.

Chapter 23

(page 428)

1. C, H, O, N.
3. By synthesis from inorganic substances in the air and in the water.
5. Nitrogen in the air is fixed by bacteria into soil nitrates, which are picked up by plants and changed to plant protein. Plant protein becomes animal protein, then waste, which is denitrified and changed back to air nitrogen.

7. Equal, less.
9. Compounds containing an acid group, COOH, and an amine group, NH_2; see text.
11. Because they contain both acid and basic groups; they contain an asymmetric carbon; L.
13. Becomes minimum.
15.

$$(CH_3)_2CH\!-\!\underset{\underset{NH_2}{|}}{CH}\!-\!\overset{\overset{O}{\|}}{C}\!-\!\boxed{OH + H}NH\!-\!CH_2COOH \longrightarrow$$

valine · · · · · · · · · · glycine

$$(CH_3)_2CHCH\!-\!\underset{\underset{NH_2}{|}}{}\overset{\overset{O}{\|}}{C}\!-\!NHCH_2COOH$$

valylglycine

$$NH_2\!-\!CH_2CO\boxed{OH} + (CH_3)_2CHCH\!-\!\underset{\underset{\boxed{H}NH}{|}}{}\overset{\overset{O}{\|}}{C}\!-\!OH \longrightarrow$$

glycine · · · · · · · · · · valine

$$NH_2CH_2\!-\!\overset{\overset{O}{\|}}{C}\!-\!NHCH\!-\!\underset{\underset{CH(CH_3)_2}{|}}{}COOH$$

glycylvaline

17. Alanine, arginine, asparagine, cysteine, glutamic acid, glutamine, glycine, histidine, isoleucine, leucine, lysine, phenylalanine, proline, serine, threonine, tyrosine, valine.
19. Proteoses, peptones, polypeptides, tripeptides, dipeptides, amino acids.
21. According to composition, function, or shape.
23. Nucleoproteins, glycoproteins, phosphoproteins, chromoproteins, lipoproteins, metalloproteins.
25. Seventy percent alcohol goes all the way through a cell before it coagulates the protein inside.
27. Egg white; otherwise the stomach would digest the antidote.
29. Xanthoproteic test: HNO_3 for protein containing a benzene ring.
    Biuret test: NaOH + $CuSO_4$ for two or more peptide linkages.

Million's test: Hg + HNO₃ for protein containing tyrosine.

Hopkins–Cole test: glyoxalic acid for protein containing tryptophan.

Ninhydrin test: ninhydrin for free amino acid or peptides.

31. A drop of unknown is placed at the bottom of a piece of filter paper suspended in a solvent. As the solvent rises in the paper, it carries the components of the unknown upward at different rates, depending upon their relative solubilities. In two-dimensional paper chromatography, the paper is then turned at right angles and placed in another solvent to repeat the process in a second direction.

33. Column is packed with an inert material coated with a thin layer of a high-boiling, nonvolatile liquid. Sample is injected into the top of the column, which is then heated as an inert carrier is passed through. The mixture in the unknown separates gradually as the more volatile components vaporize first.

35. In certain microorganisms; used as antibiotics.

37. A dipolar ion formed when the carboxyl group of an amino acid donates a hydrogen ion to the amino group of the same amino acid.

39. Because the amino acids are all of the L configuration.

41. Salt bridges, hydrogen bonds, disulfide bonds, hydrophobic bonds.

43. Storage, ferritin; transportation, hemoglobin, serum albumin; enzyme, pepsin; hormone, insulin.

Chapter 24
(page 451)

1. Enzymes increase the speed of a reaction.

3. An increase in temperature increases its rate of reaction (if temperature is too high, the enzyme ceases to function); a decrease in temperature decreases the rate.

5. Each enzyme has an optimum pH at which it functions best.

7. Increase in concentration leads to increased rate.

9. Essential for metabolism of carbohydrates, fats, proteins; pantothenic acid, adenine, ribose, phosphoric acid, mercaptoethanolamine.

11. Hydrolytic enzymes, oxidation–reduction enzymes, transferases, isomerases, ligases, lyases.

13. Enzymes that catalyze hydrolysis of esters; gastric lipase in gastric juice and pancreatic lipase in pancreatic juice; both catalyze hydrolysis of fats to fatty acids and glycerol.

15. An enzyme for the hydrolysis of nucleic acids.

17. An enzyme that catalyzes the transfer of functional groups; an enzyme that catalyzes the interconversion of optical, geometric, and structural isomers.

19. A coenzyme found in the mitrochondria; it functions in electron transport and oxidative phosphorylation.

21. With three points of attachment on an active site, only one optical isomer will fit.

23. Nonspecific inhibitors affect all enzymes in the same manner; specific inhibitors affect only one enzyme or group of enzymes.

25. Use of chemicals to destroy infectious microorganisms and cancerous cells without damaging the host cells; antibiotics such as penicillin and tetracycline and antimetabolites such as sulfanilamide.

27. Levels of some enzymes increase or decrease during disease and so blood levels are analyzed for abnormal concentrations of those enzymes.

29. Competitive inhibitors compete for the active site of the enzyme. Noncompetitive inhibitors prevent the enzyme from functioning by bonding to another part of that enzyme, not the active site.

31. In the lock-and-key theory the active site is rigid. In the induced-fit theory, the active site is flexible and can change upon binding to a substrate.

33. Key metabolic enzymes whose activity can be changed by molecules other than the substrate; to ensure that biologic processes remain coordinated at all times.

Chapter 25
(page 466)

1. To produce simple molecules that can be absorbed into the bloodstream through the intestinal walls.

3. In the mouth, stomach, and small intestine.

5. 1 to 2, presence of HCl.

7. A higher-than-normal acid condition in the stomach; may be indicative of gastric ulcers, hypertension, or gastritis.

9. Pepsinogen is an inactive form of the enzyme pepsin.

11. It stimulates the pancreas to release the pancreatic juice into the small intestine.

13. A substance secreted by the parietal cells in the walls of the stomach; vitamin B_{12} must react with the intrinsic factor before it can be absorbed. A lack of the intrinsic factor is associated with pernicious anemia.

15. Trypsinogen, for hydrolysis of protein; chymotrypsinogen, for hydrolysis of protein; carboxypeptidase, for hydrolysis of polypeptides and dipeptides; steapsin, for hydrolysis of fat; amylopsin, for hydrolysis of starch to maltose.

17. In the liver, in the gallbladder.
7.8 to 8.6; bile salts, bile pigments, and cholesterol.

21. From the breakdown of hemoglobin; in urine and feces.

23. Jaundice is produced, causing yellow pigmentation of the skin.

25. Digestion of carbohydrates begins in the mouth with ptyalin, which acts on starch; digestion is completed in the small intestine through the acid of enzymes in the pancreatic and intestinal juices; absorbed through villi in the small intestine.

27. Digestion of protein begins in the stomach through the action of pepsin; it is continued in the small intesine through the action of trypsin and chymotrypsin; absorbed through villi of the small intestine.

29. Serves as a place for reabsorption of water and some salts and for formation of feces.

31. Action of bacteria on waste products.

33. An intestinal enzyme that changes trypsinogen to trypsin in the small intestines.

Chapter 26

(page 493)

1. Glucose.

3. Synthesis of glycogen from glucose; in muscles and liver.

5. Point at which blood sugar spills over into urine.

7. By oxidation; by conversion to glycogen; by conversion to fat; through the action of hormones.

9. $A—P{\sim}P{\sim}P \rightarrow A—P{\sim}P + phosphate$;
7600 cal per mole.

11. Glycogenesis occurs primarily in the liver and the muscles. First, glucose is changed to glucose 6-P under the influence of the enzyme glucoki-

nase, the hormone insulin, and ATP. Glucose 6-P is changed to glucose 1-P through the action of the enzyme phosphoglucomutase. Glucose 1-P reacts with UTP to form UDPG under the influence of the enzyme UDPG pyrophosphorylase. Finally, UDPG under the influence of the enzyme glycogen synthetase and insulin and cyclic AMP forms glycogen.

13. No; because the enzyme glucose 6-phosphatase is found in the liver but not in the muscle.

15. It supplies most of the body's energy for anaerobic muscular contraction.

17. No, only one-fifth of it.

19. By conversion of blood glucose to muscle glycogen.

21. It provides five-carbon sugars for synthesis of nucleic acids and $NADP^+$.

23. It is the main fuel of the cycle.

25. Most of the energy is produced in the Krebs cycle.

27. Formation of glucose from noncarbohydrate substances; high-protein diet, decreased carbohydrate diet, starvation, severe diabetes.

29. Because the reaction of pyruvic acid to acetyl CoA is not reversible.

31. During periods of emotional stress; it promotes glycogenolysis.

33. An enzyme that binds oxygen in the electron transport system, then reduces it with electrons received from other cytochromes, and finally converts the O_2 to H_2O; it contains the elements Fe and Cu; it is essential to life because it controls a critical step in the body's metabolic process.

Chapter 27

(page 507)

1. Through lacteals of villi of small intestine into lymphatics to thoracic duct to bloodstream to liver.

3. Glycerol is oxidized to dihydroxyacetone phosphate, which is part of the glycolysis sequence and so becomes part of carbohydrate metabolism.

5. Acetyl CoA; acetyl CoA and propionyl CoA.

7. Carbohydrate, 4 kcal; fat, 9 kcal.   9. 129

11. Liver.

13. In adipose tissue under skin and around internal organs; reserve supply of fuel, support for internal organs, shock absorber for internal organs, insulation.

15. Conversion of glucose to fat; in the mitochondria and the cytoplasm.
17. They form part of the framework of cell membranes.
19. In lecithin a 1,2-diglyceride reacts with cytidine diphosphate choline; in cephalin, a 1,2-diglyceride reacts with cytidine diphosphate ethanolamine.
21. It is a precursor for vitamin D, sex hormones, and hormones of the adrenal cortex.
23. Gallstones.

Chapter 28

(page 522)

1. To build new tissue; to replace old tissue; to form hemoglobin, enzymes, and hormones; to produce energy.
3. When intake of nitrogen equals output.
5. Starvation, malnutrition, prolonged fever, wasting illness.
7. One that does not contain all the essential amino acids.
9. Process of removing α-amino group of an amino acid to form an α-keto acid; in the liver.
11. A reaction whereby an amino group from an amino acid is transferred to a keto acid; to manufacture amino acids the body needs.
13. In the liver.
15. Removal of a carboxyl group from an amino acid; primary amine.
17. A blue-green pigment from hemoglobin produced in reticuloendothelial cells of liver, spleen, and bone marrow.
19. Jaundice is caused by accumulation of excess bilirubin in blood due to excessive hemolysis, a blocked bile duct, or damage to liver cells as in hepatitis, cirrhosis, and cancer.
21. Hyperalimentation is the administration of glucose and hydrolyzed protein intravenously; used for patients with malnutrition.

Chapter 29

(page 535)

1. Through the lungs, skin, intestines, and kidneys.
3. A coil of capillaries in the kidneys; a structure around the glomerulus in the kidneys.

5. Blood pressure.
7. 600 to 2500 mL; weather conditions, drugs, muscular activity.
9. A decreased output of urine; high fever and certain kidney diseases.
11. Excessive water intake; diuretics and high-protein diet.
13. 4.7 to 8.0; diet.
15. Precipitation of calcium phosphate.
17. Urea: from metabolism of protein: H_2NCNH_2.
19.
21. Deposits of uric acid and urates in joints and tissues; leukemia, severe liver disease.
23. Number of milligrams of creatinine excreted within a 24-hr period per kilogram of body weight.
25. From arginine, methionine, and glycine.
27. A condition in which abnormal amounts of creatine occur in the urine.
29. Increases in hyperparathyroidism; decreases in hypoparathyroidism and renal diseases.
31. Nephritis, nephrosis, severe heart disease.
33. Sample of urine is heated and acidified with dilute acetic acid. Protein forms a cloud. Without acidification, phosphates give a false test.
35. Lactose and galactose during pregnancy; pentoses after a meal containing plums, grapes, or cherries.
37. Add sodium nitroprusside and then NH_4OH until alkaline. A red-pink color is positive.
39. Obstruction to flow of bile from liver.

Chapter 30

(page 569)

1. To carry oxygen, minerals, and food to cells and carbon dioxide and wastes from cells. Also to carry hormones, enzymes, blood cells, and antibodies. Blood acts as a buffer and helps regulate body temperature.
3. 5000 to 10,000 per cubic millimeter.
5. Defense against infection.

7. The percentage of each type of leukocyte present; they are involved in the clotting of the blood.

9. Blood serum is blood plasma without the fibrinogen for clotting.

11. Increases during fever and pregnancy; decreases in diarrhea and hemorrhaging.

13. A conjugated protein.

15. Chlorophyll a has Mg in its center while hemoglobin has Fe, and both have different side chains; cytochrome c has Fe in its center like hemoglobin but has sulfur bonding to protein.

17. Lack of vitamin B_{12} or the intrinsic factor.

19. Hemorrhaging.

21. Albumin causes most of the colloid osmotic pressure of the blood and draws water back into the bloodstream.

23. α and β-Globulins aid in transportation of carbohydrates, lipids, and metal ions; γ-globulins contain the antibodies.

25. When a blood vessel is cut, the tissues liberate thromboplastin; in the presence of thromboplastin and Ca^{2+}, prothrombin is changed to thrombin, which in turn acts on fibrinogen to change it to fibrin, the clot.

27. They are poisonous.

29. A clot formed in a blood vessel; paralysis or death.

31. CO_2 is carried as H_2CO_3, as $NaHCO_3$, and as $HHbCO_2$.

33. Since blood cells can carry only a limited amount of HCO_3^- ion, much of that ion diffuses out into the plasma while Cl^- comes in; this is the chloride shift. The reverse takes place in the lungs.

35. Bicarbonate, phosphate, protein buffers.

37. $HCl + KHCO_3 \longrightarrow KCl + H_2CO_3$
$NaOH + H_2CO_3 \longrightarrow H_2O + NaHCO_3$

39. As perspiration, as urine, through lungs, through feces.

41. To maintain osmotic pressure of extracellular fluid, to control water retention in tissue spaces, to help maintain blood pressure, to help maintain acid–base balance, to regulate irritability of nerve and muscle tissue.

43. Deficient water intake, excessive water output, poor kidney excretion, rapid administration of sodium salts, hyperactivity of adrenal cortex.

45. Too low an intake or too great an output of potassium, or sudden shift of K^+ from extracellular to intracellular fluids; general feeling of being ill, lack of energy, muscular weakness, apathy, dizziness on rising, cramps, numbness of fingers and toes.

47. A hypoactive parathyroid may cause hypocalcemia; a hyperactive parathyroid may cause hypercalcemia.

49. In the formation of HCl in the stomach and in the transportation of oxygen by the blood.

51. $\dfrac{3.6 \times 10 \times 2}{24} = 3$ mEq/L

53. pCO_2 and HCO_3^-; pH.

55. To enhance the excretion of HCO_3^-.

57. A drop in plasma protein caused by protein deficiency; swollen abdomen and extremities.

Chapter 31

(page 589)

1. Hematopoietic.

3. Protein and/or polysaccharides.

5. Macrophages, T cells, B cells.

7. Lymphokines.

9. Autoimmunity, immunologic tolerance, immunosuppressions, aging, and AIDS.

11. Set of genes associated with the immune response and transplanted organs.

13. Molecules that promote attachment of phagocytic cell; most effective is the antibody to the target particle.

15. Granulocytes and macrophages.

17. Macrophage-processed antigens and MHC gene products.

19. Th—help B cells produce immunoglobulin; Ts—provide immune tolerance; CTL—attract and destroy target cells such as transplanted tissues, tumor cells, and virally infected cells; T_{dh}—involved in allergic hypersensitivity reactions and in defense against viruses, fungi, and other organisms that replicate intracellularly.

21. Structure of the H chain.

23. Identical copies of an antibody that consists of only one H chain and one L chain; can be used for antitumor therapy, immunosuppression, fertility control, and drug toxicity reversal.

25. Survival depends on how the host reacts to the transplant.

27. (1) Hyperacute—if immediate, due to preformed antibodies, no treatment; if accelerated, that is, 1 to 5 days, due to T cells and does not usually respond to treatment. (2) Acute—cellular: after 14 days, due to T_{dh}, responds to

antirejection therapy; humoral: after 7 days, due to antibodies, not very responsive to antirejection therapy. (3) Chronic—takes months to years, due to both T cells and antibodies; treatment sometimes effective.

29. The complement system is a series of plasma enzyme reactions that punch holes (lyse) in antigens coated by IgM or IgA.

31. A severe hemolytic disease that develops when an Rh-negative mother carries an Rh-positive child. It can be prevented by injecting the mother with Rh immunoglobulin within 72 hours after delivery of an Rh-incompatible baby.

33. Myasthenia gravis, diabetes, autoimmune hemolytic anemia, systemic lupus, erythrematosis, rheumatoid arthritis, pernicious anemia rheumatic fever, multiple sclerosis.

Chapter 32

(page 616)

1. Both are carried by the bloodstream; both are required in small amounts; neither furnishes energy; vitamins are supplied in food, hormones are produced by the body.

3. Fish liver oils, butter, milk.

5. Provitamin A; it is changed in the body into two molecules of vitamin A.

7. In the liver; absorbed with aid of bile.

9. Light changes rhodopsin to *trans*-retinal and opsin. *trans*-Retinal is changed to 11-*cis*-retinal, and back to rhodopsin to begin the cycle anew.

11. Because it is produced by exposure of the skin to ultraviolet rays of the sun.

13. One thousand and 800 retinol equivalents for male and female, respectively; 400 IU vitamin D.

15. Rickets, osteomalacia, osteoporosis; calcification of soft tissues.

17. Green leafy vegetables, liver, eggs, cheese, edible fats and oils; insoluble in water, stable to heat, destroyed by acid and alkaline solutions, unstable to light and oxidants; necessary for production of prothrombin for blood-clotting process.

19. Lack of appetite, arrested growth, loss of weight, beriberi.

21. Liver, kidney, heart, yeast, peanuts, wheat germ; yeast, liver, egg yolk, germ of grain, and seeds.

23. Egg yolk, yeast, kidney and lean meats, skimmed milk, broccoli, sweet potatoes, molasses; soluble in water, insoluble in fat solvents, stable in acid and alkaline solutions; causes degeneration in adrenal cortex and a failure in reproduction.

25. Fresh fruits and vegetables, dry cereals, milk, meats, eggs; soluble in water and alcohol, insoluble in fat solvents, strong reducing agent rapidly destroyed by heat; functions in oxidation–reduction reactions in cellular respiration, in formation of intercellular substance that cements tissues together.

27. By acting as competitive inhibitor.

29. When a nerve cell is stimulated, it releases acetylcholine, which stimulates the adjacent nerve cell to release acetylcholine. This process continues until the impulse reaches the brain. After the impulse has passed, acetylcholine is regenerated to be used again.

Chapter 33

(page 644)

1. Both are necessary in small amounts and both are highly specific, but hormones are produced in the body whereas vitamins are not.

3. The pyloric mucosa; stimulates secretion of HCl and produces greater motility in the stomach.

5. Insulin and glucagon.

7. It is a protein and would be digested.

9. Increases rate of oxidation of glucose, facilitates glycogenesis, and increases synthesis of fatty acids, protein, and RNA.

11. No; it temporarily reduces blood sugar level.

13. Increases blood sugar level.

15. Increased blood sugar level, sugar in urine, ketone bodies in blood, acidosis.

17. An Individual who has retarded growth because of absence or failure of development of thryoid gland; cured by administration of thyroid hormones before adulthood.

19. Increase in size of thyroid and neck.

21. It influences the metabolism of calcium and phosphorus.

23. Glucocorticoids; affect metabolism of carbohydrates, fats, and protein; corticosterone, cortisone, cortisol. Mineralocorticoids; affect transportation of electrolytes and distribution of water; aldosterone. Androgens or estrogens; affect secondary sex characteristics; dehydroepiandrosterone.

25. Excessive loss of NaCl, low blood pressure, low body temperature, hypoglycemia, elevation of serum potassium, muscular weakness, nausea, loss of appetite, brownish pigmentation of skin.
27. Epinephrine, relaxes smooth muscles and elevates blood pressure; norepinephrine, raises blood pressure.
29. Dwarfs develop normally but do not grow in size; they are not retarded but may be sexually retarded.
31. Vasopressin, stimulates kidney to reabsorb water; oxytocin, contracts muscles of uterus.
33. Neurohormones; stimulate and inhibit release of hormones.
35. Follicles of ovary; corpus luteum.
37. Its hormones are now known to be controlled by the hypothalamus.
39. By binding to opiate receptors in the brain.
41. By triggering impulses that stimulate the release of endorphins and enkephalins.
43. Acts as a chemical messenger to regulate enzyme activity; involved in action of F prostaglandins.

Chapter 34

(page 666)

1. Chromosomes contain genes, and genes are made of DNA, which holds the hereditary information.
3.
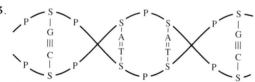
5. The DNA unwinds, and each half acts as a template for nucleotides to collect on and form a new chain, which then coils up to form two new chains of DNA.
7. In RNA, adenine bonds to uracil; in DNA, adenine bonds to thymine.

9. Cystic fibrosis.   11. See page 656.
13. A synthetic RNA composed only of uracil was substituted for mRNA in a system. The polypeptide formed contained only phenylalanine. Thus, since the RNA had the code UUU, the corresponding code of AAA on DNA specifies phenylalanine.
15. A miscopying of the genetic code or omission or change in sequence of the letters of the code.
17. (a) Normal red blood cells have a concave shape; sickle cells have a sickle shape.
    (b) Hemoglobin and hemoglobin-S differ only in the number 6 amino acid of a 146-amino acid chain.
19. If it is deficient in an enzyme, the body might lack the compound produced through the action of that enzyme.

Chapter 35

(page 677)

1. As cardiac cells die, their intracellular enzymes leak into the plasma.
3. Creatine kinase and AST peak at day 1; LD peaks at day 3.
5. Urine color, odor, and specific gravity.
7. Large meal, metabolic or respiratory alkalosis, ingestion of large amounts of stomach antacids; ammonia-generating bacteria.
9. Nitrogen metabolism.
11. Hyperglycemia, usually diabetes mellitus.
13. Excess insulin, deficiency of cortisol, ACTH, or growth hormone, glycogen storage diseases, severe liver damage, starvation.
15. Chylomicrons, VLDL, LDL, HDL.
17. Hyperthyroidism, liver disease, starvation, certain genetic diseases.
19. pH 7.37 to 7.44; pCO_2 34 to 45 mm Hg (men), 31 to 42 mm Hg (women); pO_2 85 to 95 mm Hg.
21. Heart.

Answers to Practice Tests

Chapter Number	Test Question Number									
	1	2	3	4	5	6	7	8	9	10
1	c	a	d	c	b	b	a	b	d	b
2	c	d	c	a	b	c	a	a	c	d
3	b	a	b	c	d	b	b	b	a	a
4	c	b	c	b	a	d	d	d	d	c
5	b	a	a	b	a	a	c	a	c	b
6	b	b	a	a	b	c	a	c	c	c
7	b	a	b	a	a	b	d	b	d	a
8	d	a	c	c	d	a	c	d	b	d
9	b	c	a	c	a	b	c	c	d	b
10	d	d	a	b	a	c	a	a	b	a
11	c	a	c	b	b	d	b	b	b	c
12	c	a	d	a	c	b	d	b	a	b
13	b	d	b	a	d	a	a	d	c	d
14	b	c	b	d	b	a	d	c	d	b
15	d	a	a	c	d	d	d	a	b	d
16	b	d	b	b	b	a	c	a	b	a
17	a	a	b	a	a	b	a	b	b	b
18	c	d	a	c	c	c	b	b	a	b
19	b	b	c	d	c	c	b	c	c	d
20	b	a	b	d	d	a	c	b	a	d
21	d	d	a	b	d	c	d	b	c	b
22	b	c	c	a	d	d	d	a	b	a
23	b	c	d	b	d	a	a	b	c	d
24	c	a	c	c	d	a	a	b	d	b
25	c	d	c	a	d	d	b	c	a	d
26	b	b	a	d	d	d	c	b	b	b
27	a	c	c	d	d	b	d	c	d	b
28	d	b	d	d	b	d	b	a	b	c
29	a	d	d	c	b	a	c	d	a	b
30	d	b	b	c	c	b	c	b	a	a
31	c	a	b	d	a	b	d			
32	c	d	c	a	d	a	d	a	d	a
33	b	d	d	a	a	b	a	c	a	b
34	a	d	b	c	d	c	d	d	c	a
35	c	d	b	d	a	b